# BASIC ELECTRICITY

# By THE BUREAU OF NAVAL PERSONNEL

Basic Electricity
By The Bureau of Naval Personnel

Print ISBN 13: 978-1-4209-7101-9

This edition copyright © 2020. Digireads.com Publishing.

Cover Image: a detail of "Seamless background of electrical circuit of radio device", (resistance, transistor, diode, capacitor, inductor) in green tone, by Vertyr / Shutterstock Images.

Please visit *www.digireads.com*

# CONTENTS

| Chapter | | Page |
|---|---|---|

## ACTIVE DUTY ADVANCEMENT REQUIREMENTS

| REQUIREMENTS * | E1 to E2 | E2 to E3 | #† E3 to E4 | # E4 to E5 | † E5 to E6 | † E6 to E7 | † E7 to E8 | † E8 to E9 |
|---|---|---|---|---|---|---|---|---|
| SERVICE | 4 mos. service— or completion of recruit training. | 6 mos. as E-2. | 6 mos. as E-3. | 12 mos. as E-4. | 24 mos. as E-5. | 36 mos. as E-6. 8 years total enlisted service. | 36 mos. as E-7. 8 of 11 years total service must be enlisted. | 24 mos. as E-8. 10 of 13 years total service must be enlisted. |
| SCHOOL | Recruit Training. | | Class A for PR3, DT3, PT3. AME 3, HM 3 | | | Class B for AGC MUC, MNC. | | |
| PRACTICAL FACTORS | Locally prepared check-offs. | Records of Practical Factors, NavPers 1414/1, must be completed for E-3 and all PO advancements. | | | | | | |
| PERFORMANCE TEST | | | Specified ratings must complete applicable performance tests before taking examinations. | | | | | |
| ENLISTED PERFORMANCE EVALUATION | As used by CO when approving advancement. | | Counts toward performance factor credit in advancement multiple. | | | | | |
| EXAMINATIONS ** | Locally prepared tests. | See below. | Navy-wide examinations required for all PO advancements. | | | | Navy-wide, selection board. | |
| NAVY TRAINING COURSE (INCLUDING MILITARY REQUIREMENTS) | | | Required for E-3 and all PO advancements unless waived because of school completion, but need not be repeated if identical course has already been completed. See NavPers 10052 (current edition). | | | | Correspondence courses and recommended reading. See NavPers 10052 (current edition). | |
| AUTHORIZATION | Commanding Officer | | U.S. Naval Examining Center | | | Bureau of Naval Personnel | | |

\* All advancements require commanding officer's recommendation .

† 1 year obligated service required for E-5 and E-6; 2 years for E-6, E-7, E-8 and E-9.

# Military leadership exam required for E-4 and E-5.

\*\* For E-2 to E-3, NAVEXAMCEN exams or locally prepared tests may be used.

# INACTIVE DUTY ADVANCEMENT REQUIREMENTS

| REQUIREMENTS * | E1 to E2 | E2 to E3 | E3 to E4 | E4 to E5 | E5 to E6 | E6 to E7 | E8 | E9 |
|---|---|---|---|---|---|---|---|---|
| TOTAL TIME IN GRADE | 4 mos. | 6 mos. | 15 mos. | 18 mos. | 24 mos. | 36 mos. | 36 mos. | 24 mos. |
| TOTAL TRAINING DUTY IN GRADE † | 14 days | 14 days | 14 days | 14 days | 28 days | 42 days | 42 days | 28 days |
| PERFORMANCE TESTS | | | Specified ratings must complete applicable performance tests before taking examination. | | | | | |
| DRILL PARTICIPATION | Satisfactory participation as a member of a drill unit. | | | | | | | |
| PRACTICAL FACTORS (INCLUDING MILITARY REQUIREMENTS) | Record of Practical Factors, NavPers 1414/1, must be completed for all advancements. | | | | | | | |
| NAVY TRAINING COURSE (INCLUDING MILITARY REQUIREMENTS) | Completion of applicable course or courses must be entered in service record. | | | | | | | |
| EXAMINATION | Standard Exam | Standard Exam or Rating Training. | | Standard Exam required for all PO Advancements. | | Standard Exam, Selection Board. Also pass Mil. Leadership Exam for E-4 and E-5. | | |
| AUTHORIZATION | Commanding Officer | | | U.S. Naval Examining Center | | Bureau of Naval Personnel | | |

\* Recommendation by commanding officer required for all advancements.

† Active duty periods may be substituted for training duty.

STUDY GUIDE CHART

| Chapter | AB | AD | AE | AM | AO | AQ | AS | AT | AW | AX | CE | CM | DS | EM | ET | FT | GM | IC | MN | MT | RD | RM | ST | TD | TM |
|---|---|---|---|---|---|---|---|---|---|---|---|---|---|---|---|---|---|---|---|---|---|---|---|---|---|
| 1 | E-4 | E-4 | E-4 | E-4 | E-4 | E-4 | E-4 | E-4 | E-4 | E-4 | E-4 |  | E-4 |  | E-4 | E-4 | E-4 |  | E-4 | E-4 | E-4 | E-4 | E-4 | E-4 | E-4 |
| 2 | E-4 | E-4 | E-4 | E-4 | E-4 | E-4 | E-4 | E-4 | E-4 | E-4 |  |  | E-4 | E-4 | E-4 | E-4 | E-4 |  | E-4 | E-4 | E-4 |  |  | E-4 | E-4 |
| 3 | E-4 | E-4 | E-4 | E-4 | E-4 | E-4 | E-4 | E-4 | E-4 | E-4 | E-4 | E-4 | E-4 | E-4 | E-4 | E-4 | E-4 | E-4 | E-4 | E-4 | E-5 | E-4 | E-4 | E-4 | E-4 |
| 4 | E-4 | E-4 | E-4 | E-4 | E-4 | E-4 | E-4 | E-4 | E-4 | E-4 | E-4 |  | E-4 | E-4 | E-4 | E-4 | E-4 | E-4 | E-4 | E-4 | E-5 | E-4 | E-4 | E-4 | E-4 |
| 5 | E-4 | E-4 | E-4 | E-4 | E-4 | E-4 | E-4 | E-4 | E-4 | E-4 | E-4 |  | E-4 | E-4 | E-4 | E-4 | E-4 | E-4 | E-4 | E-4 | E-5 | E-4 | E-4 | E-4 | E-4 |
| 6 | E-4 | E-4 | E-4 | E-4 | E-4 | E-4 | E-4 | E-4 | E-4 | E-4 | E-4 |  | E-4 | E-4 | E-4 | E-4 | E-4 | E-4 | E-4 | E-4 | E-5 | E-4 | E-4 | E-4 | E-4 |
| 7 | E-4 | E-4 | E-4 | E-4 | E-4 | E-4 | E-4 | E-4 | E-4 | E-4 | E-4 |  | E-4 | E-4 | E-4 | E-4 | E-4 | E-4 | E-4 | E-4 | E-4 | E-4 |  | E-4 | E-4 |
| 8 |  |  | E-4 | E-4 | E-4 | E-4 | E-4 | E-4 | E-4 | E-4 | E-4 |  | E-4 | E-4 | E-4 | E-4 | E-4 | E-4 | E-4 | E-4 | E-5 | E-4 | E-4 | E-4 | E-4 |
| 9 |  |  | E-4 | E-4 | E-4 | E-4 | E-4 | E-4 | E-4 | E-4 | E-4 |  | E-4 | E-4 | E-4 | E-4 | E-4 | E-4 | E-4 | E-4 | E-4 | E-4 | E-4 | E-4 | E-4 |
| 10 |  |  | E-4 | E-4 | E-4 | E-4 | E-4 | E-4 | E-4 | E-4 | E-4 |  | E-4 | E-4 | E-4 | E-4 | E-4 | E-4 | E-4 | E-4 | E-4 | E-4 | E-4 | E-4 | E-4 |
| 11 |  |  | E-4 | E-4 | E-4 | E-4 | E-4 | E-4 | E-4 | E-4 | E-4 |  | E-4 | E-4 | E-4 | E-4 | E-4 | E-4 | E-4 | E-4 | E-4 | E-4 | E-4 | E-4 | E-4 |
| 12 |  |  | E-4 | E-4 | E-4 | E-4 | E-4 | E-4 | E-4 | E-4 | E-4 |  | E-4 | E-4 | E-4 | E-4 | E-4 | E-4 | E-4 | E-4 | E-4 | E-4 | E-4 | E-4 | E-4 |
| 13 |  |  | E-4 | E-4 | E-4 | E-4 | E-4 | E-4 | E-4 | E-4 | E-4 |  | E-4 | E-4 | E-4 | E-4 | E-4 | E-5 | E-4 | E-4 | E-4 | E-4 | E-4 | E-4 | E-4 |
| 14 |  |  | E-4 | E-4 | E-4 | E-4 | E-4 | E-4 | E-4 | E-4 | E-4 |  | E-4 | E-4 | E-4 | E-4 | E-4 | E-4 | E-4 | E-4 | E-4 | E-4 | E-4 | E-4 |  |
| 15 |  | E-4 | E-4 |  |  | E-4 | E-4 | E-4 | E-4 | E-4 | E-4 |  | E-4 | E-4 | E-4 | E-4 | E-4 | E-4 | E-4 | E-4 | E-4 |  | E-4 | E-4 | E-4 |
| 16 |  |  | E-4 |  |  | E-4 | E-4 | E-4 |  | E-4 | E-4 |  |  | E-4 | E-4 | E-4 | E-4 | E-4 | E-4 | E-4 |  |  | E-4 | E-4 | E-4 |
| 17 |  |  | E-4 |  |  | E-4 | E-4 | E-4 |  | E-4 | E-4 |  |  | E-4 | E-4 | E-4 | E-4 | E-4 | E-4 | E-4 |  |  | E-4 | E-4 | E-4 |
| 18 |  |  | E-4 | E-4 | E-4 | E-4 | E-4 | E-4 |  | E-4 | E-4 |  |  | E-4 | E-4 | E-4 | E-4 | E-4 |  | E-4 |  |  | E-4 | E-4 | E-4 |
| 19 |  |  | E-4 | E-4 | E-4 | E-4 | E-4 | E-4 |  | E-4 | E-4 |  |  | E-4 | E-4 | E-4 | E-4 | E-4 | E-4 | E-4 |  |  | E-4 | E-4 | E-4 |
| 20 |  | E-5 | E-5 | E-5 |  | E-5 | E-4 | E-5 |  | E-5 |  |  | E-5 | E-4 | E-4 | E-4 | E-5 | E-5 | E-5 | E-5 |  |  | E-5 | E-5 |  |
| 21 |  | E-5 | E-5 | E-4 |  | E-4 | E-4 | E-4 |  | E-5 |  |  | E-4 |  | E-5 | E-4 | E-5 | E-4 | E-5 | E-5 |  |  | E-5 | E-4 |  |

# CHAPTER 1

# SAFETY

In the performance of his normal duties, the technician is exposed to many potentially dangerous conditions and situations. No training manual, no set of rules or regulations, no listing of hazards can make working conditions completely safe. However, it is possible for the technician to complete a full career without serious accident or injury. Attainment of this goal requires that he be aware of the main sources of danger, and that he remain constantly alert to those dangers. He must take the proper precautions and practice the basic rules of safety. He must be safety conscious at all times, and this safety consciousness must become second nature to him.

Much pertinent safety information is contained in Rate Training Manuals. Of particular worth is the Standard First Aid Training Course, NavPers 10081-B. In addition, directives concerning safety are published by all major commands on those specific hazards and procedures falling under the cognizance of those commands. The Chief of Naval Operations has issued a listing of specific safety precautions compiled by the Navy Department. This publication cross references safety directives by subject matter and by the identifying designation.

The purpose of this chapter is to indicate some of the major hazards encountered in the normal working conditions of the technician, and to indicate some of the basic precautions that must be observed. Although many of these hazards and precautions are general and apply to all personnel, some of them are peculiar or especially applicable to personnel concerned with electrical and electronic maintenance.

Most accidents which occur in noncombat operations can be prevented if the full cooperation of personnel is gained, and if care is exercised to eliminate unsafe acts and conditions. In the following paragraphs, some general safety rules are listed. These rules apply to personnel in all types of activities, and each individual should strictly observe the following precautions as applicable to his work or duty:

1. Report any unsafe condition or any equipment or material which he considers to be unsafe.

2. Warn others whom he believes to be endangered by known hazards or by failure to observe safety precautions.

3. Wear or use available protective clothing or equipment of the type approved for safe performance of his work or duty.

4. Report any injury or evidence of impaired health occurring in the course of work or duty.

5. Exercise, in the event of any unforseen hazardous occurrence, such reasonable caution as is appropriate to the situation.

## ELECTRICAL SAFETY PRECAUTIONS

Safety precautions in this chapter are not intended to replace information given in instructions or maintenance manuals. If at any time there is doubt as to what steps and procedures to follow, consult the leading petty officer.

## EFFECTS OF ELECTRIC SHOCK

The amount of current that may pass through the body without danger depends on the individual and the current quantity, type, path, and length of contact time.

Body resistance varies from 1,000 to 500,000 ohms for unbroken, dry skin. (Resistance and its unit of measurement are discussed later in this manual.) Resistance is lowered by moisture and high voltage, and is highest with dry skin and low voltage. Breaks, cuts, or burns may lower body resistance. A current of 1 milliampere can be felt and will cause a person to avoid it. (The term milliampere is discussed later in this manual; however, for this discussion it is sufficient to define milliampere as a very small amount of current or 1/1,000 of an ampere. Current as low as 5 milliamperes can be dangerous. If the palm of the hand makes contact with the conductor, a current of about 12 milliamperes will tend to cause the hand muscles to contract, freezing the body to the conductor. Such a shock may or may not cause serious damage, depending on the contact time and your physical condition, particularly the condition of your heart. A current of only 25 milliamperes has been known to be fatal; 100 milliamperes is likely to be fatal.

Due to the physiological and chemical nature of the human body, five times more direct current than alternating current is needed to freeze the same body to a conductor. Also, 60-hertz (cycles per second) alternating current is about the most dangerous frequency. This is normally used in residential, commercial, and industrial power.

The damage from shock is also proportional to the number of vital organs transversed, especially the percentage of current that reaches the heart.

Currents between 100 and 200 milliamperes are lethal. Ventricular fibrillation of the heart occurs when the current through the body approaches 100 milliamperes. Ventricular fibrillation is the uncoordinated actions of the walls of the heart's ventricles. This in turn causes the loss of the pumping action of the heart. This fibrillation will usually continue until some force is used to restore the coordinations of the heart's actions.

Severe burns and unconsciousness are also produced by currents of 200 milliamperes or higher. These currents usually do not cause death if the victim is given immediate attention. The victim will usually respond if rendered resuscitation in the form of artificial respiration. This is due to the 200 milliamperes of current clamping the heart muscles which prevents the heart from going into ventricular fibrillation.

When a person is rendered unconscious by a current passing through the body, it is impossible to tell how much current caused the unconsciousness. Artifical respiration must be applied immediately if breathing has stopped.

## GENERAL SAFETY PRECAUTIONS

Because of the possibility of injury to personnel, the danger of fire, and possible damage to material, all repair and maintenance work on electrical equipment should be performed only by duly authorized personnel.

When any electrical equipment is to be overhauled or repaired, the main supply switches or cutout switches in each circuit from which power could possibly be fed should be secured in the open position and tagged. The tag should read, "This circuit was ordered open for repairs and shall not be closed except by direct order of...." (usually the person directly in charge of the repairs). After the work has been completed, the tag (or tags) should be removed by the same person.

The covers of fuse boxes and junction boxes should be kept securely closed except when work is being done. Safety devices such as interlocks, overload relays, and fuses should never be altered or disconnected except for replacements. Safety or protective devices should never be changed or modified in any way without specific authorization.

The interlock switch is ordinarily wired in series with the power-line leads to the electronic power supply unit, and is installed on the lid or door of the enclosure so as to break the circuit when the lid or door is opened. A true interlock switch is entirely automatic in action; it does not have to be manipulated by the operator. Multiple interlock switches, connected in series, may be used for increased safety. One switch may be installed on the access door of a device, and another on the cover or the power-supply section. Complex interlock systems are provided when several separate circuits must be opened for safety.

Because electrical and electronic equipment may have to be serviced without deenergizing the circuits, interlock switches are constructed so that they can be disabled by the technician. However, to minimize the danger of disabling them accidentally, they are generally located in such a manner that a certain amount of manipulation is necessary in order to disable them.

Fuses should be removed and replaced only after the circuit has been deenergized. When a fuse blows, it should be replaced only with a fuse of the same current and voltage ratings. When possible, the circuit should be carefully checked before making the replacement, since the burned-out fuse is often the result of circuit fault.

## HIGH VOLTAGE SAFETY PRECAUTIONS

It is human nature to become careless with routine procedures. To illustrate the results of unsafe practices and to reemphasize the need for good safety habits, particularly around high voltage or high current circuits, consider the following incident.

A technician was electrocuted while attempting to bypass an interlock circuit in the vicinity of high voltages on a piece of electrical equipment. This was the direct result of violating a basic safety practice and indirectly an individual lack of equipment knowledge.

Many pieces of electrical equipment employ voltages which are dangerous and may be fatal if contacted. Practical safety precautions have been incorporated into electrical systems; when the most basic rules of safety are ignored, the built-in protection becomes useless.

The following rules are basic and should be followed at all times by all personnel when working with or near high voltage circuits:

1. CONSIDER THE RESULT OF EACH ACT—there is absolutely no reason for an individual to take chances that will endanger his life or the lives of others.

2. KEEP AWAY FROM LIVE CIRCUITS—do not change parts or make adjustments inside the equipment with high voltages on.

3. DO NOT SERVICE ALONE—always service equipment in the presence of another person capable of rendering assistance of first aid in an emergency.

4. DO NOT TAMPER WITH INTERLOCKS—do not depend on interlocks for protection; always shut down equipment. Never remove, short circuit, or tamper with interlocks except to repair the switch.

5. DO NOT GROUND YOURSELF—make sure you are not grounded when adjusting equipment or using measuring equipment. Use only one hand when servicing energized equipment. Keep the other hand behind you.

6. Do not energize equipment if there is any evidence of water leakage; repair the leak and wipe up the water before energizing.

These rules, teamed with the idea that voltage shows no favoritism and that personal caution is your greatest safeguard, may prevent serious injury or even death.

## WORKING ON ENERGIZED CIRCUITS

Insofar as is practicable, repair work on energized circuits should not be undertaken. When repairs on operating equipment must be made because of emergency conditions, or when such repairs are considered to be essential, the work should be done only by experienced personnel, and if possible, under the supervision of a senior petty officer of the assigned shop. Every known safety precaution should be carefully observed. Ample light for good illumination should be provided; the worker should be insulated from ground with some suitable nonconducting material such as several layers of dry canvas, dry wood, or a rubber mat of approved construction. The worker should, if possible, use only one hand in accomplishing the necessary repairs. Helpers should be stationed near the main switch or the circuit breaker so that the equipment can be deenergized immediately in case of emergency. A man qualified in first aid for electric shock should stand by during the entire period of the repair.

## GROUNDING OF EQUIPMENT

A poor safety ground, or one that is wired incorrectly, is more dangerous than no ground at all. The poor ground is dangerous because it does not offer full protection, while the user is lulled into a false sense of security. The incorrectly wired ground is a hazard because one of the line wires and the safety ground are transposed, making the shell of the tool "hot" the instant the plug is connected. Thus the unwary user is trapped, unless by pure chance the safety ground is connected to the grounded side of the line on a single-phase grounded system, or no grounds are present on an ungrounded system. In this instance the user again goes blithely along using the tool until he encounters a receptacle which has its wires transposed or a ground appears on the system.

Because there is no absolutely foolproof method of insuring that all tools are safely grounded (and because of the tendency of the average technician to ignore the use of the grounding wire), the old method of using a separate external grounding wire has been discontinued. Instead, a 3-wire, standard, color-coded cord with a polarized plug and a ground pin is required. In this manner, the safety ground is made a part of the connecting cord and plug. Since the polarized plug can be connected only to a mating receptacle, the user has no choice but to use the safety ground.

All new tools, properly connected, use the green wire as the safety ground. This wire is attached to the metal case of the tool at one end and to the polarized grounding pin in the connector at the other end. It normally carries no current, but is used only when the tool insulation fails, in which case it short circuits the electricity around the user to ground and protects him from shock. The green lead must never be mixed with the black or white leads which are the true current-carrying conductors.

Check the resistance of the grounding system with a low reading ohmmeter to be certain that the grounding is adequate (less than 1 ohm is acceptable). Ohmmeter and its use is discussed in chapter 15. If the resistance indicates greater than 1 ohm, use a separate ground strap.

Some old installations are not equipped with receptacles that will accept the grounding plug. In this event, use one of the following methods:

1. Use an adapter fitting.

2. Use the old type plug and bring the green ground wire out separately.

3. Connect an independent safety ground line. When using the adapter, be sure to connect the ground lead extension to a good ground. (Do not use the center screw which holds the cover plate on the receptacle.) Where the separate safety ground leads are externally connected to a ground, be certain to first connect the ground and then plug in the tool. Likewise, when disconnecting the tool, first remove the line plug and then disconnect the safety ground. The safety ground is always connected first and removed last.

## BATTERY SAFETY PRECAUTIONS

The principal hazard in connection with batteries is the danger of acid burns when refilling or when handling them. These burns can be prevented by the proper use of eyeshields, rubber gloves, rubber aprons, and rubber boots with nonslip soles. Rubber boots and apron need be worn only when batteries are being refilled. It is a good practice, however, to wear the eyeshield whenever working around batteries to prevent the possibility of acid burns of the eyes. Wood slat floorboards, if kept in good condition, are helpful in preventing slips and falls as well as electric shock from the high-voltage side of charging equipment.

Another hazard is the danger of explosion due to the ignition of hydrogen gas given off during battery charging operation. This is especially true where the accelerated charging method is used. Open flames or smoking is prohibited in the battery charging room, and the charging rate should be held at a point

that will prevent the rapid liberation of hydrogen gas. Manufacturers' recommendations as to the charging rates for various size batteries should be closely followed and a shop exhaust system should be used.

Particular care should be taken by technicians to prevent short circuits while batteries are being charged, tested, or handled. Hydrogen gas, which is accumulated while charging, is highly explosive and a spark from a shorted circuit could easily ignite the gas, causing serious damage to personnel and equipment.

Extreme caution should be exercised when installing and removing batteries. The nature of battery construction is such that the batteries are heavy for their size and are somewhat awkward to handle. These characteristics dictate the importance of using proper safety precautions. There is the possibility of acid causing damage to equipment or injury to personnel and the danger of an explosion that may be caused from the gas that is produced as the battery is charged. Follow the prescribed safety precautions in working with batteries.

## PRECAUTIONS WITH CHEMICALS

Volatile fluids are liquids which evaporate rapidly at relatively low temperatures. In this discussion, the term will be used to include pressurized gases escaping from their containers. In safety considerations, two main types of fluids must be considered: First, those which result in explosive vapors; and second, those which are toxic.

### Explosive Vapors

Fuels, alcohol, painting materials and supplies, insulating varnish, certain cleaning supplies, and many industrial gases produce potentially explosive vapors when allowed to accumulate in closed spaces. The hazards relating to these materials are associated with the FLASHPOINT of the liquid. This is the lowest temperature at which the liquid gives off vapor which accumulates near the surface in sufficient quantity to form a combustible mixture with air. Although liquid oxygen does not have a flashpoint, the explosive effect is the same. Almost any material that will oxidize becomes highly explosive when in the presence of liquid oxygen.

The flashpoint varies with specific materials. Personnel working with volatile materials should become faimilar with the characteristics of the particular materials with which they are associated. They should know the flash-point and also the concentration which constitutes a combustible mixture.

A general safety rule regarding explosive vapors is to always provide adequate ventilation so as to prevent accumulation of vapors, or to dispel accumulated vapors prior to operating electrical equipment in that space. The presence or absence of an odor of gasoline or other flammable or explosive vapors is not a reliable indicator of flammability, since the ability to detect odors varies between observers.

Toxic Vapors

Some liquids, upon evaporating, produce vapors which are harmful to personnel. Carbon tetrachloride is an effective solvent, an excellent cleaning material, and an excellent fire extinguisher—but it produces vapors so toxic that its use for any purpose is specifically prohibited by Navy directives. Many other materials are either prohibited or very rigidly limited for use. In most instances, the precautions are listed on the container. These precautions must at all times be rigidly adhered to and enforced. Safe substitutable materials should be used whenever permissible.

Safety considerations require that all personnel make themselves familiar with the hazards involved in the use of all materials. Toxic materials and vapors may be easily detectable, or they may be almost completely undetectable; they may act very slowly, or they may act almost instantly; they may cause discomfort, temporary damage, permanent injury, or even death; they may or may not be explosive in addition to being toxic. Toxic vapors may produce headache, dizziness, nausea, and a general feeling of illness. They may cause a gradual loss of interest, energy, or awareness. They may result in unconsciousness. They may cause temporary or permanent damage to the respiratory system, eyes, or skin. They may cause paralysis or death. Avoid prolonged exposure to vapors not known to be safe.

CONFINED SPACES

When personnel are working in confined spaces, adequate ventilation must be provided. This includes oxygen for normal breathing, cooling to prevent heat exhaustion, air movement and exchange to prevent hazardous accumulations of vapors, and an additional or alternate source of ventilation for use in the event of emergency. When a worker is sent into a confined space for any reason, provisions should be made in advance for his rescue in the event of accident or emergency. These provisions include the use of safety lines for locating the worker and for retrieving him from the space. Some means of communication must be established so that the conditions existing inside and outside the space may be made known to personnel concerned. A safety man must be provided to keep a constant check on the condition of the space and the worker, and he must be prepared to sound the alarm for additional help or to render assistance to the worker in the confined space, as required.

Fumes tend to collect in confined spaces, so the condition of the space should be checked prior to entry. The worker should also check conditions as he enters and monitor them during his stay. He should maintain constant communication with his safety man and inform him of any abnormal conditions that may exist.

Equipment used by personnel working in confined spaces is a matter of considerable importance. Enough light should be provided so that he can see clearly what he is doing. The light provided should be insulated so that it does not present a shock hazard (confined spaces are usually quite warm, and a safety light produces additional heat, so perspiration may become a serious problem). When possible, explosion-proof equipment should be used in confined spaces, and protective clothing should be used if toxic fumes are known or suspected to exist within the space.

PREVENTING ELECTRICAL FIRES

General cleanliness of the work area and of electrical apparatus is essential for the prevention of electrical fires. Oil, grease, and carbon dust can be ignited by electrical arcing. Therefore, electrical and electronic equipment should be kept absolutely clean and free of all such deposits.

Wiping rags and other flammable waste material must always be placed in tightly closed metal containers, which must be emptied at the end of the day's work.

Containers holding paints, varnishes, cleaners, or any volatile solvents should be kept tightly closed when not in actual use. They must be stored in a separate building or in a fire-resisting room which is well-ventilated and where they will not be exposed to excessive heat or to the direct rays of the sun.

## FIREFIGHTING

In case of electrical fires, the following steps should be taken:

1. Deenergize the circuit.
2. Call the Fire Department.
3. Control or extinguish the fire, using the correct type of fire extinguisher.
4. Report the fire to the appropriate authority.

For combating electrical fires, use a $CO_2$ (carbon dioxide) fire extinguisher and direct it toward the base of the flame. Carbon tetrachloride should never be used for firefighting since it changes to phosgene (a poisonous gas) upon contact with hot metal, and even in open air this creates a hazardous condition. The application of water to electrical fires is dangerous; the foam-type fire extinguishers should not be used since the foam is electrically conductive.

In cases of cable fires in which the inner layers of insulation or insulation covered by armor are burning, the only positive method of preventing the fire from running the length of the cable is to cut the cable and separate the two ends. All power to the cable should be secured and the cable should be cut with a wooden handled ax or insulated cable cutter. Keep clear of the ends after they have been cut.

When selenium rectifiers (discussed later in this manual) burn out, fumes of selenium dioxide are liberated which cause an overpowering stench. The fumes are poisonous and should not be breathed. If a rectifier burns out, deenergize the equipment immediately and ventilate the area. Allow the damaged rectifier to cool before attempting any repairs. If possible, move the equipment containing it out of doors. Do not touch or handle the defective rectifier while it is hot since a skin burn might result, through which some of the selenium compound could be absorbed.

Fires involving wood, paper, cloth, or explosives should be fought with water. Water works well on them. Therefore, advantage is taken of its inexpensiveness, availability, and safety in handling.

A steady stream of water does not work in extinguishing fires involving substances like oil, gasoline, kerosene, or paint, because these substances will float on top of the water and keep right on burning. Also, a stream of water will scatter the burning liquid and spread the fire. For this reason, foam or fog must be used in fighting such fires.

## SAFETY PRECAUTIONS WHEN USING ELECTRICAL TOOLS

As a general precaution, be sure that all tools used conform to Navy standards as to quality and type, and use them only for the purposes for which they were intended. All tools in active use should be maintained in good repair, and all damaged or nonworking tools should be turned in for repair or replacement.

When using a portable power drill, grasp it firmly during the operation to prevent it from bucking or breaking loose, thereby causing injury to yourself or damage to the tool.

Use only straight, undamaged, and properly sharpened drills. Tighten the drill securely in the chuck, using the key provided; never with wrenches or pliers. It is important that the drill be set straight and true in the chuck. The work should be firmly clamped and, if of metal a center punch should be used to score the material before the drilling operation is started.

In selecting a screwdriver for electrical work, be sure that it has a nonconducting handle. The screwdriver should not be used as a substitute for a punch or a chisel, and care should be taken that one is selected of the proper size to fit the screw.

When using a fuse puller, make certain that it is the proper type and size for the particular fuse being pulled.

The soldering iron is a fire hazard and a potential source of burns. Always assume that a soldering iron is hot; never rest the iron anywhere but on a metal surface or rack provided for that purpose. Keep the iron holder in the open to minimize the danger of fire from accumulated heat. Do not shake the iron to dispose of excess solder—a drop of hot solder may strike someone, or strike the equipment and cause a short circuit. Hold small soldering jobs with pliers or clamps.

When cleaning the iron, place the cleaning rag on a suitable surface and wipe the iron across it—do not hold the rag in the hand. Disconnect the iron when leaving the work, even for a short time—the delay may be longer than planned.

## PORTABLE POWER TOOLS

All portable power tools should be carefully inspected before being used to see that they are clean, well-oiled, and in a proper state of repair. The switches should operate normally, and the cords should be clean and free of defects. The case of all electrically driven tools should be grounded. Sparking portable electric tools should not be used in any place where flammable vapors, gases, liquids, or exposed explosives are present.

Be sure that power cords do not come in contact with sharp objects. The cords should not be allowed to kink, nor be left where they might be run over. They should not be allowed to come in contact with oil, grease, hot surfaces, or chemicals; and when damaged, should be replaced instead of being patched with tape. When unplugging power tools from receptacles, grasp the plug, not the cord.

## SHOP MACHINERY

Daily electrical work requires the use of certain pieces of shop machinery, such as a power grinder or drill press. In addition to the general precautions on the use of tools, there are a few other precautions which should be observed when working with machinery. The most important ones are:

1. Never operate a machine with a guard or cover removed.

2. Never operate mechanical or powered equipment unless throroughly familiar with the controls. When in doubt, consult the appropriate instruction or ask someone who knows.

3. Always make sure that everyone is clear before starting or operating mechanical equipment.

4. Never try to clear jammed machinery without first cutting off the source of power.

5. When hoisting heavy machinery (or equipment) by a chain fall, always keep everyone clear, and guide the hoist with lines attached to the equipment.

6. Never plug in electric machinery without insuring that the source voltage is the same as that called for on the nameplate of the machine.

### ELECTRICAL HAZARDS

Every person who works with electrical equipment should be constantly alert to the hazards of the equipment to which he may be exposed, and also be capable of rendering first aid to injured personnel. The installation, operation, and maintenance of electrical equipment requires enforcement of a stern safety code. Carelessness on the part of the operator or the maintenance technician can result in serious injury or death due to electric shock, falls, burns, flying objects, etc. After an accident has happened, investigation almost invariably shows that it could have been prevented by the exercising of simple safety precautions and procedures with which the personnel should have been familiar.

Each man concerned with electrical equipment should make it his personal responsibility to read and become thoroughly familiar with the safety practices and procedures contained in applicable safety directives, manuals, and other publications, and in equipment technical manuals prior to performing work on electrical equipment. It is the individual's responsibility to identify and eliminate unsafe conditions and unsafe acts which cause accidents.

### SHOCK

Electric shock is a jarring, shaking sensation resulting from contact with electric circuits or from the effects of lightning. The victim usually feels that he has received a sudden blow; if the voltage and resulting current is sufficiently high, the victim may become unconscious. Severe burns may appear on the skin at the place of contact; muscular spasm may occur, causing the victim to clasp the apparatus or wire which caused the shock and be unable to turn it loose.

The following procedures is recommended for rescue and care of shock victims:

1. Remove the victim from electrical contact at once, but do not endanger yourself. This can be accomplished by: (1) Throwing the switch if it is nearby; (2) cutting the cable or wires to the apparatus, using an ax with a wooden handle while taking care to protect your eyes from the flash when the wires are severed; (3) using a dry stick, rope, belt, coat, blanket, or any other nonconductor of electricity, to drag or push the victim to safety.

2. Determine whether the victim is breathing. Keep him lying down in a comfortable position and loosen the clothing about his neck, chest, and abdomen so that he can breath freely. Protect him from exposure to cold, and watch him carefully.

3. Keep him from moving about. In this condition, the heart is very weak, and any sudden muscular effort or activity on the part of the patient may result in heart failure.

4. Do not give stimulants or opiates. Send for a medical officer at once and do not leave the patient until he has adequate medical care.

5. If the victim is not breathing, it will be necessary to apply artificial respiration without delay, even though he may appear to be lifeless.

DO NOT STOP ARTIFICIAL RESPIRATION UNTIL MEDICAL AUTHORITY PRONOUNCES THE VICTIM BEYOND HELP.

For complete coverage on administering artificial respiration and treating burns, refer to Standard First Aid Training Course, NavPers 10081-B.

## BURNS AND WOUNDS

In administering first aid for burns, the objectives are to relieve the pain, to make the patient as comfortable as possible, to prevent infection, and to guard against shock which often accompanies burns of a serious nature. Detailed information on the proper method for treating various types of burns is contained in the Standard First Aid Training Course.

Any break in the skin is dangerous because it allows germs to enter the wound. Although infection may occur in any wound, it is of particular danger in wounds which do not bleed freely, wounds in which torn tissue or skin falls back into place and so prevents the entrance of air, and wounds which involve crushing of tissues.

Minor wounds should be washed immediately with soap and clean water, dried, and painted with a mild, nonirritating antiseptic. Benzalkonium chloride tincture, which is now found in most Navy first aid kits, is a good antiseptic; it does not irritate the skin but it effectively destroys many of the microorganisms which cause infection. (Benzalkonium chloride is commonly sold under the name of Zephiran or Zepherin chloride.) Apply a dressing if necessary.

Larger wounds should be treated only by medical personnel. Merely cover the wound with a dry sterile compress and fasten the compress in place with a bandage.

Since the saving of a person's life often depends on prompt and correct first aid treatment, all personnel should become thoroughly familiar with the material in the Standard First Aid Training Course.

## SAFETY PRECAUTIONS

Take time to be safe when working on electrical circuits and equipment. Carefully study the schematics and wiring diagrams of the entire system, noting what circuits must be deenergized in addition to the main power supply. Remember that electrical equipments frequently have more than one source of power. Be certain that ALL power sources are deenergized before servicing the equipment. Do not service any equipment with the power on unless it is necessary.

It must be borne in mind that deenergizing main supply circuits by opening supply switches will not necessarily "kill" all circuits in a given piece of equipment. A source of danger that has often been neglected or ignored—sometimes with tragic results—is the inputs to electrical equipment from other sources, such as synchros, remote control circuits, etc. For example, turning off the antenna safety switches will disable the antenna, but it may not turn off the antenna synchro voltages from other sources. Moreover, the rescue of a victim shocked by the power input from a remote source is often hampered because of the time required to determine the source of power and turn it off. Therefore, turn off ALL power inputs before working on equipment.

Remember that the 120-volt power supply voltage is not low, relatively harmless voltage, but is the voltage that has caused more deaths in the Navy than any other.

Do NOT work with high voltage circuits alone; have another person (safety observer), who is qualified in first aid for electrical shock, present at all times. The man stationed nearby should also know the circuits and switches controlling the equipment, and should be given instructions to pull the switch immediately if anything unforeseen happens.

Always be aware of the nearness of high voltage lines or circuits. Use rubber gloves where applicable, and stand on approved rubber matting. Remember, not all so-called rubber mats are good insulators.

Equipment containing metal parts, such as brushes and brooms, should not be used in an area within 4 feet of high voltage circuits or any electric wiring having exposed surfaces.

Inform remote stations as to the circuit on which work is being performed.

Keep clothing, hands, and feet dry if at all possible. When it is necessary to work in wet or damp locations, use a dry platform or wooden stool to sit or stand on, and place a rubber mat or other nonconductive material on top of the wood. Use insulated tools and insulated flashlights of the molded type when required to work on exposed parts.

Do not wear loose or flapping clothing. The use of thin-soled shoes with metal plates or hobnails is prohibited. Safety shoes with nonconducting soles should be worn if available. Flammable articles, such as celluloid cap visors, should not be worn.

When working on an electrical apparatus, technicians should first remove all rings, wristwatches, bracelets, ID chains and tags, and similar metal items. Care should be taken that the clothing does not contain exposed zippers, metal buttons, or any type of metal fastener.

Do NOT work on energized circuits unless absolutely necessary. Be sure to take time to lock out (or block out) the switch and tag it. Locks for this purpose should be readily available; if a lock cannot be obtained, remove the fuse and tag it.

Use one hand when turning switches on or off. Keep the doors to switch and fuse boxes closed except when working inside or replacing fuses. Use a fuse puller to remove cartridge fuses after first making certain that the circuit is dead.

All supply switches or cutout switches from which power could possible be fed should be secured in the OPEN (safety) position and tagged. The tag should read "THIS CIRCUIT WAS ORDERED OPEN FOR REPAIRS AND SHALL NOT BE CLOSED EXCEPT BY DIRECT ORDER ------" (the person making, or directly in charge of, repairs).

Never short out, tamper with, or block open an interlock switch.

Keep clear of exposed equipment; when it is necessary to work on it, use one hand only as much as possible.

Warning signs and suitable guards should be provided to prevent personnel from coming into accidental contact with high voltages.

Avoid reaching into enclosures except when absolutely necessary; when reaching into an enclosure, use rubber blankets to prevent accidental contact with the enclosure.

Do not use bare hands to remove hot parts from their holders. Use asbestos gloves if necessary.

Use a shorting stick, similar to the one shown in figure 1-1 to discharge all high voltage charges. Before a worker touches a capacitor or any part of a circuit which is known or likely to be connected to a capacitor (whether the circuit is deneegized or disconnected entirely), he should short circuit the terminals to make sure that the capacitor is completely discharged. Grounded shorting sticks should be permanently attached to workbenches where electrical equipment using high voltages are regularly serviced.

Make certain that the equipment is properly grounded. Ground all test equipment to the equipment under test.

Turn off the power before connecting alligator clips to any circuit.

When measuring circuits over 300 volts, do not hold the test prods.

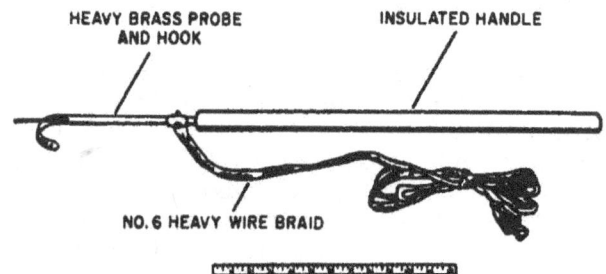

AT.14

Figure 1-1.—Shorting stick.

SAFETY EDUCATION

Safety is an all-hand responsibility. It is the job of every person in the Navy to exercise precaution to insure that people will not be injured or killed, or equipment damaged or ruined. Safety information is presented in many different ways — for example: (1) Written material, as given in this chapter; (2) safety bulletins; (3) lectures; (4) movies; (5) courses in first aid; and (6) posters.

Every shop in which you work should emphasize safety. One of the ways in which this can be done is through the use of posters. The Navy makes available numerous safety posters. Some of these are general in nature and some relate to specific types of work. These posters should be placed in a conspicuous area and as new ones are printed they should replace the older ones.

Four of the current posters that relate to shop and electrical safety are depicted in figure 1-2.

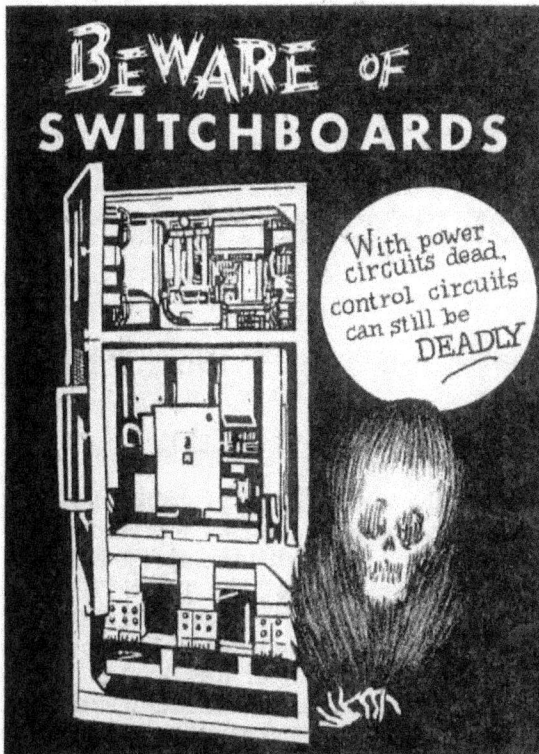

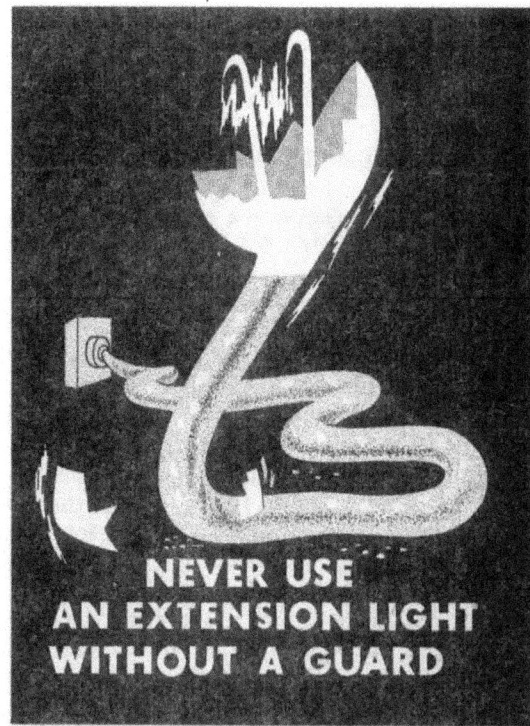

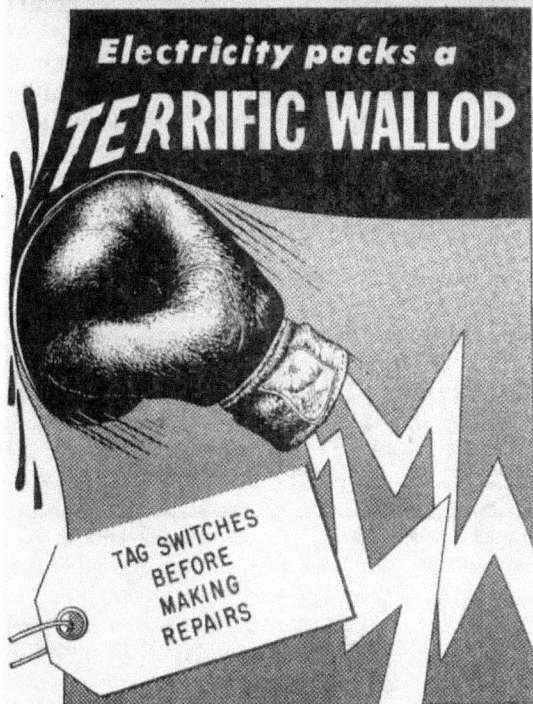

Figure 1-2.—Safety posters.

BE.1

# CHAPTER 2

# FUNDAMENTAL CONCEPTS OF ELECTRICITY

The word "electric" is actually a Greek-derived word meaning AMBER. Amber is a translucent (semitransparent) yellowish mineral, which, in the natural form, is composed of fossilized resin. The ancient Greeks used the words "electric force" in referring to the mysterious forces of attraction and repulsion exhibited by amber when it was rubbed with a cloth. They did not understand the fundamental nature of this force. They could not answer the seemingly simple question, "What is electricity?". This question is still unanswered. Though electricity might be defined as "that force which moves electrons," this would be the same as defining an engine as "that force which moves an automobile." The effect has been described, not the force.

Presently little more is known than the ancient Greeks knew about the fundamental nature of electricity, but tremendous strides have been made in harnessing and using it. Elaborate theories concerning the nature and behavior of electricity have been advanced, and have gained wide acceptance because of their apparent truth and demonstrated workability.

From time to time various scientists have found that electricity seems to behave in a constant and predictable manner in given situations, or when subjected to given conditions. These scientists, such as Faraday, Ohm, Lenz, and Kirchhoff, to name only a few, observed and described the predictable characteristics of electricity and electric current in the form of certain rules. These rules are often referred to as "laws." Thus, though electricity itself has never been clearly defined, its predictable nature and easily used form of energy has made it one of the most widely used power sources in modern time. By learning the rules, or laws, applying to the behavior of electricity, and by understanding the methods of producing, controlling, and using it, electricity may be "learned" without

ever having determined its fundamental identity.

## THE MOLECULE

One of the oldest, and probably the most generally accepted, theories concerning electric current flow is that it is comprised of moving electrons. This is the ELECTRON THEORY. Electrons are extremely tiny parts, or particles, of matter. To study the electron, therefore, the structural nature of matter itself must first be studied. (Anything having mass and inertia, and which occupies any amount of space, is composed of matter.) To study the fundamental structure or composition of any type of matter, it must be reduced to its fundamental fractions. Assume the drop of water in figure 2-1 (A) was halved again and again. By continuing the process long enough, the smallest particle of water possible—the molecule—would be obtained. All molecules are composed of atoms.

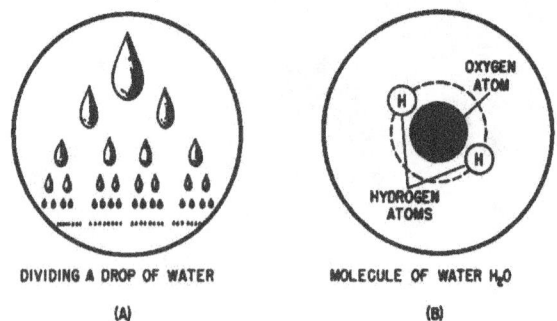

DIVIDING A DROP OF WATER
(A)

MOLECULE OF WATER $H_2O$
(B)

BE.2

Figure 2-1.—Matter is made up of molecules.

A molecule of water ($H_2O$) is composed of one atom of oxygen and two atoms of hydrogen, as represented in figure 2-1 (B). If the molecule of water were further subdivided, there would remain only unrelated atoms of oxygen and hydrogen, and the water would no longer exist as such. This example illustrates the following fact—the molecule is the smallest particle to which a substance can be reduced and still be called by the same name. This applies to all substances—liquids, solids, and gases.

When whole molecules are combined or separated from one another, the change is generally referred to as a PHYSICAL change. In a CHEMICAL change the molecules of the substance are altered such that new molecules result. Most chemical changes involve positive and negative ions and thus are electrical in nature. All matter is said to be essentially electrical in nature.

### THE ATOM

In the study of chemistry it soon becomes apparent that the molecule is far from being the ultimate particle into which matter may be subdivided. The salt molecule may be decomposed into radically different substances—sodium and chlorine. These particles that make up molecules can be isolated and studied separately. They are called ATOMS.

The atom is the smallest particle that makes up that type of material called an ELEMENT. The element retains its characteristics when subdivided into atoms. More than 100 elements have been identified. They can be arranged into a table of increasing weight, and can be grouped into families of material having similar properties. This arrangement is called the PERIODIC TABLE OF THE ELEMENTS.

The idea that all matter is composed of atoms dates back more than 2,000 years to the Greeks. Many centuries passed before the study of matter proved that the basic idea of atomic structure was correct. Physicists have explored the interior of the atom and discovered many subdivisions in it. The core of the atom is called the NUCLEUS. Most of the mass of the atom is concentrated in the nucleus. It is comparable to the sun in the solar system, around which the planets revolve. The nucleus contains PROTONS (positively charged particles) and NEUTRONS which are electrically neutral.

Most of the weight of the atom is in the protons and neutrons of the nucleus. Whirling around the nucleus are one or more smaller particles of negative electric charge. THESE ARE THE ELECTRONS. Normally there is one proton for each electron in the entire atom so that the net positive charge of the nucleus is balanced by the net negative charge of the electrons whirling around the nucleus. THUS THE ATOM IS ELECTRICALLY NEUTRAL.

The electons do not fall into the nucleus even though they are attracted strongly to it. Their motion prevents it, as the planets are prevented from falling into the sun because of their centrifugal force of revolution.

The number of protons, which is usually the same as the number of electrons, determines the kind of element in question. Figure 2-2 shows a simplified picture of several atoms of different materials based on the conception of planetary electrons describing orbits about the nucleus. For example, hydrogen has a nucleus consisting of 1 proton, around which rotates 1 electron. The helium atom has a nucleus containing 2 protons and 2 neutrons with 2 electrons encircling the nucleus. Near the other extreme of the list of elements is curium (not shown in the figure), an element discovered in the 1940's, which has 96 protons and 96 electrons in each atom.

The Periodic Table of the Elements is an orderly arrangement of the elements in ascending atomic number (number of planetary electrons) and also in atomic weight (number of protons and neutrons in the nucleus). The various kinds of atoms have distinct masses or weights with respect to each other. The element most closely approaching unity (meaning 1) is hydrogen whose atomic weight is 1.008 as compared with oxygen whose atomic weight is 16. Helium has an atomic weight of approximately 4, lithium 7, fluorine 19, and neon 20, as shown in figure 2-2.

Figure 2-3 is a pictorial summation of the discussion that has just been presented. Visible matter, at the left of the figure, is broken down first to one of its basic molecules, then to one of the molecule's atoms. The atom is then further reduced to its subatomic particles—the protons, neutrons, and electrons. Subatomic particles are electric in nature. That is, they are the particles of matter most affected by an electric force. Whereas the whole molecule or a whole atom is electrically neutral, most subatomic particles are not neutral (with the exception of the neutron). Protons are inherently positive, and electrons are inherently negative. It is these inherent characteristics which make subatomic particles sensitive to electric force.

When an electric force is applied to a conducting medium, such as copper wire, electrons in the outer orbits of the copper atoms are forced

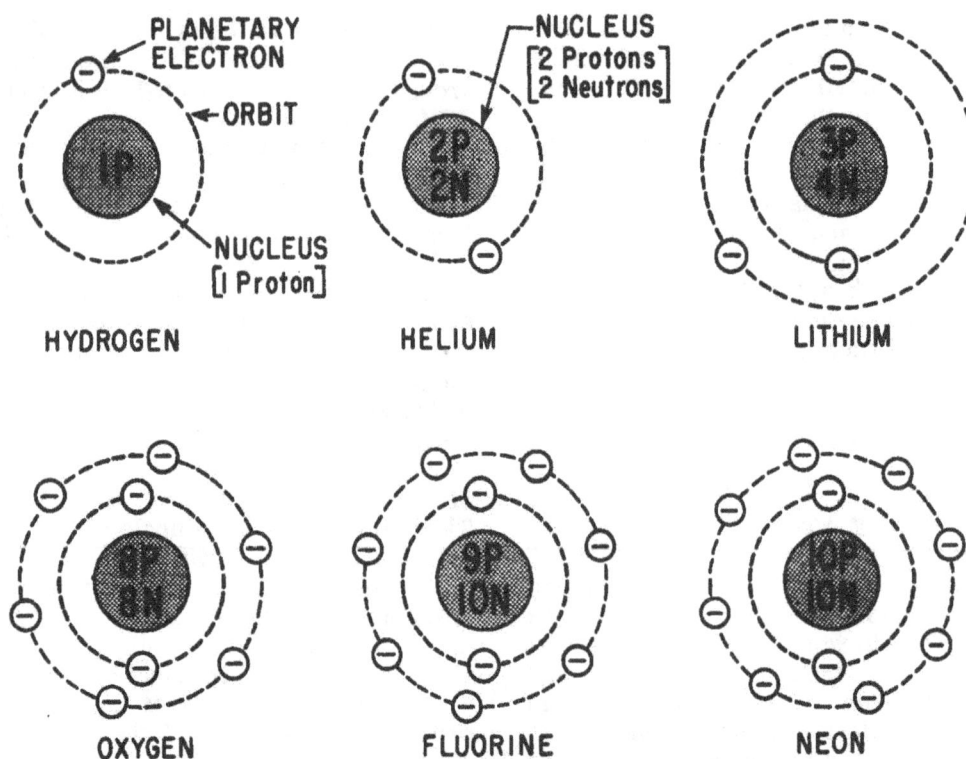

Figure 2-2.—Atomic structure of elements.

BE.3

out of orbit and impelled along the wire. The direction of electron movement is determined by the direction of the impelling force. The protons do not move, mainly because they are extremely heavy. The proton of the lightest element, hydrogen, is approximately 1,850 times heavier than its electron. Thus, it is the relatively light electron that is moved by electricity.

When an orbital electron is removed from an atom it is called a FREE ELECTRON. Some of the electrons of certain metallic atoms are so loosely bound to the nucleus that they are comparatively free to move from atom to atom. Thus, a very small force or amount of energy will cause such electrons to be removed from the atom and become free electrons. It is these free electrons that constitute the flow of an electric current in electrical conductors.

If the internal energy of an atom is raised above its normal state, the atom is said to be EXCITED. Excitation may be produced by causing the atoms to collide with particles that are impelled by an electric force. In this way, energy is transferred from the electric source to the atom. The excess energy absorbed by an atom may become sufficient to cause loosely bound

outer electrons to leave the atom against the force that acts to hold them within. An atom that has thus lost or gained one or more electrons is said to be IONIZED. If the atom loses electrons it becomes positively charged and is referred to as a POSITIVE ION. Conversely, if the atom gains electrons, it becomes negatively charged and is referred to as a NEGATIVE ION. An ion may then be defined as a small particle of matter having a positive or negative charge.

## CONDUCTORS, SEMICONDUCTORS, AND INSULATORS

Substances that permit the free motion of a large number of electrons are called CONDUCTORS. Copper wire is considered a good conductor because it has many free electrons. Electrical energy is transferred through conductors by means of the movement of free electrons that migrate from atom to atom inside the conductor. Each electron moves a very short distance to the neighboring atom where it replaces one or more electrons by forcing them out of their orbits. The replaced electrons repeat the process in other nearby atoms until the

MATTER

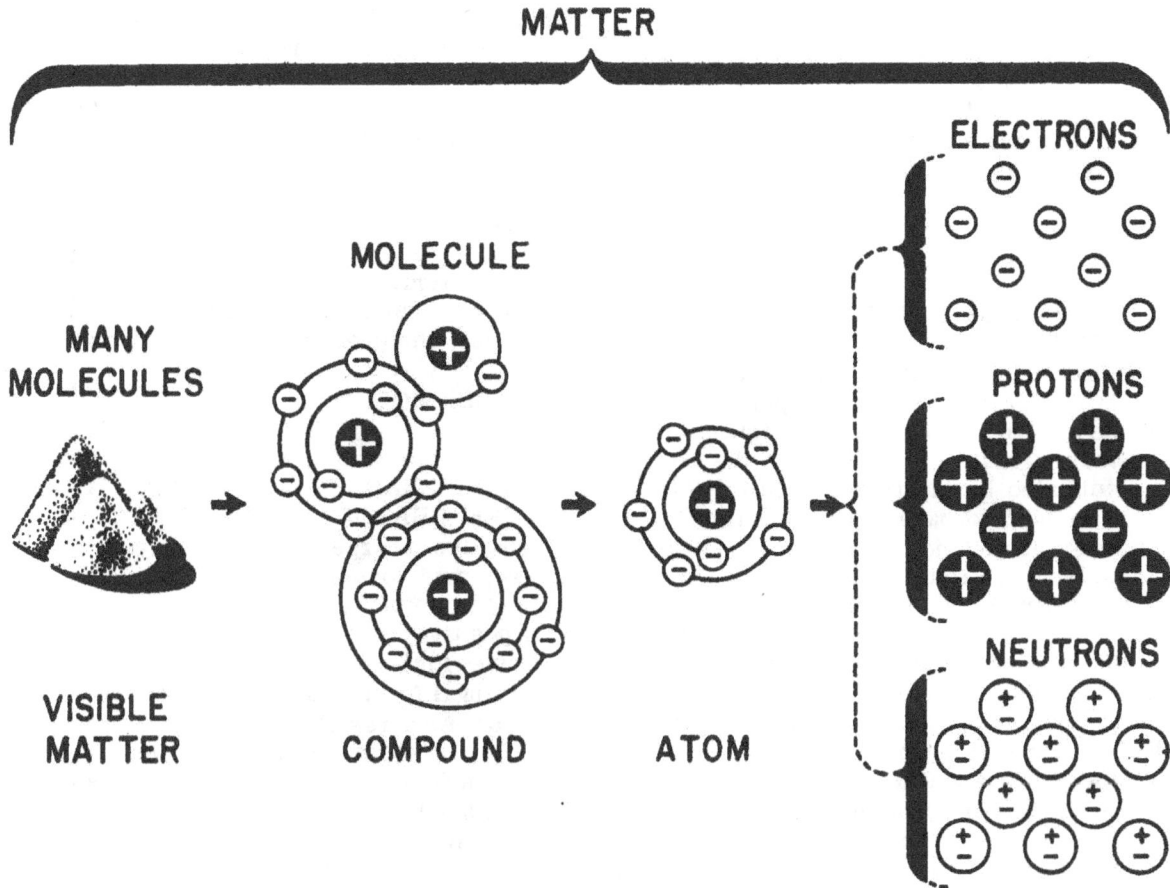

BE.4

Figure 2-3.—Breakdown of visible matter to electric particles.

movement is transmitted throughout the entire length of the conductor. The greater the number of electrons that can be made to move in a material under the application of a given force the better are the conductive qualities of that material. A good conductor is said to have a low opposition or low resistance to the current (electron) flow. Among the most commonly known metals used as conductors are silver, copper, and aluminum. The best conductor is silver, copper, and aluminum, in that order. However, copper is used more extensively because it is less expensive. A material's ability to conduct electricity also depends on its dimensions.

In contrast to good conductors, some substances such as rubber, glass, and dry wood have very few free electrons. In these materials large amounts of energy must be expended in order to break the electrons loose from the influence of the nucleus. Substances containing very few free electrons are called POOR CONDUCTORS, NONCONDUCTORS, or INSULATORS. Actually, there is no sharp dividing line between conductors and insulators, since electron motion is known to exist to some extent in all matter. Electricians simply use the best conductors as wires to carry current and the poorest conductors as insulators to prevent the current from being diverted from the wires.

Listed below are some of the best conductors and best insulators arranged in accordance with their respective abilities to conduct or to resist the flow of electrons.

| Conductors | Insulators |
| --- | --- |
| Silver | Dry air |
| Copper | Glass |
| Aluminum | Mica |
| Zinc | Rubber |
| Brass | Asbestos |
| Iron | Bakelite |

14

A semiconductor is a material that is neither a good conductor nor a good insulator. Germanium and silicon are substances that fall into this category. These materials, due to their peculiar crystalline structure, may under certain conditions act as conductors; under other conditions, as insulators. As the temperature is raised, however, a limited number of electrons become available for conduction.

## STATIC ELECTRICITY

In a natural, or neutral state, each atom in a body of matter will have the proper number of electrons in orbit around it. Consequently, the whole body of matter comprised of the neutral atoms will also be electrically neutral. In this state, it is said to have a "zero charge," and will neither attract nor repel other matter in its vicinity. Electrons will neither leave nor enter the neutrally charged body should it come in contact with other neutral bodies. If, however, any number of electrons are removed from the atoms of a body of matter, there will remain more protons than electrons, and the whole body of matter will become electrically positive. Should the positively charged body come in contact with another body having a normal charge, or having a negative (too many electrons) charge, an electric current will flow between them. Electrons will leave the more negative body and enter the positive body. This electron flow will continue until both bodies have equal charges.

When two bodies of matter have unequal charges, and are near one another, an electric force is exerted between them because of their unequal charges. However, since they are not in contact, their charges cannot equalize. The existence of such an electric force, where current cannot flow, is referred to as static electricity. "Static" means "not moving." This is also referred to as an ELECTROSTATIC FORCE.

One of the easiest ways to create a static charge is by the friction method. With the friction method, two pieces of matter are rubbed together and electrons are "wiped off" one onto the other. If materials that are good conductors are used, it is quite difficult to obtain a detectable charge on either. The reason for this is that equalizing currents will flow easily in and between the conducting materials. These currents equalize the charges almost as fast as they are created. A static charge is easier to obtain by rubbing a hard nonconducting material

against a soft, or fluffy, nonconductor. Electrons are rubbed off one material and onto the other material. This is illustrated in figure 2-4.

When the hard rubber rod is rubbed in the fur, the rod accumulates electrons. Since both fur and rubber are poor conductors, little equalizing current can flow, and an electrostatic charge is built up. When the charge is great enough, equalizing currents will flow regardless of the material's poor conductivity. These currents will cause visible sparks, if viewed in darkness, and produce a cracking sound.

## CHARGED BODIES

One of the fundamental laws of electricity is that LIKE CHARGES REPEL EACH OTHER and UNLIKE CHARGES ATTRACT EACH OTHER. A positive charge and negative charge, being unlike, tend to move toward each other. In the atom the negative electrons are drawn toward the positive protons in the nucleus. This attractive force is balanced by the electron's centrifugal force caused by its rotation about the nucleus. As a result, the electrons remain in orbit and are not drawn into the nucleus. Electrons repel each other because of their like negative charges, and protons repel each other because of their like positive charges.

The law of charged bodies may be demonstrated by a simple experiment. Two pith (paper pulp) balls are suspended near one another by threads, as shown in figure 2-5.

If the hard rubber rod is rubbed to give it a negative charge, and then held against the right-hand ball in part (A), the rod will impart a negative charge to the ball. The right-hand ball will be charged negative with respect to the left-hand ball. When released, the two balls will be drawn together, as shown in figure 2-5 (A). They will touch and remain in contact until the left-hand ball acquires a portion of the negative charge of the right-hand ball, at which time they will swing apart as shown in figure 2-5 (C). If, positive charges are placed on both balls (fig. 2-5 (B)), the balls will also be repelled from each other.

## COULOMB'S LAW OF CHARGES

The amount of attracting or repelling force which acts between two electrically charged bodies in free space depends on two things—(1) their charges, and (2) the distance between them. The relationship of charge and distance to

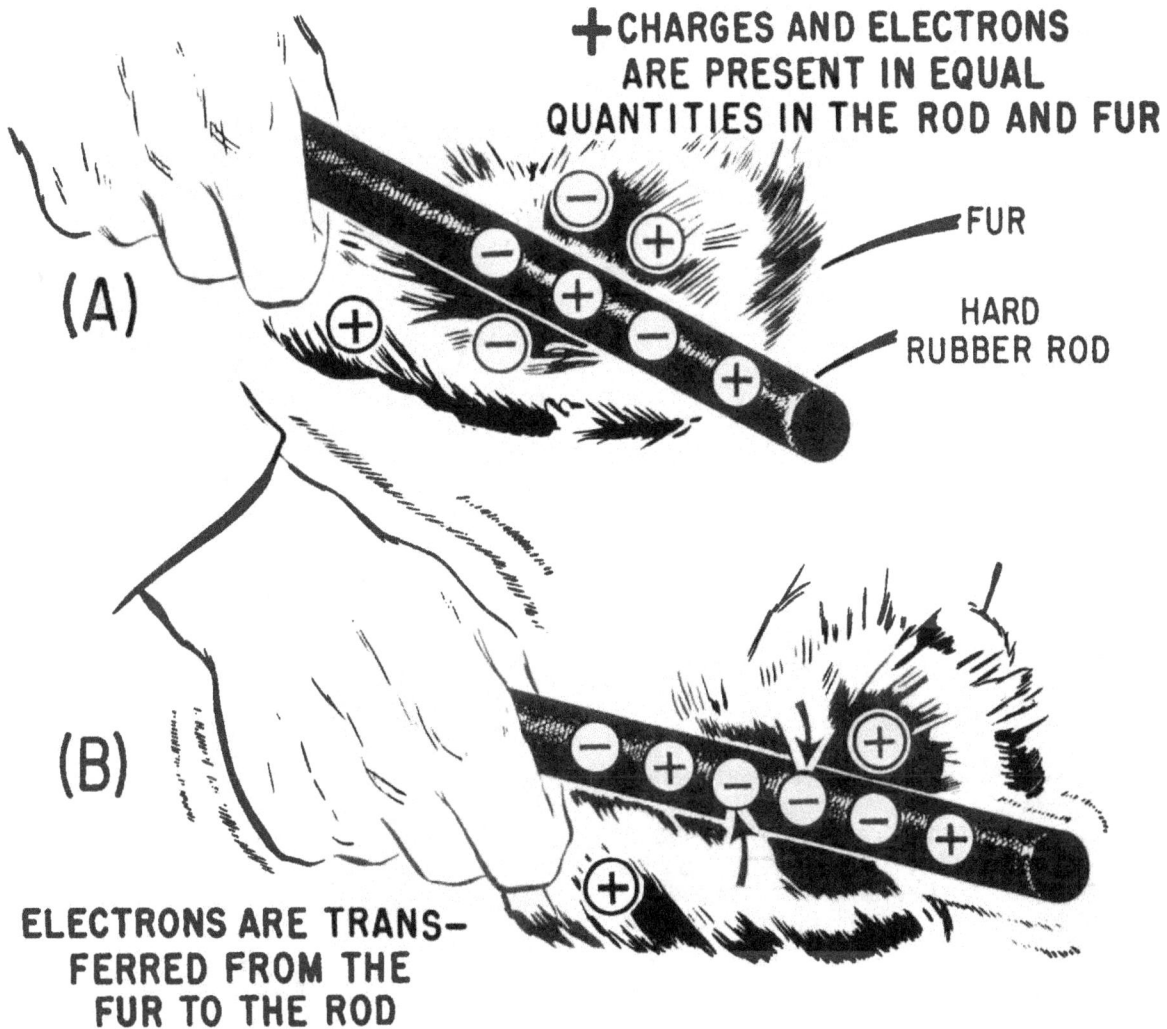

**+ CHARGES AND ELECTRONS ARE PRESENT IN EQUAL QUANTITIES IN THE ROD AND FUR**

FUR

HARD RUBBER ROD

(A)

(B)

**ELECTRONS ARE TRANS- FERRED FROM THE FUR TO THE ROD**

BE.5

Figure 2-4.—Producing static electricity by friction.

electrostatic force was first discovered and written by a French scientist named Charles A. Coulomb. Coulomb's Law states that CHARGED BODIES ATTRACT OR REPEL EACH OTHER WITH A FORCE THAT IS DIRECTLY PROPORTIONAL TO THE PRODUCT OF THEIR CHARGES, AND IS INVERSELY PROPORTIONAL TO THE SQUARE OF THE DISTANCE BETWEEN THEM.

ELECTRIC FIELDS

The space between and around charged bodies in which their influence is felt is called an ELECTRIC FIELD OF FORCE. The electric field is always terminated on material objects and extends between positive and negative charges. It can exist in air, glass, paper, or a vacuum. ELECTROSTATIC FIELDS and DIELECTRIC FIELDS are other names used to refer to this region of force.

Fields of force spread out in the space surrounding their point of origin and, in general, DIMINISH IN PROPORTION TO THE SQUARE OF THE DISTANCE FROM THEIR SOURCE.

The field about a charged body is generally represented by lines which are referred to as ELECTROSTATIC LINES OF FORCE. These lines are imaginary and are used merely to represent the direction and strength of the field. To avoid confusion, the lines of force exerted by a positive charge are always shown leaving the

16

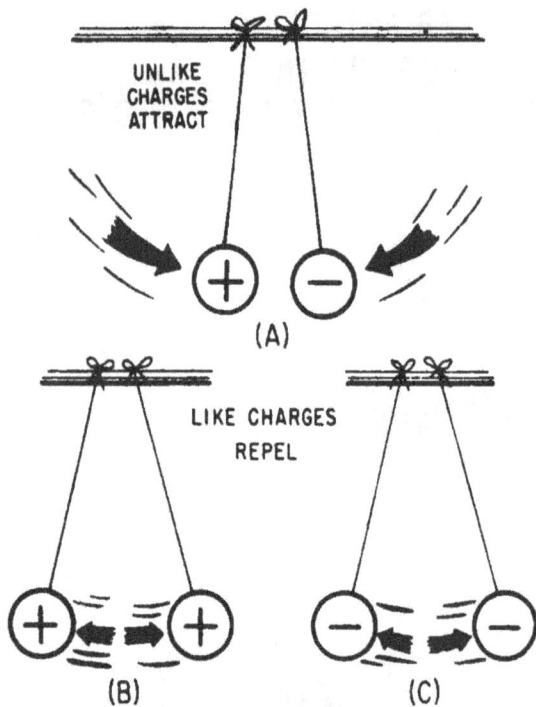

BE.6

Figure 2-5.—Reaction between charged bodies.

charge, and for a negative charge they are shown as entering. Figure 2-6 illustrates the use of lines to represent the field about charged bodies.

Figure 2-6 (A) represents the repulsion of like-charged bodies and their associated fields. Part (B) represents the attraction between unlike-charged bodies and their associated fields.

MAGNETISM

A substance is said to be a magnet if it has the property of magnetism—that is, if it has the power to attract such substances as iron, steel, nickel, or cobalt, which are known as MAGNETIC MATERIALS. A steel knitting needle, magnetized by a method to be described later, exhibits two points of maximum attraction (one at each end) and no attraction at its center. The points of maximum attraction are called MAGNETIC POLES. All magnets have at least two poles. If the needle is suspended by its middle so that it rotates freely in a horizontal plane about its center, the needle comes to rest in a approximately north-south line of direction. The same pole will always point to the north, and the other will always point toward the south. The magnetic

pole that points northward is called the NORTH POLE, and the other the SOUTH POLE.

A MAGNETIC FIELD exists around a simple bar magnet. The field consists of imaginary lines along which a MAGNETIC FORCE acts. These lines emanate from the north pole of the magnet, and enter the south pole, returning to the north pole through the magnet itself, thus forming closed loops.

A MAGNETIC CIRCUIT is a complete path through which magnetic lines of force may be established under the influence of a magnetizing force. Most magnetic circuits are composed largely of magnetic materials in order to contain the magnetic flux. These circuits are similar to the ELECTRIC CIRCUIT, which is a complete path through which current is caused to flow under the influence of an electromotive force.

Magnets may be conveniently divided into three groups.

1. NATURAL MAGNETS, found in the natural state in the form of a mineral called magnetite.

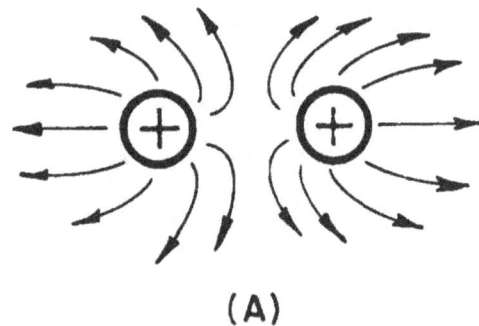

(A)

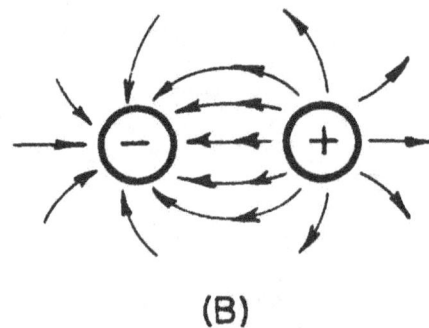

(B)

BE.7

Figure 2-6.—Electrostatic lines of force.

2. PERMANENT MAGNETS, bars of hardened steel (or some form of alloy such as alnico) that have been permanently magnetized.

3. ELECTROMAGNETS, composed of soft-iron cores around which are wound coils of insulated wire. When an electric current flows through the coil, the core becomes magnetized. When the current ceases to flow, the core loses most of its magnetism.

Permanent magnets and electromagnets are sometimes called ARTIFICIAL MAGNETS to further distinguish them from natural magnets, and are discussed in this manual under the heading of "Artificial Magnets."

## NATURAL MAGNETS

For many centuries it has been known that certain stones (magnetite, $Fe_3O_4$) have the ability to attract small pieces of iron. Because many of the best of these stones (natural magnets) were found near Magnesia in Asia Minor, the Greeks called the substance MAGNETITE, or MAGNETIC.

Before this, ancient Chinese observed that when similar stones were suspended freely, or floated on a light substance in a container of water, they tended to assume a nearly north-and-south position. Probably Chinese navigators used bits of magnetite floating on wood in a liquid-filled vessel as crude compasses. At that time it was not known that the earth itself acts like a magnet, and these stones were regarded with considerable superstitious awe. Because bits of this substance were used as compasses they were called LOADSTONES (or lodestones), which means "leading stones."

Natural magnets are also found in the United States, Norway, and Sweden. A natural magnet, demonstrating the attractive force at the poles, is shown in figure 2-7 (A).

## ARTIFICIAL MAGNETS

Natural magnets no longer have any practical value because more powerful and more conveniently shaped permanent magnets can be produced artificially. Commercial magnets are made from special steels and alloys—for example, alnico, made principally of aluminum, nickel, and cobalt. The name is derived from the first two letters of the three principal elements of which it is composed. An artificial magnet is shown in figure 2-7 (B).

An iron, steel, or alloy bar can be magnetized by inserting the bar into a coil of insulated

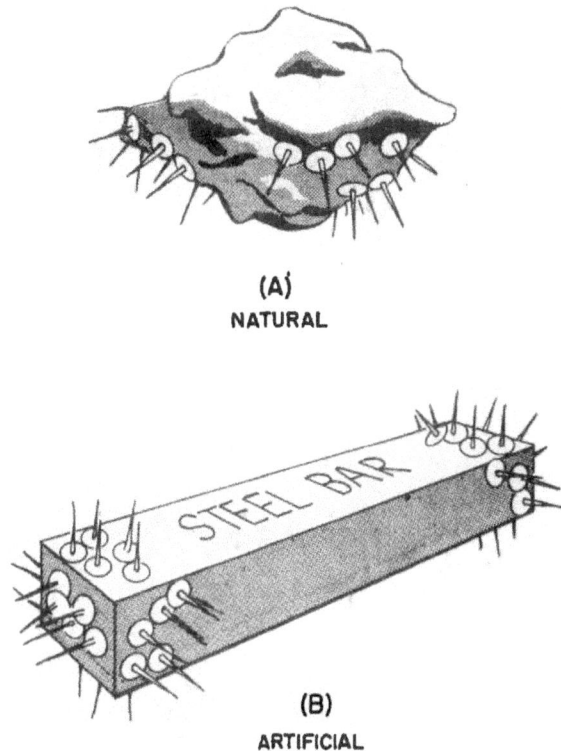

(A)
NATURAL

(B)
ARTIFICIAL

BE.8

Figure 2-7.—(A) Natural magnet;
(B) artificial magnet.

wire and passing a heavy direct current through the coil, as shown in figure 2-8 (A). This aspect of magnetism is treated later in the chapter. The same bar may also be magnetized if it is stroked with a bar magnet, as shown in figure 2-8 (B). It will then have the same magnetic property that the magnet used to induce the magnetism has— namely, there will be two poles of attraction, one at either end. This process produces a permanent magnet by INDUCTION—that is, the magnetism is induced in the bar by the influence of the stroking magnet.

Artificial magnets may be classified as "permanent" or "temporary" depending on their ability to retain their magnetic strength after the magnetizing force has been removed. Hardened steel and certain alloys are relatively difficult to magnetize and are said to have a LOW PERMEABILITY because the magnetic lines of force do not easily permeate, or distribute themselves readily through the steel. (Permeability is a measure of the relative ability of a substance to conduct magnetic lines of force as compared with

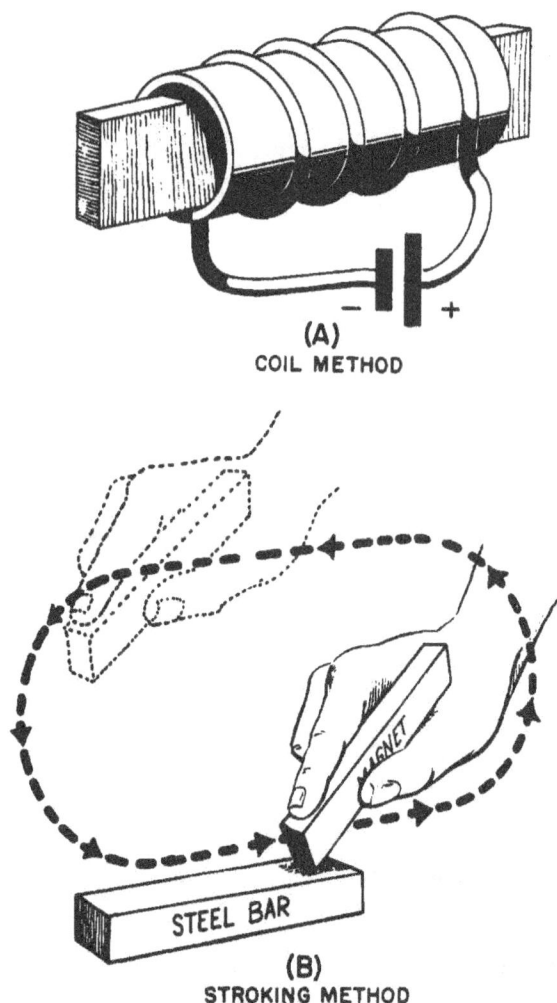

(A)
COIL METHOD

(B)
STROKING METHOD

STEEL BAR

BE.9

Figure 2-8.—Methods of producing
artificial magnets.

air. It is discussed in greater detail later in this manual.) Once magnetized, however, these materials retain a large part of their magnetic strength and are called PERMANENT MAGNETS. Permanent magnets are used extensively in electric instruments, meters, telephone receivers, permanent-magnet loudspeakers, and magnetos. Conversely, substances that are relatively easy to magnetize—such as soft iron and annealed silicon steel—are said to have a HIGH PERMEABILITY. Such substances retain only a small part of their magnetism after the magnetizing force is removed and are called TEMPORARY MAGNETS. Silicon steel and similar materials are used in transformers where the magnetism is constantly changing and in generators and motors where the strength of the fields can be readily changed.

The magnetism that remains in a temporary magnet after the magnetizing force is removed is called RESIDUAL MAGNETISM. The fact that temporary magnets retain even a small amount of magnetism is an important factor in the build-up of voltage in self-excited d-c generators.

NATURE OF MAGNETISM

A popular theory of magnetism considers the molecular alinement of the material. This is known as Weber's Theory. This theory assumes all magnetic substances to be composed of tiny molecular magnets. All unmagnetized materials have the magnetic forces of its molecular magnets neutralized by adjacent molecular magnets thereby eliminating any magnetic effect. A magnetized material will have most of its molecular magnets lined up so that the north pole of each molecule points in one direction, and the south pole faces the opposite direction. A material with its molecules thus alined will then have one effective north pole, and one effective south pole. An illustration of Weber's Theory is shown in figure 2-9 (A) where a steel bar is magnetized by stroking. When a steel bar is stroked several times in the same direction by a magnet, the magnetic force from the north pole of the magnet causes the molecules to aline themselves. The polarity of the magnet formed is dependent upon the direction of the magnetizing force as it is brought over the random magnetic molecules.

Some justification of Weber's Theory occurs when a magnet is split in half. It is found that each half possess both a north and a south magnetic pole as shown in figure 2-9 (B). The polarities of the poles are in the same respective directions as the poles of the original magnet. If a magnet is further divided into small parts, it will be found that each part, down to its last molecule, will all have similar north and south poles. Each part would exhibit its own magnetic properties.

Further support of Weber's Theory comes from the fact that when a bar magnet is held out of alinement with the earth's magnetic field and repeatedly jarred or heated, the molecular alinement is disarranged and the material becomes demagnetized. For example, measuring devices which make use of permanent magnets become inaccurate when subjected to severe jarring or exposure to opposing magnetic fields.

19

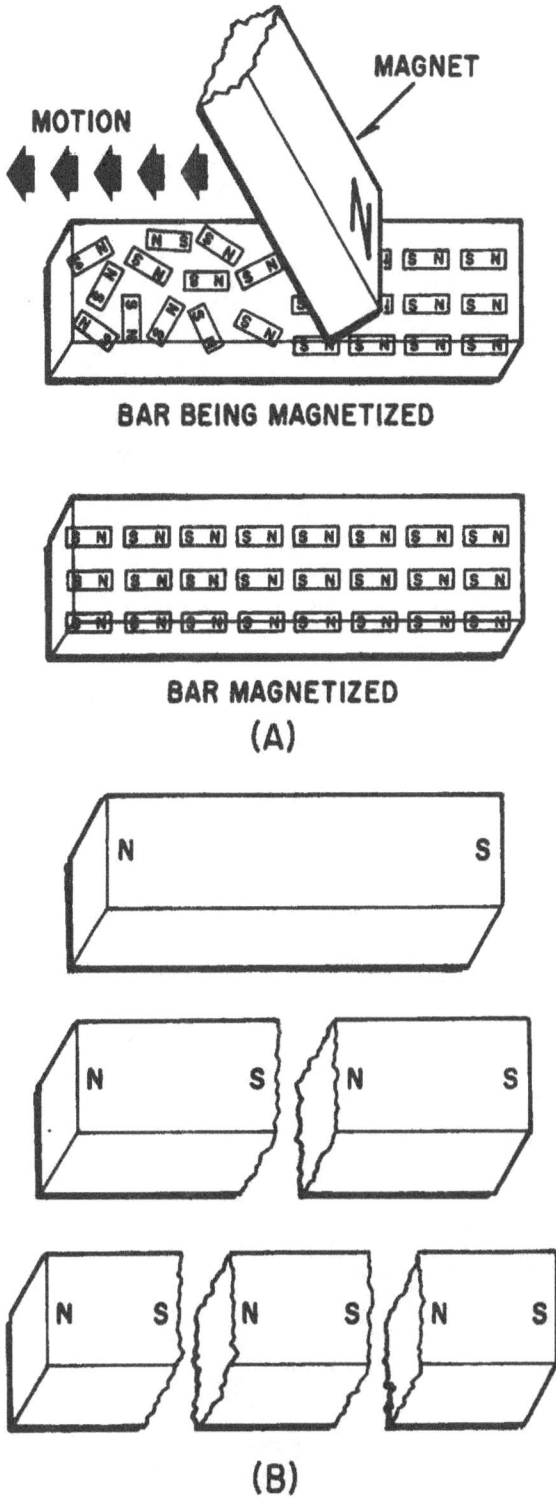

BAR BEING MAGNETIZED

BAR MAGNETIZED

(A)

(B)

BE.10

Figure 2-9.—(A) Molecular magnets;
(B) broken magnets.

## DOMAIN THEORY

A more modern theory of magnetism is based on the electron spin principle. From the study of atomic structure it is known that all matter is composed of vast quantities of atoms, each atom containing one or more orbital electrons. The electrons are considered to orbit in various shells and subshells depending upon their distance from the nucleus. The structure of the atom has previously been compared to the solar system, wherein the electrons orbiting the nucleus correspond to the planets orbiting the sun. Along with their orbital motion about the sun, these planets also revolve on their axes. It is believed that the electron also revolves on its axis as it orbits the nucleus of an atom.

It has been experimentally proven that an electron has a magnetic field about it along with an electric field. The effectiveness of the magnetic field of an atom is determined by the number of electrons spinning in each direction. If an atom has equal numbers of electrons spinning in opposite directions, the magnetic fields surrounding the electrons cancel one another, and the atom is unmagnetized. However, if more electrons spin in one direction than another, the atom is magnetized. An atom such as iron with an atomic number of 26 has 26 protons in the nucleus and 26 revolving electrons orbiting its nucleus. If 13 electrons are spinning in a clockwise direction and 13 electrons are spinning in a counterclockwise direction, the opposing magnetic fields will be neutralized. When more than 13 electrons spin in either direction, the atom is magnetized. An example of a magnetized atom of iron is shown in figure 2-10. Note that in this specific illustration the electrons magnetic fields in all except the M shell neutralize each other. As illustrated in the diagram, there exists 15 electrons spinning in one direction and only 11 electrons spinning in an opposite direction. Therefore, the unopposed magnetic fields of 4 electrons will cause this iron atom to become an infinitely small magnet.

When a number of such atoms are grouped together to form an iron bar, there is an interaction between the magnetic forces of various atoms. The small magnetic force of the field surrounding an atom affects adjacent atoms, thus producing a small group of atoms with parallel magnetic fields. This group of from $10^{14}$ to $10^{15}$ magnetic atoms, having their magnetic poles oriented in the same direction, is known as a DOMAIN. Throughout a domain

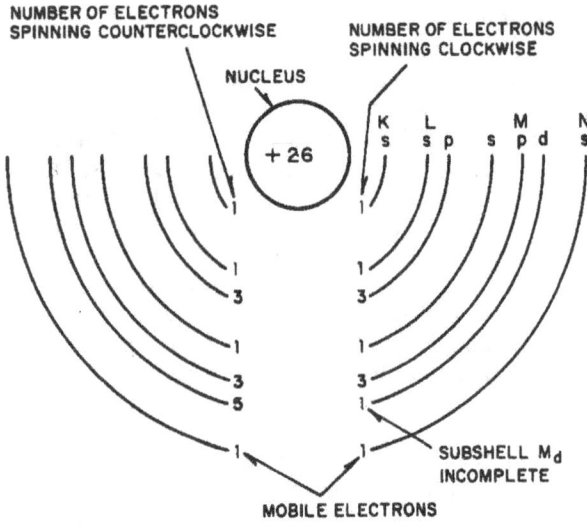

NUMBER OF ELECTRONS SPINNING COUNTERCLOCKWISE

NUMBER OF ELECTRONS SPINNING CLOCKWISE

NUCLEUS

+26

SUBSHELL $M_d$ INCOMPLETE

MOBILE ELECTRONS

BE.11

Figure 2-10.—Iron atom.

there is an intense magnetic field without the influence of any external magnetic field. Since about 10 million tiny domains can be contained in 1 cubic millimeter, it is apparent that every magnetic material is made up of a large number of domains. The domains in any substance are always magnetized to saturation but are randomly orientated throughout a material. Thus, the strong magnetic field of each domain is neutralized by opposing magnetic forces of other domains. When an external field is applied to a magnetic substance the domains will line up with the external field. Since the domains themselves are naturally magnetized to saturation, the magnetic strength of a magnetized material is determined by the number of domains alined by the magnetizing force.

## MAGNETIC FIELDS AND LINES OF FORCE

If a bar magnet is dipped into iron filings, many of the filings are attracted to the ends of the magnet, but none are attracted to the center of the magnet. As mentioned previously, the ends of the magnet where the attractive force is the greatest are called the POLES of the magnet. By using a compass, the line of direction of the magnetic force at various points near the magnet may be observed. The compass needle itself is a magnet. The north end of the compass needle always points toward the south pole, S, as shown in figure 2-11 (A), and thus the sense of direction (with respect to the polarity of the bar magnet) is also indicated. At the center, the compass

needle points in a direction that is parallel to the bar magnet.

When the compass is placed successively at several points in the vicinity of the bar magnet the compass needle alines itself with the field at each position. The direction of the field is indicated by the arrows and represents the direction in which the north pole of the compass needle will point when the compass is placed in this field. Such a line along which a compass needle alines itself is called a MAGNETIC LINE OF FORCE. (This magnetic line of force does not actually exist but is an imaginary line used to illustrate and describe the pattern of the magnetic field. As mentioned previously, the magnetic lines of force are assumed to emanate from the north pole of a magnet, pass through the surrounding space, and enter the south pole. The lines of force then pass from the south pole to the north pole inside the magnet to form a closed loop. Each line of force forms an independent closed loop and does not merge with or cross other lines of force. The lines of force between the poles of a horseshoe magnet are shown in figure 2-11 (B).

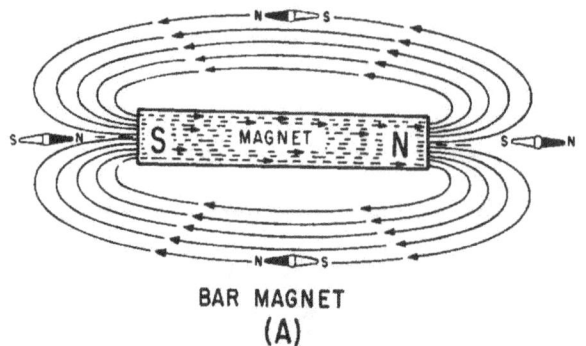

BAR MAGNET
(A)

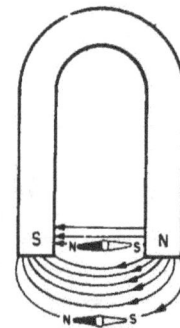

HORSESHOE MAGNET
(B)

BE.12

Figure 2-11.—Magnetic lines of force.

21

Although magnetic lines of force are imaginary, a simplified version of many magnetic phenomena can be explained by assuming the magnetic lines to have certain real properties. The lines of force can be compared to rubberbands which stretch outward when a force is exerted upon them and contract when the force is removed. The characteristics of magnetic lines of force can be described as follows:

1. Magnetic lines of force are continuous and will always form closed loops.

2. Magnetic lines of force will never cross one another.

3. Parallel magnetic lines of force traveling in the same direction repel one another. Parallel magnetic lines of force traveling in opposite directions tend to unite with each other and form into single lines traveling in a direction determined by the magnetic poles creating the lines of force.

4. Magnetic lines of force tend to shorten themselves. Therefore, the magnetic lines of force existing between two unlike poles cause the poles to be pulled together.

5. Magnetic lines of force pass through all materials, both magnetic and nonmagnetic.

The space surrounding a magnet, in which the magnetic force acts, is called a MAGNETIC FIELD. Michael Faraday was the first scientist to visualize the magnet field as being in a state of stress and consisting of uniformly distributed lines of force. The entire quantity of magnetic lines surrounding a magnet is called MAGNETIC FLUX. Flux in a magnetic circuit corresponds to current in an electric circuit.

The number of lines of force per unit area is called FLUX DENSITY and is measured in lines per square inch or lines per square centimeter. Flux density is expressed by the equation

$$B = \frac{\Phi}{A},$$

where B is the flux density, $\Phi$ (Greek phi) is the total number of lines of flux, and A is the cross-sectional area of the magnetic circuit. If A is in square centimeters, B is in lines per square centimeter, or GAUSS. The terms FLUX and FLOW of magnetism are frequently used in textbooks. However, magnetism itself is not thought to be a stream of particles in motion, but is simply a field of force exerted in space.

A visual representation of the magnetic field around a magnet can be obtained by placing a plate of glass over a magnet and sprinkling iron filings onto the glass. The filings arrange themselves in definite paths between the poles. This arrangement of the filings shows the pattern of the magnetic field around the magnet, as in figure 2-12.

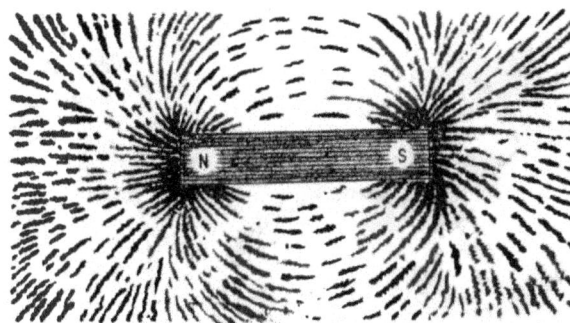

BE.13

Figure 2-12.—Magnetic field pattern around a magnet.

The magnetic field surrounding a symmetrically shaped magnet has the following properties:

1. The field is symmetrical unless disturbed by another magnetic substance.

2. The lines of force have direction and are represented as emanating from the north pole and entering the south pole.

## LAWS OF ATTRACTION AND REPULSION

If a magnetized needle is suspended near a bar magnet, as in figure 2-13, it will be seen that a north pole repels a north pole and a south pole repels a south pole. Opposite poles, however, will attract each other. Thus, the first two laws of magnetic attraction and repulsion are:

1. LIKE magnetic poles REPEL each other.

2. UNLIKE magnetic poles ATTRACT each other.

The flux patterns between adjacent UNLIKE poles of bar magnets, as indicated by lines, are shown in figure 2-14 (A). Similar patterns for adjacent LIKE poles are shown in figure 2-14 (B). The lines do not cross at any point and they act as if they repel each other.

Figure 2-15 shows the flux pattern (indicated by lines) around two bar magnets placed close together and parallel with each other. Figure 2-15 (A) shows the flux pattern when opposite

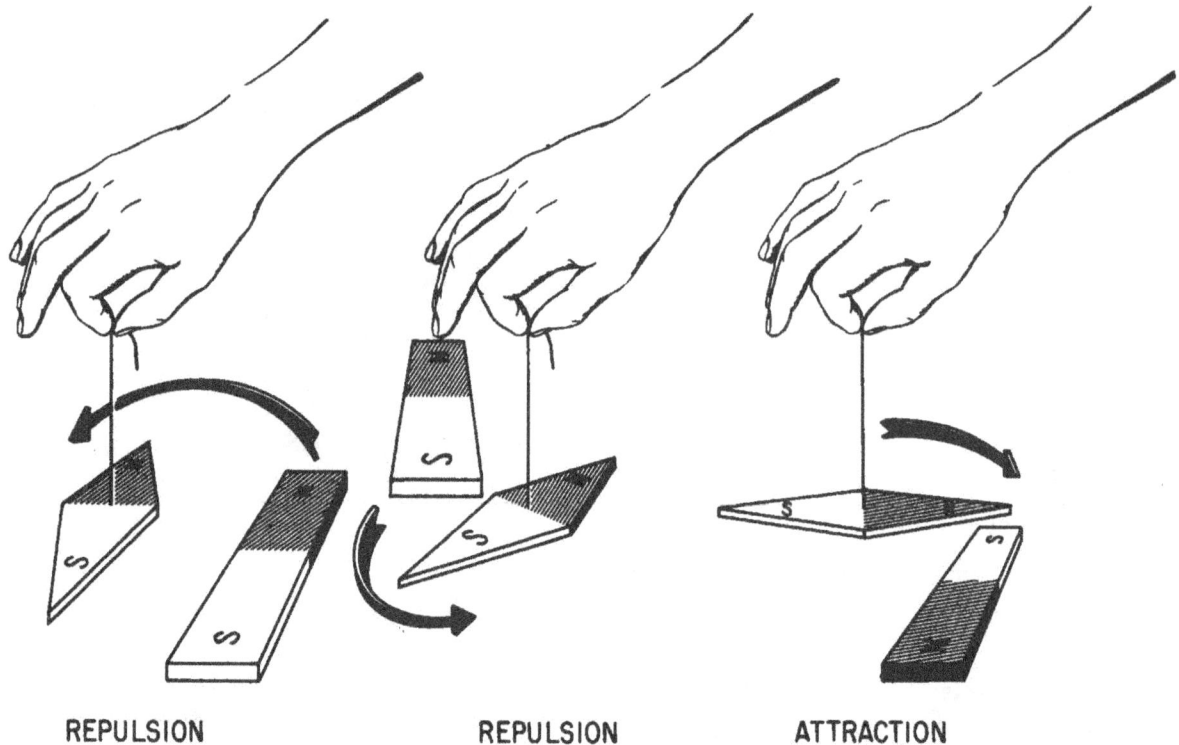

REPULSION          REPULSION          ATTRACTION

BE.14

Figure 2-13.—Laws of attraction and repulsion.

poles are adjacent; and figure 2-15 (B) shows the flux pattern when like poles are adjacent.

The THIRD LAW of magnetic attraction and repulsion states in effect that the force of attraction or repulsion existing between two magnetic poles decreases rapidly as the poles are separated from each other. Actually, the force of attraction or repulsion varies directly as the product of the separate pole strengths and inversely as the square of the distance separating the magnetic poles, provided the poles are small enough to be considered as points. For example, if the distance between two north poles is increased from 2 feet to 4 feet, the force of repulsion between them is decreased to one-fourth of its original value. If either pole strength is doubled, the distance remaining the same, the force between the poles will be doubled.

THE EARTH'S MAGNETISM

As has been stated, the earth is a huge magnet; and surrounding the earth is the magnetic field produced by the earth's magnetism. The magnetic polarities of the earth are as indicated in figure 2-16. The geographic poles are also shown at each end of the axis of rotation of the earth. The magnetic axis does not coincide with the geographic axis, and therefore the magnetic and geographic poles are not at the same place on the surface of the earth.

The early users of the compass regarded the end of the compass needle that points in a northerly direction as being a north pole. The other end was regarded as a south pole. On some maps the magnetic pole of the earth towards which the north pole of the compass pointed was designated a north magnetic pole. This magnetic pole was obviously called a north pole because of its proximity to the north geographic pole.

When it was learned that the earth is a magnet and that opposite poles attract, it was necessary to call the magnetic pole located in the northern hemisphere a SOUTH MAGNETIC POLE and the magnetic pole located in the southern hemisphere a NORTH MAGNETIC POLE. The matter of naming the poles was arbitrary. Therefore, the polarity of the compass needle that points toward the north must be opposite to the polarity of the earth's magnetic pole located there.

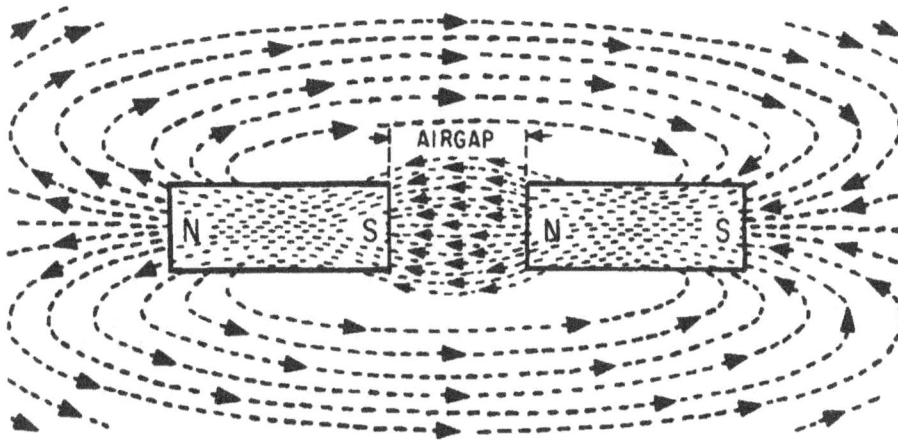

UNLIKE POLES ATTRACT

(A)

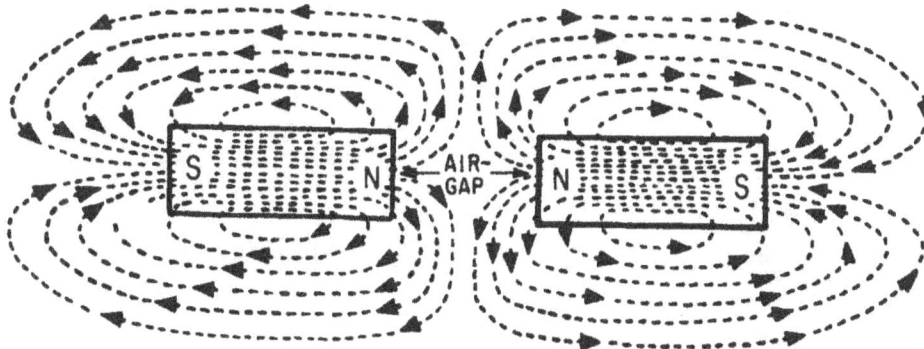

LINES OF FORCE

LIKE POLES REPEL

(B)

BE.15

Figure 2-14.—Lines of force between unlike and like poles.

As has been stated, magnetic lines of force are assumed to emanate from the north pole of a magnet and to enter the south pole as closed loops. Because the earth is a magnet, lines of force emanate from its north magnetic pole and enter the south magnetic pole as closed loops. The compass needle alines itself in such a way that the earth's lines of force enter at its south pole and leave at its north pole. Because the north pole of the needle is defined as the end that points in a northerly direction it follows that the magnetic pole in the vicinity of the north geographic pole is in reality a south magnetic pole, and vice versa.

Because the magnetic poles and the geographic poles do not coincide, a compass will not (except at certain positions on the earth) point in a true (geographic) north-south direction—that is, it will not point in a line of direction that passes through the north and south

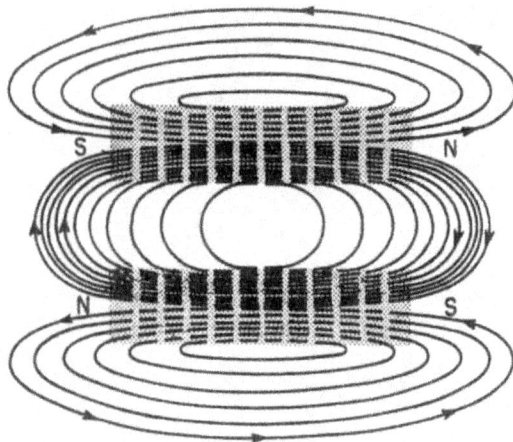

FLUX PATTERN—ATTRACTION

(A)

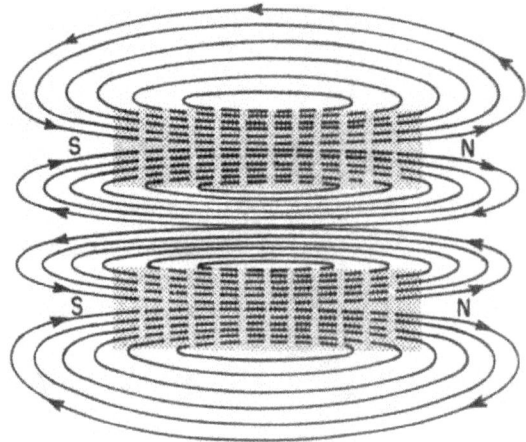

FLUX PATTERN—REPULSION

(B)

BE.16

Figure 2-15.—Flux patterns of adjacent parallel bar magnets.

geographic poles, but in a line of direction that makes an angle with it. This angle is called the angle of VARIATION OR DECLINATION.

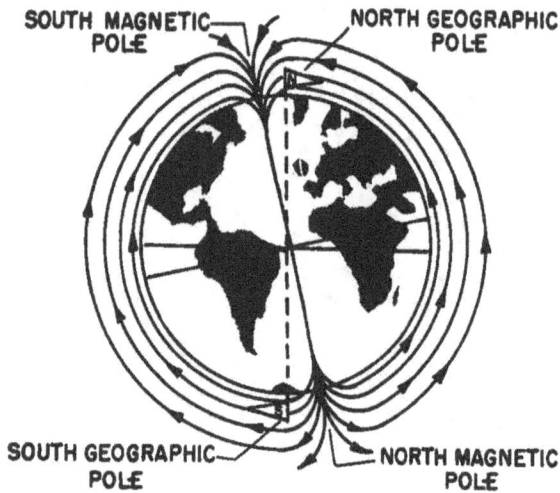

BE.17

Figure 2-16.—Earth's magnetic poles.

MAGNETIC SHIELDING

There is not a known INSULATOR for magnetic flux. If a nonmagnetic material is placed in a magnetic field, there is no appreciable change in flux—that is, the flux penetrates the nonmagnetic material. For example, a glass plate placed between the poles of a horseshoe magnet will have no appreciable effect on the field although glass itself is a good insulator in an electric circuit. If a magnetic material (for example, soft iron) is placed in a magnetic field, the flux may be redirected to take advantage of the greater permeability of the magnetic material as shown in figure 2-17. Permeability, as discussed earlier, is the quality of a substance which determines the ease with which it can be magnetized.

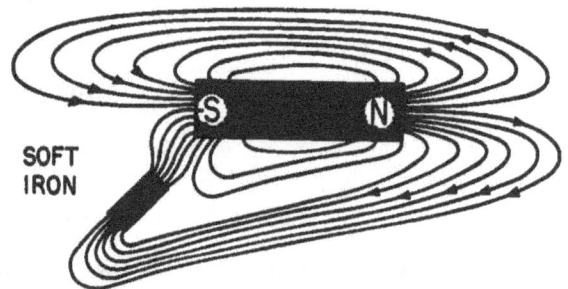

BE.18

Figure 2-17.—Effects of a magnetic substance in a magnetic field.

The sensitive mechanism of electric instruments and meters can be influenced by stray magnetic fields which will cause errors in their readings. Because instrument mechanisms

25

cannot be insulated against magnetic flux, it is necessary to employ some means of directing the flux around the instrument. This is accomplished by placing a soft-iron case, called a MAGNETIC SCREEN OR SHIELD, about the instrument. Because the flux is established more readily through the iron (even though the path is longer) than through the air inside the case, the instrument is effectively shielded, as shown by the watch and soft-iron shield in figure 2-18.

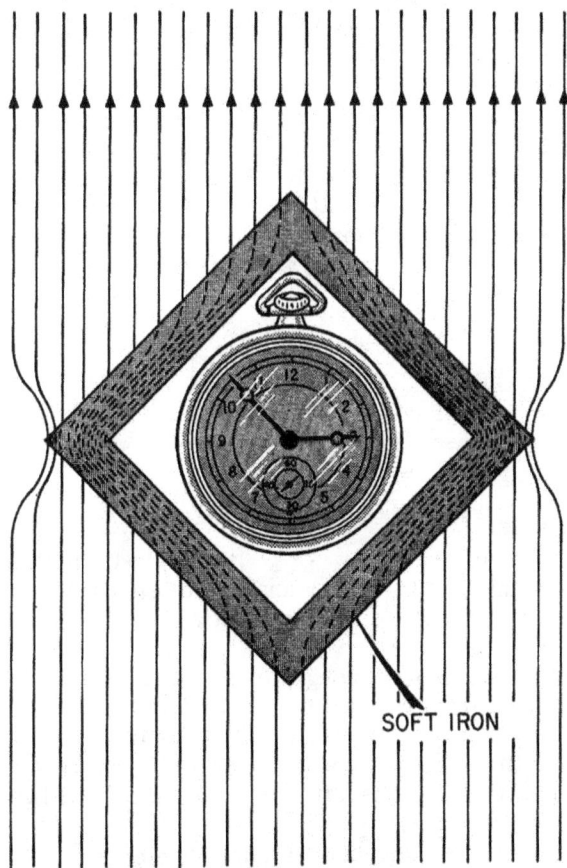

SOFT IRON

BE.19

Figure 2-18.—Magnetic shield.

## MAGNETIC MATERIALS

Early magnetic studies classified materials merely as being magnetic and nonmagnetic. Present studies classify materials into one of three groups; namely, paramagnetic, diamagnetic, and ferromagnetic.

PARAMAGNETIC materials are those that become only slightly magnetized even though under the influence of a strong magnetic field.

This slight magnetization is in the same direction as the magnetizing field. Materials of this type are aluminum, chromium, platinum, and air.

DIAMAGNETIC materials can also be only slightly magnetized when under the influence of a very strong field. These materials, when slightly magnetized, are magnetized in a direction opposite to the external field. Some diamagnetic materials are copper, silver, gold, and mercury.

Paramagnetic and diamagnetic materials have a very low permeability. Paramagnetic materials have a permeability slightly greater than one; diamagnetic materials have a permeability less than one. Because of the difficulty in obtaining some magnetization of paramagnetic and diamagnetic materials, these materials are considered for all practical purposes as nonmagnetic materials.

The most important group of materials for applications of electricity and electronics are the FERROMAGNETIC MATERIALS. Ferromagnetic materials are those which are relatively easy to magnetize such as iron, steel, cobalt, Alnico, and Permalloy, the latter two being alloys. Alnico consists primarily of aluminum, nickel, and cobalt. These new alloys can be very strongly magnetized with Alnico capable of obtaining a magnetic strength great enough to lift five hundred times its own weight.

Ferromagnetic materials all have a high permeability. However, as previously discussed, a material, such as steel used to make a permanent magnet, is considered to have a relatively low permeability in comparison to other ferromagnetic materials.

## MAGNETIC SHAPES

Because of the many uses of magnets, they are found in various shapes and sizes. However, magnets usually come under three general classifications; namely, bar magnets, horseshoe magnets, and ring magnets.

The bar magnet is most often used in schools and laboratories for studying the properties and effects of magnetism. In the preceding test material, the bar magnet proved very helpful in demonstrating magnetic effects.

Another type of magnet is the ring magnet used for computer memory cores. A common application for a temporary ring magnet would be the shielding of electrical instruments as previously discussed.

The shape of the magnet most frequently used in electrical or electronic equipment is called the horseshoe magnet. A horseshoe magnet is similar to a bar magnet but is bent in the shape of a horseshoe. The horseshoe magnet provides much more magnetic strength than a bar magnet of the same size and material because of the closeness of the magnetic poles. The magnetic strength from one pole to the other is greatly increased due to the concentration of the magnetic field in a smaller area. Electrical measuring devices quite frequently use horseshoe type magnets.

## CARE OF MAGNETS

A piece of steel that has been magnetized can lose much of its magnetism by improper handling. If it is jarred or heated, there will be a disalinement of its domains resulting in the loss of some of its effective magnetism. Had this steel formed the horseshoe magnet of a meter, the meter would no longer be operable or would give inaccurate readings. Therefore, care must be exercised when handling instruments containing magnets. Severe jarring or subjecting the instrument to high temperature will damage the device.

A magnet may also become weakened from loss of flux. Thus, when storing magnets one should always try to avoid excess leakage of magnetic flux. A horseshoe magnet should always be stored with a keeper, a soft iron bar used to join the magnetic poles. By use of the keeper while the magnet is being stored, the magnetic flux will continuously circulate through the magnet and not leak off into space.

When storing bar magnets, the same principle must be remembered. Therefore, bar magnets should always be stored in pairs with a north pole and a south pole placed together. This provides a complete path for the magnetic flux without any flux leakage.

The study of electricity and magnetism and how they interact with each other is given more thorough coverage in later chapters in this manual. The discussion of magnetism up to this point has been mainly intended to clarify terms and meanings, such as "polarity," "fields," "lines of force," etc. Only one fundamental relationship between magnetism and electricity is discussed in this chapter. This relationship pertains to magnetism as used to generate a voltage and it is discussed in the material that follows.

### DIFFERENCE IN POTENTIAL

The force that causes free electrons to move in a conductor as an electric current may be referred to as follows:

1. Electromotive force (emf).
2. Voltage.
3. Difference in potential.

When a difference in potential exists between two charged bodies that are connected by a conductor, electrons will flow along the conductor. This flow is from the negatively charged body to the positively charged body until the two charges are equalized and the potential difference no longer exists.

An analogy of this action is shown in the two water tanks connected by a pipe and valve in figure 2-19. At first the valve is closed and all the water is in tank A. Thus, the water pressure across the valve is at maximum. When the valve is opened, the water flows through the pipe from A to B until the water level becomes the same in both tanks. The water then stops flowing in the pipe, because there is no longer a difference in water pressure between the two tanks.

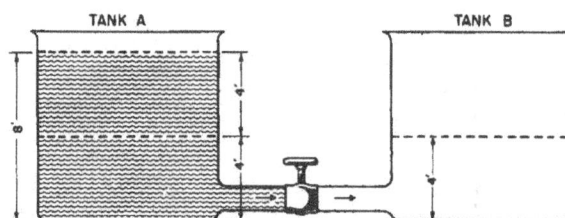

BE.20

Figure 2-19.—Water analogy of electric difference in potential.

Current flow through an electric circuit is directly proportional to the difference in potential across the circuit, just as the flow of water through the pipe in figure 2-19 is directly proportional to the difference in water level in the two tanks.

A fundamental law of current electricity is that the CURRENT IS DIRECTLY PROPORTIONAL TO THE APPLIED VOLTAGE; that is, if the voltage is increased, the current is increased. If the voltage is decreased, the current is decreased.

### PRIMARY METHODS OF PRODUCING A VOLTAGE

Presently, there are six commonly used methods of producing electromotive force (emf). Some of these methods are much more widely used than others. The following is a list of the six most common methods of producing electromotive force.

1. FRICTION.—Voltage produced by rubbing two materials together.

2. PRESSURE (Piezoelectricity).—Voltage produced by squeezing crystals of certain substances.

3. HEAT (Thermoelectricity).—Voltage produced by heating the joint (junction) where two unlike metals are joined.

4. LIGHT (Photoelectricity).—Voltage produced by light striking photosensitive (light sensitive) substances.

5. CHEMICAL ACTION.—Voltage produced by chemical reaction in a battery cell.

6. MAGNETISM.—Voltage produced in a conductor when the conductor moves through a magnetic field, or a magnetic field moves through the conductor in such a manner as to cut the magnetic lines of force of the field.

## VOLTAGE PRODUCED BY FRICTION

This is the least used of the six methods of producing voltages. Its main application is in Van de Graf generators, used by some laboratories to produce high voltages. As a rule, friction electricity (often referred to as static electricity) is a nuisance. For instance, a flying aircraft accumulates electric charges from the friction between its skin and the passing air. These charges often interfere with radio communication, and under some circumstances can even cause physical damage to the aircraft. Most individuals are familiar with static electricity and have probably received unpleasant shocks from friction electricity upon sliding across dry seat covers or walking across dry carpets, and then coming in contact with some other object.

## VOLTAGE PRODUCED BY PRESSURE

This action is referred to as piezoelectricity. It is produced by compressing or decompressing crystals of certain substances. To study this form of electricity, the meaning of the word "crystal" must first be understood. In a crystal, the molecules are arranged in an orderly and uniform manner. A substance in its crystallized state and its noncrystallized state is shown in figure 2-20.

For the sake of simplicity, assume that the molecules of this particular substance are spherical (ball-shaped). In the noncrystallized state, in (A), note that the molecules are arranged irregularly. In the crystallized state, (B), the molecules are arranged in a regular and uniform manner. This illustrates the major physical difference between crystal and noncrystal forms of matter. Natural crystalline matter is rare; an example of matter that is crystalline in its natural form is diamond, which is crystalline carbon. Most crystals are manufactured.

Crystals of certain substances, such as Rochelle salt or quartz, exhibit peculiar electrical characteristics. These characteristics, or effects, are referred to as "piezoelectric." For instance, when a crystal of quartz is

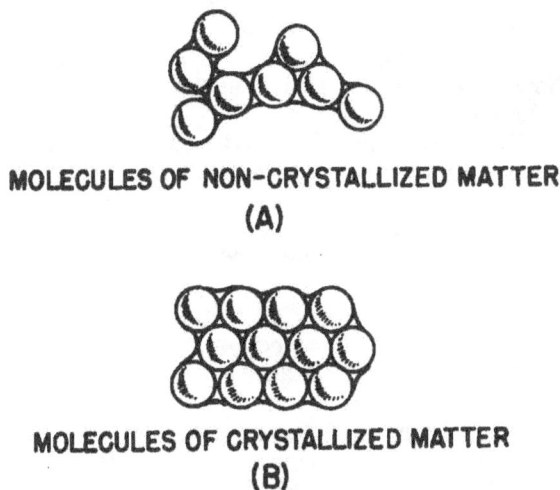

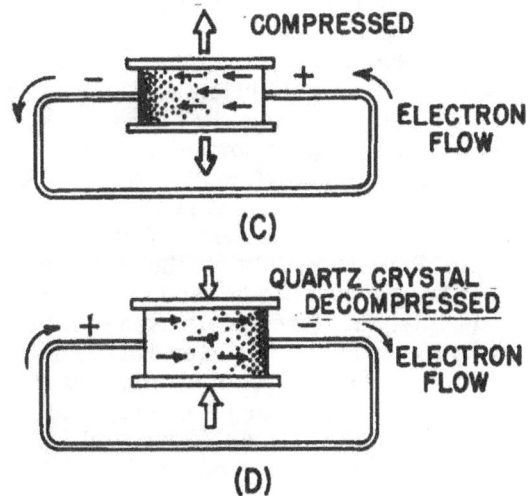

MOLECULES OF NON-CRYSTALLIZED MATTER
(A)

MOLECULES OF CRYSTALLIZED MATTER
(B)

COMPRESSED

ELECTRON FLOW
(C)

QUARTZ CRYSTAL DECOMPRESSED

ELECTRON FLOW
(D)

BE.21

Figure 2-20.—(A) Noncrystallized structure; (B) crystallized structure; (C) compression of a crystal; (D) decompression of a crystal.

compressed, as in figure 2-20 (C), electrons tend to move through the crystal as shown. This tendency creates an electric difference of potential between the two opposite faces of the crystal. (The fundamental reasons for this action are not known. However, the action is predictable, and therefore useful.) If an external wire is connected while the pressure and emf are present, electrons will flow. If the pressure is held constant, the electron flow will continue until the charges are equalized. When the force is removed, the crystal is decompressed, and immediately causes an electric force in the opposite direction (D). Thus, the crystal is able to convert mechanical force, either pressure or tension, to electrical force.

The power capacity of a crystal is extremely small. However, they are useful because of their extreme sensitivity to changes of mechanical force or changes in temperature. Due to other characteristics not mentioned here, crystals are most widely used in communication equipment.

## VOLTAGE PRODUCED BY HEAT

When a length of metal, such as copper, is heated at one end, electrons tend to move away from the hot end toward the cooler end. This is true of most metals. However, in some metals, such as iron, the opposite takes place and electrons tend to move TOWARD the hot end. These characteristics are illustrated in figure 2-21. The negative charges (electrons) are moving through the copper away from the heat and through the iron toward the heat. They cross from the iron to the copper at the hot junction, and from the copper through the current meter to the iron at the cold junction. This device is generally referred to as a thermocouple.

Thermocouples have somewhat greater power capacities than crystals, but their capacity is still very small if compared to some other sources. The thermoelectric voltage in a thermocouple depends mainly on the difference in temperature between the hot and cold junctions. Consequently, they are widely used to measure temperature, and as heat-sensing devices in automatic temperature control equipment. Thermocouples generally can be subjected to much greater temperatures than ordinary thermometers, such as the mercury or alcohol types.

## VOLTAGE PRODUCED BY LIGHT

When light strikes the surface of a substance, it may dislodge electrons from their orbits around the surface atoms of the substance. This occurs because light has energy, the same as any moving force.

Some substances, mostly metallic ones, are far more sensitive to light than others. That is, more electrons will be dislodged and emitted from the surface of a highly sensitive metal, with a given amount of light, than will be emitted from a less sensitive substance. Upon losing electrons, the photosensitive (light sensitive) metal becomes positively charged, and an electric force is created. Voltage produced in this manner is referred to as "a photoelectric voltage."

The photosensitive materials most commonly used to produce a photoelectric voltage are various compounds of silver oxide or copper oxide. A complete device which operates on the photoelectric principle is referred to as a "photoelectric cell." There are many sizes and types of photoelectric cells in use, each of which serves the special purpose for which it was designed. Nearly all, however, have some of the basic features of the photoelectric cells shown in figure 2-22.

The cell (fig. 2-22 (A)) has a curved light-sensitive surface focused on the central anode. When light from the direction shown strikes the sensitive surface, it emits electrons toward the anode. The more intense the light, the greater is the number of electrons emitted. When a wire is connected between the filament and the back, or dark side, the accumulated electrons will flow to the dark side. These electrons will eventually pass through the metal of the reflector and replace the electrons leaving the light-sensitive surface. Thus, light energy is converted to a flow of electrons, and a usable current is developed.

The cell (fig. 2-22 (B)) is constructed in layers. A base plate of pure copper is coated with light-sensitive copper oxide. An additional semitransparent layer of metal is placed over the copper oxide. This additional layer serves two purposes:

1. It is EXTREMELY thin to permit the penetration of light to the copper oxide.

2. It also accumulates the electrons emitted by the copper oxide.

An externally connected wire completes the electron path, the same as in the reflector type cell. The photocell's voltage is utilized as needed by connecting the external wires to some other device, which amplifies (enlarges) it to a usable level.

BE.22

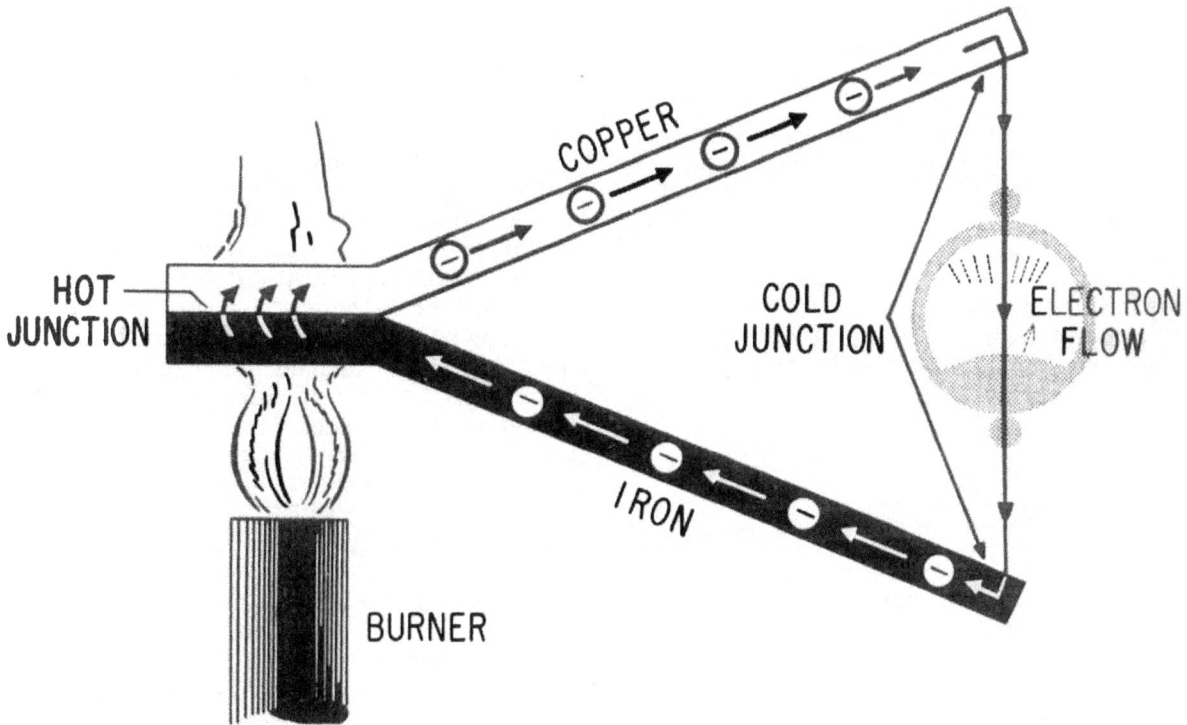

Figure 2-21.—Voltage produced by heat.

Figure 2-22.—Voltage produced by light.

BE.23

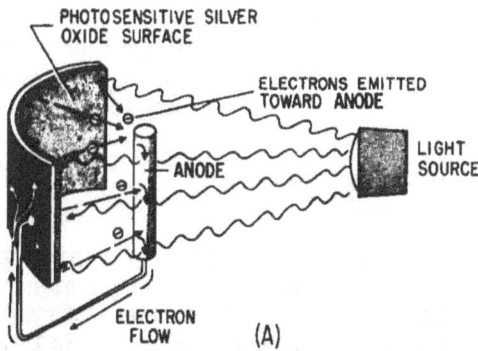

A photocell's power capacity is very small. However, it reacts to light-intensity variations in an extremely short time. This characteristic makes the photocell very useful in detecting or accurately controlling a great number of processes or operations. For instance, the photoelectric cell, or some form of the photoelectric principle, is used in television cameras, automatic manufacturing process controls, door openers, burglar alarms, and so forth.

## VOLTAGE PRODUCED BY CHEMICAL ACTION

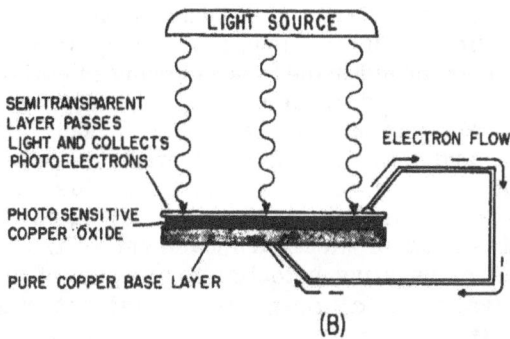

Up to this point, it has been shown that electrons may be removed from their parent atoms and set in motion by energy derived from a source of friction, pressure, heat, or light. In general, these forms of energy do not alter the molecules of the substances being acted upon. That is, molecules are not usually added, taken away, or split-up when subjected to these four forms of energy. Only electrons are involved.

When the molecules of a substance are altered, the action is referred to as CHEMICAL. For instance, if the molecules of a substance combines with atoms of another substance, or

30

gives up atoms of its own, the action is chemical in nature. Such action always changes the chemical name and characteristics of the substance affected. For instance, when atoms of oxygen from the air come in contact with bare iron, they merge with the molecules of iron. This iron is "oxidized." It has changed chemically from iron to iron oxide, or "rust." Its molecules have been altered by chemical action.

In some cases, when atoms are added to or taken away from the molecules of a substance, the chemical change will cause the substance to take on an electric charge. The process of producing a voltage by chemical action is used in batteries and is explained in chapter 3.

## VOLTAGE PRODUCED BY MAGNETISM

Magnets or magnetic devices are used for thousands of different jobs. One of the most useful and widely employed applications of magnets is in the production of vast quantities of electric power from mechanical sources. The mechanical power may be provided by a number of different sources, such as gasoline or diesel engines, and water or steam turbines. However, the final conversion of these source energies to electricity is done by generators employing the principle of electromagnetic induction. These generators, of many types and sizes, are discussed in later chapters of this manual. The important subject to be discussed here is the fundamental operating principle of ALL such electromagnetic-induction generators.

To begin with, there are three fundamental conditions which must exist before a voltage can be produced by magnetism. They are as follows:

1. There must be a CONDUCTOR, in which the voltage will be produced.

2. There must be a MAGNETIC FIELD in the conductor's vicinity.

3. There must be relative motion between the field and the conductor. The conductor must be moved so as to cut across the magnetic lines of force, or the field must be moved so that the lines of force are cut by the conductor.

In accordance with these conditions, when a conductor or conductors MOVE ACROSS a magnetic field so as to cut the lines of force, electrons WITHIN THE CONDUCTOR are impelled in one direction or another. Thus, an electric force, or voltage, is created.

In figure 2-23, note the presence of the three conditions needed for creating an induced voltage:

1. A magnetic field exists between the poles of the C-shaped magnet.

2. There is a conductor (copper wire).

3. There is relative motion. The wire is moved back and forth ACROSS the magnetic field.

In figure 2-23 (A), the conductor is moving TOWARD the front of the page. This occurs because of the magnetically induced emf acting on the electrons in the copper. The right-hand end becomes negative, and the left-hand end positive. The conductor is stopped (B), motion is eliminated (one of the three required conditions), and there is no longer an induced emf. Consequently, there is no longer any difference in potential between the two ends of the wire. The conductor at (C) is moving away from the front of the page. An induced emf is again created. However, note carefully that the REVERSAL OF MOTION has caused a REVERSAL OF DIRECTION in the induced emf.

If a path for electron flow is provided between the ends of the conductor, electrons will leave the negative end and flow to the positive end. This condition is shown in part (D). Electron flow will continue as long as the emf exists. In studying figure 2-23, it should be noted that the induced emf could also have been created by holding the conductor stationary and moving the magnetic field back and forth.

The more complex aspects of power generation by use of mechanical motion and magnetism are discussed in later chapters of this manual under the heading "Generators."

## ELECTRIC CURRENT

The drift or flow of electrons through a conductor is called electric current or electron flow. During the early study of electricity, electric current was erroneously assumed to be a movement of electrons from a positive potential to a negative potential. This assumption, termed conventional current flow, is a concept that became entrenched in the minds of many scientists, technicians, and writers. Consequently, conventional current flow is indicated in many textbooks, and this concept of electron movement should be realized. However, since this early concept, it has been positively determined that the direction of electron movement is from a region of negative potential to a region of less negative potential or more positive potential. Various terms may be used in this manual and other textbooks to describe current flow. The terms current, current flow, electron flow,

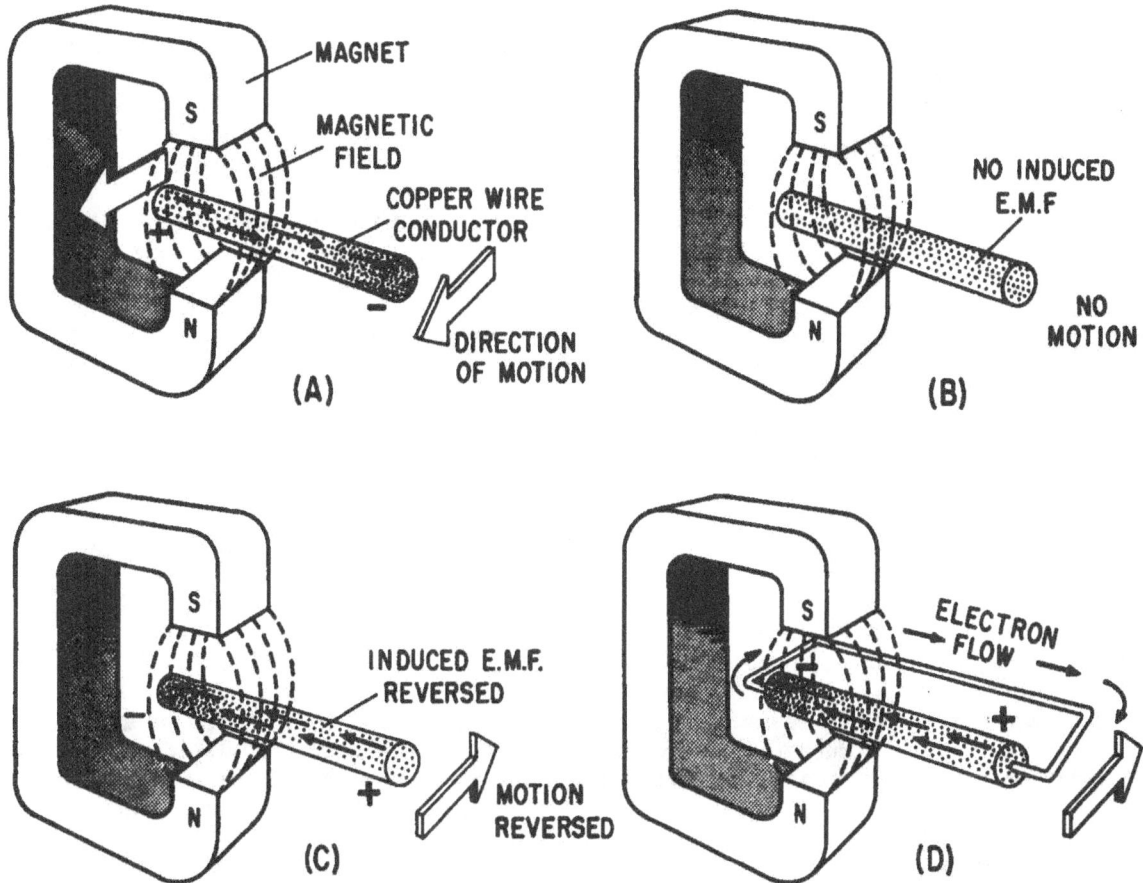

Figure 2-23.—Voltage produced by magnetism.

BE.24

electron current, etc., may be used to describe the same phenomenon; however, the reader should realize that regardless of the term used, the movement of electrons will be from a negative potential to a positive potential.

Electric current is generally classified into two general types—direct current and alternating current. Direct current flows in the same direction whereas an alternating current periodically reverses direction. These two types of current are discussed in greater detail later in this manual.

In order to determine the amount (number) of electrons flowing in a given conductor, it is necessary to adopt a unit of measurement of current flow. The term AMPERE is used to define the unit of measurement of the rate at which current flows (electron flow). The symbol for current flow is I. Current flow is measured in amperes. The abbreviation for ampere

is amp. One ampere may be defined as the flow of $6.28 \times 10^{18}$ electrons per second past a fixed point in a conductor.

A unit quantity of electricity is moved through an electric circuit when 1 ampere of current flows for 1 second of time. This unit is equivalent to $6.28 \times 10^{18}$ electrons, and is called the COULOMB. The coulomb is to electricity as the gallon is to water. The symbol for the coulomb is Q. The rate of flow of current in amperes and the quantity of electricity moved through a circuit are related by the common factor of time. Thus, the quantity of electric charge, in coulombs, moved through a circuit is equal to the product of the current in amperes, I, and the duration of flow in seconds, t. Expressed as an equation, Q = It.

For example, if a current of 2 amperes flows through a circuit for 10 seconds the quantity of electricity moved through the circuit is 2 x 10, or 20 coulombs. Conversely, current flow may be expressed in terms of coulombs and time in seconds. Thus, if 20 coulombs are moved through a circuit in 10 seconds, the average current flow is $\frac{20}{10}$, or 2 amperes. Note that the current flow in amperes implies the rate of flow of coulombs per second without indicating either coulombs or seconds. Thus a current flow of 2 amperes is equivalent to a rate of flow of 2 coulombs per second. Frequently, the ampere is much too large a unit for practical utilization. Therefore, the milliampere (ma), one-thousandth of an ampere (or the microampere, one-millionth of an ampere), is used. The device used to measure current is called an ammeter and is discussed later in this manual.

## RESISTANCE

Every material offers some resistance, or opposition, to the flow of electric current through it. Good conductors, such as copper, silver, and aluminum, offer very little resistance. Poor conductors, or insulators, such as glass, wood, and paper, offer a high resistance to current flow.

The size and type of material of the wires in an electric circuit are chosen so as to keep the electrical resistance as low as possible. In this way, current can flow easily through the conductors, just as water flows through the pipe between the tanks in figure 2-19. If the water pressure remains constant the flow of water in the pipe will depend on how far the valve is opened. The smaller the opening, the greater the opposition to the flow, and the smaller will be the rate of flow in gallons per second.

In the electric circuit, the larger the diameter of the wires, the lower will be their electrical resistance (opposition) to the flow of current through them. In the water analogy, pipe friction opposes the flow of water between the tanks. This friction is similar to electrical resistance. The resistance of the pipe to the flow of water through it depends upon (1) the length of the pipe, (2) the diameter of the pipe, and (3) the nature of the inside walls (rough or smooth). Similarly, the electrical resistance of the conductors depends upon (1) the length of the wires, (2) the diameter of the wires, and (3) the material of the wires (copper, aluminum, etc.).

Temperature also affects the resistance of electrical conductors to some extent. In most conductors (copper, aluminum, iron, etc.) the resistance increases with temperature. Carbon is an exception. In carbon the resistance decreases as temperature increases. Certain alloys of metals (manganin and constantan) have resistance that does not change appreciably with temperature.

The relative resistance of several conductors of the same length and cross section is given in the following list with silver as a standard of 1 and the remaining metals arranged in an order of ascending resistance:

| | |
|---|---|
| Silver . . . . . . . . . . . . . . | 1.0 |
| Copper . . . . . . . . . . . . . | 1.08 |
| Gold . . . . . . . . . . . . . . . | 1.4 |
| Aluminum . . . . . . . . . . . | 1.8 |
| Platinum . . . . . . . . . . . . | 7.0 |
| Lead . . . . . . . . . . . . . . | 13.5 |

The resistance in an electrical circuit is expressed by the symbol R. Manufactured circuit parts containing definite amounts of resistance are called RESISTORS. Resistance (R) is measured in OHMS. One ohm is the resistance of a circuit element, or circuit, that permits a steady current of 1 ampere (1 coulomb per second) to flow when a steady emf of 1 volt is applied to the circuit.

## CONDUCTANCE

Electricity is a study that is frequently explained in terms of opposites. The term that is exactly the opposite of resistance is conductance. Conductance (G) is the ability of a material to pass electrons. The unit of conductance is the Mho, which is ohm spelled backwards. Whereas the symbol used to represent resistance is the Greek letter omega ($\Omega$), the symbol used to represent conductance is the Greek letter omega upside down ($\mho$). The relationship that exists between resistance and conductance is the reciprocal. A reciprocal of a number is obtained by dividing the number into one. In terms of resistance and conductance:

$$R = \frac{1}{G}$$
$$G = \frac{1}{R}$$

If the resistance of a material is known, dividing its value into one will give its conductance. Similarly, if the conductance is known, dividing its value into one will give its resistance.

# CHAPTER 3

# BATTERIES

Batteries are widely used as sources of direct-current electrical energy in automobiles, boats, aircraft, ships, portable electric/electronic equipment, and lighting equipment. In some instances, they are used as the only source of power; while in others, they are used as a secondary or standby power source.

A battery consists of a number of cells assembled in a common container and connected together to function as a source of electrical power.

## CELL

A cell is a device that transforms chemical energy into electrical energy. The simplest cell, known as either a galvanic or voltaic cell, is shown in figure 3-1. It consists of a piece of carbon (C) and a piece of zinc (Zn) suspended in a jar that contains a solution of water ($H_2O$) and sulfuric acid ($H_2SO_4$).

The cell is the fundamental unit of the battery. A simple cell consists of two strips, or electrodes, placed in a container that holds the electrolyte.

## ELECTRODES

The electrodes are the conductors by which the current leaves or returns to the electrolyte. In the simple cell, they are carbon and zinc strips that are placed in the electrolyte; while in the dry cell (fig. 3-2), they are the carbon rod in the center and the zinc container in which the cell is assembled.

## EI ECTROLYTE

The electrolyte is the solution that acts upon the electrodes which are placed in it. The electrolyte may be a salt, an acid, or an alkaline solution. In the simple galvanic cell and in the automobile storage battery, the electrolyte is in a liquid form; while in the dry cell, the electrolyte is a paste.

## PRIMARY CELL

A primary cell is one in which the chemical action eats away one of the electrodes, usually the negative. When this happens, the electrode must be replaced or the cell must be discarded. In the galvanic type cell, the zinc electrode and the liquid solution are usually replaced when this happens. In the case of the dry cell, it is usually cheaper to buy a new cell. Some primary cells have been developed to the state where they can be recharged.

## SECONDARY CELL

A secondary cell is one in which the electrodes and the electrolyte are altered by the chemical action that takes place when the cell delivers current. These cells may be restored to their original condition by forcing an electric current through them in the opposite direction to that of discharge. The automobile storage battery is a common example of the secondary cell.

## BATTERY

As was previously mentioned, a battery consists of two or more cells placed in a common container. The cells are connected in series, in parallel, or in some combination of series and parallel, depending upon the amount of voltage and current required of the battery. The connection of cells in a battery is discussed in more detail later in this chapter.

## BATTERY CHEMISTRY

If a conductor is connected externally to the electrodes of a cell, electrons will flow under

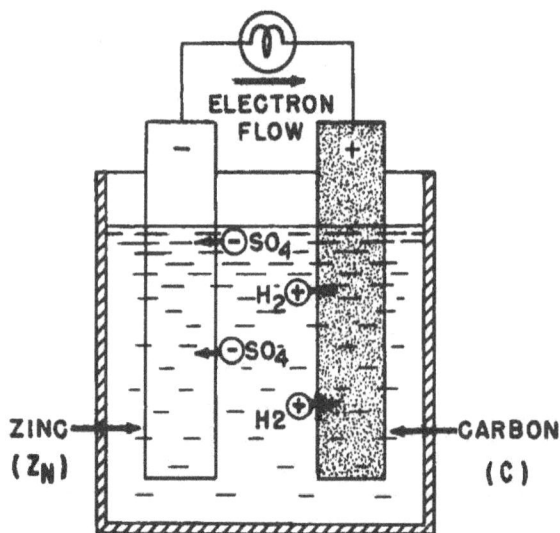

BE.25

Figure 3-1.—Simple voltaic cell.

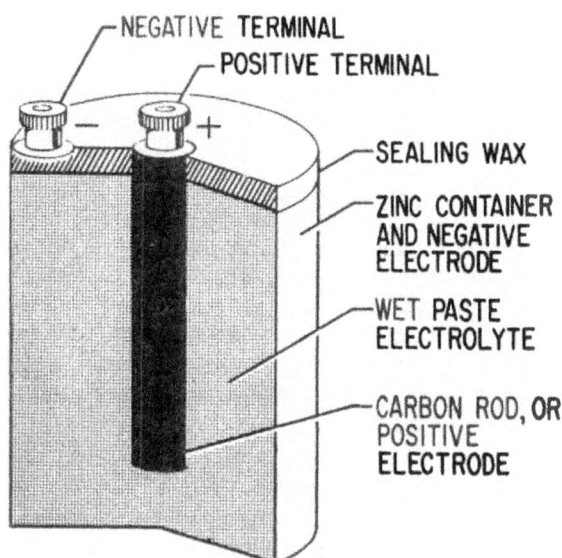

BE.26

Figure 3-2.—Dry cell, cross-sectional view.

the influence of a difference in potential across the electrodes from the zinc (negative) through the external conductor to the carbon (positive), returning within the solution to the zinc. After a short period of time, the zinc will begin to waste away because of the "burning" action

of the acid. If zinc is surrounded by oxygen, it will burn (become oxidized) as a fuel. In this respect, the cell is like a chemical furnace in which energy released by the zinc is transformed into electrical energy rather than heat energy.

The voltage across the electrodes depends upon the materials from which the electrodes are made and the composition of the solution. The difference of potential between carbon and zinc electrodes in a dilute solution of sulfuric acid and water is about 1.5 volts.

The current that a primary cell may deliver depends upon the resistance of the entire circuit, including that of the cell itself. The internal resistance of the primary cell depends upon the size of the electrodes, the distance between them in the solution, and the resistance of the solution. The larger the electrodes and the closer together they are in solution (without touching), the lower the internal resistance of the primary cell and the more current it is capable of supplying to a load.

When current flows through a cell, the zinc gradually dissolved in the solution and the acid is neutralized. A chemical equation is sometimes used to show the chemical action that takes place. The symbols in the equation represent the different materials that are used. The symbol for carbon is C and zinc Zn. The equation is quantitative and equates the number of parts of the materials used before and after the zinc is oxidized. As stated previously in chapter 2, all matter is composed of atoms and molecules, with the atom being the smallest part of an element and the molecule the smallest part of a compound.

A compound is a chemical combination of two or more elements in which the physical properties of the compound are different from those of the elements comprising it. For instance, a molecule of water ($H_2O$) is composed of two atoms of hydrogen ($H_2$) and one atom of oxygen (O). Ordinarily, hydrogen and oxygen are gases; but when combined, as stated above, they form water, which normally is a liquid. However, sulfuric acid ($H_2SO_4$) and water ($H_2O$) form a mixture (not a compound), because the identity of both liquids is preserved when they are in solution together.

When a current flows through a primary cell having carbon and zinc electrodes and a dilute solution of sulfuric acid and water, the chemical reaction that occurs can be expressed as

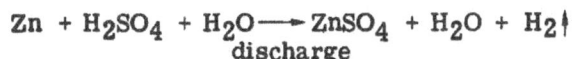

$$Zn + H_2SO_4 + H_2O \longrightarrow ZnSO_4 + H_2O + H_2\uparrow$$
$$\text{discharge}$$

The expression indicates that as current flows, a molecule of zinc combines with a molecule of sulfuric acid to form a molecule of zinc sulfate ($ZnSO_4$) and a molecule of hydrogen ($H_2$). The zinc sulfate dissolves in the solution and the hydrogen appears as gas bubbles around the carbon electrode. (A gas is designated by the arrow pointing upward in the equation.) As current continues to flow, the zinc is gradually consumed and the solution changes to zinc sulfate and water. The carbon electrode does not enter into the chemical changes taking place but simply provides a return path for the current.

In the process of oxidizing the zinc, the solution breaks up into positive and negative ions that move in opposite directions through the solution (fig. 3-1). The positive ions are hydrogen ions that appear around the carbon electrode (positive terminal). They are attracted to it by the free electrons from the zinc that are returning to the cell by way of the external load and the positive carbon terminal. The negative ions are $SO_4$ ions that appear around the zinc electrode. Positive zinc ions enter the solution around the zinc electrode and combine with the negative $SO_4$ ions to form zinc sulfate ($ZnSO_4$), a grayish-white substance that dissolves in water. At the same time that the positive and negative ions are moving in opposite directions in the solution, electrons are moving through the external circuit from the negative zinc terminal, through the load, and back to the positive carbon terminal. When the zinc is used up, the voltage of the cell is reduced to zero. There is no appreciable difference in potential between zinc sulfate and carbon in a solution of zinc sulfate and water.

Polarization

The chemical action that occurs in the cell (fig. 3-1) while the current is flowing causes hydrogen bubbles to form on the surface of the positive carbon electrode in great numbers until the entire surface is surrounded. This action is called polarization. Some of these bubbles rise to the surface of the solution and escape into the air. However, many of the bubbles remain until there is no room for any more to be formed.

The hydrogen tends to set up an electromotive force in the opposite direction to that of the cell, thus increasing the effective internal resistance, reducing the output current, and lowering the terminal voltage.

A cell that is heavily polarized has no useful output. There are several ways to prevent polarization from occurring or to overcome it after it has occurred. The very simplest method might be to remove the carbon electrode and wipe off the hydrogen bubbles. When the electrode is replaced in the electrolyte, the emf and current are again normal. This method is not practicable because polarization occurs rapidly and continuously in the simple voltaic cell. A commercial form of voltaic cell, known as the dry cell, employs a substance rich in oxygen as a part of the positive carbon electrode, which will combine chemically with the hydrogen to form water, $H_2O$. One of the best depolarizing agents used is manganese dioxide ($MnO_2$), which supplies enough free oxygen to combine with all of the hydrogen so that the cell is practically free from polarization.

The chemical action that occurs may be expressed as

$$2MnO_2 + H_2 \longrightarrow Mn_2O_3 + H_2O$$

The manganese dioxide combines with the hydrogen to form water and a lower oxide of manganese. Thus the counter emf of polarization does not exist in the cell, and the terminal voltage and output current are maintained normal.

Local Action

When the external circuit is opened, the current ceases to flow, and theoretically all chemical action within the cell stops. However, commercial zinc contains many impurities, such as iron, carbon, lead, and arsenic. These impurities form many small cells within the zinc electrode in which current flows between the zinc and its impurities. Thus the zinc is oxidized even though the cell itself is an open circuit. This wasting away of the zinc on open circuit is called local action. For example, a small local cell exists on a zinc plate containing impurities of iron, as shown in figure 3-3. Electrons flow between the zinc and iron and the solution around the impurity becomes ionized. The negative $SO_4$ ions combine with the positive $Zn$ ions to form $ZnSO_4$. Thus the acid is depleted in solution and the zinc consumed.

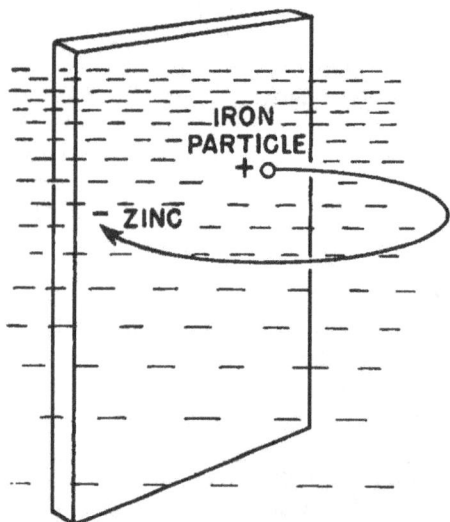

BE.27

Figure 3-3.—Local action on zinc electrode.

Local action may be prevented by using pure zinc (which is not practical), by coating the zinc with mercury, or by adding a small percentage of mercury to the zinc during the manufacturing process. The treatment of the zinc with mercury is called amalgamating (mixing) the zinc. Since mercury is 13.6 times as heavy as an equal volume of water, small particles of impurities having a lower relative weight than that of mercury will rise (float) to the surface of the mercury. The removal of these impurities from the zinc prevents local action. The mercury is not readily acted upon by the acid; and even when the cell is delivering current to a load, the mercury continues to act on the impurities in the zinc, causing them to leave the surface of the zinc electrode and float to the surface of the mercury. This process greatly increases the life of the primary cell.

## TYPES OF BATTERIES

The development of new and different types of batteries in the past decade has been so rapid that it is virtually impossible to have a complete knowledge of all of the various types currently being developed or now in use. A few recent developments are the silver-zinc, nickel-zinc, nickel-cadmium, silver-cadmium, magnesium-magnesium perchlorate, mercury, thermal, and water-activated batteries.

The lead-acid battery has been in service for a relatively long period of time; however, there are still various improvements being incorporated into the battery to improve its efficiency and life span. The material presented in this chapter, though not all inclusive, provides the reader with a knowledge of various types of batteries.

## DRY (PRIMARY) CELL

The dry cell is so called because its electrolyte is not in a liquid state. Actually, the electrolyte is a moist paste. If it should become dry, it would no longer be able to transform chemical energy to electrical energy. The name dry cell, therefore, is not strictly correct in a technical sense.

### Construction of the Dry Cell

The construction of a common type of dry cell is shown in figure 3-4. The internal parts of the cell are located in a cylindrical zinc container. This zinc container serves as the negative electrode of the cell. The container is lined with a nonconducting material, such as blotting paper, to insulate the zinc from the paste. A carbon electrode is located in the center, and it serves as the positive terminal of the cell. The paste is a mixture of several substances. Its composition may vary, depending on its manufracturer. Generally, however, the paste will contain some combination of the following substances: ammonium chloride (sal ammoniac), powdered coke, ground carbon, maganese dioxide, zinc chloride, graphite, and water.

This paste, which is packed in the space between the carbon and the blotting paper, also serves to hold the carbon electrode rigid in the center of the cell. When packing the paste in the cell, a small expansion space is left at the top. The cell is then sealed with asphalt-saturated cardboard.

Binding posts are attached to the elctrodes so that wires may be conveniently connected to the cell.

Since the zinc container is one of the electrodes, it must be protected with some insulating material. Therefore, it is common practice for the manufacturer to encluse the cells in cardboard containers.

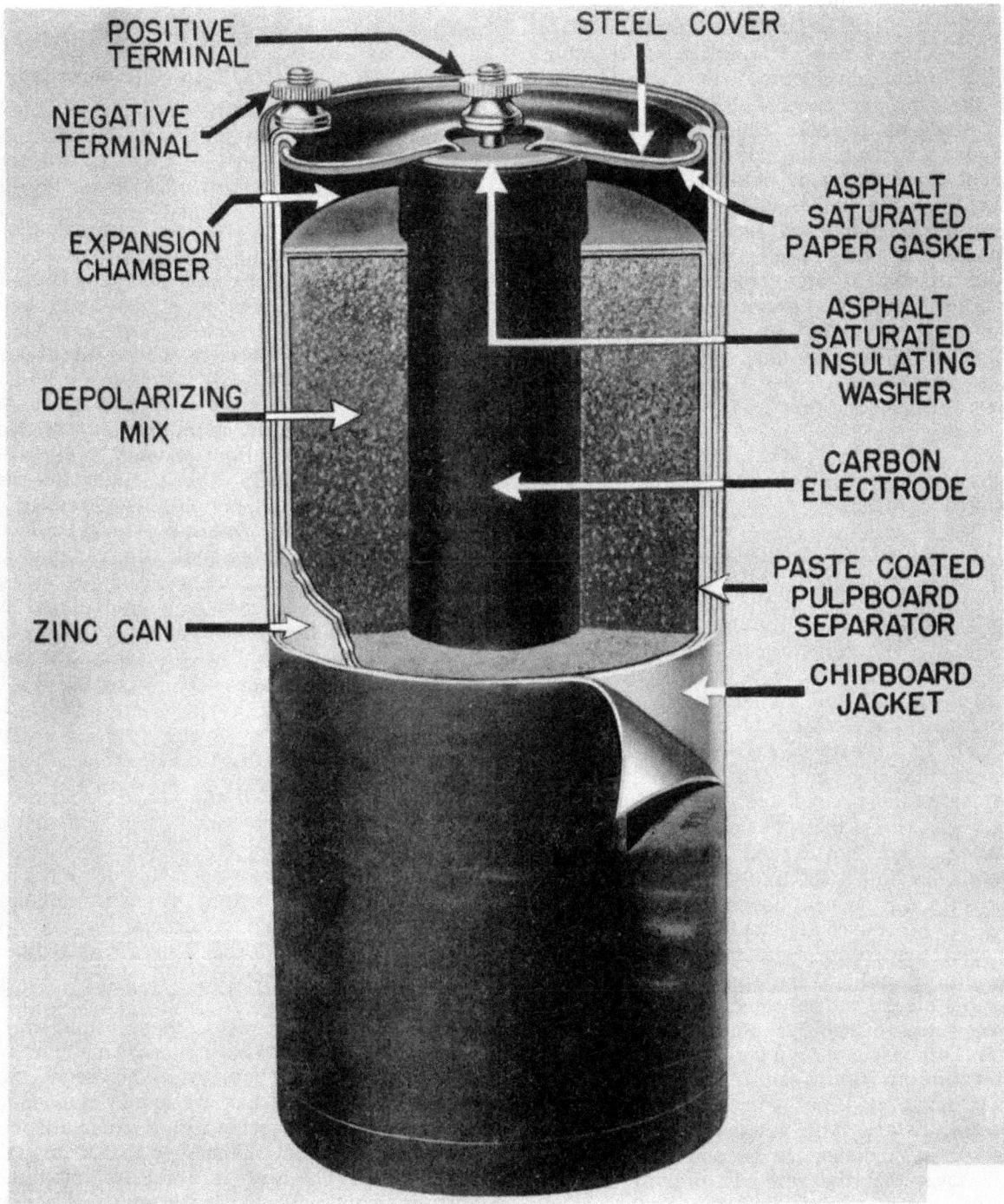

POSITIVE TERMINAL

NEGATIVE TERMINAL

EXPANSION CHAMBER

DEPOLARIZING MIX

ZINC CAN

STEEL COVER

ASPHALT SATURATED PAPER GASKET

ASPHALT SATURATED INSULATING WASHER

CARBON ELECTRODE

PASTE COATED PULPBOARD SEPARATOR

CHIPBOARD JACKET

BE.28

Figure 3-4.—Cutaway view of the general-purpose dry cell.

## Chemical Action of the Dry Cell

The dry cell (fig. 3-4) is fundamentally the same as the simple voltaic cell (wet cell) described earlier, as far as its internal chemical action is concerned. The action of the water and the ammonium chloride in the paste, together with the zinc and carbon electrodes, produces the voltage of the cell. The manganese dioxide is added to reduce the polarization when line current flows and zinc chloride reduces local action when the cell is idle. The blotting paper (paste coated pulpboard separator) serves two purposes, one being to keep the paste from making actual contact with the zinc container and the other being to permit the electrolyte to filter through to the zinc slowly. The cell is sealed at the top to keep air from entering and drying the electrolyte. Care should be taken to prevent breaking this seal.

## Rating of the Standard Size Cell

One of the popular sizes in general use is the standard, or No. 6, dry cell. It is approximately 2 1/2 inches in diameter and 6 inches in length. The voltage is about 1 1/2 volts when new but decreases as the cell ages. When the open-circuit voltage falls below 0.75 to 1.2 volts (depending upon the circuit requirements), the cell is usually discarded. The amount of current that the cell can deliver, and still give satisfactory service, depends upon the length of time that the current flows. For instance, if a No. 6 cell is to be used in a portable radio, it is likely to supply current constantly for several hours. Under these conditions, the current should not exceed 1/8 ampere, the rated constant-current capacity of a No. 6 cell. If the same cell is required to supply current only occasionally, for only short periods of time, it could supply currents of several amperes without undue injury to the cell. As the time duration of each discharge decreases, the interval of time between discharges increases, the allowable amount of current available for each discharge becomes higher, up to the amount that the cell will deliver on short circuit.

The short-circuit current test is another means of evaluating the condition of a dry cell. A new cell, when short circuited through an ammeter, should supply not less than 25 amperes. A cell that has been in service should supply at least 10 amperes if it is to remain in service.

## Rating of the Unit Size Cell

A popular size of dry cell, the size D, is 1 3/8 inches in diameter and 2 3/4 inches in length. It is also known as the unit cell. The size D cell voltage is 1.5 volts when new. A discharged cell may expand, allowing the electrolyte to leak and cause corrosion. Some manufacturers place a steel jacket around the zinc container to prevent this action.

## Shelf Life

A cell that is not being used (sitting on the shelf) will gradually deteriorate because of slow internal chemical actions (local action) and changes in moisture content. However, this deterioration is usually very slow if cells are properly stored. Highgrade cells of the larger sizes should have a shelf life of a year or more. Smaller size cells have a proportionately shorter shelf life, ranging down to a few months for the very small sizes. If unused cells are stored in a cool place, their shelf life will be greatly increased; therefore, to minimize deterioration, they should be stored in refrigerated spaces ($10°$ F to $35°$ F) that are not dehumidified (dry).

## MERCURY CELLS

With the advent of the space program and the development of small transceivers and miniaturized equipment, a power source of miniaturized size was needed. Such equipment requires a small battery which is capable of delivering maximum electrical capacity per unit volume while operating in varying temperatures and at a constant discharge voltage. The mercury battery, which is one of the smallest batteries, meets these requirements.

Present mercury batteries are manufactured in three basic structures. The wound anode type (fig. 3-5) (A)) has its anode composed of a corrugated zinc strip with a paper absorbent wound in an offset manner so that it protrudes at one end. The zinc is amalgamated (mixed) with mercury (10 percent), and the paper is impregnated with the electrolyte which causes it to swell and produce a positive contact pressure.

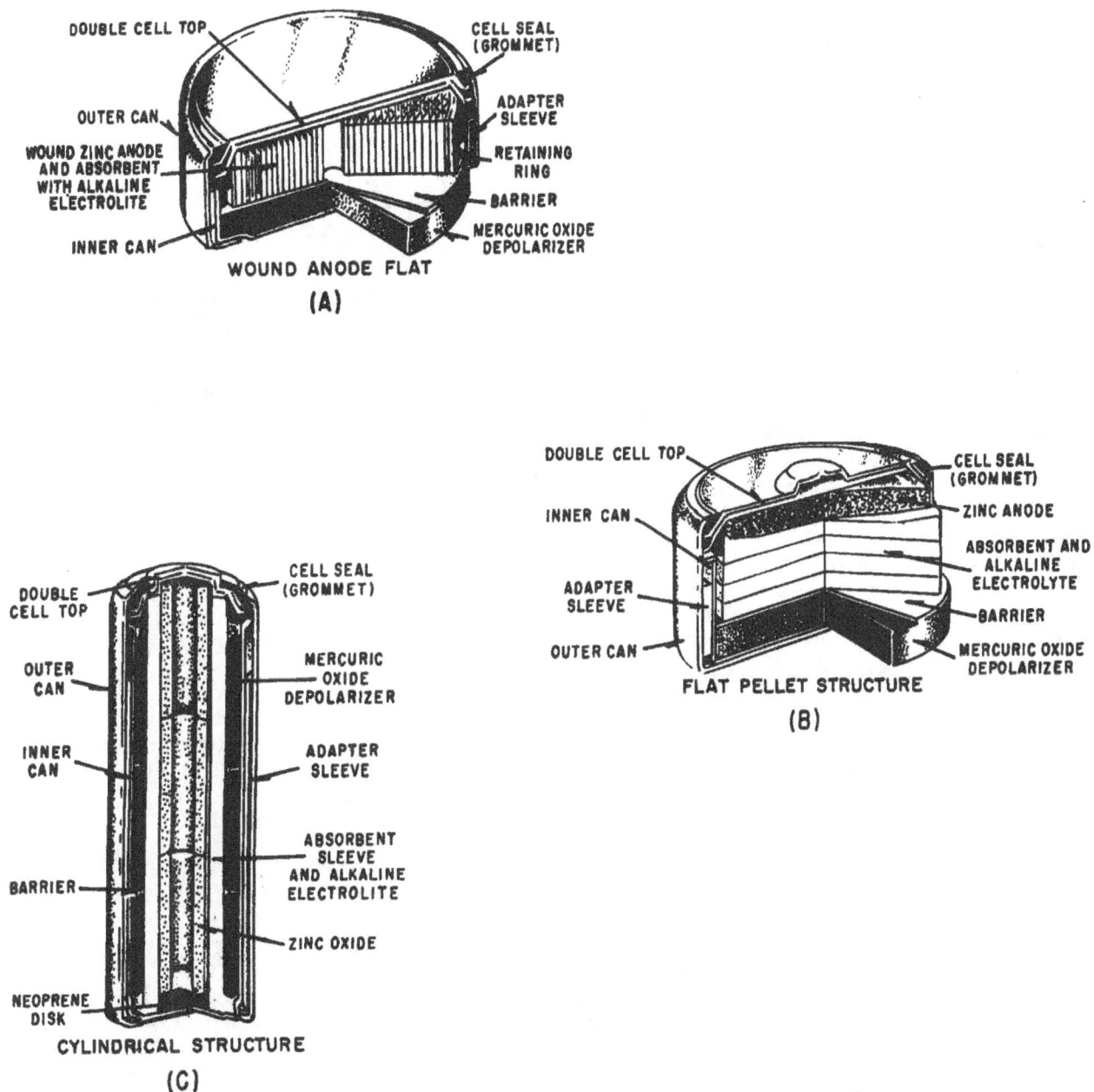

**DOUBLE CELL TOP** · **CELL SEAL (GROMMET)** · **OUTER CAN** · **ADAPTER SLEEVE** · **WOUND ZINC ANODE AND ABSORBENT WITH ALKALINE ELECTROLITE** · **RETAINING RING** · **BARRIER** · **INNER CAN** · **MERCURIC OXIDE DEPOLARIZER**

WOUND ANODE FLAT
(A)

DOUBLE CELL TOP · CELL SEAL (GROMMET) · INNER CAN · ZINC ANODE · ADAPTER SLEEVE · ABSORBENT AND ALKALINE ELECTROLYTE · OUTER CAN · BARRIER · MERCURIC OXIDE DEPOLARIZER

FLAT PELLET STRUCTURE
(B)

DOUBLE CELL TOP · CELL SEAL (GROMMET) · OUTER CAN · MERCURIC OXIDE DEPOLARIZER · INNER CAN · ADAPTER SLEEVE · BARRIER · ABSORBENT SLEEVE AND ALKALINE ELECTROLITE · ZINC OXIDE · NEOPRENE DISK

CYLINDRICAL STRUCTURE
(C)

BE.29

Figure 3-5.—Mercury cells.

In the pressed powder cells (fig. 3-5 (B) and (C)), the zinc powder is preamalgamated prior to pressing into shape; its porosity allows electrolyte impregnation with oxidation in depth when current is discharged. A double can structure is used in the larger sized cells. The space between the inner and outer containers provide passage for any gas generated by an improper chemical balance or impurities present within the cell. The construction is such that, if excessive gas pressures are experienced, the compression of the upper part of the grommet by internal pressure allows the gas to escape into the space between the two cans. A paper tube surrounds the inner can so that any liquid carried by discharging gas will be absorbed, maintaining a leak resistant structure. Release of excessive gas pressure automatically reseals the cell.

NOTE: Mercury batteries have been known to explode with considerable force when shorted. Caution should be exercised to insure that the battery is not accidentally shorted.

The overall chemical action by which the mercury cell produces electricity is given by the following chemical formula:

$$Zn + H_2O + HgO \longrightarrow ZnO + H_2O + Hg$$

This action, the same as in other type cells, is a process of oxidation. The alkaline electrolyte is in contact with the zinc electrode. The zinc oxidizes (Zn changes to ZnO), thus taking atoms of oxygen from water molecules in the electrolyte. This leaves positive hydrogen ions, which move toward the mercuric oxide pellet, causing polarization. These hydrogen ions take oxygen from the mercuric oxide (thus changing HgO to Hg). Where one molecule of water is destroyed at the negative electrode, one molecule is produced at the positive electrode, maintaining the net amount of water constant. By absorbing oxygen, the zinc electrode accumulates excess electrons, making it negative. By giving up oxygen, the mercuric oxide electrode loses electrons, making it positive. In the discharged state, the negative electrode is zinc oxide, and the positive electrode is ordinary mercury.

## RESERVE CELL

A reserve cell is one in which the elements are kept dry until the time of use; the electrolyte is then admitted and the cell starts producing current. In theory, this means that a reserve cell should be able to be stored for an indefinite period of time before it is activated.

One new reserve cell (fig. 3-6) is the alkaline manganese cell of the standard D size (flashlight battery). This reserve cell exhibits a high efficiency over a wide temperature range and is capable of momentary high current pulses in the range of 12 to 15 amperes.

The reserve cell is manufactured in a dry state, the electrolyte being contained in a plastic vial within the cell. When stored in this manner, the cell has a shelf life capability of over 10 years. To activate the cell, the activating mechanism is rotated 35° in either direction. This releases a spring-loaded plunger which breaks the plastic vial of electrolyte. Continued

BE. 30

Figure 3-6.—Reserve cell.

rotation permits the activating mechanism to be removed and discarded, resulting in a D size cell. A safety device is incorporated to prevent accidental activation during handling and transit.

Activation time is approximately 2 seconds when the cell is not under load. When under a 4-ohm load, the activation time (to reach a 1.35-volt level) is less than 5 seconds at 70° F and less than 30 seconds at 30° F.

The cell has been designed so that it is not position-sensitive during either the activation or the discharge period, and after activation can be handled and used as a standard D cell. After activation, shelf life of the reserve cell is approximately 2 years less than that of the standard alkaline manganese cell.

Reserve cells are used for emergency lighting and communications equipment, military ordnance devices, and in any other use where long storage ability is of prime importance.

## COMBINING CELLS

In many cases, a battery-powered device may require more electrical energy than one cell can provide. The device may require either a higher voltage or more current, and in some cases both. Under such conditions it is necessary to combine, or interconnect, a sufficient number of cells to meet the higher requirements. Cells connected in series provide a higher voltage, while cells connected in parallel provide a higher current capacity. To provide adequate power when both voltage and current requirements are greater than the capacity of one cell, a combination series-parallel network of cells must be interconnected.

## Series Connected Cells

Assume that a load requires a power supply with a potential of 6 volts and a current capacity of 1/8 ampere. Since a single cell normally supplies a potential of only 1.5 volts, more than one cell is obviously needed. To obtain the higher potential, the cells are connected in series as shown in figure 3-7 (A).

In a series hookup, the negative electrode of the first cell is connected to the positive electrode of the second cell, the negative electrode of the second to the positive of the third, etc. The positive electrode of the first cell and negative electrode of the last cell then serve as the power takeoff terminals of the battery. In this way, the potential is boosted 1.5 volts by each cell in the series line. There are four cells, so the output terminal voltage is 1.5 x 4 = 6 volts. When connected to the load, 1/8 ampere flows through the load and each cell of the battery. This is within the capacity of each cell. Therefore, only four series-connected cells are needed to supply this particular load.

## Parallel Connected Cells

In this case, assume an electrical load requires only 1.5 volts, but will draw 1/2 ampere of current. (Assume that a cell will supply only 1/8 ampere.) To meet this requirement, the cells are connected in parallel, as shown in figure 3-8 (A). In a parallel connection, all positive cell electrodes are connected to one line, and all negative electrodes are connected to the other. No more than one cell is connected between the lines at any one point; so the potential between the lines is the same as that of one cell, or 1.5 volts. However, each cell may contribute its maximum allowable current of 1/8 ampere to the line. There are four cells, so the total line current is 1/8 x 4 = 1/2 ampere. Hence, four cells in parallel have enough capacity to supply a load requiring 1/2 ampere at 1.5 volts.

## Series-Parallel Connected Cells

Figure 3-9 depicts a battery network supplying power to a load requiring both a voltage and current greater than one cell can provide. To provide the required 4.5 volts, groups of three 1.5-volt cells are connected in series. To provide the required 1/2 ampere of current, four series groups are connected in parallel, each supplying 1/8 ampere of current.

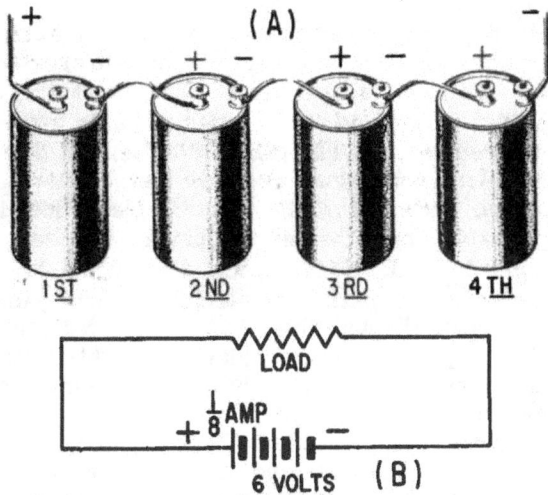

BE. 31

Figure 3-7.—(A) Pictorial view of series connected cells; (B) schematic of series connection.

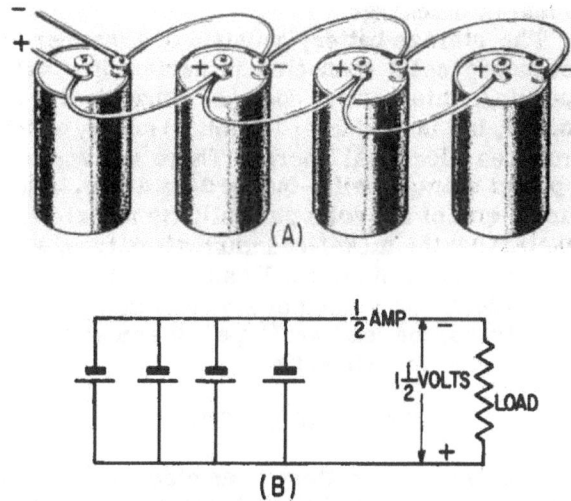

BE. 32

Figure 3-8.—(A) Pictorial view of parallel-connected cells; (B) schematic of parallel correction.

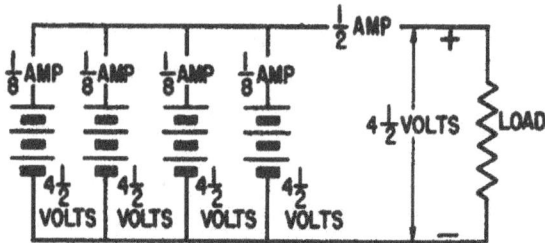

BE. 33
Figure 3-9.—Series-parallel connected cells.

## SECONDARY (WET) CELLS

Secondary cells function on the same basic chemical principles as primary cells. They differ mainly in that they may be recharged, whereas the primary cell is not normally recharged. (As mentioned earlier, some primary cells have been developed to the state where they may be recharged.) Some of the materials of a primary cell are consumed in the process of changing chemical energy to electrical energy. In the secondary cell, the materials are merely transferred from one electrode to the other as the cell discharges. Discharged secondary cells may be restored (charged) to their original state by forcing an electric current from some other source through the cell in the opposite direction to that of discharge.

The storage battery consists of a number of secondary cells connected in series. Properly speaking, this battery does not store electrical energy, but is a source of chemical energy which produces electrical energy. There are various types of storage cells—the lead-acid type, which has an emf of 2.2 volts per cell; the nickel-iron alkali type; the nickel-cadmium alkali type, with an emf of 1.2 volts per cell; and the silver-zinc type, which has an emf of 1.5 volts per cell. Of these types, the lead-acid type is the most widely used, and is described first.

## LEAD-ACID BATTERY

The lead-acid battery is an electrochemical device for storing chemical energy until it is released as electrical energy. Active materials within the battery react chemically to produce a flow of direct current whenever current consuming devices are connected to the battery terminal posts. This current is produced by chemical reaction between the active material of the plates (electrodes) and the electrolyte (sulfuric acid). The lead-acid battery is used extensively throughout the world. The parts of a lead-acid battery are illustrated in figure 3-10 and are discussed in the following paragraphs.

## BATTERY CONSTRUCTION

A lead-acid battery consists of a number of cells connected together, the number needed depending upon the voltage desired. Each cell produces approximately 2 volts.

A cell consists of a hard rubber, plastic, or bituminous material compartment into which is placed the cell element, consisting of two types of lead plates, known as positive and negative plates. (See fig. 3-11.) These plates are insulated from each other by suitable separators (usually made of plastic, rubber, or glass) and submerged in a sulfuric acid solution (electrolyte).

There are a variety of plates used in the lead-acid battery—pasted plates, spun-lead (Plante) plates, Gould plates, and ironclad plates. These plates are each designed to fulfill a specific purpose. The most commonly used plates (the pasted plates) are discussed briefly.

The pasted plates are formed by applying lead-oxide pastes to a grid (fig. 3-12) made of lead-antimony alloy. The grid is designed to give the plates mechanical strength, hold the active material in place, and provide adequate conductivity for the electric current created by the chemical action. The active material (lead oxide) is applied to the grids in paste form and allowed to dry. The plates are then put through an electrochemical process that converts the active material of the positive plates into lead peroxide, and that of the negative plates into sponge lead. This action is caused by immersing the plates in electrolyte and passing a current through them in the proper direction. This type of plate is relatively light in weight compared to the other plates that are more rugged and durable in construction.

After the plates have been formed, they are built into positive and negative groups. The negative group of plates always has one more plate than the positive group so that both sides of the positive plates are acted upon chemically. This keeps the expansion and contraction that takes place in the positive plates the same on both sides and prevents buckling. These groups

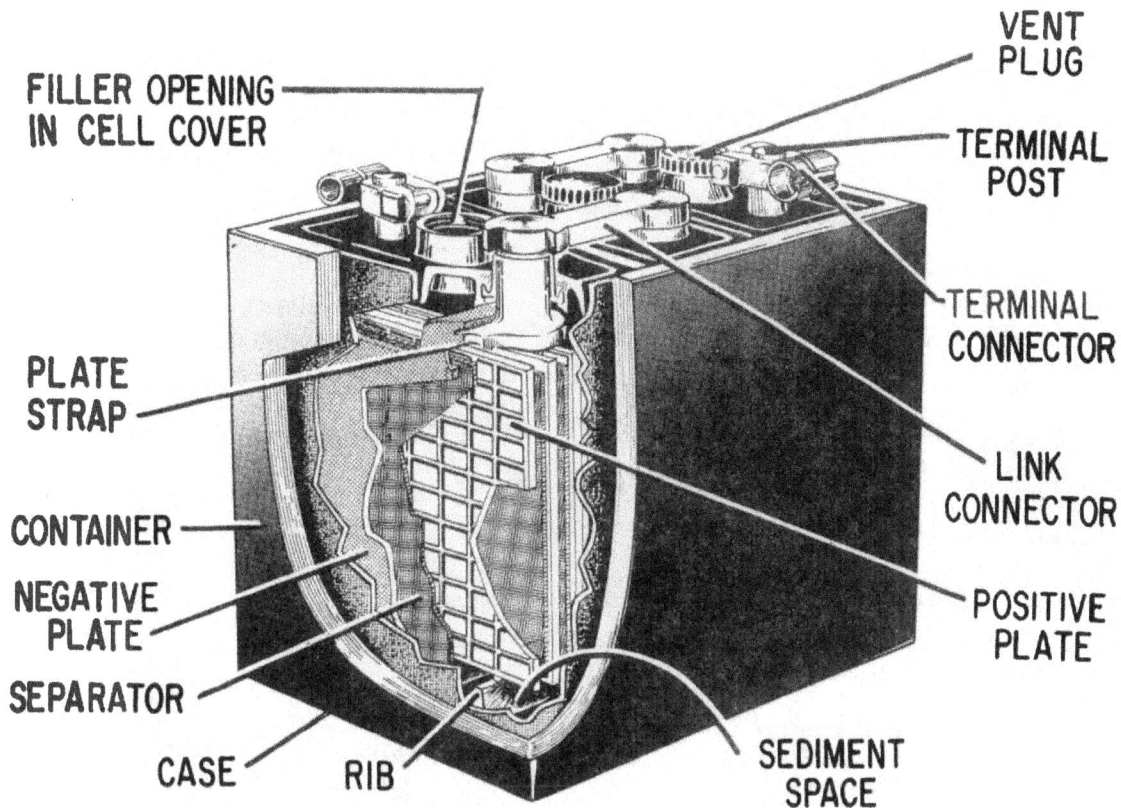

BE.34

Figure 3-10.—Lead-acid battery construction.

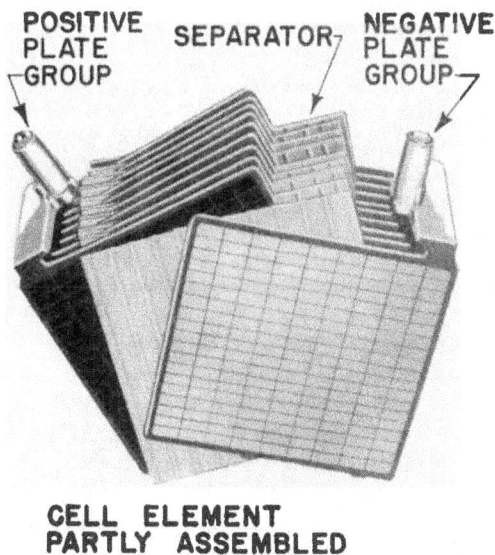

CELL ELEMENT
PARTLY ASSEMBLED

BE.35

Figure 3-11.—Plate arrangement.

are then assembled together with separators to become cell elements. The separators are grooved vertically on one side and smooth on the other. The grooved side is placed next to the positive plate to permit free circulation of the electrolyte around the active material.

The positive plates which are lead peroxide and the negative plates which are spongy lead are referred to as the active material of the battery. However, these materials alone in a container will cause no chemical action unless there is a path for interaction between them. To provide this path for interaction and to carry the electric current within the battery are the functions of the electrolyte.

A battery container is the receptacle for the cells that make up the battery. Most containers are made from hard rubber, plastic, or bituminous composition that is resistant to acid and mechanical shock, and is able to withstand extreme weather conditions. Most batteries are assembled in a one-piece container with compartments for each individual cell. The bottom of the container has ribs

BE.36

Figure 3-12.—Grid structure.

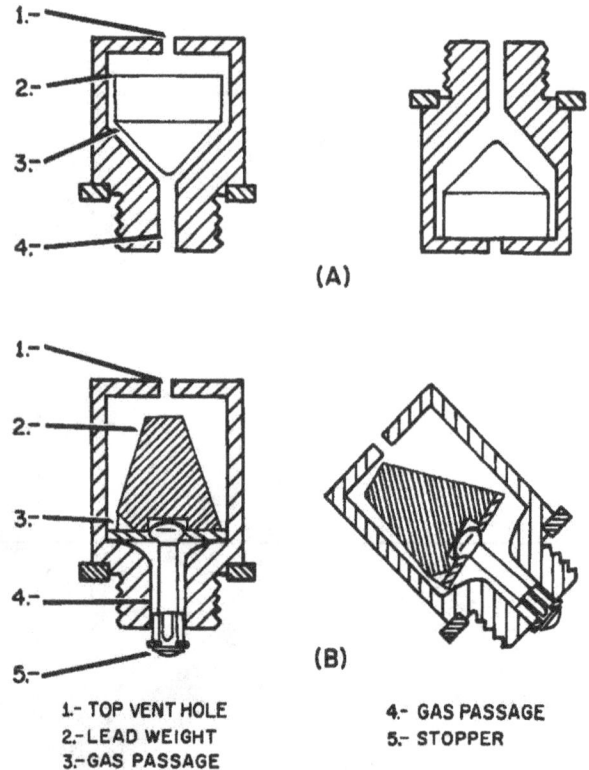

1.- TOP VENT HOLE  4.- GAS PASSAGE
2.-LEAD WEIGHT   5.- STOPPER
3.-GAS PASSAGE

BE.37

Figure 3-13.—Nonspill vent plugs.

molded into it to provide support for the elements and a sediment space for the flakes of active material that drop off the plates during the life of the battery.

The battery or cell covers and the battery container are usually made of the same material. The cell covers provide openings for the two-element terminals and a vent plug.

Cell connectors are used to connect the cell of a battery in series. The element in each cell is placed so that the negative terminal of one cell is physically located adjacent to the positive terminal of the next cell; they are connected both physically and electrically by a cell connector. Connectors must be of sufficient size to carry the current demands of the battery without over-heating.

Vent plugs are made of various designs to function in conjunction with the cover vent openings to permit the escape of gases that form within the cells while preventing leakage or loss of the electrolyte. A typical vent plug used in an automobile battery is depicted in figure 3-10.

Some batteries utilize a nonspill type of vent plug which makes it possible to place the battery in any position without loss of the electrolyte. (See fig. 3-13.) This type vent plug has found wide use in aircraft.

Sealing compound, generally made of a bituminous substance, is used to form a seal between the cell cover and the container. The compound is an acid resistant material that must conform to rigid vibration and heat standards. This insures that the sealing compound does not melt or flow at summer temperatures and does not crack at winter temperatures. Batteries with a polystyrene jar use a polystyrene cement as a sealer.

The terminals of a lead-acid battery are normally distinguishable from one another by their physical size and the marking by the manufacturer. The positive terminal marked (+) is slightly larger than the negative terminal marked (-).

BATTERY OPERATION

In its charged condition, the active materials in the lead-acid battery are lead peroxide (used as the positive plate) and sponge lead (used as the negative plate). The electrolyte is a mixture of sulfuric acid and water. The strength (acidity) of the electrolyte is measured in terms of its specific gravity. Specific gravity is the ratio of the weight of a given volume

45

of electrolyte to an equal volume of pure water. Concentrated sulfuric acid has a specific gravity of about 1.830; pure water has a specific gravity of 1.000. The acid and water are mixed in a proportion to give the specific gravity desired. For example, an electrolyte with a specific gravity of 1.210 requires roughly one part of concentrated acid to four parts of water.

In a fully charged battery, the positive plates are pure lead peroxide and the negative plates are pure lead. Also, in a fully charged battery, all acid is in the electrolyte so that the specific gravity is at its maximum value. The active materials of both the positive and negative plates are porous, and have absorptive qualities similar to a sponge.

The pores are therefore filled with the battery solution (electrolyte) in which they are immersed. As the battery discharges, the acid in contact with the plates separates from the electrolyte. It forms a chemical combination with the active material of the plate, changing it to lead sulfate. Thus, as the discharge continues, lead sulfate forms on the plates, and more acid is taken from the electrolyte. The water content of the electrolyte becomes progressively higher; that is, the ratio of water to acid increases. As a result, the specific gravity of the electrolyte will gradually decrease during discharge.

When the battery is being charged, the reverse takes place. The acid held in the sulfated plate material is driven back into the electrolyte; further charging cannot raise its specific gravity any higher. When fully charged, the material of the positive plates is again pure lead peroxide and that of the negative plates is pure lead.

Electrical energy is derived from a cell when the plates react with the electrolyte. As a molecule of sulfuric acid separates, part of it combines with the negative sponge lead plates. Thus, it makes the sponge lead plates negative, and at the same time forms lead sulfate. The remainder of the sulfuric acid molecule, lacking electrons, has thus become a positive ion. The positive ions migrate through the electrolyte to the opposite (lead peroxide) plates, and take electrons from them. This action neutralizes the positive ions, forming ordinary water. It also makes the lead peroxide plates positive, by taking electrons from them. Again, lead sulfate is formed in the process.

The action just described is represented in more detail by the following chemical equation:

discharging

$$Pb + PbO_2 + 2H_2SO_4 \rightleftharpoons 2PbSO_4 + 2H_2O$$

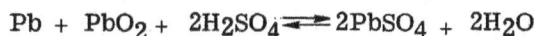

The left side of the expression represents the cell in the charged condition, and the right side represents the cell in the discharged condition.

In the charged condition, the positive plate contains lead peroxide, $PbO_2$; the negative plate is composed of sponge lead, $Pb$; and the solution contains sulfuric acid, $H_2SO_4$. In the discharged condition, both plates contain lead sulfate, $PbSO_4$, and the solution contains water, $H_2O$. As the discharge progresses, the acid content of the electrolyte becomes less and less because it is used in forming lead sulfate, and the specific gravity of the electrolyte decreases. A point is reached where so much of the active material has been converted into lead sulfate that the cell can no longer produce sufficient current to be of practical value. At this point the cell is said to be discharged (fig. 3-14 (C)). Since the amount of sulfuric acid combining with the plates at any time during discharge is in direct proportion to the ampere-hours (product of current in amperes and time in hours) of discharge, the specific gravity of the electrolyte is a guide in determining the state of discharge of the lead-acid cell.

If the discharged cell is properly connected to a direct-current charging source (the voltage of which is slightly higher than that of the cell), current will flow through the cell, in the opposite direction to that of discharge, and the cell is said to be charging (fig. 3-14 (D)). The effect of the current will be to change the lead sulfate on both the positive and negative plates back to its original active form of lead peroxide and sponge lead, respectively. At the same time, the sulfate is restored to the electrolyte with the result that the specific gravity of the electrolyte increases. When all the sulfate has been restored to the electrolyte, the specific gravity will be maximum. The cell is then fully charged and is ready to be discharged again.

It should always be remembered that the addition of sulfuric acid to a discharged lead-acid cell does not recharged the cell. Adding acid only increases the specific gravity of the electrolyte and does not convert the lead sulfate on the plates back into active material (sponge lead and lead peroxide), and consequently does

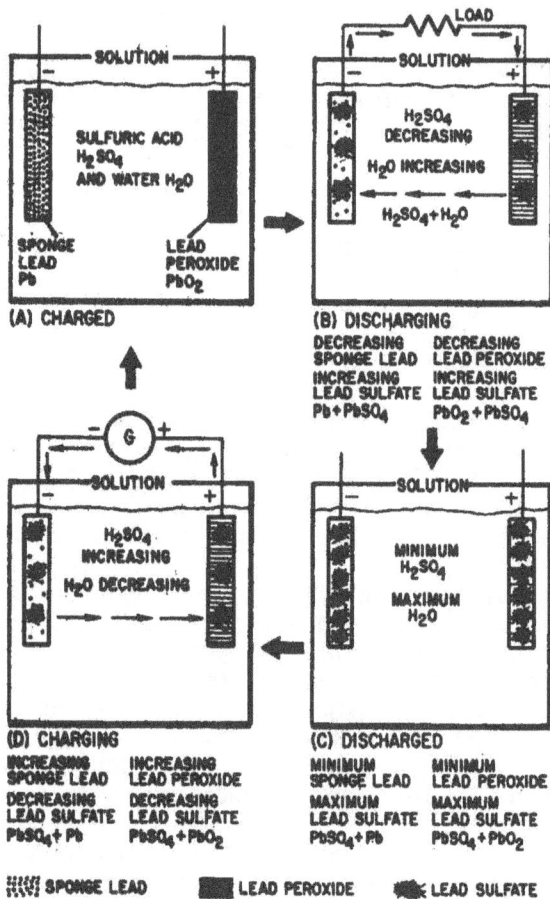

SPONGE LEAD | LEAD PEROXIDE | LEAD SULFATE

BE.38

Figure 3-14.—Chemical action in lead-acid cell.

not bring the cell back to a charged condition. A charging current must be passed through the cell to do this.

As a cell charge nears completion, hydrogen gas ($H_2$) is liberated at the negative plate and oxygen gas ($O_2$) is liberated at the positive plate. This action occurs because the charging current is greater than the amount that is necessary to reduce the small remaining amount of lead sulfate on the plates. Thus, the excess current ionizes the water in the electrolyte. This action is necessary to assure full charge to the cell.

## SPECIFIC GRAVITY

The ratio of the weight of a certain volume of liquid to the weight of the same volume of water is called the specific gravity of the liquid. The specific gravity of pure water is 1.000. Sulfuric acid has a specific gravity of 1.830; thus sulfuric acid is 1.830 times as heavy as water. The specific gravity of a mixture of sulfuric acid and water varies with the strength of the solution from 1.000 to 1.830.

As a storage battery discharges, the sulfuric acid is depleted and the electrolyte is gradually converted into water. This action provides a guide in determining the state of discharge of the lead-acid cell. The electrolyte that is usually placed in a lead-acid battery has a specific gravity of 1.350 or less. Generally, the specific gravity of the electrolyte in standard storage batteries is adjusted between 1.210 and 1.220. On the other hand the specific gravity of the electrolyte in submarine batteries when charged is from 1.250 to 1.265, while in aircraft batteries when fully charged it is from 1.285 to 1.300.

### Hydrometer

The specific gravity of the electrolyte is measured with a hydrometer. In the syringe type hydrometer (fig. 3-15), part of the battery electrolyte is drawn up into a glass tube by means of a rubber bulb at the top.

The hydrometer float consists of a hollow glass tube weighted at one end and sealed at both ends. A scale calibrated in specific gravity is laid off axially along the body (stem) of the tube. The hydrometer float is placed inside the glass syringe and the electrolyte to be tested is drawn up into the syringe, thus immersing the hydrometer float into the solution. When the syringe is held approximately in a vertical position, the hydrometer float will sink to a certain level in the electrolyte. The extent to which the hydrometer stem protudes above the level of the liquid depends upon the specific gravity of the solution. The reading on the stem at the surface of the liquid is the specific gravity of the electrolyte in the syringe.

The Navy uses two types of hydrometer bulbs, or floats, each having a different scale. The type-A hydrometer is used with submarine batteries and has two different floats with scales from 1.960 to 1.240 and from 1.120 to 1.300. The type-B hydrometer is used with portable storage batteries and aircraft batteries. It has a scale from 1.100 to 1.300.

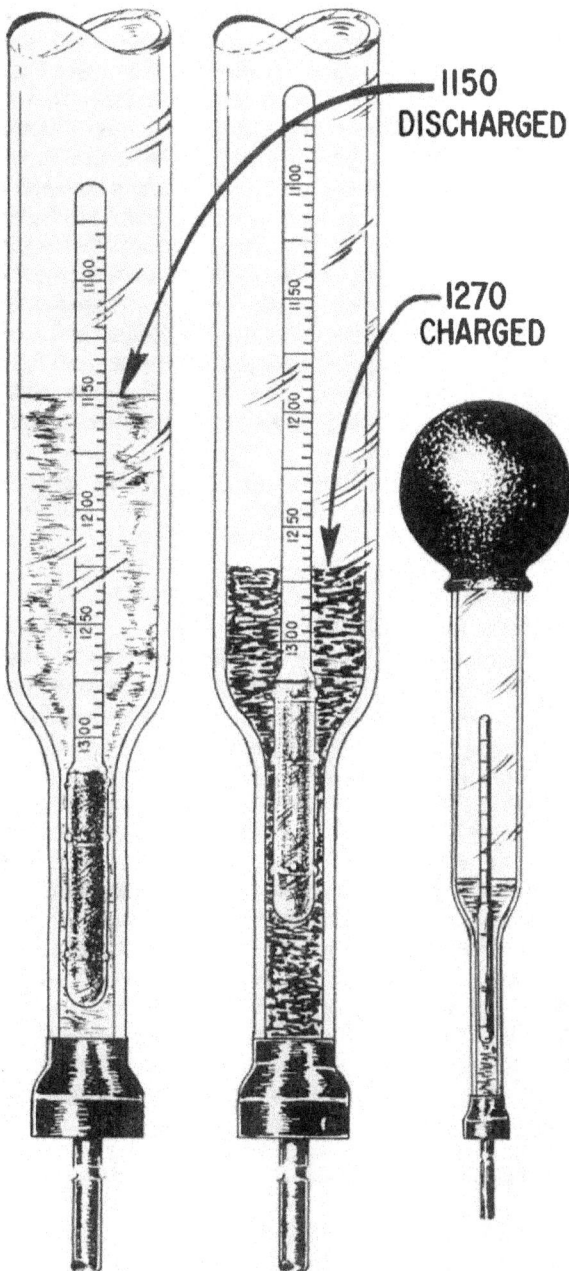

1150
DISCHARGED

1270
CHARGED

BE.39

Figure 3-15.—Type-B hydrometer.

CAUTION: Hydrometers should be flushed daily with fresh water to prevent inaccurate readings. Storage battery hydrometers must not be used for any other purpose.

## Corrections

The specific gravity of the electrolyte is affected by its temperature. The electrolyte expands and becomes less dense when heated and its specific gravity reading is lowered. On the other hand, the electrolyte contracts and becomes denser when cooled and its specific gravity reading is raised. In both cases the electrolyte may be from the same fully charged storage cell. Thus, the effect of temperature is to distort the readings.

Most standard storage batteries use $80°$ F as the normal temperature to which specific gravity readings are corrected. To correct the specific gravity reading of a storage battery, add 4 points to the reading for each $10°$ F above $80°$ and subtract 4 points for each $10°$ F below $80°$. The electrolyte in a cell should be at the normal level when the reading is taken. If the level is below normal, there will not be sufficient fluid drawn into the tube to cause the float to rise. If the level is above normal there is too much water, the electrolyte is weakened, and the reading is too low. A hydrometer reading is inaccurate if taken immediately after water is added, because the water tends to remain at the top of the cell. When water is added, the battery should be charged for at least an hour to mix to electrolyte before a hydrometer reading is taken.

## Adjusting Specific Gravity

Only authorized personnel should add acid to a battery. Acid with a specific gravity above 1.350 is never added to a battery.

If the specific gravity of a cell is more than it should be, it can be reduced to within limits by removing some of the electrolyte and adding distilled water. The battery is charged for 1 hour to mix the solution, and then hydrometer readings are taken. The adjustment is continued until the desired true readings are obtained.

## MIXING ELECTROLYTES

The electrolyte of a fully charged battery usually contains about 38 percent sulfuric acid by weight, or about 27 percent by volume. In preparing the electrolyte, distilled water and sulfuric acid are used. New batteries may be delivered with containers of concentrated sulfuric acid of 1.830 specific gravity or electrolyte of 1.400 specific gravity, both of which

48

must be diluted with distilled water to make electrolyte of the proper specific gravity. The container used for diluting the acid should be made of glass, earthenware, rubber, or lead.

When mixing electrolyte, ALWAYS POUR ACID INTO WATER—never pour water into acid. Pour the acid slowly and cautiously to prevent excessive heating and splashing. Stir the solution continuously with a nonmetallic rod to mix the heavier acid with the lighter water and to keep the acid from sinking to the bottom. When concentrated acid is diluted, the solution becomes very hot.

## TREATMENT OF ACID BURNS

If acid or electrolyte from a lead-acid battery comes into contact with the skin, the affected area should be washed as soon as possible with large quantities of fresh water, after which a salve such as petrolatum, boric acid, or zinc ointment should be applied. If none of these salves are available, clean lubricating oil will suffice. When washing, large amounts of water should be used, since a small amount of water might do more harm than good in spreading the acid burn.

Acid spilled on clothing may be neutralized with dilute ammonia or a solution of baking soda and water.

## CAPACITY

The capacity of a battery is measured in ampere-hours. As mentioned before, the ampere-hour capacity is equal to the product of the current in amperes and the time in hours during which the battery is supplying this current. The ampere-hour capacity varies inversely with the discharge current. The size of a cell is determined generally by its ampere-hour capacity. The capacity of a cell depends upon many factors, the most important of these are as follows:

1. The area of the plates in contact with the electrolyte.

2. The quantity and specific gravity of the electrolyte.

3. The type of separators.

4. The general condition of the battery (degree of sulfating, plates buckled, separators warped, sediment in bottom of cells, etc.).

5. The final limiting voltage.

## RATING

Storage batteries are rated according to their rate of discharge and ampere-hour capacity. Most batteries, except aircraft and some used for radio and sound systems, are rated according to a 20-hour rate of discharge—that is, if a fully charged battery is completely discharged during a 20-hour period, it is discharged at the 20-hour rate. Thus if a battery can deliver 20 amperes continuously for 20 hours, the battery has a rating of 20 x 20, or 400 ampere-hours. Thus the 20-hour rating is equal to the average current that a battery is capable of supplying without interruption for an interval of 20 hours. (NOTE: Aircraft batteries are rated according to a 1-hour rate of discharge.) Some other ampere-hour ratings used are 6-hour and 10-hour ratings.

All standard batteries deliver 100 percent of their available capacity if discharged in 20 hours or more, but they will deliver less than their available capacity if discharged at a faster rate. The faster they discharge, the less ampere-hour capacity they have.

The low-voltage limit, as specified by the manufacturer, is the limit beyond which very little useful energy can be obtained from a battery. For example, at the conclusion of a 20-hour discharge test on a battery, the closed-circuit voltmeter reading is about 1.75 volts per cell and the specific gravity of the electrolyte is about 1.060. At the end of a charge, its closed-circuit voltmeter reading, while the battery is being charged at the finishing rate, is between 2.4 and 2.6 volts per cell. The specific gravity of the electrolyte corrected to 80°F is between 1.210 and 1.220. In climates of 40°F and below authority may be granted to increase the specific gravity to 1.280. Other batteries, of higher normal specific gravity, may also be increased.

## TEST DISCHARGE

The test discharge is the best method of determining the capacity of a battery. Most battery switchboards are provided with the necessary equipment for giving test discharges. If proper equipment is not available, a tender, repair ship, or shore station may make the test. To determine the battery capacity, a battery is normally given a test discharge once every 6 months. Test discharges are also given whenever any cell of a battery after charge cannot be brought within 10 points of full charge, or when one or more cells is found to have less than normal voltage after an equalizing charge.

A test discharge must always be preceded by an equalizing charge. Immediately after the equalizing charge, the battery is discharged at its 20-hour rate until either the total battery voltage drops to a value equal to 1.75 times the number of cells in series or the voltage of any individual cell drops to 1.65 volts, whichever occurs first. The rate of discharge should be kept constant throughout the test discharge. Because standard batteries are rated at the 20-hour capacity, the discharge rate for a 200 ampere-hour battery is 200/20, or 10 amperes. If the temperature of the electrolyte at the beginning of the charge is not exactly 80° F, the time duration of the discharge must be corrected for the actual temperature of the battery.

A battery of 100-percent capacity discharges at its 20-hour rate for 20 hours before reaching its low-voltage limit. If the battery or one of its cells reaches the low-voltage limit before the 20-hour period has elapsed, the discharge is discontinued immediately and the percentage of capacity is determined from the equation

$$C = \frac{H_a}{H_t} \times 100$$

where C is the percentage of ampere-hour capacity available, $H_a$ the total hours of discharge, and $H_t$ the total hours for 100-percent capacity. The date for each test discharge should be recorded on the storage battery record sheet.

For example, a 200-ampere-hour 6-volt battery delivers an average current of 10 amperes for 20 hours. At the end of this period the battery voltage is 5.25 volts. On a later test the same battery delivers an average current of 10 amperes for only 14 hours. The discharge was stopped at the end of this time because the voltage of the middle cell was found to be only 1.65 volts. The percentage of capacity of the battery is now $\frac{14}{20} \times 100$, or 70 percent. Thus, the ampere-hour capacity of this battery is reduced to 0.7 x 200 = 140 ampere hours.

STATE OF CHARGE

After a battery is discharged completely from full charge at the 20-hour rate, the specific

gravity has dropped about 150 points to about 1.060. The number of points that the specific gravity drops per ampere-hour can be determined for each type of battery. For each ampere-hour taken out of a battery a definite amount of acid is removed from the electrolyte and combined with plates.

For example, if a battery is discharged from full charge to the low-voltage limit at the 20-hour rate and if 100 ampere-hours are obtained with a specific gravity drop of 150 points, there is a drop of $\frac{150}{100}$, or 1.5 points per ampere-hour delivered. If the reduction is specific gravity per ampere-hour is known, the drop in specific gravity for this battery may be predicted for any number of ampere-hours delivered to a load. For example, if 70 ampere-hours are delivered by the battery at the 20-hour rate or any other rate or collection of rates, the drop in specific gravity is 70 x 1.5, or 105 points.

Conversely, if the drop in specific gravity per ampere-hour and the total drop in specific gravity are known, the ampere-hours delivered by a battery may be determined. For example, if the specific gravity of the previously considered battery is 1.210 when the battery is fully charged and 1.150 when it is partly discharged, the drop in specific gravity is 1,210 - 1,150, or 60 points, and the number of ampere-hours taken out of the battery is $\frac{60}{1.5}$, or 40 ampere-hours. Thus the number of ampere-hours expended in any battery discharge can be determined from the following items.

1. The specific gravity when the battery is fully charged.
2. The specific gravity after the battery has been discharged.
3. The reduction is specific gravity per ampere-hour.

Voltage alone is not a reliable indication of the state of charge of a battery except when the voltage is near the low-voltage limit on discharge. During discharge the voltage falls. The higher the rate of discharge the lower will be the terminal voltage. Open-circuit voltage is of little value because the variation between full charge and complete discharge is so small—only about 0.1 volt per cell. However, abnormally low voltage does indicate injurious sulfation or some other serious deterioration of the plates.

## TYPES OF CHARGES

The following types of charges may be given to a storage battery, depending upon the condition of the battery:
1. Initial charge.
2. Normal charge.
3. Equalizing charge.
4. Floating charge.
5. Fast charge.

### Initial Charge

When a new battery is shipped dry, the plates are in an uncharged condition. After the electrolyte has been added it is necessary to convert the plates into the charged condition. This is accomplished by giving the battery a long low-rate initial charge. The charge is given in accordance with the manufacturer's instructions, which are shipped with each battery. If the manufacturer's instructions are not available, reference should be made to the detailed instruction in current directives.

### Normal Charge

A normal charge is a routine charge that is given in accordance with the nameplate data during the ordinary cycle of operation to restore the battery to its charged condition. The following steps should be observed:
1. Determine the starting and finishing rate from the nameplate data.
2. Add water, as necessary, to each cell.
3. Connect the battery to the charging panel and make sure the connections are clean and tight.
4. Turn on the charging circuit and set the current through the battery at the value given as the starting rate.
5. Check the temperature and specific gravity of pilot cells hourly.
6. When the battery begins to gas freely, reduce the charging current to the finishing rate.

A normal charge is complete when the specific gravity of the pilot cell, corrected for temperature, is within 5 points (0.005) of the specific gravity obtained on the previous equalizing charge.

### Equalizing Charge

An equalizing charge is an extended normal charge at the finishing rate. It is given periodically to insure that all the sulfate is driven from the plates and that all the cells are restored to a maximum specific gravity. The equalizing charge is continued until the specific gravity of all cells, corrected for temperature, shows no change for a 4-hour period. Readings of all cells are taken every half hour.

### Floating Charge

A battery may be maintained at full charge by connecting it across a charging source that has a voltage maintained within the limits of from 2.13 to 2.17 volts per cell of the battery. In a floating charge, the charging rate is determined by the battery voltage rather than by a definite current value. The voltage is maintained between 2.13 and 2.17 volts per cell with an average as close to 2.15 volts as possible.

### Fast Charge

A fast charge is used when a battery must be recharged in the shortest possible time. The charge starts at a much higher rate than is normally used for charging. It should be used only in an emergency as this type charge may be harmful to the battery.

## CHARGING RATE

Normally, the charging rate of Navy storage batteries is given on the battery nameplate. If the available charging equipment does not have the desired charging rates, the nearest available rates should be used. However, the rate should never be so high that violent gassing occurs. NEVER ALLOW THE TEMPERATURE OF THE ELECTROLYTE IN ANY CELL TO RISE ABOVE 125°F.

## CHARGING TIME

The charge must be continued until the battery is fully charged. Frequent readings of specific gravity should be taken during the charge. These readings should be corrected to 80°F and compared with the reading taken before the battery was placed on charge. If the rise in specific gravity in points per ampere-hour is known, the approximate time in hours required to complete the charge is as follows:

$$\frac{\text{rise in specific gravity in points to complete charge}}{\text{rise in specific gravity in points per ampere-hour} \times \text{charging rate in amperes}}$$

## GASSING

When a battery is being charged, a portion of the energy is dissipated in the electrolysis of the water in the electrolyte. Thus, hydrogen is released at the negative plates and oxygen at the positive plates. These gases bubble up through the electrolyte and collect in the air space at the top of the cell. If violent gassing occurs when the battery is first placed on charge, the charging rate is too high. If the rate is not too high steady gassing which develops as the charging proceeds, indicates that the battery is nearing a fully charged condition. A mixture of hydrogen and air can be dangerously explosive. No smoking, electric sparks, or open flames should be permitted near charging batteries.

## NICKEL-CADMIUM BATTERIES

The nickel-cadmium batteries are far superior to the lead-acid type. Some are physically and electrically interchangeable with the lead-acid type, while some are sealed units which use standard plug and receptacle connections which are used on other electrical components. These batteries generally require less maintenance than lead-acid batteries throughout their service life in regard to the adding of electrolyte or water.

The nickel-cadmium and lead-acid batteries have capacities that are comparable at normal discharge rates, but at high discharge rates the nickel-cadmium battery can:

1. Be charged in a short time.
2. Deliver a large amount of power.
3. Stay idle in any state of charge for an indefinite time and keep a full charge when stored for a long time.
4. Be charged and discharged any number of times without any appreciable damage.

5. The individual cells may be replaced if a cell wears out; the rest of the cells do not have to be replaced.

Due to their superior capabilities, nickel-cadmium batteries are being used extensively in many military applications that require a battery with a high discharge rate. A prime example is the aircraft storage battery.

Some lead-acid batteries are equipped with the same quick-disconnect receptacle and plug used on nickel-cadmium batteries. In distinguishing a lead-acid battery from a nickel-cadmium battery or a silver-zinc battery, the nameplate of each battery should be checked, since the physical appearance could be the same. (See fig. 3-16.)

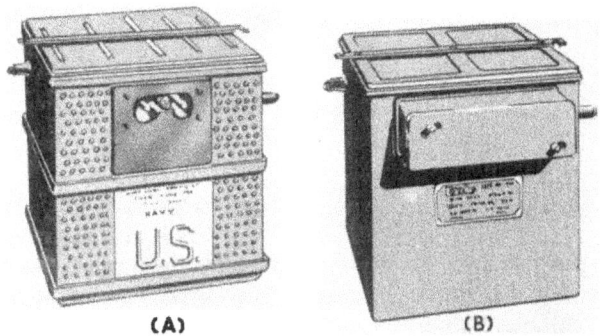

AE.98

Figure 3-16.—(A) Nickel-cadmium battery; (B) lead-acid battery.

The nickel-cadmium battery plates are constructed of nickel powder sintered to a nickel wire screen. The active materials (nickel-hydroxide on the positive plate and cadmium-hydroxide on the negative plate) are electrically bonded to the basic plate structure. The separators are constructed of plastic, nylon cloth, or a special type of cellophane, and assembled as a cell core with plates. (See fig. 3-17.)

The construction of the sintered-plate cell is accomplished by a powder metallurgy process. Carbonyl nickel powder is lightly compressed in a mold and then is subjected either to a temperature of about 1,600 F in a sintering furnace or to a sudden heavy electric current. Either process causes the individual grains of nickel to weld at their points of contact, producing a porous plaque which is approximately 80 percent open holes and 20 percent solid nickel. The plaques are then impregnated with

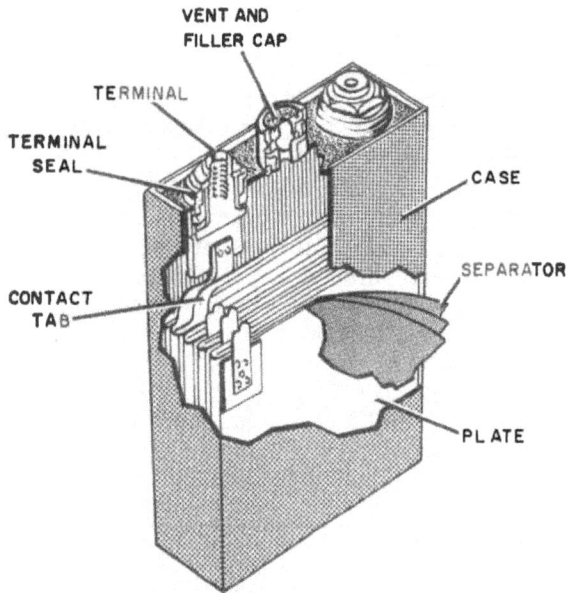

BE.40

Figure 3-17.—Nickel-cadmium cell.

active materials. They are then soaked in a solution of nickel salts to make the positive plates and in a solution of cadmium salts to make negative plates. The bath is repeated until the plaques contain the amount of active material necessary to give them the capacity desired. When the plaques are impregnated, they are classified as plates.

The electrolyte used in a nickel-cadmium battery is a 30-percent-by-weight solution of potassium hydroxide in distilled water. Chemically speaking, this is just about the exact opposite to the diluted sulfuric acid used in the lead battery. As with lead-acid batteries, there are limitations on the concentration of electrolyte solution that can be used in nickel-cadmium cells. The specific gravity of the solution should not be outside the range of 1.240 to 1.300 at 70 F. The electrolyte in the nickel-cadmium battery does not chemically react with the plates as the electrolyte does in the lead battery. It acts only as a conductor of current between plates—therefore, no flaking or shredding of active material. Consequently the plates do not deteriorate, nor does the specific gravity of the electrolyte appreciably change. For this reason, it is not possible to determine the charge state of a nickel-cadmium battery by checking the electrolyte with a hydrometer; neither can the charge be determined by a voltage test because of the inherent character-

istic that the voltage remains constant during 90 percent of the discharge cycle.

No external vent is required since gassing of this type battery is practically negligible. As a safety precaution, however, relief valves have been installed in the fill hole cap of each cell (fig. 3-18) in order to release any excess gas that is formed when the battery is charged improperly.

AE.100

Figure 3-18.—Relief valves in negative post of cells.

CHARGE DETERMINATION

At the present time, a simple method of determining the exact state of charge has not been developed. In an aircraft battery, the only practical method is to measure the open circuit battery voltage. If the open circuit voltage exceeds 1.4 volts per cell and the battery voltage is 26.0 volts (normally 28.0 volts) or more, it can be assumed that the battery is fully charged. If the cell voltage is less than 1.0 volt under load, or the battery voltage measures 25 volts or less, it can be assumed that the battery has expended over 90 percent of its capacity and is in need of charging.

In a battery shop, either of two methods may be used for determining the state of charge of a nickel-cadmium battery; namely, the constant potential method or the discharge method. The constant potential method consists of connecting a constant potential of 28.5 ± 0.3 volts across the battery and observing the charging current. If the current falls to 3 amperes or less within 5 minutes, the battery is charged.

The discharge method consists of placing a 15-ampere load across the battery for 5 minutes. If the voltage does not drop below 22 volts during the discharge period, the battery may be returned to service after being recharged.

The available ampre-hour capacity cannot be accurately determined. Therefore, it is recommended that any battery whose charge is unknown or subject to doubt be discharged to or beyond the manufacturer's set end point of 1.0 or 1.1 volts, and then be recharged in accordance with the appropriate instructions. This process will prevent possible damage to the cells from overcharge.

## CHARGING

Nickel-cadmium batteries should preferably be charged at an ambient temperature of 70° to 80°F. Never allow a battery on charge to exceed 100°F as this may cause overcharging and gassing. In the battery shop a thermometer should be placed between the central cells in such a manner that the bulb of the thermometer is located below the top of the cell. Whenever the temperature of battery is 100°F or higher, the battery should not be charged.

The rate of charging of a nickel-cadmium battery is dependent upon two factors, the first being the charging voltage, and the second being the temperature of the battery. In hot weather where the ground air temperature approaches 90°F or higher, the battery can be adequately charged at 27 volts. In mild ground air temperatures ranging from 35° to 85°F, the battery can be satisfactorily charged at 27.5 volts. In cold, subfreezing weather the battery requires a charging voltage of 28.5 volts.

The nickel-cadmium battery was designed and constructed to operate without gassing of the cells. The charging voltage should be maintained below the gassing voltage (approximately 29.4 volts at 80°F) so that the life of the battery is prolonged. Therefore, on constant potential charging in the battery shop, the voltage should be set at 28 volts or less. Under no circumstances should this voltage exceed 28.5 volts.

If the battery has never been placed in service, follow the manufacturer's instructions accompanying the battery for the initial charge. If possible, the battery should be charged by the constant potential method.

For constant potential charging, maintain the battery at 28 volts for 4 hours, or until the current drops below 3 amperes. Do not allow battery temperature to exceed 100°F.

For constant current charging, start the charge at 10 to 15 amperes and continue until the voltage reaches 28.5 volts—then reduce the current to 4 amperes and continue charging until the battery voltage reaches 28.5 volts, or until the battery temperature exceeds 100°F and the voltage begins to decline.

Never add electrolyte unless the battery is fully charged. Allow the fully charged battery to stand for a period of 3 or 4 hours before distilled water can be added to bring the electrolyte to the proper level. A hydrometer or syringe can be used for introduction of the distilled water—just enough to cover the top of the plates. The battery solution is then recycled to stir the water and prevent it from freezing during cold-weather operation.

## SAFETY PRECAUTIONS

The electrolyte used in nickel-cadmium batteries is potassium hydroxide (KOH). This is a highly corrosive alkaline solution, and should be handled with the same degree of caution as sulfuric acid ($H_2SO_4$). Personnel should always wear rubber gloves, a rubber apron, and protective goggles when handling and servicing these batteries. If the electrolyte is spilled on the skin or clothing, the exposed area should be rinsed immediately with water, or if available, vinegar, lemon juice, or boric acid solution. If the face or eyes are affected, treat as above and report immediately for medical examination and treatment.

The battery shop used for nickel-cadmium batteries should be a separate isolated shop from the lead-acid battery shop.

## SILVER-ZINC BATTERIES

Silver-zinc batteries are used largely in military applications and in some industrial applications where their unique characteristics are sufficiently important to justify their comparatively high cost.

The silver-zinc battery was developed for one major and one secondary purpose. The major purpose was to secure a large quantity of electrical power for emergency operations. The secondary purpose was to permit a design weight savings in new batteries. A lightweight, silver-zinc battery provides as much electrical capacity as a much larger lead-acid or nickel-cadmium battery.

Operational silver-zinc batteries have a nominal operating voltage of 24 volts, obtained with sixteen 1.5 volt cells. Cell electrolytic levels should be monitored and adjusted periodically. The other required operations that might be considered maintenance are the normal recharging of the battery and keeping the top surfaces of the cells reasonably clean.

## CHARACTERISTICS

Because of its extremely low internal resistance, the silver-zinc battery is capable of discharge rates of up to 30 times its ampere-hour rating. The low internal resistance (as low as 0.0003 ohms per cell) is due primarily to the excellent conductivity of its plates, the close plate spacing (possible because small amounts of electrolyte may be used successfully), and the fact that the composition (and therefore the conductivity) of the electrolyte does not change during discharge. The internal conductivity of the battery increases during discharge as the positive plates are changed from oxides of silver (fair conductors) to metallic silver.

The high electrical capacity per unit of space and weight is a result of the close plate spacing, the large degree to which the active plate materials are utilized, and the absence of heavy supporting grids in the plate. Silver-zinc batteries are capable of producing as much as six times more energy per unit of weight and volume than other types. Silver-zinc cells have been built with capacities ranging from tenths of ampere-hours to thousands of ampere-hours.

Good voltage regulation is provided by the relatively constant voltage discharge characteristic of the silver-zinc battery. Terminal voltage is essentially constant throughout most of the discharge when discharged at rates higher than 2- or 3-hour rate.

Silver-zinc batteries have a maximum service cycle life which is less than that of other types, but their life expectancy compares favorably with that of other types of batteries that are designed for maximum capacity per unit of space and weight such as nickel-cadmium batteries.

## OPERATION

The construction and electrochemical reactions of the silver-zinc battery are somewhat similar to those of the nickel-cadmium type. When in the fully charged condition, the positive plates are composed of silver oxide and the negative plates of zinc. As the battery discharges, the positive plates are reduced to metallic silver and the negative plates are oxidized. Thus, when the battery is discharging, electrons are flowing out of the cathode (negative plates) and into the anode (positive plates) by way of the external circuit.

The electrolyte, potassium hydroxide in aqueous solution, exists as potassium (K) and hydroxide (OH) ions, which serve only to conduct the electric charge between the plates. Thus, the electronic or metallic conduction in the external circuit is balanced by the ionic or electrolytic conduction through the electrolyte, so as to maintain the net charge transfer into and out of each electrode the same.

As with other types of alkaline cells, and unlike lead-acid cells, the electrolyte does not take part in the chemical transformations and therefore its specific gravity does not change with the state of charge of the cell. As long as the plates are covered, the electrical capacity of the battery is independent of the amount of electrolyte present.

In general, silver-zinc batteries require maintenance which is similar in many respects to that which is required of the lead-acid batteries. Testing the open circuit voltage of the battery is the method by which its state of charge is determined.

A silver-zinc battery tester or a voltmeter which reads accurately to 0.1 volt should be used to test the open circuit voltage of the battery. If the reading is below 25.6 volts, remove the battery cover and inspect the top of the battery for corrosion or damaged cells. If any damage is evident, remove and replace the battery.

## CHARGING

The silver-zinc battery is usually shipped in a dry condition. Only the special electrolyte furnished in the filling kit provided with each new battery should be used. Some battery types may use electrolytes containing special additives; other electrolytes, if used, may degrade the battery. The electrolyte must be kept in a closed alkali-resistant container, or it will absorb carbon dioxide from the air and deteriorate. The filling kit contains detailed instructions for filling and should be followed in detail. (NOTE: Batteries that will not be used within 30 days should be stored in the dry state.)

Silver-zinc batteries are sensitive to excessive voltage during charging and may be damaged if the voltage exceeds 2.05 volts per cell. Precaution must therefore be taken to insure that the charging equipment is adjusted accurately to cut off the current at 28.7 volts.

Where charging is not monitored automatically or periodically, a voltage cut off system must be used which will interrupt the charging current when the voltage rises to 28.7 volts.

If possible, charging should be performed at an ambient temperature of 60° to 90°F, and the battery temperature during charging should not exceed 150°F as measured at the intercell connections.

While silver-zinc batteries do not generate any harmful gases during normal charge and discharge operations, they do generate both oxygen and hydrogen gases during excessive overcharging. All vent caps and spong-rubber plugs must be removed from the vent hole during charging operations. If electrolyte is forced from the vent holes or if excessive gassing is evident, it is an indication of overheating, and the charging should be interrupted for 8 hours to allow the battery to cool. After charging, the batteries should be allowed to stand idle at least 8 hours.

The level of the electrolyte of each cell of the battery should be checked after charging and the level adjusted by either removing any excess electrolyte or by adding distilled water if low.

The safety precautions relating to silver-zinc batteries are the same as those which have been presented for the nickel-cadmium batteries.

## SILVER-CADMIUM BATTERY

One of the recent developments in storage batteries is the silver-cadmium battery. Generally, the most important requirements for evaluating and designing a battery are for high energy density, good voltage regulation, long shelf life, repeatable number of cycles, and long service life expectancy. The silver-cadmium battery is designed to offer the overall maximum performance in all of these expectations.

The silver-cadmium battery has more than twice the wet shelf life of the silver-zinc battery. The long shelf life plus the good voltage regulation makes the silver-cadmium battery a highly desirable addition to the family of electric storage batteries. Limitations include lower cell voltage than other rechargeable batteries and high initial cost.

# CHAPTER 4

# SERIES D-C CIRCUITS

## SIMPLE ELECTRIC CIRCUIT

Whenever two unequal charges are connected by a conductor, a complete pathway for current flow exist. Current will flow from the negative to the positive charge. This was illustrated in chapter 2.

An electric circuit is a completed conducting pathway, consisting not only of the conductor, but including the path through the voltage source. Current flows from the positive terminal through the source, emerging at the negative terminal. As an example, a lamp connected by conductors across a dry cell forms a simple electric circuit. (Refer to fig. 4-1.)

Current flows from the negative (-) terminal of the battery through the lamp to the positive (+) battery terminal, and continues by going through the battery from the positive (+) terminal to the negative (-) terminal. As long as this pathway is unbroken, it is a closed circuit and current will flow. However, if the path is broken at ANY point, it is an open circuit and no current flows. (See fig. 4-1 (B).)

Current flow in the external circuit is the movement of electrons in the direction indicated by the arrows (from the negative terminal through the lamp to the positive terminal). (See fig. 4-1 (A).) Current flow in the internal battery circuit is the simultaneous movement in opposite directions of positive hydrogen ions toward the positive terminal of the battery and negative ions toward the negative terminal.

## SCHEMATIC REPRESENTATION

A SCHEMATIC is a diagram in which symbols are used for the various components instead of pictures. These symbols are used in an effort to make the diagrams easier to draw and easier to understand. In this respect, schematic symbols aid the technician in the same way that shorthand aids the stenographer. In previous chapters the schematic symbols for cells and resistances were presented. These symbols will now be used to discuss the circuits of figure 4-1.

A schematic diagram of a basic circuit is shown in figure 4-2. The battery is designated by the letter symbols $E_{bb}$, the light bulb in the circuit is labeled $R_1$. Since in reality the light bulb element is nothing more than a wire wound resistor, the conventional resistor symbol is used for the bulb in this discussion. It should be noted, however, that the light bulb has its own specific schematic symbol and is not normally drawn as a resistor. The standard symbol used for a light bulb is discussed at a later time when need arises.

In studies of electricity and electronics many circuits are analyzed which consist mainly of specially designed resistive components. As previously stated, these components are called resistors. Throughout the remaining analysis of the basic circuit, the resistive component will be a physical resistor. However, the resistive component could be any one of several electrical devices.

A closed loop of wire (conductor) is not necessarily a circuit. A source of voltage must be included to make it an electric circuit. In any electric circuit where electrons move around a closed loop, current, voltage, and resistance are present. The physical pathway for current flow is actually the circuit. Its resistance controls the amount of current flow around the circuit. By knowing any two of the three quantities, such as voltage and current, the third (resistance) may be determined. This is done mathematically by the use of Ohm's LAW.

## OHM'S LAW

In the early part of the 19th century Georg Simon Ohm proved by experiment that a precise

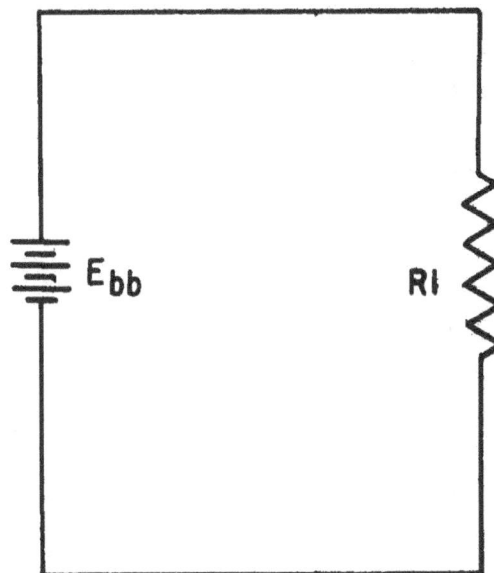

BE.42
Figure 4-2.—Schematic diagram of a basic circuit.

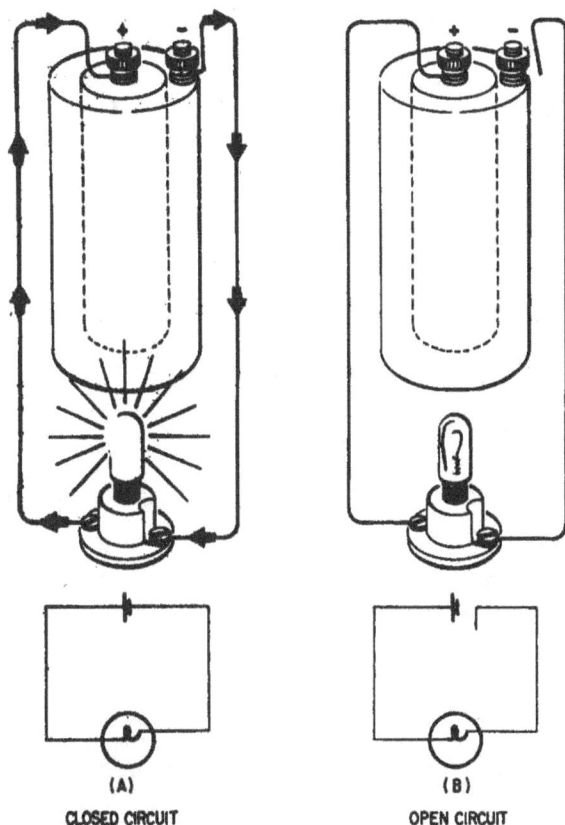

BE.41
Figure 4-1.—(A) Simple electric circuit (closed); (B) simple electric circuit (open).

relationship exists between current, voltage, and resistance. This relationship is called Ohm's Law and is stated as follows:

The current in a circuit is DIRECTLY proportional to the applied voltage and INVERSELY proportional to the circuit resistance. Ohm's Law may be expressed as an equation:

$$I = \frac{E}{R} \qquad (4\text{-}1)$$

Where: I = current in amperes

E = voltage in volts

R = resistance in ohms

If any two of the quantities in equation (4-1) are known, the third may be easily found. For example, Figure 4-3 shows a circuit containing a resistance of 1.5 ohms and a source voltage

of 1.5 volts. How much current flows in the circuit?

Given: E = 1.5 volts

R = 1.5 ohms

I = ?

Solution: $I = \frac{E}{R}$

$I = \frac{1.5}{1.5}$

I = 1 ampere

To observe the effect of source voltage on circuit current, the above problem will be solved again using double the previous source voltage.

Given: E = 3 volts

R = 1.5 ohms

I = ?

Solution: $I = \frac{E}{R}$

$I = \frac{3}{1.5}$

I = 2 amperes

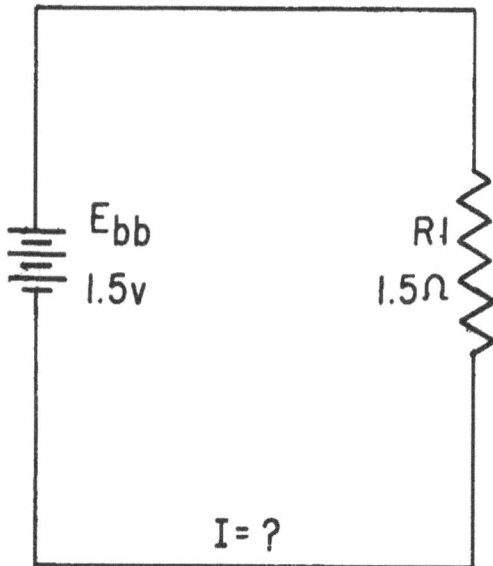

BE.43

Figure 4-3.—Determining current in a basic circuit.

Notice that as the source voltage doubles, the circuit current also doubles.
Circuit current is directly proportional to applied voltage and will change by the same factor that the voltage changes.

To verify the statement that current is inversely proportional to resistance, assume the resistor in figure 4-3 to have a value of 3 ohms.

Given:
$$E = 1.5 \text{ volts}$$
$$R = 3 \text{ ohms}$$
$$I = ?$$

Solution:
$$I = \frac{E}{R}$$
$$I = \frac{1.5}{3}$$
$$I = 0.5 \text{ ampere}$$

Comparing this current of 0.5 ampere for the 3 ohm resistor, to the 1 ampere of current obtained with the 1.5-ohm resistor, shows that doubling the resistance will reduce the current to one half the original value. Circuit current is inversely proportional to the circuit resistance.

In many circuit applications current is known and either the voltage or the resistance will be the unknown quantity. To solve a problem in which current and resistance are known, the basic formula for Ohm's Law must be transposed to solve for E as follows:

Basic equation: $I = \dfrac{E}{R}$        (4-1)

Multiply both sides of the equation by R.

$$IR = \frac{E}{\not{R}} \not{R}$$
$$IR = E$$
$$E = IR \qquad (4-2)$$

To transpose the basic formula when resistance is unknown:

Basic equation: $I = \dfrac{E}{R}$        (4-1)

Multiply both sides of the equation by R.

$$IR = \frac{E}{\not{R}} \not{R}$$
$$IR = E$$

Divide both sides of the equation by I.

$$\frac{\not{I}R}{\not{I}} \quad \frac{E}{I}$$
$$R = \frac{E}{I} \qquad (4-3)$$

Example: What voltage is required to properly light a lamp having a resistance of 10 ohms and a current rating of 1 ampere?

First draw a circuit like figure 4-4 including all the given information.

Given:
$$R = 10 \text{ ohms}$$
$$I = 1 \text{ ampere}$$
$$E = ?$$

Solution:
$$E = IR$$
$$E = 1 \times 10$$
$$E = 10 \text{ volts}$$

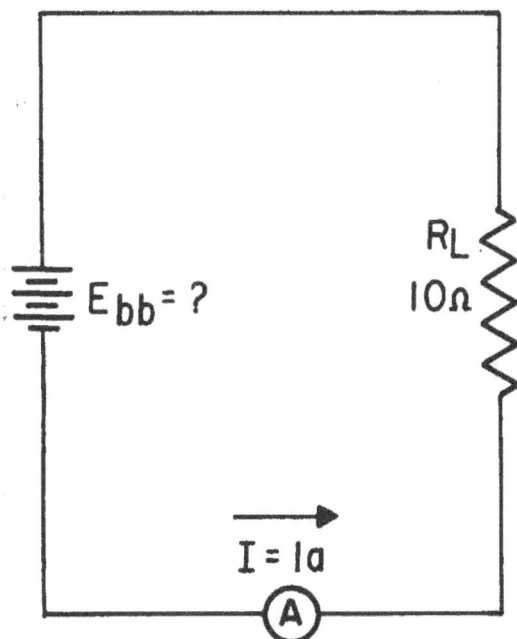

BE.44

Figure 4-4.—Determining voltage in a basic circuit.

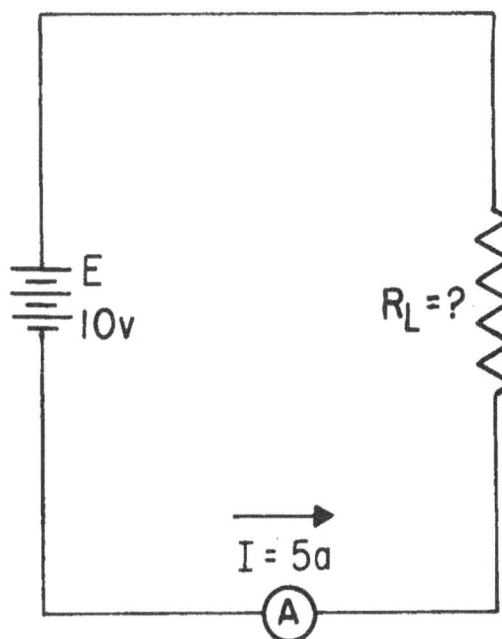

BE.45

Figure 4-5.—Determining resistance in a basic circuit.

Example: When a 10 volt source is connected to a circuit, the circuit draws 5 amperes of current from the source. How much resistance is contained in the circuit?

Given:     E = 10 volts

           I = 5

           R = ?

Draw and label a circuit like figure 4-5.

Solution:   $R = \frac{E}{I}$     (4-3)

           $R = \frac{10}{5}$

           R = 2 ohms

Although the three equations representing Ohm's Law are fairly simple, they are perhaps the most important of all electrical equations. These three equations and the law they represent must be thoroughly understood before continuing on to more advanced theory.

## GRAPHICAL ANALYSIS

One of the most valuable methods of inquiry available to the technician is that of graphical analysis. No other method provides a more convenient or more rapid way to observe the characteristics of an electrical device.

The first step in constructing a graph consists of obtaining a table of data from which the graph will evolve. The information in the table can be obtained experimentally by taking laboratory measurements on the device under examination, or can be obtained theoretically through a series of computations. The latter method will be used here.

Let us assume that the characteristics of the circuit shown in figure 4-6 are to be investigated using Ohm's Law and graphical methods. Since there are three variables (E, I, and R) under consideration, there are three unique graphs that may be constructed.

In constructing any graph of electrical quantities, it is standard practice to vary one quantity in a specified way, and note the changes which occur in a second quantity. The quantity which is intentionally varied is called the independent variable and is plotted on the X-AXIS. The second quantity which changes as a result of changes in the first quantity is called the dependent variable and is plotted on the Y-AXIS. Any other quantities involved are held constant.

60

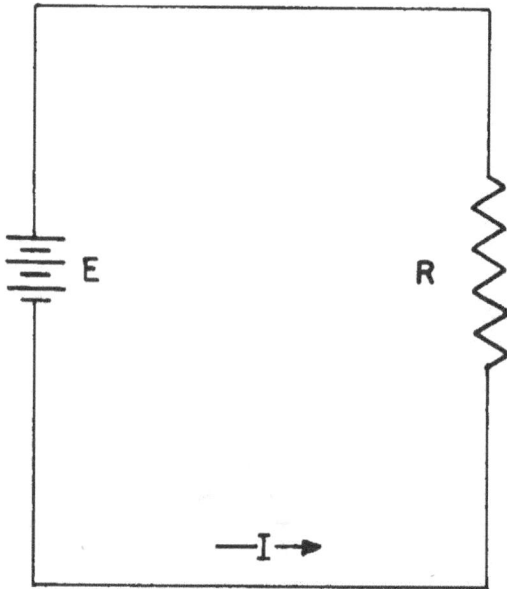

BE.46

Figure 4-6.—Three variables in a series circuit.

In the circuit of figure 4-6 the resistance will remain fixed (constant) and the voltage (independent variable) will be varied. The resulting changes in current (dependent variable) will then be graphed.

To aid in compiling the data, a table of values is completed as shown in figure 4-7. This table shows R to be held constant at 10 ohms as E is varied from 0 to 20 volts in 5-volt steps. Through the use of Ohm's Law the value of current in column two of the table can be calculated for each value of voltage in column one. When the table is complete, the information it contains can be used to construct the graph in figure 4-7. For example, when the voltage applied to the 10-ohm resistor is 10 volts, the current is 1 ampere. These values of current and voltage determine a point on the graph. When all the points have been plotted, a smooth curve is drawn through the points. This curve is called the volt-ampere characteristic for the 10-ohm resistor.

Through the use of this curve the value of current through the resistor can be quickly determined for any value of voltage between 0 and 20 volts.

A very important characteristic of a fixed resistor is illustrated by the graph in figure 4-7. Since the volt-ampere characteristic curve is a straight line, it shows that equal changes of voltage across a resistor produce equal changes in current through the resistor. Because of this straight line characteristic, the fixed resistor is called a linear device.

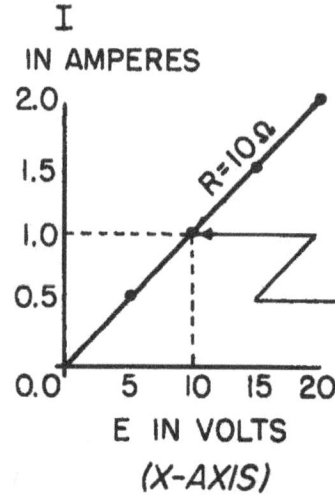

BE.47

Figure 4-7.—Volt-ampere characteristic.

This graph illustrates an important characteristic of the basic law—namely, the current varies directly with the applied voltage if the resistance is constant.

If the voltage across a load is maintained at a constant value, the current through the load will depend solely upon the effective resistance of the load. For example, with a constant voltage of 12 volts and a resistance of 12 ohms, the current will be $\frac{12}{12}$, or 1 ampere.

If the resistance is halved, the current will be doubled; if the resistance is doubled, the current will be halved. In other words, the current will vary inversely with the resistance.

If the resistance of the load is reduced in steps of 2 ohms starting at 12 ohms and continuing to 2 ohms, the current through the load becomes $\frac{12}{10}$ = 1.2 amperes; $\frac{12}{8}$ = 1.5 amperes; $\frac{12}{6}$ = 2 amperes; and so forth. The relation between current and resistance in this example is expressed as a graph (fig. 4-8), whose equation is $I = \frac{12}{R}$. The numerator of the fraction represents a constant value of 12 volts in this example. As R approaches a small value the current approaches a very large value. The example illustrates a second equally important relation in Ohm's law—namely, that the current varies inversely with the resistance.

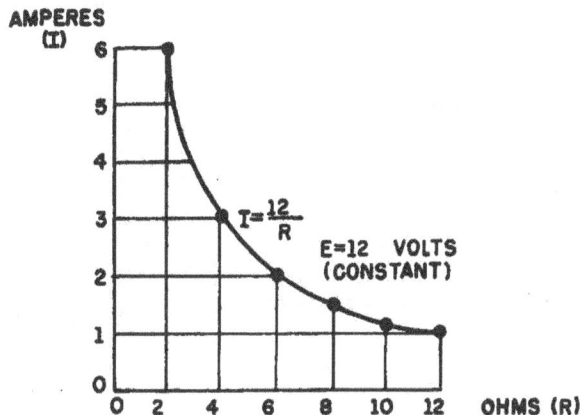

Figure 4-8.—Relation between current and resistance.

If the current through the load is maintained constant at 5 amperes, the voltage across the load will depend upon the resistance of the load and will vary directly with it. The relation between voltage and resistance is shown in the graph of figure 4-9. Values of resistance are plotted horizontally along the X-axis to the right of the origin, and corresponding values of voltage are plotted vertically along the Y-axis above the origin. The graph is a straight line having the equation E = 5R. The coefficient 5 represents the assumed current of 5 amperes which is constant in this example. Thus, a third important relation is illustrated—namely, that the voltage across a device varies directly with the effective resistance of the device provided the current through the device is maintained constant.

## APPLYING OHM'S LAW

Equation (4-1) may be transposed to solve for the resistance if the current and voltage are known, or to solve for the voltage if the current and resistance are known. Thus, $R = \dfrac{E}{I}$ and E = IR. For example, if the voltage across a device is 50 volts and the current through it is 2 amperes, the resistance of the device will be

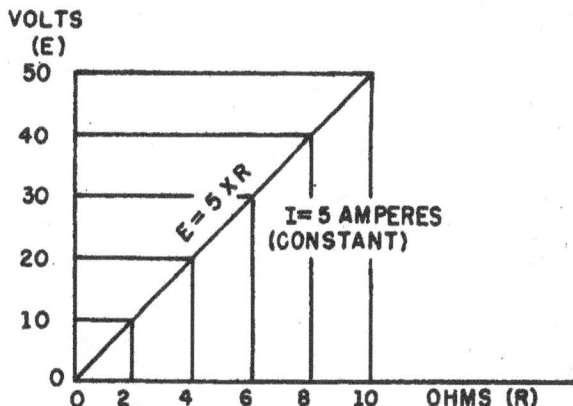

Figure 4-9.—Graph of voltage vs. resistance with constant current.

$\dfrac{50}{2}$, or 25 ohms. Also, if the current through a wire is 3 amperes and the resistance of the wire is 0.5 ohm, the voltage drop across the wire will be 3 x 0.5, or 1.5 volts.

Equation (4-1) and its transpositions may be obtained readily with the aid of figure 4-10. The circle containing E, I, and R is divided into two parts with E above the line and IR below it. To determine the unknown quantity, first cover that quantity with a finger. The location of the remaining uncovered letters in the circle will indicate the mathematical operation to be performed. For example, to find I, cover I with a finger. The uncovered letters indicate that E is to be divided by R, or $I = \dfrac{E}{R}$. To find E, cover E. The result indicates that I is to be multiplied by R, or E = IR. To find R, cover R. The result indicates that E is to be divided by I, or $R = \dfrac{E}{I}$.

The beginning student is cautioned not to rely wholly on the use of this diagram when transposing simple formulas but rather to use it to supplement his knowledge of the algebraic method. Algebra is a basic tool in the solution of electrical problems and the importance of knowing how to use it should not be underemphasized or bypassed after the student has learned a shortcut method such as the one indicated in this figure.

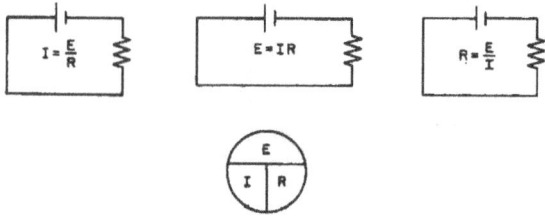

BE.50

Figure 4-10.—Ohm's law in diagram form.

## ELECTRIC POWER AND ENERGY

### POWER

Power, whether electrical or mechanical, pertains to the rate at which work is being done. Work is done whenever a force causes motion. If a mechanical force is used to lift or move a weight, work is done. However, force exerted WITHOUT causing motion, such as the force of a compressed spring acting between two fixed objects, does not constitute work.

Previously, it was shown that voltage is electrical force, and that voltage forces current to flow in a closed circuit. However, when voltage exists between two points, but current cannot flow, no work is done. This is similar to the spring under tension that produced no motion. When voltage causes electrons to move, work is done. The instantaneous RATE at which this work is done is called the electric power rate, and its measure is the watt.

A total amount of work may be done in different lengths of time. For example, a given number of electrons may be moved from one point to another in 1 second or in 1 hour, depending on the RATE at which they are moved. In both cases, total work done is the same. However, when the work is done in a short time, the wattage, or INSTANTANEOUS POWER RATE is greater than when the same amount of work is done over a longer period of time.

As stated, the basic unit of power is the WATT, and it is equal to the voltage across a circuit multiplied by current through the circuit. This represents the rate at any given instant at which work is being done in moving electrons through the circuit. The symbol P indicates electrical power. Thus, the basic power formula is $P = E \times I$. E is the voltage across and I is the current through the resistor or circuit whose power is being measured. The amount of power will change when either voltage or current, or both voltage and current change. This relation is shown with the graph and simple circuit in figure 4-11.

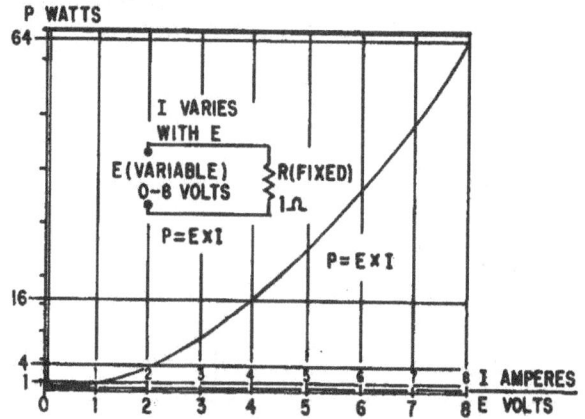

BE.51

Figure 4-11.—Graph of power related to changing voltage and current.

The resistance is 1 ohm, and does not change. Voltage E is increased in steps of 1 volt from 0 to 8. By applying Ohm's law, the current I is determined for each step of voltage. For instance, when E is 1 volt, the current I is

$$I = \frac{E}{R} \qquad (4\text{-}1)$$

$$I = \frac{1}{1}$$

$$I = 1 \text{ ampere}$$

Power P, in watts, is determined by applying the basic power formula $P = E \times I$. When E is 1 volt, I is 1 ampere, so P is

$$P = E \times 1$$

$$P = 1 \times 1$$

$$P = 1 \text{ watt}$$

However, when E is 2 volts, and I is 2 amperes, P becomes:

$$P = E \times 1$$

$$P = 2 \times 2$$

$$P = 4 \text{ watts}$$

It is important to note that when voltage E was doubled from 1 volt to 2 volts, power P doubled TWICE from 1 watt to 4 watts. This occurred because the doubling of voltage caused a doubling of current, therefore power was doubled twice. This is shown as follows:

$$P = E \times I$$

$$P = E \times 2 \times I \times 2$$

$$P = (1 \times 2) \times (1 \times 2)$$

$$P = 2 \times 2$$

$$P = 4 \text{ watts.}$$

This shows that the power in a circuit of fixed resistance is caused to change at a SQUARE rate by changes in applied voltage. Thus, the basic power formula $P = E \times I$ may also be $P = \dfrac{E^2}{R}$. To further illustrate the square-rate relation between power and voltage, note on the graph that power is the square of voltage (when resistance is 1 ohm). For instance, when E is 2 volts, P is 4 watts. When E is doubled to 4 volts, P is 16 watts, and when E is redoubled to 8 volts, P becomes 64 watts. When resistance is any value other than 1 ohm, power will not be the exact square of voltage in quantity but it will still vary at a square rate. That is, regardless of the value of resistance, so long as it is fixed, when voltage doubles, power doubles twice. Also, when the voltage is halved, power is halved twice.

Another important relation may be seen by studying figure 4-11. Thus far power has been calculated with voltage and current ($P = E \times I$), and with voltage and resistance ($P = \dfrac{E^2}{R}$). Referring to figure 4-11, note that power also varies as the square of current just as it does with voltage. Thus, another formula for power, with current and resistance as its factors, is $P = I^2 R$. Note that resistance R is a divisor in one formula ($P = \dfrac{E^2}{R}$), but is a multiplier in the other ($P = I^2 R$). This is true because of substitutions in the original formula $P = E \times I$. That is, the

Ohm's law equivalent of I is $\dfrac{E}{R}$. If this equivalent is substituted for I in the power formula $P = E \times I$, the results are as follows:

$$P = E \times I \qquad (4\text{-}4)$$

$$P = E \times \frac{E}{R}$$

$$P = \frac{E^2}{R} \qquad (4\text{-}5)$$

In addition, the Ohm's law equivalent of E is $I \times R$. If this equivalent is substituted for E, the power formula becomes

$$P = E \times I$$

$$P = (I \times R) \times I$$

$$P = I^2 R \qquad (4\text{-}6)$$

In the foregoing discussion, and in figure 4-11, it was shown how variations of the voltage impressed across a fixed resistance caused variations in the circuit current and power. The following discussion refers to figure 4-12. In this circuit, the voltage E is fixed at 10 volts, and the resistance R is the variable factor. (The arrow through the resistance means it is variable).

When the resistance R is set at 1 ohm, the current I is 10 amperes, and the power is

$$P = I^2 R \qquad (4\text{-}6)$$

$$P = (10^2) \times R$$

$$P = 100 \times 1$$

$$P = 100 \text{ watts}$$

When the resistance is doubled to 2 ohms, this same calculation will show that power is halved to 50 watts, as the graph shows. Subsequent redoubling of the resistance to 4 and then to 8 ohms causes the power to be halved each time to 25 and 12-1/2 watts, respectively. Conversely, you should note the relation when starting with resistance at 10 ohms, with 10 watts of power. If resistance is halved to 5 ohms, power is doubled to 20 watts.

In figures 4-11 and 4-12, current and power were caused to vary as a function of voltage, in one case, and of resistance in the other. In figure 4-13, however, current is held constant.

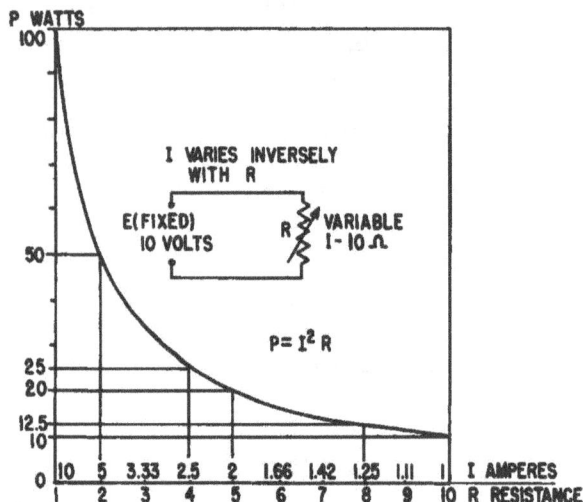

BE.52

Figure 4-12.—Graph of power related to changing resistance and current.

This is done by raising or lowering voltage and resistance equally and directly with each other. The resulting variations in power are linear with those changes. That is, power is changed step-for-step with voltage. The changing resistance maintains a constant current despite changes of voltage. At any point on the graph, voltage divided by resistance is 1 ampere. Had the current been allowed to vary, power would have changed at a curved, or square rate, instead of linearly as on the graph.

Up to this point, four of the most important basic electrical quantities have been discussed. These are E, I, R, and P. It is of fundamental importance that you thoroughly understand the interrelation of these quantities. You should understand how any one of these quantities either controls or is controlled by the others in an electrical circuit. These relations are further explained in the treatment that follows. You should compare each statement carefully with its associated formula. Check each formula for correctness by applying it to the graphs in figures 4-11, 4-12, or 4-13. (The appropriate figure number is indicated after each of the following statements)

　1. Power, as related to E and I: P = EI (fig. 4-11).

This formula states that P is the product of E multiplied by I, regardless of their individual values. If either E or I varies, P varies

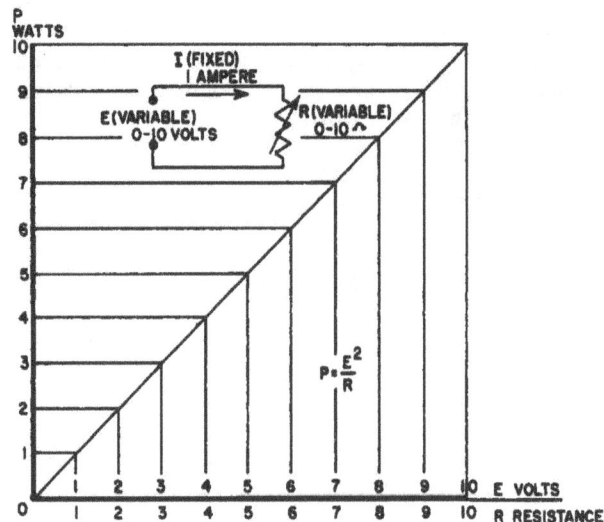

BE.53

Figure 4-13.—Graph of power related to changing voltage and resistance.

proportionally. If both E and I vary, P varies at a square rate.

　2. Power, as related to I and R: $P = I^2R$ (fig. 4-12).

This formula states that if R is held constant, and I is varied, P varies as the square of I, because I appears as a squared quantity $(I^2)$ in the formula. Also, if I is held constant and R is varied, P varies directly and proportionally to R, because R is a multiplier in the formula.

　3. Power, as related to E and R: $P = \dfrac{E^2}{R}$ (fig. 4-13).

This formula states that if R is held constant as E is varied, P varies as the square of E, because E appears as a squared quantity $(E^2)$ in the formula. Also, if E is held constant and R is varied, P varies inversely but proportionally to R, because R is a divisor in the formula.

In the preceding paragraph, P was expressed in terms of alternate pairs of the other three basic quantities E, I, and R. In practice, you should be able to express any one of the three basic quantities, as well as P, in terms of any two of the others. Figure 4-14 is a summary of twelve basic formulas you should know. The four quantities E, I, R, and P are at the center of the figure.

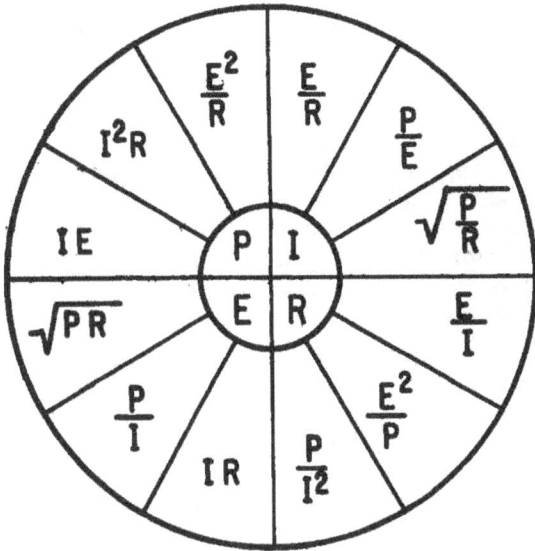

BE.54

Figure 4-14.—Summary of basic formulas.

Adjacent to each quantity are three segments. Note that in each segment, the basic quantity is expressed in terms of two other basic quantities, and no two segments are alike.

## RATING OF ELECTRICAL DEVICES BY POWER

Electrical lamps, soldering irons, and motors are examples of electrical devices that are rated in watts. The wattage rating of a device indicates the rate at which the device converts electrical energy into another form of energy, such as light, heat, or motion.

For example, a 100-watt lamp will produce a brighter light than a 75-watt lamp, because it converts more electrical energy into light.

Electric soldering irons are of various wattage ratings, with the high-wattage irons changing more electrical energy to heat than those of low-wattage ratings.

If the normal wattage rating is exceeded, the equipment or device will overheat and probably be damaged. For example, if a lamp is rated 100 watts at 110 volts and is connected to a source of 220 volts, the current through the lamp will double. This will cause the lamp to use four times the wattage for which it is rated, and it will burn out quickly.

## POWER CAPACITY OF ELECTRICAL DEVICES

Rather than indicate a device's ability to do work, its wattage rating may indicate the device's operating limits. These power limits generally are given as the maximum or minimum safe voltages and currents to which a device may be subjected. However, in cases where a device is not limited to any specific operating voltage, its limits are given directly in watts.

### Resistors

A resistor is an example of such a device. It may be used in circuits with widely different voltages, depending on the desired current. However, the resistor has a maximum current limitation for each voltage applied to it. The product of the resistor's voltage and current at any time must not exceed a certain wattage.

Thus, resistors are rated in watts, in addition to their ohmic resistance. Resistors of the same resistance value are available in different wattage values. Carbon resistors, for example, are commonly made in wattage ratings of 1/3, 1/2, 1, and 2 watts. (See fig. 4-15.) The larger the physical size of a carbon resistor, the higher its wattage rating, since a larger amount of material will absorb and give up heat more easily.

When resistors of wattage ratings greater than 2 watts are needed, wire-wound resistors

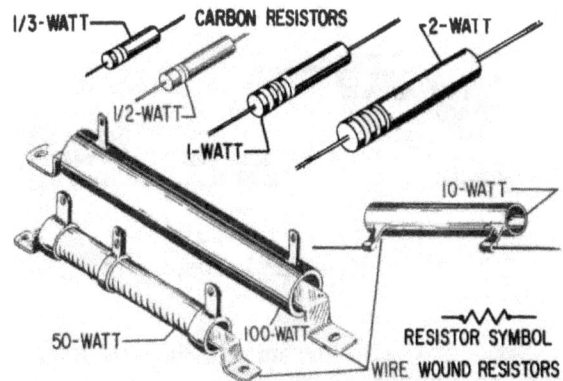

BE.55

Figure 4-15.—Resistors of different wattage ratings.

are used. Such resistors are made in ranges between 5 and 200 watts, with special types being used for power in excess of 200 watts.

## Fuses

When current passes through a resistor, electric energy is transformed into heat, which raises the temperature of the resistor. If the temperature becomes too high, the resistor may be damaged. The metal wire in a wound resistor may melt, opening the circuit and interrupting current flow. This effect is used to an advantage in fuses.

Fuses are actually metal resistors with very low resistance values. They are designed to "blow out" and thus open a circuit when current exceeds the fuse's rated value.

When the power consumed by the fuse raises the temperature of its metal too high, the metal melts, or "blows." In service, fuses are generally connected as shown in figure 4-16.

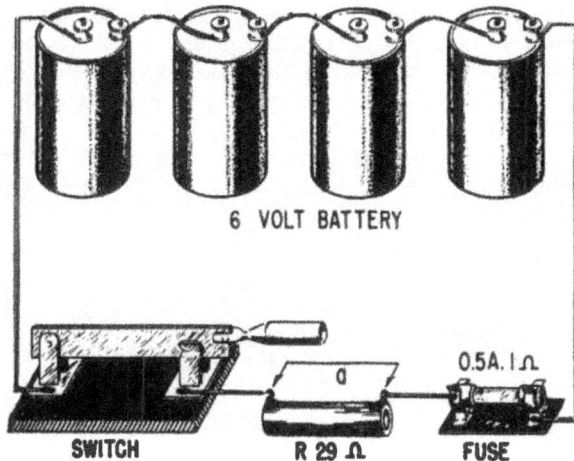

6 VOLT BATTERY

SWITCH          R 29 Ω          FUSE

0.5A.1Ω

BE.56

Figure 4-16.—Simple fused circuit.

Note that all current flowing through the load resistance of 29 ohms must also flow through the half-ampere fuse. Under normal conditions, the total resistance would be $29 + 1 = 30$ ohms. If the switch were closed, the current I would be

$$I = \frac{E}{R}$$

$$I = \frac{6}{30}$$

$$I = 0.2 \text{ ampere}$$

This would be less than the rated current of the fuse, and it would not open. However, if the short conductor (a, fig. 4-16) were connected, the load resistance would be bypassed, or "shorted out." Only the fuse resistance of one ohm would remain in the circuit. Current would then be

$$I = \frac{E}{R}$$

$$I = \frac{6}{1}$$

$$I = 6 \text{ amperes}$$

Six amperes of current will cause the half-ampere fuse to open very quickly, because its wattage rating is greatly exceeded.

There are a great number of different types and sizes of fuses presently in use. Figure 4-17 shows three of the most common types. Fuses are discussed in greater detail in chapter 14 of this manual.

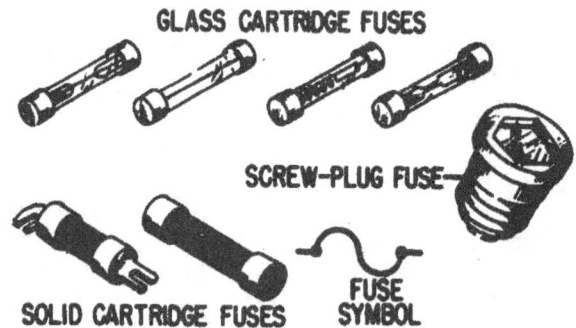

GLASS CARTRIDGE FUSES

SCREW-PLUG FUSE

SOLID CARTRIDGE FUSES          FUSE SYMBOL

BE.57

Figure 4-17.—Typical fuse types.

## ENERGY

Energy is defined as the ability to do work. Energy is expended when work is done, because it takes energy to maintain a force when that force acts through a distance. The total energy expended to do a certain amount of work is equal to the working force multiplied by the distance through which the force moved to do the work. This is the mechanical definition.

In electricity, total energy expended is equal to the rate at which work is done, multiplied by the length of time the rate is measured. Essentially, energy W is equal to power P times time t.

An equation for energy is derived by multiplying both sides of equation (4-4) by the common factor of time, t, and equating the expression to the energy, W, as

$$W = Pt$$

$$W = EIt. \qquad (4-7)$$

Similarly, both sides of equations (4-5) and (4-6) may be multiplied by the time factor, t, and equated to the energy, W, as

$$W = \frac{E^2}{R} t, \qquad (4-8)$$

and

$$W = I^2Rt. \qquad (4-9)$$

In the energy equations (4-7), (4-8), and (4-9), E is in volts and I in amperes. If t is expressed in hours, W will be in watt-hours.

If t is expressed in seconds, W will be in watt-seconds or joules (1 joule is equal to 1 watt-second). Since $Q = It$ (where Q is in coulombs, I in amperes, and t in seconds), it is possible to substitute Q for It in equation (4-7) with the resulting expression for energy. Thus

$$W = QE \qquad (4-10)$$

where W is the energy in joules or watt-seconds, Q is the quantity in coulombs, and E is in volts. (As explained in chapter 2, Q is the symbol for coulombs. This is the measure of a quantity of electrons. The Q is to electricity as the gallon is to water.)

Electrical energy is bought and sold in units of kilowatt-hours ($3,600 \times 10^3$ joules), and is totalized in large central generating stations in terms of megawatt-hours ($3,600 \times 10^6$ joules). For example, if the average demand over a 10-hour period is 70 megawatts, the total energy delivered is 70 x 10, or 700 megawatt-hours. This amount of energy is equivalent to 700 x 1,000 = 700,000 kilowatt-hours, or 700 x 3,600 $x\ 10^6 = 2,520,000 \times 10^6$ joules. The most

practical unit to use depends in part upon the magnitude of the quantity of energy involved, and in this example the megawatt-hour is appropriate.

## SERIES CIRCUIT CHARACTERISTICS

As previously mentioned, an electric circuit is a complete path through which electrons can flow from the negative terminal of the voltage source, through the connecting wires or conductors, through the load or loads, and back to the positive terminal of the voltage source. A circuit is thus made up of a voltage source, the necessary connecting conductors, and the effective load.

If the circuit is arranged so that the electrons have only ONE possible path, the circuit is called a SERIES CIRCUIT. Therefore, a series circuit is defined as a circuit that contains only one path for current flow. Figure 4-18 shows a series circuit having several lamps.

## RESISTANCE

Referring to figure 4-18, the current in a series circuit, in completing its electrical path, must flow through each lamp inserted into the circuit. Thus, each additional lamp offers added resistance. In a series circuit, THE TOTAL CIRCUIT RESISTANCE ($R_T$) IS EQUAL TO THE SUM OF THE INDIVIDUAL RESISTANCES.

As an equation:
$$R_T = R_1 + R_2 + R_3 \cdots R_n \qquad (4-11)$$

NOTE: The subscript n denotes any number of additional resistances that might be in the equation.

Example: Three resistors of 10 ohms, 15 ohms, and 30 ohms are connected in series across a battery whose emf is 110 volts (fig. 4-19). What is the total resistance?

Given:

$$R_1 = 10 \text{ ohms}$$

$$R_2 = 15 \text{ ohms}$$

$$R_3 = 30 \text{ ohms}$$

$$R_T = ?$$

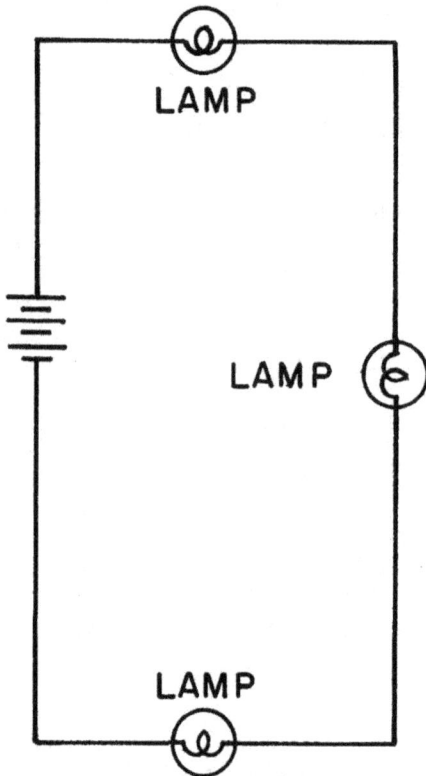

SERIES CIRCUIT

BE.58

Figure 4-18.—Series circuit.

Solution: $\quad R_T = R_1 + R_2 + R_3$

$$R_T = 10 + 15 + 30$$

$$R_T = 55 \text{ ohms}$$

In some circuit applications, the total resistance is known and the value of a circuit resistor has to be determined. Equation (4-11) can be transposed to solve for the value of the unknown resistance.

Example: The total resistance of a circuit containing three resistors is 40 ohms (fig. 4-20). Two of the circuit resistors are 10 ohms each. Calculate the value of the third resistor.

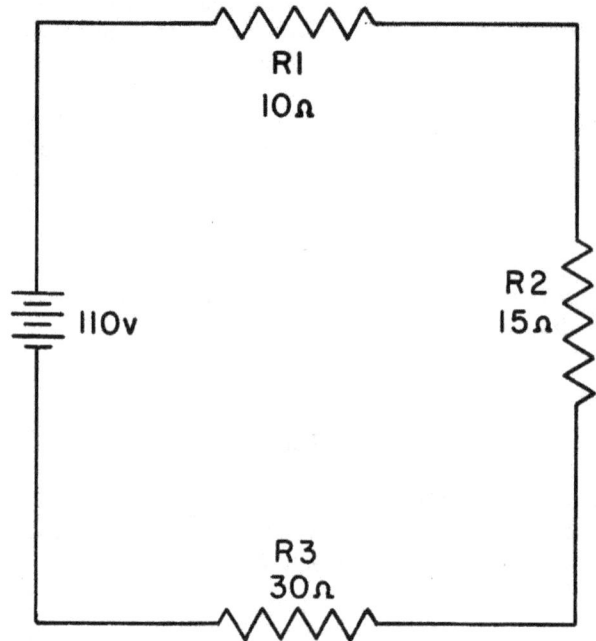

BE.59

Figure 4-19.—Solving for total resistance in a series circuit.

Given: $\quad\quad R_T = 40 \text{ ohms}$

$$R_1 = 10 \text{ ohms}$$

$$R_2 = 10 \text{ ohms}$$

$$R_3 = ?$$

Solution: $\quad R_T = R_1 + R_2 + R_3 \quad\quad (4\text{-}11)$

Subtracting $(R_1 + R_2)$ from both sides of the equation

$$R_3 = R_T - R_1 - R_2$$

$$R_3 = 40 - 10 - 10$$

$$R_3 = 40 - 20$$

$$R_3 = 20 \text{ ohms}$$

CURRENT

Since there is but one path for current in a series circuit, the same current must flow

69

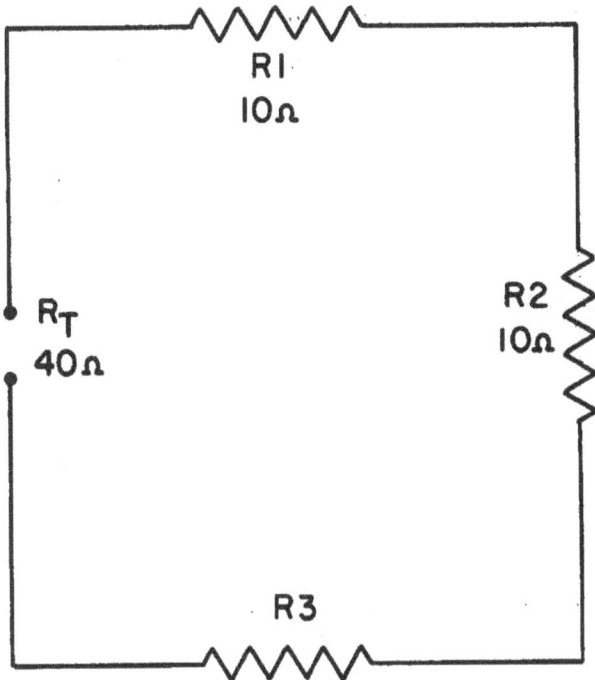

BE.60

Figure 4-20.—Calculating the value of one resistance in a series circuit.

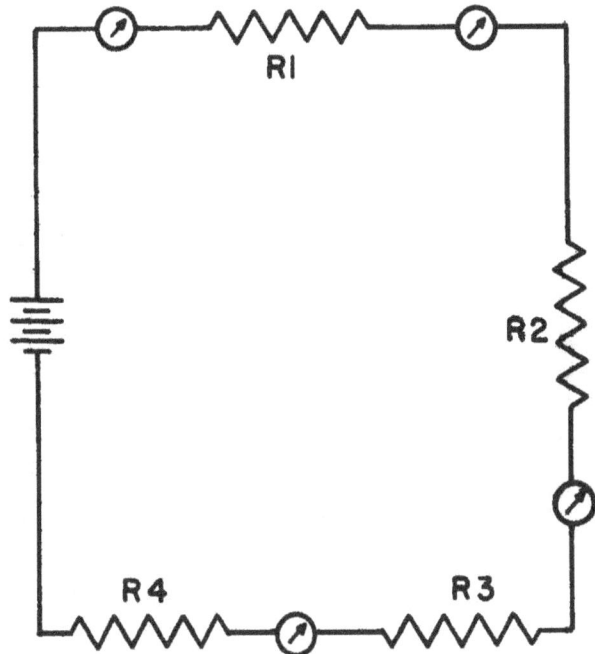

BE.61

Figure 4-21.—Current in a series circuit.

through each part of the circuit. To determine the current throughout a series circuit, only the current through one of the parts need be known.

The fact that the same current flows through each part of a series circuit can be verified by inserting ammeters into the circuit at various points as shown in figure 4-21. If this were done, each meter would be found to indicate the same value of current.

## VOLTAGE

As stated previously, the voltage drop across the resistor in the basic circuit is the total voltage across the circuit and is equal to the applied voltage. The total voltage across a series circuit is also equal to the applied voltage, but consists of the sum of two or more individual voltage drops. In any series circuit the SUM of the resistor voltage drops must equal the source voltage. This statement can be proven by an examination of the circuit shown in figure 4-22. In this circuit a source potential ($E_T$) of 20 volts is impressed across a series circuit

consisting of two 5 ohm resistors. The total resistance of the circuit is equal to the sum of the two individual resistances, or 10 ohms. Using Ohm's Law the circuit current may be calculated as follows:

$$I = \frac{E_T}{R_T}$$

$$I = \frac{20}{10}$$

$$I = 2 \text{ amperes}$$

Knowing the size of the resistors to be 5 ohms each, and the current through the resistors to be 2 amperes, the voltage drops across the resistors can be calculated. The voltage ($E_1$) across $R_1$ is therefore:

$$E_1 = IR_1$$

$$E_1 = 2 \text{ amperes x 5 ohms}$$

$$E_1 = 10 \text{ volts}$$

Since $R_2$ is the same ohmic value as $R_1$ and carries the same current, the voltage drop

70

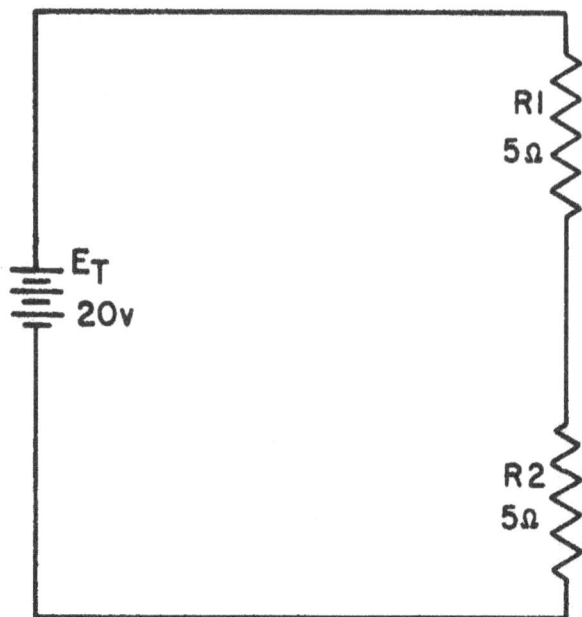

BE.62

Figure 4-22.—Calculating total resistance in a series circuit.

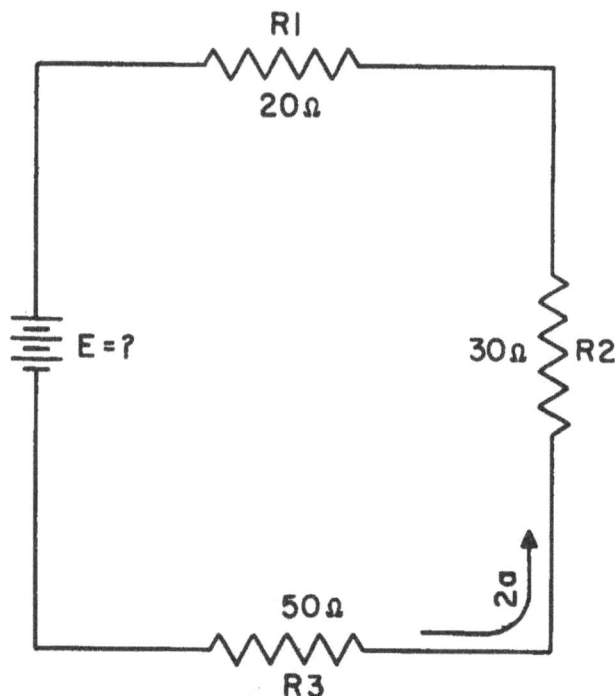

BE.63

Figure 4-23.—Solving for applied voltage in a series circuit.

across $R_2$ is also equal to 10 volts. Adding these two 10 volt drops together gives a total drop of 20 volts exactly equal to the applied voltage. For a series circuit then:

$$E_T = E_1 + E_2 + E_3 \ldots E_n \quad (4\text{-}12)$$

Example: A series circuit consists of three resistors having values of 20 ohms, 30 ohms, and 50 ohms respectively. Find the applied voltage if the current through the 30 ohm resistor is 2 amperes.

To solve the problem, a circuit diagram is first drawn and labeled as shown in figure 4-23.

Given:

$R_1$ = 20 ohms

$R_2$ = 30 ohms

$R_3$ = 50 ohms

I   = 2 amperes

Solution: Since the circuit involved is a series circuit, the same 2 amperes of current flows

through each resistor. Using Ohm's Law, the voltage drops across each of the three resistors can be calculated and are:

$$E_1 = 40 \text{ volts}$$

$$E_2 = 60 \text{ volts}$$

$$E_3 = 100 \text{ volts}$$

Once the individual drops are known they can be added to find the total or applied voltage:

$$E_T = E_1 + E_2 + E_3 \quad (4\text{-}12)$$

$$E_T = 40v + 60v + 100v$$

$$E_T = 200 \text{ volts}$$

NOTE: In using Ohm's Law, the quantities used in the equation MUST be taken from the SAME part of the circuit. In the above example the voltage across $R_2$ was computed using the current through $R_2$ and the resistance of $R_2$.

It must be emphasized that the potential difference across a resistor remains constant, for it is a measure of the amount of energy required to move a unit charge from one point to another. As long as the source produces electric energy as rapidly as it is consumed in a resistance, the potential difference across the resistance will remain at a constant voltage. The value of this voltage is determined by the applied voltage and the proportional relationship of circuit resistances. The voltage drops that occur in a series circuit are in direct proportions to the resistance across which they appear. This is a result of having the same current flow through each resistor. Thus, the larger the resistor the larger will be the voltage drop across it.

## POWER

Each of the resistors in a series circuit consumes power which is dissipated in the form of heat. Since this power must come from the source, the total power must be equal in amount to the power consumed by the circuit resistances. In a series circuit the total power is equal to the SUM of the powers dissipated by the individual resistors. Total power ($P_T$) is thus equal to:

$$P_T = P_1 + P_2 + P_3 + \dots P_n \qquad (4\text{-}13)$$

Example: A series circuit consists of three resistors having values of 5 ohms, 10 ohms, and 15 ohms. Find the total power dissipation when 120 volts is applied to the circuit. (See figure 4-24.)

given:

$$R_1 = 5 \text{ ohms}$$

$$R_2 = 10 \text{ ohms}$$

$$R_3 = 15 \text{ ohms}$$

$$E = 120 \text{ volts}$$

Solution: The total resistance is found first.

$$R_T = R_1 + R_2 + R_3 \qquad (4\text{-}11)$$

$$R_T = 5 + 10 + 15$$

$$R_T = 30 \text{ ohms}$$

Using total resistance and the applied voltage, the circuit current is calculated.

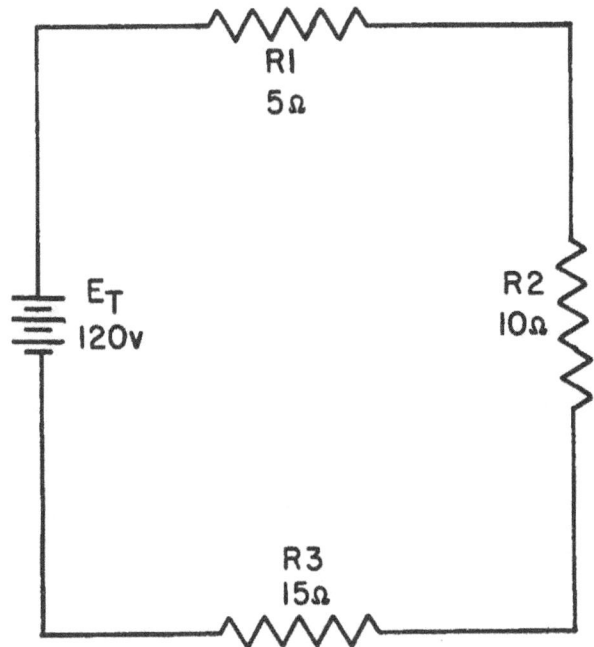

BE. 64

Figure 4-24.—Solving for total power in a series circuit.

$$I = \frac{E_T}{R_T}$$

$$I = \frac{120}{30}$$

$$I = 4 \text{ amperes}$$

Using the power formulas, the individual power dissipations can be calculated. For resistor $R_1$:

$$P_1 = I^2 R_1 \qquad (4\text{-}6)$$

$$P_1 = (4)^2 5$$

$$P_1 = 80 \text{ watts}$$

For $R_2$:

$$P_2 = I^2 R_2 \qquad (4\text{-}6)$$

$$P_2 = (4)^2 10$$

$$P_2 = 160 \text{ watts}$$

For $R_3$:

$$P_3 = I^2 R_3 \qquad (4\text{-}6)$$

$$P_3 = (4)^2 15$$

$$P_3 = 240 \text{ watts}$$

To obtain total power:

$$P_T = P_1 + P_2 + P_3 \qquad (4\text{-}13)$$

$$P_T = 80 + 160 + 240$$

$$P_T = 480 \text{ watts}$$

To check the answer the total power delivered by the source can be calculated:

$$P_{source} = I_{source} \times E_{source} \qquad (4\text{-}4)$$

$$P_{source} = 4 \text{ a. } \times 120 \text{ v.}$$

$$P_{source} = 480 \text{ watts}$$

Thus the total power is equal to the sum of the individual power dissipations.

### RULES FOR SERIES D-C CIRCUITS

The important factors governing the operation of a series circuit are listed below. These factors have been set up as a group of rules so that they may be easily studied. These rules must be completely understood before the study of more advanced circuit theory is undertaken.

1. The same current flows through each part of a series circuit.

2. The total resistance of a series circuit is equal to the sum of the individual resistances.

3. The total voltage across a series circuit is equal to the sum of the individual voltage drops.

4. The voltage drop across a resistor in a series circuit is proportional to the size of the resistor.

5. The total power dissipated in a series circuit is equal to the sum of the individual power dissipations.

## GENERAL CIRCUIT ANALYSIS

To establish a procedure for solving series circuits, the following sample problem will be solved.

Example: Three resistors of 5 ohms, 10 ohms, and 15 ohms are connected across a battery rated at 90 volts terminal voltage. Completely solve the circuit (fig. 4-25).

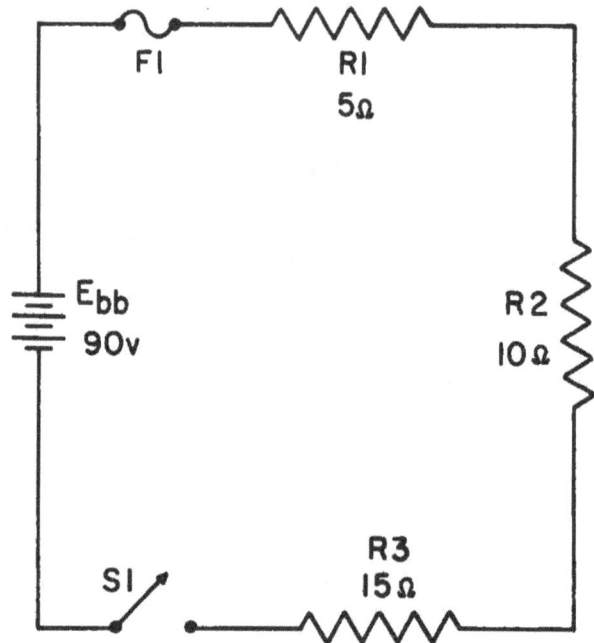

BE.65

Figure 4-25.—Solving for various values in a series circuit.

In solving the circuit the total resistance will be found first. Next, the circuit current will be calculated. Once the current is known the voltage drops and power dissipations can be calculated.

The total resistance is:

$$R_T = R_1 + R_2 + R_3$$

$$R_T = 5 \text{ ohms} + 10 \text{ ohms} + 15 \text{ ohms}$$

$$R_T = 30 \text{ ohms}$$

By Ohm's Law the current is:

$$I = \frac{E_{bb}}{R_T}$$

$$I = \frac{90}{30}$$

$$I = 3 \text{ amperes}$$

The voltage ($E_1$) across $R_1$ is:

$$E_1 = IR_1$$

$$E_1 = 3 \text{ amperes} \times 5 \text{ ohms}$$

$$E_1 = 15 \text{ volts}$$

The voltage ($E_2$) across $R_2$ is:

$$E_2 = IR_2$$

$$E_2 = 3 \text{ amperes} \times 10 \text{ ohms}$$

$$E_2 = 30 \text{ volts}$$

The voltage ($E_3$) across $R_3$ is:

$$E_3 = IR_3$$

$$E_3 = 3 \text{ amperes} \times 15 \text{ ohms}$$

$$E_3 = 45 \text{ volts}$$

The power dissipated in $R_1$ is:

$$P_1 = I \times E_1$$

$$P_1 = 3 \text{ amperes} \times 15 \text{ volts}$$

$$P_1 = 45 \text{ watts}$$

The power dissipated in $R_2$ is:

$$P_2 = I \times E_2$$

$$P_2 = 3 \text{ amperes} \times 30 \text{ volts}$$

$$P_2 = 90 \text{ watts}$$

The power dissipated in $R_3$ is:

$$P_3 = I \times E_3$$

$$P_3 = 3 \text{ amperes} \times 45 \text{ volts}$$

$$P_3 = 135 \text{ watts}$$

The total power dissipated is:

$$P_T = E_T \times I$$

$$P_T = 90 \text{ volts} \times 3 \text{ amperes}$$

$$P_T = 270 \text{ watts}$$

Example: Four resistors $R_1$ = 10 ohms, $R_2$ = 10 ohms, $R_3$ = 50 ohms, and $R_4$ = 30 ohms are connected in series across a battery. The current through the circuit is 0.5 amperes. (See figure 4-26.)

a. What is the battery voltage?
b. What is the voltage across each resistor?
c. What is the power expended in each resistor?
d. What is the total power?

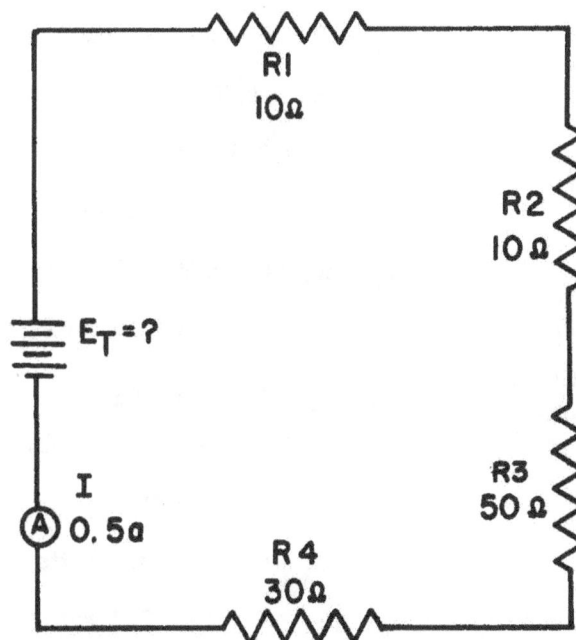

BE.66

Figure 4-26.—Computing series circuit values.

Given:

$$R_1 = 10 \text{ ohms}$$

$$R_2 = 10 \text{ ohms}$$

$$R_3 = 50 \text{ ohms}$$

$$R_4 = 30 \text{ ohms}$$

$$I = 0.5 \text{ ampere}$$

Find:

$$E_1 = ? \qquad P_1 = ?$$

$$E_2 = ? \qquad P_2 = ?$$

$$E_3 = ? \qquad P_3 = ?$$

$$E_4 = ? \qquad P_4 = ?$$

$$E_T = ? \qquad P_T = ?$$

Solution:

(a) $R_T = R_1 + R_2 + R_3 + R_4$

$R_T = 10 + 10 + 50 + 30 = 100 \text{ ohms}$

$E_T = IR_T = 0.5 \times 100 = 50 \text{ volts}$

(b) $E_1 = IR_1 = 0.5 \times 10 = 5 \text{ volts}$

$E_2 = IR_2 = 0.5 \times 10 = 5 \text{ volts}$

$E_3 = IR_3 = 0.5 \times 50 = 25 \text{ volts}$

$E_4 = IR_4 = 0.5 \times 30 = 15 \text{ volts}$

Check: $E_T = E_1 + E_2 + E_3 + E_4$

$E_T = 5 + 5 + 25 + 15 = 50 \text{ volts}$

(c) Power consumed in $R_1$ is:

$P_1 = IE_1 = 0.5 \times 5 = 2.5 \text{ watts}$

$P_2 = IE_2 = 0.5 \times 5 = 2.5 \text{ watts}$

$P_3 = IE_3 = 0.5 \times 25 = 12.5 \text{ watts}$

$P_4 = IE_4 = 0.5 \times 15 = 7.5 \text{ watts}$

(d) Total power

$$P_T = P_1 + P_2 + P_3 + P_4$$

$$P_T = 2.5 + 2.5 + 12.5 + 7.5 = 25 \text{ watts}$$

Check:

$$P_T = I_T^2 R_T = 0.5^2 \times 100 = 25 \text{ watts}$$

or:

$$P_T = I_T E_T = 0.5 \times 50 = 25 \text{ watts}$$

or:

$$P_T = \frac{E_T^2}{R_T} = \frac{50^2}{100} = \frac{2500}{100} = 25 \text{ watts}$$

An important fact to keep in mind when applying Ohm's Law to a series circuit is to consider whether the values used are component values or total values. When the information available enables the use of Ohm's Law to find total resistance, total voltage and total current, total values must be inserted into the formula.

To find total resistance:

$$R_T = \frac{E_T}{I_T}$$

To find total voltage:

$$E_T = I_T \times R_T$$

To find total current:

$$I_T = \frac{E_T}{R_T}$$

NOTE: $I_T$ is equal to I in a series circuit. However, the distinction between $I_T$ and I in the formula should be noted. The reason being that future circuits may have several currents, and it will be necessary to differentiate between $I_T$ and other currents.

To compute any quantity (E, I, R, or P) associated with a single given resistor, the values used in the formula must be obtained from that particular resistor. For exampler, to find the value of an unknown resistance, the voltage across and the current through that particular resistor must be used.

To find the value of a resistor:

$$R = \frac{E_R}{I_R}$$

To find the voltage drop across a resistor:

$$E_R = I_R \times R$$

To find current through a resistor:

$$I_R = \frac{E_R}{R}$$

## KIRCHHOFF'S VOLTAGE LAW

In 1847 Kirchhoff extended the use of Ohm's Law by developing a simple concept concerning the voltages contained in a series circuit loop. Kirchhoff's Law is stated as follows.

The algebraic sum of the instantaneous emf's and voltage drops around any closed circuit loop is zero.

Through the use of Kirchhoff's Law, circuit problems can be solved which would be difficult and often impossible with only a knowledge of Ohm's Law. When the law is properly applied, an equation can be set up for a closed loop and the unknown circuit values may be calculated.

### POLARITY OF VOLTAGE

To apply Kirchhoff's Voltage Law, the meaning of voltage POLARITY must be understood. In the circuit shown in figure 4-27 the current is seen to be flowing in a counterclockwise direction due to the arrangement of the battery source $E_{bb}$. Notice that the end of resistor $R_1$ into which the current flows is marked NEGATIVE ( - ). The end of $R_1$ at which the current leaves is marked POSITIVE (+). These polarity markings are used to show that the end of $R_1$ into which the current flows is at a higher negative potential than is the end of the resistor at which the current leaves. Point A is thus more negative than point B.

Point C, which is at the same potential as point B, is labeled negative. This is to indicate that point C, though positive with respect to point A, is more negative than point D. To say a point is positive (or negative), without stating what it is positive IN RESPECT TO, has no meaning.

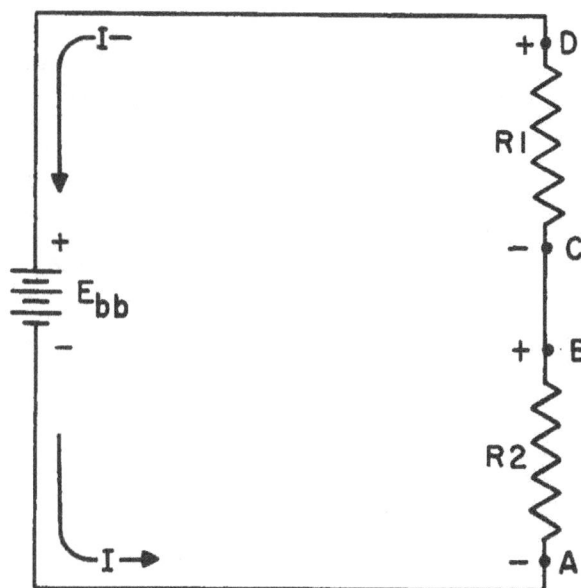

BE.67

Figure 4-27.—Voltage polarities.

Kirchhoff's Voltage Law can be written as an equation as shown below:

$$E_a + E_b + E_c + \ldots E_n = 0 \qquad (4\text{-}14)$$

where $E_a$, $E_b$, etc., are the voltage drops and emf's around any closed circuit loop. To set up the equation for an actual circuit, the following procedure is used.

1. Assume a direction of current through the circuit. (Correct direction desirable but not necessary.)

2. Using assumed direction of current, assign polarities to all resistors through which the current flows.

3. Place correct polarities on any source included in the circuit.

4. Starting at any points in the circuit, trace around the circuit writing down the magnitude and polarity of the voltage across each component in succession. The polarity used is the sign AFTER the component is passed through. Stop

when reaching the point at which the trace was started.

5. Place these voltages with their polarities into equation (4-14) and solve for the desired quantity.

Example: Three resistors are connected across a 50 volt source. What is the voltage across the third resistor if the voltage drops across the first two resistors are 25 volts and 15 volts?

Solution: A diagram is first drawn as shown in figure 4-28. Next a direction of current is assumed as shown. Using this current, the polarity markings are placed at each end of each resistor and also on the terminals of the source. Starting at point A, trace around the circuit in the direction of current flow recording the voltage and polarity of each component. Starting at point A these voltages would be as follows:

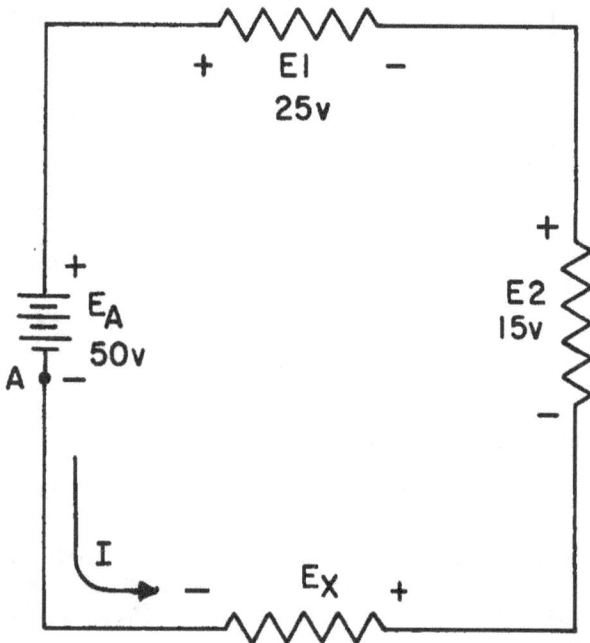

BE.68

Figure 4-28.—Determining unknown voltage in a series circuit.

Basic formula:

$$E_a + E_b + E_c \ldots E_n = 0 \qquad (4\text{-}14)$$

From the circuit:

$$(+E_x) + (+E_2) + (+E_1) + (-E_A) = 0$$

Substituting values from circuit:

$$E_x + 15 + 25 + 50 = 0$$

$$E_x - 10 = 0$$

$$E_x = 10 \text{ volts}$$

Thus, the unknown voltage ($E_x$) is found to be 10 volts.

Using the same idea as above, a problem can be solved in which the current is the unknown quantity.

Example: A circuit having a source voltage of 60 volts contains three resistors of 5 ohms, 10 ohms, and 15 ohms. Find the circuit current.

Solution: Draw and label the circuit (fig. 4-29). Establish a direction of current flow and assign polarities. Next, starting at any point, point (A) will be chosen in this example; write out the loop equation.

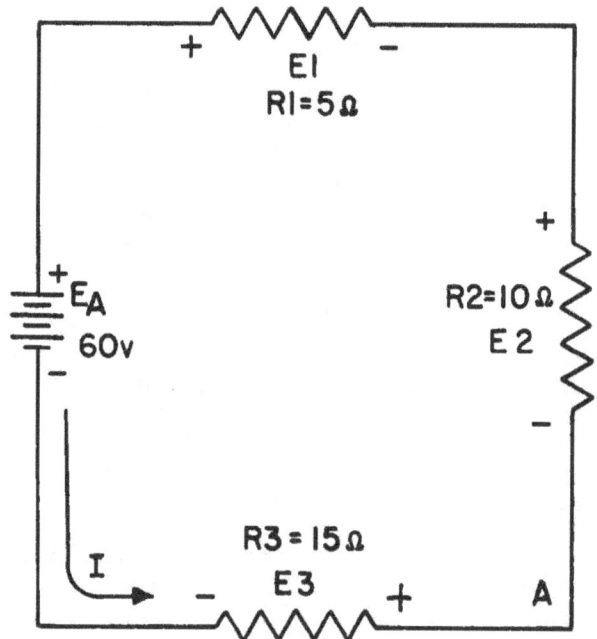

BE.69

Figure 4-29.—Correct direction of assumed current.

Basic equation:

$$E_a + E_b + E_c + \ldots E_n = 0 \qquad (4\text{-}14)$$

$$+ E_2 + E_1 - E_A + E_3 = 0$$

Since $E = IR$, by substitution:

$$IR_2 + IR_1 - E_A + IR_3 = 0$$

Substituting values:

$$10I = 5I - 60 + 15I = 0$$

Combining like terms:

$$30I - 60 = 0$$

$$30I = 60$$

$$I = 2 \text{ amperes}$$

Since the current obtained in the above calculations is a positive 2 amperes, the assumed direction of current was correct. To show what happens if the incorrect direction of current is assumed, the problem will be solved as before but with the opposite direction of current. The circuit is redrawn showing the new direction of current and new polarities in figure 4-30.

Solution:

$$E_a + E_b + E_c + \ldots E_n = 0 \qquad (4\text{-}14)$$

Starting at point (A):

$$- E_2 - E_1 - E_A - E_3 = 0$$

$$-IR_2 - IR_1 - E_A - IR_3 = 0$$

$$10I - 5I - 60 - 15I = 0$$

$$-30I - 60 = 0$$

$$-30I = 60$$

$$I = -2 \text{ amperes}$$

Notice that the AMOUNT of current is the SAME as before. Its polarity, however, is NEGATIVE. The negative polarity simply indicates the wrong direction of current was assumed. Should it be necessary to use this current in further calculations on the circuit, the negative polarity should be retained in the calculations.

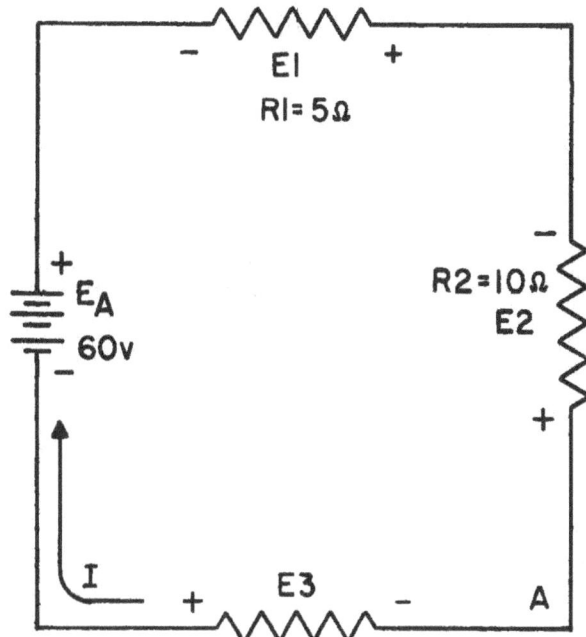

BE.70

Figure 4-30.—Incorrect direction of assumed current.

## SERIES AIDING AND OPPOSING SOURCES

In many practical applications a circuit may contain more than one source. Sources of emf that cause current to flow in the same direction are considered to be SERIES AIDING and their voltages add. Sources of emf that would tend to force current in opposite directions are said to be SERIES OPPOSING, and the effective source voltage is the difference between the opposing voltages. When two opposing sources are inserted into a circuit, current flow would be in a direction determined by the larger source. Examples of series aiding and opposing sources are shown in figure 4-31.

## MULTIPLE SOURCE SOLUTIONS

A simple solution may be obtained for a multiple source circuit through the use of Kirchhoff's Voltage Law. In applying this method, the exact same procedure is used for the multiple source as was used above for the single source circuit. This is demonstrated by the following problem.

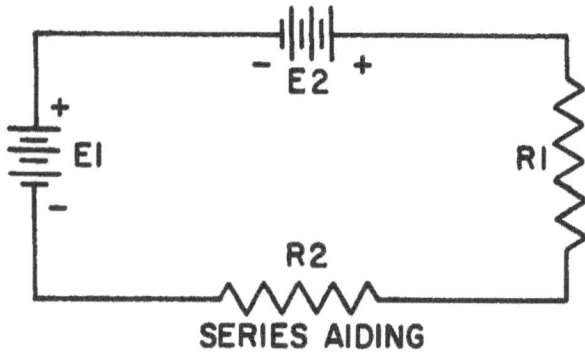

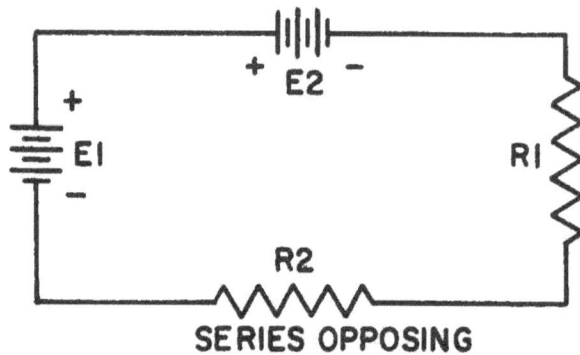

SERIES AIDING

SERIES OPPOSING

BE. 71

Figure 4-31.—Aiding and opposing sources.

Example: Using Kirchhoff's Voltage equation, find the amount of current in the circuit shown in figure 4-32.

Solution: As before, a direction of current flow is assumed and polarity signs are placed on the drawing. The loop equation will be starting at point A.

Basic equation:

$$E_a + E_b + E_c + \dots E_n = 0 \qquad (4\text{-}14)$$

From the circuit:

$$E_{bb2} + E_1 - E_{bb1} + E_{bb3} + E_2 = 0$$

$$+ 20 + 60I - 180 + 40 + 20I = 0$$

Combining like terms:

$$+ 80I - 120 = 0$$

$$80I = 120$$

$$I = 1.5 \text{ amperes}$$

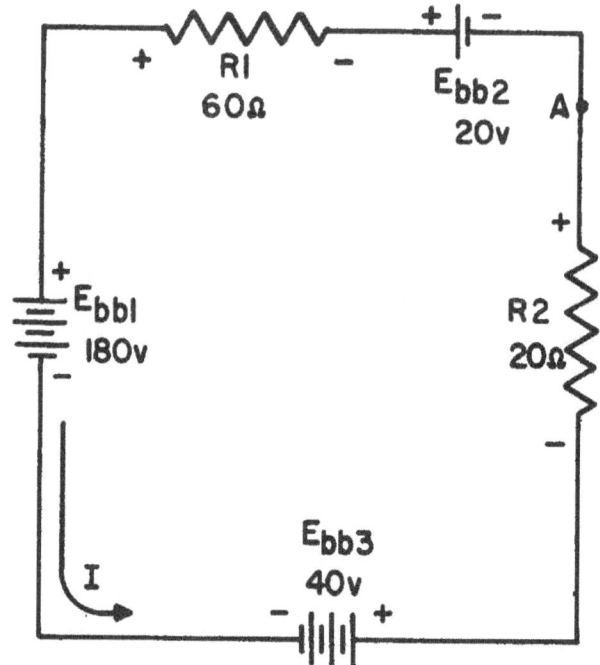

BE.72

Figure 4-32.—Solving for circuit current using Kirchhoff's voltage equation.

## VOLTAGE REFERENCES

### REFERENCE POINT

A reference point is an arbitrarily chosen point to which all other points are compared. In series circuits, any point can be chosen as a reference and the electrical potential at all other points can be determined in reference to the initial point. In the example of figure 4-33 point A shall be considered as the reference. Each series resistor in the illustrated circuit is of equal value; therefore, the applied voltage is equally distributed across each resistor. The potential at point B is 25 volts more positive than A. Points C and D are 50 volts and 75 volts respectively more positive than point A.

79

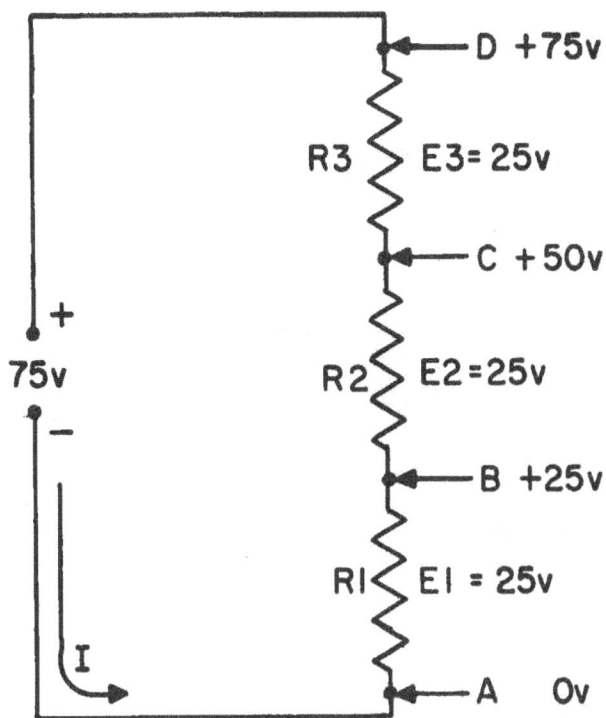

BE.73

Figure 4-33.—Reference points in a series circuit.

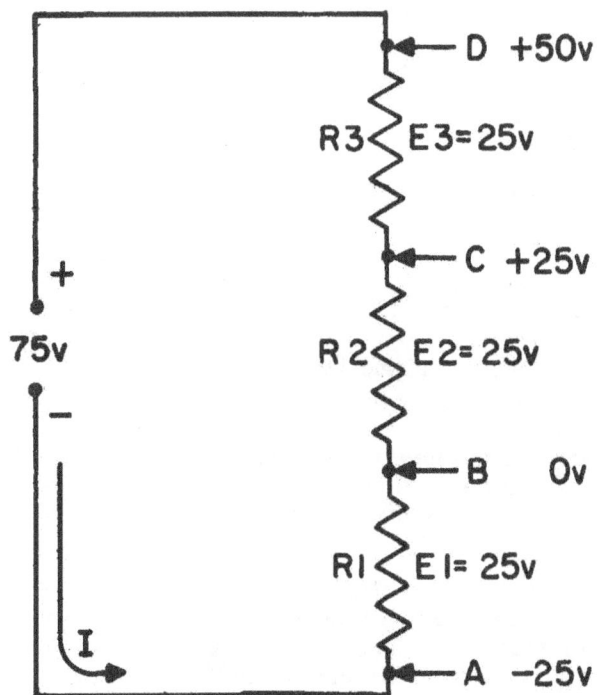

BE.74

Figure 4-34.—Determining potentials with respect to a reference point.

If point B is used as the reference as in figure 4-34, point D would be positive 50 volts in respect to the new reference point B. The former reference point A is 25 volts negative in respect to point B.

GROUND

As in the previous circuit illustration, the reference point of a circuit is always considered to be at zero potential. Since the earth (ground) is said to be at a zero potential, the term GROUND is used to denote a common electrical point of zero potential. In figure 4-35, point A is the zero reference or ground and is symbolized as such.

Point C is 75 volts positive and point B is 25 volts positive in respect to ground.

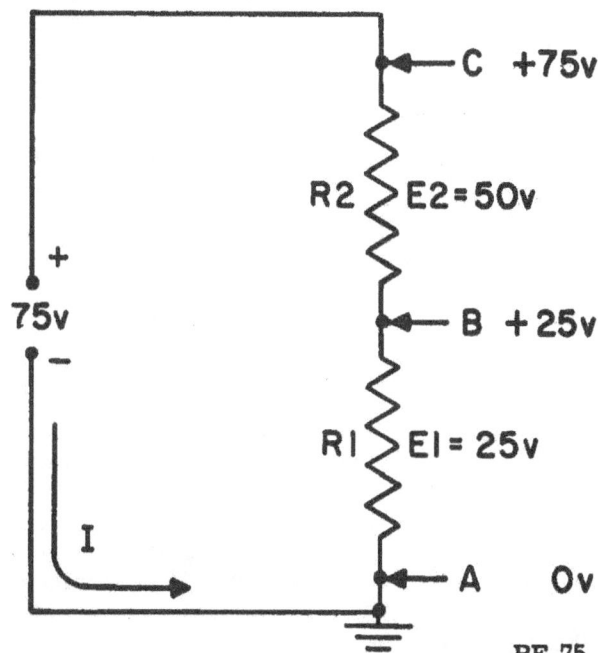

BE.75

Figure 4-35.—Use of ground symbols.

In many electrical/electronic equipments, the metal chassis is the common ground for the many electrical circuits. The value of ground is noted when considering its contribution to economy, simplification of schematics, and ease of measurement. When completing each electrical circuit, common points of a circuit at zero potential are connected directly to the metal chassis thereby eliminating a large amount of connecting wire. The electrons pass through the metal chassis (conductor) to reach other points of the circuit. An example of a grounded circuit is illustrated in figure 4-36.

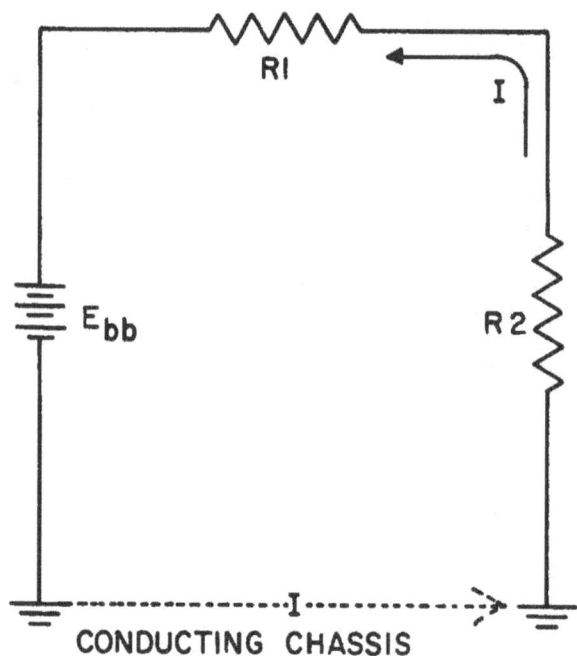

BE.76

Figure 4-36.—Ground used as a conductor.

Most voltage measurements used to check proper circuit operation in electronic equipment are taken in respect to ground. One meter lead is attached to ground and the other meter lead is moved to various test points.

## OPEN AND SHORT CIRCUITS

A circuit is said to be OPEN when a break exists in a complete conducting pathway. Although an open occurs any time a switch is thrown to deenergize a circuit, an open may also develop accidentally due to abnormal circuit conditions. To restore a circuit to proper operation, the open must be located and its cause determined.

Sometimes an open can be located visually by a close inspection of the circuit components. Defective components, such as burned out resistors and fuses can usually be discovered by this method. Others such as a break in wire covered by insulation, or the melted element of an enclosed fuse, are not visible to the eye. Under such conditions, the understanding of an open's effect on circuit conditions enables a technician to make use of a voltmeter or ohmmeter to locate the open component.

In figure 4-37, the series circuit consists of two resistors and a fuse. Notice the effects on circuit conditions when the fuse opens.

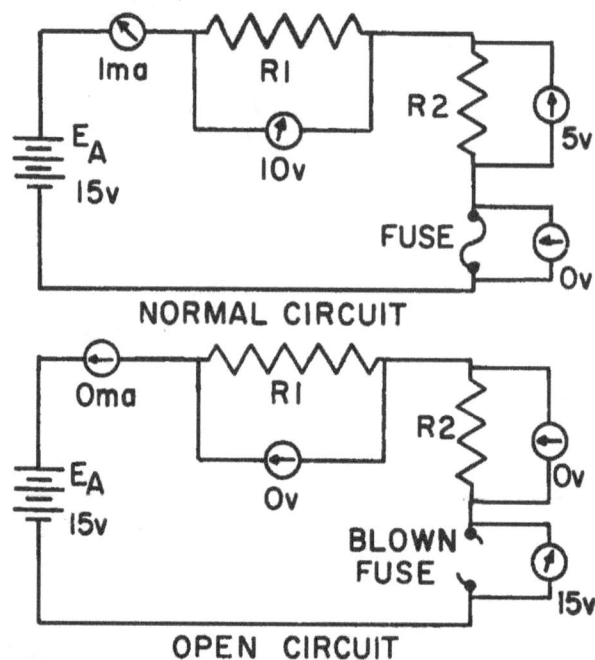

BE.77

Figure 4-37.—Normal and open circuit conditions.

Current ceases to flow; therefore, there is no longer a voltage drop across the resistors. Each end of the open conducting path becomes an extension of the battery terminals and the voltage felt across the open is equal to the applied voltage.

An open circuit, such as found in figure 4-37 could also have been located with an ohmmeter. However, when using an ohmmeter to check a circuit, it is important to first deenergize the circuit. The reason being that an ohmmeter has its own power source and would be damaged if connected to an energized circuit.

The ohmmeter used to check a series circuit would indicate the ohmic value of each resistance it is connected across. The open circuit due to its almost infinite resistance would cause no deflection on the ohmmeter as indicated by the illustration, figure 4-38.

A SHORT CIRCUIT is an accidental path of low resistance which passes an abnormal amount of current. A short circuit exists whenever the resistance of the circuit or the resistance of a part of a circuit drops in value to almost zero ohms. A short often occurs as a result of improper wiring or broken insulation.

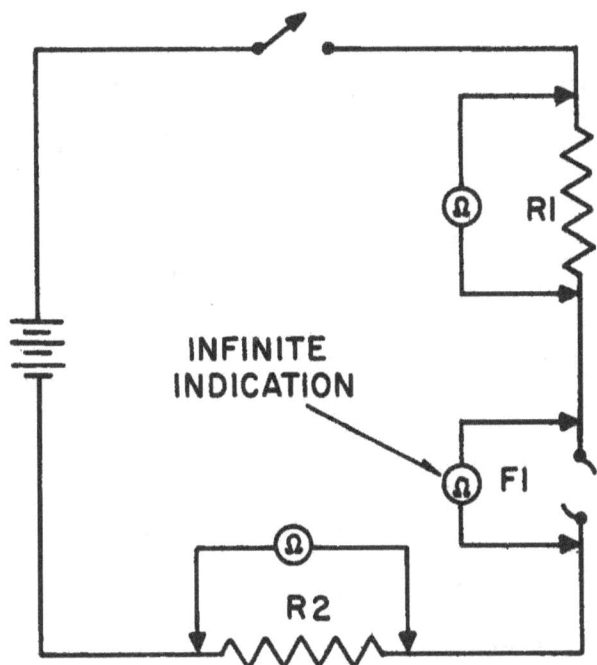

BE.78

Figure 4-38.—Ohmmeter readings in a series circuit.

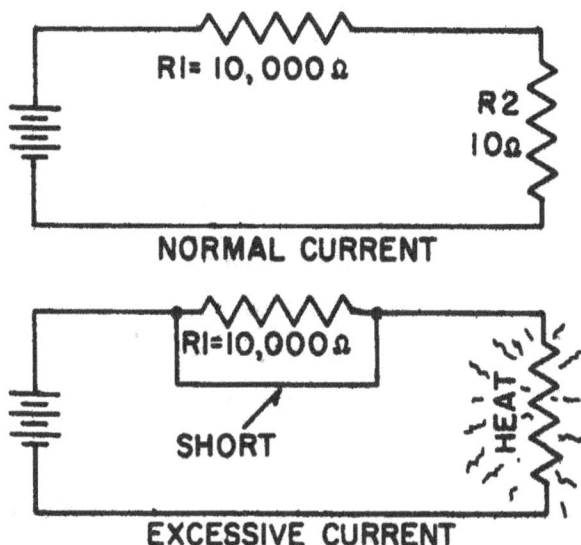

BE.79

Figure 4-39. —Normal and short circuit conditions.

In figure 4-39 a short is caused by improper wiring. Note the effect on current flow. Since the resistor has in effect been replaced with a piece of wire, practically all the current flows through the short and very little current flows through the resistor. Electrons flow through the short, a path of almost zero resistance and complete the circuit by passing through the 10-ohm resistor and the battery. The amount of current flow increases greatly because its resistive path has decreased from 10,010 ohms to 10 ohms. Due to the excessive current through the 10-ohm resistor, the increased heat dissipated by the resistor will destroy the component.

## EFFECT OF SOURCE RESISTANCE ON VOLTAGE, POWER, AND EFFICIENCY

All sources of emf have some internal resistance that acts in series with the load resistance. The source resistance is generally indicated in circuit diagrams as a separate resistor connected in series with the source. Both the voltage and power made available to the load may be increased if the resistance of the source is reduced.

The effects of source resistance, $R_s$, on load voltage may be illustrated by the use of figure 4-40. In figure 4-40 (A), the circuit is open, and therefore a voltmeter connected across the battery will read the open-circuit voltage. In the case of a dry cell, the open-circuit voltage is 1.5 volts. In figure 4-40 (B), the cell is short-circuited through the ammeter, and a current of 30 amperes flows from the source. In this case the voltage of the cell is developed across the internal resistance of the cell. The internal resistance of the cell is therefore,

$$R_s = \frac{E_s}{I} = \frac{1.5}{30} = 0.05 \text{ ohm.}$$

If a load, $R_L$, of 0.10 ohm is connected to the circuit, as shown in figure 4-40 (C), the current, I, becomes

$$I = \frac{E_s}{R_t} = \frac{1.5}{0.15} = 10 \text{ amperes.}$$

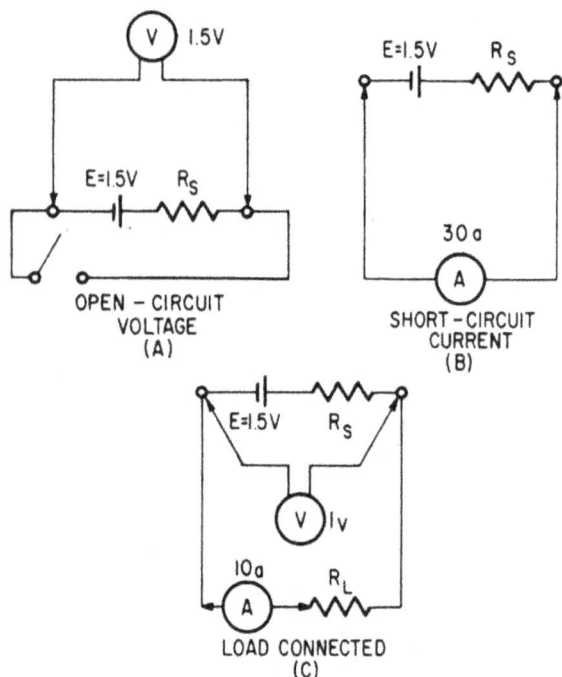

BE.80

Figure 4-40.—Effect of source resistance on load voltage.

The voltage available at the load is

$$E_L = IR_L = 10 \times 0.1 = 1 \text{ volt.}$$

The voltage absorbed across the internal resistance of the cell is

$$IR_S = 10 \times 0.05 = 0.5 \text{ volt.}$$

Thus the effect of the internal resistance is to decrease the terminal voltage from 1.5 volts to 1 volt when the cell delivers 10 amperes to the load.

The effect of the source resistance on the power output of a d-c source may be shown by an analysis of the circuit in figure 4-41 (A). When the variable load-resistor, $R_L$, is set at the zero ohms position (equivalent to a short circuit) the current is limited only by the internal resistance, $R_S$, of the source. The short-circuit current, I, is determined as

$$I = \frac{E_S}{R_S} = \frac{100}{5} = 20 \text{ amperes}$$

This is the maximum current that may be drawn from the source. The terminal voltage across the short circuit is zero and all the voltage is absorbed within the terminal resistance of the source.

If the load resistance, $R_L$, is increased (the internal resistance remaining the same), the current drawn from the source will decrease. Consequently, the voltage drop across the internal resistance will decrease. At the same time, the terminal voltage applied across the load will increase and will approach a maximum as the current approaches zero.

The MAXIMUM POWER TRANSFER THEOREM says in effect that maximum power is transferred from the source to the load when the resistance of the load is equal to the internal resistance of the source. This theorem is illustrated in the tabular chart and the graph of figure 4-41 (B) and (C). When the load resistance is 5 ohms, thus matching the source resistance, the maximum power of 500 watts is developed in the load.

The efficiency of power transfer (ratio of output to input power) from the source to the load increases as the load resistance is increased. The efficiency approaches 100 percent as the load resistance approaches a relatively large value compared with that of the source, since less power is lost in the source. The efficiency of power transfer is only 50 percent at the maximum power transfer resistance of 5 ohms and approaches zero efficiency at relatively low values of load resistance compared with that of the source.

Thus the problem of high efficiency and maximum power transfer is resolved as a compromise somewhere between the low efficiency of maximum power transfer and the high efficiency of the high-resistance load. Where the amounts of power involved are large and the efficiency is important, the load resistance is made large relative to the source resistance so that the losses are kept small. In this case the efficiency will be high. Where the problem of matching a source to a load is of paramount importance, as in communications circuits, a strong signal may be more important than a high percentage of efficiency. In such cases, the efficiency of transmission will be only about 50 percent. However, the power of transmission will be the maximum of which the source is capable of supplying.

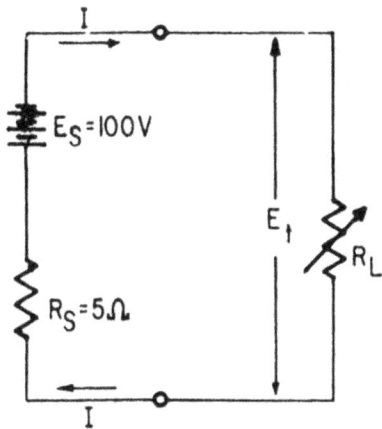

$E_S$ = OPEN-CIRCUIT VOLTAGE OF SOURCE

$R_S$ = INTERNAL RESISTANCE OF SOURCE

$E_t$ = TERMINAL VOLTAGE

$R_L$ = RESISTANCE OF LOAD

$P_L$ = POWER USED IN LOAD

I = CURRENT FROM SOURCE

% EFF. = PERCENTAGE OF EFFICIENCY

(A)
CIRCUIT AND SYMBOL DESIGNATIONS

| $R_L$ | $E_t$ | I | $P_L$ | %EFF. |
|---|---|---|---|---|
| 0 | 0 | 20 | 0 | 0 |
| 1 | 16.6 | 16.6 | 275.6 | 16.6 |
| 2 | 28.6 | 14.3 | 409 | 28.6 |
| 3 | 37.5 | 12.5 | 468.8 | 37.5 |
| 4 | 44.4 | 11.1 | 492.8 | 44.4 |
| 5 | 50 | 10 | 500 | 50 |
| 6 | 54.5 | 9.1 | 495.4 | 54.5 |
| 7 | 58.1 | 8.3 | 482.2 | 58.1 |
| 8 | 61.6 | 7.7 | 474.3 | 61.6 |
| 9 | 63.9 | 7.1 | 453.7 | 63.9 |
| 10 | 66 | 6.6 | 435.6 | 66 |
| 20 | 80 | 4 | 320 | 80 |
| 30 | 87 | 2.9 | 252 | 87 |
| 40 | 88 | 2.2 | 193.6 | 88 |
| 50 | 91 | 1.82 | 165 | 91 |

(B)
CHART

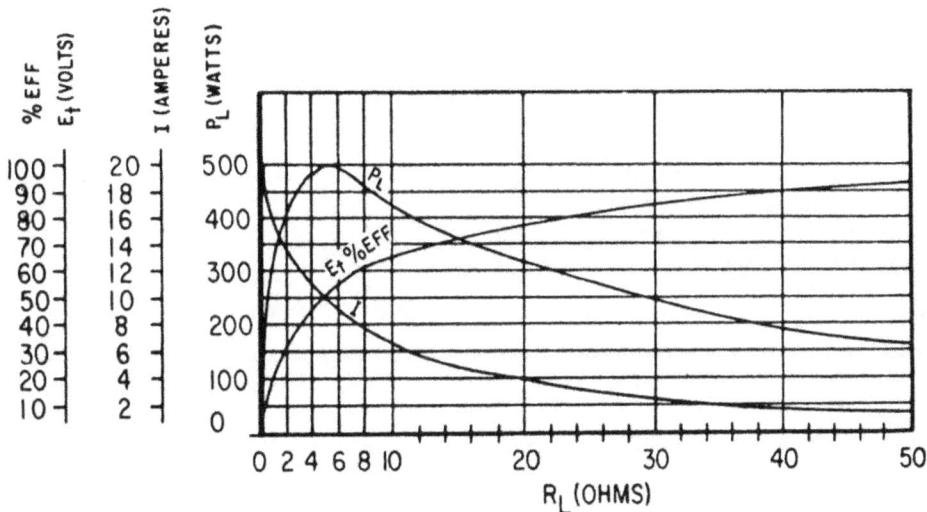

(C)
GRAPH

BE.81

Figure 4-41.—Effect of source resistance on power output.

# CHAPTER 5

# PARALLEL D-C CIRCUITS

An adequate understanding of modern electrical equipment requires a progressive development in the study of typical electrical circuits. In stepping-stone fashion, the discussion of series d-c circuits will now be followed by a consideration of the characteristics of parallel d-c circuits. It will be shown how the principles applied to series circuits can be used to determine the reactions of such quantities as voltage, current, and resistance in parallel and series-parallel circuits.

Along with the progressive introduction of electrical theories and circuit characteristics comes a corresponding progression in the use of mathematical equations and problem solving methods. A basic knowledge of powers of ten, fractions, fractional equations, and the use of simultaneous equations is required for the comprehension of material presented in this chapter.

## PARALLEL CIRCUIT CHARACTERISTICS

A parallel circuit is defined as one having more than one current path connected to a common voltage source. Parallel circuits, therefore, must contain two or more load resistances which are not connected in series. An example of a basic parallel circuit is shown in figure 5-1.

Commencing at the voltage source ($E_{bb}$) and tracing counterclockwise around the circuit, two complete and separate paths can be identified in which current can flow. One path is traced from the source through resistance $R_1$ and back to the source; the other, from the source through resistance $R_2$ and back to the source.

## VOLTAGE

You have seen that the source voltage in a series circuit divides proportionately across each resistor in the circuit. In a parallel circuit (fig. 5-1), the same voltage is present across all the resistors of a parallel group. This voltage is equal to the applied voltage ($R_{bb}$). The foregoing statement can be expressed in equation form as

$$E_{bb} = E_{R1} = E_{R2} = E_{Rn} \qquad (5-1)$$

Voltage measurements taken across the resistors of a parallel circuit, as illustrated by figure 5-2, verify the above equation. Each voltmeter indicates the same amount of voltage. Notice that the voltage across each resistor is the same as the applied voltage.

Example. Assume that the current through a resistor of a parallel circuit is known to be 4.5 milliamperes (ma) and the value of the resistor is 30,000 ohms. Determine the potential across the resistor. The circuit is shown in figure 5-3.

Given:

$$R_2 = 30K$$

$$I_{R2} = 4.5 \text{ milliamperes}$$

Find:

$$E_{R2} = ?$$

$$E_{bb} = ?$$

Solution: Select proper equation.

$$E = IR \qquad (4-2)$$

Substitute known values:

$$E_{R2} = I_{R2} \times R_2$$

$$E_{R2} = 4.5 \text{ milliamperes} \times 30,000 \text{ ohms}$$

Express in powers of ten:

$$E_{R2} = (4.5 \times 10^{-3}) \times (30 \times 10^3)$$

$$E_{R2} = 4.5 \times 30$$

Resultant:

$$E_{R2} = 135 \text{ v}$$

Therefore:

$$E_{bb} = 135 \text{ v}$$

Having determined the voltage across one resistor ($R_2$) in a parallel circuit, the value of the source voltage ($E_{bb}$) and the potentials across any other resistors that may be connected in parallel with it are known (equations 5-1).

## CURRENT DIVISION

The current in a circuit . . . . is inversely proportional to the circuit resistance. This fact, obtained from Ohm's law, establishes the relationship upon which the following discussion is developed.

A single current flows in a series circuit. Its value is determined in part by the total resistance of the circuit. However, the source current in a parallel circuit divides among the available paths in relation to the value of the resistors in the circuit. Ohm's law remains unchanged. For a given voltage, current varies inversely with resistance.

The behavior of current in parallel circuits will be shown by a series of illustrations using example circuits with different values of resistance for a given value of applied voltage.

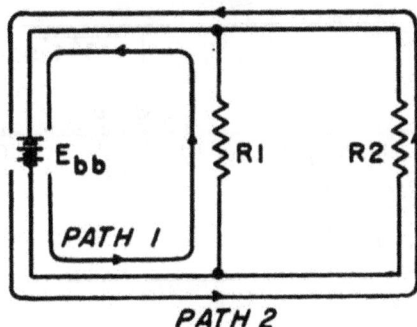

BE.82

Figure 5-1.—Example of a basic parallel circuit.

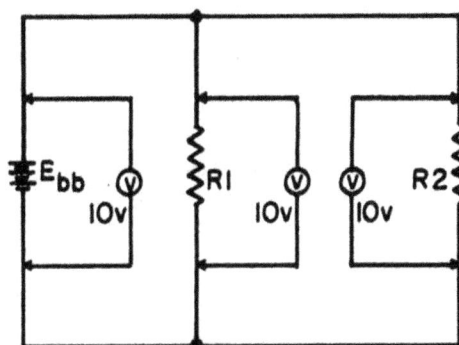

BE.83

Figure 5-2.—Voltage comparison in a parallel circuit.

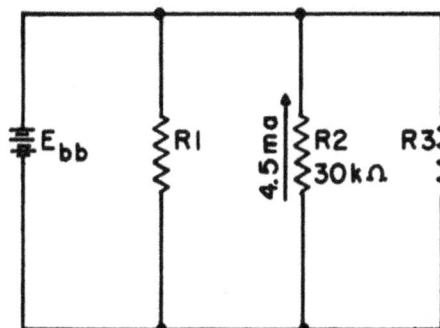

BE.84

Figure 5-3.—Example problem parallel circuit.

Part (A) of figure 5-4 shows a basic series circuit. Here the total current must pass through the single resistor. The amount of current is determined as

$$I_t = \frac{E_{bb}}{R_1} = \frac{50}{10} = 5 \text{ amperes}$$

Part (B) of figure 5-4 shows the same resistor ($R_1$) with a second resistor ($R_2$) of equal value connected in parallel across the voltage source. Applying the proper equation from Ohm's law, the current flow through each resistor is seen to be the same as through the single resistor in part (A). These individual currents are determined as follows:

$$I_{R1} = \frac{E_{bb}}{R_1} = \frac{50}{10} = 5 \text{ amperes}$$

$$I_{R2} = \frac{E_{bb}}{R_2} = \frac{50}{10} = 5 \text{ amperes}$$

However, it is apparent that if 5 amperes of current flows through each of the two resistors, there must be a total current of 10 amperes drawn from the source. The distribution of current in the simple parallel circuit shown in figure 5-4 (B) is as follows:

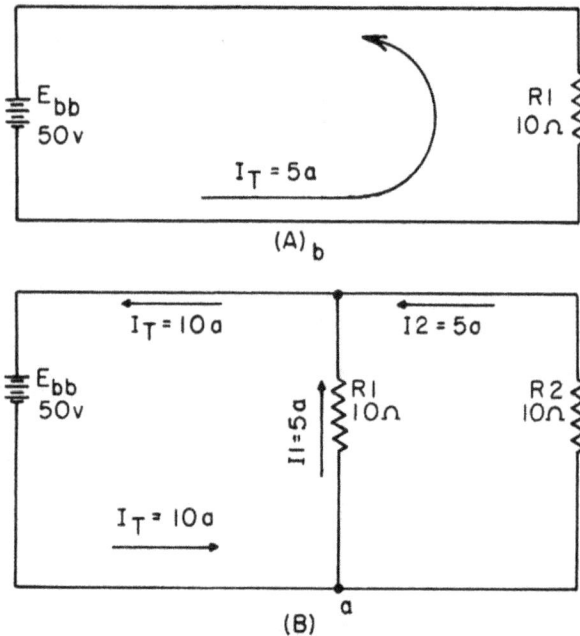

BE.85

Figure 5-4.—Analysis of current in parallel circuit.

The total current of 10 amperes leaves the negative terminal of the battery and flows to point a. Since point a is a connecting point for the two resistors, it is called a junction. At junction a the total current divides into two smaller currents of 5 amperes each. These two currents flow through their respective resistors and rejoin at junction b. The total current then flows from junction b back to the positive terminal of the source. Thus, the source supplies a total current of 10 amperes and each of the two equal resistors carries one-half the total current.

Each individual current path in the circuit of figure 5-4 (B) is referred to as a branch. Each branch will carry a current that is a portion of the total current. Two or more branches form a network.

From the foregoing observations, the characteristics of current in a parallel circuit can be expressed in terms of the following general equation

$$I_t = I_1 + I_2 + \ldots I_n \qquad (5\text{-}2)$$

The analysis of current in parallel circuits is continued with the use of the following example circuits.

Compare part (A) of figure 5-5 with part (B) of the preceding example circuit in figure 5-4. Notice that doubling the value of the second branch resistor ($R_2$) has no effect on the current in the first branch ($I_{R1}$), but does reduce its own branch current ($I_{R2}$) to one-half its original value. The total circuit current drops to a value equal to the sum of the branch currents. These facts are verified as follows:

$$I_1 = \frac{E_{bb}}{R_1} = \frac{50}{10} = 5 \text{ amperes}$$

$$I_2 = \frac{E_{bb}}{R_2} = \frac{50}{20} = 2.5 \text{ amperes}$$

$$I_t = I_1 + I_2$$

$$I_t = 5 + 2.5 = 7.5 \text{ amperes}$$

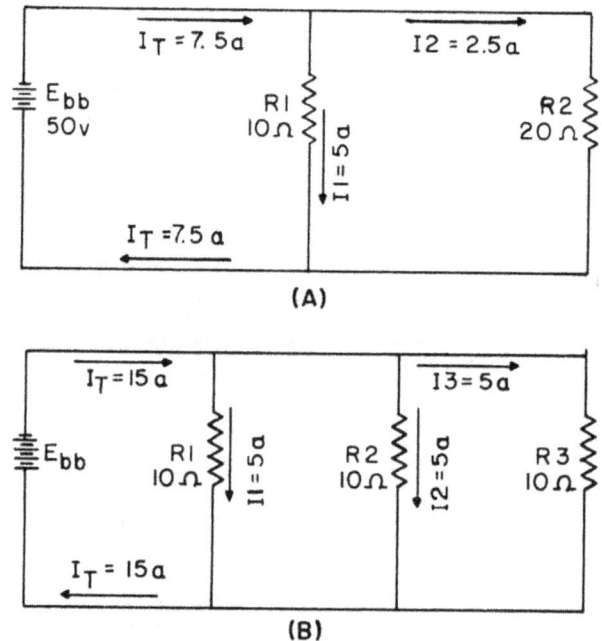

BE.86

Figure 5-5.—Current behavior in parallel circuits.

Now compare the two circuits of figure 5-5. Notice that the sum of the ohmic values of the resistors in both circuits is equal and that the applied voltage is the same value. However, the total current in part (B) is twice the amount in part (A). It is apparent, therefore, that the manner in which resistors are connected in a circuit, as well as their actual ohmic value, affects the total current flow. This phenomenon will be illustrated in more detail in the discussion of resistance. The amount of current flow in the branch circuits and the total current in the circuit (fig. 5-5 (B)), are determined as follows:

$$I_1 = \frac{E_{bb}}{R_1} = \frac{50}{10} = 5 \text{ amperes}$$

$$I_2 = \frac{E_{bb}}{R_2} = \frac{50}{10} = 5 \text{ amperes}$$

$$I_3 = \frac{E_{bb}}{R_3} = \frac{50}{10} = 5 \text{ amperes}$$

$$I_t = I_1 + I_2 + I3$$

$$I_t = \frac{E_{bb}}{R_1} + \frac{E_{bb}}{R_2} + \frac{E_{bb}}{R_3}$$

$$I_t = \frac{50}{10} + \frac{50}{10} + \frac{50}{10} = 15 \text{ amperes}$$

The division of current in a parallel network follows a definite pattern. This pattern is described by Kirchhoff's current law which is stated as follows:

The algebraic sum of the currents entering and leaving any junction of conductors is equal to zero. This law can be stated mathematically as

$$I_a + I_b + \ldots\ldots + I_n = 0 \qquad (5-3)$$

where $I_a$, $I_b$, etc., are the currents entering and leaving the junction. Currents entering the junction are assumed to be positive, and currents leaving the junction are considered negative. When solving a problem using equation (5-3), the currents must be placed into the equation with the proper polarity signs attached.

Example. Solve for the value of $I_3$ in figure 5-6.

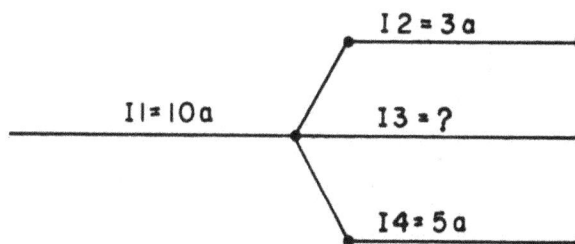

BE.87

Figure 5-6.—Circuit for example problem.

Solution: First the currents are given proper signs.

$$I_1 = + 10 \text{ amperes}$$

$$I_2 = - 3 \text{ amperes}$$

$$I_3 = \quad ? \text{ amperes}$$

$$I_4 = - 5 \text{ amperes}$$

these currents are placed into equation (5-3) with the proper signs as follows:

Basic equation:

$$I_a + I_b + \ldots\ldots I_n = 0 \qquad (5-3)$$

Substitution:

$$I_1 + I_2 + I_3 + I_4 = 0$$
$$(+10) + (-3) + (I_3) + (-5) = 0$$

Combining like terms:

$$I_3 + 2 = 0$$

$$I_3 = -2 \text{ amperes}$$

thus, $I_3$ has a value of 2 amperes, and the negative sign shows it to be a current leaving the junction.

Example. Using figure 5-7, solve for the magnitude and direction of $I_3$:

Solution:

$$I_a + I_b + \ldots\ldots I_n = 0 \qquad (5-3)$$
$$I_1 + I_2 + I_3 + I_4 = 0$$

88

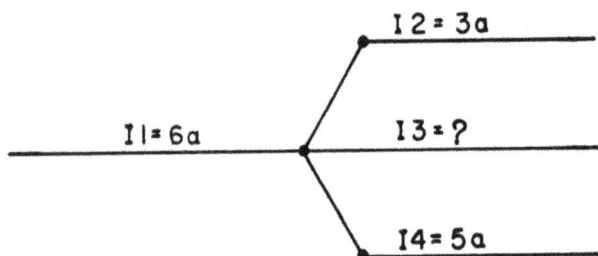

BE. 88

Figure 5-7.—Circuit for example problem.

$$(+6a) + (-3a) + (I_3) + (-5a) = 0$$
$$I_3 - 2a = 0$$
$$I_3 = 2 \text{ amperes}$$

Thus, $I_3$ is 2 amperes, and its positive sign shows it to be a current entering the junction.

## PARALLEL RESISTANCE

The preceding discussion of current introduced certain principles involving the characteristics and effects of resistance in parallel circuits. A detailed explanation of the characteristics of parallel resistances will be considered in this section. The explanation will commence with a simple parallel circuit. Various methods used to determine the total resistance in parallel circuits will be described.

In the example diagram (fig. 5-8), two cylinders of conductive material having a resistance value of 10 ohms each are connected across a 5-volt battery. A complete circuit consisting of two parallel paths is formed and current will flow as shown.

Computing the individual currents shows that there is one-half an ampere of current flowing through each resistance. Accordingly, the total current flowing from the battery to the junction of the resistors, and returning from the resistors to the battery, is equal to 1 ampere. The total resistance of the circuit can be determined by substituting total values of voltage and current into the following equation. This equation is derived from Ohm's law.

$$R_t = \frac{E_t}{I_t}$$
$$R_t = \frac{5}{1} = 5 \text{ ohms}$$

This computation shows the total resistance to be 5 ohms, one-half the value of either of the two resistors.

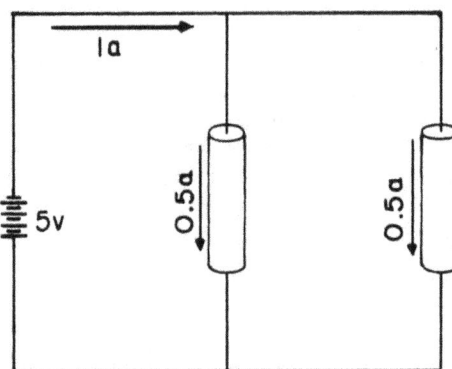

BE.89

Figure 5-8.—Two equal resistors connected in parallel.

Since the total resistance of this parallel circuit is smaller than either of the two resistors, the term "total resistance" does not mean the sum of the individual resistor values. The total resistance of resistors in parallel is also referred to as equivalent resistance. In many texts the terms total resistance and equivalent resistances are used interchangeably.

There are several methods used to determine the equivalent resistance of parallel circuits. The most appropriate method for a particular circuit depends on the number and value of the resistors. For the circuit described above, the following simple equation is used:

$$R_{eq} = \frac{R}{N}$$

where

$R_{eq}$ = equivalent parallel resistance

$R$ = ohmic value of one resistor

$N$ = number of resistors

This equation is valid for any number of equal value parallel resistors.

An understanding of why the equivalent resistance of two parallel resistors is smaller than the resistance of either of the two resistors can be gained by an examination of figure 5-8. The two 10-ohm cylinders have fixed equal volumes. If the cylinders were combined into one cylinder as shown in figure 5-9, the volume would double. If the same length is retained and the volume is doubled, the cross-sectional area will double. When the cross-sectional area of a material is increased, the resistance is decreased proportionately.

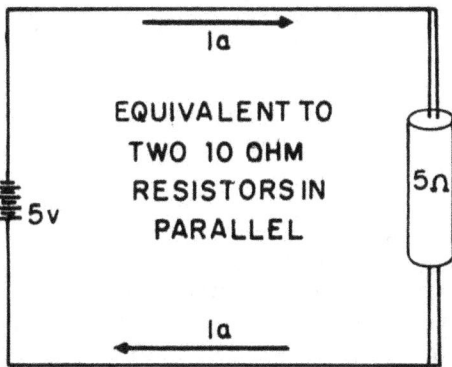

BE. 90

Figure 5-9.—Equivalent parallel circuit.

Since, in this case, the cross-sectional area is two times the original area, the resistance is one-half the former value. Therefore, when two equal value resistors are connected in parallel, they present a total resistance equivalent to a single resistor of one-half the value of either of the original resistors.

Example. Four 40-ohm resistors are connected in parallel. What is their equivalent resistance?

Solution:

$$R_{eq} = \frac{R}{N} = \frac{40}{4} = 10 \text{ ohms}$$

Circuits containing parallel resistance of unequal value will now be considered. Refer to example circuit in figure 5-10.

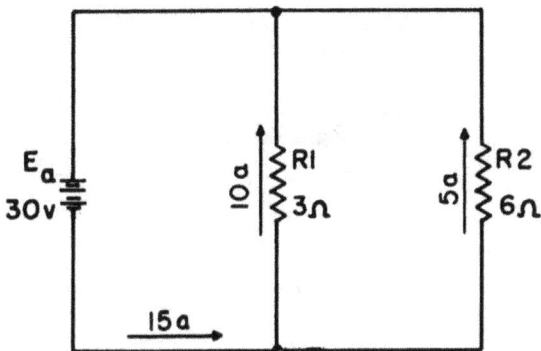

BE. 91

Figure 5-10.—Example circuit with unequal parallel resistors.

Given:

$$R_1 = 3\Omega, R_2 = 6\Omega, E_a = 30v$$

Known:

$I_1 = 10$ amperes, $I_2 = 5$ amperes, $I_t = 15$ amperes.

Determine:

Solution:

$$R_{eq} = ?$$

$$R_{eq} = \frac{E_a}{I_t}$$

$$R_{eq} = \frac{30}{15} = 2 \text{ ohms}$$

Notice that the equivalent resistance of two ohms is less than the value of either branch resistor. In parallel circuits the equivalent resistance will always be smaller than the resistance of any branch.

## RECIPROCAL METHOD

Many circuits are encountered in which resistors of unequal value are connected in parallel. It is therefore desirable to develop a formula which can be used to compute the equivalent resistance of two or more unequal parallel resistors. This equation can be derived as follows:

Given:

$$I_t = I_1 + I_2 + \ldots \ldots I_n \qquad (5-2)$$

Substituting $\frac{E}{R}$ for I gives:

$$\frac{E_t}{R_t} = \frac{E_1}{R_1} + \frac{E_2}{R_2} + \ldots \ldots \frac{E_n}{R_n}$$

Since in a parallel circuit $E_t = E_1 = E_2 = E_n$

$$\frac{E}{R_t} = \frac{E}{R_1} + \frac{E}{R_2} + \ldots \ldots \frac{E}{R_n}$$

Dividing both sides by E:

$$\frac{E}{E R_t} = \frac{E}{E R_1} + \frac{E}{E R_2} + \ldots \ldots \frac{E}{E R_n}$$

$$\frac{1}{R_t} = \frac{1}{R_1} + \frac{1}{R_2} + \ldots \ldots \frac{1}{R_n}$$

90

Taking the reciprocal of both sides:

$$\frac{1}{\frac{1}{R_t}} = \frac{1}{\frac{1}{R_1} + \frac{1}{R_2} + \ldots \frac{1}{R_n}}$$

Simplifying:

$$R_t = \frac{1}{\frac{1}{R_1} + \frac{1}{R_2} + \ldots \ldots \frac{1}{R_n}}$$

This formula is called "the reciprocal of the sum of the reciprocals" and is the one normally used to solve for the equivalent resistance of a number of parallel resistors.

Example. Given three parallel resistors of 20 ohms, 30 ohms, and 40 ohms; find the equivalent resistance using the reciprocal equation. (See fig. 5-11.)

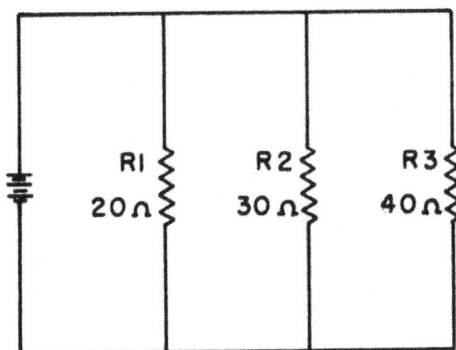

BE.92

Figure 5-11.—Example parallel circuit with unequal branch resistors.

Solution:

Select the proper equation:

$$R_{eq} = \frac{1}{\frac{1}{R_1} + \frac{1}{R_2} + \frac{1}{R_3}}$$

Substitute:

$$R_{eq} = \frac{1}{\frac{1}{20} + \frac{1}{30} + \frac{1}{40}}$$

Find LCD:

$$R_{eq} = \frac{1}{\frac{6}{120} + \frac{4}{120} + \frac{3}{120}} = \frac{1}{\frac{13}{120}}$$

Invert:

$$R_{eq} = \frac{120}{13} = 9.23 \text{ ohms}$$

Some parallel circuit problems can be solved more conveniently by considering the ease with which current can flow. The degree to which a circuit permits or conducts current is called the conductance (G) of the circuit. The unit of conductance is the MHO, which is ohms spelled backwards. The conductance of a circuit is the reciprocal of the resistance. The conductance can therefore be found using the following formula:

$$G = \frac{1}{R}$$

also:

$$R = \frac{1}{G}$$

In a parallel circuit, the total conductance is equal to the sum of the individual branch conductances. As an equation:

$$G_t = G_1 + G_2 + \ldots \ldots G_n \qquad (5\text{-}6)$$

Example: Determine the equivalent (total) resistance of the circuit shown in the preceding example (fig. 5-11), using the conductance method.

Solution:

$$G_1 = \frac{1}{R_1} = \frac{1}{20} = 0.050 \text{ mho}$$

$$G_2 = \frac{1}{R_2} = \frac{1}{30} = 0.033 \text{ mho}$$

$$G_3 = \frac{1}{R_3} = \frac{1}{40} = 0.025 \text{ mho}$$

$$G_t = G_1 + G_2 + G_3 \qquad (5\text{-}6)$$

$$G_t = 0.050 + 0.033 + 0.025 = 0.108 \text{ mho}$$

Since:

$$R_t = \frac{1}{G_t}$$

$$R_t = \frac{1}{0.108} = 9.25 \text{ ohms}$$

The value of equivalent resistance determined by the conductance method is almost identical to the value determined by the reciprocal of the sum of the reciprocals methods.

## PRODUCT OVER THE SUM

A convenient formula for finding the equivalent resistance of two parallel resistors can be derived from equation (5-5) as shown below:

$$R_t = \cfrac{1}{\cfrac{1}{R_1} + \cfrac{1}{R_2}} \qquad (5-5)$$

Finding the LCD:

$$R_t = \cfrac{1}{\cfrac{R_2 + R_1}{R_1 \times R_2}}$$

Taking the reciprocal:

$$R_t = \frac{R_1 \times R_2}{R_1 + R_2}$$

This equation, called the product over the sum formula, is used so frequently it should be committed to memory.

Example. What is the equivalent resistance of a 20 ohm and a 30 ohm resistor connected in parallel?

Given:

$$R_1 = 20$$

$$R_2 = 30$$

Find:

$$R_{eq} = ?$$

Solution:

$$R_t = \frac{R_1 \times R_2}{R_1 + R_2}$$

$$R_t = \frac{20 \times 30}{20 + 30}$$

$$R_t = 12 \text{ ohms}$$

## PARALLEL CIRCUIT REDUCTION

In the study of electricity, it is often necessary to resolve a complex circuit into a simpler

form. Any complex circuit consisting of resistances can be reduced to a basic equivalent circuit containing the source and total resistance. This process is called reduction to an equivalent circuit. An example of circuit reduction is shown in figure 5-12.

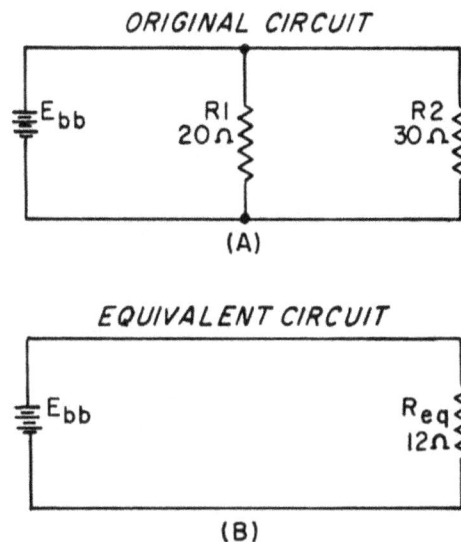

*ORIGINAL CIRCUIT*

(A)

*EQUIVALENT CIRCUIT*

(B)

BE.93

Figure 5-12.—Parallel circuit with equivalent circuit.

The circuit shown in figure 5-12 (A) is reduced to the simple circuit shown in (B).

## COMPUTING TOTAL POWER

Power computations in a parallel circuit are essentially the same as those used for the series circuit. Since power dissipation in resistors consists of a heat loss, power dissipations are additive regardless of how the resistors are connected in the circuit. The total power dissipated is equal to the sum of the powers dissipated by the individual resistors. Like the series circuit, the total power consumed by the parallel circuit is

$$P_t = P1 + P2 + \ldots P_n \qquad (4-13)$$

Example. Find the total power consumed by the circuit in figure 5-13.

Solution:

$$P_{R1} = E_{bb} \times I_{R1}$$

$$P_{R1} = 50 \times 5$$

$P_{R1} = 250 \text{ w}$

$P_{R2} = E_{bb} \times I_{R2}$

$P_{R2} = 50 \times 2$

$P_{R2} = 100 \text{ w}$

$P_{R3} = E_{bb} \times I_{R3}$

$P_{R3} = 50 \times 1$

$P_{R3} = 50 \text{ w}$

$P_t = P_1 + P_2 + P_3$

$P_t = 250 + 100 + 50$

$P_t = 400 \text{ w}$

Note that the power dissipated in the branch circuits is determined in the same manner as the power dissipated by individual resistors in a series circuit. The total power ($P_t$) is then obtained by summing up the powers dissipated in the branch resistors using equation (4-13).

Since, in the example shown in figure 5-13, the total current is known, the total power could be determined by the following method:

$P_t = E_{bb} \times I_t$

$P_t = 50v \times 8a$

$P_t = 400 \text{ w}$

Rules for solving parallel d-c circuits are as follows:

1. The same voltage exists across each branch of a parallel circuit and is equal to the source voltage.
2. The current through a branch of a parallel network is inversely proportional to the amount of resistance of the branch.
3. The total current of a parallel circuit is equal to the sum of the currents of the individual branches of the circuit.
4. The total resistance of a parallel circuit is equal to the reciprocal of the sum of the reciprocals of the individual resistances of the circuit.
5. The total power consumed in a parallel circuit is equal to the sum of the power consumption of the individual resistances.

## TYPICAL PROBLEMS IN PARALLEL CIRCUITS

Problems involving the determination of resistance, voltage, current, and power in a parallel circuit are solved as simply as in a series circuit. The procedure is the same— (1) draw a circuit diagram, (2) state the values given and the values to be found, (3) state the applicable equations, and (4) substitute the given values and solve for the unknown.

For example, the parallel circuit of figure 5-14 consists of 2 branches (a and b). Branch a consists of 3 lamps in parallel. Their ratings are $L_1 = 50$ watts, $L_2 = 25$ watts, and $L_3 = 75$ watts. Branch b also has 3 lamps in parallel with ratings of $L_4 = 150$ watts, $L_5 = 200$ watts, and $L_6 = 250$ watts. The source voltage is 100 volts.

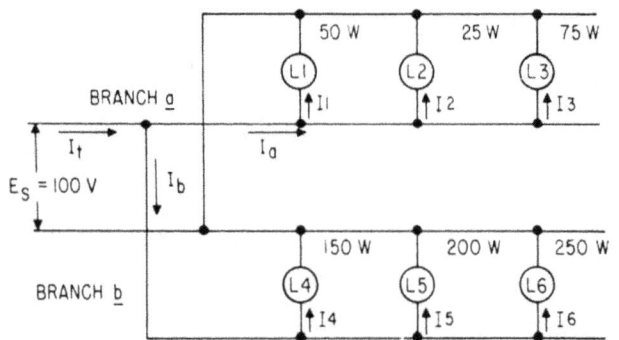

BE.94

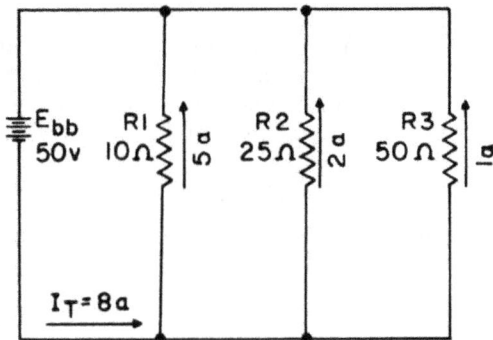

Figure 5-13.—Example parallel circuit.

BE.95

Figure 5-14.—Typical parallel circuit.

Problem:

1. Find the current in each lamp.
2. Find the resistance of each lamp.
3. Find the current in branch a.
4. Find the current in branch b.
5. Find the total circuit current.
6. Find the total circuit resistance.
7. Find the total power supplied to the circuit.
8. Check 7 by a separate calculation.

Solution:

1. The current in $L_1$ is $I = \dfrac{P}{E} = \dfrac{50}{100} =$

   0.50 ampere.

   The current in $L_2$ is $\dfrac{25}{100} = 0.25$ ampere.

   The current in $L_3$ is $\dfrac{75}{100} = 0.75$ ampere.

   The current in $L_4$ is $\dfrac{150}{100} = 1.50$ amperes.

   The current in $L_5$ is $\dfrac{200}{100} = 2.00$ amperes.

   The current in $L_6$ is $\dfrac{250}{100} = 2.5$ amperes.

2. The resistance of $L_1$ is $R = \dfrac{E}{I} = \dfrac{100}{0.5} =$

   200 ohms.

   The resistance of $L_2$ is $\dfrac{100}{0.25} = 400$ ohms.

   The resistance of $L_3$ is $\dfrac{100}{0.75} = 133$ ohms.

   The resistance of $L_4$ is $\dfrac{100}{1.5} = 66.7$ ohms.

   The resistance of $L_5$ is $\dfrac{100}{2.0} = 50$ ohms.

   The resistance of $L_6$ is $\dfrac{100}{2.5} = 40$ ohms.

3. The current in branch a is

   $I_1 + I_2 + I_3 = 0.5 + 0.25 + 0.75 = 1.5$ amperes

4. The current in branch b is

   $I_4 + I_5 + I_6 = 1.5 + 2.0 + 2.5 = 6.0$ amperes

5. The total circuit current is

   $I_a + I_b = 1.5 + 6.0 = 7.5$ amperes

6. The total circuit resistance is

   $R_t = \dfrac{E}{I_t} = \dfrac{100}{7.5} = 13.3$ ohms

7. The total power supplied to the circuit is:

   $50w + 25w + 75w + 150w + 200w + 250w$

   $= 750$ watts

8. The total power is also equal to

   $P_t = EI_t = 100 \times 7.5 = 750$ watts

## SERIES-PARALLEL COMBINATIONS

In the preceding discussions, series and parallel d-c circuits have been considered separately. However, the technician will seldom encounter a circuit that consists solely of either type of circuit. Most circuits consist of both series and parallel elements. A circuit of this type will be referred to as a combination circuit. The solution of a combination circuit is simply a matter of application of the laws and rules discussed prior to this point.

SOLVING A COMBINATION CIRCUIT

At least three resistors are required to form a combination circuit. Two basic series-parallel circuits are shown in figure 5-15. In figure 5-15 (A), $R_1$ is connected in series with the parallel combination made up of $R_2$ and $R_3$.

The total resistance $(R_t)$ of figure 5-15 (A) is determined in two steps. First, the equivalent resistance of the parallel combination of $R_2$ and $R_3$ is determined as follows:

$$R_{2,3} = \frac{R_2 R_3}{R_2 + R_3} = \frac{3 \times 6}{3 + 6} = \frac{18}{9} = 2 \text{ ohms}$$

The sum of $R_{2,3}$ and $R_1$ — that is, $R_t$ — is

$$R_t = R_{2,3} + R_1 = 2 + 2 = 4 \text{ ohms}$$

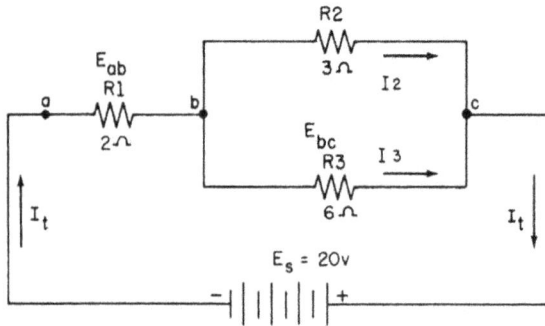

R1 IN SERIES WITH PARALLEL COMBINATION OF R2 AND R3

(A)

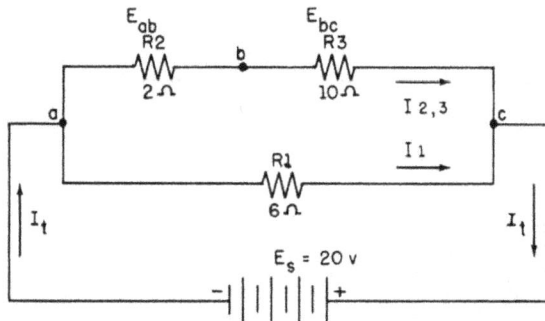

R1 IN PARALLEL WITH THE SERIES COMBINATION OF R2 & R3

(B)

BE.96

Figure 5-15.—Compound circuits—series-parallel connections.

If the total resistance ($R_t$) and the source voltage ($E_s$) are known, the total current ($I_t$) may be determined by Ohm's law. Thus, in figure 5-15 (A),

$$I_t = \frac{E_s}{R_t} = \frac{20}{4} = 5 \text{ amperes}$$

If the values of the various resistors and the current through them are known, the voltage drops across the resistors may be determined by Ohm's law. Thus,

and
$$E_{ab} = I_t R_1 = 5 \times 2 = 10 \text{ volts}$$

$$E_{bc} = I_t R_{2,3} = 5 \times 2 = 10 \text{ volts}$$

According to Kirchhoff's voltage law, the sum of the voltage drops around the closed circuit is equal to the source voltage. Thus,

or
$$E_{ab} + E_{bc} = E_s$$

$$10 + 10 = 20 \text{ volts}$$

If the voltage drop ($E_{bc}$) across $R_{2,3}$—that is, the drop between points b and c—is known, the current through the individual branches may be determined as follows:

$$I_2 = \frac{E_{bc}}{R_2} = \frac{10}{3} = 3.333 \text{ amperes}$$

and

$$I_3 = \frac{E_{bc}}{R_3} = \frac{10}{6} = 1.666 \text{ amperes}$$

According to Kirchhoff's current law, the sum of the currents flowing in the individual parallel branches is equal to the total current. Thus,

$$I_2 + I_3 = I_t$$

or

$$3.333 + 1.666 = 5 \text{ amperes (approx)}$$

The total current flows through $R_1$; and at point b it divides between the two branches in inverse proportion to the resistance of each branch. Twice as much current goes through $R_2$ as through $R_3$ because $R_2$ has one-half the resistance of $R_3$. Thus, 3.333, or two-thirds of 5 amperes flows through $R_2$; and 1.666, or one-third of 5 amperes flows through $R_3$.

In figure 5-15 (B), $R_1$ is in parallel with series combination of $R_2$ and $R_3$. The total resistance ($R_t$) is determined in two steps. First, the sum of the resistance of $R_2$ and $R_3$—that is, $R_{2,3}$—is determined as follows:

$$R_{2,3} = R_2 + R_3 = 2 + 10 = 12 \text{ ohms}$$

Second, the total resistance ($R_t$) is the result of combining $R_{2,3}$ in parallel with $R_1$, or

$$R_t = \frac{R_{2,3} R_1}{R_{2,3} + R_1} = \frac{12 \times 6}{12 + 6} = 4 \text{ ohms}$$

If the total resistance ($R_t$) and the source voltage ($E_s$) are known, the total current ($I_t$) may be determined by Ohm's law. Thus, in figure 5-15 (B),

$$I_t = \frac{E_s}{R_t} = \frac{20}{4} = 5 \text{ amperes}$$

A portion of the total current flows through the series combination of $R_2$ and $R_3$ and the remainder flows through $R_1$. Because current varies inversely with the resistance, two-thirds of the total current flows through $R_1$ and one-third flows through the series combination of $R_2$ and $R_3$, since $R_1$ is one-half of $R_2 + R_3$.

The source voltage ($E_S$) is applied between points a and c, and therefore the current $I_1$ through $R_1$ is

$$I_1 = \frac{E_S}{R_1} = \frac{20}{6} = 3.333 \text{ amperes}$$

and the current, $I_{2,3}$, through $R_{2,3}$ is

$$I_{2,3} = \frac{E_s}{R_{2,3}} = \frac{20}{12} = 1.666 \text{ amperes}$$

According to Kirchhoff's current law the sum of the individual branch currents is equal to the total current, or

$$I_t = I_1 + I_{2,3}$$

$$5 = 3.333 + 1.666 \text{ (approx)}$$

Combination circuits may be made up of a number of resistors arranged in numerous series and parallel combinations. In more complicated circuits, special theorems, rules, and formulas are used. These are based on Ohm's Law and provide faster solutions for particular applications. Series formulas are applied to the series parts of the circuit, and parallel formulas are applied to the parallel parts. For example, in figure 5-16, the total resistance ($R_t$) may be obtained in three logical steps.

First, $R_3$, $R_4$, and $R_5$ in figure 5-16 (A), are in series (there is only one path for current) and may be combined in figure 5-16 (B), to give the resistance, $R_s$, of the three resistors. Thus,

$$R_s = R_3 + R_4 + R_5 = 5 + 9 + 10 = 24 \text{ ohms}$$

and it is now in parallel with $R_2$ (because they both receive the same voltage).

The combined resistance of $R_s$ in parallel with $R_2$ is

$$R_{s,2} = \frac{R_2 R_s}{R_2 + R_s} = \frac{8 \times 24}{8 + 24} = 6 \text{ ohms}$$

as in the figure 5-16 (C).

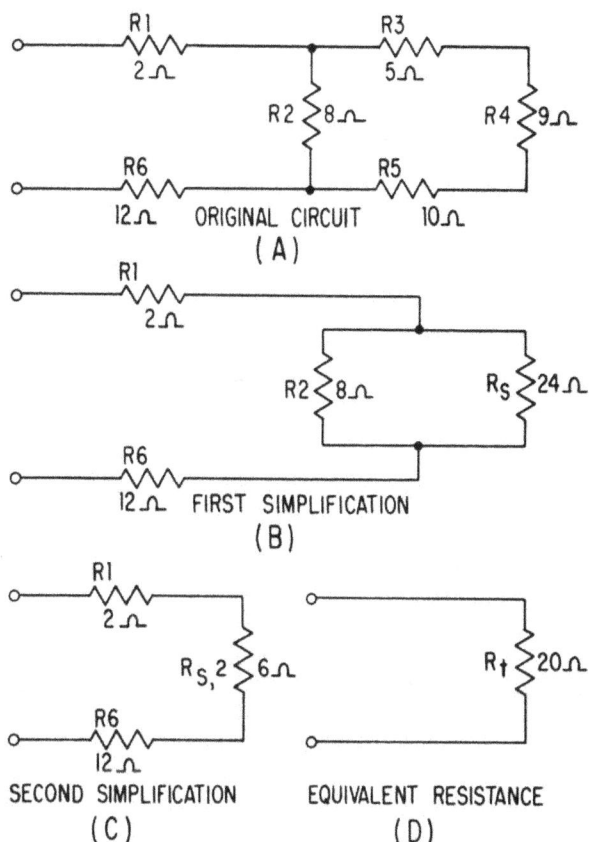

BE.97

Figure 5-16.—Solving total resistance in a compound circuit.

Third, the total resistance ($R_t$) is determined by combining resistors $R_1$ and $R_6$ with $R_{s,2}$ as

$$R_t = R_1 + R_6 + R_{s,2} = 2 + 12 + 6 = 20 \text{ ohms}$$

Other compound circuits may be solved in a similar manner. For example, in figure 5-17, the total resistance ($R_t$) may be found by simplifying the circuit in successive steps beginning with the resistance, $R_1$ and $R_2$. Thus,

$$R_{1,2} = \frac{R_1 R_2}{R_1 + R_2} = \frac{3 \times 6}{3 + 6} = \frac{18}{9} = 2 \text{ ohms}$$

and it is in series with $R_3$.

The resistances, $R_{1,2}$ and $R_3$, are added to give the resultant resistance, $R_{1,2,3}$. Thus,

$$R_{1,2,3} = R_{1,2} + R_3 = 2 + 4 = 6 \text{ ohms}$$

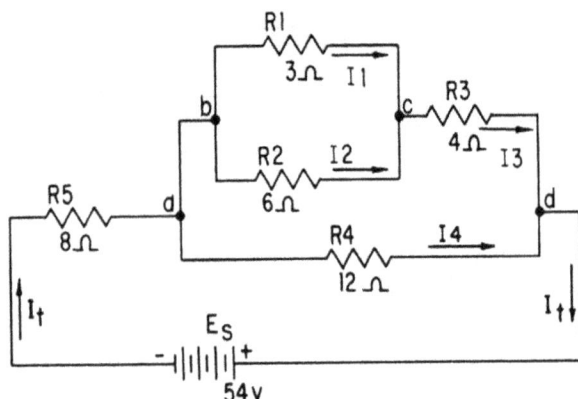

BE.98

Figure 5-17.—Compound circuit for solving resistance, voltage, current, and power.

$R_{1,2,3}$ is in parallel with $R_4$. The combined resistance, $R_{1,2,3,4}$, is determined as follows:

$$R_{1,2,3,4} = \frac{R_{1,2,3} R_4}{R_{1,2,3} + R_4} = \frac{6 \times 12}{6 + 12} = \frac{72}{18} = 4 \text{ ohms}$$

This equivalent resistance is in series with $R_5$. Thus, the total resistance $(R_t)$ of the circuit is

$$R_t = R_{1,2,3,4} + R_5 = 4 + 8 = 12 \text{ ohms}$$

By Ohm's Law, the line current $(I_t)$ is

$$I_t = \frac{E_s}{R_t} = \frac{54}{12} = 4.5 \text{ amperes}$$

The line current flows through $R_5$ and therefore the voltage drop, $E_5$, across $R_5$ is

$$E_5 = I_t R_5 = 4.5 \times 8 = 36 \text{ volts}$$

According to Kirchhoff's voltage law, the sum of the voltage drops around the circuit is equal to the source voltage; accordingly, the voltage between points a and d is

$$E_{ad} = E_s - E_5 = 54 - 36 = 18 \text{ volts}$$

The current through $R_4$ is

$$I_4 = \frac{E_4}{R_4} = \frac{18}{12} = 1.5 \text{ amperes}$$

The resistance, $R_{1,2,3}$, of parallel resistors $R_1$ and $R_2$ in series with resistor $R_3$ is 6 ohms.

$E_{ad}$ is applied across 6 ohms; therefore the current, $I_3$, through $R_3$ is

$$I_3 = \frac{E_{ad}}{R_{1,2,3}} = \frac{18}{6} = 3 \text{ amperes}$$

The voltage drop, $E_3$, across $R_3$ is

$$E_3 = I_3 R_3 = 3 \times 4 = 12 \text{ volts}$$

and the voltage across the parallel combination of $R_1$ and $R_2$—that is, $E_{bc}$—is

$$E_{bc} = I_{1,2} R_{1,2} = 3 \times 2 = 6 \text{ volts}$$

where $I_{1,2}$ is the current through the parallel combinations of $R_1$ and $R_2$. By Kirchhoff's current law, $I_{1,2}$ is equal to $I_3$. The current, $I_1$ through $R_1$ is

$$I_1 = \frac{E_{bc}}{R_1} = \frac{6}{3} = 2 \text{ amperes}$$

and the current, $I_2$, through $R_2$ is

$$I_2 = \frac{E_{bc}}{R_2} = \frac{6}{6} = 1 \text{ ampere}$$

The preceding computations may be checked by the application of Kirchhoff's voltage and current law to the entire circuit. Briefly, the sum of the voltage drops around the circuit is equal to the source voltage. Voltage $E_5$ across $R_5$ is 36 volts and voltage $E_{ad}$ across $R_4$ is 18 volts—that is,

$$E_s = E_5 + E_{ad}$$

or 
$$54 = 36 + 18 \text{ volts}$$

Likewise, the voltage drop, $E_{bc}$, across the parallel combination of $R_1$ and $R_2$ plus the voltage drop, $E_3$, across $R_3$ should be equal to the voltage across points a and d. $E_{bc}$ is 6 volts and $E_3$ is 12 volts. Therefore,

$$E_{ad} = E_{bc} + E_3 = 6 + 12 = 18 \text{ volts}$$

Kirchhoff's current law says in effect that the sum of the branch currents is equal to the line current, $I_t$. The line current is 4.5 amperes, and therefore the sum of $I_4$ and $I_3$ should be 4.5 amperes, or

$$I_t = I_4 + I_3 = 1.5 + 3 = 4.5 \text{ amperes}$$

The power consumed in a circuit element is determined by one of the three power formulas. For example, in figure 5-17 the power, $P_1$ consumed in $R_1$ is

$$P_1 = I_1 E_{bc} = 2 \times 6 = 12 \text{ watts}$$

the power $P_2$ consumed in $R_2$ is

$$P_2 = I_2 E_{bc} = 1 \times 6 = 6 \text{ watts}$$

the power $P_3$ consumed in $R_3$ is

$$P_3 = I_3 E_3 = 3 \times 12 = 36 \text{ watts}$$

the power $P_4$ consumed in $R_4$ is

$$P_4 = I_4 E_4 = 1.5 \times 18 = 27 \text{ watts}$$

and the power $P_5$ consumed in $R_5$ is

$$P_5 = I_5 E_5 = 4.5 \times 36 = 162 \text{ watts}$$

The total power $P_t$, consumed is

$$P_t = P_1 + P_2 + P_3 + P_4 + P_5$$

$$= 12 + 6 + 36 + 27 + 162$$

$$= 243 \text{ watts}$$

The total power is also equal to the total current multiplied by the source voltage, or

$$P_t = I_t E_s = 4.5 \times 54 = 243 \text{ watts}$$

EFFECTS OF SOURCE RESISTANCE

The parallel circuits discussed up to this point have been explained and solved without considering the internal resistance of the source. Every known source possesses resistance. In a battery the resistance is partially due to the opposition offered to the movement of current through the electrolyte. A schematic representation of source resistance is shown in figure 5-18.

The internal resistance of the battery is labeled ($R_i$) and is always shown schematically connected in series with the source. Under load conditions this internal resistance will have a voltage drop across it and must be considered as part of the external circuit. The voltage at battery terminals A and B will always be less than the generated voltage of the battery since a portion of the generated voltage will be dropped across the internal resistance of the battery.

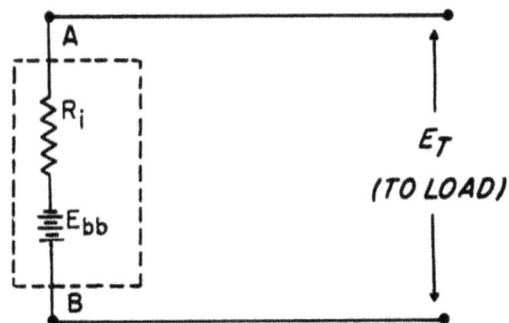

BE.99

Figure 5-18.—Battery with internal resistance.

The presence of internal resistance results in (1) a diminished voltage supplied to the components that comprise the load, (2) a decrease in total current, and (3) an increase in total resistance. The power dissipated by the circuit is also affected. The effect of internal resistance on the circuit is analyzed using the example circuit shown in figure 5-19.

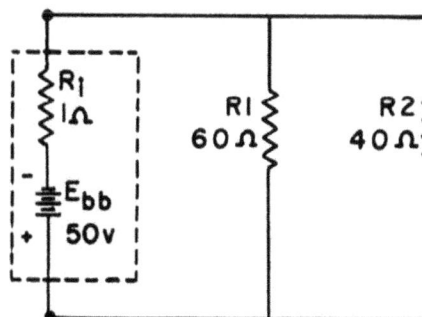

BE.100

Figure 5-19.—Effect of source resistance on a parallel circuit.

The circuit shown in figure 5-19 can no longer be classified as a parallel circuit because there is a series resistance to be considered. The circuit is solved in the following manner.

Determine $R_{eq}$ for the parallel network:

$$R_{eq} = \frac{R_1 \times R_2}{R_1 + R_2} = \frac{60 \times 40}{60 + 40} = \frac{2400}{100} = 24 \text{ ohms}$$

Reduce to an equivalent circuit (fig. 5-20).

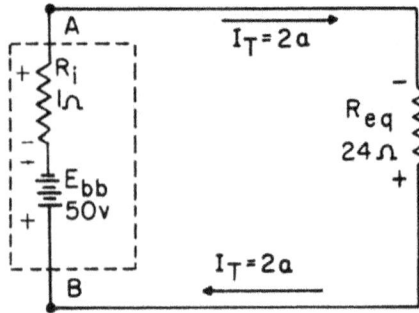

BE.101

Figure 5-20.—Equivalent circuit.

Compute the total series resistance:

$$R_t = R_i + R_{eq}$$

$$R_t = 1 + 24 = 25 \text{ ohms}$$

Compute total current:

$$I_t = \frac{E_{bb}}{R_t} = \frac{50 \text{ v}}{25} = 2 \text{ amperes}$$

Determine voltage drop across $R_{eq}$:

$$E_{R_{eq}} = I_t \times R_{eq}$$

$$E_{R_{eq}} = 2 \times 24$$

$$E_{R_{eq}} = 48 \text{ volts}$$

Find voltage drop across $R_i$:

$$E_{R_i} = I_t \times R_i$$

$$E_{R_i} = 2a \times 1 \text{ ohm}$$

$$E_{R_i} = 2 \text{ volts}$$

Determine power dissipated by load resistors:

$$P_{R_{eq}} = I_t \times E_{R_{eq}}$$

$$P_{R_{eq}} = 2a \times 48 \text{ v}$$

$$P_{R_{eq}} = 96 \text{ w}$$

Determine power dissipated by source resistance:

$$P_{R_i} = I_t \times E_{R_i}$$

$$P_{R_i} = 2a \times 2 \text{ v}$$

$$P_{R_i} = 4 \text{ w}$$

Determine total power dissipation:

$$P_t = P_{R_{eq}} + P_{R_i}$$

$$P_t = 96w + 4w$$

$$P_t = 100 \text{ w}$$

Circuit efficiency is determined by the following formula:

$$\text{Percent Eff} = \frac{P_o}{P_{in}} \times 100$$

where

Percent Eff = percent of efficiency

$P_o$ = power supplied to load device

$P_{in}$ = power supplied by the source

For the circuit of figure 5-20 the percent Efficiency is:

$$\text{Percent Eff} = \frac{P_o}{P_{in}} \times 100$$

$$\text{Percent Eff} = \frac{96}{96 + 4} \times 100$$

$$\text{Percent Eff} = \frac{96w}{100w} \times 100 = 96 \text{ percent}$$

From this efficiency relationship, we may conclude that the source resistance does affect the total power dissipated by the equivalent (load) resistance. The source resistance also affects the transfer of power. As stated in the preceding chapter, maximum transfer of power occurs when the circuit is 50 percent efficient, or when there is an equal amount of voltage dropped across the load and the source resistance.

OPEN AND SHORT CIRCUITS

In comparing the effects of an open in series and parallel circuits, the major difference to be noted is that an open in a parallel circuit would not necessarily disable the entire circuit; i.e., the current flow would not be reduced to zero, unless the open condition existed at some point electrically common to all other parts of the circuit.

A short circuit in a parallel network has an effect similar to a short in a series circuit. In general, the short will cause an increase in current and the possibility of component damage regardless of the type of circuit involved.

Opens and shorts, alike, if occurring in a branch circuit of a parallel network, will result in an overall change in the equivalent resistance. This can cause undesirable effects in other parts of the circuit due to the corresponding change in the total current flow.

To prevent damage to equipment due to a short circuit, a fuse or overload relay is normally placed in the circuit in series with the more sensitive components or in series with the source. The effects of a short circuit occurring in a fused network is shown in figure 5-21 and is explained as follows:

In figure 5-21, with the switch in position one (as shown), a value of current flows that does not exceed the rated current capacity of the fuse. If the switch is thrown to position two, the straight wire conductor will be in parallel with the load resistors. The equivalent resistance of the straight wire and the resistors, all connected in parallel, will be less than the resistance of the straight wire. This follows from the fact that the total resistance of a parallel circuit is always less than the smallest resistance in the branch. Since a complete path still exists to permit current flow, and the equivalent resistance is effectively zero, the current will rise rapidly until the current capacity of the fuse is reached. The fuse will then open the circuit causing the current to stop flowing. A short usually causes components to fail in a circuit which is not properly fused, or otherwise protected. The failure may take the form of a burned-out resistor, damaged source, or a fire in the circuit components and wiring.

## VOLTAGE DIVIDER

In practically all electronic devices, such as radio receivers and transmitters, certain design requirements recur again and again. For instance, a typical radio receiver may require a number of different voltages at various points in its circuitry. In addition, all the various voltages must be derived from a single primary power supply. The most common method of meeting these requirements is by the use of a

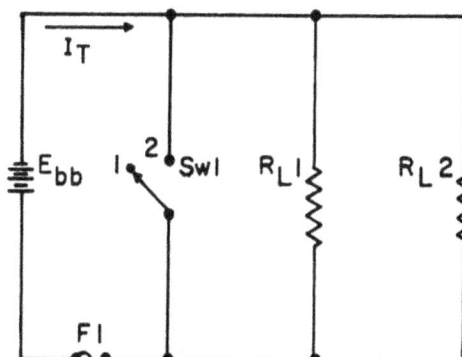

BE. 102

Figure 5-21.—Example of a circuit protected from shorts by a fuse.

voltage-divider network. A typical voltage divider consists of two or more resistors connected in series across the primary power supply. The primary voltage $E_s$ must be as high or higher than any of the individual voltages it is to supply. As the primary voltage is dropped by successive steps through the series resistors, any desired fraction of the original voltage may be "tapped off" to supply individual requirements. The values of the series resistors to be used is dictated by the voltage drops required.

If the total current flowing in the divider circuit is affected by the loads placed on it, then the voltage drops of each divider resistor will also be affected. When a voltage divider is being designed, the maximum current drawn by the loads will determine the value of the resistors that form the voltage divider. Normally, the resistance values chosen for the divider will permit a current equal to 10 percent of the total current drawn by the external loads. This current which does not flow through any of the load devices is called bleeder current.

A voltage divider circuit is shown in figure 5-22. The divider is connected across a 270-volt source and supplies three loads simultaneously— 10 ma (1 milliampere is 0.001 ampere) at 90 volts, between terminal 1 and ground; 5 ma at 150 volts, between terminal 2 and ground; and 30 ma at 180 volts, between terminal 3 and ground. The current in resistor A is 15 ma. The current, voltage, resistance, and power of the 4 resistors are to be determined.

Kirchhoff's law of currents applied to terminal 1 indicates that the current in resistor B

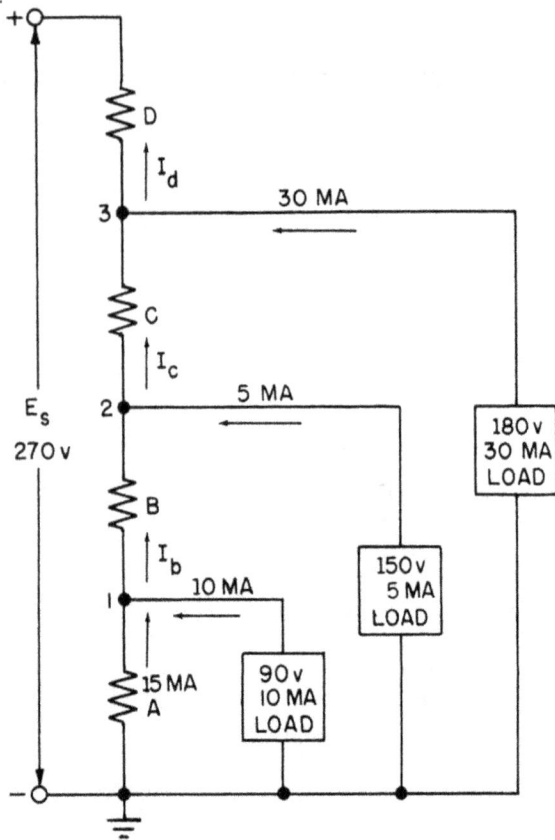

BE.103

Figure 5-22.—Voltage divider, to determine R and P.

is equal to the sum of 15 ma from resistor A and 10 ma from the 90-volt load. Thus,

$$I_b = 15 + 10 = 25 \text{ ma}$$

Similarly,

$$I_c = 25 + 5 = 30 \text{ ma}$$

and

$$I_d = 30 + 30 = 60 \text{ ma}$$

Kirchhoff's voltage law indicates that the voltage across resistor A is 90 volts; the voltage across B is

$$E_b = 150 - 90 = 60 \text{ volts}$$

the voltage across C is

$$E_c = 180 - 150 = 30 \text{ volts}$$

and the voltage across D is

$$E_d = 270 - 180 = 90 \text{ volts}$$

Before solving for the various resistances, it should be recalled that in the formula, $R = \dfrac{E}{I}$, R will be in ohms if E is in volts and I is in amperes. In many electronic circuits, particularly those being considered, it is just as valid and considerably simpler to let R be in thousands of ohms (k-ohms), E in volts, and I in milli-amperes. In the following formulas this convention will be followed.

Applying Ohm's law to determine the resistances—

resistance of A if $R_a = \dfrac{E_a}{I_a} = \dfrac{90}{15} = 6$ k-ohms

resistance of B if $R_b = \dfrac{E_b}{I_b} = \dfrac{60}{25} = 2.4$ k-ohms

resistance of C is $R_c = \dfrac{E_c}{I_c} = \dfrac{30}{30} = 1$ k-ohm

resistance of D is $R_d = \dfrac{E_d}{I_d} = \dfrac{90}{60} = 1.5$ k-ohms

The power absorbed by

resistor A is $P_a = E_a I_a = 90 \times 0.015 = 1.35$ watts

resistor B is $P_b = E_b I_b = 60 \times 0.025 = 1.50$ watts

resistor C is $P_c = E_c I_c = 30 \times 0.030 = 0.90$ watt

resistor D is $P_d = E_d I_d = 90 \times 0.060 = 5.40$ watts

The total power absorbed by the 4 resistors is

$$1.35 + 1.50 + 0.90 + 5.40 = 9.15 \text{ watts}$$

The power absorbed by the load connected to

terminal 1 is $P_1 = E_1 I_1 = 90 \times 0.010 = 0.90$ watt

terminal 2 is $P_2 = E_2 I_2 = 150 \times 0.005 = 0.75$ watt

terminal 3 is $P_3 = E_3 I_3 = 180 \times 0.030 = 5.4$ watts

The total power supplied to the 3 loads is

$$0.90 + 0.75 + 5.4 = 7.05 \text{ watts}$$

The total power supplied to the entire circuit including the voltage divider and the 3 loads is

$$9.15 + 7.05 = 16.2 \text{ watts}$$

This value is checked as

$$P_t = E \times I_t = 270 \times 0.060 = 16.2 \text{ watts}$$

In figure 5-23 the voltage divider resistances are given and the current in $R_5$ is to be found. The load current in $R_1$ is 6 ma; the current in $R_2$ is 4 ma; and the current in $R_3$ is 10 ma. The source voltage is 510 volts. Kirchhoff's current law may be applied at the junctions a, b, c, and d to determine expressions for the current in resistors $R_4$, $R_5$, $R_6$, and $R_7$. Accordingly, the current in $R_4$ is $I + 6 + 4 + 10$, or $I + 20$; the current in $R_5$ is $I$; the current in $R_6$ is $I + 6$; the current in $R_7$ is $I + 6 + 4$, or $I + 10$.

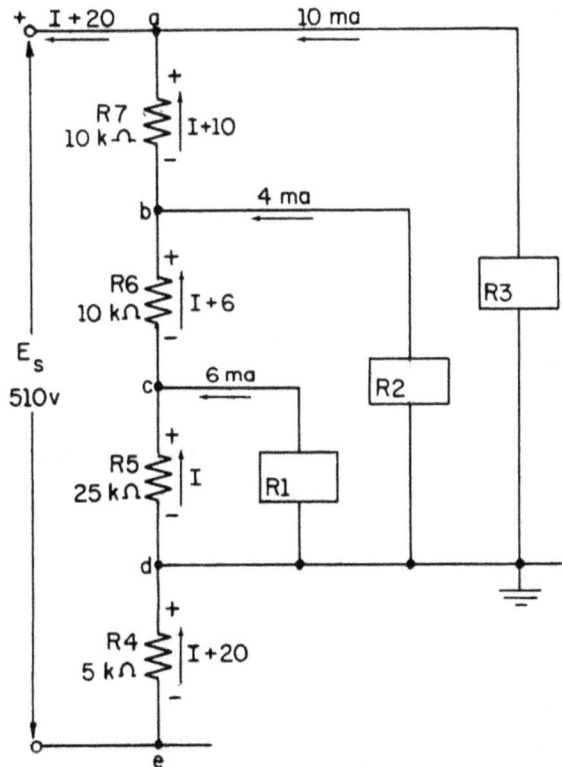

BE.104

Figure 5-23.—Voltage divider, to determine E and R.

The voltage across $R_4$ may be expressed in terms of the resistance in k-ohms and the current in milliamperes as $5(I + 20)$ volts. Similarly, the voltage across $R_5$ is equal to $25I$; the voltage across $R_6$ is $10(I + 6)$ and the voltage across $R_7$ is $10(I + 10)$. Kirchhoff's law of volt-

ages may be applied to the voltage divider to solve for the unknown current, I, by expressing the source voltage in terms of the given values of voltage, resistance, and current (both known and unknown values). The sum of the voltages across $R_4$, $R_5$, $R_6$, and $R_7$ is equal to the source voltage as follows:

$$E_4 + E_5 + E_6 + E_7 = E_s$$

$$5(I + 20) + 25I + 10(I + 6) + 10(I + 10) = 510$$

$$5I + 100 + 25I + 10I + 60 + 10I + 100 = 510$$

$$50I + 260 = 510$$

$$50I = 510 - 260$$
$$50I = 250$$
$$I = 5 \text{ ma}$$

The current of 5 ma through $R_5$ produces a voltage drop across $R_5$ of $5 \times 25$, or 125 volts. Since $R_1$ is in parallel with $R_5$, the voltage across load $R_1$ is 125 volts. The current through $R_4$ is $5 + 20$, or 25 ma and the corresponding voltage is $5 \times 25$, or 125 volts. Since point d is at ground potential, point c is 125 volts positive with respect to ground, whereas point e is 125 volts negative with respect to ground. The current in $R_6$ is $5 + 6$ or 11 ma and the voltage drop across $R_6$ is $11 \times 10$, or 110 volts. The current in $R_7$ is $5 + 10$, or 15 ma and the voltage drop is $15 \times 10$ or 150 volts. The total voltage is the sum of the voltages across the divider. Thus,

$$125 + 125 + 110 + 150 = 510$$

The power absorbed by each resistor in the voltage divider may be found by multiplying the voltage across the resistor by the current in the resistor. If the current is expressed in amperes and the emf in volts, the power will be expressed in watts. Thus the power in $R_4$ is

$$P_4 = E_4 I_4 = 125 \times 0.025 = 3.125 \text{ watts}$$

Similarly the power in $R_5$ is $125 \times 0.005 = 0.625$ watt; the power in $R_6$ is $110 \times 0.011 = 1.21$ watts; and in $R_7$ is $150 \times 0.015 = 2.25$ watts. The total power in the divider is

$$3.125 + 0.625 + 1.21 + 2.25 = 7.21 \text{ watts}$$

The voltage across load $R_1$ is the voltage across $R_5$, or 125 volts. The power in $R_1$ is

$$P_1 = E_1 I_1 = 125 \times 0.006 = 0.750 \text{ watts}$$

The voltage across load $R_2$ is equal to the sum of the voltages across $R_5$ and $R_6$. Thus,

$$E_2 = E_5 + E_6 = 125 + 110 = 235 \text{ volts}$$

The power in load $R_2$ is

$$P_2 = E_2 I_2 = 235 \times 0.004 = 0.940 \text{ watt}$$

The voltage across load $R_3$ is equal to the sum of the voltages across $R_5$, $R_6$, and $R_7$. Thus,

$$E_3 = E_5 + E_6 + E_7 = 125 + 110 + 150 = 385 \text{ volts}$$

The power in load $R_3$ is

$$P_3 = E_3 I_3 = 385 \times 0.010 = 3.85 \text{ watts}$$

The total power in the three loads is

$$0.75 + 0.94 + 3.85 = 5.54 \text{ watts}$$

and the total power supplied by the source is equal to the sum of the power absorbed by the voltage divider and the three loads, or

$$7.21 + 5.54 = 12.75 \text{ watts}$$

The total power may be checked by

$$P_t = E_t I_t = 510 \times 0.025 = 12.75 \text{ watts}$$

The resistances of load resistors $R_1$, $R_2$, and $R_3$ are determined by means of Ohm's law as follows:

$$R_1 = \frac{E_1}{I_1} = \frac{125}{6} = 20.83 \text{ k-ohms}$$

$$R_2 = \frac{E_2}{I_2} = \frac{235}{4} = 58.75 \text{ k-ohms}$$

and

$$R_3 = \frac{E_3}{I_3} = \frac{385}{10} = 38.5 \text{ k-ohms}$$

The variation of voltages and currents found in the previous examples are undesirable in a voltage divider. It must be designed to provide voltages that are as stable as possible. A voltage divider consisting of two resistors will be designed using the circuit configuration shown in figure 5-24. The supply voltage is 200 volts. It is desired to furnish voltages of 50 and 200 volts to two loads drawing 6 and 18 milliamperes

respectively. Assume bleeder current to be 10 percent of the required load current.

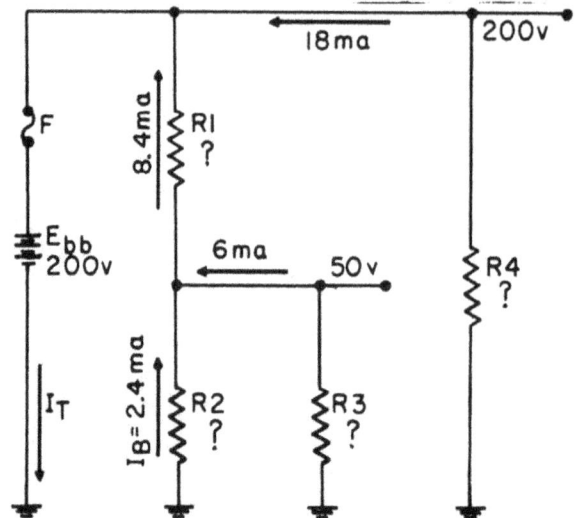

BE.105

Figure 5-24.—Example circuit for proposed voltage divider.

Total load current is specified as 24 milliamperes. The bleeder current, therefore, should be

$$I_b = 10 \text{ percent } I_L$$

$$I_b = 10 \text{ percent} \times 24 \text{ ma}$$

$$I_b = 2.4 \text{ ma}$$

The bleeder current and the current through resistor $R_3$ combine and both currents flow through $R_1$. This current value may be computed

$$I_{R1} = I_b + I_{R3}$$

$$I_{R1} = 2.4 \text{ ma} + 6 \text{ ma}$$

$$I_{R1} = 8.4 \text{ ma}$$

The total current may also be determined

$$I_t = 8.4 \text{ ma} + 18 \text{ ma}$$

$$I_t = 26.4 \text{ ma}$$

The resistance values of $R_3$ and $R_4$ must be as follows:

$$R_3 = \frac{E_{R3}}{I_{R3}} = \frac{50}{6 \times 10^{-3}} = 8.33 \text{ k-ohms}$$

$$R_4 = \frac{E_{R4}}{I_{R4}} = \frac{200}{18 \times 10^{-3}} = 11.1 \text{ k-ohms}$$

Computing for $R_1$ and $R_2$

$$R_1 = \frac{E_{R1}}{I_{R1}} = \frac{150}{8.4 \times 10^{-3}} = 17.85 \text{ k-ohms}$$

$$R_2 = \frac{E_{R2}}{I_{R2}} = \frac{50}{2.4 \times 10^{-3}} = 20.82 \text{ k-ohms}$$

## TYPICAL PROBLEMS IN SERIES-PARALLEL CIRCUITS

As seen by the preceding calculations, problems involving the determination of resistance, voltage, current, and power in a series-parallel circuit are relatively simple. The procedure is the same as for series and parallel circuits— (1) draw the circuit diagram, (2) state the values given and the values to be found, (3) state the applicable equations, and (4) substitute the given values and solve for the unknown. For an example refer to figure 5-25.

Problems:

1. Find the resistance of branch (a).
2. Find the resistance of branch (b).
3. Find the total circuit resistance.
4. Find the total circuit current.
5. Find the voltages $E_{R1}$, $E_a$, and $E_b$.
6. Find the current for branch (a) and (b).
7. Find the voltages $E_{R2}$ and $E_{R5}$.
8. Find the currents $I_1$, $I_2$, $I_3$, and $I_4$.
9. Find the voltages $E_{R3}$, $E_{R6}$, and $E_{R7}$.

10. Find the power for $R_8$, branches (a) and (b), and $R_1$.
11. Find the total circuit power.

Solutions:

1. The resistance of branch (a) $R_a$ is

$$R_a = \frac{R_3 \times R_4}{R_3 + R_4} + R_5$$

$$R_a = \frac{100 \times 100}{100 + 100} + 50$$

$$R_a = 50 + 50 = 100 \text{ ohms}$$

2. The resistance of branch (b) $R_b$ is

$$R_b = R_2 + \frac{(R_7 + R_8) R_6}{R_6 + R_7 + R_8}$$

$$R_b = 20 + \frac{(80 + 80)\ 160}{80 + 80 + 160}$$

$$R_b = 20 + 80 = 100 \text{ ohms}$$

3. The total circuit resistance $R_T$ is

$$R_T = \frac{R_a \times R_b}{R_a + R_b} + R_1$$

$$R_T = \frac{100 \times 100}{100 + 100} + 50$$

$$R_T = 50 + 50 = 100 \text{ ohms}$$

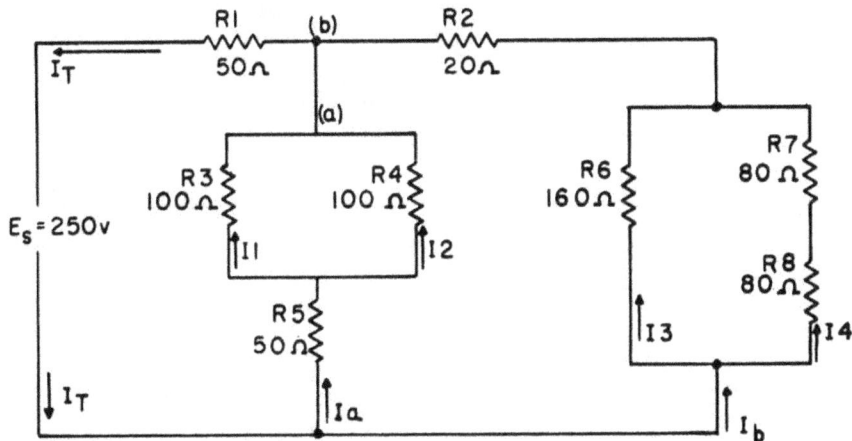

Figure 5-25.—Typical series-parallel circuit.

BE.106

104

4. The total circuit current is $= I_T = \dfrac{E}{R_T}$

$$I_T = \frac{E}{R_T} = \frac{250}{100} = 2.5 \text{ amperes}$$

5. The voltage drop of $R_1$ is

$$E_{R1} = I_1 R_1 = 2.5 \times 50 = 125 \text{ volts}$$

The voltage for (a) is

$$E_a = E_S - E_{R1} = 250 - 125 = 125 \text{ volts}$$

The voltage for (b) is

$$E_b = E_S - E_{R1} = 250 - 125 = 125 \text{ volts}$$

6. The current for branch (a) is

$$I_a = \frac{E_a}{R_a} = \frac{125}{100} = 1.25 \text{ amperes}$$

The current for branch (b) is

$$I_b = \frac{E_b}{R_b} = \frac{125}{100} = 1.25 \text{ amperes}$$

7. The voltage drop across $R_2$ is

$$E_{R2} = I_b R_2 = 1.25 \times 20 = 25 \text{ volts}$$

The voltage drop across $R_5$ is

$$E_{R5} = I_a R_5 = 1.25 \times 50 = 62.5 \text{ volts}$$

8. The current $I_1$ is

$$I_1 = \frac{E_{R3}}{R_3} = \frac{62.5}{100} = 0.625 \text{ ampere}$$

The current $I_2$ is

$$I_2 = I_a - I_1 = 1.25 - 0.625 = 0.625 \text{ ampere}$$

The current $I_3$ is

$$I_3 = \frac{E_{R6}}{R_6} = \frac{100}{160} = 0.625 \text{ ampere}$$

The current for $I_4$ is

$$I_4 = I_b - I_3 = 1.25 - 0.625 = 0.625 \text{ ampere}$$

9. The voltage drop across $R_3$ is

$$E_{R3} = I_1 R_3 = 0.625 \times 100 = 62.5 \text{ volts}$$

The voltage drop across $R_6$ is

$$E_{R6} = I_3 R_6 = 0.625 \times 160 = 100 \text{ volts}$$

The voltage drop across $R_8$ is

$$E_{R8} = I_4 R_8 = 0.625 \times 80 = 50 \text{ volts}$$

10. The power consumed by $R_8$ is

$$P_{R8} = I_4 E_{R8} = 0.625 \times 50 = 31.25 \text{ watts}$$

The power consumed by branch (a) is

$$P_a = I_a E_a = 1.25 \times 125 = 156.25 \text{ watts}$$

The power consumed by branch (b) is

$$P_b = I_b E_b = 1.25 \times 125 = 156.25 \text{ watts}$$

The power consumed by $R_1$ is

$$P_{R1} = I_{R1} E_{R1} = 2.5 \times 125 = 312.5 \text{ watts}$$

11. The total power consumed by the circuit is

$$P_T = P_{R1} + P_a + P_b = 312.5 + 156.25 + 156.25$$
$$= 625 \text{ watts}$$

or $\quad P_T = E I_T = 250 \times 2.5 = 625 \text{ watts}$

# CHAPTER 6

# NETWORK ANALYSIS OF D-C CIRCUITS

## SPECIAL NETWORK TECHNIQUES

The circuit solutions studied up to this point have been accomplished mainly through the use of formulas derived from Ohm's law. Like many other fields of science, electronics has its share of special shortcut methods. These methods, however, must be reserved until enough background theory has been presented to make their use worthwhile. This chapter will therefore be devoted to methods of solution which either simplify circuit calculations or solve circuits which cannot be solved by ordinary methods.

## LOOP ANALYSIS

Loop analysis is a valuable method of circuit analysis in which Kirchhoff's voltage law is the key to the solution. In this method, two or more equations are formed which are then solved simultaneously.

Figure 6-1 shows a network containing five resistors and a source. This network will not actually be solved but is included so that the terms and procedures can be defined. In solving this circuit by loop analysis, three currents are assumed as shown. Each current is arbitrarily assigned a counterclockwise direction (the true direction is unimportant at this time). Each of the three closed current paths is called a mesh. The individual circuit components (resistors) which form the meshes are called elements.

To solve the circuit in figure 6-1, three equations are formed, one for each mesh. The number of equations required is always equal to the number of meshes. These equations are then solved simultaneously.

## TYPICAL SOLUTION

In this section the combination circuit shown in figure 6-2 will be solved using the loop method of analysis. Notice that the given circuit is a two mesh circuit and therefore two equations are required for the solution.

To begin the solution, a current circulating in a counterclockwise direction is assumed in each mesh. In addition, it is assumed that these currents cause voltage drops across the circuit resistors. Polarities are then assigned to each resistor, according to the direction of current flowing through the resistor. Notice that resistor $R_2$ carries two currents which flow in opposite directions. A separate set of polarity signs is used for the voltage drops caused by each of these currents. After assigning currents and polarities to the circuit, the voltage equations can be formed.

Recalling Kirchhoff's statement that the algebraic sum of the emf's and voltage drops around any closed loop is equal to zero, an equation can be formed for loop ABEF of figure 6-2. This equation is written starting at point A and tracing around the loop in the direction of assumed current. The polarities used for each voltage drop in the equation are those found following the component traced through. For example, in tracing through $R_1$ from E to F the positive polarity sign is used. (For those needing a review of Kirchhoff's voltage equations, refer to chapter 5.) Therefore, the equation for loop ABEF is

$$\text{ABEF:} \quad 20I_1 - 20I_2 + 8I_1 - 40 = 0 \qquad (1)$$

$$\text{Simpliying:} \quad 28I_1 - 20I_2 = 40 \qquad (2)$$

Notice that in passing through $R_2$, two voltage drops of opposite polarity are encountered. Both of these drops ($20I_1$ and $-20I_2$) must be included in the equation along with their proper polarities.

## ELEMENTS

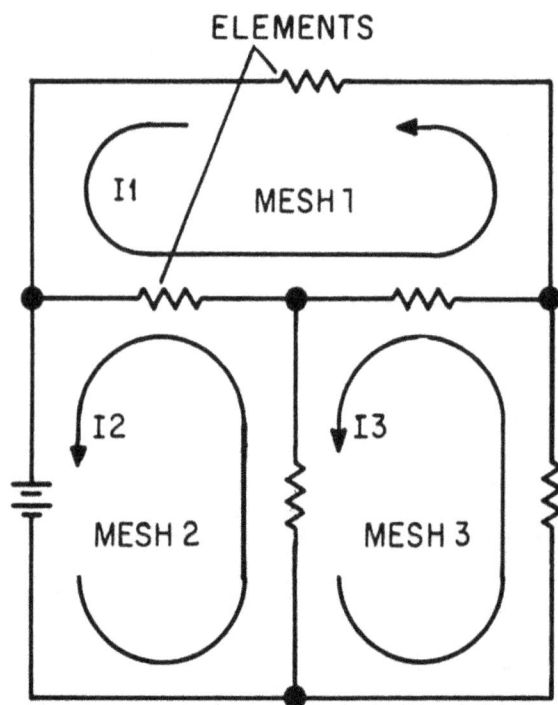

BE.107

Figure 6-1.—A three mesh network.

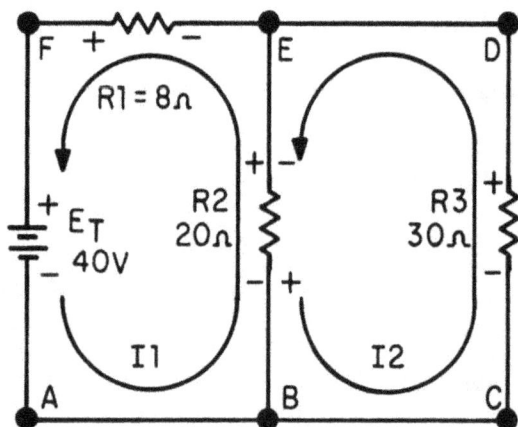

BE.108

Figure 6-2.—Example circuit.

Next, an equation is written for loop BCDE.

This equation is

BCDE:  $+ 30I_2 + 20I_2 - 20I_1 = 0$    (3)

Simplifying:  $20I_1 + 50I_2 = 0$    (4)

Again notice that two opposing voltage drops ( $+ 20I_2$ and $- 20I_1$) were included for $R_2$. Equations (2) and (4), repeated below, are now solved simultaneously to obtain I1 and I2.

$$28I_1 - 20I_2 = 40 \qquad (2)$$

$$- 20I_1 + 50I_2 = 0 \qquad (4)$$

In order to eliminate $I_2$, equation (2) is multiplied by 5 and equation (4) is multiplied by 2. The resulting equations (5) and (6) are then added.

$$140I_1 - 100I_2 = 200 \qquad (5)$$

$$- 40I_1 + 100I_2 = 0 \qquad (6)$$

$$\overline{\phantom{00}100I_1 \qquad\quad = 200}$$

Dividing both sides by 100:

$$I_1 = \frac{200}{100}$$

$$I_1 = 2 \text{ amperes}$$

To obtain the value of $I_2$, 2 amperes is now substituted into equation (4) and in place of $I_1$.

$$- 20 \times 2 + 50I_2 = 0$$

$$- 40 + 50I_2 = 0$$

$$50I_2 = 40$$

$$I_2 = 0.8 \text{ ampere}$$

Thus, current $I_1$ is 2 amperes and current $I_2$ is 0.8 ampere. The voltage drops can now be evaluated using these currents and Ohm's law as follows:

$$E_{R1} = I_1 R_1$$

$$E_{R1} = 2 \times 8$$

$$E_{R1} = 16 \text{ volts}$$

Notice that the actual current through $R_2$ is the algebraic sum of the two opposing currents. Therefore, the voltage across $R_2$ is

$$E_{R2} = (I_1 - I_2) R_2$$

$$E_{R2} = 1.2 \times 20$$

$$E_{R2} = 24 \text{ volts}$$

$$E_{R3} = I_2R_3$$

$$E_{R3} = 0.8 \times 30$$

$$E_{R3} = 24 \text{ volts}$$

## MULTIPLE SOURCE CIRCUITS

Quite frequently, networks containing more than one source must be solved. Although a circuit of this type may look complicated, the solution is no more difficult than the one discussed previously. In fact, the same method of solution is used in both single and multiple source circuits.

Figure 6-3 shows a multiple source circuit which will be used in the example solution. In the diagram a counterclockwise current has been assumed in each mesh and polarities assigned accordingly. Note that $R_2$ has two opposing voltage drops, one for each current. The voltage equation for loop ABEF is

$$+5 + 20I_1 - 20I_2 + 20 + 5I_1 - 50 = 0 \qquad (7)$$

Simplifying: $\qquad 25I_1 - 20I_2 = 25 \qquad (8)$

Loop BCDE:

$$-100 + 7I_2 + 20 + 10I_2 + 20I_2 - 20I_1 - 5 = 0 \qquad (9)$$

Simplifying: $\qquad -20I_1 + 37I_2 = 85 \qquad (10)$

Multiply (8) by 4: $\quad 100I_1 - 80I_2 = 100 \qquad (11)$
Multiply (10) by 5: $-100I_1 + 185I_2 = 425 \qquad (12)$

Add: $\qquad\qquad\qquad +105I_2 = 525$

$$I_2 = \frac{525}{105}$$

$$I_2 = 5 \text{ amperes}$$

Substitute (13) in (8)

$$25I_1 - 20(5) = 25$$

$$25I_1 - 100 = 25$$

$$25I_1 = 125$$

$$I_1 = 5 \text{ amperes} \qquad (14)$$

Had one of the above currents been negative, this would have indicated an incorrectly assumed direction of current flow. The magnitude

of current would be correct. Now that the currents are known, the voltage drops across the resistors can be calculated.

$$E_{R1} = I_1R_1$$

$$E_{R1} = 25 \text{ volts}$$

$$E_{R2} = (I_1-I_2)R_2$$

$$E_{R2} = 0 \text{ volts}$$

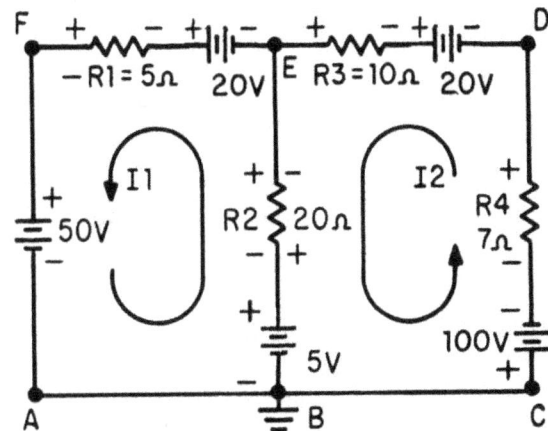

BE.109

Figure 6-3.—Multiple source circuit.

Notice that equal and opposite currents flow through $R_2$ and no voltage drop occurs across it.

$$E_{R3} = I_2R_3$$

$$E_{R3} = 50 \text{ volts}$$

$$E_{R4} = I_2R_4$$

$$E_{R4} = 35 \text{ volts}$$

## EQUIVALENT CIRCUITS

In the analysis of many circuits, the solution is primarily concerned with the computation of values for load current and load voltage. In most cases additional calculations are required as intermediate steps in the solution. The following discussion will show how many of these intermediate steps can be eliminated.

Let us now examine the load resistor ($R_L$) shown in figure 6-4. This resistor is connected to a switch so that it can be connected across the terminals of either of the two circuits. When the switch is in the position illustrated, $R_L$ is connected to circuit A.

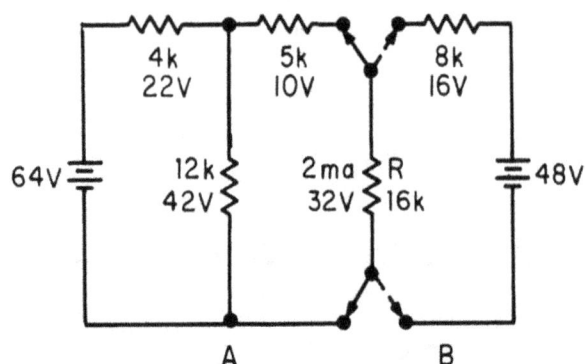

BE.110

Figure 6-4.—Equivalent circuits.

By using a lengthy series of steps, the values of load current (2 ma) and load voltage (32 v) can be computed. Using the methods studied up to this time, computations of total resistance, total current, and various voltage drops would be required as intermediate steps of the solution. (An alternate method would be loop analysis.)

Keeping in mind the values of load voltage and current determined for circuit A, assume the switch to be in the position which places $R_L$ across the terminals of circuit B. In this simple series circuit the load voltage and current can be easily determined. Notice that again the load voltage is 32 volts and the load current is 2 milliamperes. Since the load voltage and current are the same regardless of which circuit is used, as far as the load is concerned circuit B could be substituted for circuit A. Any circuit that may be substituted for another circuit is called an equivalent circuit. The purpose of this section is to show how a simple equivalent circuit can be developed for any complex circuit. This simple equivalent circuit is then used for the calculations instead of the original complex circuit. It should not be necessary to emphasize the hours of work that can be saved by utilizing an equivalent circuit.

### THEVENIN'S THEOREM

one of the most valuable equivalent circuits is one known as Thevenin's equivalent circuit.

This circuit is derived from Thevenin's Theorem, stated as follows:

Thevenin's Theorem: Any linear network of impedance and sources, if viewed from any two points in the network, can be replaced by an equivalent impedance $Z_{th}$ in series with an equivalent voltage source $E_{th}$. (The term impedance means any opposition to current flow and in d-c circuits will be taken to mean resistance.)

According to this theorem, any linear d-c circuit regardless of its complexity can be replaced by a Thevenin's equivalent shown in figure 6-5.

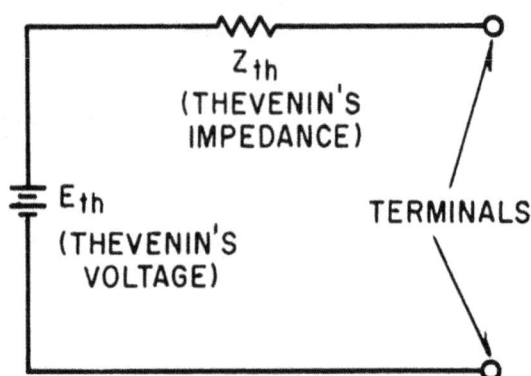

BE.111

Figure 6-5.—Thevenin's equivalent circuit.

The process whereby a Thevenin's equivalent circuit is developed for a given network is best illustrated by an example.

Assume that the circuit in figure 6-6 (A) is to be used to develop a Thevenin's equivalent circuit. Basically the problem consists of finding values of $E_{th}$ and $Z_{th}$. These quantities can be found using the following procedure.

### APPLICATION OF THEVENIN'S THEOREM

1. Disconnect the section of the circuit considered as the load ($R_L$ in fig. 6-6 (A)).

2. By measurement or calculation, determine the voltage that would appear between the load terminals with the load disconnected (terminals X and Y). This open circuit voltage is called Thevenin's voltage ($E_{th}$).

3. Replace each source within the circuit by its internal impedance. (A constant voltage

source such as a battery is replaced with a short, while a constant current source is replaced with an open. (See fig. 6-6 (B).)

4. By measurement or calculation, determine the impedance (resistance) the load would see, looking back into the network from the load terminals. This is Thevenin's impedance ($Z_{th}$). (See fig. 6-6 (B).)

5. Draw the equivalent circuit consisting of $R_L$ and $Z_{th}$ in series, connected across source $E_{th}$ (fig. 6-6 (C). Solve for the load current and voltage.

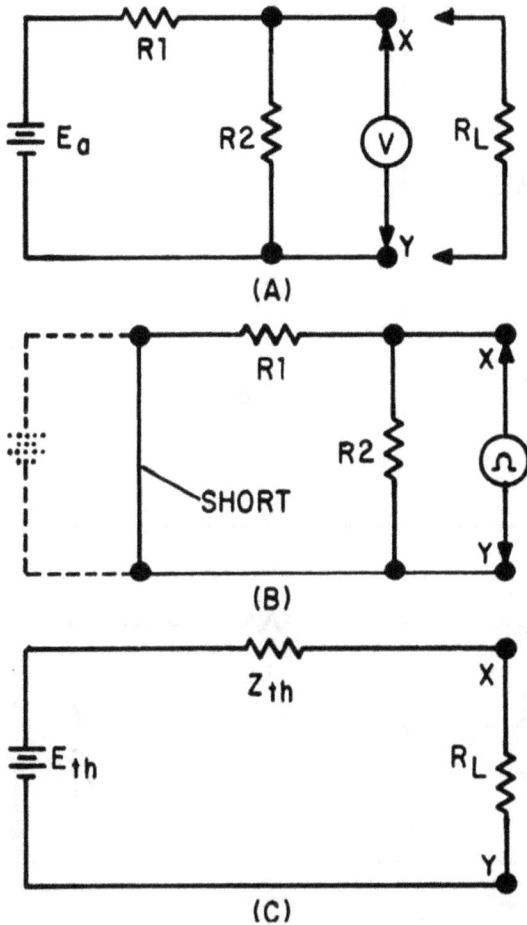

(A)

(B)

(C)

BE.112

Figure 6-6.—Developing a Thevenin's equivalent.

## A TYPICAL SOLUTION

In the following solution, the load voltage will be computed using loop analysis. The circuit will then be solved a second time using a Thevenin's equivalent circuit. The two methods can thus be compared as to results and ease of computation.

Figure 6-7 shows a three mesh circuit for which the output voltage $E_L$ is to be determined. Currents are assumed in each mesh and appropriate polarities are assigned. The voltage equations are then written as follows:

$$\text{ABGH:} \qquad + 10I_1 - 10I_2 - 110 = 0 \qquad (15)$$

$$10I_1 - 10I_2 = 110 \qquad (16)$$

$$\text{BCFG:} + 30I_2 - 30I_3 + 20I_2 + 10I_2 - 10I_1 = 0 \,(17)$$

$$- 10I_1 + 60I_2 - 30I_3 = 0 \,(18)$$

$$\text{CDEF:} \qquad + 10I_3 + 30I_3 - 30I_2 = 0 \,(19)$$

$$- 30I_2 + 40I_3 = 0 \,(20)$$

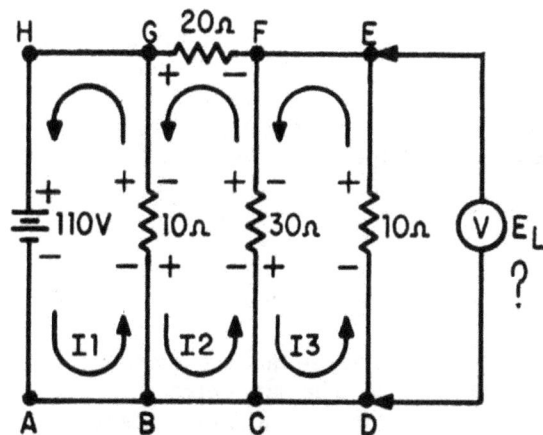

BE.113

Figure 6-7.—Example circuit.

Adding (16) to (18):

$$10I_1 - 10I_2 \qquad = 110$$

$$-10I_1 + 60I_2 - 30I_3 = 0$$

$$\overline{\qquad 50I_2 - 30I_3 = 110} \qquad (21)$$

Multiplying (20) by 3: $\quad -90I_2 + 120I_3 = 0$

Multiplying (21) by 4: $\quad 200I_2 - 120I_3 = 440$

Adding: $\qquad\qquad \overline{\quad 110I_2 \qquad\quad = 440} \qquad (22)$

$$I_2 = 4 \text{ amperes} \qquad (23)$$

110

Substituting (23) into (20)

$$-30(4) + 40I_3 = 0$$

$$-120 + 40I_3 = 0$$

$$40I_3 = 120$$

$$I_3 = 3 \text{ amperes} \qquad (24)$$

Substituting (23) into (16)

$$10I_1 - 10(4) = 110$$

$$10I_1 - 40 = 110$$

$$10I_1 = 150$$

$$I_1 = 15 \text{ amperes} \qquad (25)$$

Once all three currents are known, the voltages are computed and added for each loop as a check. If the computed voltages around each loop add up to zero, the computed currents are correct. The output voltage $E_L$ is then

$$E_L = I_3 R_L$$

$$E_L = 30 \text{ volts}$$

Most people will agree that the above solution is long and tedious. The same circuit will now be solved through the use of a Thevenin's equivalent circuit. The procedure is as follows:

1. Disconnect the load as shown in (A) of figure 6-8.

2. Compute $E_{th}$, the no load (open circuit) voltage between terminals E and D. This voltage is the same as the voltage across $R_3$. Since 110 volts are applied to $R_2$ and $R_3$ in series, the voltage across $R_3$ is three-fifths of 110 volts or 66 volts.

3. Replace the source with its internal impedance. In this case the source is a battery (constant voltage source) and is replaced with a short as in (B) of figure 6-8. This shorts out both the battery terminals and R1, resulting in the circuit shown in (C) of figure 6-8.

NOTE: The internal resistance of a flashlight cell is approximately 0.005 ohm. Thus, the internal resistance of a battery is usually considered to be 0 ohms.

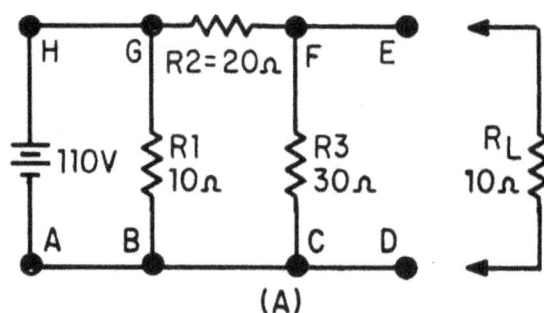

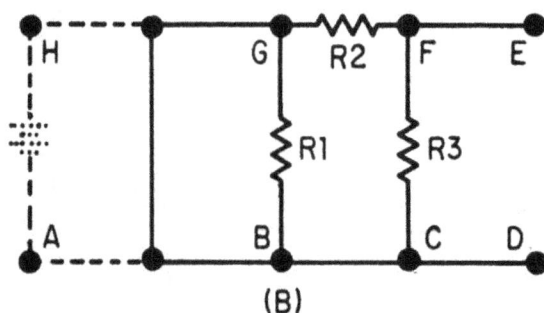

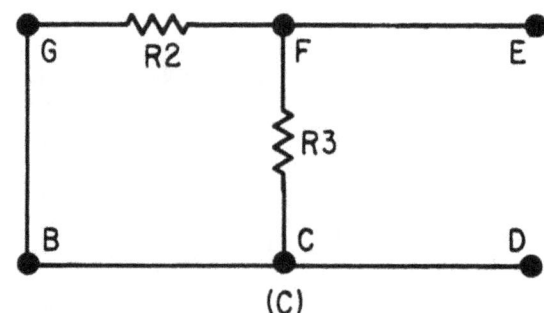

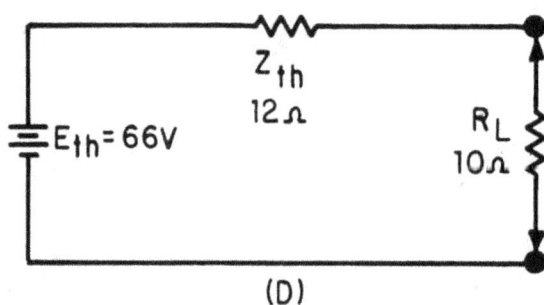

(A)

(B)

(C)

(D)

BE.114

Figure 6-8.—Evolution of the equivalent circuit.

4. Determine the resistance ($Z_{th}$) the load would see "looking back" into the network from terminals E and D. Notice that in (C) of the figure, two separate paths exist between terminals E and D. One of these paths is through $R_2$ and the other is through $R_3$, indicating the two resistors to be in parallel. Since $R_2$ and $R_3$ are in parallel, this resistance $Z_{th}$ is

$$Z_{th} = \frac{R_2 R_3}{R_2 + R_3}$$

$$Z_{th} = \frac{20 \times 30}{20 + 30}$$

$$Z_{th} = 12 \text{ ohms}$$

5. Draw the equivalent circuit and connect the load resistance as in (D) of figure 6-8, including the values obtained for $E_{th}$ and $Z_{th}$. Using Ohm's law, solve for the load current and voltage.

$$I_L = \frac{E_{th}}{Z_{th} + R_L}$$

$$I_L = \frac{66}{22}$$

$$I_L = 3 \text{ amperes}$$

Notice that the load currents ($I_3$ in the loop analysis and $I_L$ in Thevenin's equivalent) are identical.

$$E_L = I_L R_L$$

$$E_L = 3 \times 10$$

$$E_L = 30 \text{ volts}$$

This is the same value of voltage found by loop analysis. At this point one should stop and compare the labor required by each method in order to reach a solution. Once the steps used in applying Thevenin's theorem are learned, this method is by far the simpler of the two methods.

VOLTAGE DIVIDER EQUATION

As an aid to the application of Thevenin's equivalent circuit, an equation can be derived which will yield the load voltage in one simple calculation. This equation (the derivation will be left as a problem for the student) is:

$$E_L = \frac{E_{th} R_L}{Z_{th} + R_L} \qquad (26)$$

To illustrate the use of equation (26) the Thevenin equivalent will be developed for the circuit in figure 6-9. As before, the load circuit will be opened to determine $E_{th}$, the open circuit load

voltage. With the switch in figure 6-9 open, no load current flows through $R_L$ or $R_3$. The voltage between X and Y is therefore the same as the voltage across $R_2$. Applying the voltage divider formula this voltage is found to be:

$$E_{R2} = \frac{E_a R_2}{R_1 + R_2}$$

$$E_{R2} = \frac{100 \times 30}{50}$$

$$E_{R2} = 60 \text{ volts}$$

Since this is the open circuit load voltage, this voltage is $E_{th}$.

$$E_{th} = 60 \text{ volts}$$

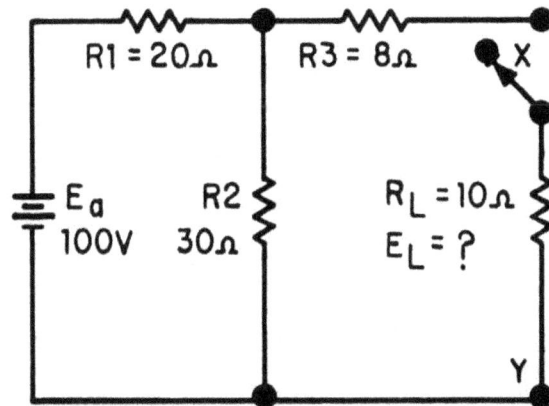

Figure 6-9.—Example circuit.

Next the source is replaced with a short circuit. This places $R_1$ and $R_2$ in parallel and the impedance $Z_{th}$ looking back into the network is:

$$Z_{th} = R_3 + \frac{R_1 R_2}{R_1 + R_2}$$

$$Z_{th} = 8 + \frac{20 \times 30}{50}$$

$$Z_{th} = 20 \text{ ohms}$$

The Thevenin's equivalent circuit is drawn and the load connected as in figure 6-10. In one step the output voltage across the load can be found as follows:

$$E_L = \frac{E_{th} R_L}{Z_{th} + R_L}$$

$$E_L = \frac{60 \times 10}{30}$$

$$E_L = 20 \text{ volts}$$

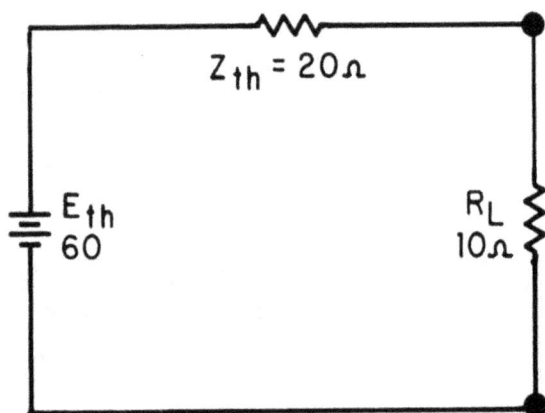

BE.116

Figure 6-10.—Thevenin's equivalent for figure 6-9.

Using Thevenin's equivalent circuit and the voltage divider equation as tools, certain complex circuits can be quickly solved. This technique should not be forgotten as it will be a time saving aid in future chapters.

## NORTON'S THEOREM

Another important theorem that can be used as an aid in solving complex circuits is called Norton's Theorem. This theorem is similar to Thevenin's and is stated as follows:

Norton's Theorem: Any linear network of impedance and sources, if viewed from any two points in the network, can be replaced by an equivalent impedance $Z_{th}$ in shunt with an equivalent current source $I_n$.

This equivalent circuit is shown in figure 6-11. Notice that the impedance $Z_{th}$ is placed in shunt

(parallel) with a constant current source. The impedance $Z_{th}$ is the same impedance used in Thevenin's equivalent circuit.

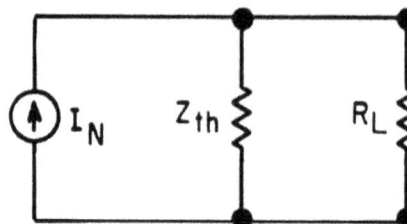

BE.117

Figure 6-11.—Norton's equivalent circuit.

To illustrate the application of Norton's theorem, a Norton's equivalent circuit will be developed for the network shown in figure 6-12. The quantities $Z_{th}$ and $I_n$ can be found as follows:

1. Disconnect the section of the circuit considered as the load ($R_L$ in fig. 6-12 (A)).

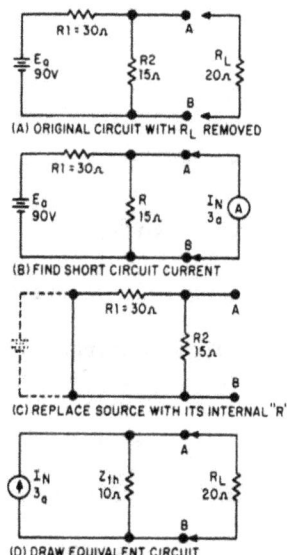

(A) ORIGINAL CIRCUIT WITH $R_L$ REMOVED

(B) FIND SHORT CIRCUIT CURRENT

(C) REPLACE SOURCE WITH ITS INTERNAL "R"

(D) DRAW EQUIVALENT CIRCUIT

BE.118

Figure 6-12.—Evolution of Norton's equivalent circuit.

2. By measurement or calculation determine the current that would flow through a wire connected between the load terminals (A and B of fig. 6-12 (B)). This short-circuit load current is Norton's current ($I_n$).

3. Remove short from load terminals. Replace each source within the network with its internal impedance (the battery in fig. 6-12 (C) is replaced with a short).

4. By measurement or calculation, determine the impedance $Z_{th}$ looking back into the network. (Same as for Thevenin's equivalent.)

5. Draw the equivalent circuit consisting of $R_L$, $Z_{th}$, and source $I_n$ all in parallel (fig. 6-12 (C)) and then solve for the desired quantities.

As an aid to the application of Norton's theorem, a current divider equation can be derived which will yield the load current in one calculation. Using the notation from Norton's equivalent circuit this equation is

$$I_L = \frac{I_n Z_{th}}{Z_{th} + R_L} \tag{27}$$

The current through the short ($I_n$) in figure 6-12 (B) is 3 amperes, since the short effectively places $R_1$ directly across the source. With $I_n$ equal to 3 amperes and $Z_{th}$ equal to 10 ohms, the load current $I_L$ is

$$I_L = \frac{I_n Z_{th}}{Z_{th} + R_L} \tag{27}$$

$$I_L = \frac{3 \times 10}{10 + 20}$$

$$I_L = 1 \text{ ampere}$$

Should it be desired to develop the Thevenin's equivalent circuit, $E_{th}$ can be easily determined. The Thevenin's and Norton's equivalent circuits are closely related such that

$$E_{th} = I_n Z_{th} \tag{28}$$

Therefore, $E_{th}$ for figure 6-12 is

$$E_{th} = I_n Z_{th}$$

$$E_{th} = 3 \times 10$$

$$E_{th} = 30 \text{ volts}$$

Thus, if either equivalent circuit is known it is a simple matter to convert to the other. Usually a Norton's equivalent circuit is used when the load current is desired, and a Thevenin's equivalent circuit is used when the load voltage is required.

## BRIDGE CIRCUITS

### SIMPLE RESISTANCE BRIDGE

A resistance bridge circuit in its simplest form is shown in figure 6-13. Two identical resistors, $R_1$ and $R_2$, are connected in parallel across a 20-volt power source. The network becomes a bridge circuit when a cross connection or "bridge" is placed between the two resistors.

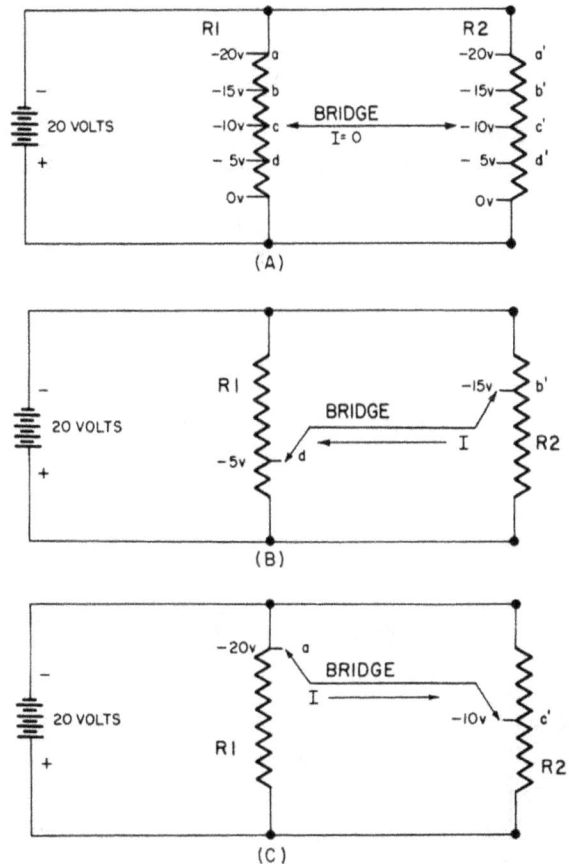

BE.119

Figure 6-13.—Simple resistance bridges.

The voltage across both resistors is dropped at the same rate, because the resistors are identical. Therefore, points a-a', b-b', c-c', and d-d' are at equal potentials. If the bridge is connected between points of equal potential, as in figure 6-13 (A), no current will flow through the bridge. However, if the bridge is connected between points of unequal potential, current will flow from the more negative to the less negative end, as shown in figure 6-13 (B). In (B), current flows right-to-left from b' to d. In (C), current flows left-to-right from a to c'. Thus, it can be seen that the direction of bridge current is controlled by the relative potential of the two ends of the bridge resistor. When the bridge is across equal potentials, and no current flows, it is said to be balanced. When it is across unequal potentials, and current flows, it is said to be unbalanced. The bridge may be unbalanced in either or both of two ways—(1) by connecting the bridge to unequal potentials, or (2) by using resistors of unequal values, thus causing an unbalanced condition.

## UNBALANCED RESISTANCE BRIDGE

Figure 6-14 shows an unbalanced bridge using unequal resistors. The two parallel legs are $R_1$-$R_4$ and $R_3$-$R_5$. $R_2$ is the bridge. The current and voltage drop of each resistor, and the source voltage $E_s$, are to be determined.

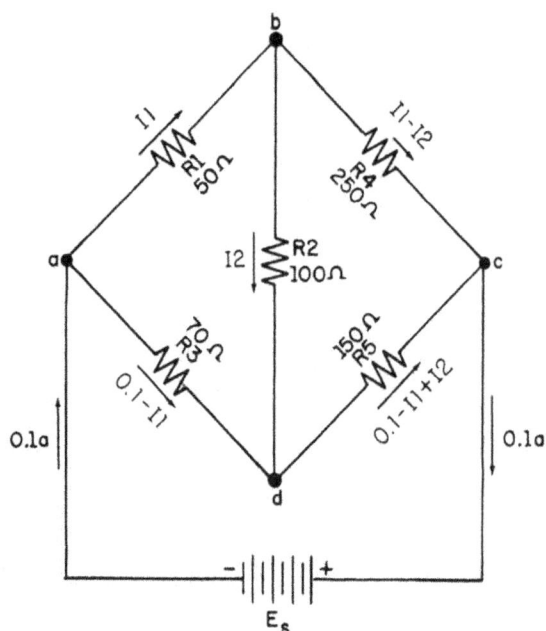

BE.120

Figure 6-14.—Unbalanced resistance bridge.

The current of 0.1 ampere flowing into junction a divides into two parts. The part flowing through R1 is indicated as $I_1$ and the part through R3 is 0.1 - I1. Similarly, at junction b, $I_1$ divides, part flowing through $R_2$ and the remainder through $R_4$. The part through $R_2$ is designated $I_2$, and the part through $R_4$ is $I_1 - I_2$.

The direction of current through $R_2$ may be assumed arbitrarily.

If the solution indicates a positive value for $I_2$, the assumed direction is proved to be correct. The current through $R_4$ is $I_1 - I_2$. At junction d the currents may be analyzed in a similar manner. Current $I_2$ through $R_2$ joins current 0.1 - $I_1$, from $R_3$; and the current through $R_5$ is 0.1 - $I_1$ + $I_2$.

The unknown currents, $I_1$ and $I_2$, may be determined by establishing two voltage equations in which they both appear. These equations are solved for $I_1$ and $I_2$ in terms of the given values of current and resistance. The first voltage equation is developed by tracing clockwise around the closed circuit containing resistors $R_1$, $R_2$, and $R_3$. The trace starts at junction a, and proceeds to b, to d, and back to a. The algebraic sum of the voltages around this circuit is zero. These voltages are expressed in terms of resistance and current. Going from a to b, the voltage drop is in the direction of the arrow and is equal to $-50I_1$; the drop across $R_2$, going from b to d, is $-100I_2$; and the voltage from d to a (in opposite direction to the arrow) is $+ 70(0.1 - I_1)$. Thus,

$$-50I_1 - 100I_2 + 70(0.1 - I_1) = 0$$

Multiplying both sides by -1,

$$50I_2 + 100I_2 - 70(0.1 - I_1) = 0$$

and transposing and simplifying,

$$120I_1 + 100I_2 = 7 \qquad (29)$$

The second voltage equation is established by tracing clockwise around the circuit, which includes resistors $R_4$, $R_5$, and $R_2$. Starting at junction b, the trace proceeds to c, to d, and returns to b. The voltage across $R_4$, from b to c, is $-250(I_1 - I_2)$; the voltage across $R_5$, from c to d, is $+ 150(0.1 - I_1 + I_2)$; and the voltage across $R_2$, from d to b, is $+ 100I_2$. Thus,

$$-250(I_1 - I_2) + 150(0.1 - I_1 + I_2) + 100I_2 = 0$$

from which,

$$400I_1 - 500I_2 = 15 \qquad (30)$$

Equations (29) and (30) may be solved simultaneously by multiplying equation (29) by the factor 5 and then adding the equations to eliminate $I_2$ as follows:

$$400I_1 - 500I_2 = 15$$

$$600I_1 + 500I_2 = 35$$

$$1,000I_1 = 50$$

$$I_1 = 0.05 \text{ ampere}$$

Substituting the value of 0.05 for $I_1$ in equation (29) and solving for $I_2$,

$$120(0.05) + 100I_2 = 7$$

$$100I_2 = 1$$

$$I_2 = 0.01 \text{ ampere}$$

Thus the current in $R_1$ is $I_1$ = 0.05 ampere. The current in $R_2$ is $I_2$ = 0.01 ampere. The current in $R_3$ is $0.1 - I_1 = 0.1 - 0.05$, or 0.05 ampere. The current in $R_4$ is $I_1 - I_2$ = 0.05 - 0.01, or 0.04 ampere. The current in $R_5$ is $0.1 - I_1 + I_2$ or 0.1 - 0.05 + 0.01 = 0.06 ampere. The voltages $E_1$, $E_2$, $E_3$, $E_4$, and $E_5$ are as follows:

$E_1$ across $R_1$ is $I_1R_1$ = 0.05 x 50 = 2.5 volts.

$E_2$ across $R_2$ is $I_2R_2$ = 0.01 x 100 = 1.0 volt.

$E_3$ across $R_3$ is $(0.1 - I_1)R_3$=0.05 x 70=3.5 volts.

$E_4$ across $R_4$ is $(I_1 - I_2)R_4$=0.04 x250=10 volts.

$E_5$ across $R_5$ is $(0.1 - I_1+I_2)R_5$=0.06 x 150 =

9.0 volts.

The source voltage $E_s$ is equal to the sum of voltages across $R_3$ and $R_5$ or $R_1$ and $R_4$. Thus,

$$E_s = E_1 + E_4$$

$$= 2.5 + 10 = 12.5 \text{ volts}$$

and

$$E_s = E_3 + E_5$$

$$= 3.5 + 9 = 12.5 \text{ volts}$$

The voltage across $R_2$ is the difference in the voltages across $R_3$ and $R_1$. It is also the difference in the voltages across $R_4$ and $R_5$.

WHEATSTONE BRIDGE

A type of circuit that is widely used for precision measurements of resistance is the Wheatstone bridge. The circuit diagram of a Wheatstone bridge is shown in figure 6-15 (A). $R_1$, $R_2$, and $R_3$ are precision variable resistors, and $R_x$ is the resistor whose unknown value of resistance is to be determined. After the bridge has been properly balanced, the unknown resistance may be determined by means of a simple formula. The galvanometer, G, is inserted across terminals b and d to indicate the condition of balance. When the bridge is properly balanced there is no difference in potential across terminals bd, and the galvanometer deflection, when the switch is closed, will be zero.

The operation of the bridge is explained in a few logical steps. When the switch to the battery is closed, electrons flow from the negative terminal of the battery to point a. Here the current divides, as it would in any parallel circuit, a part of it passing through $R_1$ and $R_2$ and the remainder passing through $R_3$ and $R_x$. The two currents, labeled $I_1$ and $I_2$, unite at point c and return to the positive terminal of the battery. The value of $I_1$ depends on the sum of resistances $R_1$ and $R_2$, and the value of $I_2$ depends on the sum of resistances $R_3$ and $R_x$. In each case, according to Ohm's law, the current is inversely proportional to the resistance.

$R_1$, $R_2$, and $R_3$ are adjusted so that when the galvanometer switch is closed there will be no deflection of the needle. When the galvanometer shows no deflection there is no difference of potential between points b and d. This means that the voltage drop $(E_1)$ across $R_1$, between points a and b, is the same as the voltage drop $(E_3)$ across $R_3$, between points a and d. By similar reasoning, the voltage drops across $R_2$ and $R_x$—that is, $E_2$ and $E_x$—are also equal. Expressed algebraically,

$$E_1 = E_3$$

or

$$I_1R_1 = I_2R_3$$

and

$$E_2 = E_x$$

or

$$I_1R_2 = I_2R_x$$

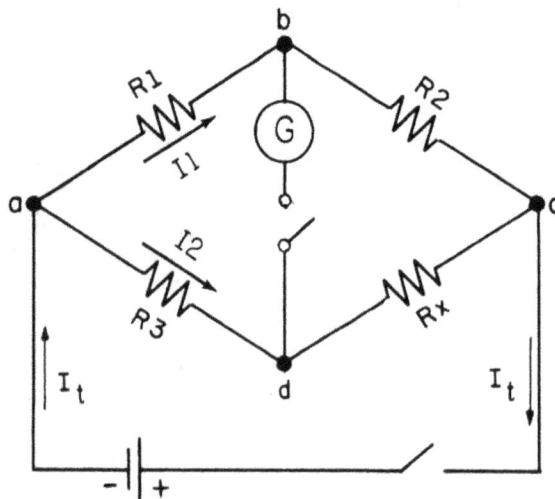

**(A)**

SCHEMATIC WHEATSTONE-BRIDGE CIRCUIT

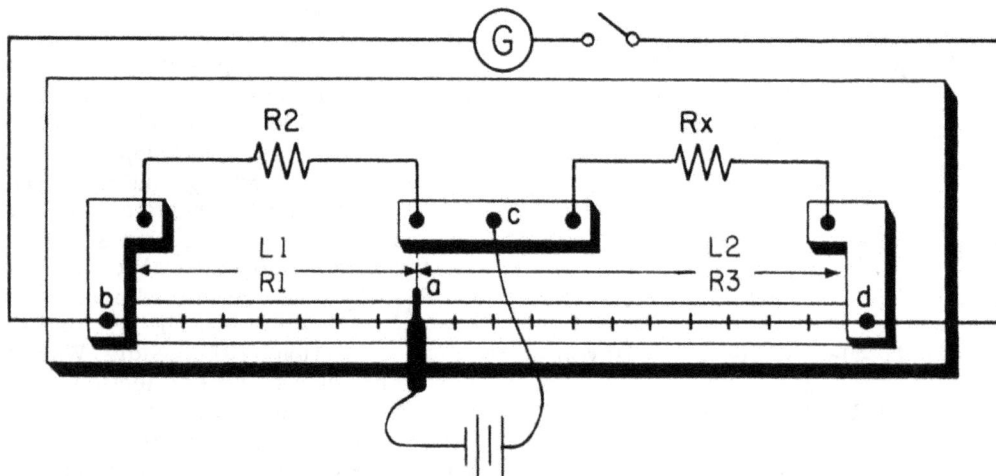

**(B)**

SLIDE-WIRE BRIDGE

Figure 6-15.—Wheatstone bridge circuit.

BE.121

Dividing the voltage drops across $R_1$ and $R_3$ by the respective voltage drops across $R_2$ and $R_x$,

$$\frac{I_1 R_1}{I_1 R_2} = \frac{I_2 R_3}{I_2 R_x}$$

Simplifying,

$$\frac{R_1}{R_2} = \frac{R_3}{R_x}$$

Therefore,

$$R_x = \frac{R_2 R_3}{R_1}$$

The resistance values of $R_1$, $R_2$, and $R_3$ are readily determined from the markings on the standard resistors, or from the calibrated dials if a dial-type bridge is used.

The Wheatstone bridge may be of the slide-wire type, as shown in figure 6-15 (B). In the slide-wire circuit, the slide-wire (b to d), corresponds to $R_1$ and $R_3$ of figure 6-15 (A). The wire may be an alloy of uniform cross section; for example German silver or nichrome, having a resistance of about 100 ohms. Point a is established where the slider contacts the wire. The bridge is balanced by moving the slider along the wire.

The equation for solving for $R_x$ in the slide-wire bridge of figure 6-15 (B), is similar to the one used for solving for $R_x$ in figure 6-15 (A). However, in the slide-wire bridge the length $L_1$ corresponds to the resistance $R_1$, and the length $L_2$ corresponds to the resistance $R_3$. Therefore, $L_1$ and $L_2$ may be substituted for $R_1$ and $R_3$ in the equation. The resistance of $L_1$ and $L_2$ varies uniformly with slider movement because in a wire of uniform cross section the resistance varies directly with the length; therefore the ratio of the resistances equals to the corresponding ratio of the lengths. Substituting $L_1$ and $L_2$ for $R_1$ and $R_3$

$$R_x = \frac{L_2 R_2}{L_1}$$

A meter stick is mounted underneath the slide-wire and $L_1$ and $L_2$ are easily read in centimeters. For example, if a balance is obtained when $R_2$ = 150 ohms, $L_1$ = 25 cm. and $L_2$ = 75 cm., the unknown resistance is

$$R_x = \frac{75}{25} \times 150 = 450 \text{ ohms}$$

## PARALLEL SOURCES SUPPLYING A COMMON LOAD

The circuit shown in figure 6-16 illustrates two sources of emf, $E_{S1}$ and $E_{S2}$, having internal resistances of 2 and 2.5 ohms respectively, connected in parallel, and supplying a 5-ohm load. Neglecting the resistance of the lead wires, it is desired to determine the current delivered by each source to the load, the load current, and the load voltage.

The problem may be solved by establishing two voltage equations in which the voltages are expressed in terms of the unknown currents $I_1$ and $I_2$, the known resistances, and the known

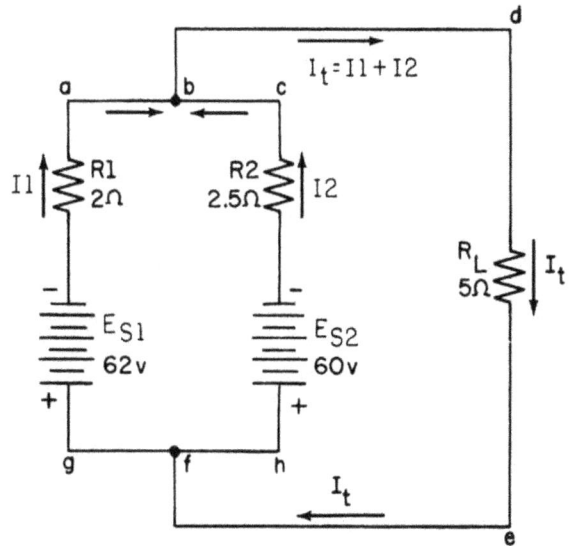

BE.122

Figure 6-16.—Parallel sources supplying a common load.

voltages. The equations are then solved simultaneously as in previous examples, to eliminate one of the unknown currents. The other unknown current is solved by substitution.

The first voltage equation is established by starting at point g and tracing clockwise around circuit gabdefg. The total load current is equal to the sum of the source currents, $I_1 + I_2$. The first voltage equation is,

$$62 - 2I_1 - 5(I_1 + I_2) = 0$$

from which,

$$7I_1 + 5I_2 = 62 \tag{31}$$

The second voltage equation is established by starting at point h and tracing around circuit hcbdefh. Thus,

$$60 - 2.5I_2 - 5(I_1 + I_2) = 0$$

from which,

$$5I_1 + 7.5I_2 = 60 \tag{32}$$

118

$I_2$ is eliminated by multiplying equation (31) by 1.5 and subtracting equation (32) from the result, as follows:

$$10.5I_1 + 7.5I_2 = 93$$
$$5.0I_1 + 7.5I_2 = 60$$
$$5.5I_1 = 33$$
$$I_1 = 6 \text{ amperes}$$

Substituting this value in equation (31),

$$7 \times 6 + 5I_2 = 62$$

from which

$$I_2 = 4 \text{ amperes}$$

The load current is $I_1 + I_2 = 6 + 4 = 10$ amperes. Thus source $E_{s1}$ supplies 6 amperes and source $E_{s2}$ supplies 4 amperes.

The load voltage is equal to the voltage developed across terminals f and b and is equal to the difference in a given source voltage and the internal voltage absorbed across the coresponding source resistance.

The statement applies equally to either source since both are in parallel with the load. In terms of source $E_{s1}$

$$E_{fb} = E_{s1} - I_1 R_1$$
$$= 62 - 6 \times 2$$
$$= 50 \text{ volts}$$

and in terms of source $E_{s2}$

$$E_{fb} = E_{s2} - I_2 R_2$$
$$= 60 - 4 \times 2.5$$
$$= 50 \text{ volts}$$

A further check on the load voltage is to express this value in terms of the load current and the load resistance as follows:

$$E_{fb} = (I_1 + I_2) R_L$$
$$= (6 + 4)(5)$$
$$= 50 \text{ volts}$$

## DISTRIBUTION CIRCUITS

### TWO-WIRE DISTRIBUTION CIRCUITS

Up to this point the voltage drop and the power lost in the line wires connecting the load and the source have been neglected. When the load is located at some distance from the source, the line resistance becomes an appreciable part of the total circuit resistance and the voltage and power lost in the line become significant even with moderate loads.

In figure 6-17, load M draws 7 amperes through terminals b and e, and the parallel group of lamps draws 5 amperes through terminals c and d. The current in line wires bc and de is 5 amperes. The line current in wires ab and ef is 5 + 7 = 12 amperes.

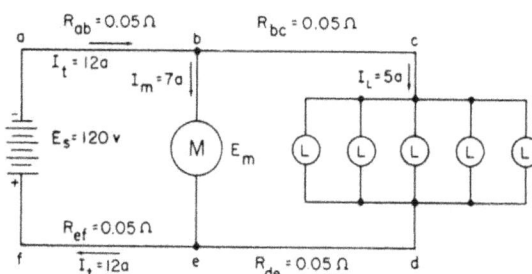

BE.123

Figure 6-17.—Simple two-wire distribution circuit.

The voltage source supplies a constant potential of 120 volts between points a and f. The resistance of line wires ab and ef is 2 x 0.05 = 0.1 ohm. The voltage drop across line wires ab and ef is 12 x 0.1 = 1.2 volts. The voltage drop across line wires bc and de is 5 x 0.1 = 0.5 volt. The voltage across M is 120 - 1.2 = 118.8 volts and the voltage across the five lamps is 118.8 - 0.5 = 118.3 volts.

The power dissipated in line wires ab and ef is equal to $(12)^2 \times 0.1 = 14.4$ watts. The power absorbed by line wires bc and de is $(5)^2 \times 0.1 = 2.5$ watts. The total power absorbed by the line wires is 14.4 + 2.5 = 16.9 watts. The power delivered to load M is 118.8 x 7 = 831.6 watts, and to the 5 lamps is 118.3 x 5 = 591.5 watts. The total power supplied to the entire circuit is equal to 16.9 + 831.6 + 591.5 = 1,440 watts and is equal to the product of the total applied voltage and the total current. Thus,

$$P_t = E_t I_t = 120 \times 12 = 1,440 \text{ watts}$$

## THREE-WIRE DISTRIBUTION CIRCUITS

Three-wire distribution circuits transmit power at 240 volts and utilize it at 120 volts. The direct-current 3-wire system includes a positive feeder, a negative feeder, and a neutral wire, as shown in figure 6-18 (A). The loads are connected between the negative feeder and the neutral, and between the positive feeder and the neutral. When the loads are unbalanced (unequal), the neutral wire carries a current equal to the difference in the currents in the negative and positive feeders.

In the example of figure 6-18 (A), load $L_1$ draws 10 amperes, load $L_2$ draws 4 amperes, and the neutral wire carries a current of 10 - 4 = 6 amperes. The direction of flow of the current in the neutral wire is always the same as that of the smaller of the currents in the positive and negative feeders. Thus the flow is to the left in the lower (positive) wire and also in the neutral. The current in the upper (negative) wire is 10 amperes and in the lower wire it is 4 amperes. The algebraic sum of the currents entering and leaving junction c is equal to zero. Thus,

$$+ 10 - 6 - 4 = 0$$

To find load voltage, $E_1$, a voltage equation is established in which $E_1$ is expressed in terms of the source voltage, $E_{s1}$, and the IR drops in the negative feeder and neutral wire. The algebraic sum of the voltages around the circuit fabcf, is equal to zero. Starting at f and proceeding clockwise,

$$+120 - 10 \times 0.5 - E_1 - 6 \times 0.5 = 0$$

$$E_1 = 112 \text{ volts}$$

Thus the voltage across load $L_1$ is 112 volts. This voltage is less than the source voltage by an amount equal to the sum of the voltage drops in the negative (5 volts) and the neutral (3 volts) wires.

To find load voltage $E_2$ a voltage equation is established in which $E_2$ is expressed in terms of the source voltage, $E_{s2}$, and the IR drops in the positive feeder and the neutral wire. The algebraic sum of the voltages around the circuit efcde is zero. Starting at e and proceeding clockwise,

$$+120 + (6 \times 0.5) - E_2 - (4 \times 0.5) = 0$$

$$E_2 = 121 \text{ volts}$$

In tracing the circuit from f to c, note that the direction is against the arrow representing current flow, and therefore that the IR drop of (6 x 0.5) volts is preceded by a plus sign. The load voltage, $E_2$, is 121 volts, which is 1 volt higher than the source voltage, $E_{s2}$. The total source voltage ($E_{s1} + E_{s2}$) is 240 volts and the total load voltage ($E_1 + E_2$) is 112 + 121 = 233 volts. This value is also equal to the difference between the total source voltage and the sum of the voltage drops in the positive and negative feeders, or 240 - (2 + 5) = 233 volts.

When the loads are balanced on the positive and negative sides of the 3-wire system, the neutral current is zero and the currents in the outside wires (positive and negative) are equal. When the loads are unbalanced the neutral wire carries the unbalanced current. The voltage on the heavily loaded side falls while the voltage on the lightly loaded side rises. The lower the resistance of the neutral wire, the less imbalance in voltage there will be for a given unbalanced load.

A more complicated 3-wire circuit is shown in figure 6-18 (B). The source voltage is 120 volts between each outside wire and the neutral, or center, wire. Load currents in the upper side of the system are indicated as 10, 4, and 8 amperes respectively for loads 1, 2, and 3. In the lower side of the system, the load currents are 12 and 6 amperes respectively for loads 4 and 5. In order to determine the various load voltages it is necessary to find the currents in each outside wire and in the neutral wire. The resistances of these wires are indicated, and therefore the voltage drops and the load voltages may be calculated after the currents are determined.

To find the currents in the various sections of the wires, it is best to start at the load farthest removed from the source. The polarities of the sources are such that electrons flow out of the negative terminal at n and return to the positive terminal at b.

Currents flowing toward a junction are assumed to be positive, and those flowing away from a junction are assumed to be negative. Applying Kirchhoff's current law at junction h, the

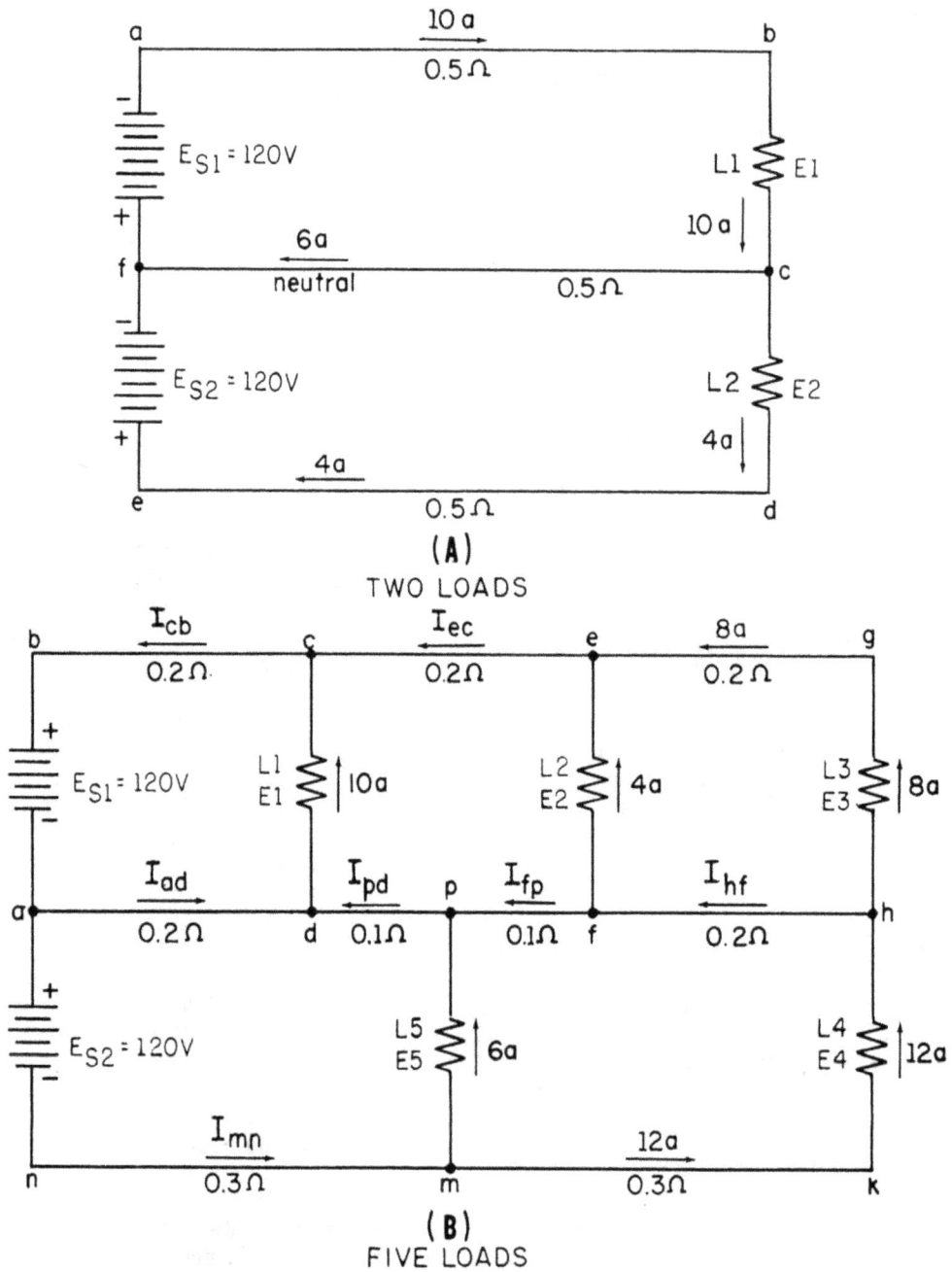

Figure 6-18.—Three-wire distribution circuits.

BE.124

neutral current, $I_n$ (flowing from h to f) is determined as

$$12 - 8 - I_{hf} = 0$$

$$I_{hf} = 4 \text{ amperes}$$

Applying the same rule successively to junctions f, e, p, m, d, and c, it follows that at junction f,

$$4 - 4 - I_{fp} = 0$$

$$I_{fp} = 0 \text{ amperes}$$

at junction e,

$$4 + 8 - I_{ec} = 0$$

$$I_{ec} = 12 \text{ amperes}$$

at junction p,

$$6 + 0 - I_{pd} = 0$$

$$I_{pd} = 6 \text{ amperes}$$

at junction m,

$$+ I_{mn} - 6 - 12 = 0$$

$$I_{mn} = 18 \text{ amperes}$$

at junction d,

$$I_{ad} + 6 - 10 = 0$$

$$I_{ad} = 4 \text{ amperes}$$

at junction c,

$$- I_{cb} + 10 + 12 = 0$$

$$I_{cb} = 22 \text{ amperes}$$

Thus, $E_{s1}$ supplies 22 amperes and $E_{s2}$ supplies 18 amperes. The electron flow in all parts of the lower wire is outward from the source, and the electron flow in all parts of the upper wire is back toward the source. The current in the neutral wire is always equal to the difference in the currents in the two outside wires, and the electron flow is in the direction of the smaller of these two currents. Thus in figure 6-18 (B), the neutral current in section ad is 4 amperes, which is the difference between 18 amperes and 22 amperes; and it is in the direction of the smaller current in section nm. The neutral current in section pd is 6 amperes, which is the difference between 18 amperes and 12 amperes; and it is in the same direction as the 12 amperes in section ec. The neutral current in section fp is zero because the current in each outside wire in that section is 12 amperes. The neutral current in section hf is 4 amperes, which is the difference between 12 amperes and 8 amperes, and it is in the direction of the smaller outside current in section ge.

In order fo find the voltages across the loads in figure 6-18 (B), Kirchhoff's voltage law is applied to the various individual circuits. Thus to find the voltage, $E_1$, across load $L_1$, the algebraic sum of the voltages around the circuit abcda is equated to zero, and E1 is then readily determined. Starting at a,

$$- 120 + (22 \times 0.2) + E_1 + (4 \times 0.2) = 0$$

$$E1 = 114.8 \text{ volts}$$

To find load voltage $E_2$, the algebraic sum of the voltages around circuit dcefpd is set equal to zero. Starting at d,

$$- 114.8 + (12 \times 0.2) + E_2$$

$$+ (0 \times 0.1) - (6 \times 0.1) = 0$$

$$E_2 = 113 \text{ volts}$$

To find load voltage $E_3$, the algebraic sum of the voltages around loop feghf is set equal to zero. Starting at f,

$$- 113 + (8 \times 0.2) + E_3 - (4 \times 0.2) = 0$$

$$E_3 = 112.2 \text{ volts}$$

To find load voltage $E_4$, the algebraic sum of the voltages around loop nadpfhkmn is set equal to zero. Loop mpfhkm cannot be used because it would contain two unknown voltages, $E_5$ and $E_4$. Starting at n,

$$- 120 - (4 \times 0.2) + (6 \times 0.1) + (0 \times 0.1)$$

$$+ (4 \times 0.2) + E_4 + (12 \times 0.3) + (18 \times 0.3) = 0$$

$$E_4 = 110.4 \text{ volts}$$

To find load voltage $E_5$, the algebraic sum of the voltages around loop nadpmn is set equal to zero. Starting at n,

$$- 120 - (4 \times 0.2) + (6 \times 0.1)$$

$$+ E_5 + (18 \times 0.3) = 0$$

$$E_5 = 114.8 \text{ volts}$$

In each case, the equations used contain one unknown; and thus a simple solution is quickly obtained. It is necessary that the path traced include a completely closed loop and that all but one of the voltages within that loop be known. Simple transposition of the resulting equation gives the desired voltage.

# CHAPTER 7

# ELECTRICAL CONDUCTORS AND WIRING TECHNIQUES

Since all electrical circuits utilize conductors of one type of another, it is essential that you know the basic physical features and electrical characteristics of the most common types of conductors.

As stated previously, any substance that permits the free motion of a large number of electrons is classed as a conductor. A conductor may be made from many different types of metals, but only the most commonly used types of materials will be discussed in this chapter.

To compare the resistance and size of one conductor with that of another, a standard or unit size of conductor must be established. A convenient unit of linear measurement, as far as the diameter of a piece of wire is concerned, is the mil (0.001 of an inch); and a convenient unit of wire length is the foot. The standard unit of size in most cases is the MIL-FOOT; that is, a wire will have unit size if it has diameter of 1 mil and a length of 1 foot. The resistance in ohms of a unit conductor of a given substance is called the specific resistance, or specific resistivity, of the substance.

Gage numbers are a further convenience in comparing the diameter of wires. The gage commonly used is the American wire gage (AWG), formerly the Brown and Sharpe (B and S) gage.

## MIL

### SQUARE MIL

The square mil is a convenient unit of cross-sectional area for square or rectangular conductors. A square mil is the area of a square, the sides of which are 1 mil, as shown in figure 7-1 (A). To obtain the cross-sectional area in square mils of a square conductor, square one side measured in mils. To obtain the cross-sectional area in square mils of a rectangular conductor, multiply the length of one side by that of the other, each length being expressed in mils.

For example, find the cross-sectional area of a large rectangular conductor 3/8 inch thick and 4 inches wide. The thickness may be expressed in mils as 0.375 x 1,000 = 375 mils, and the width as 4 x 1,000, or 4,000 mils. The cross-sectional area is 375 x 4,000, or 1,500,000 square mils.

### CIRCULAR MIL

The circular mil is the standard unit of wire cross-sectional area used in American and English wire tables. Because the diameters of round conductors, or wires, used to conduct electricity may be only a small fraction of an inch, it is convenient to express these diameters in mils, to avoid the use of decimals. For example, the diameter of a wire is expressed as 25 mils instead of 0.025 inch. A circular mil is the area of a circle having a diameter of 1 mil, as shown in figure 7-1 (B). The area in circular mils of a round conductor is obtained by squaring the diameter measured in mils. Thus, a wire having a diameter of 25 mils has an area of $25^2$ or 625 circular mils. By way of comparison, the basic formula for the area of a circle is $A = \pi R^2$ and in this example the area in square inches is

$$A = \pi R^2 = 3.14(0.0125)^2 = 0.00049 \text{ sq. in.}$$

If D is the diameter of a wire in mils, the area in square mils is

$$A = \pi \left( \frac{D}{2} \right)^2 = \frac{3.1416}{4} D^2 = 0.7854D^2 \text{ sq. mils.}$$

123

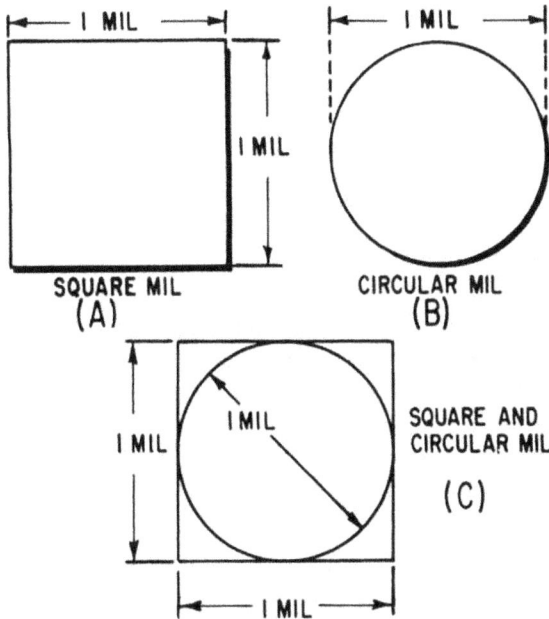

BE.125

Figure 7-1.—(A) Square mil; (B) circular mil; (C) comparison of circular to square mil.

Therefore, a wire 1 mil in diameter has an area of

$$A = 0.7854 \times 1^2 = 0.7854 \text{ sq. mils,}$$

which is equivalent to 1 circular mil. The cross-sectional area of a wire in circular mils is therefore determined as

$$A = \frac{0.7854D^2}{0.7854} = D^2 \text{ circular mils,}$$

where D is the diameter in mils. Thus, the constant $\frac{\pi}{4}$ is eliminated from the calculation.

In comparing square and round conductors it should be noted that the circular mil is a smaller unit of area than the square mil, and therefore there are more circular mils than square mils in any given area. The comparison is shown in figure 7-1 (C). The area of a circular mil is equal to 0.7854 of a square mil. Therefore, to determine the circular-mil area when the square-mil area is given, divide the area in square mils by 0.7854. Conversely, to determine the square-mil area when the circular-mil area is given, multiply the area in circular mils by 0.7854.

For example, a No. 12 wire has a diameter of 80.81 mils. What is (1) its area in circular mils and (2) its area in square mils?

Solution:

(1)  $A = D^2 = 80.81^2 = 6,530$ circular mils

(2)  $A = 0.7854 \times 6,530 = 5,128.7$ square mils

A rectangular conductor is 1.5 inches wide and 0.25 inch thick. (1) What is its area in square mils? (2) What size of round conductor in circular mils is necessary to carry the same current as the rectangular bar?

Solution:

(1) 1.5″ = 1.5   x 1,000 = 1,500 mils

0.25″= 0.25   x 1,000 = 250 mils

A = 1,500  x  250  = 375,000 sq. mils

(2) To carry the same current, the cross-sectional area of the rectangular bar and the cross-sectional area of the round conductor must be equal. There are more circular mils than square mils in this area, and therefore

$$A = \frac{375,000}{0.7854} = 477,000 \text{ circular mils}$$

A wire in its usual form is a slender rod or filament of drawn metal. In large sizes, wire becomes difficult to handle, and its flexibility is increased by stranding. The strands are usually single wires twisted together in sufficient numbers to make up the necessary cross-sectional area of the cable. The total area in circular mils is determined by multiplying the area of one strand in circular mils by the number of strands in the cable.

CIRCULAR-MIL-FOOT

A circular-mil-foot, as shown in figure 7-2, is actually a unit of volume. It is a unit conductor 1 foot in length and having a cross-sectional area of 1 circular mil. Because it is considered a unit conductor, the circular-mil-foot is useful in making comparisons between wires that are made of different metals. For example, a basis of comparison of the RESISTIVITY (to be treated later) of various substances may be made by determining the resistance of a circular-mil-foot of each of the substances.

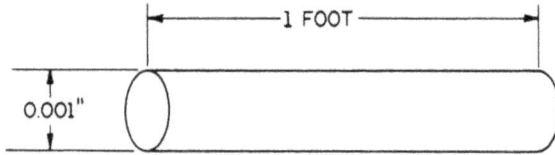

BE.126

Figure 7-2.—Circular-mil-foot.

In working with certain substances it is sometimes more convenient to employ a different unit volume. Accordingly, unit volume may also be taken as the centimeter cube; and specific resistance becomes the resistance offered by a cube-shaped conductor 1 cm. long and 1 sq. cm. in cross-sectional area. The inch cube may also be used. The unit of volume employed is given in tables of specific resistances.

## SPECIFIC RESISTANCE OR RESISTIVITY

Specific resistance, or resistivity, is the resistance in ohms offered by unit volume (the circular-mil-foot) of a substance to the flow of electric current. Resistivity is the reciprocal of conductivity. A substance that has a high resistivity will have a low conductivity, and vice versa.

Thus, the specific resistance of a substance is the resistance of a unit volume of that substance. Many tables of specific resistance are based on the resistance in ohms of a volume of the substance 1 foot long and 1 circular mil in cross-sectional area. The temperature at which the resistance measurement is made is also specified. If the kind of metal of which a conductor is made is known, the specific resistance of the metal may be obtained from a table. The specific resistances of some common substances are given in table 7-1.

The resistance of a conductor of uniform cross section varies directly as the product of the length and the specific resistance of the conductor and inversely as the cross-sectional area of the conductor. Therefore the resistance of a conductor may be calculated if the length, cross-sectional area, and specific resistance

Table 7-1.—Specific resistance.

| Substance | Specific resistance at 20° C. | |
| | Centimeter cube (microhms) | Circular-mil-foot (ohms) |
| --- | --- | --- |
| Silver . . . . . . | 1.629 | 9.8 |
| Copper (drawn). | 1.724 | 10.37 |
| Gold . . . . . . . | 2.44 | 14.7 |
| Aluminum . . . | 2.828 | 17.02 |
| Carbon (amorphous.) | 3.8 to 4.1 | . . . . . . . . |
| Tungsten . . . . | 5.51 | 33.2 |
| Brass . . . . . . | 7.0 | 42.1 |
| Steel (soft) . . . | 15.9 | 95.8 |
| Nichrome . . . | 109.0 | 660.0 |

of the substance are known. Expressed as an equation, the resistance, R in ohms, of a conductor is

$$R = \rho \frac{L}{A}$$

where $\rho$ (Greek rho) is the specific resistance in ohms per circular-mil-foot, L the length in feet and A the cross-sectional area in circular mils.

For example, what is the resistance of 1,000 feet of copper wire having a cross-sectional area of 10,400 circular mils (No. 10 wire), the wire temperature being 20° C?

Solution:

The specific resistance, from table 7-1, is 10.37. Substituting the known values in the preceding equation, the resistance, R, is determined as

$$R = \rho \frac{L}{A} = 10.37 \times \frac{1,000}{10,400} = 1 \text{ ohm, approximately}$$

If R, $\rho$, and A are known, the length may be determined by a simple mathematical transposition. This is of value in many applications. For example, when it is desired to locate a ground in a telephone line, special test equipment is used that operates on the principle that the resistance of a line varies directly with its length. Thus, the distance between the test point and a fault can be computed accurately.

As has been mentioned in preceding chapters, conductance (G) is the reciprocal of resistance.

When R is in ohms, the conductance is expressed in mhos. Where resistance is opposition to flow, conductance is the ease with which the current flows. Conductance is mhos is equivalent to the number of amperes flowing in a conductor per volt of applied emf. Expressed in terms of the specific resistance, length, and cross section of a conductor,

$$G = \frac{A}{\rho L}$$

The conductance, G, varies directly as the cross-sectional area, A, and inversely as the specific resistance, $\rho$ and the length, L. When A is in circular mils, $\rho$ is in ohms per circular-mil-foot, L is in feet, and G is in mhos.

The relative conductance of several substances is given in table 7-2.

Table 7-2.—Relative conductance.

| Substance | Relative conductance (Silver = 100%) |
|---|---|
| Silver | 100 |
| Copper | 98 |
| Gold | 78 |
| Aluminum | 61 |
| Tungsten | 32 |
| Zinc | 30 |
| Platinum | 17 |
| Iron | 16 |
| Lead | 15 |
| Tin | 9 |
| Nickel | 7 |
| Mercury | 1 |
| Carbon | 0.05 |

## WIRE MEASURE

### RELATION BETWEEN WIRE SIZES

Wires are manufactured in sizes numbered according to a table known as the American wire gage (AWG). As may be seen in table 7-3, the wire diameters become smaller as the gage numbers become larger. The largest wire size shown in the table is 0000 (read "4 naught"), and the smallest is number 40. Larger and smaller sizes are manufactured but are not commonly used by the Navy. The ratio of the diameter of one gage number to the diameter of the next higher gage number is a constant, 1.123. The cross-sectional area varies as the square of the diameter. Therefore, the ratio of the cross section of one gage number to that of the next higher gage number is the square of 1.123, or 1.261. Because the cube of 1.261 is very nearly 2, the cross-sectional area is approximately halved, or doubled, every three gage numbers. Also because 1.261 raised to the 10th power is very nearly equal to 10, the cross-sectional area is increased or decreased 10 times every 10 gage numbers.

A No. 10 wire has a diameter of approximately 102 mils, a cross-sectional area of approximately 10,400 circular mils, and a resistance of approximately 1 ohm per 1,000 feet. From these facts it is possible to estimate quickly the cross-sectional area and the resistance of any size copper wire without referring directly to a wire table.

For example, to estimate the cross-sectional area and the resistance of 1,000 feet of No. 17 wire, the following reasoning might be employed. A No. 17 wire is 3 sizes removed from a No. 20 wire and therefore has twice the cross-sectional area of a No. 20 wire. A No. 20 wire is 10 sizes removed from a No. 10 wire and therefore has one-tenth the cross section of a No. 10 wire. Therefore, the cross-sectional area of a No. 17 wire is 2 x 0.1 x 10,000 = 2,000 circular mils. Since resistance varies inversely with the cross-sectional area, the resistance of a No. 17 wire is 10 x 1 x 0.5 = 5 ohms per 1,000 feet.

A wire gage is shown in figure 7-3. It will measure wires ranging in size from number 0 to number 36. The wire whose size is to be measured is inserted in the smallest slot that will just accommodate the bare wire. The gage number corresponding to that slot indicates the wire size. The slot has parallel sides and should not be confused with the semicircular opening at the end of the slot. The opening simply permits the free movement of the wire all the way through the slot.

### STRANDED WIRES AND CABLES

A WIRE is a slender rod or filament of drawn metal. The definition restricts the term to what would ordinarily be understood as "solid wire." The word "slender" is used because the length of a wire is usually large in comparison with the diameter. If a wire is covered with insulation, it is properly called an insulated wire. Although the term "wire" properly refers to the metal, it is generally understood to include the insulation.

Table 7-3.—Standard annealed solid copper wire.

(American wire gage--B & S)

| Gage number | Diameter (mils) | Cross section | | Ohms per 1,000 ft. | | Ohms per mile 25° C. (=77° F.) | Pounds per 1,000 ft. |
| | | Circular mils | Square inches | 25° C. (=77° F.) | 65° C. (=149° F.) | | |
|---|---|---|---|---|---|---|---|
| 0000 | 460.0 | 212,000.0 | 0.166 | 0.0500 | 0.0577 | ·0.264 | 641.0 |
| 000 | 410.0 | 168,000.0 | .132 | .0630 | .0727 | .333 | 508.0 |
| 00 | 365.0 | 133,000.0 | .105 | .0795 | .0917 | .420 | 403.0 |
| 0 | 325.0 | 106,000.0 | .0829 | .100 | .116 | .528 | 319.0 |
| 1 | 289.0 | 83,700.0 | .0657 | .126 | .146 | .665 | 253.0 |
| 2 | 258.0 | 66,400.0 | .0521 | .159 | .184 | .839 | 201.0 |
| 3 | 229.0 | 52,600.0 | .0413 | .201 | .232 | 1.061 | 159.0 |
| 4 | 204.0 | 41,700.0 | .0328 | .253 | .292 | 1.335 | 126.0 |
| 5 | 182.0 | 33,100.0 | .0260 | .319 | .369 | 1.685 | 100.0 |
| 6 | 162.0 | 26,300.0 | .0206 | .403 | .465 | 2.13 | 79.5 |
| 7 | 144.0 | 20,800.0 | .0164 | .508 | .586 | 2.68 | 63.0 |
| 8 | 128.0 | 16,500.0 | .0130 | .641 | .739 | 3.38 | 50.0 |
| 9 | 114.0 | 13,100.0 | .0103 | .808 | .932 | 4.27 | 39.6 |
| 10 | 102.0 | 10,400.0 | .00815 | 1.02 | 1.18 | 5.38 | 31.4 |
| 11 | 91.0 | 8,230.0 | .00647 | 1.28 | 1.48 | 6.75 | 24.9 |
| 12 | 81.0 | 6,530.0 | .00513 | 1.62 | 1.87 | 8.55 | 19.8 |
| 13 | 72.0 | 5,180.0 | .00407 | 2.04 | 2.36 | 10.77 | 15.7 |
| 14 | 64.0 | 4,110.0 | .00323 | 2.58 | 2.97 | 13.62 | 12.4 |
| 15 | 57.0 | 3,260.0 | .00256 | 3.25 | 3.75 | 17.16 | 9.86 |
| 16 | 51.0 | 2,580.0 | .00203 | 4.09 | 4.73 | 21.6 | 7.82 |
| 17 | 45.0 | 2,050.0 | .00161 | 5.16 | 5.96 | 27.2 | 6.20 |
| 18 | 40.0 | 1,620.0 | .00128 | 6.51 | 7.51 | 34.4 | 4.92 |
| 19 | 36.0 | 1,290.0 | .00101 | 8.21 | 9.48 | 43.3 | 3.90 |
| 20 | 32.0 | 1,020.0 | .000802 | 10.4 | 11.9 | 54.9 | 3.09 |
| 21 | 28.5 | 810.0 | .000636 | 13.1 | 15.1 | 69.1 | 2.45 |
| 22 | 25.3 | 642.0 | .000505 | 16.5 | 19.0 | 87.1 | 1.94 |
| 23 | 22.6 | 509.0 | .000400 | 20.8 | 24.0 | 109.8 | 1.54 |
| 24 | 20.1 | 404.0 | .000317 | 26.2 | 30.2 | 138.3 | 1.22 |
| 25 | 17.9 | 320.0 | .000252 | 33.0 | 38.1 | 174.1 | 0.970 |
| 26 | 15.9 | 254.0 | .000200 | 41.6 | 48.0 | 220.0 | 0.769 |
| 27 | 14.2 | 202.0 | .000158 | 52.5 | 60.6 | 277.0 | 0.610 |
| 28 | 12.6 | 160.0 | .000126 | 66.2 | 76.4 | 350.0 | 0.484 |
| 29 | 11.3 | 127.0 | .0000995 | 83.4 | 96.3 | 440.0 | 0.384 |
| 30 | 10.0 | 101.0 | .0000789 | 105.0 | 121.0 | 554.0 | 0.304 |
| 31 | 8.9 | 79.7 | .0000626 | 133.0 | 153.0 | 702.0 | 0.241 |
| 32 | 8.0 | 63.2 | .0000496 | 167.0 | 193.0 | 882.0 | 0.191 |
| 33 | 7.1 | 50.1 | .0000394 | 211.0 | 243.0 | 1,114.0 | 0.152 |
| 34 | 6.3 | 39.8 | .0000312 | 266.0 | 307.0 | 1,404.0 | 0.120 |
| 35 | 5.6 | 31.5 | .0000248 | 335.0 | 387.0 | 1,769.0 | 0.0954 |
| 36 | 5.0 | 25.0 | .0000196 | 423.0 | 488.0 | 2,230.0 | 0.0757 |
| 37 | 4.5 | 19.8 | .0000156 | 533.0 | 616.0 | 2,810.0 | 0.0600 |
| 38 | 4.0 | 15.7 | .0000123 | 673.0 | 776.0 | 3,550.0 | 0.0476 |
| 39 | 3.5 | 12.5 | .0000098 | 848.0 | 979.0 | 4,480.0 | 0.0377 |
| 40 | 3.1 | 9.9 | .0000078 | 1,070.0 | 1,230.0 | 5,650.0 | 0.0299 |

A CONDUCTOR is a wire or combination of wires not insulated from one another, suitable for carrying an electric current.

A STRANDED CONDUCTOR is a conductor composed of a group of wires or of any combination of groups of wires. The wires in a stranded conductor are usually twisted together.

A CABLE is either a stranded conductor (single-conductor cable) or a combination of conductors insulated from one another

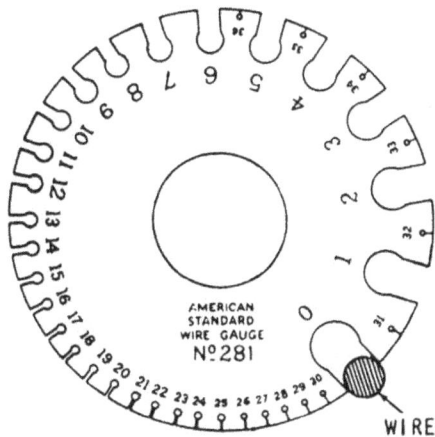

Figure 7-3.—Wire gage.

(multiple-conductor cable). The term cable is a general one, and in practice it is usually applied only to the larger sizes of conductors. A small cable is more often called a stranded wire or cord. Cables may be bare or insulated. The insulated cables may be sheathed (covered) with lead, or protective armor.

Figure 7-4 shows some of the different types of wire and cables used in the Navy.

Conductors are stranded mainly to increase their flexibility. The arrangement of the wire strands in concentric-layer cables is as follows:

The first layer of strands around the center is made up of 6 conductors; the second layer is made up of 12 conductors; the third layer is made up of 18 conductors; and so on. Thus, standard cables are composed of 1, 7, 19, 37, and so forth strands.

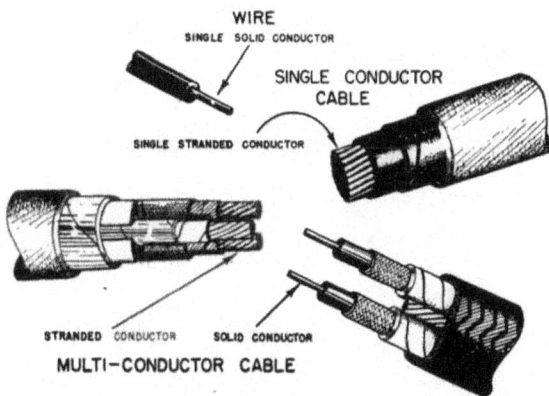

Figure 7-4.—Conductors.

The overall flexibility may be increased by further stranding of the individual strands.

Figure 7-5 shows a typical 37-strand cable. It also shows how the total circular-mil cross-sectional area of a stranded cable is determined.

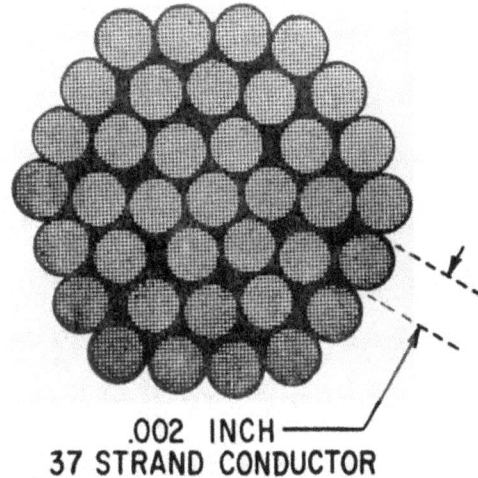

.002 INCH
37 STRAND CONDUCTOR

DIAMETER OF EACH STRAND = .002 INCH
DIAMETER OF EACH STRAND, MILS = 2 MILS
CIRCULAR MIL AREA OF
EACH STRAND = $D^2$ = 4 CM
TOTAL CM AREA OF
CONDUCTOR = 4 x 37 = 148 CM

BE.129

Figure 7-5.—Stranded conductor.

## FACTORS GOVERNING THE SELECTION OF WIRE SIZE

Several factors must be considered in selecting the size of wire to be used for transmitting and distributing electric power.

One factor is the allowable power loss ($I^2R$ loss) in the line. This loss represents electrical energy converted into heat. The use of large conductors will reduce the resistance and therefore the $I^2R$ loss. However, large conductors are more expensive initially than small ones; they are heavier and require more substantial supports.

A second factor is the permissible voltage drop (IR drop) in the line. If the source maintains a constant voltage at the input to the line, any variation in the load on the line will cause a variation in line current, and a consequent variation in the IR drop in the line. A wide variation in the IR drop in the line causes poor voltage regulation at the load. The obvious remedy is to reduce either $I$ or $R$. A reduction in load current lowers the amount of power being transmitted, whereas a reduction in line resistance increases the size and weight of conductors required. A compromise is generally reached whereby the voltage variation at the load is within tolerable limits and the weight of line conductors is not excessive.

A third factor is the current-carrying ability of the line. When current is drawn through the line, heat is generated. The temperature of the line will rise until the heat radiated, or otherwise dissipated, is equal to the heat generated by the passage of current through the line. If the conductor is insulated, the heat generated in the conductor is not so readily removed as it would be if the conductor were not insulated. Thus, the protect the insulation from too much heat, the current through the conductor must be maintained below a certain value. Rubber insulation will begin to deteriorate at relatively low temperatures. Varnished cloth insulation retains its insulating properties at higher temperatures; and other insulation— for example, asbestos or silicon—is effective at still higher temperatures.

Electrical conductors may be installed in locations where the ambient (surrounding) temperature is relatively high; in which case the heat generated by external sources constitutes an appreciable part of the total conductor heating. Due allowance must be made for the influence of external heating on the allowable conductor current and each case has its own specific limitations. The maximum allowable operating temperature of insulated conductors is specified in tables and varies with the type of conductor insulation being used.

Tables have been prepared by the National Board of Fire Underwriters giving the safe current ratings for various sizes and types of conductors covered with various types of insulation. For example, the allowable current-carrying capacities of copper conductors at not over 30°C (86° F) room temperature for single conductors in free air are given in table 7-4.

## COPPER VS. ALUMINUM CONDUCTORS

Although silver is the best conductor, its cost limits its use to special circuits where a substance with high conductivity is needed.

The two most generally used conductors are copper and aluminum. Each has characteristics that make its use advantageous under certain circumstances. Likewise, each has certain disadvantages.

Copper has a higher conductivity; it is more ductile (can be drawn out), has relatively high tensile strength, and can be easily soldered. It is more expensive and heavier than aluminum.

Table 7-4.—Current-carrying capacities (in amperes) of single copper

conductors at ambient temperature of below 30°C

| Size | Rubber or thermoplastic | Thermoplastic asbestos, var-cam, or asbestos var-cam | Impregnated asbestos | Asbestos | Heat-Resistant or Moisture Resistant |
|------|------|------|------|------|------|
| 0000 | 300 | 385 | 475 | 510 | 370 |
| 000 | 260 | 330 | 410 | 430 | 320 |
| 00 | 225 | 285 | 355 | 370 | 275 |
| 0 | 195 | 245 | 305 | 325 | 235 |
| 1 | 165 | 210 | 265 | 280 | 205 |
| 2 | 140 | 180 | 225 | 240 | 175 |
| 3 | 120 | 155 | 195 | 210 | 150 |
| 4 | 105 | 135 | 170 | 180 | 130 |
| 6 | 80 | 100 | 125 | 135 | 100 |
| 8 | 55 | 70 | 90 | 100 | 70 |
| 10 | 40 | 55 | 70 | 75 | 55 |
| 12 | 25 | 40 | 50 | 55 | 40 |
| 14 | 20 | 30 | 40 | 45 | 30 |

Although aluminum has only about 60 percent of the conductivity of copper, its lightness makes possible long spans, and its relatively large diameter for a given conductivity reduces corona—that is, the discharge of electricity from the wire when it has a high potential. The discharge is greater when smaller diameter wire is used than when larger diameter wire is used. However, aluminum conductors are not easily soldered, and aluminum's relatively large size for a given conductance does not permit the economical use of an insulation covering.

A comparison of some of the characteristcs of copper and aluminum is given in table 7-5.

Table 7-5.—Characteristics of copper and aluminum.

| Characteristics | Copper | Aluminum |
|---|---|---|
| Tensile strength (lb/in$^2$). | 55,000 | 25,000 |
| Tensile strength for same conductivity (lb). | 55,000 | 40,000 |
| Weight for same conductivity (lb). | 100 | 48 |
| Cross section for same conductivity (C.M.). | 100 | 160 |
| Specific resistance ($\Omega$/mil ft). | 10.6 | 17 |

## TEMPERATURE COEFFICIENT

The resistance of pure metals—such as silver, copper, and aluminum—increases as the temperature increases. However, the resistance of some alloys—such as constantan and manganin—changes very little as the temperature changes. Measuring instruments use these alloys because the resistance of the circuits must remain constant if accurate measurements are to be achieved.

In table 7-1 the resistance of a circular-mil-foot of wire (the specific resistance) is given at a specific temperature, 20°C in this case. It is necessary to establish a standard temperature because, as has been stated, the resistance of pure metals increase with an increase in temperature; and a true basis of comparison cannot be made unless the resistances of all the substances being compared are measured at the same temperature. The amount of increase in the resistance of a 1-ohm sample of the conductor per degree rise in temperature above 0°C is called the TEMPERATURE COEFFICIENT OF RESISTANCE. For copper, the value is approximately 0.00427 ohm. For pure metals, the temperature coefficient of resistance ranges between 0.003 and 0.006 ohm.

Thus, a copper wire having a resistance of 50 ohms at an initial temperature of 0°C will have an increase in resistance of 50 x 0.00427, or 0.214 ohms for the entire length of wire for each degree of temperature rise above 0° C. At 20°C the increase in resistance is approximately 20 x 0.214, or 4.28 ohms. The total resistance at 20°C is 50 + 4.28, or 54.28 ohms.

## CONDUCTOR INSULATION

To be useful and safe, electric current must be forced to flow only where it is needed. It must be "channeled" from the power source to a useful load. In general, current-carrying conductors must not be allowed to come in contact with one another, their supporting hardware, or personnel working near them. To accomplish this, conductors are coated or wrapped with various materials. These materials have such a high resistance that they are, for all practical purposes, nonconductors. They are generally referred to as insulators or insulating material.

Because of the expense of insulation and its stiffening effect, together with the great variety of physical and electrical conditions under which the conductors are operated, only the necessary minimum of insulation is applied for any particular type of cable designed to do a specific job. Therefore there is a wide variety of insulated conductors available to meet the requirements of any job.

Two fundamental properties of insulation materials (for example, rubber, glass, asbestos, and plastic) are insulation resistance and dielectric strength. These are entirely different and distinct properties.

INSULATION RESISTANCE is the resistance to current leakage through and over the surface of insulation materials. Insulation resistance can be measured by means of a megger without damaging the insulation, and information so obtained serves as a useful guide in appraising the general condition of insulation. However, the data obtained in this manner may not give a true picture of the condition of the insulation. Clean-dry insulation having cracks or other faults may show a high value of insulation resistance but would not be suitable for use.

DIELECTRIC STRENGTH is the ability of the insulator to withstand potential difference and is usually expressed in terms of the voltage at which the insulation fails because of the electrostatic stress. Maximum dielectric strength values can be measured by raising the voltage of a TEST SAMPLE until the insulation breaks down.

## RUBBER

One of the most common types of insulation is rubber. The voltage that may be applied to a rubber-covered pair of conductors (twisted pair) is dependent on the thickness and the quality of the rubber covering. Other factors being equal, the thicker the insulation the higher may be the applied voltage. Figure 7-6 shows two types of rubber-covered wire. One is a single, solid conductor, and the other is a 2-conductor cable in which each stranded conductor is covered with rubber insulation. In each case the rubber serves the same purpose—to confine the current to its conductor.

It may be seen from the enlarged cross-sectional view that a thin coating of tin separates the copper conductor from the rubber insulation. If the thin coating of tin were not used, chemical action would take place and the rubber would become soft and gummy where it makes contact with the copper. When small, solid, or stranded conductors are used, a winding of cotton threads is applied between the conductors and the rubber insulation.

## PLASTICS

Plastic has become one of the more common types of material used as insulation for electrical conductors. It has good insulating, flexibility, and moisture resistant qualities under various conditions. There are various types of plastics used as insulating material, thermoplastic being one of the most common. With the use of thermoplastic the conductor temperature can be higher than with some other types of insulating materials without damage to the insulating quality of the material.

## VARNISHED CAMBRIC

Heat is developed when current flows through a wire, and when a large amount of current flows, considerable heat may be developed. The heat can be dissipated if air is circulated freely around the wire. If a cover of insulation is used, the heat is not removed so readily and the temperature may reach a high value.

Rubber is a good insulator at relatively low voltage as long as the temperature remains low. Too much heat will cause even the best grade of rubber insulation to become brittle and crack. VARNISHED CAMBRIC insulation will stand much higher temperatures than rubber insulation. Varnished cambric is cotton cloth that has been coated with an insulating varnish. Figure 7-7 shows some of the detail of a cable covered with varnished cambric insulation. The varnished cambric is in tape form and is wound around the conductor in layers. An oily compound is applied between each layer of the tape. This compound prevents water from seeping through the insulation. It also acts as a lubricant between the layers of tape, so they will slide over each other when the cable is bent.

This type of insulation is used on high-voltage conductors associated with switch gear in substations and power houses and other locations subjected to high temperatures. It is also used on high-voltage generator coils and leads, and also on transformer leads because it is unaffected by oils or grease and because it has a high dielectric strength. Varnished cambric and paper insulation for cables are the two types of insulating materials most widely used at voltages above 15,000 volts, but such cables are always lead-covered to keep the moisture out.

## ASBESTOS

Even varnished cambric may break down when the temperature goes above 85°C (185° F). When the combined effects of a high ambient (surrounding) temperature and a high internal temperature due to large current flow through the wire makes the total temperature of the wire go above 85°C, ASBESTOS insulation is used.

Asbestos is a good insulation for wires and cables used under very high temperature conditions. It is fire resistant and does not change with age. One type of asbestos-covered wire is shown in figure 7-8. It consists of a stranded copper conductor covered with felted asbestos, which is, in turn, covered with asbestos braid. This type of wire is used in motion-picture projectors, arc lamps, spotlights, heating element leads, and so forth.

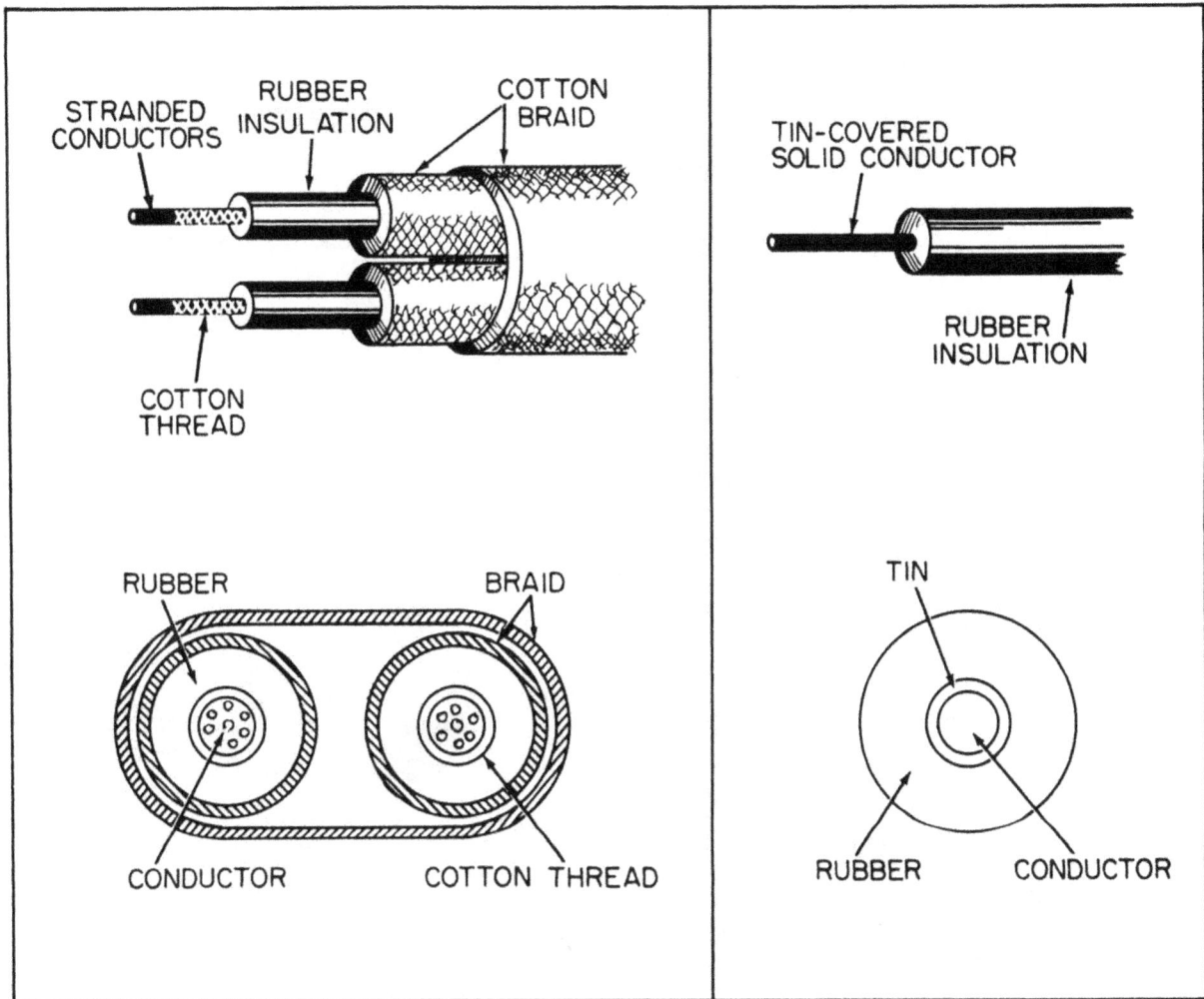

Figure 7-6.—Rubber insulation.

BE.130

Another type of asbestos-covered cable is shown in figure 7-9. It serves as leads for motors and transformers that sometimes must operate in hot, wet locations. The varnished cambric covers the inner layer of felted asbestos and prevents moisture from reaching the innermost layer of asbestos. Asbestos loses its insulating properties when it becomes wet, and will in fact become a conductor. The varnished cambric prevents this from happening because it resists moisture. Although this insulation will withstand some moisture, it should not be used on conductors that may at times be partly immersed in water, unless the insulation is protected with an outer lead sheath.

PAPER

Paper has little insulation value alone, but when impregnated with a high grade of mineral oil, it serves as a satisfactory insulation for high-voltage cables. The oil has a high dielectric strength, and tends to prevent breakdown of paper insulation when the paper is thoroughly saturated with it. The thin paper tape is wrapped in many layers around the conductors, and it is then soaked with oil.

The 3-conductor cable shown in figure 7-10 consists of paper insulation on each conductor with a spirally wrapped nonmagnetic metallic tape over the insulation. The space between

132

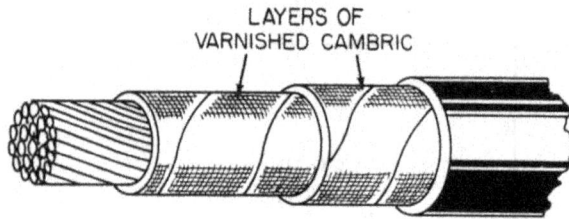

BE.131

Figure 7-7.—Varnished cambric insulation.

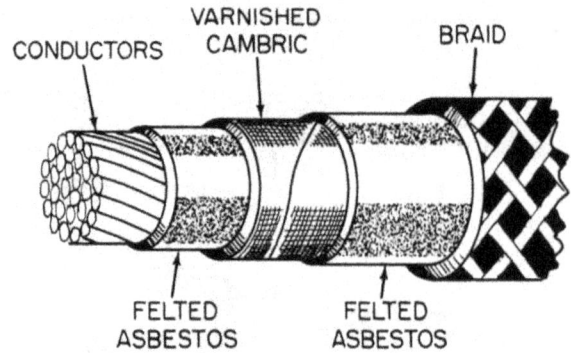

BE.132

Figure 7-8.—Asbestos insulation.

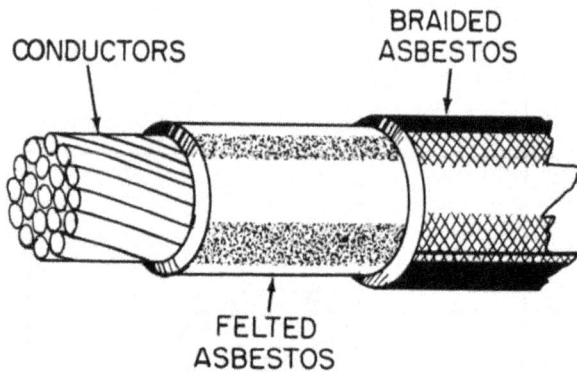

BE.133

Figure 7-9.—Asbestos and varnished cambric insulation.

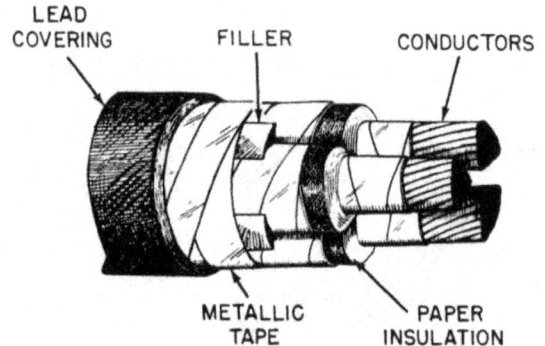

BE.134

Figure 7-10.—Paper-insulated power cables.

conductors is filled with a suitable spacer to round out the cable and another nonmagnetic metal tape is used to secure the entire cable, and then a lead sheath is applied over all. This type of cable is used on voltages from 10,000 volts to 35,000 volts.

## SILK AND COTTON

In certain types of circuits—for example communications circuits—a large number of conductors are needed, perhaps as many as several hundred. Figure 7-11 shows a cable containing many conductors, each insulated from the others by silk and cotton threads. The use of silk and cotton as insulation keeps the size of the cable small enough to be handled easily. The silk and cotton threads are wrapped around the individual conductors in reverse directions, and the covering is then impregnated with a special wax compound.

Because the insulation in this type of cable is not subjected to high voltage, thin layers of silk and cotton are used.

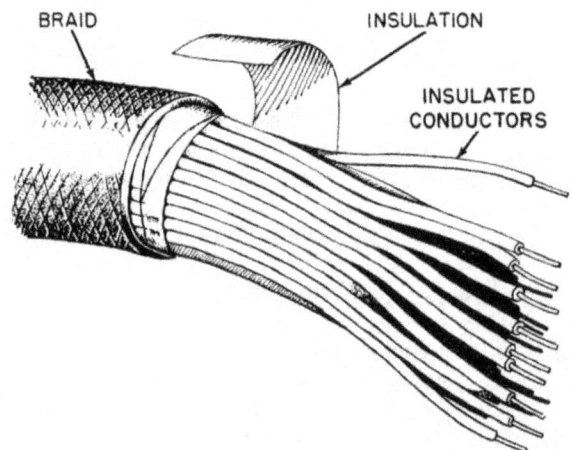

BE.135

Figure 7-11.—Silk and cotton insulation.

## ENAMEL

The wire used on the coils of meters, relays, small transformers, and so forth, is called MAGNET WIRE. This wire is insulated with an enamel coating. The enamel is a synthetic compound of cellulose acetate (wood pulp and magnesium). In the manufacture, the bare wire is passed through a solution of the hot enamel and then cooled. This process is repeated until the wire acquires from 6 to 10 coatings. Enamel has a higher dielectric strength than rubber for equal thickness. It is not practical for large wires because of the expense and because the insulation is readily fractured when large wires are bent.

Figure 7-12 shows an enamel-coated wire. Enamel is the thinnest insulating coating that can be applied to wires. Hence, enamel-insulated magnet wire makes smaller coils. Enameled wire is sometimes covered with one or more layers of cotton covering to protect the enamel from nicks, cuts, or abrasions.

### CONDUCTOR PROTECTION

Wires and cables are generally subject to abuse. The type and amount of abuse depends on how and where they are installed and the manner in which they are used. Cables buried directly in the ground must resist moisture, chemical action, and abrasion. Wires installed in buildings must be protected against mechanical injury and overloading. Wires strung on crossarms on poles are kept far enough apart so that they do not touch; but snow, ice, and strong winds necessitate the use of conductors having high tensile strength and substantial supporting frame structures.

Generally, except for overload transmission lines, wires or cables are protected by some form of covering. The covering may be some type of insulator like rubber or plastic. Over this an outer covering of fibrous braid may be applied. If conditions require, a metallic outer covering may be used. The type of outer covering used depends on how and where the wire or cable is to be used.

### FIBROUS BRAID

Cotton, linen, silk, rayon, and jute are types of FIBROUS BRAIDS. They are used for outer covering under conditions where the wires or cables are not exposed to heavy mechanical injury. Interior wiring for lights or power is usually done with impregnated cotton, braid-covered, rubber-insulated wire. Generally, the wire will be further protected by a flame-resistant nonmetallic outer covering or by a flexible or rigid conduit.

Figure 7-13 shows a typical building wire. In this instance two braid coverings are used for extra protection. The outer braid is soaked with a compound that resists moisture and flame.

Impregnated cotton braid is used as a covering for outdoor overhead conductors to afford protection against abrasion. For example,

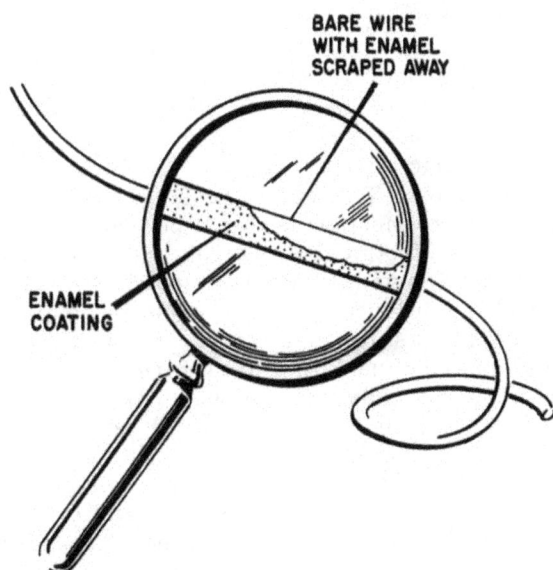

BE.136

Figure 7-12.—Enamel insulation.

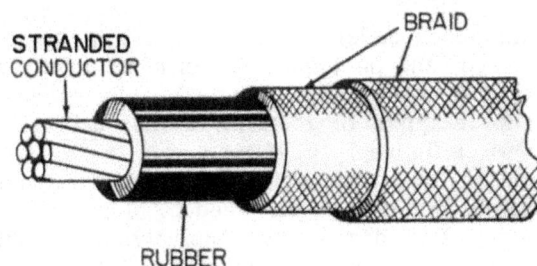

BE.137

Figure 7-13.—Fibrous braid covering.

the service wires from the transformer secondary mains to the service entrance and also the high-voltage primary mains to the transformer are protected in this manner.

## LEAD SHEATH

Subway-type cables or wires that are continually subjected to water must be protected by a watertight cover. This watertight cover is made either of a continuous lead jacket or a rubber sheath molded around the cable.

Figure 7-14 is an example of a lead-sheathed cable used in power work. The cable shown is a stranded 3-conductor type. Each conductor is insulated and then wrapped with a layer of rubberized tape. The conductors are twisted together and fillers or rope are added to form a rounded core. Over this is wrapped a second layer of tape called the SERVING, and finally the lead sheath is molded around the cable.

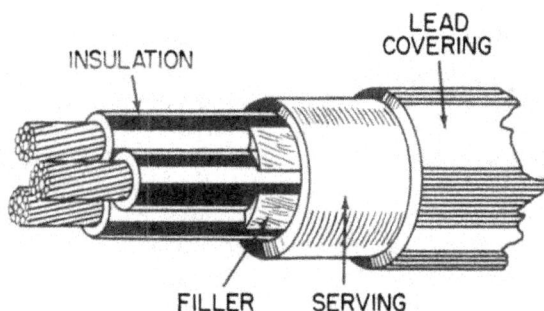

BE.138
Figure 7-14.—Lead sheathed cable.

## METALLIC ARMOR

Metallic armor provides a tough protective covering for wires or cables. The type, thickness, and kind of metal used to make the armor depend on the use of the conductors, the circumstances under which the conductors are to be used, and on the amount of rough treatment that is to be expected.

Four types of metallic armor cables are shown in figure 7-15.

Wire braid armor is used wherever light, flexible protection is needed. This type of armor is used almost exclusively aboard ship. The individual wires that are woven together to form the metal braid may be made of steel, copper, bronze, or aluminum. Besides mechanical protection, the wire braid also presents a

static shield. This is important in radio work aboard ship to prevent interference from stray fields.

When cables are buried directly in the ground, they might be injured from two sources—moisture, and abrasion. They are protected from moisture by a lead sheath, and from abrasion by steel tape or interlocking armor covers. The steel tape covering, as shown in figure 7-15, is wrapped around the cable and then covered with a serving of jute. It is known as Parkway cable. The interlocking armor covering can withstand impacts better than steel tape. Interlocking armor has other uses besides underground work. In wiring the interior of buildings, interlocking armor-covered wire (BX cable) without the lead sheath is frequently used.

Armor wire is the best type of covering to withstand severe wear and tear. Underwater leaded cable usually has an outer armor wire cover.

All wires and cables do not have the same type of protective covering. Some coverings are designed to withstand moisture, others to withstand mechanical strain, and so forth. A cable may have a combination of each type, each doing its own job.

## CONDUCTOR SPLICES AND TERMINAL CONNECTIONS

Conductor splices and connections are an essential part of any electric circuit. When conductors join each other, or connect to a load, splices or terminals must be used. It is important that they be properly made, since any electric circuit is only as good as its weakest link. The basic requirement of any splice or connection is that it be both mechanically and electrically as strong as the conductor or device with which it is used. High-quality workmanship and materials must be employed to insure lasting electrical contact, physical strength, and insulation (if required). The most common methods of making splices and connections in electric cables will now be discussed.

The first step in making a splice is preparing the wires or conductor. Insulation must be removed from the end of the conductor and the exposed metal cleaned. In removing the insulation from the wire, a sharp knife is used in much the same manner as in sharpening a pencil. That is, the knife blade is moved at a

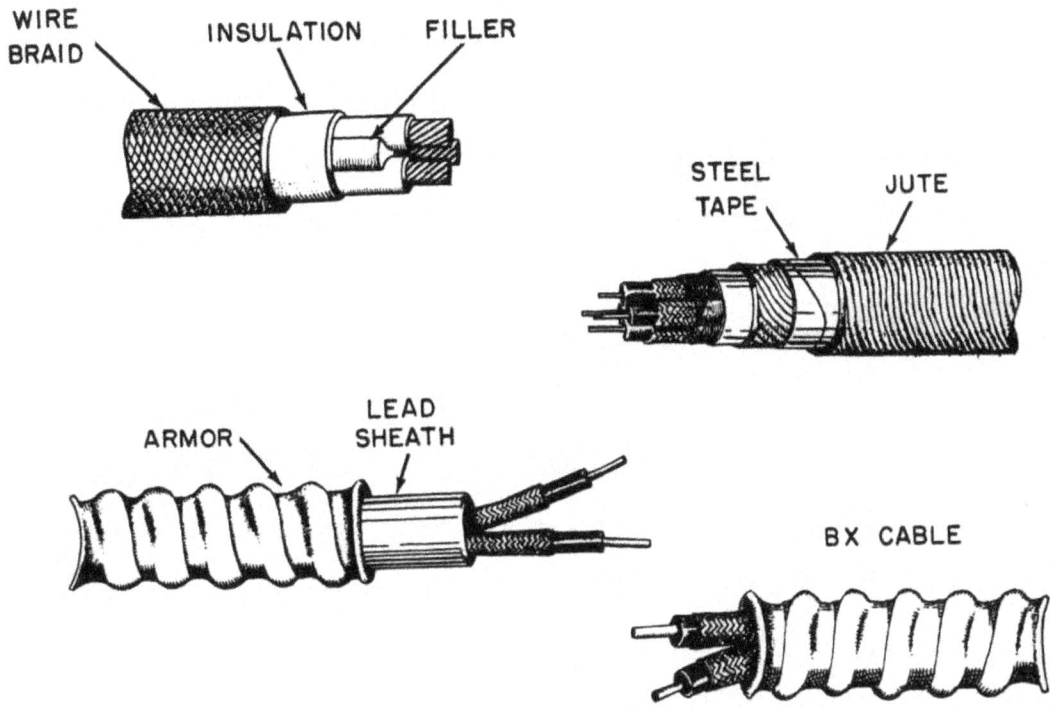

Figure 7-15.—Metallic armor.

small angle with the wire to avoid "nicking" the wire. This produces a taper on the cut insulation, as shown in figure 7-16. The insulation may also be removed by using a plier-like hand-operated wire stripper. After the insulation is removed, the bare wire ends should then be scraped bright with the back of a knife blade or rubbed clean with fine sandpaper.

WESTERN UNION SPLICE

Small, solid conductors may be joined together by a simple connection known as the WESTERN UNION SPLICE. In most instances the wires may be twisted together with the fingers and the ends clamped into position with a pair of pliers.

Figure 7-17 shows the steps in making a Western Union splice. First, the wires are prepared for splicing by removing sufficient insulation and cleaning the conductor. Next, the wires are brought to a crossed position and a long twist or bend is made in each wire. Then one of the wire ends is wrapped four of five times around the straight portion of the wire. The other end wire is wrapped in a similar manner. Finally, the ends of the wires should

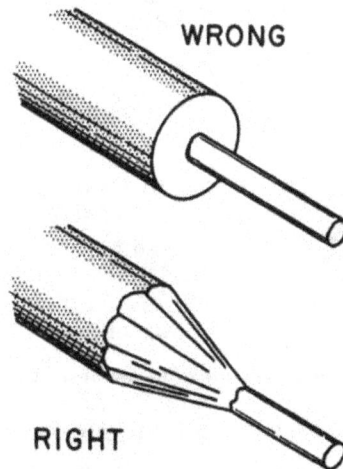

Figure 7-16.—Removing insulation from a wire.

be pressed down as close as possible to the straight portion of the wire this prevents the sharp ends from puncturing the tape covering that is wrapped over the splice.

136

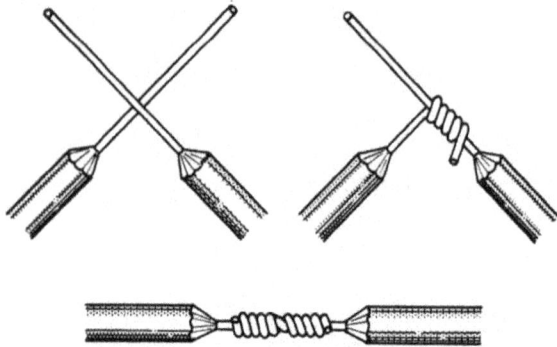

BE.141

Figure 7-17.—Western Union splice.

## STAGGERED SPLICE

Joining small, multiconductor cables together presents somewhat of a problem. Each conductor must be spliced and taped; and if the splices are directly opposite each other, the overall size of the joint becomes large and bulky. A smoother and less bulky joint may be made by staggering the splices.

Figure 7-18 shows how a 2-conductor cable is joined to a similar cable by means of the staggered splice. Care should be exercised to ensure that a short wire is connected to a long wire, and that the sharp ends are clamped firmly down on the conductor.

## RATTAIL JOINT

Wiring that is installed in buildings is usually placed inside long lengths of steel pipe (conduit). Whenever branch circuits are required, junction or pull boxes are inserted in the conduit. One type of splice that is used for branch circuits is the rettail joint shown in figure 7-19.

The ends of the conductors to be joined are stripped of insulation. The wires are then twisted to form the rattail effect.

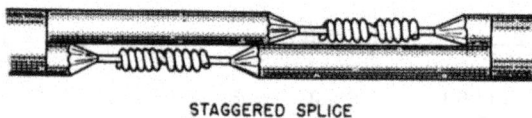

STAGGERED SPLICE

BE.142

Figure 7-18.—Staggered splice.

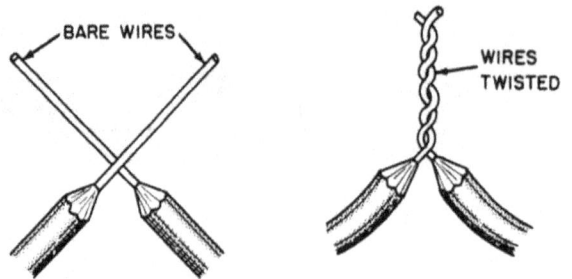

BE.143

Figure 7-19.—Rattail joint.

## FIXTURE JOINT

A fixture joint is used to connect a light fixture to the branch circuit of an electrical system where the fixture wire is smaller in diameter than the branch wire. Like the rattail joint, it will not stand much mechanical strain.

The first step is to remove the insulation from the wires to be joined. Figure 7-20 shows the steps in making a fixture joint.

After the wires are prepared, the fixture wire is wrapped a few times around the branch wire, as shown in the figure. The wires are not twisted, as in the rattail joint. The end of the branch wire is then bent over the completed turns. The remainder of the bare fixture wire

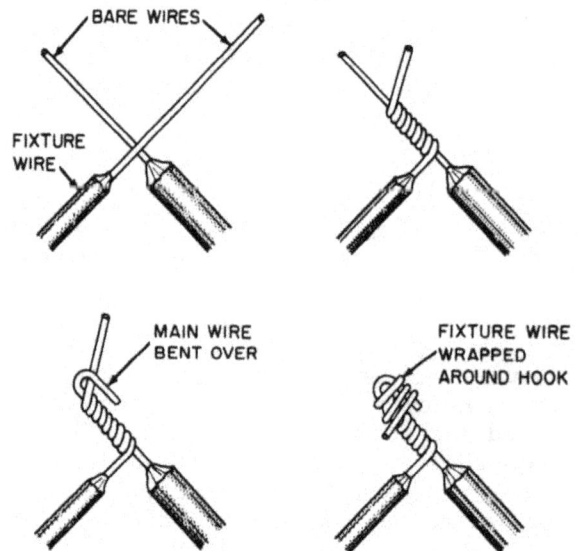

BE.144

Figure 7-20.—Fixture joint.

137

is then wrapped over the bent branch wire. Soldering and taping completes the job.

## KNOTTED TAP JOINT

All of the splices considered up to this point are known as BUTTED splices. Each was made by joining the FREE ends of the conductors together. Sometimes, however, it is necessary to join a conductor to a CONTINUOUS wire, and such a junction is called a TAP joint.

The main wire, to which the branch wire is to be tapped, has about one inch of insulation removed. The branch wire is stripped of about three inches of insulation. The steps in making the tap are shown in figure 7-21.

The branch wire is crossed over the main wire, as shown in figure, with about three-fourths of the bare portion of the branch wire extending above the main wire. The end of the branch wire is bent over the main wire, brought under the main wire, around the branch wire, and then over the main wire to form a knot. It is then wrapped around the main conductor in short, tight turns and the end is trimmed off.

The knotted tap is used where the splice is subject to strain or slip. When there is no mechanical strain, the knot may be eliminated.

### SOLDERING EQUIPMENT

Soldering operations are a vital part of electrical/electronics maintenance procedures. It is a manual skill which can and must be learned by all personnel who work in the field of electricity. Practice is required to develop proficiency in the techniques of soldering; however, practice serves no useful purpose unless it is founded on a thorough understanding of basic principles. This discussion is devoted to providing information regarding some important aspects of soldering operations.

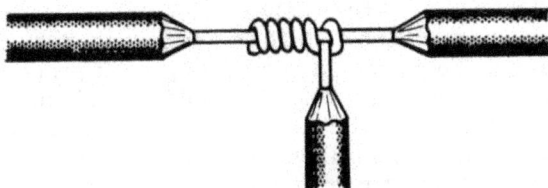

**KNOTTED TAP JOINT**

BE.145

Figure 7-21.—Knotted tap joint.

Both the solder and the material to be soldered must be heated to a temperature which allows the solder to flow. If either is heated inadequately, "cold" solder joints result. Such joints do not provide either the physical strength or the electrical conductivity required. Appreciably exceeding the flow point temperature, however, is likely to cause damage to the parts being soldered. Various types of solder flow at different temperatures. In soldering operations it is necessary to select a solder that will flow at a temperature low enough to avoid damage to the part being soldered, or to any other part or material in the immediate vicinity.

The duration of high heat conditions is almost as important as the temperature. Insulation and many other materials in electrical equipment are susceptible to damage from heat. They are damaged if exposed to excessively high temperatures, or deteriorate if exposed to less drastically elevated temperatures for prolonged periods. The time and temperature limitations depend on many factors—the kind and amount of metal involved, the degree of cleanliness, the ability of the material to withstand heat, and the heat transfer and dissipation characteristics of the surroundings.

### SOLDER

The three grades of solder generally used for electrical work are 40-60, 50-50, and 60-40 solder. The first figure is the percentage of tin, while the other is the percentage of lead. The higher the percentage of tin content, the lower the temperature required for melting. Also, the higher the tin content, the easier the flow, less the time required to harden, and generally the easier it is to do a good soldering job.

In addition to the solder, there must be flux to remove any oxide film on the metals being joined, otherwise they cannot fuse. The flux enables the molten solder to wet the metals so the solder can stick. The two types of flux are acid flux and rosin flux. Acid flux is more active in cleaning metals but is corrosive. Rosin is always used for the light soldering work in making wire connections. Generally, the rosin is in the hollow core of solder intended for electrical work, so that a separate flux is unnecessary. Such rosin-core solder is the type generally used. It should be noted, though, that the flux is not a substitute for cleaning the metals to be soldered. The metal must be shiny clean for the solder to stick.

## SOLDERING PROCESS

Cleanliness is a prime perequisite for efficient, effective soldering. Solder will not adhere to dirty, greasy, or oxidized surfaces. Heated metals tend to oxidize rapidly, and the oxide must be removed prior to soldering. Oxides, scale, and dirt can be removed by mechanical means (such as scraping or cutting with an abrasive) or by chemical means. Grease or oil films can be removed by a suitable solvent. Cleaning should be accomplished immediately prior to the actual soldering operation.

Items to be soldered should normally be tinned before making mechanical connection. When the surface has been properly cleaned, a thin, even coating of flux may be placed over the surface to be tinned to prevent oxidation while the part is being heated to soldering temperature. Rosin core solder is usually preferred in electrical work, but a separate rosin flux may be used instead. Separate rosin flux is frequently used when tinning wires in cable fabrication. Tinning is the coating of the material to be soldered with a light coat of solder.

The tinning on a wire should extend only far enough to take advantage of the depth of the terminal or receptacle. Tinning or solder on wires subject to flexing causes stiffness, and may result in breakage.

The tinned surfaces to be joined should be shaped and fitted, then mechanically joined to make good mechanical and electrical contact. They must be held still with no relative movement of the parts. Any motion between parts will likely result in a poor solder connection.

## SOLDERING TOOLS

### Soldering Irons

All high quality irons operate in the temperature range of 500° to 600° F. Even the little 25-watt midget irons produce this temperature. The important difference in iron sizes is not temperature, but thermal inertia (the capacity of the iron to generate and maintain a satisfactory soldering temperature while giving up heat to the joint to be soldered). Although it is not practical to try to solder a heavy metal box with the 25-watt iron, that iron is quite suitable for replacing a half-watt resistor in a printed circuit. An iron with a rating as large as 150 watts would be satisfactory for use on a printed circuit, provided that suitable soldering techniques are used. One advantage of using a small iron for small work is that it is light and easy to handle and has a small tip which is easily inserted into close places. Also, even though its temperature is high, it does not have the capacity to transfer large quantities of heat.

Some irons have built-in thermostats. Others are provided with thermostatically controlled stands. These devices control the temperature of the soldering iron, but are a source of trouble. A well-designed iron is self-regulating by virtue of the fact that the resistance of its element increases with rising temperature, thus limiting the flow of current. For critical work, it is convenient to have a variable transformer for fine adjustment of heat; but for general-purpose work, no temperature regulation is needed.

### Soldering Gun

The soldering gun has gained great popularity in recent years because it heats and cools rapidly. It is especially well adapted to maintenance and troubleshooting work where only a small part of the technician's time is spent actually soldering. A soldering iron, if kept hot constantly, oxidizes rapidly and is therefore difficult to keep clean.

A transformer in the soldering gun supplies approximately 1 volt at high current to a loop of copper which acts as the tip. It heats to soldering temperature in 3 to 5 seconds, but may overheat to the point of incandescence if left on over 30 seconds. The gun is operated with a finger switch so that the gun heats only while the switch is depressed.

Since the gun normally operates only for short periods at a time, it is comparatively easy to keep clean and well tinned; thus, little oxidation is allowed to form. However, the tip is made of pure copper, and is susceptible to pitting which results from the dissolving action of the solder.

Tinning of the tip is always desirable unless it has already been done. The gun or iron should always be kept tinned in order to permit proper heat transfer to the work to be soldered. Tinning also provides adequate control of the heat to prevent thermal spillover to nearby materials. Tinning of the tip of a gun may be somewhat more difficult than tinning the tip of an iron. Maintaining the proper tining on either type, however, may be made easier by tinning with silver solder. The temperature at which the bond is formed between the copper tip and the silver solder is considerably higher than with lead-tin solder. This tends to decrease the pitting action of the solder on the copper tip.

Pitting of the tip indicates the need for retinning, after first filing away a portion of the tip. Retinning too often results in using up the tip too fast.

Overheating can easily occur when using the gun to solder delicate wiring. With practice, however, the heat can be accurately controlled by pulsing the gun on and off with its trigger. For most jobs, even the LOW position of the trigger overheats the soldering gun after 10 seconds; the HIGH position is used only for fast heating and for soldering heavy connections.

Heating and cooling cycles tend to loosen the nuts or screws which retain the replaceable tips on soldering irons or guns. When the nut on a gun is loosened, the resistance of the tip connection increases, and the temperature of the connection is increased. Continued loosening may eventually cause an open circuit. Therefore, the nut should be tightened periodically.

### Resistance Soldering

A time-controlled resistance soldering set is now available. The set consists of a transformer that supplies 3 or 6 volts at high current to stainless steel or carbon tips. The transformer is turned ON by a foot switch and OFF by an electronic timer. The timer can be adjusted for as long as 3 seconds soldering time. This set is especially useful for soldering cables to plugs and similar connectors—even the smallest types available.

In use, the double-tip probes of the soldering unit are adjusted to straddle the connector cup to be soldered. One pulse of current heats it for tinning and, after the wire is inserted, a second pulse of current completes the job. Since the soldering tips are hot only during the brief period of actual soldering, burning of wire insulation and melting of connector inserts are greatly minimized.

The greatest difficulty with this device is keeping the probe tips free of rosin and corrosion. A cleaning block is mounted on the transformer case for this purpose. Some technicians prefer fine sandpaper for cleaning the double tips. CAUTION: Do not use steel wool. It is dangerous when used around electrical equipment.

### Pencil Iron and Special Tips

An almost indispensable item is the pencil type soldering iron with an assortment of tips (fig. 7-22). Miniature soldering irons, with wattage ratings of less than 40 watts are easy to used and are recommended. In an emergency, larger irons can be converted and used on subminiature equipment as described later in this section.

One type of iron is equipped with several different tips that range from one-fourth to one-half inch in size (diameter) and are of various shapes. This feature makes it adaptable to a variety of jobs. Unlike most tips which are held in place by setscrews, these tips have threads and screw into the barrel. This feature provides excellent contact with the heating element, thus improving heat transfer efficiency. A pad of "antiseize" compound is supplied with each iron. This compound is applied to the threads each time a tip is installed in the iron, thereby enabling the tip to be easily removed when another is to be inserted.

A special feature of this iron is the soldering pot that screws in like a tip and holds about a thimbleful of solder. It is useful for tinning the ends of large numbers of wires.

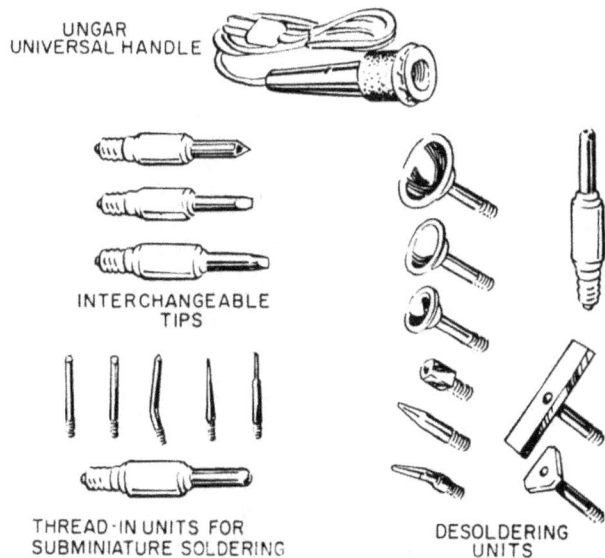

BE.146
Figure 7-22.—Pencil iron kit with special tips.

The interchangeable tips are of various sizes and shapes for specific applications. Extra tips may be obtained and shaped to serve special purposes. The thread-in units are useful in soldering subminiature items. The desoldering units are specifically designed for performing special and individual functions.

Another advantage of the pencil soldering iron is its possible use as an improvised light source for inspections. Simply remove the soldering tip and insert a 120-volt, 6-watt, type 6S6, candelabra screwbase lamp bulb into the socket.

If leads, tabs, or small wires are bent against a board or terminal, slotted tips may be used to simultaneously melt the solder and straighten the leads.

A hollow tip, which fits over a pin terminal, may be used to desolder and resolder wiring at cables or feed-through terminals.

Many miniature components have multiple connections, all of which must be desoldered to permit removal of the component in one operation. These connections may be desoldered individually by heating each connection and brushing away the solder. With this method, particular care must be taken to insure that loose solder does not stick to other parts of become lodged where it may cause a short circuit. A more efficient method is to use the specially shaped desoldering units (fig. 7-22). Select the proper size and shape tip that will contact all terminals to be desoldered—and nothing else. Do not permit the tip to remain in contact with the terminals too long at one time.

If no suitable tip is available for a particular operation, an improvised tip may be made. Wrap a length of copper wire around one of the regular tips and bend the wire into the proper shape for the purpose. This method also serves to reduce tip temperature when a larger iron must be used on miniature components. (See fig. 7-23.)

In connection with the discussion of soldering tools and devices, the selection of solder and flux is also critical. A small diameter rosin core solder with a high tin-lead ratio (60/40) is normally preferred in miniature circuits where heat is critical.

### Soldering Aids

Several devices other than the soldering iron and its tips are required in soldering miniature circuits. Several of these (brushes, probes, scrapers, knives, etc.) have been mentioned previously.

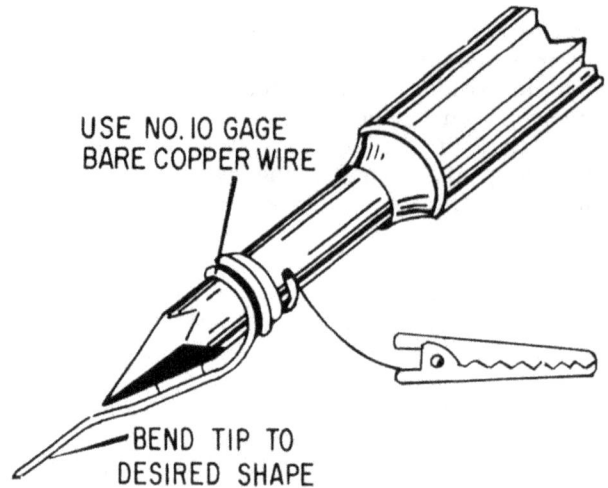

USE NO. 10 GAGE BARE COPPER WIRE

BEND TIP TO DESIRED SHAPE

AT.74

Figure 7-23.—Improvised tip to reduce tip temperature.

Some type of thermal shunt is essential in all soldering operations which involve heat-sensitive components. Pliers, tweezers, or hemostats may be used for some applications, but their effectiveness is limited. A superior heat shunt, as shown in figure 7-24, permits soldering the leads of component parts without overheating the part itself.

For maximum effectiveness, any protective coating should be removed before applying the heat shunt. The shunt should be attached carefully to prevent damage to the leads, terminals, or component parts. The shunt should be clipped to the lead, between the joint and the part being protected. As the joint is heated, the shunt absorbs the excess heat before it can reach the part and cause damage.

A small piece of beeswax may be placed between the protected unit and the heat shunt. When the beeswax begins to melt, the temperature limit has been reached. The heat source

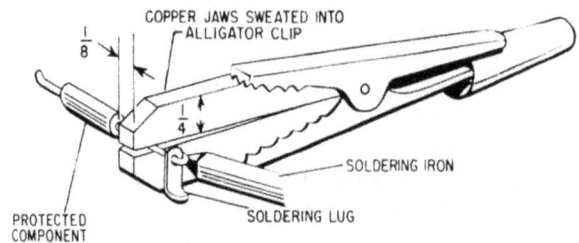

COPPER JAWS SWEATED INTO ALLIGATOR CLIP

$\frac{1}{8}$

$\frac{1}{4}$

SOLDERING IRON

PROTECTED COMPONENT

SOLDERING LUG

AE.47

Figure 7-24.—Heat shunt.

should be removed immediately, but the shunt should be left in place.

Premature removal of the heat shunt permits the unrestricted flow of heat from the melted solder into the component. The shunt should be allowed to remain in place until it cools to room temperature. A clip-on type shunt is preferred because it requires positive action to remove the shunt, but does not require that the technician maintain pressure to hold it in place.

Another invaluable soldering aid is the "solder sucker" syringe. One type is shown in figure 7-25. Its purpose is to "suck up" excess solder (and incidentally the excess heat) from a joint. The only requirements of an efficient solder sucker are a controllable source of vacuum (squeeze bulb), a solder receiver, and a tip. The tip must be able to withstand the heat of molten solder. Teflon is ideal, but may be difficult to acquire. A silicon rubber-covered Fiberglas sleeving with an inner diameter of 0.162 inch and the bulb from a medicine dropper makes a suitable syringe. (The glass or plastic tip of the medicine dropper cannot withstand the heat.)

## SOLDER CONNECTIONS

Frequent arguments occur in electrical shops concerning the proper method of making soldered connections to terminals and binding posts. For

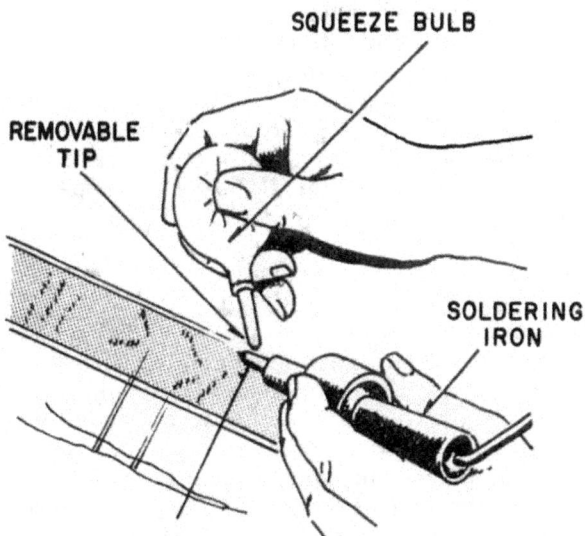

Figure 7-25.—Solder sucker.

AT.75

many years it was considered necessary to wrap the lead tightly around the terminal, so as to provide maximum mechanical support and strength. General Specification for Soldering Process states that the parts to be joined shall be held together in such a manner that the parts shall not move in relation to one another during the soldering process. The joint must not be disturbed until the solder has completely solidified.

Electronics Laboratories tested many standard capacitors and resistors soldered to terminals of various types. The joints were then subjected to vibrations far in excess of those normally encountered in electrical and electronic equipment. The connections were made with various degrees of wrapping around the terminals, with main reliance for physical strength being placed on the solder. As a result of these tests and others conducted by other organizations, the joints illustrated in figure 7-26 are recommended. Wrappings of three-eighths to three-fourths turn are usually recommended so that the joint need not be held during the application and cooling of the solder.

Excessive wrappings of leads results in increased heat requirements, more strain on parts, greater difficulty of inspection, greater difficulty of assembly and disassembly of the joints, and increased danger of breaking the parts or terminals during desoldering operations. Insufficient wrapping may result in poor solder joints due to movement of the lead during the soldering operation.

The areas to be joined must be heated to or slightly above the flow temperature of the solder. The application of heat must be carefully controlled to prevent damage to components of the assembly, insulation, or nearby materials. Solder is then applied to the heated area. Only enough solder should be used to make a satisfactory joint. Heavy fillets or beads must be avoided.

Solder should not be melted with the soldering tip and allowed to flow onto the joint. The joint should be heated and the solder applied to the joint. When the joint is adequately heated, the solder will flow evenly. Excessive temperature tends to carbonize flux, thus hindering the soldering operation.

No liquid should be used to cool a solder joint. By using the proper tools and soldering technique, a joint should not become so hot that rapid cooling is needed.

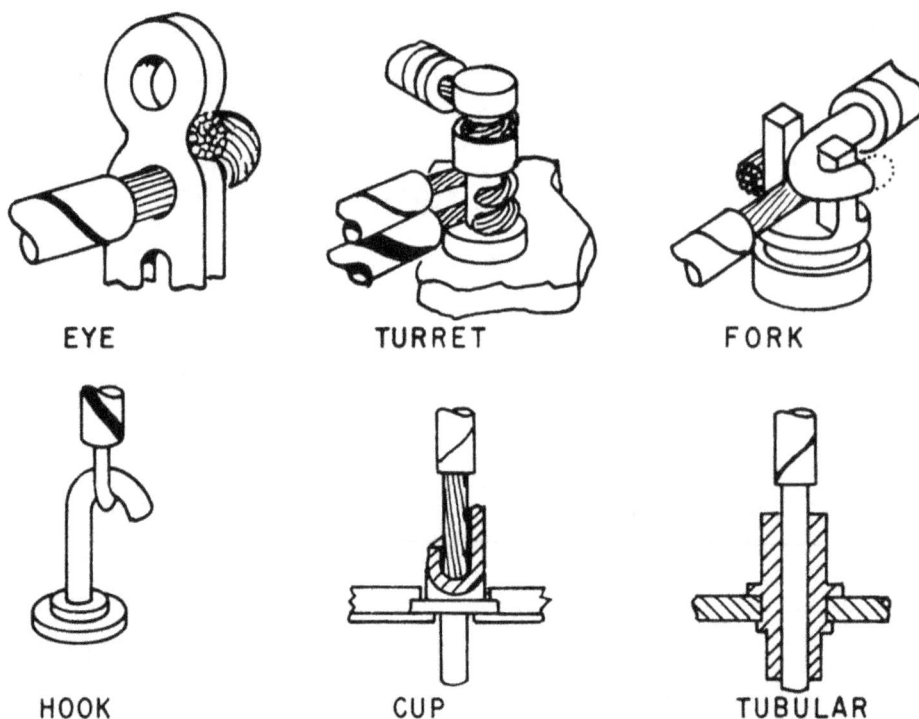

EYE          TURRET          FORK

HOOK          CUP          TUBULAR

AT.262

Figure 7-26.—Wrapping of terminals for soldering.

If, for any reason, a satisfactory joint is not initially obtained, the joint must be taken apart, the surfaces cleaned, excess solder removed, and the entire soldering operation (except tinning) repeated.

After the joint has cooled, all flux residues should be removed. Any flux residue remaining on the surface of electrical contacts may collect dirt and promote arcing at a later time. This cleaning is necessary even when rosin-core solder is used.

Connections should never be soldered or desoldered while equipment power is on or while the circuit is under test. Always discharge any capacitors in the circuit prior to any soldering operation.

SOLDER SPLICERS

The solder-type splicer is essentially a short piece of metal tube. Its inside diameter is just large enough to allow the tip of a stranded conductor to be inserted in either end, after the conductor tip has been stripped of insulation. This type of splicer is shown in figure 7-27.

The splicer is first heated and filled with solder. While still molten, the solder is then poured out, leaving the inner surfaces tinned. When the conductor tips are stripped, the length of exposed strands should be long enough so that the insulation butts against the splicer when the conductors are tinned and fully inserted. (See fig. 7-27 (B).) When heat is applied to the connection and the solder melts, excess solder will be squeezed out through the vents. This must be cleaned away. After the splice has cooled, insulating material must be wrapped or tied over the joint.

SOLDER TERMINAL LUGS

In addition to being joined or spliced to one another, conductors are often connected to other objects, such as motors and switches. Since this is where a length of conductor ends (terminates), such connections are referred to as terminal points. In some cases, it is allowable to bend the end of the conductor into a small "eye" and put it around a terminal binding post. Where a mounting screw is used, the screw is passed through the eye. The conductor tip which forms the eye should be bent as

143

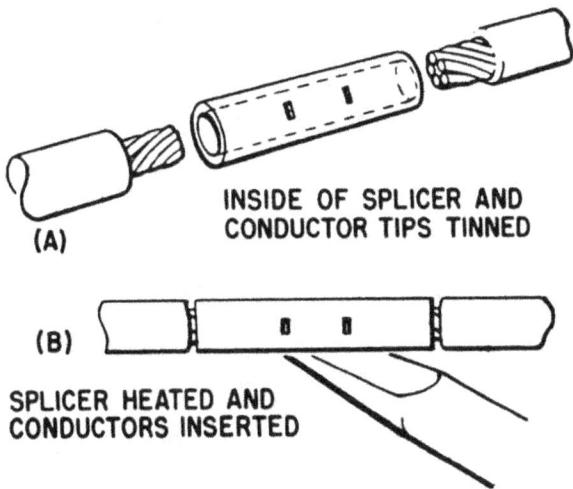

INSIDE OF SPLICER AND
CONDUCTOR TIPS TINNED

(A)

(B)

SPLICER HEATED AND
CONDUCTORS INSERTED

BE. 147

Figure 7-27.—Steps in using solder splicer.

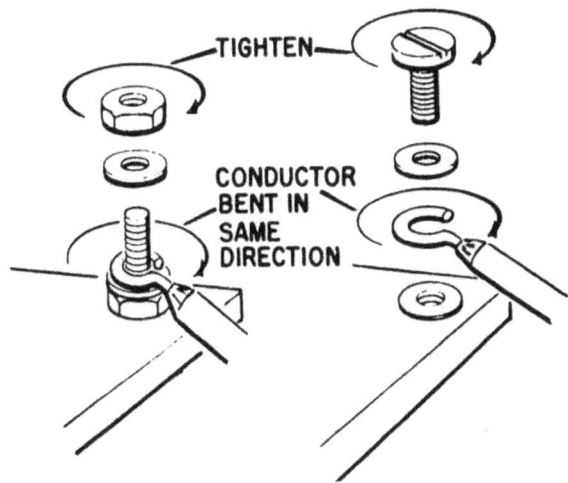

TIGHTEN

CONDUCTOR
BENT IN
SAME
DIRECTION

BE. 148

Figure 7-28.—Conductor terminal connection.

shown in figure 7-28. Note that when the screw or binding nut is tightened, it also tends to tighten the conductor eye.

This method of connection is sometimes not desirable. When design requirements are more rigid, terminal connections are made by using special hardware devices called terminal lugs. There are terminal lugs of many different sizes and shapes, but all are essentially the same as the type shown in figure 7-29.

Each type of lug has a barrel (sleeve) which is wedged, crimped, or soldered to its conductor. There is also a tongue with a hold or slot in it to receive the terminal post or screw. When mounting a solder-type terminal lug to a conductor, first tin the inside of the barrel. The conductor tip is stripped and also tinned, then inserted in the preheated lug. When mounted, the conductor insulation should butt against the lug barrel, so that there is no exposed conductor.

## SOLDERLESS CONNECTORS

Splicers and terminal lugs which do not require solder are more widely used than those which do require solder. Solderless connectors are attached to their conductors by means of several different devices, but the principle of each is essentially the same. They are all squeezed (crimped) tightly onto their conductors. They afford adequate electrical contact, plus great mechanical strength. In addition, solder-

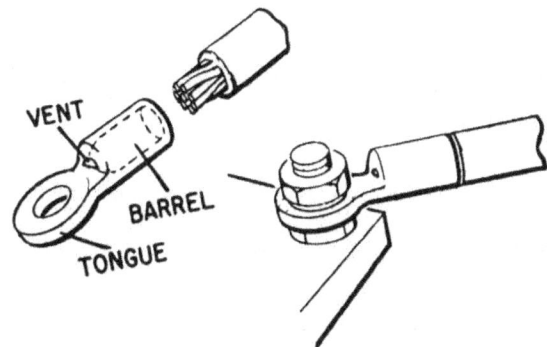

VENT

BARREL

TONGUE

BE. 149

Figure 7-29.—Solder-type terminal lug.

less connectors are easier to mount correctly because they are free from the most common problems of solder connector mounting; namely, cold solder joints, burned insulation, and so forth.

Solderless connectors are made in a great variety of sizes and shapes, and for many different purposes. Only a few are discussed here.

144

## SOLDERLESS SPLICERS

Three of the most common types solderless splicers, classified according to their methods of mounting, are the split-sleeve, split-tapered-sleeve, and crimp-on splicers.

### Split-Sleeve Splicer

A split-sleeve splicer is shown in figure 7-30. To connect this splicer to its conductor, the stripped conductor tip is first inserted between the split-sleeve jaws. Using a tool designed for that purpose, the slide ring is forced toward the end of the sleeve. The sleeve jaws are closed tightly on the conductor, and the slide ring holds them securely.

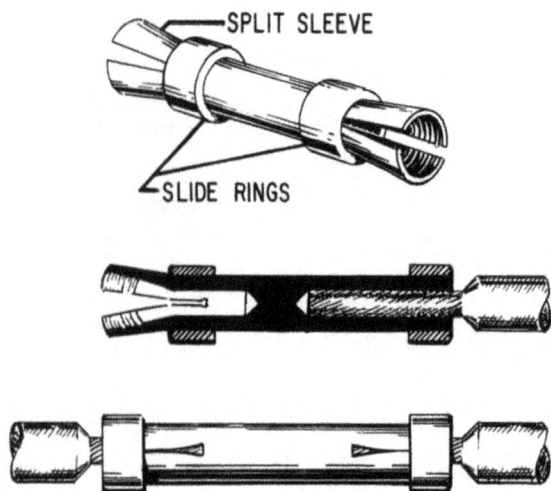

Figure 7-30.—Split-sleeve splicer.

BE. 150

### Split-Tapered-Sleeve Splicer

A cross-sectional view of a split-tapered-sleeve splicer is shown in figure 7-31 (A). To mount this type of splicer, the conductor is stripped and inserted in the split-tapered sleeve. The threaded sleeve is turned or screwed into the tapered bore of the body. As the sleeve is turned in, the split segments are squeezed tightly around the conductor by the narrowing bore. The finished splice (fig. 7-31 (B)), must be covered with insulation.

### Crimp-On Splicer

The crimp-on splicer (fig. 7-32) is the simplest of the splicers discussed. The type

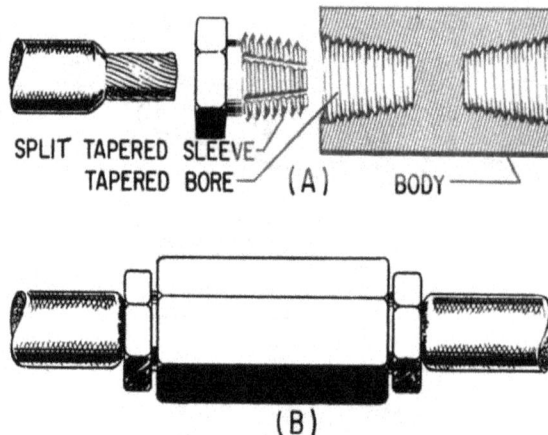

Figure 7-31.—Split-tapered-sleeve splice.

BE. 151

shown is preinsulated, though uninsulated types are manufactured. These splicers are mounted with a special plier-like hand-crimping tool designed for that purpose. The stripped conductor tips are inserted in the splicer, which is then squeezed tightly closed. The insulating sleeve grips the outer insulated conductor, and the metallic internal splicer grips the bare conductor strands.

## SOLDERLESS TERMINAL LUGS

Solderless terminal lugs are used more widely than solder terminal lugs. They afford adequate electrical contact, plus great mechanical strength. In addition, solderless lugs are easier to attach correctly, because they are free from the most common problems of solder terminal lugs; namely, cold solder joints, burned insulation, and so forth. There are many sizes and shapes of these lugs, each intended for a different type of service of conductor size. Only a few are discussed here.

These are classified according to their method of mounting. They are the split-tapered-sleeve (wedge-type), split-tapered-sleeve (threaded-type), and crimp-on.

### Split-Tapered-Sleeve Terminal Lug (Wedge)

This type lug is shown in figure 7-33. It is commonly referred to as a "wedge-on," because of the manner in which it is secured to a

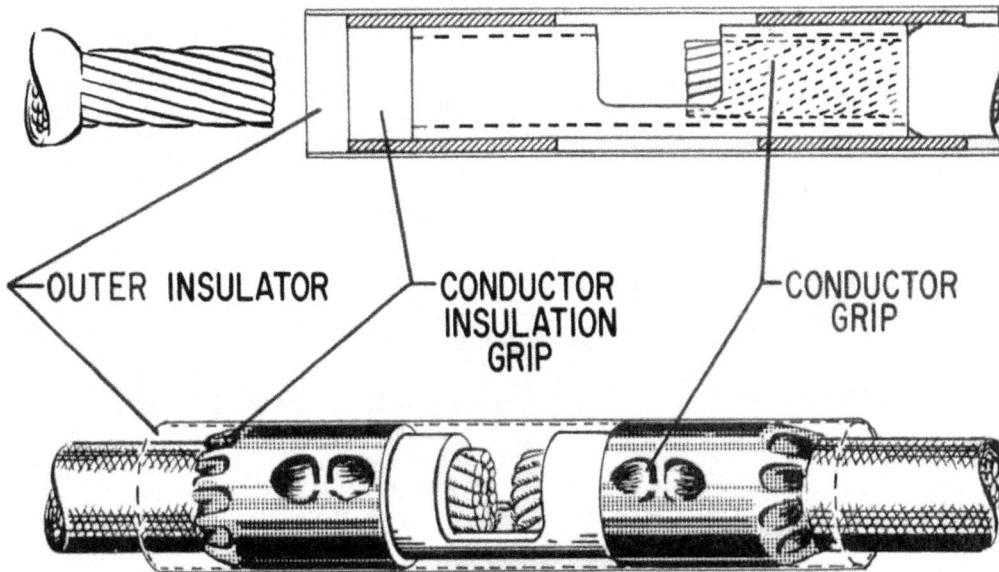

BE.152

Figure 7-32.—Crimp-on splicer.

conductor. The stripped conductor is inserted through the hole in the split sleeve. When the sleeve is forced or "wedged" down into the barrel, its tapered segments are squeezed tightly around the conductor.

Split-Tapered-Sleeve
Terminal Lug (Threaded)

This lug (fig. 7-34) is attached to a conductor in exactly the same manner as a split-sleeve splicer. The segments of the threaded split sleeve squeezes tightly around the conductor as it is turned into the tapered bore of the barrel. For this reason, the lug is commonly referred to as a "screw-wedge."

Crimp-On Terminal Lug

The crimp-on lug is shown in figure 7-35. This lug is simply squeezed or "crimped" tightly onto a conductor. This is done by using the same tool used with the crimp-on splicer. The lug shown is preinsulated, but uninsulated types are manufactured. When mounted, both the conductor and its insulation are gripped by the lug.

TAPING A SPLICE

The final step in completing a splice or joint is the placing of insulation over the bare

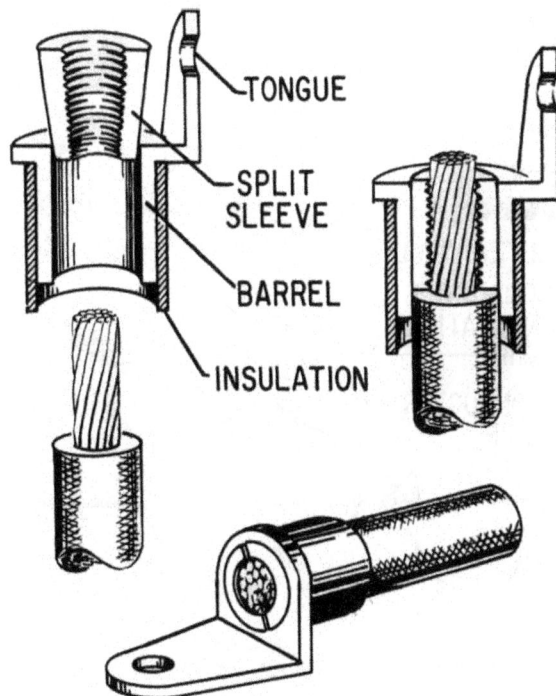

BE. 153

Figure 7-33.—Split-tapered-sleeve terminal lug (wedge type).

146

## RUBBER TAPE

Latex (rubber) tape is a splicing compound. It is used where the original insulation was rubber. The tape is applied to the splice with a light tension so that each layer presses tightly against the one underneath it. This pressure causes the rubber tape to blend into a solid mass. When the application is completed, an insulation similar to the original has been restored.

Between each layer of latex tape, when it is in roll form, there is a layer of paper or treated cloth. This layer prevents the latex from fusing while still on the roll. The paper or cloth is peeled off and discarded before the tape is applied to the splice.

Figure 7-36 shows the correct way to cover a splice with rubber insulation. The rubber splicing tape should be applied smoothly and under tension so that there will be no air spaces between the layers. In putting on the first layer, start near the middle of the joint instead of the end. The diameter of the completed insulated joint should be somewhat greater that the overall diameter of the original cable, including the insulation.

## FRICTION TAPE

Putting rubber tape over the splice means that the insulation has been restored to a great degree. It is also necessary to restore the protective covering. Friction tape is used for this purpose; it also affords a minor degree of electrical insulation.

Friction tape is a cotton cloth that has been treated with a sticky rubber compound. It comes in rolls similar to rubber tape except that no paper or cloth separator is used. Friction tape is applied like rubber tape; however, it does not stretch.

The friction tape should be started slightly back on the original braid covering. Wind the tape so that each turn overlaps the one before it; and extend the tape over onto the braid covering at the other end of the splice. From this point a second layer is wound back along the splice until the original starting point is reached. Cutting the tape and firmly pressing down the end complete the job. When proper care is taken, the splice can take as much abuse as the rest of the wire.

Weatherproof wire has no rubber insulation, just a braid covering. In that case, no rubber tape is necessary, only friction tape need be used.

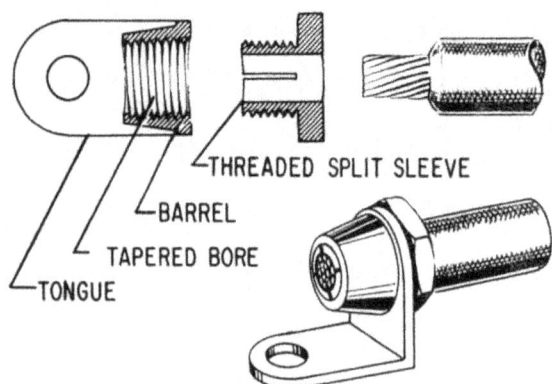

BE. 154

Figure 7-34.—Split-tapered-sleeve terminal lug (threaded).

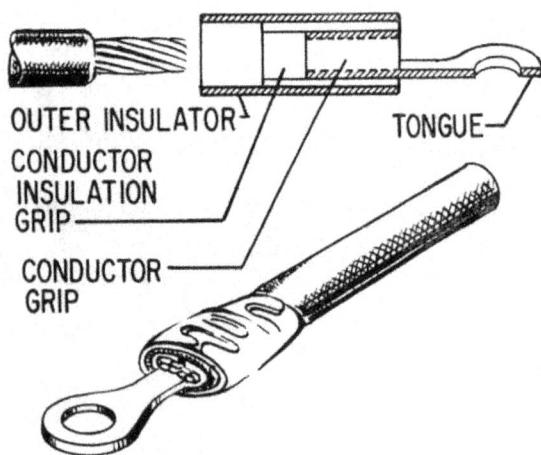

BE. 155

Figure 7-35.—Crimp-on terminal lug.

wire. The insulation should be of the same basic substance as the original insulation. Usually a rubber splicing compound is used.

## PLASTIC ELECTRICAL TAPE

Plastic electrical tape has come into wide use in recent years. It has certain advantages over rubber and friction tape. For example, it will withstand higher voltages for a given thickness. Single thin layers of certain commercially available plastic tape will stand several thousand volts without breaking down. However, to provide an extra margin of safety, several layers are usually wound over the splice. Because the tape is very thin, the extra layers add only a very small amount of bulk; but at the same time the added protection, normally furnished by friction tape, is provided by the additional layers of plastic tape. In the choice of plastic tape, the factor of expense must be balanced against the other factors involved.

Plastic electric tape normally has a certain amount of stretch so that it easily conforms to the contour of the splice without adding unnecessary bulk. The lack of bulkiness is especially important in some junction boxes where space is at a premium.

For high temperatures—for example, above 175° F.—a special type of tape backed with glass cloth is used.

BE. 156

Figure 7-36.—Applying rubber tape.

# CHAPTER 8

# ELECTROMAGNETISM AND MAGNETIC CIRCUITS

The fundamental theories concerning simple magnets and magnetism were discussed in chapter 2 of this manual. Those discussions dealt mainly with forms of magnetism that were not related directly to electricity—permanent magnets for instance. Only brief mention was made of those forms of magnetism having direct relation to electricity (such as "producing electricity with magnetism"). This chapter resumes the study of magnetism where chapter 2 left off. Therefore, it may be necessary for you to review parts of chapter 2 from time to time. This chapter begins the more advanced study of magnetism as it is affected by electric current flow, and the closely related study of electricity as it is affected by magnetism. This general subject area is most often referred to as electromagnetism.

Magnetism and basic electricity are so closely related that one cannot be studied at length without involving the other. This close fundamental relationship will be continually borne out in other chapters of this manual, such as in the study of generators, transformers, and motors. The technician, to be proficient in electricity, must become familiar with such general relationships that exist between magnetism and electricity as follows:

1. Electric current flow will always produce some form of magnetism.

2. Magnetism is by far the most commonly used means for producing or using electricity.

3. The peculiar behavior of electricity under certain conditions is caused by magnetic influences.

## MAGNETIC FIELD AROUND A CURRENT-CARRYING CONDUCTOR

In 1819 Hans Christian Oersted, a Danish physicist, found that a definite relation exists between magnetism and electricity. He discovered that an electric current is accompanied by certain magnetic effects and that these effects obey definite laws. If a compass is place in the vicinity of a current-carrying conductor, the needle alines itself at right angles to the conductor, thus indicating the presence of a magnetic force. The presence of this force can be demonstrated by passing an electric current through a vertical conductor which passes through a horizontal piece of cardboard, as illustrated in figure 8-1. The magnitude and direction of the force are determined by setting a compass at various points on the cardboard and noting the deflection.

The direction of the force is assumed to be the direction the north pole of the compass points. These deflections show that a magnetic field exists in circular form around the conductor. When the current flows upward, the field direction is clockwise, as viewed from the top, but if the polarity of the supply is reversed so that the current flows downward, the direction of the field is counterclockwise.

The relation between the direction of the magnetic lines of force around a conductor and the direction of current flow along the conductor may be determined by means of the left-hand rule for a conductor. If the conductor is grasped in the left hand with the thumb extended in the direction of electron flow ( - to +), the fingers will point in the direction of the magnetic lines of force. This is the same direction in which the north pole of a compass would point if the compass was placed in the magnetic field.

Arrows generally are used in electric diagrams to denote the direction of current flow along the length of wire. Where cross sections of wire are shown, a special view of the arrow is used. A cross-sectional view of a conductor

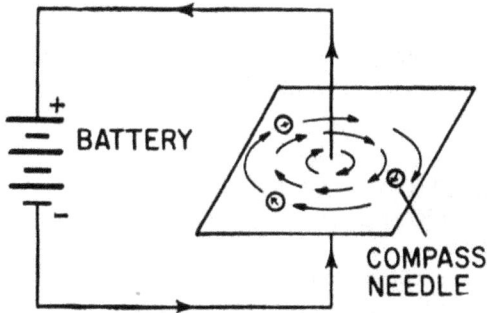

BE. 157

Figure 8-1.—Magnetic field around a
current-carrying conductor.

that is carrying current toward the observer is
illustrated in figure 8-2 (A). The direction of
current is indicated by a dot, which represents
the head of the arrow. A conductor that is
carrying current away from the observer is
illustrated in figure 8-2 (B). The direction of
current is indicated by a cross, which represents the tail of the arrow.

When two parallel conductors carry current
in the same direction, the magnetic fields tend
to encircle both conductors, drawing them together with a force of attraction, as shown in
figure 8-3 (A). Two parallel conductors carrying currents in opposite directions are shown in

figure 8-3 (B). The field around one conductor
is opposite in direction to the field around the
other conductor. The resulting lines of force
are crowded together in the space between the
wires, and tend to push the wires apart. Therefore, two parallel adjacent conductors carrying
currents in the same direction attract each other
and two parallel conductors carrying currents
in opposite directions repel each other.

MAGNETIC FIELD OF A COIL

The magnetic field around a current-carrying
wire exists at all points along its length. The
field consists of concentric circles in a plane
perpendicular to the wire. (See fig. 8-1.) When

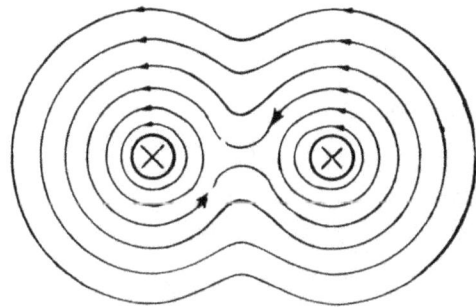

CURRENTS FLOWING IN
THE SAME DIRECTION
(A)

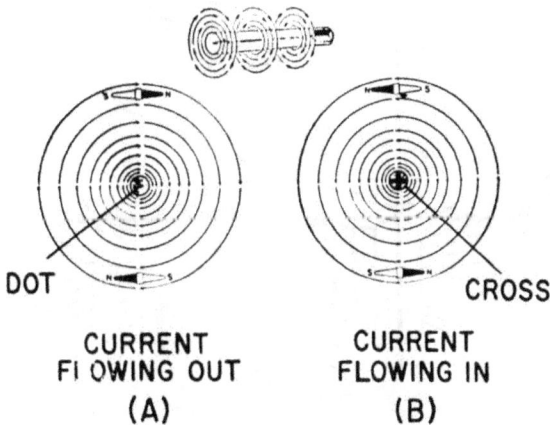

DOT                              CROSS

CURRENT                    CURRENT
FLOWING OUT             FLOWING IN
(A)                               (B)

BE. 158

Figure 8-2.—Magnetic field around a
current-carrying conductor, detailed view.

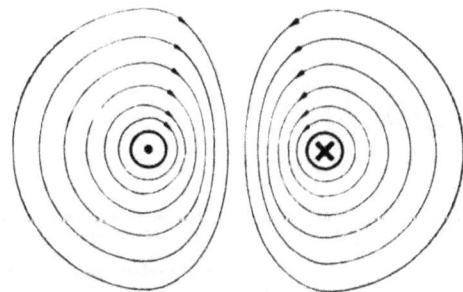

CURRENTS FLOWING IN
OPPOSITE DIRECTIONS
(B)

BE. 159

Figure 8-3.—Magnetic field around two
parallel conductors.

150

this straight wire is wound around a core, as shown in figure 8-4(A), it becomes a coil and the magnetic field assumes a different shape. Part (A) is a partial cutaway view which shows the construction of a simple coil. Part (B) is a complete cross-sectional view of the same coil. The two ends of the coil are identified as a and b. When current is passed through the coiled conductor, as indicated, the magnetic field of each turn of wire links with the fields of adjacent turns, as explained in connection with figure 8-3 (A). The combined influence of all the turns produces a two-pole field similar to that of a simple bar magnet. One end of the coil will be a north pole and the other end will be a south pole.

## POLARITY OF AN ELECTROMAGNETIC COIL

In figure 8-2, it was shown that the direction of the magnetic field around a straight conductor depends on the direction of current flow through that conductor. Thus, a reversal of current flow through a conductor causes a reversal in the direction of the magnetic field that is produced. It follows that a reversal of the current flow through a coil also causes a reversal of its two-pole field. This is true because that field is the product of the linkage between the individual turns of wire on the coil. Therefore, if the field of each turn is reversed, it follows that the total field (coil's field) is also reversed.

When the direction of electron flow through a coil is known, its polarity may be determined by use of the left-hand rule for coils. This rule is illustrated in figure 8-5, and is stated as follows: Grasping the coil in the left hand, with the fingers "wrapped around" in the direction of electron flow, the thumb will point toward the north pole.

## STRENGTH OF AN ELECTRO-MAGNETIC FIELD

The strength, or intensity, of a coil's field depends on a number of factors. The major factors are listed below. All of these factors are discussed under headings that follow.

1. The number of turns of conductor.
2. The amount of current flow through the coil.
3. The ratio of the coil's length to its width.
4. The type of material in the core.

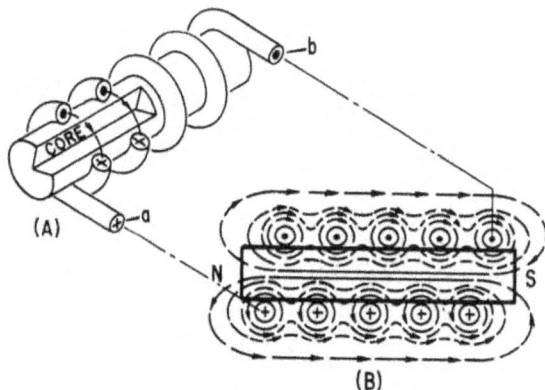

BE. 160

Figure 8-4.—Magnetic field produced by a current-carrying coil.

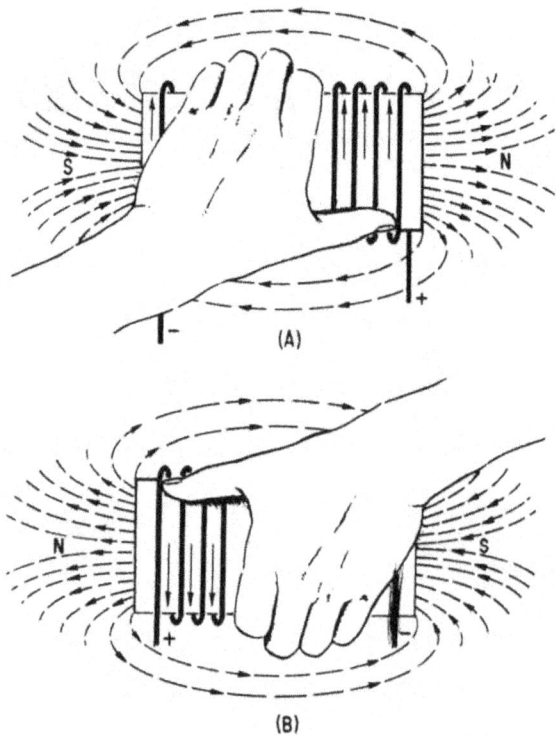

BE. 161

Figure 8-5.—Left-hand rule for coil polarity.

## MAGNETIC CIRCUITS

Many electrical devices depend upon magnetism in one or more forms for their operation. To have these devices function efficiently, engineers work out intricate designs for the required magnetic conditions. The magnets designed to do a particular job must have the required strength, and must be provided with paths, or circuits, of suitable shapes and materials.

A magnetic circuit is defined as the path (or paths) taken by the magnetic lines of force leaving a north pole, passing through the entire circuit and returning to the south pole. A magnetic circuit may be a series or parallel circuit or any combination.

## OHM'S LAW EQUIVALENT FOR MAGNETIC CIRCUITS

The law of current flow in the electric circuit is similar to the law for the establishing of flux in the magnetic circuit.

Ohm's law for electric circuits states that the current is directly proportional to the applied voltage and inversely proportional to the resistance offered by the circuit. Expressed mathematically,

$$I = \frac{E}{R}$$

Rowland's law for magnetic circuits states, in effect, that the number of lines of magnetic flux in maxwells ($\Phi$) is directly proportional to the magnetomotive force in gilberts (F) and inversely proportional to the reluctance ($\mathcal{R}$) offered by the circuit. The unit of reluctance sometimes used is the REL, $\mathcal{R}$ . Expressed mathematically.

$$\Phi = \frac{F}{\mathcal{R}}$$

The similarity of Ohm's law and Rowland's law is apparent. However, the units used in the expression for Rowland's law need to be explained.

The magnetic flux, $\Phi$, (phi) is similar to current in the Ohm's law formula, and comprises the total number of lines of force existing in the magnetic circuit. The maxwell is the unit of flux—that is, 1 line of force is equal to 1 maxwell. However, the maxwell is often referred to as simply a line of force, line of induction, or line.

The magnetomotive force, F, or mmf, comparable to electromotive force in the Ohm's law formula, is the force that produces the flux in the magnetic circuit. The practical unit of magnetomotive force is the ampere-turn. Another unit of magnetomotive force sometimes used is the gilbert, designated by the capital letter, F. The gilbert is the magnetomotive force required to establish 1 maxwell in a magnetic circuit having 1 unit of reluctance (1 rel). The magnetomotive force in gilberts is expressed in terms of ampere-turns as

$$F = 1.257 \ IN$$

where F is in gilberts, I is in amperes, and N is the number of complete turns of wire encircling the circuit.

The unit of intensity of magnetizing force per unit of length is designated as H, and is sometimes expressed as gilberts per centimeter of length. Expressed mathematically,

$$H = \frac{1.257 \ IN}{1}$$

where 1 is the length in centimeters.

The reluctance, $\mathcal{R}$ , similar to resistance in the Ohm's law formula, is the opposition offered by the magnetic circuit to the passage of magnetic flux. The unit of reluctance, symbol $\mathcal{R}$ pronounced REL, is the reluctance of 1 centimeter-cube of air. The reluctance of a magnetic substance varies directly as the length of the flux path and inversely as the cross-sectional area and the permeability, $\mu$, of the substance. Expressed mathematically,

$$\mathcal{R} = \frac{1}{\mu A}$$

where 1 is the length in centimeters, and A is the cross-sectional area in square centimeters.

Permeability, designated by the Greek letter mu, $\mu$, is treated under a separate heading. However, it is defined here to permit a fuller interpretation of Rowland's law and also a practical application of this law. Permeability is a measure of the relative ability of a substance to conduct magnetic lines of force as compared with air. The permeability of air is taken as 1. Permeability is indicated as the ratio of the flux density in lines per square centimeter (gauss, B) to the intensity of the magnetizing force in gilberts per centimeter of length, indicated by H. Expressed mathematically,

$$\mu = \frac{B}{H}$$

Another term used in magnetic circuits is permeance. Permeance, indicated by the symbol P, is the reciprocal of reluctance—that is,

$$P = \frac{1}{\Re}$$

Values of B, H, and $\mu$ for common magnetic substances are given in table 8-1.

Flux density B is expressed as

$$B = \frac{\Phi}{A} = \frac{20,000}{4} = 5,000 \text{ lines/cm.}^2$$

and from table 8-1 the corresponding value of H for cast steel is 3.9. The formula for H has previously been given as

$$H = \frac{1.257 \text{ IN}}{1}$$

Table 8-1.—B, H, and $\mu$ for common magnetic material.

| B | Sheet steel | | Cast steel | | Wrought iron | | Cast iron | |
|---|---|---|---|---|---|---|---|---|
| | H | $\mu$ | H | $\mu$ | H | $\mu$ | H | $\mu$ |
| 3,000 | 1.3 | 2,310 | 2.8 | 1,070 | 2.0 | 1,500 | 5.0 | 600 |
| 4,000 | 1.6 | 2,500 | 3.4 | 1,177 | 2.5 | 1,600 | 8.5 | 471 |
| 5,000 | 1.9 | 2,630 | 3.9 | 1,281 | 3.0 | 1,666 | 14.5 | 347 |
| 6,000 | 2.3 | 2,605 | 4.5 | 1,332 | 3.5 | 1,716 | 24.0 | 250 |
| 7,000 | 2.6 | 2,700 | 5.1 | 1,371 | 4.0 | 1,750 | 38.5 | 182 |
| 8,000 | 3.0 | 2,666 | 5.8 | 1,380 | 4.5 | 1,778 | 60.0 | 133 |
| 9,000 | 3.5 | 2,570 | 6.5 | 1,382 | 5.0 | 1,800 | 89.0 | 101 |
| 10,000 | 3.9 | 2,560 | 7.5 | 1,332 | 5.6 | 1,782 | 124.0 | 80.6 |
| 11,000 | 4.4 | 2,500 | 9.0 | 1,222 | 6.5 | 1,692 | 166.0 | 66.4 |
| 12,000 | 5.0 | 2,400 | 11.5 | 1,042 | 7.9 | 1,520 | 222.0 | 54.1 |
| 13,000 | 6.0 | 2,166 | 16.0 | 813 | 10.0 | 1,300 | 290.0 | 44.8 |
| 14,000 | 9.0 | 1,558 | 21.5 | 651 | 15.0 | 934 | 369.0 | 38.0 |
| 15,000 | 15.5 | 970 | 32.0 | 469 | 25.0 | 600 | .... | ... |
| 16,000 | 27.0 | 594 | 49.0 | 327 | 49.0 | 327 | .... | ... |
| 17,000 | 52.5 | 324 | 74.0 | 230 | 93.0 | 183 | .... | ... |
| 18,000 | 92.0 | 196 | 115.0 | 156 | 152.0 | 118 | .... | ... |
| 19,000 | 149.0 | 127 | 175.0 | 108 | 229.0 | 83 | .... | ... |
| 20,000 | 232.0 | 86 | 285.0 | 70 | .... | .... | .... | ... |

*B = flux density in lines per square centimeter; H = gilberts per centimeter of length; $\mu$ = permeability; $\mu = \frac{B}{H}$.

Permeance is like conductance in electric circuits, and is defined as the property of a magnetic circuit that permits lines of magnetic flux to pass through the circuit.

A comparison of the units, symbols, and equations used in applying Ohm's law to electric circuits and Rowland's law to magnetic circuits is given in table 8-2.

As a practical application of Rowland's law, let it be required to find the ampere-turns (IN) necessary to produce 20,000 lines of flux in a cast steel ring having a cross-sectional area of 4 square centimeters and an average length of 20 centimeters, as shown in figure 8-6.

from which

$$\text{IN} = \frac{H1}{1.257}$$

Substituting 3.9 for H and 20 for 1 in the preceding equation,

$$\text{IN} = \frac{3.9 \times 20}{1.257} = 62 \text{ ampere-turns}$$

Table 8-2.—Comparison of electric and magnetic circuits.

|  | Electric circuit | Magnetic circuit |
|---|---|---|
| Force......... | Volt, E, or e.m.f. | Gilberts, F, or m.m.f. |
| Flow ......... | Ampere, I | Flux, $\Phi$, in maxwells |
| Opposition...... | Ohms, R | Reluctance, $\mathcal{R}$, or rels |
| Law......... | Ohm's law, $I = \dfrac{E}{R}$ | Rowland's law, $\Phi = \dfrac{F}{\mathcal{R}}$ |
| Intensity of force . | Volts per cm. of length | $H = \dfrac{1.257IN}{\ell}$, gilberts per centimeter of length. |
| Density........ | Current density— for example, amperes per cm.$^2$. | Flux density, $B$.—for example, lines per cm.$^2$, or gausses. |

## PROPERTIES OF MAGNETIC MATERIALS

### PERMEABILITY

When an annealed sheet steel core is used in an electromagnet it produces a stronger magnet than if a cast iron core is used. This is true because annealed sheet steel is more readily acted upon by the magnetizing force of the coil than is hard cast iron. Therefore, soft sheet steel is said to have greater permeability because the magnetic lines are established more easily in it than in cast iron. The ratio of the flux produced by a coil when the core is iron (or some other substance) to the flux produced when the core is air is called the permeability of the iron (or whatever substance is used), the current in the coil being the same in each case.

The permeability of a substance is thus a measure of the relative ability to conduct magnetic lines of force, or its magnetic conductivity. The permeability of air is 1. The permeability of nonmagnetic materials, such as wood, aluminum, copper, and brass is essentially unity, or the same as for air.

Magnetization curves for the four magnetic materials listed in table 8-1 are given in figure 8-7.

The permeability of magnetic materials varies with the degree of magnetization, being smaller for high values of flux density, as is indicated in table 8-1 and figure 8-8.

### HYSTERESIS

The simplest method of illustrating the property of hysteresis is by graphical means such as the hysteresis loop shown in figure 8-9.

In this figure the magnetizing force is indicated in gilberts per centimeter of length along the plus and minus H axis, and the flux density is indicated in gausses along the plus and minus B axis. The intensity of the magnetizing force, H, applied by means of a current-carrying coil of wire around the sample of magnetic material, is varied uniformly through one cycle of operation, starting at zero. The force, H, is increased in the positive direction (current flowing in a given direction through the coil) to 11 gilberts per centimeter. During this time the flux density, B, increases from zero to 14,000 at point A. If H is decreased to zero, the descending curve of flux density does not return to zero via its rise path. Instead, it returns to point

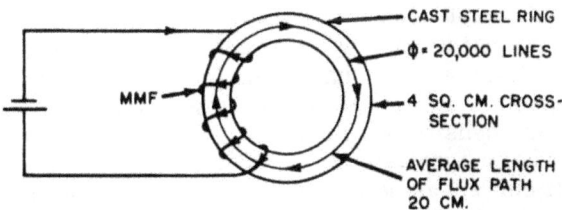

CAST STEEL RING

$\Phi$ = 20,000 LINES

4 SQ. CM. CROSS-SECTION

AVERAGE LENGTH OF FLUX PATH 20 CM.

MMF

BE. 162

Figure 8-6.—Determining ampere-turns in a magnetic circuit.

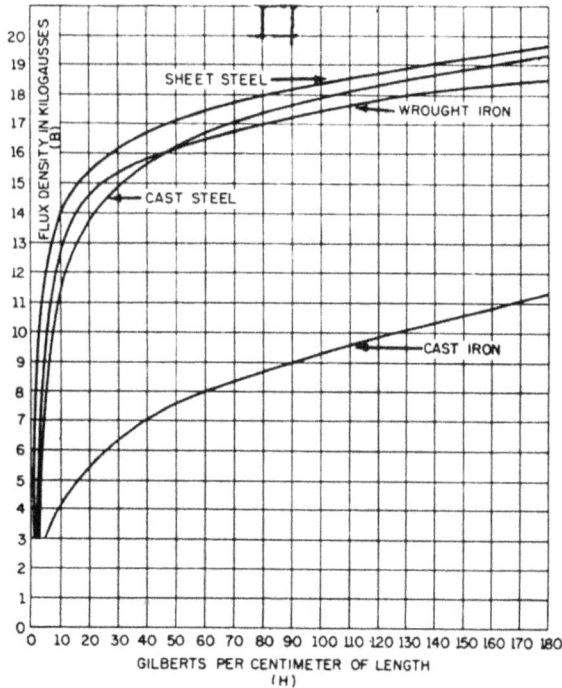

BE. 163

Figure 8-7.—Magnetization curves for four magnetic materials.

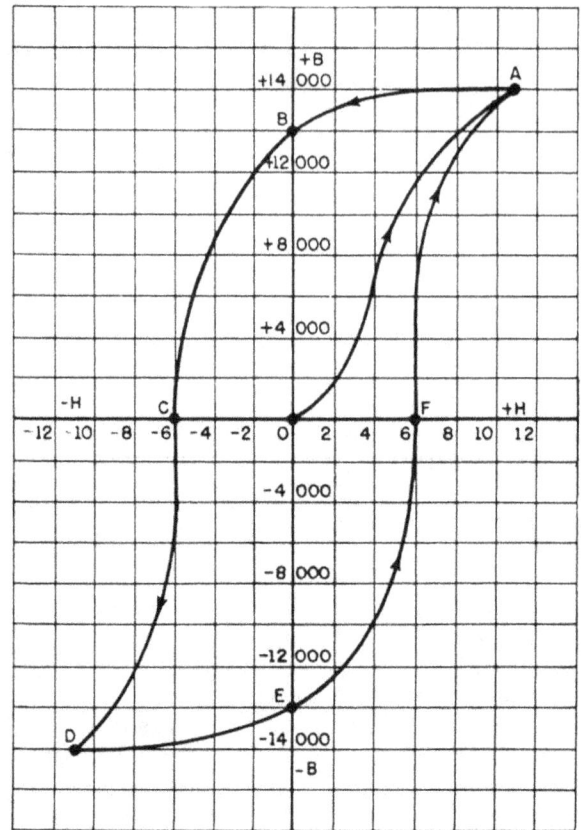

BE. 165

Figure 8-9.—Hysteresis loop.

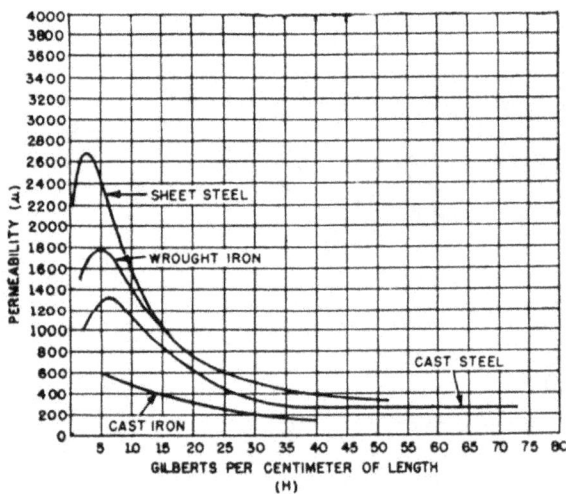

BE. 164

Figure 8-8.—Permeability curves.

B, where the flux density is 13,000. The magnetic flux indicated by the length of line OB represents the retentivity of the magnetic substance.

Retentivity is the ability of a magnetic substance to retain its magnetism after the magnetizing force has been removed. Retentivity is most apparent in hard steel and is least apparent in soft iron.

The value of the residual, or remaining, magnetism when H has been reduced to zero, depends on the substance used and the degree of flux density attained. In this example the residual magnetism is 13,000 gausses.

If current is now sent through the coil in the opposite direction, so that the intensity of the magnetizing force becomes -H, the force will have to be increased to point C before the residual magnetism is reduced to zero. The magnetizing force, OC, necessary to reduce the

155

residual magnetism to zero is called the coercive force. In this example the coercive force is 6 gilberts per centimeter.

If the magnetizing force is continued to -11 gilberts per centimeter, the curve descends from C to D, magnetizing the sample of magnetic material with the opposite polarity. If the magnetizing force is reduced again to zero, the flux density is reduced to point E. The magnetic flux indicated by the length of line OE represents the retentivity of the magnetic substance, as did line OB. The residual magnetism is again 13,000 gausses.

If the current through the coil is again reversed (sent through in the original direction), the magnetization curve moves to zero when the magnetising force is increased to point F.

Thus, when the magnetizing force goes through a complete cycle, the resulting magnetization likewise goes through a complete cycle.

From the foregoing analysis it is apparent that hysteresis is the property of a magnetic substance that causes the magnetization to lag behind the force that produces it. The lag of magnetization behind the force that produces it is caused by molecular friction. Energy is needed to move the molecules (or domains) through a cycle of magnetization. If the magnetization is reversed slowly, the energy loss may be negligible. However, if the magnetization is reversed rapidly, as when commercial alternating current is used, considerable energy may be disipated. If the molecular friction is great, as when hard steel is used, the losses may be very great. Another factor that determines hysteresis loss is the maximum density of the flux established in the magnetic material.

A comparison of the hysteresis loops for annealed steel and hard steel is shown in figure 8-10. The area within each loop is a measure of the hysteresis energy loss per cycle of operation. Thus, as shown in the figure, more energy is dissipated in molecular friction in hard steel than in annealed steel. It is therefore important that substances having low hysteresis loss be used for transformer cores and similar a-c applications.

## ELECTROMAGNETS

An electromagnet is composed of a coil of wire wound around a core of soft iron. When direct current flows through the coil the core will become magnetized with the same polarity

that the coil (solenoid) would have without the core. If the current is reversed, the polarity of both the coil and the soft-iron core is reversed.

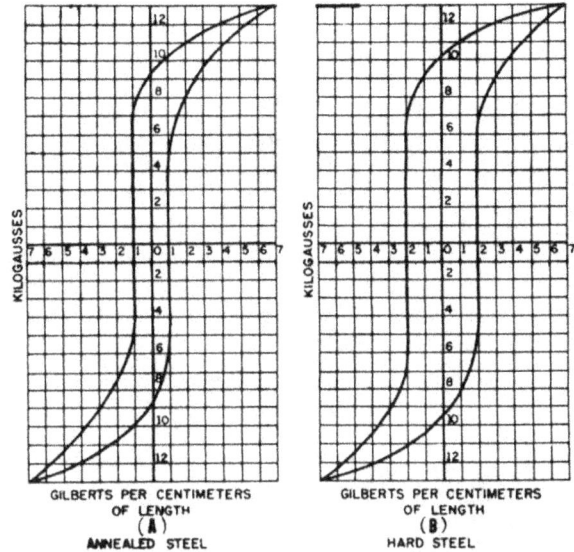

BE.166

Figure 8-10.—Comparison of hysteresis loops.

The polarity of the electromagnet is determined by the left-hand rule in the same manner that the polarity of the solenoid in figure 8-5 was determined. If the coil is grasped in the left hand in such a way that the fingers curve around the coil in the direction of electron flow ( - to + ), the thumb will point in the direction of the north pole.

The addition of the soft-iron core does two things for the current-carrying coil, or solenoid. First, the magnetic flux is increased because the soft-iron core is more permeable than the air core; second, the flux is more highly concentrated. The permeability of soft iron is many times that of air, and therefore the flux density is increased considerably when a soft-iron core is inserted in the coil.

The magnetic field around the turns of wire making up the coil influences the molecules in the iron bar causing, in effect, the individual molecular magnets, or domains, to line up in the direction of the field established by the coil. Essentially the same effect is produced in a soft-iron bar when it is under the influence of a permanent magnet.

The magnetomotive force resulting from the current flow around the coil does not increase the magnetism that is inherent in the iron core, it merely reorientates the "atomic" magnets that were present before the magnetizing force was applied. If substantial numbers of the tiny magnets are orientated in the same direction, the core is said to be magnetized.

When soft iron is used, most of the atomic magnets return to what amounts to a miscellaneous orientation upon removal of the magnetizing current, and the iron is said to be demagnetized, if hard steel is used, more of them will remain in alinement with the direction of the flux produced by the flow of current through the coil, and the metal is said to be a permanent magnet. Soft iron and other magnetic materials having high permeability and low retentivity are generally used in electromagnets.

It is known from experience that a piece of soft iron is attracted to either pole of a permanent magnet. A soft-iron bar is likewise attracted by a current-carrying coil, if the coil and bar are orientated as in figure 8-11. As shown in the figure, the lines of force extend through the soft iron and magnetize it. Because unlike poles attract, the iron bar is pulled toward the coil. If the bar is free to move, it will be drawn into the coil to a position near the center where the field is the strongest.

The solenoid-and-plunger type of magnet in various forms is employed extensively aboard ships and aircraft. These are used to operate the feeding mechanism of carbon-arc searchlights; to open circuit-breakers automatically when the load current becomes excessive; to close switches for motorboat starting; to fire guns; and to operate flood valves, magnetic brakes, and many other devices.

The armature-type of electromagnet also has extensive applications. In this type of magnet the coil is wound on and insulated from the iron core. The core is not movable. When current flows through the coil the iron core becomes magnetized and causes a pivoted soft-iron armature located near the electromagnet, to be attracted toward it. This type of magnet is used in door bells, relays, circuit breakers, telephone receivers, and so forth.

APPLICATIONS

Electric Bell

The electric bell is one of the most common devices employing the electromagnet. A simple

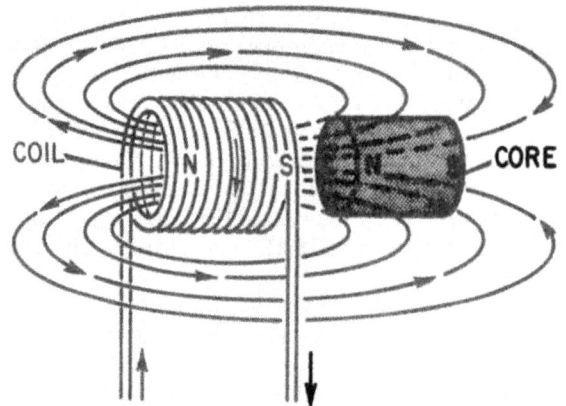

BE.167

Figure 8-11.—Solenoid with iron core.

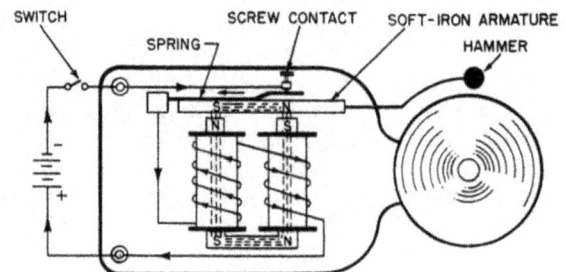

BE.168

Figure 8-12.—Electric bell.

electric bell is shown in figure 8-12. Its operation is explained as follows:

1. When the switch is closed, current flows from the negative terminal of the battery, through the contact points, the spring, the two coils, and back to the positive terminal of the battery.

2. The cores are magnetized, and the soft-iron armature (magnetized by induction) is pulled down, thus causing the hammer to strike the bell.

3. At the instant the armature is pulled down, the contact is broken, and the electromagnet loses its magnetism. The spring pulls the armature up so that contact is reestablished, and the operation is repeated. The speed with which the hammer is moved up and down depends on the

stiffness of the spring and the mass of the moving element.

The magnetomotive forces of the two coils are in series aiding and therefore the magnetization of the core is increased over that produced by one coil alone.

## Circuit Breaker

A circuit breaker, like a fuse, protects a circuit against overloading caused by excessive loads or short circuits. In this device the winding of an electromagnet is connected in series with the load circuit to be protected and with the switch contact points. The principle of operation is shown in figure 8-13. Excessive current through the magnet winding causes the switch to be tripped, and the circuit to both breaker and load is opened by a spring. When the circuit fault has been cleared, the circuit is closed again by manually resetting the circuit breaker. Circuit breakers and their applications are discussed in greater detail in chapter 14 of this manual.

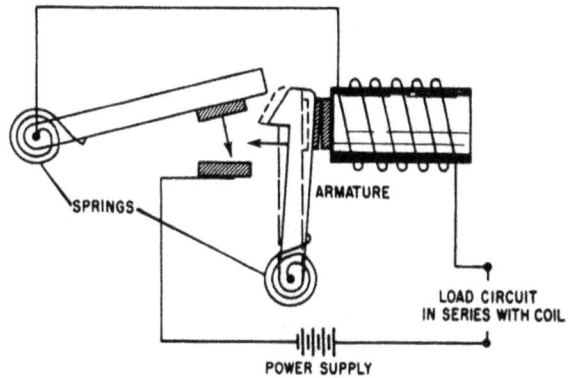

BE.169

Figure 8-13.—Magnetic circuit breaker.

Many more applications of electromagnets are discussed throughout this manual. Their applications to generators, motors, voltage regulators, reverse current relays, and servomechanisms will be covered.

# CHAPTER 9

# INTRODUCTION TO ALTERNATING-CURRENT ELECTRICITY

The first faltering steps of scientific achievement in the field of electricity were performed with crude, and for the most part, homemade apparatus. Great men, such as George Simon Ohm, had to fabricate nearly all the laboratory equipment used in their experiments. The only convenient source of electrical energy available to these early scientists was the voltaic cell, invented some years earlier. Due to the fact that cells and batteries were the only sources of power available, some of the early electrical devices were designed to operate from DIRECT CURRENT.

When the use of electricity became widespread, certain disadvantages in the use of direct current became apparent. In a direct-current system the supply voltage must be generated at the level required by the load. To operate a 240-volt lamp for example, the generator must deliver 240 volts. A 120-volt lamp could not be operated from this generator by any convenient means. A resistor could be placed in series with the 120-volt lamp to drop the extra 120 volts, but the resistor would waste an amount of power equal to that consumed by the lamp.

Another disadvantage of direct-current systems is the large amount of power lost due to the resistance of the transmission wires used to carry current from the generating station to the consumer. This loss could be greatly reduced by operating the transmission line at very high voltage and low current. This is not a practical solution in a d-c system, however, since the load would also have to operate at high voltage. As a result of the difficulties encountered with direct current, practically all modern power distribution systems use a type of current known as ALTERNATING CURRENT (a.c.). In an alternating-current system, the current flows first in one direction then reverses and flows in the opposite direction.

Unlike d-c voltage, a-c voltage can be stepped up or down by a device called a TRANSFORMER. This permits the transmission lines to be operated at high voltage and low current for maximum efficiency. Then at the consumer end the voltage is stepped down to whatever value the load requires by using a transformer. Due to its inherent advantages and versatility, alternating current has replaced direct current in all but a few commercial power distribution systems.

Many other types of current and voltage exist in addition to direct current and voltage. If a graph is constructed showing the magnitude of d-c voltage across the terminals of a battery with respect to time it would appear as in figure 9-1 (A). The d-c voltage is shown to have a constant amplitude. Some voltages go through periodic changes in amplitude like those shown in figure 9-1 (B). The pattern which results when these changes in amplitude with respect to time are plotted on graph paper is known as a WAVEFORM. Figure 9-1 (B) shows some of the common electrical waveforms. Of those illustrated, the sine wave will be dealt with most often.

## BASIC A-C GENERATOR

An alternating-current generator converts mechanical energy into electrical energy. It does this by utilizing the principle of electromagnetic induction. In the study of magnetism, it was shown that a current-carrying conductor produces a magnetic field around itself. It is also true that a changing magnetic field may produce an emf in a conductor. If a conductor lies in a magnetic field, and either the field or conductor moves, an emf is induced in the conductor. This effect is called electromagnetic induction.

## CYCLE

Figure 9-2 shows a suspended loop of wire (conductor) being rotated (moved) in a counter-

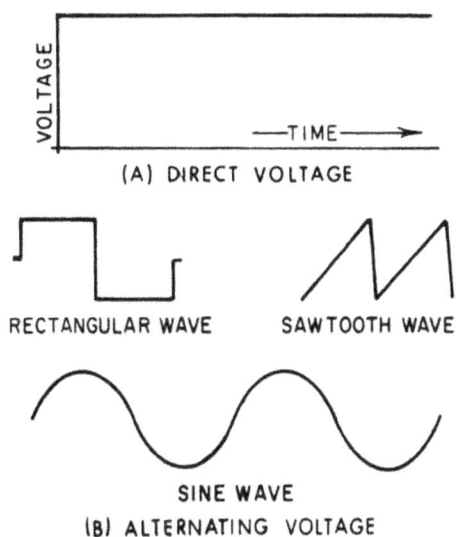

(A) DIRECT VOLTAGE

RECTANGULAR WAVE          SAWTOOTH WAVE

SINE WAVE
(B) ALTERNATING VOLTAGE

BE.170

Figure 9-1.—Voltage waveforms.

clockwise direction through the magnetic field between the poles of a permanent magnet. For ease of explanation, the loop has been divided into a dark and a light half. Notice that in part (A), the dark half is moving along (parallel to) the lines of force. Consequently, it is cutting none of these lines. The same is true of the light half, moving in the opposite direction. Since the conductors are cutting no lines of force, no emf is induced. As the loop rotates toward the position shown in part (B), it cuts more and more lines of force per second because it is cutting more directly across the field (lines of force) as it approaches the position shown in (B). At position (B) the induced voltage is greatest because the conductor is cutting directly across the field.

As the loop continues to be rotated toward the position shown in part (C), it cuts fewer and fewer lines of force per second. The induced voltage decreases from its peak value. Eventually, the loop is once again moving in a plane parallel to the magnetic field, and no voltage (zero voltage) is induced. The loop has now been rotated through half a circle (one alternation, or 180°). The sine curve shown in the lower part of the figure shows the induced voltage at every instant of rotation of the loop. Notice that this curve contains 360°, or two alternations. Two alternations represent one complete cicle of rotation.

The direction of current flow during the rotation from (B) to (C), when a closed path is provided across the ends of the conductor loop, can be determined by using the LEFT-HAND RULE FOR GENERATORS. The left-hand rule is applied as follows: Extend the left hand so that the THUMB points in the direction of conductor movement, and the FOREFINGER points in the direction of magnetic flux (north to south). By pointing the MIDDLE FINGER 90 degrees from the forefinger, it will point in the direction of current flow within the conductor.

Applying the left-hand rule to the dark half of the loop in part (B), the direction of current flow can be determined and is depicted by the heavy arrow. Similarly, direction of current flow through the light half of the loop can be determined. The two induced voltages add together to form one total emf. When the loop is further rotated to the position shown in part (D), the action is reversed. The dark half is moving up instead of down, and the light half is moving down instead of up. By applying the left-hand rule once again, it is readily apparent that the direction of the induced emf and its resulting current have reversed. The voltage builds up to maximum in this new direction, as shown by the sine-wave tracing. The loop finally returns to it original position (part E), at which point voltage is again zero. The wave of induced voltage has gone through one complete cycle.

(As mentioned previously, the cycle is two complete alternations in a period of time. Recently, the hertz (Hz) has been designated to be used in lieu of cycles per second. While it may seem confusing to the reader that in one place a cycle is used to designate two alternations per period of time and in another instance a hertz is used to designate two alternations per second, the key to determine which is used is the time factor. One hertz is one cycle per second. Therefore, throughout this manual a cycle is used when no specific time element is involved and a hertz (Hz) is used when the time element is measured in seconds.)

If the loop is rotated at a steady rate, and if the strength of the magnetic field is uniform, the number of hertz and the voltage will remain at fixed values. Continuous rotation will produce a series of sine-wave voltage cycles, or, in other words, an a-c voltage. In this way mechanical energy is converted into electrical energy.

The rotating loop in figure 9-2 is called an armature. The armature may have any number of loops or coils.

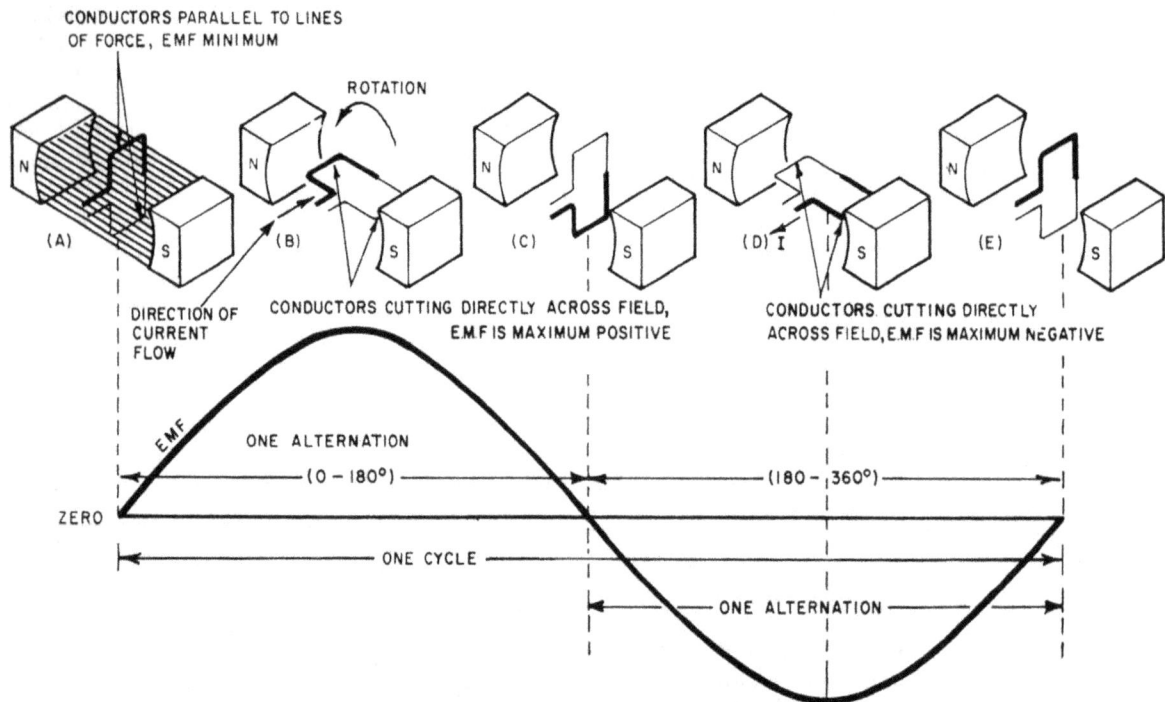

BE.171

Figure 9-2.—Basic alternating-current generator.

## FREQUENCY

The frequency (f) of an alternating current or voltage is the number of complete cycles occurring in each second of time. Hence, the speed of rotation of the loop determines the frequency. For a single loop rotating in a two-pole field you can see that each time the loop makes one complete revolution the current reverses direction twice. A single hertz will result if the loop makes one revolution each second. If it makes two revolutions per second, the output frequency will be 2 hertz. In other words, the frequency of a two-pole generator happens to be the same as the number of revolutions per second. As the speed is increased, the frequency is increased.

If an alternating-current generator has four pole pieces, as in figure 9-3, every complete mechanical revolution of the armature will produce 2 a-c cycles. When the dark half of the loop passes between poles S1 and N2, a voltage is induced which causes current to flow into the slipring attached to the dark end of the loop. When the dark half passes between N2 and S2, the induced voltage reverses direction. Another reversal occurs when it passes between S2 and

N1. The voltage at the sliprings reverses direction FOUR TIMES during each revolution. In other words, 2 cycles of a a-c voltage are generated for each mechanical revolution. If each revolution lasts 1 second, the frequency of the output is 2 hertz. The more poles that are added, the higher the frequency per revolution becomes. To find the output frequency of any a-c generator, the following formula can be used:

$$f = \frac{P \times rpm}{120}$$

where f is frequency in hertz, rpm is revolutions per minute, and P is the number of poles.

A generator made to deliver 60 hertz, having two field poles, would need an armature designed to rotate at 3,600 rpm. If it had four field poles, it would need an armature designed to rotate at 1,800 rpm. In either case, frequency would be the same. In actual practice, a generator designed for low-speed operation generally has a greater number of pole pieces, while a high-speed machine will have relatively fewer pole pieces, if both are to deliver power at the same frequency.

161

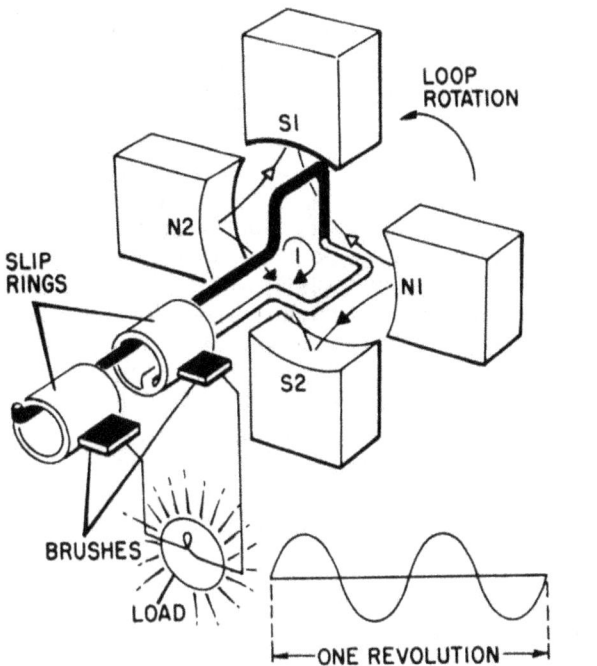

Figure 9-3.—Four-pole basic a-c generator.

## PERIOD

An individual cycle of any sine wave represents a finite amount of TIME. Figure 9-4 shows 2 cycles of a sine wave which has a frequency of 2 hertz (Hz). Since 2 cycles occur each second, 1 cycle must require one-half second of time. The time required to complete 1 cycle of a waveform is called the PERIOD of the wave. In this example the period is one-half second.

Each cycle of the waveform in figure 9-4 is seen to consist of two pulse shaped variations in voltage. The pulse which occurs during the time the voltage is postive is called the POSITIVE ALTERNATION. The pulse which occurs during the time the voltage is negative is called the NEGATIVE ALTERNATION. For a sine wave these two alternations will be identical in size and shape, and opposite in polarity.

The period of a wave is inversely proportional to its frequency. Thus, the higher the frequency (greater number of Hz), the shorter the period. In terms of an equation:

$$t = \frac{1}{f}$$

where t = period in seconds
  f = frequency in hertz

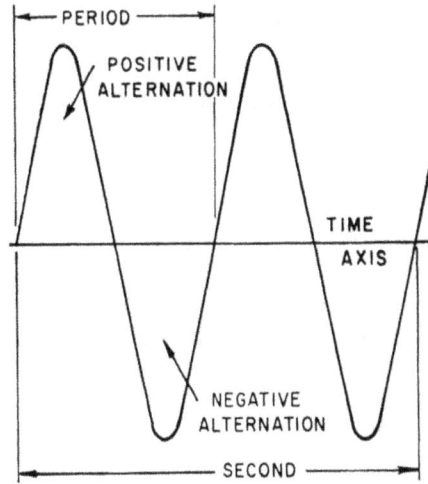

Figure 9-4.—Period of a sine wave.

## GENERATED VOLTAGE

Alternating-current generators are usually constant-potential machines because they are driven at constant speed and have fixed magnetic field strength for a given load. The effective voltage, E, generated by a single-winding (phase) a-c generator is related to the total field strength per pole, $\Phi$; the frequency f; and the total number of active conductors, N, in the armature winding; as indicated in the following equation:

$$E = 2.22\Phi fN10^{-8}$$

For example, if $\Phi = 2.5 \times 10^6$, f = 60 hertz,

and N = 96 conductors, the voltage generated is

$$E = 2.22 \times 2.5 \times 10^6 \times 60 \times 96 \times 10^{-8}$$

$$E = 320 \text{ volts}$$

The factors of poles (P) and speed (S in rpm), which appear in the equation for generated voltage in multipolar series-wound d-c generators, do not appear in the formula for the voltage generated in each phase of the a-c generator. This is because they are replaced by the equivalent factor of frequency ($f = \frac{PS}{120}$).

The length of active conductor extending under a pole does not appear in the equation directly because it is included in the factor of total magnetic flux per pole. The longer the active conductor, the more flux there will be for each pole, since the pole length and conductor length are assumed to be the same. For example, if an active conductor length is doubled, the pole length is doubled, the flux per pole is doubled, and the generated voltage is doubled.

## ANALYSIS OF A SINE WAVE

### VECTORS DEFINED

An alternating current or voltage is one in which the direction changes periodically. The electron movement is first in one direction, then in the other. The variation is of sine waveform. Straight lines drawn to scale, called VECTORS, are used in solving problems involving sine-wave currents and voltages.

A simple vector is a straight line used to denote the magnitude and direction of a given quantity. Magnitude is denoted by the length of the line, drawn to scale, and direction is indicated by an arrow at one end of the line, together with the angle that the vector makes with a horizontal reference vector.

For example, if a certain point B (fig. 9-5) lies 1 mile east of point A, the direction and distance from A to B can be shown as vector e by using a scale of approximately 1/2 inch = 1 mile.

Vectors may be rotated like the spokes of a wheel to generate angles. Positive rotation is counterclockwise and generates positive angles. Negative rotation is clockwise and generates negative angles.

The vertical projection (dotted line in fig. 9-6) of a ROTATING VECTOR may be used to represent the voltage at any instant. Vector $E_m$ represents the maximum voltage induced in a conductor rotating at uniform speed in a 2-pole

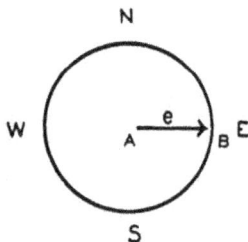

field (points 3 and 9). The vector is rotated counterclockwise through one complete revolution (360°). The point of the vector describes a circle. A line drawn from the point of the vector perpendicular to the horizontal diameter of the circle is the vertical projection of the vector.

The circle also describes the path of the conductor rotating in the bi-polar field. The vertical projection of the vector represents the voltage generated in the conductor at any instant corresponding to the position of the rotation vector as indicated by angle $\theta$. Angle $\theta$ represents selected instants at which the generated voltage is plotted. The sine curve plotted at the right of the figure represents successive values of the a-c voltage induced in the conductor as it moves at uniform speed through the 2-pole field because the instantaneous values of rotationally induced voltage are proportional to the sine of the angle $\theta$ that the rotating vector makes with the horizontal.

### EQUATION OF SINE WAVE OF VOLTAGE

The sine curve is a graph of the equation

$$e = E_m \sin \theta,$$

where e is the instantaneous voltage, $E_m$ is the maximum voltage, $\theta$ is the angle of the generator armature, and "sine" is one of the trigonometric functions shown in figure 9-6.

The instantaneous voltage, e, depends on the sine of the angle. It rises to a maximum positive value as the angle reaches 90°. This occurs because the conductor cuts directly across the flux at 90°. It falls to zero at 180°, because the conductor cuts no lines of flux at 180°. It reaches a negative peak at 270°, and becomes zero again at 360°. For example, when $\theta = 60°$ and $E_m = 100$ volts, e = 100 x sin 60° 86.6 volts. When $\theta = 240°$, e = 100 sin 240° = -86.6 volts.

Another form of the trigonometric equation for a sine wave of voltage involves the angular velocity of its rotating vector. The term "angular velocity" refers to the number of degrees (angles) through which a voltage vector rotates per second. Angular velocity is symbolized by the Greek letter omega ($\omega$). In practice, $\omega$ is generally given in terms of RADIANS per second, rather than degrees per second. A radian is a segment of the circumference of a circle. This segment is always exactly equal in length to the RADIUS of that circle. There are $\pi$ (3.14)

BE. 174

Figure 9-5.—Designating direction and distance by vectors.

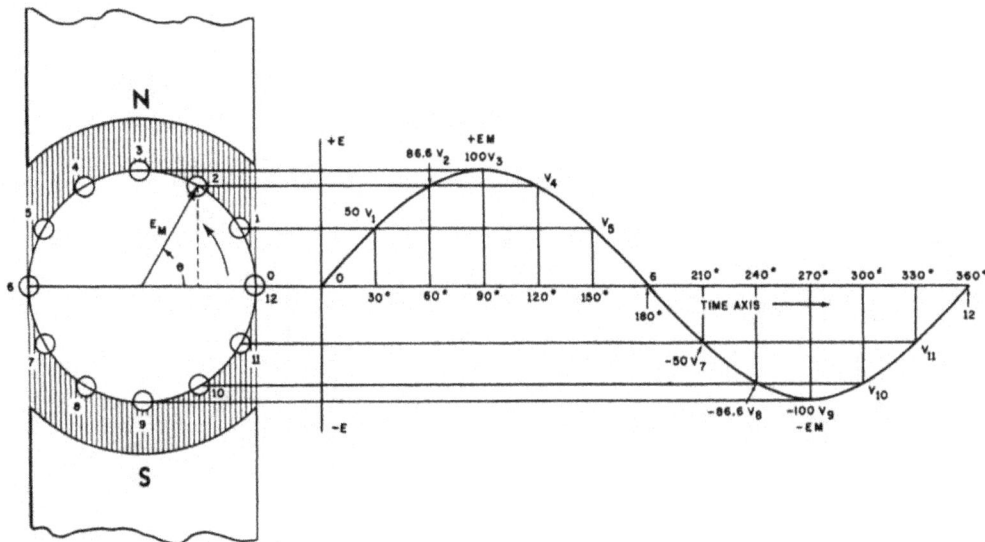

BE. 175

Figure 9-6.—Generation of sine-wave voltage.

radians in half a circle, and $2\pi$ (6.28) radians in the circumference of a complete circle. Therefore, when a voltage vector makes one complete revolution, describing one complete circle, it traverses $2\pi$ or 6.28 radians. In terms of degrees of rotation, one radian is $360°/6.28$ or $57.32°$.

The number of radians per second traversed by an alternating-voltage vector is closely related to its frequency, because either may be used to express the other. Since one vector revolution equals one complete cycle, then each cycle equals 6.28 radians of vector travel; that is, an alternating voltage whose frequency is 60 hertz would be said to have an angular velocity of 6.28 x 60, or roughly 377 radians per second. Written as a formula involving angular velocity, the term $2\pi f$ may be replaced by the simpler symbol $\omega$, if convenient. As previously stated, another form of the trigonometric equation for a sine wave of voltage involves the angular velocity of the generating vector.

The equation is

$$e = E_m \text{ x sin } \omega t.$$

It is used to determine the voltage of a rotating vector at some given instant of time. The start-ing reference, or "time zero," is usually when the voltage vector is at zero. The equation time factor t is the elapsed time from time zero, and is the exact instant at which the voltage is to be determined. To determine the exact angular position at any instant, multiply the angular velocity ($\omega$) of the vector by the time elapsed (t). By dividing elapsed time t by the period for 1 cycle, and multiplying by $360°$, the angular position of the voltage vector may be determined for any instant. Multiplying $E_m$ by the sine of that instantaneous angle will yield the instantaneous voltage e.

For example, consider a 60-Hz voltage whose peak value is 100 volts. To determine the voltage 0.00139 second from the zero point the equation would be:

$$\sin \omega t = \sin 2\pi ft = \sin 360(60)(0.00139) = \sin 30°$$

$$= 0.5$$

The instantaneous voltage e is

$$e = E_m \text{ x sin } 30°$$

$$e = 100 \text{ x } 0.5$$

$$e = 50 \text{ volts.}$$

There are four important values associated with sine waves of voltage or current:

Instantaneous-designated as e or i; maximum-designated as $E_m$ or $I_m$; average-designated as $E_{avg}$ or $I_{avg}$; and effective-designated as E or I.

The INSTANTANEOUS value may be any value between zero and maximum depending on the instant chosen, as indicated by the equation $e = E_m \sin \omega t$.

The MAXIMUM value of voltage is reached twice each cycle and is the greatest value of instantaneous voltage generated during each cycle.

The ratio of the instantaneous value of voltage to the maximum value is equal to the sine of the angle corresponding to that instant.

PEAK AMPLITUDE

One of the most frequently measured characteristics of a sine wave is its amplitude. Unlike d-c measurement, the amount of alternating current or voltage present in a circuit can be measured in various ways. In one method of measurement, the maximum amplitude of either the positive or the negative alternation is measured. The value of current or voltage obtained is called the PEAK VOLTAGE or the PEAK CURRENT. To measure the peak value of current or voltage, an oscilloscope or a special meter (peak reading meter) must be used. The peak value of a sine wave is illustrated in figure 9-7.

PEAK-TO-PEAK AMPLITUDE

A second method of indicating the amplitude of a sine wave consists of determining the total voltage or current between the positive and negative peaks. This value of current or voltage is called the PEAK-TO-PEAK VALUE (fig. 9-7). Since both alternations of a pure sine wave are identical, the peak-to-peak value is twice the peak value. Peak-to-peak voltage is usually measured with an oscilloscope, although some voltmeters have a special scale calibrated in peak-to-peak volts.

INSTANTANEOUS AMPLITUDE

The instantaneous value of a sine wave of voltage for any angle of rotation is expressed by the formula:

$$e = E_m \times \sin \theta$$

where e = the instantaneous voltage

$E_m$ = the maximum or peak voltage

$\sin \theta$ = the sine of angle at which e is desired.

Similarly the equation for the instantaneous value of a sine wave of current would be:

$$i = I_m \times \sin \theta$$

where i = the instantaneous current

$I_m$ = the maximum or peak current

$\sin \theta$ = the sin of the angle at which i desired

EFFECTIVE OR RMS VALUE

As the use of alternating current gained popularity, it became increasingly apparent that some common basis was needed on which a.c. and d.c. could be compared. A 100-watt light bulb, for example, should work just as well on 120 volts a.c. as it does on 120 volts d.c. It can be seen, however, that a sine wave of voltage having a peak value of 120 volts would not supply the

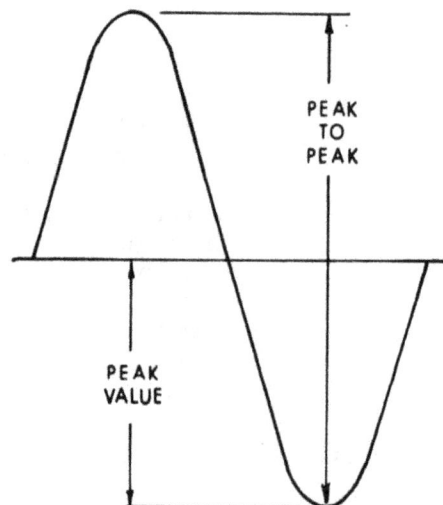

BE.176

Figure 9-7.—Peak and peak-to-peak values.

165

lamp with as much power as a steady value of 120 volts d.c.

Since the power dissipated by the lamp is a result of current flow through the lamp, the problem resolves to one of finding a MEAN alternating current ampere which is equivalent to a steady ampere of direct current.

Figure 9-8 shows a circuit in which the peak alternating current through the 10 ohm resistor is 1.414 amperes. Since the current through the resistor is changing continuously the power dissipated by the resistor will also vary. It will be minimum when the current is zero.

The variations in power throughout the cycle can best be analyzed by plotting a curve showing the instantaneous power at each point in the cycle. In the procedure to follow, the instantaneous current, the square of the instantaneous current, and the instantaneous power will be calculated in 10° steps for the first quarter of the cycle. These values are shown in table 9-1.

Table 9-1.—Instantaneous values of current and Power.

| Degrees | i | $i^2$ | P |
|---------|------|-------|-------|
| 0° | .000 | .000 | .00 |
| 10° | .245 | .060 | .60 |
| 20° | .486 | .236 | 2.36 |
| 30° | .707 | .500 | 5.00 |
| 40° | .909 | .826 | 8.26 |
| 50° | 1.083 | 1.173 | 11.73 |
| 60° | 1.225 | 1.500 | 15.00 |
| 70° | 1.329 | 1.766 | 17.66 |
| 80° | 1.393 | 1.940 | 19.40 |
| 90° | 1.414 | 2.000 | 20.00 |

Notice that at 0° the instantaneous current (i) is zero causing the power dissipated by the resistor to be zero. At 10° the instantaneous current is 0.245 amperes, the current squared is 0.060 and the power is 0.60 watt. At 90° the current has reached its maximum value of 1.414 amperes, the square of the current is 2.000 and the power dissipated is 20.00 watts.

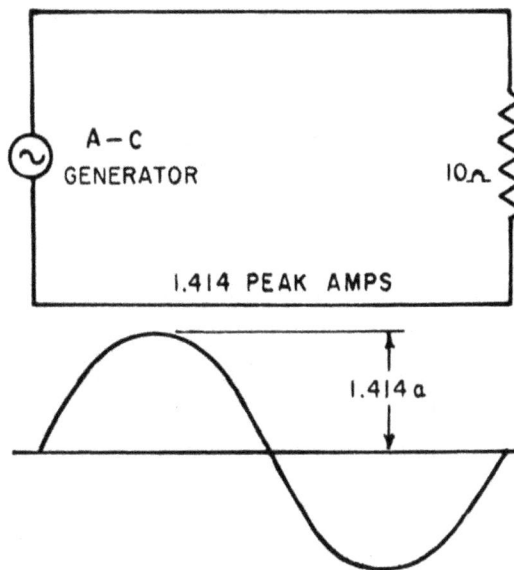

BE. 177

Figure 9-8.—Basic a-c circuit.

During the part of the sine wave of current from 90° to 180° the same values could be used as before but in a reverse order. Thus, at 100° the values of current and power would be identical to those at 80°.

Using the values of i and P from table 9-1, a graph can be constructed showing the way in which power varies throughout the cycle. This graph is shown in figure 9-9.

In this graph a sine wave of current is plotted first, using the instantaneous values from table 9-1. Next the curve representing $i^2$ and power is constructed.

Notice that the power curve has twice the frequency of the current curve, and that ALL POWER IS POSITIVE. This is due to the fact that heat is dissipated regardless of which way the current flows through the resistor.

Since all the alternations of the power curve are identical, the MEAN or AVERAGE POWER is the value HALF-WAY between the maximum and minimum values of power. Thus, the average power dissipated by the 10 ohm resistor is 10 watts, one-half the peak power. Since the curve representing power also represents current squared ($i^2$), the average or mean of the curve also lies half-way between the maximum and minimum values of $i^2$. As power is proportional to $i^2$, a d-c current having a value equal to the square root of the mean of the $i^2$ values would produce the same average power as the original sine wave of current. This mean current is called the ROOT MEAN SQUARE (RMS)

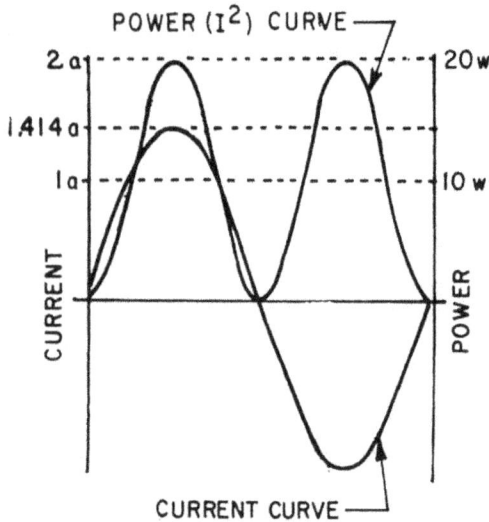

BE. 178

Figure 9-9.—Current and power curves.

current. One RMS ampere of alternating current is as effective in producing heat as one steady ampere of direct current. For this reason an RMS ampere is also callen an EFFECTIVE ampere. In figure 9-9 the peak current of 1.414 amperes produces the same amount of average power as 1 ampere of effective (RMS) current.

ANYTIME AN ALTERNATING VOLTAGE OR CURRENT IS STATED WITHOUT ANY QUALIFICATIONS, IT IS ASSUMED TO BE AN EFFECTIVE VALUE. Since effective values of a.c. are the ones generally used, most meters are calibrated to indicate effective values of voltage and current.

In many instances it is necessary to convert from effective to peak or vice-versa. Figure 9-9 shows that the peak value of a sine wave is 1.414 times the effective value and therefore:

$$E_m = E \times 1.414 \qquad (9\text{-}1)$$

where $E_m$ = maximum or peak voltage

$E$ = effective or RMS voltage

and

$$I_m = I \times 1.414$$

where

$I_m$ = maximum or peak current

$I$ = effective or RMS current.

Upon occasion it is necessary to convert a peak value of current or voltage to an effective value. The conversion factor may be derived as follows:

$$E_m = E \times 1.414 \qquad (9\text{-}1)$$

Multiplying both sides of the equation by $1/1.414$

$$E_m \times \frac{1}{1.414} = E \times 1.414 \times \frac{1}{1.414}$$

$$E_m \times \frac{1}{1.414} = E$$

Dividing 1 by 1.414

$$E = E_m \times 0.707$$

where    $E$ = the effective voltage

$E_m$ = the maximum or peak voltage

Similarly for current

$$I = I_m \times 0.707$$

where    $I$ = the effective current

$I_m$ = the maximum or peak current.

AVERAGE VALUE

The average value of a complete cycle of a sine wave is zero, since the positive alternation is idential to the negative alternation. In certain types of circuits however, it is necessary to compute the average value of one alternation. This could be accomplished by adding together a series of instantaneous values of the wave between $0°$ and $180°$, and then dividing the sum by the number of instantaneous values used. Such a computation would show one alternation of a sine wave to have an average value equal to 0.637 of the peak value. In terms of an equation:

equation:
$$E_{avg} = E_m \times 0.637$$

where

$E_{avg}$ = the average voltage of one alternation

$E_m$ = the maximum or peak voltage

similarly

$$I_{avg} = I_m \times 0.637$$

where

$I_{avg}$ = the average current in one alternation

$I_m$ = the maximum or peak current.

Figure 9-10 shows a comparison between the various values that are used to indicate the amplitude of a sine wave.

SINE WAVES IN PHASE

If a sine wave of voltage is applied to a resistance, the resulting current will also be a sine wave. This follows Ohm's law which states that the current is directly proportional to the applied voltage. Figure 9-11 shows a sine wave of voltage and the resulting sine wave of current superimposed on the same time axis. Notice that as the voltage increases in a positive direction the current increases along with it. When the voltage reverses direction, the current reverses direction. At all times the voltage and current pass through the same relative parts of their respective cycles at the same time. When two waves, such as those in figure 9-11, are precisely

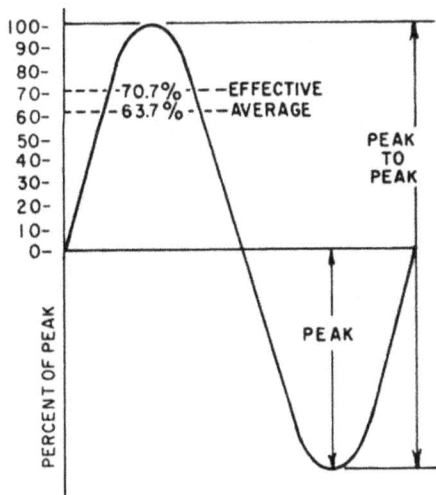

BE. 180

Figure 9-11.—Voltage and current waves in phase.

in step with one another they are said to be IN PHASE. To be in phase, the two waves must go through their maximum and minimum points at the same time and in the same direction.

In some circuits, several sine waves can be in phase with each other. Thus, it is possible to have two or more voltage drops in phase with each other and also in phase with the circuit current.

SINE WAVES OUT OF PHASE

Figure 9-12 shows a voltage wave $E_1$ considered to start at 0° (time 1). As voltage wave $E_1$ reaches its positive peak, a second voltage wave $E_2$ starts its rise (time 2). Since these waves do not go through their maximum and minimum points at the same instant of time, a PHASE DIFFERENCE exists between the two waves. The two waves are said to be out of phase. For the two waves in figure 9-12 this phase difference is 90°.

To further describe the phase relationship between two waves the terms LEAD and LAG are used. The amount by which one wave leads or lags another is measured in degrees. Referring again to figure 9-12, wave $E_2$ is seen to start 90° later in time than wave $E_1$, thus wave $E_2$ lags wave $E_1$ by 90°. This relationship could also be

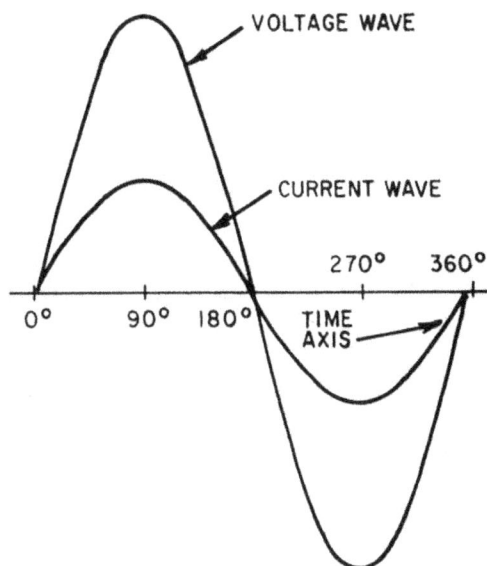

BE.179

Figure 9-10.—Various values used to indicate sine wave amplitude.

COMBINING A-C VOLTAGES

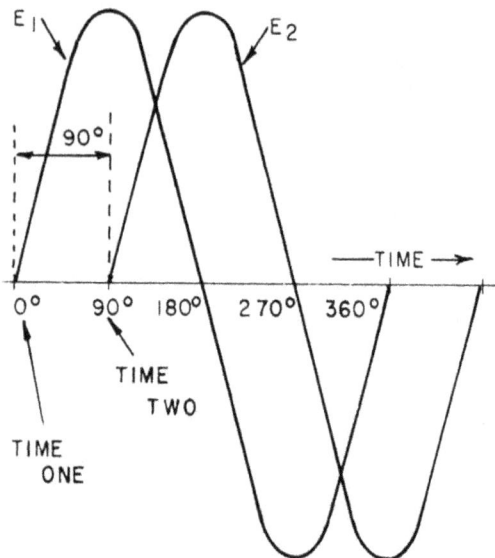

BE. 181

Figure 9-12.—Voltage waves 90° out of phase.

described by stating that wave $E_1$ leads wave $E_2$ by 90°.

It is possible for one wave to lead or lag another by any number of degrees, except 0 or 360°, in which condition the two waves are in phase. Thus, two waves may differ in phase by 45°, but two waves differing by 360° would be considered as in phase.

A phase relationship that is quite common is the one shown in figure 9-13. The two waves illustrated have a phase difference of 180°. Notice that although the waves pass through their maximum and minimum values at the same time, their instantaneous voltages are always of opposite polarity. If two such waves existed across the same component they would have a cancelling effect on each other. If the two waves are equal in amplitude the resultant wave would be zero. However, if they have different amplitudes the resultant wave would have the polarity of the larger and be the difference of the two.

To determine the phase difference between two sine waves, locate the points on the time axis where the two waves cross the time axis traveling in the same direction. The number of degrees between the crossing points is the phase difference. The wave that crosses the axis at the later time (to the right on the time axis) lags the other.

Vectors may be used to combine a-c voltages of sine waveform and of the same frequency. The angle between the vectors indicates the time difference between their positive maximum values. The length of the vectors represents either the effective value or the positive maximum values as desired.

The sine wave voltages generated in coils a and b of the simple generators (fig. 9-14 (A)) are 90° out of phase because the coils are located 90° apart on the two-pole armatures. The armatures are on a common shaft. When coil a is cutting squarely across the field, coil b is moving parallel to the field and not cutting through it. Thus, the voltage in coil a is maximum when the voltage in coil b is zero. If the frequency is 60 hertz, the time difference between the positive maximum values of these voltages is $\frac{90}{360} \times \frac{1}{60} = 0.00416$ second.

Vector $E_a$ leads vector $E_b$ by 90° in figure 9-14 (B) and sine wave a leads sine wave b by 90° in figure 9-14 (C). The curves are shown on separate axes to identify them with their respective generators; they are also projected on a common axis to show the relation between their instantaneous values. If coils a and b are connected in series and the maximum voltage generated in each coil is 10 volts, the total

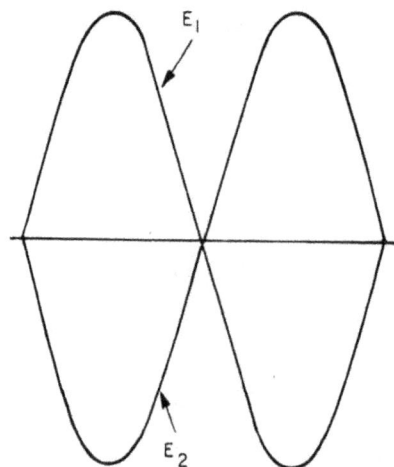

BE. 182

Figure 9-13.—Two waves 180° out of phase.

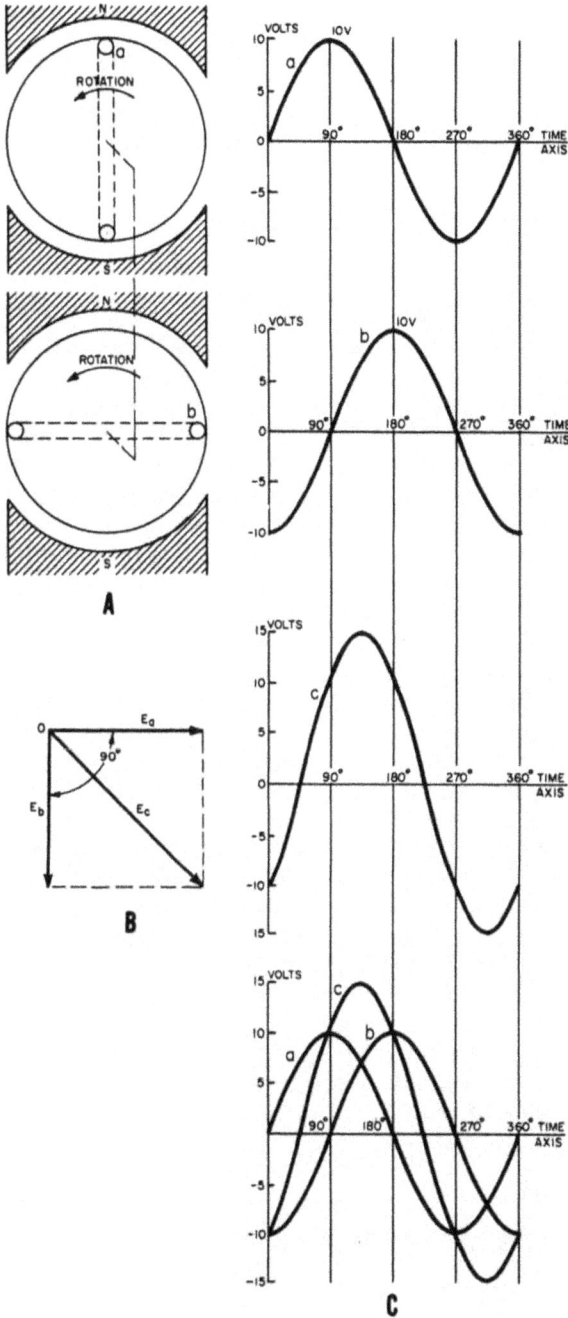

BE. 183

Figure 9-14.—Combining a-c voltages.

voltage is not 20 volts because the two maximum values of voltage do not occur at the same instant, but are separated one-fourth of a cycle.

The voltages cannot be added arithmetically because they are out of phase. These values, however, can be added vectorially.

$E_c$ in figure 9-14 (B) is the vector sum of $E_a$ and $E_b$, and is the diagonal of the parallelogram, the sides of which are $E_a$ and $E_b$. The effective voltage in each coil is 0.707 x 10 or 7.07 volts (where 10 volts is the maximum) and represents the length of the sides of the parallelogram. The effective voltage of the series combination is 7.07 x$\sqrt{2}$ or 10 volts and represents the length of the diagonal of the parallelogram.

Curve c in figure 9-14 (C) represents the sine wave variations of the total voltage $E_c$ developed in the series circuit connecting coils a and b. Voltage $E_a$ leads $E_b$ by 90°. Voltage $E_c$ lags $E_a$ by 45°, and leads $E_b$ by 45°.

Counterclockwise rotation of the vectors is considered positive rotation thus giving the sense of lead or lag. Thus, if $E_a$ and $E_b$ (fig. 9-14 (B)) are rotated counterclockwise, and their movement observed from a fixed position, $E_a$ will pass this position first, then 90° later $E_b$ will pass the position. Thus $E_b$ lags $E_a$ by 90°. If the maximum voltage in each coil is 10 volts, the maximum value of the combined voltage will be 10 x$\sqrt{2}$ or 14.4 volts. (See fig. 9-14 (C)).

The important point to be made in this discussion is that two out-of-phase voltages may be resolved into a single resultant value by the use of vectors. Though generated voltages are used to illustrate the point in this case, the same rules apply to VOLTAGE DROPS as well. Any number of out-of-phase voltages may be combined vectorially, as long as they all have the same frequency; that is, as long as they remain a fixed number of degrees apart, like the two generator loops. Their magnitudes (LENGTH of their vectors) may be different.

# CHAPTER 10

# INDUCTANCE

The study of inductance presents a very challenging but rewarding segment of electricity. It is challenging in the sense that, at first, it will seem that new concepts are being introduced. The reader will realize as this chapter progresses that these "new concepts" are merely extensions and enlargements of fundamental principles that have been acquired previously in the study of magnetism and electron physics. The study of inductance is rewarding in the sense that a thorough understanding of it will enable the reader to acquire a working knowledge of electrical circuits more rapidly and with more surety of purpose than would otherwise be possible.

Inductance is the characteristic of an electrical circuit that makes itself evident by opposing the starting, stopping, or changing of current flow. The above statement is of such importance to the study of inductance that it bears repeating in a simplified form. Inductance is the characteristic of an electrical conductor which opposes a CHANGE in current flow.

One does not have to look far to find a physical analogy of inductance. Anyone who has ever had to push a heavy load (wheelbarrow, car, etc.) is aware that it takes more work to start the load moving than it does to keep it moving. This is because the load possess the property of inertia. Inertia is the characteristic of mass which opposes a CHANGE in velocity. Therefore, inertia can hinder us in some ways and help us in others. Inductance exhibits the same effect on current in an electric circuit as inertia does on velocity of a mechanical object. The effects of inductance are sometimes desirable—sometimes undesirable.

On September 22, 1791 in Newington Butts, London, a man was born who was destined to play a great part in the laying of the foundation of the growing science of electricity. This man, Michael Faraday, started to experiment with electricity around the year 1805 while working as an apprentice bookbinder. It was in 1831 that Faraday performed experiments on magnetically coupled coils. A voltage was induced in one of the coils due to a magnetic field created by current flow in the other coil. From this experiment came the induction coil, the theory of which eventually made possible many of our modern conveniences such as the automobile, doorbell, auto radio, etc. In performing this experiment Faraday also invented the first transformer, but since alternating current had not yet been discovered the transformer had few practical applications. Two months later, based on these experiments, Faraday constructed the first direct current generator. At the same time Faraday was doing his work in England, Joseph Henry was working independently along the same lines in New York. The discovery of the property of self-induction of a coil was actually made by Henry a little in advance of Faraday and it is in honor of Joseph Henry that the unit of inductance is called the HENRY.

It was from the experiments performed by these, and many other men that the laws and theories of inductance grew.

### UNIT OF INDUCTANCE

The unit for measuring inductance, L, is the HENRY, h. An inductor has an inductance of 1 henry if an emf of 1 volt is induced in the inductor when the current through the inductor is changing at the rate of 1 ampere per second. The relation between the induced voltage, inductance, and rate of change of current with respect to time is stated mathematically as

$$E = L\,\frac{\Delta I}{\Delta t}$$

where E is the induced emf in volts, L is the inductance in henrys, and $\Delta I$ is the change in current in amperes occurring in $\Delta t$ seconds. (Delta, symbol $\Delta$, means "a change in......")

The henry is a large unit of inductance and is used with relatively large inductors. The unit employed with small inductors is the millihenry, mh. For still smaller inductors the unit of inductance is the microhenry, $\mu$h.

## SELF-INDUCTANCE

Even a perfectly straight length of conductor has some inductance. As previously explained, current in a conductor always produces a magnetic field surrounding, or linking with, the conductor. When the current changes, the magnetic field changes, and an emf is induced in the conductor. This emf is called a SELF-INDUCED EMF because it is induced in the conductor carrying the current. The direction of the induced emf has a definite relation to the direction in which the field that induces the emf varies. When the current in a circuit is increasing, the flux linking with the circuit is increasing. This flux cuts across the conductor and induces an emf in the conductor in such a direction as to oppose the increase in current and flux. This emf is sometimes referred to as counterelectromotive force (cemf). The two terms are used synonymically throughout this course. Likewise, when the current is decreasing, an emf is induced in the opposite direction and opposes the decrease in current. These effects are summarized by Lenz's law, which states that THE INDUCED EMF IN ANY CIRCUIT IS ALWAYS IN A DIRECTION TO OPPOSE THE EFFECT THAT PRODUCED IT.

The inductance is increased by shaping a conductor so that the electromagnetic field around each portion of the conductor cuts across some other portion of the same conductor. This is shown in its simplest form in figure 10-1 (A). A length of conductor is looped so that two portions of the conductor lie adjacent and parallel to one another. These portions are labeled conductor 1 and conductor 2. When the switch is closed, electron flow through the conductor establishes a typical concentric field around ALL portions of the conductor. For simplicity, however, the field is shown in a single plane that is perpendicular to both conductors. Although the field originates simultaneously in both conductors it is considered as originating in conductor 1 and its effect on conductor 2 will be noted. With increasing current, the field expands outward, cutting across a portion of

conductor 2. The resultant induced emf in conductor 2 is shown by the dashed arrow. Note that it is in OPPOSITION to the battery current and voltage, according to Lenz's law.

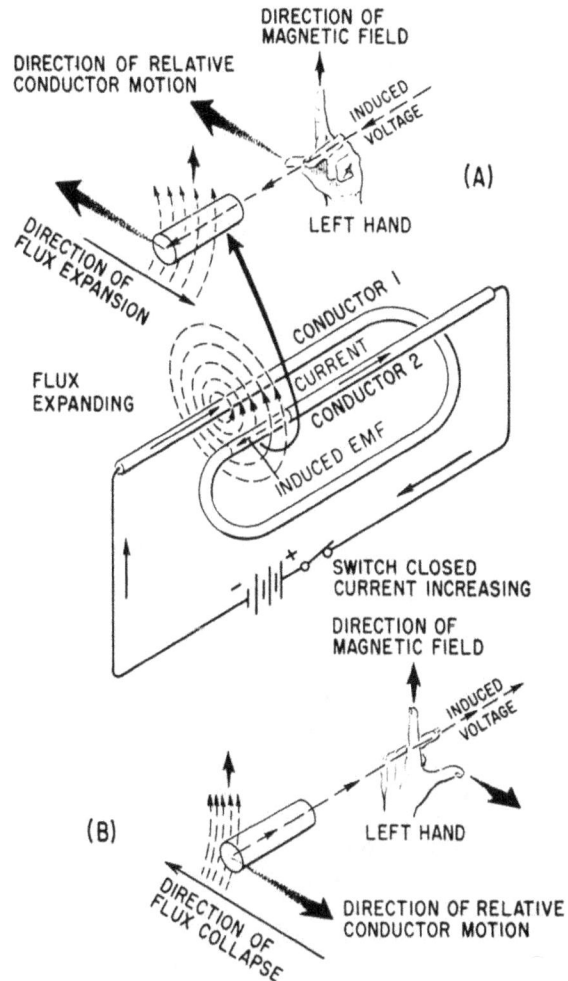

BE. 184

Figure 10-1.—Self-inductance.

The direction of this induced voltage may be determined by applying the Left-Hand Rule for Generators. This rule is applied to a portion of conductor 2 that is "lifted" and enlarged for the purpose in figure 10-1 (A). In applying this rule, the thumb of the left hand points in the direction that a conductor is moved through a field. (In this case, the field is moving, or expanding, in one direction, which is the same as the conductor moving in the opposite direction.) The index finger points in the direction of the magnetic field. The middle finger, extended as shown, will now indicate the direction of induced (generated) voltage.

In figure 10-1 (B), the same section of conductor 2 is shown, but with the switch opened and the flux collapsing. Applying the left-hand rule in this case shows that the reversal of flux MOVEMENT has caused a reversal in the direction of the induced voltage. The most important thing to note, however, is that the voltage of self-induction opposes both CHANGES in current. It delays the initial buildup of current by opposing the battery voltage, and delays the breakdown of current by exerting an induced voltage in the same direction that the battery voltage acted.

## FACTORS AFFECTING SELF-INDUCTANCE

Many things affect the self-inductance of a circuit. An important factor is the degree of linkage between the circuit conductors and its electromagnetic flux. In a straight length of conductor, there is very little flux linkage between one part of the conductor and another. Therefore, its inductance is extremely small. Conductors become much more inductive when they are wound into coils, as shown in figure 10-2. This is true because there is maximum flux linkage between the conductor turns, which lie side by side in the coil.

Inductance is further affected by the manner in which a coil is wound. The coil in figure 10-2 (A) is a poor inductor if compared to the other in the figure, because its turns are widely spaced, thus decreasing the flux linkage between its turns. Also, its lateral flux movement, indicated by the dashed arrows, does not link effectively, because there is only one layer of turns. A more inductive coil is shown in figure 10-2 (B). The turns are closely spaced, and the two layers link each other with a greater number of flux loops during all lateral flux movements. Note that nearly all turns, such as (a) are directly adjacent to four other turns (shaded), thus affording increased flux linkage.

The coil is made still more inductive by winding it in three layers, and providing a highly permeable core, as in figure 10-2 (C). The increased number of layers (cross-sectional area) improves lateral flux linkage. Note that some turns, such as (b) lie directly adjacent to six other turns (shaded). The magnetic properties of the iron core increase the total coil flux strength many times that of an air core coil of the same number of turns.

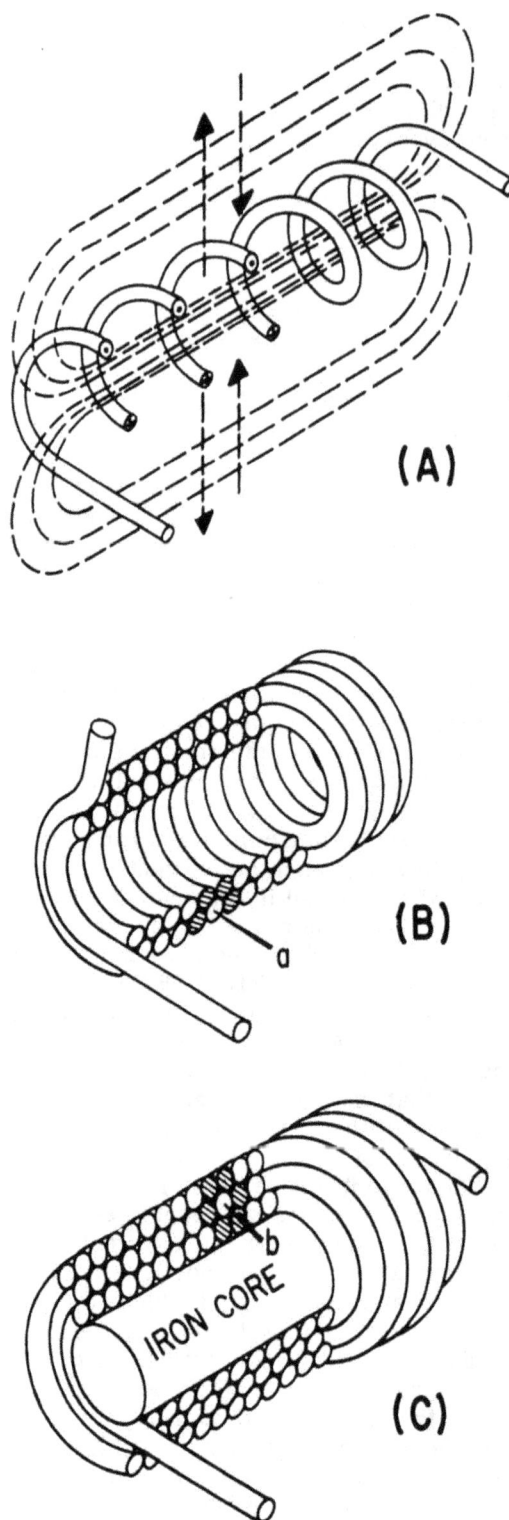

BE. 185

Figure 10-2.—Coils of various inductances.

From the foregoing, it can be seen that the primary factors controlling the inductance of a coil are (1) the number of turns of conductor, (2) the ratio of the cross-sectional area of the coil to its length, and (3) the permeability of its core material. The inductance of a coil is affected by the magnitude of current when the core is a magnetic material. When the core is air, the inductance is independent of the current. This is discussed in greater detail in the following paragraphs.

To summarize in an electric circuit, the cemf (counter electromotive force) is an induced emf or voltage. This voltage is induced in conductors of the circuit not by means of an external magnetic field, as in the case of the simple two-pole generator, but by means of the magnetic field already surrounding any conductor carrying a current. Any change in current changes the intensity of this magnetic field, and the resultant emf induced, the counter emf, is a self induced voltage. Thus, the property of a circuit which produces such an emf is called self-inductance. Actually, all elements in a circuit, including connecting wires, show some self inductance, but for all practical purposes only those elements designed to make use of this property to advantage are known as inductances or inductors. Moreover, it may be said that counter emf is present in any a-c circuit, but its effect is negligible in a circuit of moderate power, such as an electric lamp, which uses almost pure resistance as a load. But the effect of counter emf is considerable in circuits (even of very low power) which use an inductance as part of the load, such as the primary of the power transformer in an ordinary radio receiver.

## GENERATION OF COUNTER EMF, LENZ'S LAW

It will be remembered that in the case of the simple generator, motion either of the conductor or of the magnet was necessary to an induced emf. In self inductance, the equivalent of motion, that is—a change in flux density of the magnetic lines of force about a conductor, is caused by the rise or fall of the current, since, as was seen in the study of electromagnetism, the intensity of a magnetic field about a conductor is directly proportional to the current through the conductor. The force setting up the flux lines is equal to $0.4\pi NI$. Since $0.4\pi$ is a constant, the factor NI is called the ampere turns. Any change in

current changes the factor NI or ampere turns and, accordingly, the flux density. Thus, self inductance is present constantly in an a-c circuit because current is constantly changing, but it is present in a d-c circuit only at the moments of closing or opening the circuit.

That the emf self-induced in a conductor carrying current is a counter emf was deduced by H.F.E. Lenz from the principle of the conservation of energy. If the emf self-induced was not a counter emf, then an increase of current would aid the applied voltage, and this increase in applied voltage would in turn tend to increase the current. This process would continue, of course, until current reached an infinite amount, a condition not possible in the physical universe. As previously mentioned, Lenz' Law states: An induced emf always has such a direction as to oppose the action that produced it. Thus, when a current flowing through a circuit is varying in magnitude, it produces a varying magnetic field which sets up an induced emf that opposes the current change producing it. Or, it may be said that when the current in a circuit is increasing, the induced emf opposes the applied voltage and tends to keep the current from increasing; and when the current is decreasing, the induced emf aids the line voltage and tends to keep the current from decreasing.

The effect of counter emf may be observed experimentally in that an alternating current through an inductor is opposed by a force much greater than its simple d-c resistance. For example, the d-c resistance of the primary of an ordinary power transformer used in a typical radio receiver is approximately 6 ohms. As in figure 10-3, this primary is connected directly to the 120-volt, a-c outlet in the home. From Ohm's Law:

$$I = \frac{E}{R} = \frac{120}{6}$$

$$I = 20 \text{ amperes}$$

Thus, the current is calculated to be 20 amperes, but when actually measured the current is found to be approximately 1 ampere. It may be seen that some opposition other than the 6-ohm resistance is present in an a-c circuit. This opposition is the counter emf. If by mistake such a radio set is connected to a 120-volt, d-c line, the current through the primary of the transformer is 20 amperes, and the transformer burns up. Hence, Ohm's Law as stated for d-c circuits

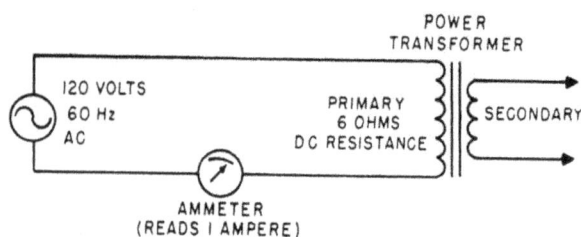

BE. 186

Figure 10-3.—Effect of cemf on current flow.

must be modified to include this effect of electrical inertia present in a-c circuits.

## MAGNITUDE OF A COUNTER EMF

The magnitude of a counter emf depends upon the same factors that govern any induced emf. In the analysis of an induced emf it was shown that the magnitude of the emf induced in a conductor of unit length depended on the number of flux lines cut per second.

This principle may be restated as Faraday's Law of electromagnetic induction: THE EMF INDUCED IN ANY CIRCUIT IS DEPENDENT UPON THE RATE OF CHANGE OF THE FLUX LINKING THE CIRCUIT. Since there is no physical movement of the conductor or of the lines of force in self inductance, the rate of change of the flux density is equivalent to movement. But, as was shown above and in the study of electromagnetism, the flux density about a conductor is directly proportional to the current in the conductor. Therefore, the magnitude of self-induced emf depends directly upon the rate of change of the current in the circuit. Thus, a rapidly changing current induces a greater counter emf than a slowly changing current. But for any a.c. the rate of change of current depends on the number of hertz or the frequency. The counter emf then depends directly upon frequency.

As was mentioned previously, the total magnitude of an induced emf depends also on the length of the conductor, since in the simple generator the length of the conductor and the number of conductors in a coil side, determine the total emf induced. Thus, a long conductor has greater counter emf induced, or has more self inductance, than a short one. If, however, a long conductor is wound on itself in the form of a coil, its self inductance is increased because

of the increase in total flux density. Such a coil, or inductance, is a solenoid, and, as was shown in the study of electromagnetism, the flux density about it may also be increased by the addition of a core material of high permeability, such as soft iron. Figure 10-4 illustrates this type of inductance.

## MEASUREMENT OF INDUCTANCE

As previously mentioned, the unit of measurement of inductance is the henry, named after Joseph Henry, the co-discoverer with Faraday of the principle of electromagnetic induction. A henry is defined as the inductance of a circuit in which a current change of 1 ampere per second causes a counter emf of 1 volt. Since the henry is defined in terms of practical units, the factor $10^{-8}$ must be used if the cemf is to be read in volts and the rate of change of the current in amperes per second. Then

$$L = \frac{0.4\pi N^2 \mu A}{l} \times 10^{-8} \text{ (henrys)}$$

where    $L$ =   self inductance of solenoid in henrys

$N$ =   number of turns of coil

$\mu$ =   permeability of core in electromagnetic units

$A$ =   cross-sectional area of core in $cm^2$

$l$ =   mean length of core in cm.

This formula reveals the following important relationships:

1. The inductance of a coil is proportional to the square of the number of turns.

2. The inductance of a coil increases directly as the permeability of the material making up the core increases.

3. The inductance of coil increases directly as the cross-sectional area of the core increases.

4. The inductance of a coil decreases as its length increases. Figure 10-5 (A) shows two coils of a fixed number of turns with different cross-sectional areas. The larger coil has a greater total flux, or less reluctance, and therefore greater inductance. Figure 10-5 (B) shows two coils of a fixed number of turns and the

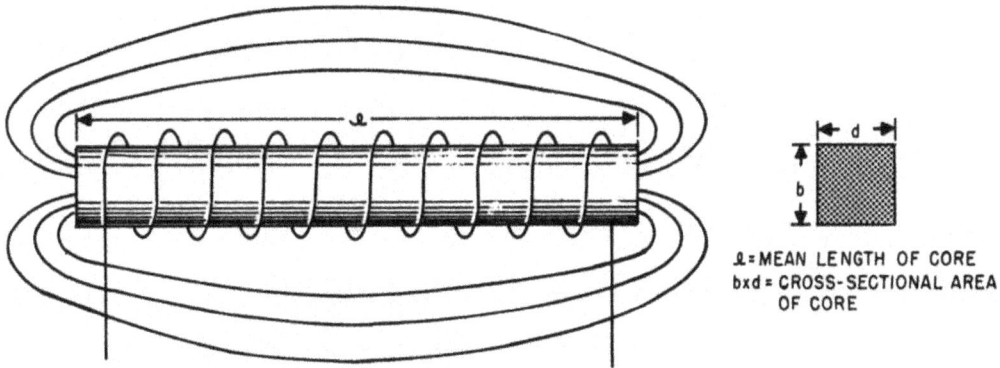

ℓ = MEAN LENGTH OF CORE
bxd = CROSS-SECTIONAL AREA
OF CORE

BE.187

Figure 10-4.—Inductance wound on iron core.

same cross-sectional area, but of different lengths. The longer coil has less total flux, or greater reluctance, and therefore, less inductance.

It should also be noted that for ferromagnetic (iron-core) inductances, the permeability, $\mu$, of the core material is not a constant, but depends on the magnitude of the magnetizing current. In a-c circuits, the current is constantly changing in magnitude and periodically in direction, and accordingly, an error is introduced in calculations of the magnitude of the inductance. In figure 10-6 (A) the relationship between the flux density B and the field intensity H is shown in the form of a hysteresis loop graph. But the ratio B/H is the definition of permeability. Therefore, it may be seen from this graph that the value of $\mu$ varies as the ratio of B to H varies for different points on the loop.

In figure 10-6 (B) a permeability curve for cast steel is shown. Permeability for this material increases to a maximum value at approximately 7,000 lines per square centimeter and then falls off as the flux density increases. This variation of $\mu$ invalidates calculations of inductance based on the formula given previously.

## MEASUREMENT OF COUNTER EMF: VOLTAGE

The formula for the magnitude of a cemf is as follows:

$$cemf = -L \frac{\Delta i}{\Delta t}$$

An examination of this formula reveals that the greater the inductance, or the faster the rate of change of the current, the greater is the cemf induced in the circuit. For example, a coil of 1-henry inductance has a current of 1 ampere flowing through it. If this current changes to 2 amperes in 1 second, the cemf will be:

$$cemf = -1 \times \frac{(2 - 1)}{1} = -1 \text{ volt}$$

If the current change remains the same but the coil used has an inductance of 10 henrys, then:

$$cemf = -10 \times \frac{(2 - 1)}{1} = -10 \text{ volts}$$

If the inductance remains, as in the first instance, at 1 henry, and the current change from 1

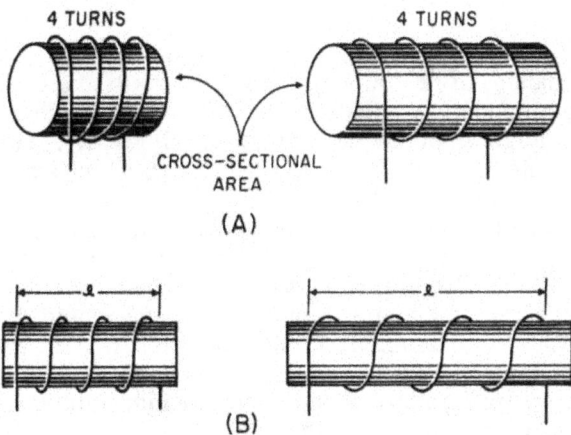

(A)

(B)

BE. 188

Figure 10-5.—Variation of inductance with size of solenoid.

(A)

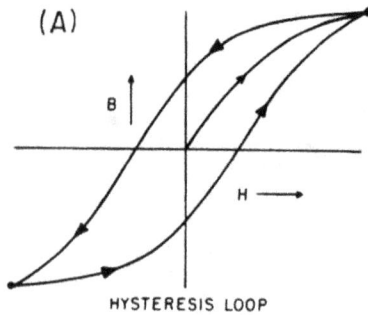

HYSTERESIS LOOP

(B)

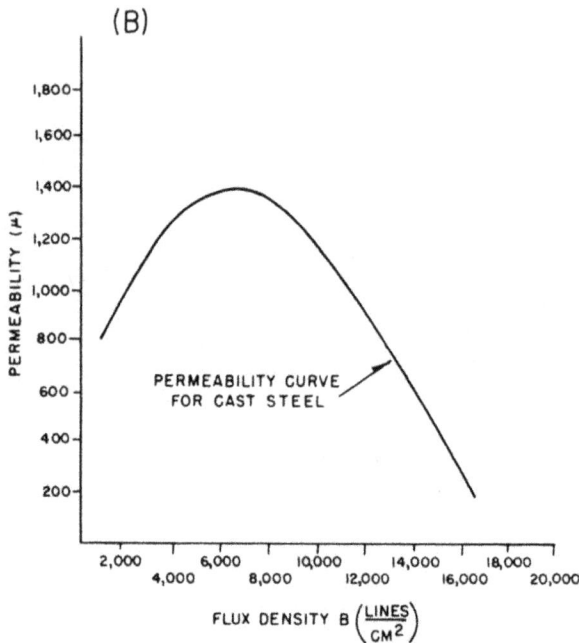

PERMEABILITY CURVE
FOR CAST STEEL

BE.189

Figure 10-6.—Variation of $\mu$ for iron-core materials.

ampere to 2 amperes takes place in one-tenth of a second, then:

$$\text{cemf} = -1 \times \frac{(2 - 1)}{1/10} = -10 \text{ volts}$$

From these examples it may be seen that a high value of opposition to the flow of current may be obtained either by increasing the inductance or the speed of the change of current in a circuit, or both. Thus, low frequency a-c circuits, because of the slow speed of change of the current, generally employ high values of inductance (iron cores) to obtain a high cemf. High-frequency a-c circuits, because of the great speed of the change of the

current, often may generate sufficient cemf with small air core inductances. Table 10-1 illustrates the rise in cemf as the rate of change of the current increases:

where $\Delta i$ (called delta i) = change in current

$\Delta t$ (called delta t) = time required for change in current.

Table 10-1.—Relationship of cemf and frequency.

| L henrys | $\Delta i$ amperes | $\Delta t$ seconds | cemf in volts |
|---|---|---|---|
| 1 | 1 | 1 | -1 |
| 1 | 1 | 1/2 | -2 |
| 1 | 1 | 1/4 | -4 |
| 1 | 1 | 1/10 | -10 |
| 1 | 1 | 1/20 | -20 |
| 1 | 1 | 1/50 | -50 |
| 1 | 1 | 1/100 | -100 |
| 1 | 1 | 1/500 | -500 |
| 1 | 1 | 1/1,000 | -1,000 |
| 1 | 1 | 1/1,000,000 | -1,000,000 |

From table 10-1 it is apparent that if a change of 1 ampere were to take place instantaneously, that is, if $\Delta t = 0$, the induced voltage e would become infinitely large. This would violate Kirchhoff's first law, which states that at any instant the applied voltage in a circuit must equal the sum of the voltage drops around the circuit. Certainly, if $\Delta t = 0$, the voltage drop across the inductance would be greater than any applied voltage could be. And by extension, it may be seen that at any instant, no matter how fast the change of current or how great the value of inductance, the induced voltage cannot be greater than the applied voltage. On the other hand, if there were no change of current, that is, if $\Delta t$ were equal to infinity, the circuit would be a d-c circuit, and e would be zero.

177

## GROWTH AND DECAY OF CURRENT
## IN AN RL SERIES CIRCUIT

If a battery is connected across a pure inductance, the current builds up to its final value at a rate that is determined by the battery voltage and the internal resistance of the battery. The current buildup is gradual because of the counter emf generated by the self-inductance of the coil. When the current starts to flow, the magnetic lines of force move out, cut the turns of wire on the inductor, and build up a counter emf that opposes the emf of the battery. This opposition causes a delay in the time it takes the current to build up to steady value. When the battery is disconnected, the lines of force collapse, again cutting the turns of the inductor and building up an emf that tends to prolong the current flow.

A voltage divider containing resistance and inductance may be connected in a circuit by means of special switch, as shown in figure 10-7 (A). Such a series arrangement is called an RL series circuit.

If switch S1, is closed (as shown), a voltage, E, appears across the divider. A current attempts to flow, but the inductor opposes this current by building up a back emf that, at the initial instant, exactly equals the input voltage, E. Because no current can flow under this condition, there is no voltage across resistor R. Figure 10-7 (B) shows that all of the voltage is impressed across L and no voltage appears across R at the instant switch Sp is closed.

As current starts to flow, a voltage, $e_r$, appears across R, and $e_L$ is reduced by the same amount. The fact that the voltage across L is reduced means that the growth current, $i_g$, is increasing and consequently $e_r$ is increasing. Figure 10-7 (B) shows that $e_L$ finally becomes zero when $i_g$ stops increasing, while $e_r$ builds up to the input voltage, E, as $i_g$ reaches its maximum value. Under steady-state conditions, only the resistor limits the size of the current.

Electrical inductance is like mechanical inertia, and the growth of current in an inductive circuit can be likened to the acceleration of a boat on the surface of the water. The boat begins to move at the instant a constant force is applied to it. At this instant its rate of change of speed (acceleration) is greatest, and all the applied force is used to overcome the inertia of the boat. After a while the speed of the boat increases (its acceleration decreases)

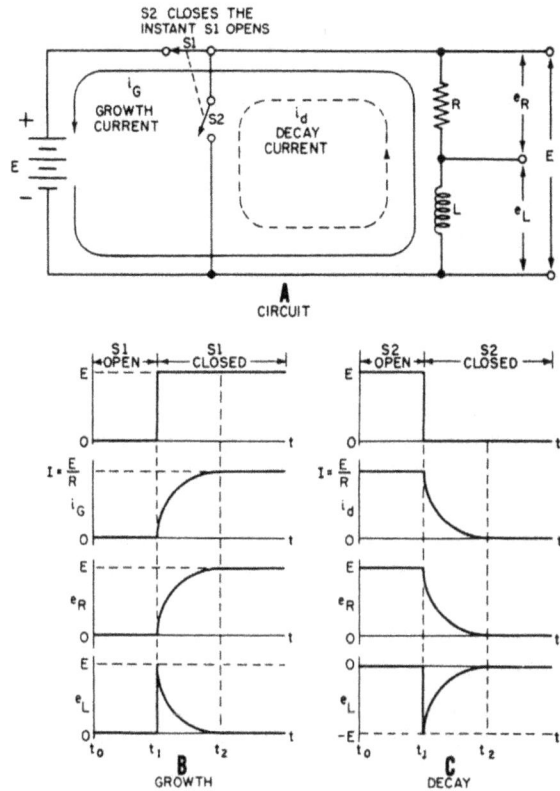

BE.190

Figure 10-7.—Growth and decay of current in an RL series circuit.

and the applied force is used up in overcoming the friction of the water against the hull. As the speed levels off and the acceleration becomes zero, the applied force equals the opposing friction force at this speed and the inertia effect disappears.

When the battery switch in the R-1 circuit of figure 10-7 (A) is closed, the rate of current increase is maximum in the inductive circuit. At this instant all the battery voltage is used in overcoming the emf of self-induction which is a maximum because the rate of change of current is maximum. Thus the battery voltage is equal to the drop across the inductor and the voltage across the resistor is zero. As time goes on more of the battery voltage appears across the resistor and less across the inductor. The rate of change of current is less and the induced emf is less. As the steady-state condition of the current flow is approached the drop across the inductor approaches zero and all of the battery voltage is used to overcome the resistance of the circuit.

Thus the voltages across the inductor and resistor change in magnitudes during the period of growth of current the same way the force applied to the boat divides itself between the inertia and friction effects. In both examples, the force is developed first across the inertia-inductive effect and finally across the friction-resistive effect.

If switch S2 is closed (source voltage E removed from the circuit), the flux that has been established around L collapses through the windings and induces a voltage, $e_L$, in L that has a polarity opposite to E and essentially equal to it in magnitude. The induced voltage, $e_L$, causes current $i_d$ to flow through R in the same direction that it was flowing when S1 was closed. A voltage, $e_r$, that is initially equal to E, is developed across R. It rapidly falls to zero as the voltage, $e_L$, across L, due to the collapsing flux, falls to zero.

L/R Time Constant

The time required for the current through an inductor to increase to 63 percent (actually, 63.2 percent) of the maximum current or to decrease to 37 percent (actually, 36.7 percent) is known as the TIME CONSTANT of the circuit. An RL circuit and its charge and discharge graphs are shown in figure 10-8. The value of the time constant in seconds is equal to the inductance in henrys divided by the circuit resistance in ohms. One set of values is given in figure 10-8 (A). $\frac{L}{R}$ is the symbol used for this time constant.

Two useful relations used in calculating $\frac{L}{R}$ time constants are as follows:

$$\frac{L(\text{in henrys})}{R(\text{in ohms})} = t \text{ (in seconds)}$$

$$\frac{L(\text{in microhenrys})}{R(\text{in ohms})} = t \text{ (in microseconds)}$$

The time constant may also be defined as the time required for the current through the inductor to grow or decay to its final value IF it continued to grow or decay at its initial rate. As may be seen in figure 10-8 (B), the slope of the dotted tangent line, ox, indicates the initial rate of current growth with respect to time. At this rate, the current would reach

its maximum value in $\frac{L}{R}$ seconds. Similarly, the slope of the dotted tangent line, YZ, indicates the initial rate of current decay with respect to time, and the decay would be completed in $\frac{L}{R}$ seconds.

The equation for the growth of current, $i_L$, through L is

$$i_L = \frac{E}{R} \left( 1 - \frac{1}{2.718^{\frac{Rt}{L}}} \right)$$

where $i_L$ is the instantaneous current through inductor L, E is the applied voltage (100 volts in this case), R is the resistance in ohms, t is the time in seconds, and L is the inductance in henrys. Figure 10-8 (B) shows a graph of this equation.

When $t = \frac{L}{R}$, the exponent $\frac{Rt}{L}$ in the preceding equation reduces to 1. Then $\frac{1}{2.718} = 0.368$. Therefore,

$$i_L = \frac{E}{R}(1-0.368) = 0.632\frac{E}{R}$$

In other words, when $t = \frac{L}{R}$, $i_L$ is equal to 63.2 percent of the ratio $\frac{E}{R}$, which is the maximum current.

When the maximum current is 10 amperes (E = 100 and R = 10), the current through L grows to 6.32 amperes in $\frac{L}{R} = \frac{10}{10}$, or 1 second.

The equation for inductor voltage, $e_L$, on growth of current is

$$e_L = E \left( \frac{1}{2.718^{\frac{Rt}{L}}} \right)$$

The graph of this equation is also shown in figure 10-8 (B). When $t = \frac{L}{R}$, $e_L = 0.368E$; that is, $e_L = 0.368 \times 100 = 36.8$ volts.

MUTUAL INDUCTANCE

MUTUAL INDUCTANCE DEFINED

Whenever two coils are located so that the flux from one coil links with the turns of the other, a change of flux in one coil will cause

179

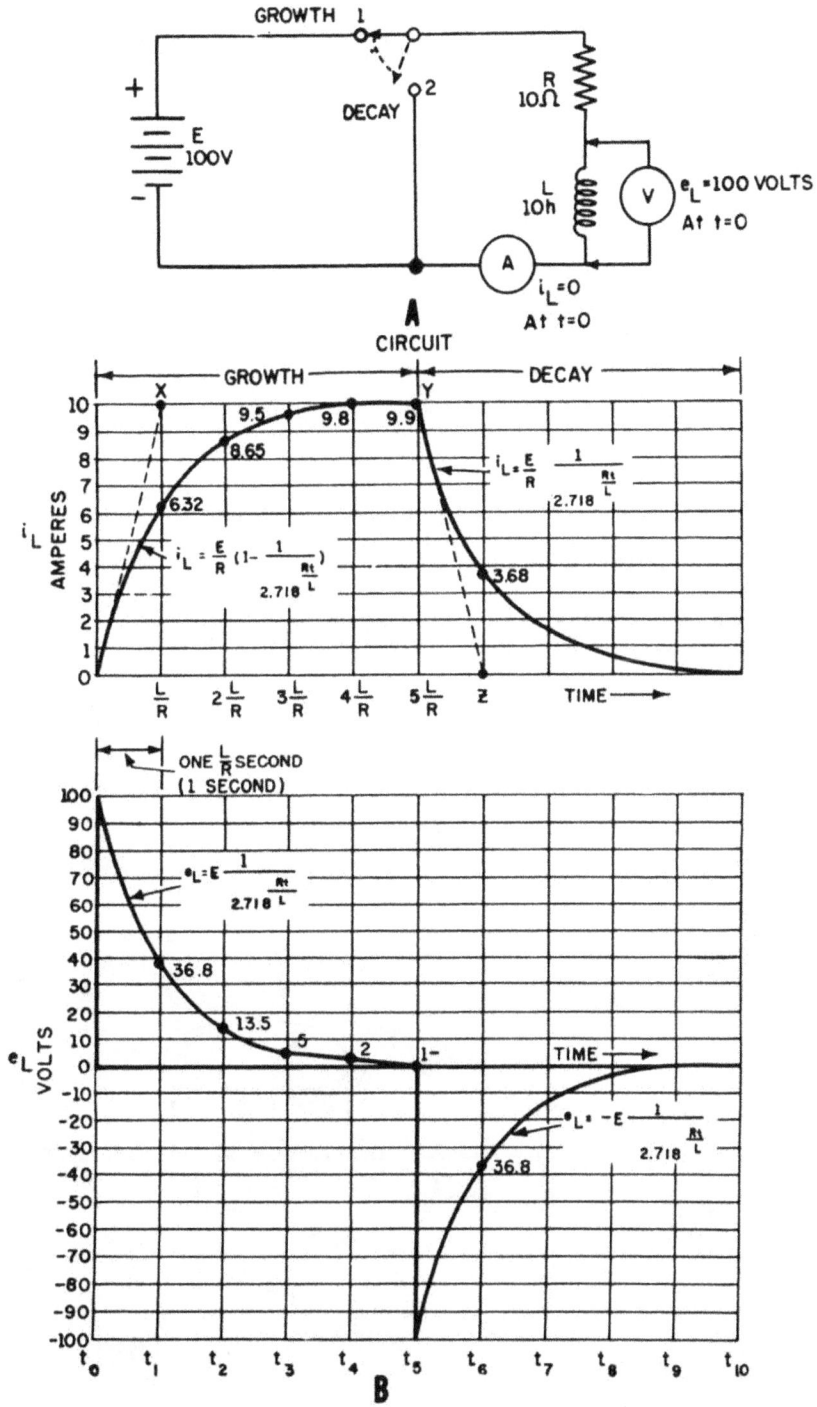

Figure 10-8.—L/R time constant.

BE.191

an emf to be induced in other coil. The two coils have MUTUAL INDUCTANCE. The amount of mutual inductance depends on the relative position of the two coils. If the coils are separated a considerable distance, the amount of flux common to both coils is small and the mutual inductance is low. Conversely, if the coils are close together so that nearly all the flux of one coil links the turns of the other mutual inductance is high. The mutual inductance can be increased greatly by mounting the coils on a common iron core.

Two coils placed close together with their axes in the same plane are shown in figure 10-9. Coil A is connected to a battery through switch S, and coil B is connected to galvanometer G. When the switch is closed (fig. 10-9 (A)), the current that flows in coil A sets up a magnetic field that links coil B, causing an induced current and a momentary deflection of galvanometer G. When the current in coil A reaches a steady value, the galvanometer returns to zero. If the switch is opened (fig. 10-9 (B)), the galvanometer deflects momentarily in the opposite direction, indicating a momentary flow of current in the opposite direction in coil B. This flow of current in coil B, which is produced by the collapsing flux of coil A.

When current flows through coil A the flux expands in coil A producing a north pole nearest coil B. Some of the flux, that expands from left to right, cuts the turns of coil B. This flux induces an emf and current in coil B that opposes the growth of current and flux in coil A. Thus the current in B tries to establish a north pole nearest coil A (like poles repel).

When the switch is opened, the magnetic field produced by coil A collapses. The collapse of the flux cuts the turns of coil B in the opposite direction and produces a south pole nearest coil A (unlike poses attract). This polarity aids the magnetism of coil A, tending to prevent the collapse of its field.

## FACTORS AFFECTING MUTUAL INDUCTANCE

The mutual inductance of two adjacent coils is dependent upon the—(1) physical dimensions of the two coils, (2) number of turns in each coil, (3) distance between the two coils, (4) relative positions of the axes of the two coils, and (5) the permeability of the cores.

If the two coils are positioned so that all the flux of one coil cuts all the turns of the

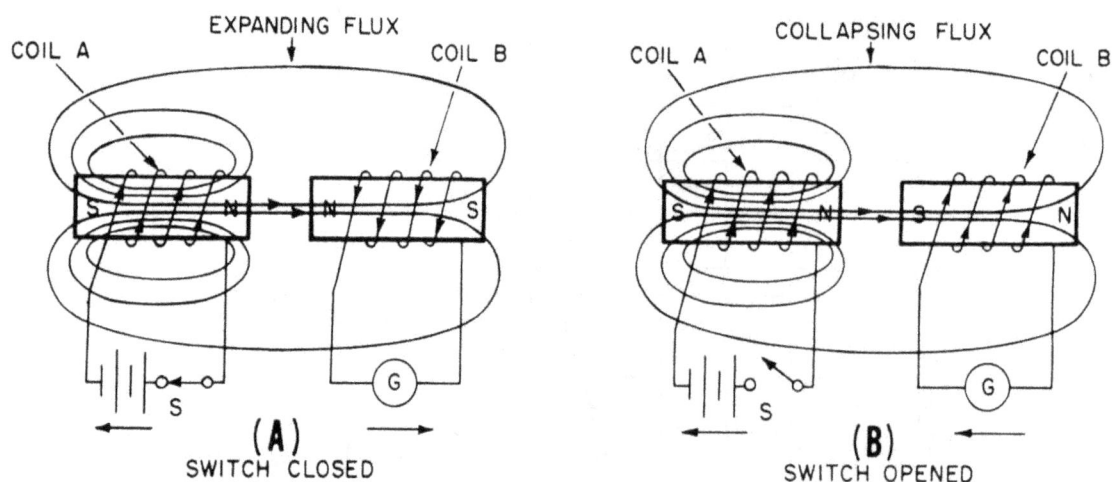

Figure 10-9.—Mutual inductance.

BE.192

181

other, the mutual inductance can be expressed as follows:

$$M = \frac{0.4\pi\mu SN_1N_2}{10^8 1}$$

where

M equals mutual inductance of coils in henrys

$N_1$ and $N_2$ equal the number of turns in coils 1 and 2 respectively

S equals area of core in square cm

$\mu$ equals permeability of core

1 equals length of core in cm

If the two coils are so positioned with respect to each other that all the flux of one coil cuts all the turns of the other, the coils have UNITY COEFFICIENT OF COUPLING. If all of the flux produced by one of the coils cuts only one-half the turns of the other coil, the coefficient of coupling is 0.5. Coefficient of coupling is designated by the letter, K. The coefficient of coupling is equal to the percentage of flux originating in one coil that cuts the other coil. It is never exactly equal to unity but it approaches this value in certain shell type transformers.

The mutual inductance between two coils, $L_1$ and $L_2$, may be expressed in terms of the inductance of each coil and the coefficient of coupling, K, as follows,

$$M = K\sqrt{L_1 L_2}$$

where M is in the same units of inductance as $L_1$ and $L_2$.

## SERIES INDUCTORS WITHOUT MAGNETIC COUPLING

When inductors are well shielded, or located far enough apart to make the effects of mutual inductance negligible, the inductance of the various inductors are added in the same manner that the resistances of resistors in series are added.

For example,

$$L_T = L_1 + L_2 + L_3 \ldots + L_N$$

where $L_t$ is the total inductance; $L_1$, $L_2$, $L_3$ are the inductances of $L_1$, $L_2$, $L_3$; and $L_N$ means that any number (N) of inductors may be used.

## SERIES INDUCTORS WITH MAGNETIC COUPLING

When two inductors in series are so arranged that the field of one links the other, the combined inductance is determined as

$$L_T = L_1 + L_2 \pm 2M$$

where $L_T$ is the total inductance, $L_1$ and $L2$ are the self inductances of $L_1$ and $L_2$ respectively, and M is the mutual inductance between the two inductors. The plus sign is used with M when the magnetomotive forces of the two inductors are aiding each other. The minus sign is used with M when the mmf's of the two inductors oppose each other. The factor 2 accounts for the influence of $L_1$ on $L_2$ and $L2$ on $L1$.

If the coils are arranged so that one can be rotated relative to the other to cause a variation in the coefficient of coupling, the mutual inductance between them can be varied. The total inductance, $L_T$, may be varied by an amount equal to 4M.

## PARALLEL INDUCTORS WITHOUT COUPLING

The total inductance, $L_T$, of inductors in parallel is calculated in the same manner that the total resistance of resistors in parallel is calculated, provided the coefficient of coupling between the coils is zero. For example,

$$\frac{1}{L_T} = \frac{1}{L_1} + \frac{1}{L_2} + \frac{1}{L_3} \ldots + \frac{1}{I_N}$$

where $L_1$, $L2$, and $L_3$ are the respective inductances of inductors $L_1$, $L_2$, and $L_3$; and $L_N$ means that any number (N) of inductors may be used.

## EFFECTS OF INDUCTANCE IN AN ELECTRIC CIRCUIT

### REACTION OF AN INDUCTOR TO FLUX CHANGE

Up to this point, the effects of inductance in an electric circuit have been considered as being controlled only by the particular inductor considered in each case. For instance, the time required for the growth and decay of current in the circuit shown in figure 10-7 is determined only by the inherent properties of the inductor L, namely, the ratio of its inductance to its resistance. When permitted to change at this inherent, or natural rate, the shaper of the L/R curve for any particular inductor will always remain the same. That is, it will always exhibit the same natural opposition to any change in current and flux, when left to its own tendencies. At this point, however, it must be pointed out that an inductor can be made to exhibit many different degrees of reaction. This is done by using external means to change the flux linking the inductor. These changes, such as variations in the applied frequency, usually do not occur at the same rate as the inductor's inherent rate. When the flux is changed RAPIDLY by some external means, the inductor's reaction is much greater than when the flux is changed gradually. That is, the inductor's self-induced emf is dependent on the RATE with respect to time at which the linking flux is changed.

The dependence of the induced voltage on the rate of change of flux with respect to time is shown by means of the simple apparatus in figure 10-10. When the switch is closed, current builds up to a maximum, and the lamp glows with its normal brilliance. If the iron core is inserted rapidly into the coil, the flux increases rapidly (because of the increased inductance of the coil), and the induced voltage opposes, according to Lenz's law, the source voltage. Therefore, less current flows through the lamp, and it dims momentarily. If the core is withdrawn rapidly from the coil, a portion of the flux that was established around the coil collapses. The resulting induced voltage opposes the decrease (again, according to Lenz's law) by aiding the source voltage, and the coil current increases. Consequently, the lamp burns brighter momentarily. The faster the iron core is moved the greater is the flux change per unit of time and the more noticeable is the effect on the lamp.

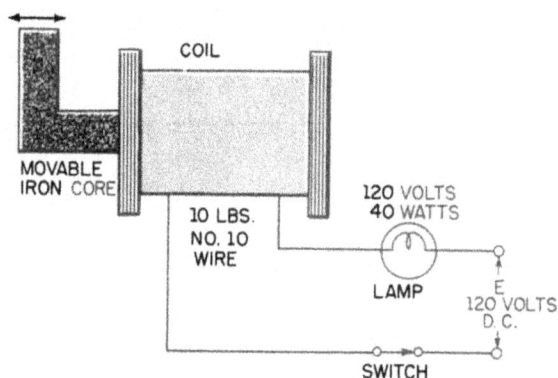

BE. 193

Figure 10-10.—Dependence of self-induced voltage on the rate of change of flux.

### EFFECTS OF INDUCTANCE IN AN A-C CIRCUIT

When a circuit containing a coil is energized with direct current, the coil's effect in the circuit is evident only when the circuit is energized, or when it is deenergized. For instance, when the switch in figure 10-11 is placed in position ① , the inductance of coil L will cause a delay in the time required for the lamps to attain normal brilliance. After they have attained normal brilliance, the inductance has no effect on the circuit as long as the switch remains closed. When the switch is opened, an electric spark will jump across the opening switch contacts. The emf which produces the spark is caused by the collapsing magnetic field cutting the turns of the inductor.

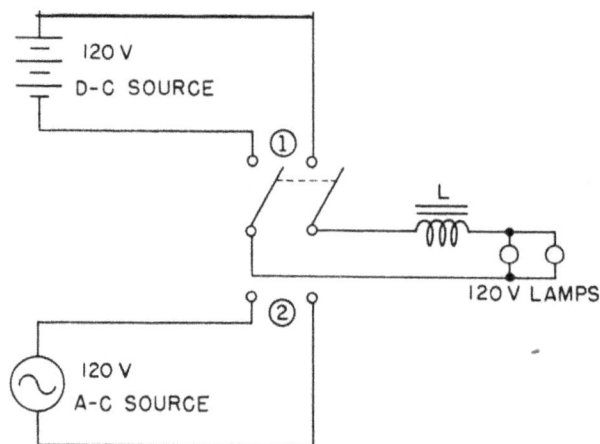

BE. 194

Figure 10-11.—Relative effects of inductance in d-c and a-c circuits.

When the inductive circuit is supplied with alternating current, however, the inductor's effect is continuous and much greater than when it was supplied with direct current. For equal applied voltages, the current through the circuit is less when a.c. is applied than when d.c. is applied, as may be demonstrated by the circuits of figure 10-11. The alternating current is accompanied by an alternating magnetic field around the coil, which cuts through the turns of the coil. This action induces a voltage in the coil that always opposes the changing current. When the switch is in position ① the lamps burn brightly on direct current but in position ② , although the effective value of the applied a-c voltage is equal to the d-c value, the lamps burn dimly because of the opposition developed across the inductance. Most of the applied voltage appears across L, with little remaining for the lamps.

## RELATION BETWEEN INDUCED VOLTAGE AND CURRENT

As stated previously, any change in current, either a rise or a fall, in a coil causes a corresponding change of the magnetic flux around the coil (fig. 10-12 (A)). If the current is sinusoidal, the induced voltage will also have the form of a sine wave. Because the current changes at its maximum rate when it is going through its zero value at $0°$, $180°$, and $360°$ (fig. 10-12 (B)), the flux change is also greatest at those times. Consequently, the self-induced voltage in the coil is at its maximum value at these instants. According to Lenz's law, the induced voltage always opposes the change in current. Thus, when the current is rising in a positive direction at $0°$, the induced emf is of opposite polarity to the impressed emf and opposes the rise in current. Later, when the current is falling toward its zero value at $180°$, the induced voltage is of the same polarity as the current and tends to keep the current from falling. Thus the induced voltage can be seen to lag the current by $90°$. The resistance of the coil is small and the principal opiosition to the current flow through the coil is the induced voltage, $E_{ind}$. The applied voltage, E, is slightly larger than $E_{ind}$ and diametrically opposed it, as indicated in the vector diagram (fig. 10-12 (C)).

The current lags the applied voltage in an inductive circuit by an angle of $90°$ and leads the induced voltage by $90°$. The induced voltage

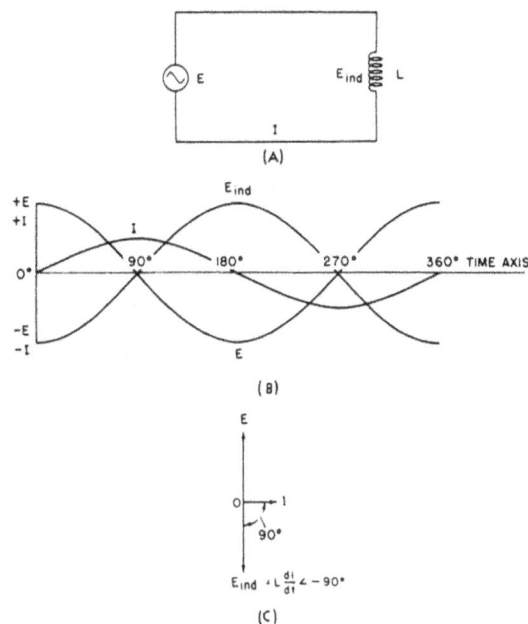

BE. 195

Figure 10-12.—Relation between induced voltage and current.

is always of opposite polarity to the applied voltage and is called a COUNTER EMF or a BACK EMF because it always opposes the change of current.

## INDUCTIVE KICK

It has been shown that an inductor in which there is a changing current becomes a source of EMF, and that the direction of this EMF is such that it tends to oppose the change in current producing it. As a result of this action the current in the inductor does not rise to its full value the instant the switch is closed, but rises at a rate which depends upon the L/R ratio. Likewise when the switch is opened, the source removed, and a short placed across the circuit, the current does not instantaneously fall to zero; it slowly decays at a rate determined by the L/R ratio of the discharge circuit.

The circuits so far considered have been composed of ideal components. However, components are not perfect, therefore no switch could be manufactured that would be capable of being moved from one position to another instantaneously. This leads to an explanation of the action of the induced voltage at the INSTANT a switch is opened. Because the magnitude of self-induced voltage can be extremely great even though the source voltage is very low, the development of high induced voltages when an inductive circuit is suddenly opened will now be studied.

Consider the circuit shown in figure 10-13 (A) consisting of a 6 volt battery, a switch, and a 30 henry coil. The resistor (R) represents the total circuit resistance, including the resistance of the coil. At the instant the switch is closed (time zero), 6 volts of CEMF will develop across the coil. This is shown in figure 10-13 (B) where $e_L$ = 6V at $t_0$.

Since at time zero the current, for all practical purposes has not started to flow, there is no voltage drop across the resistor. With an inductance of 30 henrys and a voltage of 6 volts, the initial rate of change (roc) will be:

$$roc = \frac{e}{L}$$

$$roc = \frac{6}{30}$$

$$roc = 0.2 \text{ amp/sec}$$

After the interval of time $\left(5\frac{L}{R}\right)$ the current has reached its final steady value. For the circuit of figure 10-13 (A) this will be:

$$T = \frac{L}{R}$$

$$T = \frac{30}{1}$$

$$T = 30 \text{ seconds}$$

But: steady state value equals:

$$t = 5\frac{L}{R} = 5T$$

Where: t = elapsed time, in seconds
Therefore: steady state will be reached in

$$5 \times T = t$$

$$5 \times 30 = 150 \text{ sec}$$

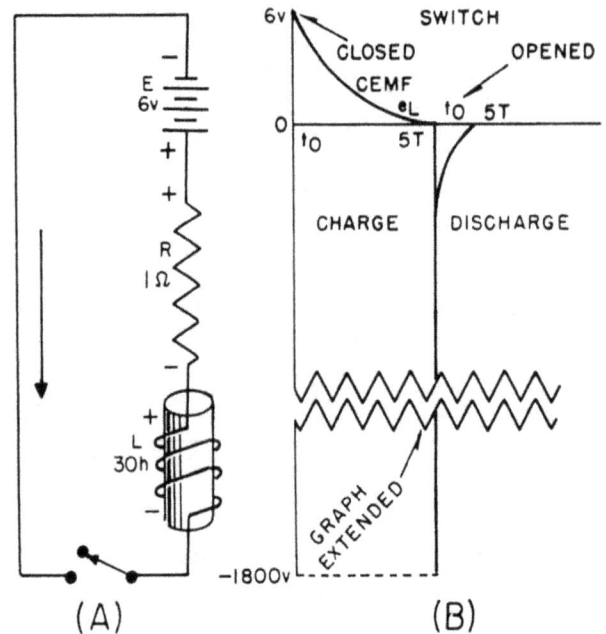

BE. 196

Figure 10-13.—Development of high induced voltage from low source.

The magnetic field around the coil is now fully established and steady. Figure 10-13 (B) shows no $e_L$ (CEMF) at $5\frac{L}{R}$ or 5T), therefore, there must be no change in current at this time.

If the switch is now opened, one side of the battery is disconnected and there is no complete path through which the current can flow. As a result, the current supplied by the battery stops immediately. However, the inductance of the coil opposes the change. Since the current supplied by the battery is no longer available to support the magnetic field, it will start to collapse. As it collapses, the magnetic lines of force cuts the turns of the coil. Whereas expanding lines of force induced an EMF with the polarity on the coil as shown in Figure 10-13 (A), collapsing lines of force induce an EMF in the coil with the opposite polarity. The inductance is therefore acting as a source of voltage attempting to maintain the same current flow in the circuit just as if the battery was still connected.

The decay time will be the same as the growth time if the same L/R ratio is maintained during decay as existed during growth. For an understanding of the change in time constant, picture for a moment, the conditions existing in the circuit shown in figure 10-14 (A). The current flowing in this circuit is:

$$I = \frac{E}{R}$$

$$I = \frac{6}{1}$$

$$I = 6 \text{ amps}$$

The voltage drop across the resistor ($E_R$) is:

$$E_{R1} = IR$$

$$E_{R1} = 6 \times 1$$

$$E_{R1} = 6 \text{ volts}$$

Now if it were possible to maintain the battery at a constant 6 volts and briefly maintain the current at 6 amperes while INSTANTANEOUSLY removing the one ohm resistor ($R_1$) and inserting a large resistance ($R_2$), as in figure 10-14 (B), the resistive voltage drop would increase:

$$E_{R2} = IR$$

$$E_{R2} = 6 \times 2000$$

$$E_{R2} = 12,000 \text{ volts}$$

The above conditions are impossible to obtain in the resistive circuit discussed and are inserted here merely to show the possibility of obtaining a high voltage pulse when exchanging a high resistance for a low resistance and MOMENTARILY maintaining the same current. These conditions can be obtained in the circuit of figure 10-13. At the exact INSTANT the switch contacts represent the insertion of an extremely high resistance (open circuit) the action of the inductance is maintaining the current at very near the value built up with the small resistance in the circuit has been increased by a very large amount, the L/R time constant of the circuit for decay is not the original 30 seconds. Due to the insertion of the high resistance, the time constant has become a fraction of a second. Transposing the following equation and substituting values will give an approximation of the voltage developed across the coil by the collapsing magnetic field.

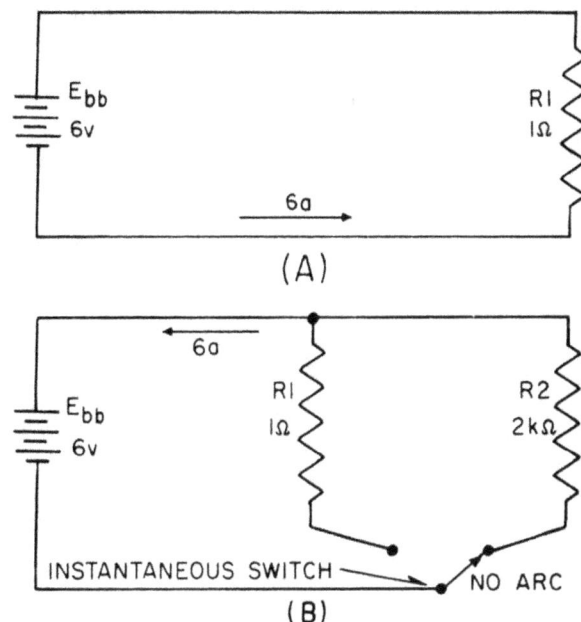

BE. 197

Figure 10-14.—Resistive demonstration circuit for high voltage pulse.

$$L = \frac{e}{\Delta i/\Delta t}$$

transposing:

$$e = L \frac{\Delta i}{\Delta t}$$

NOTE: Assuming for the purpose of illustration, that 0.00001 second ($\Delta t$) after the switch is opened the current has decreased to 5.9994 amps, a change ($\Delta i$) of 0.0006 amp:

Substituting:

$$e = L \frac{\Delta i}{\Delta t}$$

$$e = 30 \times \frac{0.0006}{0.00001}$$

$$e = 30 \times 60$$

$$e = 1800 \text{ volts}$$

The energy contained in the collapsing magnetic field must be dissipated somewhere within the circuit. The voltage developed in the conductor is sufficient to create an arc across the switch contacts. The energy in the magnetic field is dissipated in the heat of this arc. The energy expended in the arc can seriously burn an individual, damage the switch contacts, or break down the insulation of the coil. For these reasons care should be taken in the abrupt interruption of current in any inductive circuit.

The development of a large voltage pulse from a low voltage source (INDUCTIVE KICK) is not always a disadvantage, but is commonly used in the spark coil circuits (ignition system) of most gasoline engines.

# CHAPTER 11

# CAPACITANCE

Every electrical circuit, no matter how complex, is composed of no more than three basic electrical properties; resistance, inductance, and capacitance. Therefore, a thorough understanding of each of these basic properties is a necessary step toward the understanding of electrical equipment. Since resistance and inductance have been covered, the last of the basic three, capacitance, will now be discussed.

Two conductors separated by a nonconductor exhibit the property called CAPACITANCE, because this combination can store an electric charge; whereas inductance was defined as a property of a circuit which opposes a change in CURRENT. Capacitance is a property of a circuit which opposes a change in VOLTAGE. Where inductance stored energy in an ELECTROMAGNETIC field, capacitance stores energy in an ELECTROSTATIC field.

## REVIEW OF ELECTROSTATICS

In order to promote a clear understanding of capacitance, the reader should be thoroughly familiar with the theories and laws of electrostatics. For convenience, the main points of electrostatics will be briefly reviewed in this section.

When a charged body is brought into close proximity with another charged body, there is a force that causes the bodies to attract or repel one another. If the changed bodies possess the same sign of charge, a repelling force will exist between the two bodies. If they have unlike signs, there will be a force of attraction between them. The force of attraction or repulsion is caused by the electrostatic field that surrounds every charged body. If a material is charged positively, it has a deficiency of electrons. If it is charged negatively, it has an excess of electrons. The direction of the electrostatic field is represented by lines of force

drawn perpendicular to the charged surface and shown originating from the positively charged material. Each line of force is drawn in the form of an arrow and is shown pointing from positive to negative.

The force between charges is described by Coulomb's law: "The force existing between two charged bodies is directly proportional to the produce of the charges and inversely proportional to the square of the distance separating them."

If a test charge is inserted in an existing electrostatic field it will move toward one or the other of the charged areas which is causing the field to exist. The direction of movement will depend on whether the test charge is positive or negative. Previously it has been shown that a positive test charge placed in a field moves in the direction that the line of force points, from positive toward negative. In this case the test charge will be an electron and since the electron is negative it will move in a direction opposite to that of the positive charge. In other words, an electron in an electrostatic field will move AGAINST the arrow from negative toward positive. The above action is illustrated by figure 11-1.

If Coulomb's law is analyzed in connection with figure 11-1 it can be seen that the greater the distance between the electron and the positive charge the less the force of attraction. The importance of the distance between the charges creating the field will become apparent at a later time in this chapter.

One important characteristic of electrostatic lines of force is that they have the ability to pass through any known material.

## THE CAPACITOR

CAPACITANCE is defined as the property of an electrical device or circuit that tends to

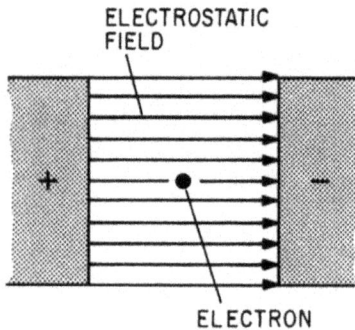

ELECTROSTATIC
FIELD

+

−

ELECTRON

BE.198

Figure 11-1.—Electron movement in an electrostatic field.

oppose a CHANGE in VOLTAGE. Capacitance
is also a measure of the ability of two conducting surfaces, separated by some form of nonconductor, to store an electric charge. For the
present time air will be used as the insulating
material between the conducting surfaces.

The device used in electrical circuits to
store a charge by virtue of an electrostatic field
is called a CAPACITOR. (The larger the capacitor, the larger the charge that can be stored.)

The simplest type of capacitor consists of
two metal plates separated by air. It has been
established that a free electron inserted in an
electrostatic field will move. The same is true,
with qualifications, if the electron is in a bound
state. The material between the two charged
surfaces of figure 11-1 (air in this case) is composed of atoms containing bound orbital electrons. Since the electrons are bound they can not
travel to the positively charged surface. Therefore, the resultant effect will be a distorting of
the electron orbits. The bound electrons will be
attracted toward the positive surface, and repelled from the negative surface. This effect is
illustrated in figure 11-2. In figure 11-2 (A),
there is no difference in charge placed across
the plates; and the structure of the atom's orbits
is undisturbed. If there is a difference in charge
across the plates as shown in figure 11-2 (B),
the orbits will be elongated in the direction of
the positive charge.

An energy is required to distort the orbits,
energy is transferred from the electrostatic
field to the electrons of each atom between the
charged plates. Since energy cannot be destroyed, the energy required to distort the
orbits can be recovered when the electron orbits
are permitted to return to their normal positions.

This effect is analogous to the storage of energy
in a stretched spring. A capacitor can thus
"store" electrical energy.

An illustration of a simple capacitor and its
schematic symbol is shown in figure 11-3. The
conductors that form the capacitor are called
PLATES. The material between the plates is
called the DIELECTRIC. In figure 11-3 (B), the
two vertical lines represent the connecting leads.
The two horizontal lines represent the capacitor
plates. Notice that the schematic symbol (B) and
the simple capacitor diagram (A) are similar in
appearance. In a practical capacitor, the parallel
plates may be constructed in various configurations (circular, rectangular, etc.); but the cross-
sectional area of the capacitor plates is
tremendously large in comparison to the cross-
sectional area of the connecting conductor. This
means that there is an abundance of free electrons available in each plate of the capacitor.
If the cross-sectional area and plate material of
the capacitor plates are the same, the number of
free electrons in each plate must be approximately the same. It should be noted that there
is a possibility of the difference in charge becoming so large as to cause ionization of the
insulating material to occur (cause bound electrons to be freed). This places a limit on the
amount of charge that can be stored in the
capacitor.

Figure 11-3 (A) depicts a capacitor in its
simplest form. It consists of two metal plates
separated by a thin layer of insulating material
(dielectric). When connected to a voltage source
(battery), the voltage forces electrons onto one
plate, making it negative, and pulls them off the
other, making it positive. Electrons cannot flow
through the dielectric. Since it takes a definite
quantity of electrons to "fill up," or charge, a
capacitor, it is said to have a CAPACITY. This

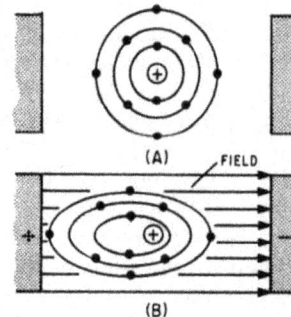

(A)

FIELD

+

−

(B)

BE.199

Figure 11-2.—Electron orbits with and without
the presence of an electric field.

189

characteristic is referred to as CAPACI-TANCE.

## DIELECTRIC MATERIALS

Various materials differ in their ability to support electric flux or to serve as dielectric material for capacitors. This phenomenon is somewhat similar to permeability in magnetic circuits. Dielectric materials, or insulators are rated in their ability to support electric flux in terms of a figure called the DIELECTRIC CONSTANT. The higher the value of dielectric constant (other factors being equal), the better is the dielectric material.

A vacuum is the standard dielectric for purposes of reference and is assigned the value of unity (or one). The dielectric constant of a dielectric material is also defined as the ratio of the capacitance of a capacitor having that particular material as the dielectric to the capacitance of the same capacitor having air as the dielectric. By way of comparison, the dielectric constant of pure water is 81; flint glass, 9.9; and paraffin paper, 3.5. The range of dielectric constants is much more restricted than is the range of permeabilities. Dielectric constants for some common materials are given in the following list:

| Material | Constant |
|---|---|
| Vacuum | 1.0000 |
| Air | 1.0006 |
| Paraffin paper | 3.5 |
| Glass | 5 - 10 |
| Mica | 3 - 6 |
| Rubber | 2.5 - 35 |
| Wood | 2.5 - 8 |
| Glycerine (15 °C) | 56 |
| Petroleum | 2 |
| Pure water | 81 |

Notice the dielectric constant for a vacuum. Since a vacuum is the standard of reference, it is assigned a constant of one; and the dielectric constants of all materials are compared to that of a vacuum. Since the dielectric constant of air has been determined experimentally to be approximately the same as that of a vacuum, the dielectric constant of AIR is also considered to be equal to one. The formula used to compute the value of capacitance using the physical factors just described is:

$$C \quad 0.2249 \ \left(\frac{KA}{d}\right) \qquad (11\text{-}1)$$

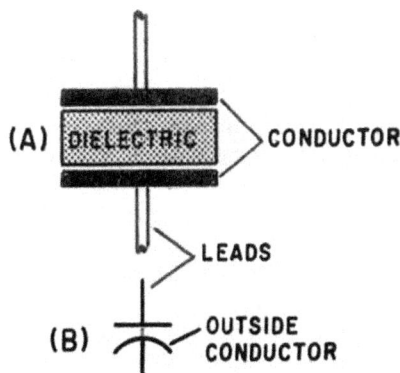

BE.200
Figure 11-3.—Capacitor and schematic symbol.

Where C = capacitance, in picofarads $(10^{-12})$
   A = area of one plate, in square inches
   d = distance between plates, in inches
   K = dielectric constant of insulating material.
0.2249 = a constant resulting from conversion from Metric to British units.

Example: Find the capacitance of a parallel plate capacitor with paraffin paper as the dielectric.

Given:     K = 3.5

      d = 0.05 inch

      A = 12 square inches

Solution:     $C = 0.2249 \left(\frac{KA}{d}\right)$      (11-1)

      $C = 0.2249 \left(\frac{3.5 \times 12}{0.05}\right)$

      C = 189 picofarads

Using equation (11-1) it is easy to visualize the effects on capacitance of the physical factors involved. It can be seen that capacitance is a direct function of the dielectric constant and the area of the capacitor plates, and an inverse function of the distance between the plates.

## UNIT OF CAPACITANCE

Capacitance is measured in a unit called the FARAD. This unit is a tribute to the memory of

Michael Faraday, a scientist who performed many early experiments with electrostatics and magnetism.

It was discovered that for a given value of capacitance, the ratio of charge deposited on one plate, to the voltage producing the movement of charge, is a constant value. This constant value is a measure of the amount of capacitance present. The symbol used to designate a capacitor is (C). The capacitance is equal to 1 farad when a voltage changing at the rate of 1 volt/per second causes a charging current of 1 amp to flow. This is expressed by the equation:

$$C = \frac{i}{\frac{\Delta e}{\Delta t}} \qquad (11\text{-}2)$$

Where C = capacitance, in farads

i = instantaneous current in amp

$\frac{\Delta e}{\Delta t}$ = rate of change of voltage, in volts, with time, in seconds

Equation (11-2) may be clearer if expressed as follows:

| Capacitance equals 1 farad when | = | Charging current of 1 amp flows |
|---|---|---|
| | | When voltage changes 1 volt in 1 second |

The farad can also be defined in terms of charge and voltage. A capacitor has a capacitance of 1 farad if it will store 1 coulomb of charge when connected across a potential of 1 volt. This relationship can be expressed mathematically as:

$$C = \frac{Q}{E}$$

Where C = capacitance in farads

Q = charge in coulombs

E = applied potential in volts

Example: What is the capacitance of two metal plates separated by one centimeter of air, if .001 coulomb of charge is stored when a potential of 200 volts is applied to the capacitor?

Given:  Q = 0.001 coulomb

E = 200 volts

Solution: $C = \frac{Q}{E}$ $\qquad (11\text{-}3)$

Converting to power of ten

$$C = \frac{10 \times 10^{-4}}{2 \times 10^2}$$

$$C = 5 \times 10^{-6}$$

$$C = 0.000005 \text{ farads}$$

Although this capacitance might appear rather small (five millionths of a farad), many electronic circuits require capacitors of much smaller value. Consequently the farad is a cumbersome unit, far too large for most applications. The MICROFARAD which is one millionth of a farad ($1 \times 10^{-6}$ farad) is a more convenient unit. The symbols used to designate microfarad are $\mu$f. In high frequency circuits even the microfarad becomes too large, and the unit MICROMICROFARAD (one millionth of a microfarad) is used. The symbols for micromicrofarads are $\mu\mu$f.

To avoid confusion and the use of double prefixes the name PICOFARAD (pf) is preferred in place of micromicrofarad. In powers of ten, 1 picofarad (or 1 micromicrofarad) is equal to $1 \times 10^{-12}$ farad.

In using equation (11-3) one must not deduce the mistaken idea that capacitance is dependent upon charge and voltage. Capacitance is determined entirely by physical factors such as plate area, plate spacing, etc.

## FACTORS AFFECTING THE VALUE OF CAPACITANCE

The capacitance of a capacitor depends on the three following factors:
1. The area of the plates.
2. The distance between the plates.
3. The dielectric constant of the material between the plates.

These three factors are related to the capacitance of a parallel-plate capacitor consisting of two plates by the formula

$$C = 0.2249\left(\frac{kA}{d}\right)$$

Where C is in picofarads, A is the area of one of the plates in square inches, d is the distance between the plates in inches, and k is the dielectric constant of the insulator separating the plates.

For example, the capacitance of a parallel-plate capacitor with an air dielectric and spacing of 0.0394 inch between the plates, each of which has an area of 15.5 square inches, is approximately

$$C = 0.225\left(\frac{1 \times 15.5}{0.0394}\right) = 88.5 \text{ picofarads}$$

From this formula it may seem that the capacitance increases when the plates are increased in area; it decreases if the spacing of the plates is increased; and it increases if the k-value is increased.

As previously mentioned, the dielectric constant, k, expresses the relative capacitance when materials other than air are used as the insulating material between the plates. For example, if mica is substituted for air as the dielectric, the capacitance increases from 3 to 6 times because the dielectric constant of mica is from 3 to 6 times greater than air (1).

If the capacitor is composed of more than two parallel plates, the capacitance is calculated by multiplying the preceding formula by N - 1, where N is the number of plates. The plates are interlaced as shown in figure 11-4, and the effect is that of increasing the capacitance of the two-plate capacitor by the factor N - 1. In the figure there are 11 plates, and the capacitance is 10 times that of a 2-plate capacitor of the same plate area, spacing, and dielectric material.

### VOLTAGE RATING OF CAPACITORS

In selecting or substituting a capacitor for use in a particular circuit, consideration must be given to (1) the value of capacitance desired and (2) the amount of voltage to which the capacitor is to be subjected. If the voltage applied across the plates is too great, the dielectric will break down and arcing will occur between the plates.

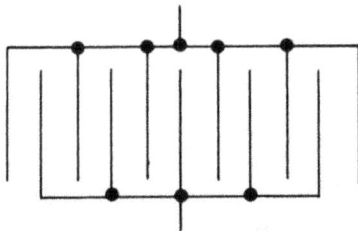

BE.201

Figure 11-4.—Capacitor multiple-plate construction.

The capacitor is then short-circuited and the possible flow of direct current through it can cause damage to other parts of the equipment. Capacitors have a voltage rating that should not be exceeded.

The working voltage of the capacitor is the maximum voltage that can be steadily applied without danger of arc-over. The working voltage depends on (1) the type of material used as the dielectric, and (2) the thickness of the dielectric.

The voltage rating of the capacitor is a factor in determining the capacitance because capacitance decreases as the thickness of the dielectric increases. A high-voltage capacitor that has a thick dielectric must have a larger plate area in order to have the same capacitance as a similar low-voltage capacitor having a thin dielectric. The voltage rating also depends on frequency because the losses, and the resultant heating effect, increase as the frequency increases.

A capacitor that may be safely charged to 500 volts d.c. cannot be safely subjected to alternating or pulsating direct voltages whose effective values are 500 volts. An alternating voltage of 500 volts (r.m.s.) has a peak voltage of 707 volts, and a capacitor to which it is applied should have a working voltage of at least 750 volts. The capacitor should be selected so that its working voltage is at least 50 percent greater than the highest voltage to be applied to it. Effective (rms) voltage and the action of capacitors in a-c circuits are described in a later chapter.

### CHARGING AND DISCHARGING A CAPACITOR

#### CHARGING

In order to better understand the action of a capacitor in conjunction with other components, the charge and discharge action of a purely capacitive circuit will be analyzed first. For ease of explanation the capacitor and voltage source used in figure 11-5 will be assumed to be perfect (no internal resistance, etc.) although this is impossible in practice.

In figure 11-5 (A) an uncharged capacitor is shown connected to a four-position switch. With the switch in position 1 the circuit is open and no voltage is applied to the capacitor. Initially each plate of the capacitor is a neutral body, and until a difference of potential is impressed across the capacitor no electrostatic field can exist between the plates.

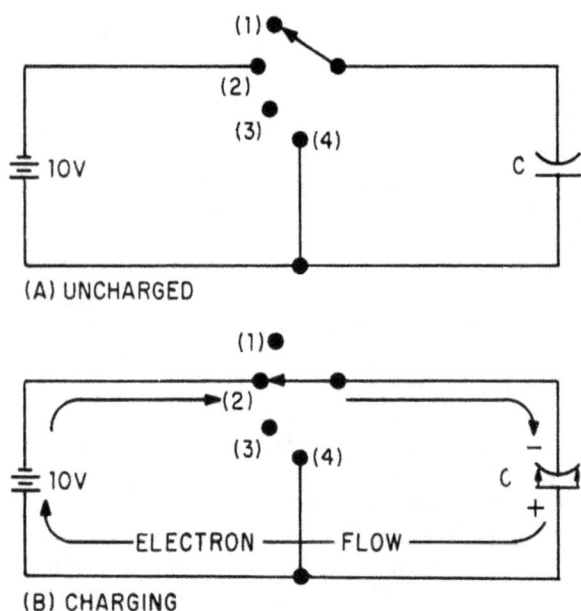

(A) UNCHARGED

(B) CHARGING

BE.202

Figure 11-5.—Charging a capacitor.

To CHARGE the capacitor the switch must be thrown to position 2 which places the capacitor across the terminals of the battery. Under the given conditions the capacitor would reach full charge instantaneously, however, the charging action will be spread out over a period of time in the following discussion so that a step-by-step analysis can be made.

At the instant the switch is thrown to position 2 (fig. 11-5 (B)) a displacement of electrons will occur simultaneously in all parts of the circuit. This electron displacement is directed away from the negative terminal and toward the positive terminal of the source. An ammeter connected in series with the source will indicate a brief surge of current as the capacitor charges.

If it were possible to analyze the motion of individual electrons in this surge of charging current, the following action would be observed. (See fig. 11-6.)

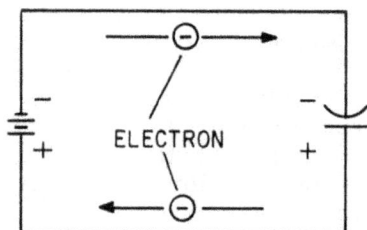

BE.203

Figure 11-6.—Electron motion during charge.

At the instant the switch is closed, the positive terminal of the battery extracts an electron from the bottom conductor and the negative terminal of the battery forces an electron into the top conductor. At this same instant an electron is forced into the top plate of the capacitor and another is pulled from the bottom plate. Thus, in every part of the circuit a clockwise DISPLACEMENT of electrons occurs in the manner of a chain reaction.

As electrons accumulate on the top plate of the capacitor and others depart from the bottom plate, a difference of potential develops across the capacitor. Each electron forced onto the top plate makes that plate more negative, while each electron removed from the bottom causes the bottom plate to become more positive. Notice that the polarity of the voltage which builds up across the capacitor is such as to oppose the source voltage. The source forces current around the circuit of figure 11-6 in a clockwise direction. The emf developed across the capacitor, however, has a tendency to force the current in a counterclockwise direction, opposing the source. As the capacitor continues to charge, the voltage across the capacitor rises until it is equal in amount to the source voltage. Once the capacitor voltage equals the source voltage, the two voltages balance one another and current ceases to flow in the circuit.

In studying the charging process of a capacitor it must be emphasized that NO current flows THROUGH the capacitor. The material between the plates of the capacitor must be an insulator.

To an observer stationed at the source or along one of the circuit conductors, the action has all the appearances of a true flow of current even though the insulating material between the plates of the capacitor prevents having a complete path. The current which appears to flow in a capacitive circuit is called DISPLACEMENT CURRENT.

To provide a better understanding of charging action, a capacitor can be compared to the mechanical system in figure 11-7. Part A of the diagram shows a metal cylinder containing a flexible rubber membrane which blocks off the cylinder. The cylinder is then filled with round balls as shown. If an additional ball is now pushed into the left hand side of the tube, the membrane will stretch and a ball will be forced out of the right hand end of the tube. To an observer who could not see inside the tube the ball would have the appearance of traveling all the way through the tube. For each ball inserted into the left hand side, one ball would leave the right hand side, although no balls actually pass all the way through the tube.

As more balls are forced into the tube it becomes increasingly difficult to force in additional balls, due to the tendency of the membrane to spring back to its original position.

If too many balls are forced into the tube, the membrane will rupture, and any number of balls can then be forced all the way through the tube.

A similar effect occurs in a capacitor when the voltage applied to the capacitor is too high. If an excessive amount of voltage is applied to a capacitor, the insulating material between the plates will break down and allow a current flow through the capacitor. In most cases this destroys the capacitor, necessitating its replacement.

When a capacitor is fully charged and the source voltage is equaled by the counter electromotive force (cem) across the capacitor, the electrostatic field between the plates of the capacitor will be maximum. Since the electrostatic field is maximum the energy stored in the dielectric will be maximum.

If the switch is now opened as shown in figure 11-8 (A), the electrons on the upper plate are isolated. Due to the intense repelling effect of these electrons, no electrons will return to the positive plate. Thus, with the switch in position 3, the capacitor will remain charged indefinitely. At this point it should be noted that the insulating dielectric material in a practical capacitor is not perfect and small leakage current will flow through the dielectric. This current will eventually dissipate the charge. A high quality capacitor may hold its charge for a month or more however.

To review briefly, when the capacitor is connected across a source, a surge of charging current will flow. This charging current develops a cemf across the capacitor which opposes the applied voltage. When the capacitor is fully charged the cemf will be equal to the applied voltage and charging current will cease. At full charge the electrostatic field between the plates is at maximum intensity and the energy stored in the dielectric is maximum. If the charged capacitor is disconnected from the source the charge will be retained for some period of time. The length of time the charge is retained depends on the amount of leakage current present. Since electrical energy is stored in the capacitor, a charged capacitor can act as a source.

DISCHARGING

To DISCHARGE a capacitor, the charges on the two plates must be neutralized. This is accomplished by providing a conducting path between the two plates (fig. 11-8 (B)). With the switch in position 4 the excess electrons on the negative plate can flow to the positive plate and neutralize its charge. When the capacitor is discharged the distorted orbits of the electrons in the dielectric return to their normal positions

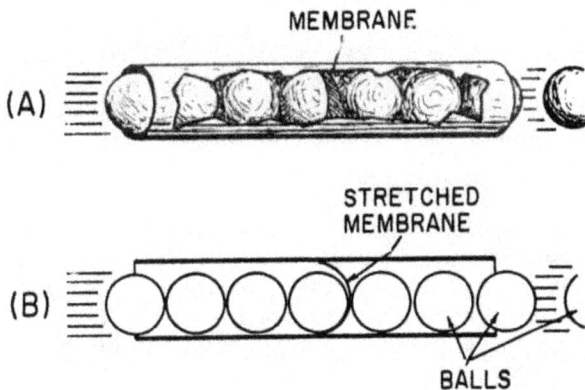

Figure 11-7.—Mechanical equivalent of a capacitor.

BE.204

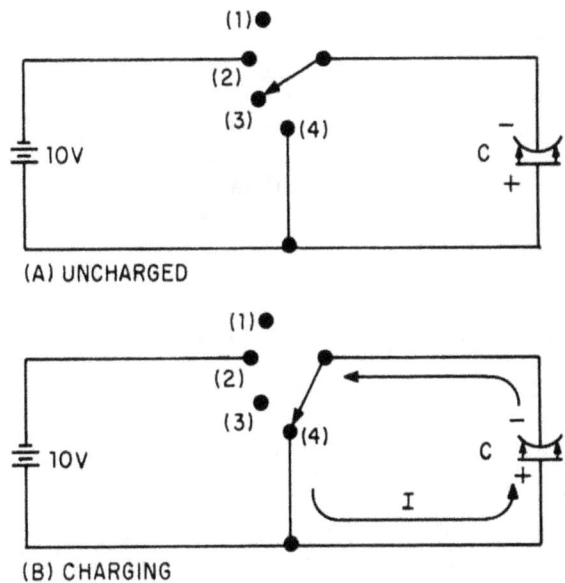

(A) UNCHARGED

(B) CHARGING

BE.205

Figure 11-8.—Discharging a capacitor.

and the stored energy is returned to the circuit. It is important to note that a capacitor does not consume power. The energy the capacitor draws from the source is recovered when the capacitor is discharged.

## CHARGE AND DISCHARGE OF AN RC SERIES CIRCUIT

Ohm's law states that the voltage across a resistance is equal to the current through it times the value of the resistance. This means that a voltage will be developed across a resistance ONLY WHEN CURRENT FLOWS through it.

A capacitor is capable of storing or holding a charge of electrons. When uncharged, both plates contain the same number of free electrons. When charged, one plate contains more free electrons than the other. The difference in the number of electrons is a measure of the charge on the capacitor. The accumulation of this charge builds up a voltage across the terminals of the capacitor, and the charge continues to increase until this voltage equals the applied voltage. The charge in a capacitor is related to the capacitance and voltage as follows:

$$Q = CE, \qquad (11-4)$$

in which Q is the charge in coulombs, C the capacitance in farads, and E the difference in potential in volts. Thus, the greater the voltage, the greater the charge on the capacitor. Unless a discharge path is provided, a capacitor keeps its charge indefinitely. Any practical capacitor, however, has some leakage through the dielectric so that the charge will gradually leak off.

A voltage divider containing resistance and capacitance may be connected in a circuit by means of a switch, as shown in figure 11-9 (A). Such a series arrangement is called an RC series circuit.

If S1 is closed, electrons flow counterclockwise around the circuit containing the battery, capacitor, and resistor. This flow of electrons ceases when C is charged to the battery voltage. At the instant current begins to flow, there is no voltage on the capacitor and the drop across R is equal to the battery voltage. The initial charging current, I, is therefore equal to $\frac{E_s}{R}$. Figure 11-9 (B) shows that at the instant the switch is closed, the entire input voltage, $E_s$, appears across R, and that the voltage across C is zero.

The current flowing in the circuit soon charges the capacitor. Because the voltage on

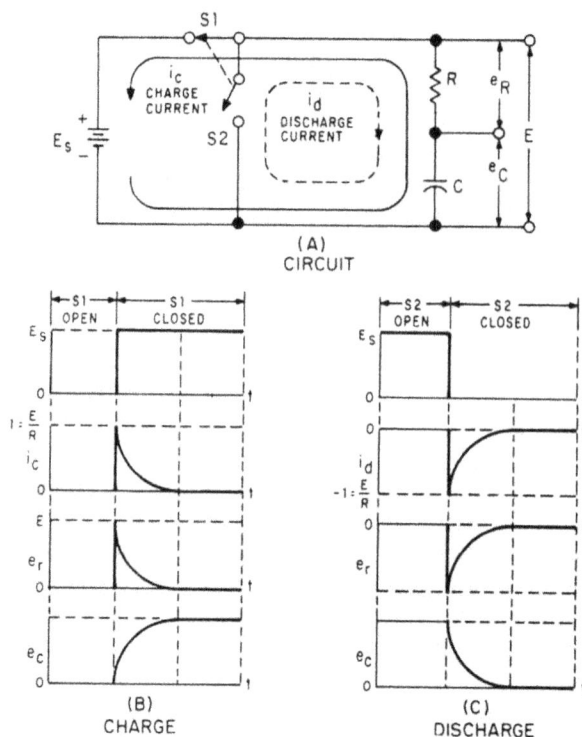

BE.206

Figure 11-9.—Charge and discharge of an RC series circuit.

the capacitor is proportional to its charge, a voltage, $e_c$, will appear across the capacitor. This voltage opposes the battery voltage—that is, these two voltages buck each other. As a result, the voltage $e_r$ across the resistor is $E_s - e_c$, and this is equal to the voltage drop $(i_cR)$ across the resistor. Because $E_s$ is fixed, $i_c$ decreases as $e_c$ increases.

The charging process continues until the capacitor is fully charged and the voltage across it is equal to the battery voltage. At this instant, the voltage across R is zero and no current flows through it. Figure 11-9 (B) shows the division of the battery voltage, $E_s$, between the resistance and capacitance at all times during the charging process.

If S2 is closed (S1 opened) in figure 11-9 (A), a discharge current, $i_d$, will discharge the capacitor. Because $i_d$ is opposite in direction to $i_c$, the voltage across the resistor will have a polarity opposite to the polarity during the charging time. However, this voltage will have the same magnitude and will vary in the same manner. During discharge the voltage across the capacitor is equal and opposite to the drop across the resistor, as shown in figure 11-9 (C).

The voltage drops rapidly from its initial value and then approaches zero slowly, as indicated in the figure.

## RC TIME CONSTANT

The time required to charge a capacitor to 63 percent (actually 63.2 percent) of maximum voltage or to discharge it to 37 percent (actually 36.8 percent) of its final voltage is known as the TIME CONSTANT of the circuit. An RC circuit with its charge and discharge graphs is shown in figure 11-10. The value of the time constant in seconds is equal to the product of the circuit resistance in ohms and its capacitance in farads, one set of values of which is given in figure 11-10 (A). RC is the symbol used for this time constant.

Some useful relations used in calculating RC time constants are as follows:

R (ohms) x C (farads) $\quad$ = t (seconds)
R (megohms) x C (microfarads) = t (seconds)
R (ohms) x C (microfarads) $\quad$ = t (microseconds)
R (megohms) x C (picofarads) = t (microseconds)

The time constant may also be defined as the time required to charge or discharge a capacitor completely IF it continues to charge or discharge at its initial rate. As may be seen in figure 11-10 (B), the slope of the dotted tangent line, OX, indicates the initial rate of charge. At this rate, the capacitor would be completely charged in RC seconds. Likewise, the slope of the dotted tangent line, YZ, indicates the initial rate of discharge with respect to time, and at this rate the capacitor would be completely discharged in RC seconds.

The equation for the rise in voltage, $e_c$, across the capacitor is

$$e_c = E \left(1 - \frac{1}{2.718^{\frac{t}{RC}}}\right)$$

where $e_c$, is the instantaneous voltage across the capacitor, E the applied voltage (100 volts in this case), t the time in seconds, R the resistance in ohms, C the capacitance in farads (0.000001 farad or 1$\mu$f in this case), and the number 2.718 the natural logarithm base. Figure 11-10 (B) shows a graph of this equation.

When t = RC, the exponent, $\frac{t}{RC}$, reduces to 1. Therefore,

$$e_c = E \left(1 - \frac{1}{2.718}\right) = 0.632E$$

In other words, when t = RC, $e_c$ is equal to 63.2 percent of E, the maximum value. When the maximum is 100 volts, as shown, the voltage across the capacitor increases to 63.2 volts in RC seconds—that is, in 10 microseconds.

The equation for the charging current, $i_c$, is

$$i_c = \frac{E}{R} \left(\frac{1}{2.718^{\frac{t}{RC}}}\right)$$

The graph of this equation is shown in figure 11-10 (B). When t = RC, $i_c$ = 0.368 x $\frac{E}{R}$; that is, when t = 10 microseconds, $i_c$ = 0.368 x $\frac{100}{10}$ = 3.68 ampere.

## UNIVERSAL TIME CONSTANT CHART

Because the impressed voltage and the values of R and C or R and L usually will be known, a universal time constant chart (fig. 11-11) can be used. Curve A is a graph of the voltage across the resistor in series with the inductor on the growth of current. Curve B is a graph of capacitor voltage on discharge, capacitor current on charge, inductor current on decay, or the voltage across the resistor in series with the capacitor on charge. The graphs of resistor voltage and current and inductor voltage on discharge are not shown because negative values would be involved.

The time scale (horizontal scale) is graduated in terms of the RC or $\frac{L}{R}$ time constants so that the curves may be used for any value of R and C or L and R. The voltage and current scales (vertical scales) are graduated in terms of the fraction of the maximum voltage or current so that the curves may be used for any value of voltage or current. If the time constant and the initial or final voltage for the circuit in question are known, the voltages across the various parts of the circuit can be obtained from the curves for any time after the switch is closed, either on charge or discharge. The same reasoning is true of the current in the circuit.

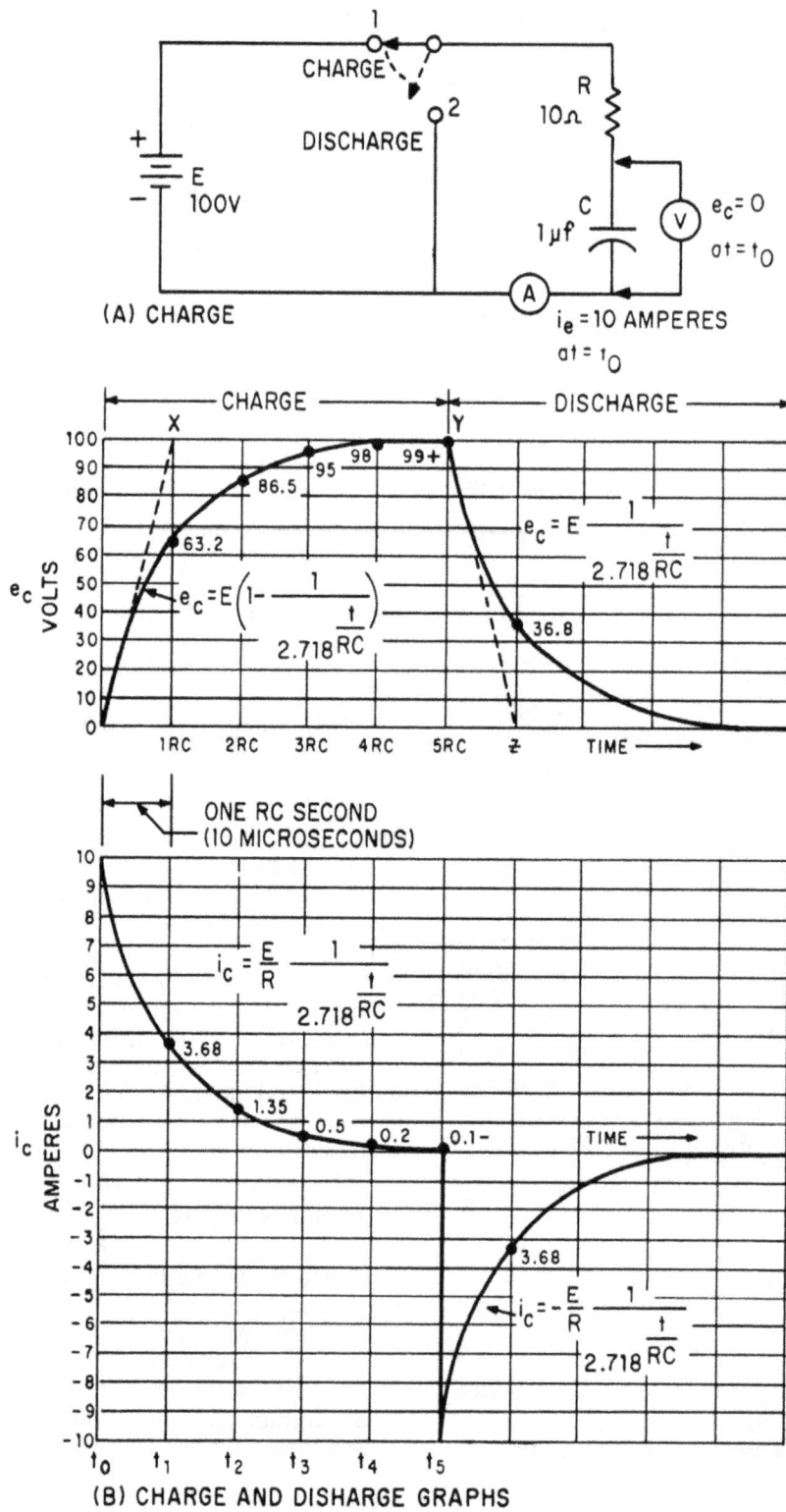

Figure 11-10.—RC time constant.

BE.207

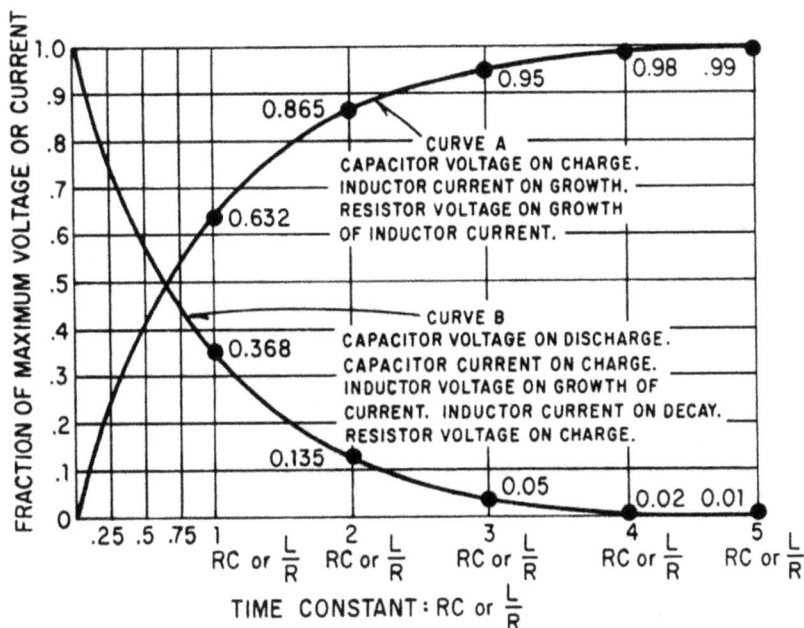

BE.208

Figure 11-11.—Universal time constant chart for RC and RL circuits.

The following problem illustrates how the universal time constant chart may be used.

A circuit is to be designed in which a capacitor must charge to one-fifth (0.2) of the maximum charging voltage in 100 microseconds (0.0001 second). Because of other considerations, the resistor must have a value of 20,000 ohms. What size of capacitor is needed?

Curve A is first consulted to determine the RC time necessary to give 0.2 of the full voltage. The time is less than 0.25 RC, approximately 0.22 RC. If 0.22 RC must be equal to 100 microseconds, one complete RC must be equal to $\frac{100}{0.22}$ = 455 microseconds, or 0.000455 second. Therefore,

$$RC = 0.000455.$$

Substituting the known value of R and solving for C,

$$C = \frac{0.000455}{20,000} = 0.000000023 \text{ farad},$$

or 0.023 microfarad.

The graphs shown in figure 11-10 are not entirely complete—that is, the charge or discharge (or the growth or decay) is not quite complete in 5 RC or 5 $\frac{L}{R}$ seconds. However, when the values reach 0.99 of the maximum (corresponding to 5 RC or 5 $\frac{L}{R}$) the graphs may be considered accurate enough for most purposes.

## CAPACITORS IN SERIES AND PARALLEL

Capacitors may be connected in series or parallel to give resultant values, which may be either the sum of the individual values (in parallel) or a value less than that of the smallest capacitance (in series).

### CAPACITORS IN SERIES

A circuit consisting of a number of capacitors in series is similar in some respects to one containing several resistors in series. In a series capacitive circuit the same DISPLACEMENT CURRENT flows through each part of the circuit and the applied voltage will divide across the individual capacitors.

Figure 11-12 shows a circuit containing a source and two series capacitors. When the switch is closed, current will flow in the direction indicated by the arrows on the diagram. Since there is only one path for current, the

198

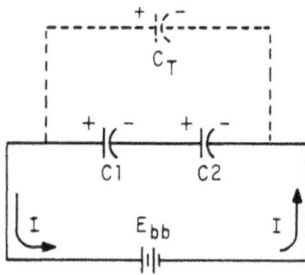

BE.209

Figure 11-12.—Series capacitive circuit.

amount of charge current in motion is the same in all parts of the circuit. This current is of brief duration and will flow only until the total voltage across the capacitors is equal to the source voltage.

Since the charge (Q) is the same in all parts of the circuit:

$$Q_t = Q_1 = Q_2 = \ldots\ldots Q_n \qquad (11\text{-}5)$$

also:

$$C = \frac{Q}{E} \qquad (11\text{-}4)$$

transposing:

$$E = \frac{Q}{C} \qquad (11\text{-}6)$$

Since the sum of the capacitor voltages must equal the source voltage (Kirchhoff's law):

$$E_t = E_1 + E_2 + \ldots E_n \qquad (11\text{-}7)$$

Substituting equation (11-6) into (11-7)

$$\frac{Q_t}{C_t} = \frac{Q_1}{C_1} + \frac{Q_2}{C_2} + \ldots\ldots \frac{Q_n}{C_n} \qquad (11\text{-}8)$$

Since by equation (11-5) all the charges are the same, dividing each term of (11-8) by $Q_t$ yields:

$$\frac{1}{C_t} = \frac{1}{C_1} + \frac{1}{C_2} + \ldots\ldots \frac{1}{C_n} \qquad (11\text{-}9)$$

Taking the reciprocal of both sides:

$$C_t = \frac{1}{\frac{1}{C_1} + \frac{1}{C_2} + \ldots\ldots \frac{1}{C_n}} \qquad (11\text{-}10)$$

Where $C_t$, $C_1$, etc., are in farads.

Equation (11-10) is the general equation used to compute the total capacitance of capacitors

connected in series. Notice the similarity between this equation and the one used to find equivalent resistance of parallel resistors. If the circuit contains only two capacitors the product over the sum formula can be used.

$$C_t = \frac{C_1 C_2}{C_1 + C_2} \qquad (11\text{-}11)$$

Where: $C_t$, $C_1$, etc., are in farads.

As might be anticipated from the above equations, the total capacitance of series connected capacitors will always be smaller than the smallest of the individual capacitors.

Example. Determine the total capacitance of a series circuit containing three capacitors of $0.01\,\mu f$, $0.25\,\mu f$, and $50,000\,pf$ respectively.

Given:

$$C_1 = 0.01\,\mu f$$
$$C_2 = 0.25\,\mu f$$
$$C_3 = 50,000\,pf$$

Solution:

$$C_t = \frac{1}{\frac{1}{C_1} + \frac{1}{C_2} + \frac{1}{C_3}} \qquad (11\text{-}10)$$

$$C_t = \frac{1}{\frac{1}{.01\mu f} + \frac{1}{.25\mu f} + \frac{1}{50,000 pf}}$$

Converting to powers of ten.

$$C_t = \frac{1}{\frac{1}{1 \times 10^{-8}} + \frac{1}{25 \times 10^{-8}} + \frac{1}{5 \times 10^{-8}}}$$

$$C_t = \frac{1}{100 \times 10^6 + 4 \times 10^6 + 20 \times 10^6}$$

$$C_t = \frac{1}{124 \times 10^6}$$

$$C_t = 0.008\mu f$$

The total capacitance of 0.008 $\mu f$ is slightly smaller than the smallest capacitor (0.01$\mu f$).

CAPACITORS IN PARALLEL

When capacitors are connected in parallel one plate of each capacitor is connected directly to one terminal of the source, while the other plate of each capacitor is connected to the other terminal of the source. In figure 11-13, since all the negative plates of the capacitors are connected together, and all the positive plates are connected together, $C_t$ appears as a capacitor with a plate area equal to the sum of all the individual plate areas. As previously mentioned, capacitance is a direct function of plate area. Connecting capacitors in parallel effectively increases plate area and thereby the capacitance.

For capacitors connected in parallel the total charge is the sum of all the individual charges.

$$Q_t = Q_1 + Q_2 + Q_3 + \ldots Q_n \qquad (11\text{-}12)$$

Transposing formula (11-4):

$$Q = CE \qquad (11\text{-}13)$$

Substitute: (11-13 into (11-12)

$$C_t E = C_1 E + C_2 E + C_3 E \qquad (11\text{-}14)$$

Divide both sides by E:

$$C_t = C_1 + C_2 + C_3 + \ldots \ldots C_n \qquad (11\text{-}15)$$

Where: all capacitances are in the same units.

Example: Determine the total capacitance in a parallel capacitive cirucit:

Given:   $C_1 = 0.03$

$C_2 = 2\mu f$

$C_3 = 0.25\mu f$

Solution:   $C_t = C_1 + C_2 + C_3 \qquad (11\text{-}14)$

$C_t = 0.03 + 2 + 0.25$

$C_t = 2.28\mu f$

## SERIES PARALLEL CONFIGURATION

If capacitors are connected in a combination of series and parallel the total capacitance is found by applying equations (11-10) and 11-14) to the individual branches.

Example. Determine the total capacitance of the circuit in figure 11-14.

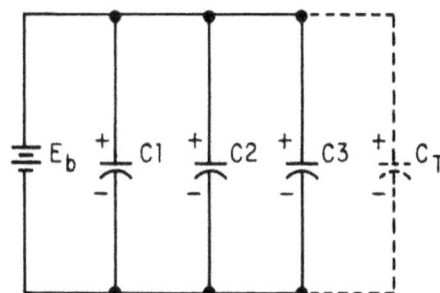

BE.210

Figure 11-13.—Parallel capacitive circuit.

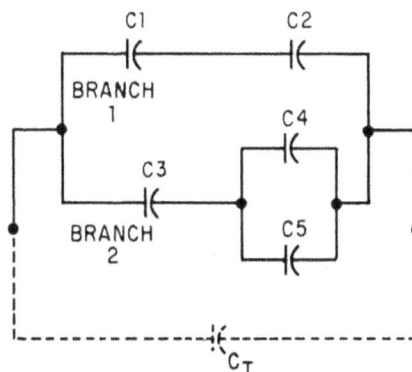

BE.211

Figure 11-14.—Series-parallel capacitance configuration.

Given:   $C_1 = 0.06\mu f$

$C_2 = 0.002\mu f$

$C_3 = 3\mu f$

$C_4 = 0.005\mu f$

$C_5 = 0.001\mu f$

Find:   $C_t = ?$

Solution: Simplify the circuit into separate branches.

BRANCH 1 - consists of a series combination $C_1$ and $C_2$. To determine total capacitance of this branch only:

Branch 1: $\quad C_{t1} = \dfrac{C_1 C_2}{C_1 + C_2}$ $\qquad$ (11-11)

$$C_{t1} = \frac{0.6 \times 0.002}{0.06 + 0.002}$$

$$C_{t1} = \frac{0.00012}{0.062}$$

$$C_{t1} = 0.00193 \mu f$$

BRANCH 2 - consists of a series-parallel combination. Solve for total capacitance of branch 2 only. The equivalent capacitance ($C_{eq}$) of the parallel combination of $C_4$ and $C_5$ must be determined first by the use of equation (11-14) for parallel configurations:

$$C_{eq} = C_4 + C_5$$

$$C_{eq} = 0.005 + 0.001$$

$$C_{eq} = 0.006 \mu f$$

Branch 2 is now reduced to an equivalent series circuit consisting of $C_3$ and the equivalent capacitance of $C_4$ and $C_5$.

The circuit figure 11-14 has been simplified considerably and appears as the circuit in figure 11-15. To illustrate the use of the

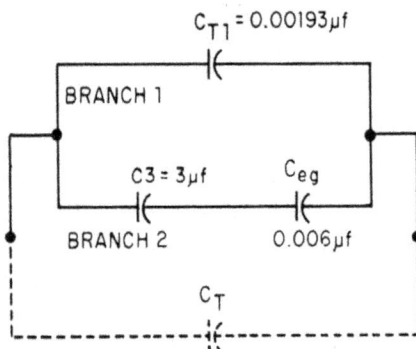

BE.212

Figure 11-15.—Simplified series-parallel capacitance configuration.

reciprocal method the total capacitance of branch 2 is determined using equation (11-10):

$$C_{t2} = \cfrac{1}{\cfrac{1}{C_3} + \cfrac{1}{C_{eq}}} \qquad (11-10)$$

$$C_{t2} = \cfrac{1}{\cfrac{1}{3} + \cfrac{1}{0.006}}$$

$$C_{t2} = \frac{1}{0.333 + 166.666}$$

$$C_{t2} = 0.00598 \mu f$$

NOTE: Equation (11-11) could also have been used to solve the previous problems.

The entire problem is now reduced to a simple parallel combination consisting of $C_{t1}$ (capacity branch 1) and $C_{t2}$ (capacity branch 2). Equation (11-15) is now used to determine the total capacitance of the original circuit.

$$C_t = C_{t1} + C_{t2}$$

$$C_t = 0.00193 + 0.00598$$

$$C_t = 0.00791 \mu f$$

CAPACITOR LOSSES

Losses occurring in capacitors may be attributed to either dielectric leakage or dielectric hysteresis. Dielectric hysteresis may be defined as an effect in a dielectric material similar to the hysteresis found in magnetic materials. It is a result of the changes in orientation of electron orbits in the dielectric because of the rapid reversals of the polarity of the line voltage. The amount of loss due to hysteresis depends upon the type of dielectric used; vacuum shows the smallest loss.

Dielectric leakage occurs in a capacitor as the result of leakage of current through the dielectric. Normally, it is assumed that the dielectric will effectively prevent the flow of current through the capacitor. However, while the resistance of the dielectric is extremely high, a minute amount of current does flow. Ordinarily, this current is so small that for all practical purposes it is considered to be of no consequence. However, if the leakage through the dielectric is abnormally high, there will be a rapid loss of charge and an overheating of the capacitor.

The power factor of a capacitor is determined by dielectric losses. If the losses are negligible and the capacitor returns the total charge to the circuit, it is considered to be a perfect capacitor with a power factor of zero. Therefore, the power factor of a capacitor is a measurement of its efficiency.

## CAPACITOR TYPES

Capacitors may be divided into two major groups—fixed and variable.

### FIXED CAPACITORS

Fixed capacitors are constructed in such manner that they possess a fixed value of capacitance which cannot be adjusted. They may be classified according to the type of material used as the dielectric, such as paper, oil, mica, and electrolyte.

A PAPER CAPACITOR is one that uses paper as its dielectric. It consists of flat thin strips of metal foil conductors, separated by the dielectric material. In this capacitor the dielectric used is waxed paper. Paper capacitors usually range in value from about 300 picofarads to about 4 microfarads. Normally, the voltage limit across the plates rarely exceeds 600 volts. Paper capacitors are sealed with wax to prevent the harmful effects of moisture and to prevent corrosion and leakage.

Many different kinds of outer covering are used for paper capacitors, the simplest being a tubular cardboard. Some types of paper capacitors are encased in a mold of very hard plastic; these types are very rugged and may be used over a much wider temperature range than the cardboard-case type. Figure 11-16 (A) shows the construction of a tubular paper capacitor; part (B) shows a completed cardboard encased capacitor.

A MICA CAPACITOR is made of metal foil plates that are separated by sheets of mica, which form the dielectric. The whole assembly is covered in molded plastic. Figure 11-17 (A) shows a cutaway view of a mica capacitor. By molding the capacitor parts into a plastic case, corrosion and damage to the plates and dielectric are prevented in addition to making the capacitor mechanically strong. Various

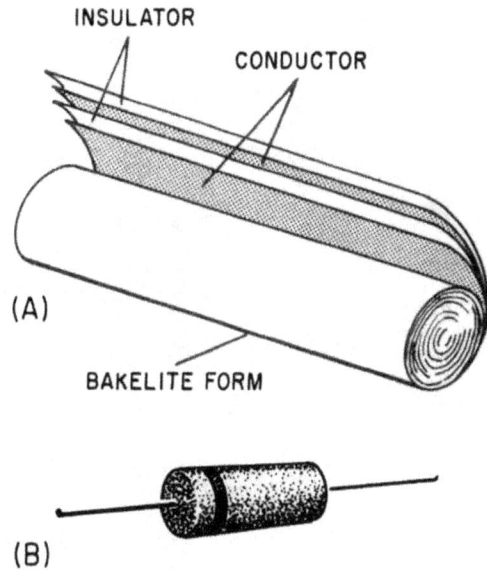

Figure 11-16.—Paper capacitor.

BE.213

types of terminals are used to connect mica capacitors into circuits; these are also molded into the plastic case.

Mica is an excellent dielectric and will withstand higher voltages than paper without allowing arcing between the plates. Common values of mica capacitors range from approximately 50 picofarads to about 0.02 microfarad. Some typical mica capacitors are shown in figure 11-17 (B).

A CERAMIC CAPACITOR is so named because of the use of ceramic dielectrics. One type of ceramic capacitor uses a hollow ceramic cylinder as both the form on which to construct the capacitor and as the dielectric material. The plates consist of thin films of metal deposited on the ceramic cylinder.

A second type of ceramic capacitor is manufactured in the shape of a disk. After leads are attached to each side of the capacitor, the capacitor is completely covered with an insulating moisture-proof coating. Ceramic capacitors usually range in value between 1 picofarad and 0.01 microfarad and may be used with voltages as high as 30,000 volts. Typical capacitors are shown in figure 11-18.

ELECTROLYTIC CAPACITORS are used where a large amount of capacitance is required. As the name implies, electrolytic capacitors contain an electrolyte. This electrolyte can be in the form of either a liquid (wet electrolytic capacitor) or a paste (dry electrolytic capacitor).

(A)

(B)

BE.214

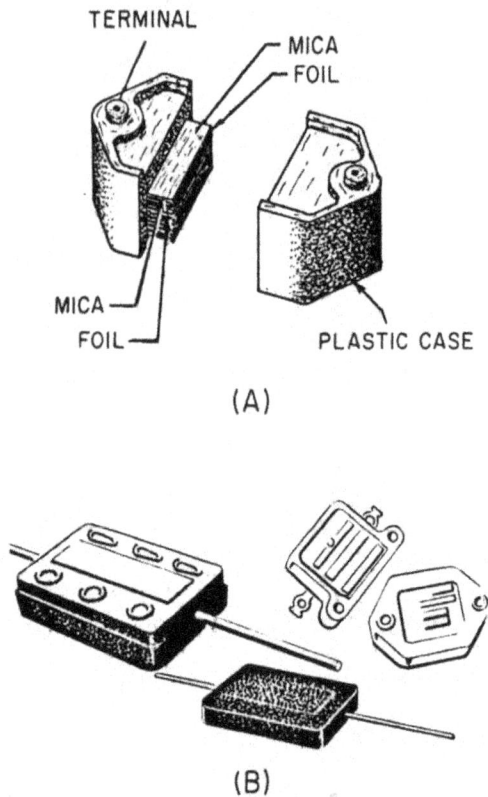

Figure 11-17.—Typical mica capacitors.

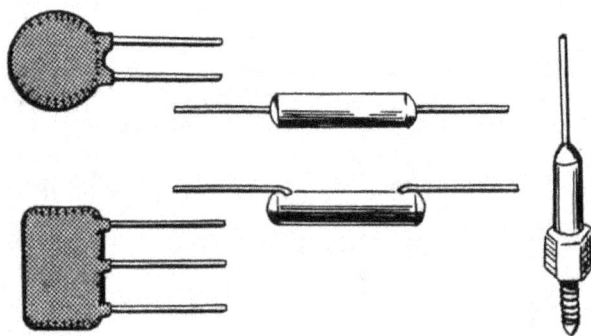

BE.215

Figure 11-18.—Ceramic capacitors.

Wet electrolytic capacitors are no longer in popular use due to the care needed to prevent spilling of the electrolyte.

Dry electrolytic capacitors consist essentially of two metal plates between which is placed the electrolyte. In most cases the capacitor is housed in a cylindrical aluminum container which acts as the negative terminal of the capacitor (fig. 11-19). The positive terminal (or terminals if the capacitor is of the multisection type) is in the form of a lug on the bottom end of the container. The size and voltage rating of the capacitor is generally printed on the side of the aluminum case.

An example of a multisection type of electrolytic capacitor is depicted in figure 11-19. The cylindrical aluminum container will normally enclose four electrolytic capacitors into one can. Each section of the capacitor is electrically independent of the other sections and one section may be defective while the other sections are still good. The can is the common negative connection with separate terminals for the positive connections identified by an embossed mark as shown in figure 11-19. The common identifying marks on electrolytic capacitors are the half moon, triangle, square, and no identifying mark. By looking at the bottom of the container and the identifying sheet pasted to the side of the container, the technician can easily identify each section.

Internally, the electrolytic capacitor is constructed similarly to the paper capacitor. The positive plate consists of aluminum foil covered with an extremely thin film of oxide which is formed by an electrochemical process. This thin oxide film acts as the dielectric of the capacitor. Next to, and in contact with the oxide, is placed

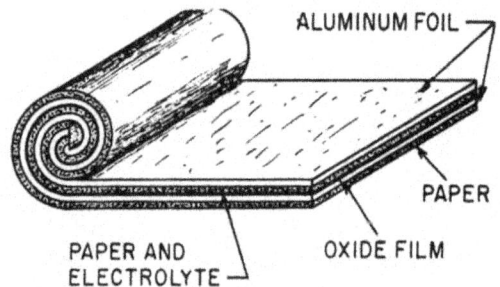

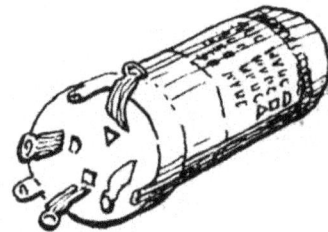

BE.216

Figure 11-19.—Construction of an electrolytic capacitor.

a strip of paper or gauze which has been impregnated with a paste-like electrolyte. The electrolyte acts as the negative plate of the capacitor. A second strip of aluminum foil is then placed against the electrolyte to provide electrical contact to the negative electrode (electrolyte). When the three layers are in place they are rolled up into a cylinder as shown in figure 11-19.

Electrolytic capacitors have two primary disadvantages in that they are POLARIZED, and they have a LOW LEAKAGE RESISTANCE. Should the positive plate be accidentally connected to the negative terminal of the source, the thin oxide film dielectric will dissolve and the capacitor will become a conductor (i.e., it will short). The polarity of the terminals is normally marked on the case of the capacitor. Since electrolytic capacitors are polarity sensitive, their use is ordinarily restricted to d-c circuits or circuits where a small a-c voltage is superimposed on a d-c voltage. Special electrolytic capacitors are available for certain a-c applications, such as motor starting capacitors. Dry electrolytic capacitors vary in size from about 4 microfarads to several thousand microfarads, and have a voltage limit of approximately 500 volts.

The type of dielectric used and its thickness govern the amount of voltage that can safely be applied to a capacitor. If the voltage applied to a capacitor is high enough to cause the atoms of the dielectric material to become ionized, an arc over will take place between the plates. If the capacitor is not self-healing, its effectiveness will be impaired. The maximum safe voltage of a capacitor is called its WORKING VOLTAGE and is indicated on the body of the capacitor. The working voltage of a capacitor is determined by the type and thickness of the dielectric. If the thickness of the dielectric is increased, the distance between the plates is also increased and the working voltage will be increased. Any change in the distance between the plates will cause a change in the capacitance of a capacitor. Because of the possibility of voltage surges (brief high amplitude pulses) a margin of safety should be allowed between the circuit voltage and the working voltage of a capacitor. The working voltage should always be higher than the maximum circuit voltage.

OIL CAPACITORS are often used in radio transmitters where high output power is desired. Oil-filled capacitors are nothing more than paper capacitors that are immersed in oil.

The oil impregnated paper has a high dielectric constant which lends itself well to the production of capacitors that have a high value. Many capacitors will use oil with another dielectric material to prevent arcing between the plates. If an arc should occur between the plates of an oil-filled capacitor, the oil will tend to reseal the hole caused by the arc. These types of capacitors are often called SELF-HEALING capacitors.

## VARIABLE CAPACITORS

Variable capacitors are constructed in such manner that their value of capacitance can be varied. A typical variable capacitor (adjustable capacitor) is the rotor/stator type. It consists of two sets of metal plates that are arranged so that the rotor plates move between the stator plates. Air is the dielectric. As the position of the rotor is changed, the capacitance value is likewise changed. This is the type capacitor used for tuning most radio receivers and it is shown in figure 11-20. Another type variable (trimmer) capacitor is shown in figure 11-21; it consists of two plates separated by a sheet of mica. A screw adjustment is used to change the distance between the plates, thereby changing the capacitance.

## COLOR CODES FOR CAPACITORS

Although the value of a capacitor may be indicated by printing on the body of a capacitor, many capacitive values are indicated by the use of a color code. The colors used to represent the numerical value of a capacitor are the same as those used to identify resistance values. There are two color coding systems that are currently in popular use: the Joint Army-Navy (JAN) and the Radio Manufacturers' Association (RMA) systems.

In each of these systems a series of colored dots (sometimes bands) is used to denote the value of the capacitor. Mica capacitors are marked with either three dots or six dots. Both systems are similar, but the six dot system contains more information about the capacitor, such as working voltage, temperature coefficient, etc. Capacitors are manufactured in various sizes and shapes. Some are small tubular resistor like devices, and others are molded rectangular flat components.

An explanation of capacitor color codes is provided in appendix IV.

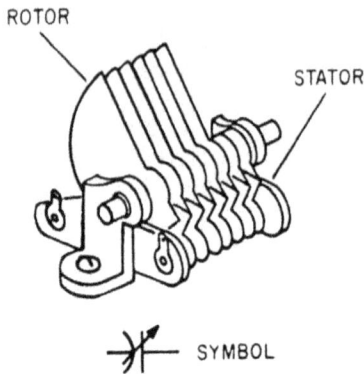

SYMBOL

BE.217

Figure 11-20.—Rotor-stator type variable capacitor.

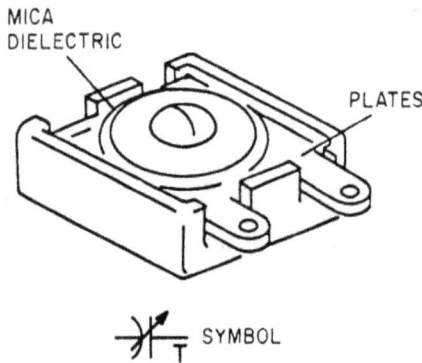

MICA DIELECTRIC

PLATES

SYMBOL

BE.218

Figure 11-21.—Trimmer capacitor.

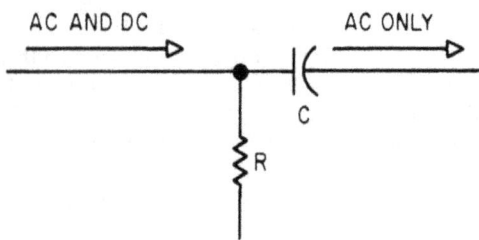

AC AND DC

AC ONLY

C

R

BE.219

Figure 11-22.—Blocking capacitor.

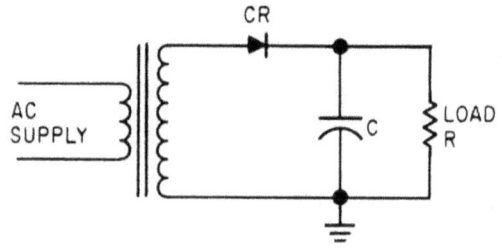

CR

AC SUPPLY

C

LOAD R

BE.220

Figure 11-23.—Filter capacitor in a power supply circuit.

## CAPACITOR APPLICATION

Capacitors are used in many ways in electrical and electronic equipment and circuits. A few of the more common applications are blocking direct current, filtering, and spark suppression.

The blocking capacitor in figure 11-22 is used in a circuit where both direct and alternating currents flow at the same time and it is necessary or desirable to pass the a.c. and block the d.c. This can be done by using a blocking capacitor as in figure 11-22. The capacitor C in the diagram allows alternating current to flow and blocks the flow of direct current.

The filter capacitor in figure 11-23 is used to maintain a steady d-c voltage by filtering out or removing undesired a-c or ripple voltages by capacitor action opposing any change in voltage. Filter capacitors are commonly used to filter power supply voltages.

Spark suppression is obtained by placing a capacitor across the contacts of relays and other movable points subject to electrical sparking when opening and closing. The capacitor minimizes the effects of the sparking and extends the life of the relay contacts or points. The buffer capacitor across the breaker points of an automobile ignition system is a fine example.

# CHAPTER 12

# INDUCTIVE AND CAPACITIVE REACTANCE

## INDUCTIVE REACTANCE

Previously, the opposition that an inductance offers to a changing current was called self-induced voltage or CEMF and was measured in volts. However, opposition to current flow is normally measured in ohms, not in volts. Since a coil reacts to a current change by generating a CEMF, a coil is said to be reactive. The opposition of a coil is therefore called reactance (X) and is measured in ohms. Since more than one kind of reactance exists, the subscript L is added to denote INDUCTIVE REACTANCE. Thus, the opposition offered by a coil to alternating current is termed inductive reactance, designated $X_L$.

## FACTORS AFFECTING INDUCTIVE REACTANCE

In the preceding chapter, it was shown that when an alternating voltage is applied to an inductor, the inductor reacts in such a way as to oppose alternating-current flow. This opposition is thus referred to in general as an inductive reactance. The amount of opposition exhibited by a coil depends on the magnitude of its self-induced voltage, $E_{ind}$. The effective value of this self-inuced voltage, or counter emf in volts, is

$$E_{ind} = 2\pi fLI$$

where $2\pi$ is a constant, I is in effective amperes, L is in henrys, and f is in hertz. The induced voltage varies directly with the frequency, the inductance, and the current.

The amount of opposition to current flow offered by an inductor is referred to as its inductive reactance. This reactance is expressed as the ratio of the inductor's counter emf to the current through the inductor, or

$$X_L = \frac{E_{ind}}{I}$$

where $X_L$ is the inductor's opposition to current flow measured in ohms, $E_{ind}$ is the inductor's counter emf in volts, and I is the inductor current in amperes. Assuming the current and counter voltage have sine waveforms, the inductive reactance $X_L$, in ohms, is

$$X_L = 2\pi fL$$

Since the inductive reactance of a coil and its counter emf are closely related, anything which influences one must also influence the other. It has been shown that the counter emf of a coil depends on its inductance and the rate of flux change around the coil. Consequently, the inductive reactance must also be affected by the same influences. The rate of flux change per unit of time depends on the FREQUENCY of the inductor current. Flux must change more rapidly at high frequencies than at low frequencies. From the foregoing it can be seen that the inductive reactance of a coil depends primarily on (1) the coil's INDUCTANCE and (2) the FREQUENCY of the current flowing through the coil.

If frequency or inductance varies, inductive reactance must also vary. A coil's inductance does not vary appreciably after the coil is manufactured, unless it is designed as a variable inductor. Thus, frequency is generally the only variable factor affecting the inductive reactance of a coil. The coil's inductive reactance will vary directly with the applied frequency.

## POWER IN AN INDUCTIVE CIRCUIT

The power in a d-c circuit is equal to the product of volts and amperes, but in an a-c circuit this is true only when the load is resistive, and has no reactance.

In a circuit possessing inductance only, the true power is zero (fig. 12-1). The current lags the applied voltage by 90°. The TRUE POWER is the average power actually consumed by the circuit, the average being taken over one complete cycle of alternating current. THE APPARENT POWER is the product of rms volts and rms amperes. Thus, in figure 12-1 (A), the apparent power is 100 x 10 = 1,000 volt-amperes. However, the power absorbed by the coil during the time the current is rising (fig. 12-1 (B)), is returned to the source during the time the current is falling, so that the average power is zero.

The product of instantaneous values of current and voltage yield the double frequency power curve P. (See fig. 12-1 (B).) The shaded areas under this curve above the X axis represent positive energy and the shaded areas below the X axis represent negative energy.

From 0° to 90° current is negative and falling; the magnetic field is collapsing and the energy in the field is being returned to the source (negative energy). The product of negative current and positive voltage is negative power.

From 90° to 180° the current is positive and rising; energy from the source is being stored in the rising magnetic field (positive energy). The product of positive current and positive voltage is equal to positive power.

From 180° to 270° current is positive and falling. Again energy is being returned to the source; the product of positive current and negative voltage is negative power.

From 270° to 360° current is negative and rising. Energy (positive) is being supplied by the source and stored in the magnetic field. The product of negative current and negative voltage is positive power.

Thus when current is rising, power is being supplied by the source and stored in the magnetic field; when current is falling, power is being returned to the source from the collapsing magnetic field. In the theoretically pure inductance shown in figure 12-1 the supplied power is equal to the returned power. Thus, average power used (true power) is zero.

The ratio of the true power to the apparent power in an a-c circuit is called the POWER FACTOR. It may be expressed as a percent or as a decimal. In the inductor of figure 12-1 (A), the power factor is $\frac{0}{1,000}$ = 0. The power factor is also equal to the cos $\theta$, where $\theta$ is the phase angle between the current and voltage. The phase angle between E and I in the inductor is

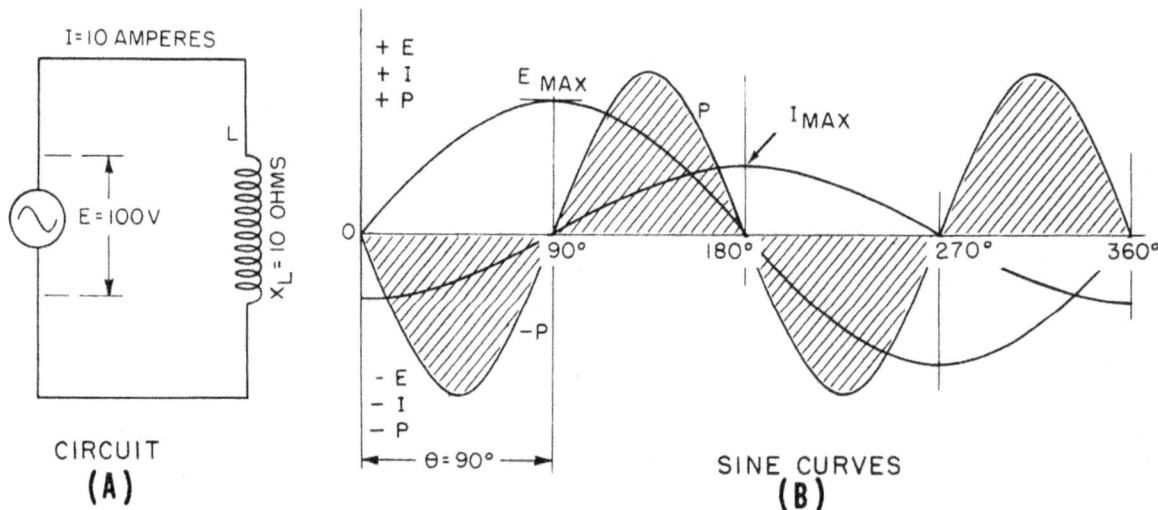

Figure 12-1.—Power in an inductive circuit.

BE. 221

$\theta = 90°$. Thus the power factor of an inductor of negligible losses is cos 90° = 0. The apparent power in a purely inductive circuit is called REACTIVE POWER and the unit of reactive power is called the VAR. This unit is derived from the first letters of the words volt-ampere-reactive.

RESISTANCE IN AN A-C CIRCUIT

In an a-c circuit containing only resistance, the current and voltage are always in phase ($\theta = 0°$). (See fig. 12-2.) The true power dissipated in heat in a resistor in an a-c circuit when sine waveforms of voltage and current are applied is equal to the product of the r.m.s. volts and the r.m.s. amperes. In the circuit of figure 12-2 (A) the power absorbed by the resistor is P = EI = 100 x 10 = 1,000 watts. The product of the instantaneous values of current and voltage (fig. 12-2 (B)) gives the power curve, P, the axis of which is displaced above the X axis by an amount that is proportional to 1,000 watts.

The true average power is

$$P = \frac{E_{max} \times I_{max}}{2} = E_{eff} \times I_{eff}$$

The power factor of a resistive circuit is cos 0° = 1, or 100 percent. The apparent power in the resistor is also equal to the true power. The reactive power in the resistive circuit is zero.

RESISTANCE AND INDUCTIVE REACTANCE IN SERIES

Because any practical inductor must be wound with wire that has resistance, it is not possible to obtain a coil without some resistance. The resistance associated with a coil may be considered as a separate resistor, R, in the series with an inductor, L, which is purely inductive and contains only inductive reactance (fig. 12-3 (A)). The resistance has been exaggerated in this example to be of the same order of magnitude as the inductive reactance of the inductor in order to simplify the trigonometric solution.

If an alternating current, $I_0$, flows through the inductor a voltage drop occurs across both the resistor and the inductor. The voltage, $E_r$, across the resistor is in phase with the current, and the voltage drop, $E_L$, across the inductor leads the current by 90° (fig. 12-3 (B)). The voltage across the resistor is assumed to be 50 volts and the voltage across the inductor, 86.6

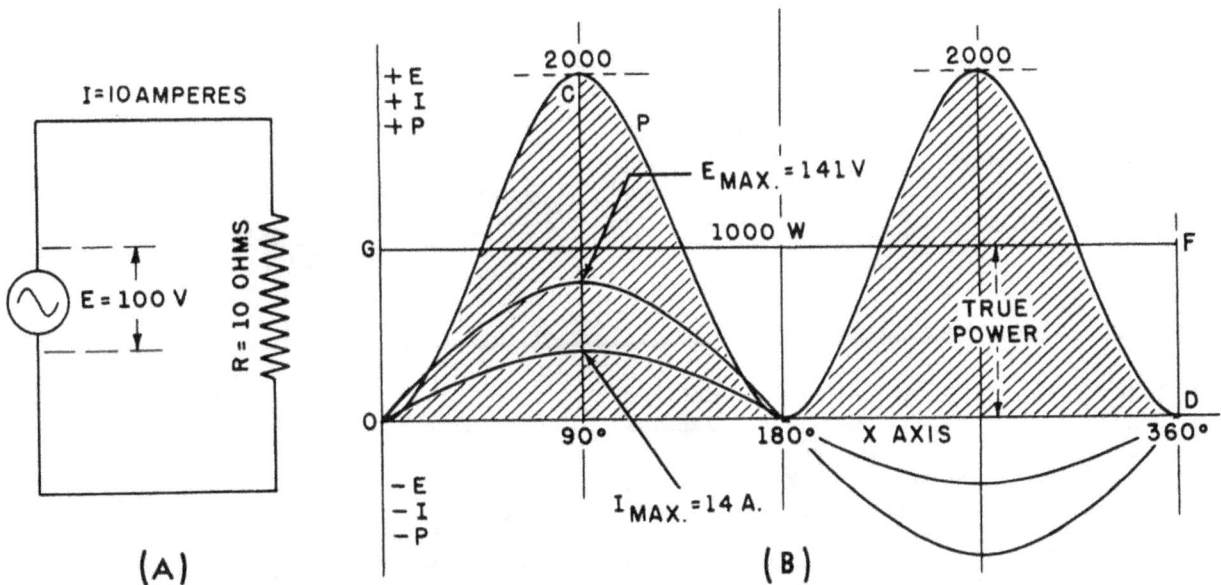

Figure 12-2.—Relation between E, I, and P in a resistive circuit.

BE.222

208

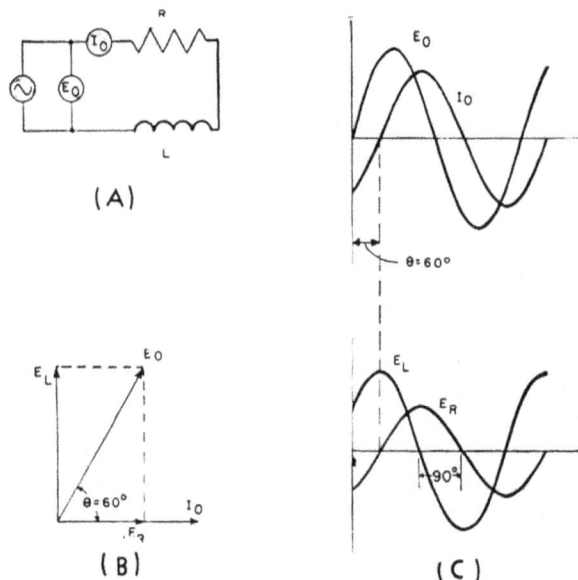

(A)

(B)

(C)

BE.223

Figure 12-3.—Resistance and inductive reactance in series.

volts. These two voltages are 90° out of phase, as indicated in the figure. The applied voltage, $E_O$, is the hypotenuse of a right traingle, the sides of which are 50 and 86.6, respectively. Thus,

$$E_O = \sqrt{50^2 + 86.6^2} = 100 \text{ volts}$$

The sine curves of circuit current, $I_O$, applied voltage, $E_O$, the voltage across the coil, $E_L$, and the voltage across the resistor, $E_r$, are shown in figure 12-3 (C).

IMPEDANCE

Impedance is the total opposition to the flow of alternating current in a circuit that contains resistance and reactance. In the case of pure inductance, inductive reactance, $X_L$ is the total opposition to the flow of current through it. In the case of pure resistance, R represents the total opposition. The combined opposition of R and $X_L$ in series or in parallel to current flow is called IMPEDANCE. The symbol for impedance is Z.

The impedance of resistance in series with inductance is

$$Z = \sqrt{R^2 + X_L^2}$$

where Z, R, and $X_L$ are the hypotenuse, base, and altitude respectively of a right triangle in which $\cos \theta = \frac{R}{Z}$, $\sin \theta = \frac{X_L}{Z}$, and $\tan \theta = \frac{X_L}{R}$. As mentioned before, $\cos \theta$ is equal to the circuit power factor; $\sin \theta$ is sometimes referred to as the reactive factor; and $\tan \theta$ is referred to as the quality, or Q, of a circuit or a circuit component. The trigonometric functions including the sine, cosine, and tangent of angles between 0° and 90° are given in appendix IX at the end of this training manual.

POWER IN A SERIES CIRCUIT
CONTAINING R and $X_L$

The true power in any circuit is the product of the applied voltage, the circuit current, and the cosine of the phase angle between them. Thus, in figure 12-4, the true power is

P = EI $\cos \theta$ = 100 x 7.07 (cos 45 ° = 0.707)

= 500 watts

The power curve is partly above the X axis (fig. 12-4 (B)) and partly below it. The axis of the power curve is displaced above the X axis an amount proportional to the true power, EI $\cos \theta$. The apparent power in this circuit is

100 x 7.07 = 707 volt-amperes

The power factor is

$$\cos \theta = \frac{\text{true power}}{\text{apparent power}} = \frac{500}{707}$$

= 0.707, or 70.7 percent

The reactive power in the LR circuit is the product of EI $\sin \theta$ where $\sin \theta$ is the reactive factor. Thus the reactive power is

100 x 7.07 (sin 45° = 0.707)= 500 VARS (lagging)

SUMMARY OF E, Z, and P
RELATIONS IN LR CIRCUIT

The relation between voltage, impedance, and power in series LR circuit with sine

209

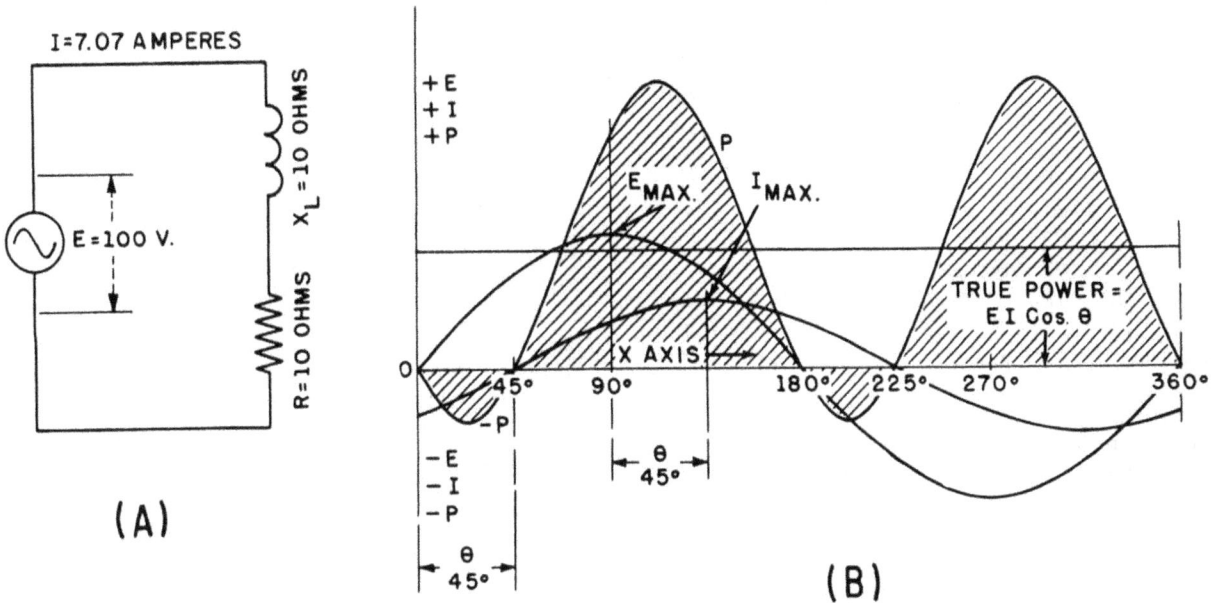

Figure 12-4.—Power in a circuit containing L and R in series.

BE.224

waveforms applied is summarized in figure 12-5. In this example the circuit contains 12 ohms of resistance in series with 16 ohms of inductive reactance (fig. 12-5 (A)). The phase relations between the applied voltage; the voltage across the resistance, R; and the voltage across the inductance, L; are shown in figure 12-5 (B). The IR drop across R is 5 x 12 = 60 volts and forms the base of the right triangle. The altitude is the $IX_L$ drop, or 5 x 16 = 80 volts, across L. The applied voltage represents the vector sum of the IR drop and the $IX_L$ drop and is the hypotenuse of the right triangle. Its magnitude is 100 volts. All voltages are effective values.

The relation between resistance, inductive reactance, and impedance is shown by the vectors of figure 12-5 (C). The resistance of 12 ohms forms the base of the right triangle. The altitude of this triangle represents the inductive reactance of the coil and has a magnitude of 16 ohms. The combined impedance of the circuit is $Z = \sqrt{12^2 + 16^2} = 20$ ohms. In this triangle $\cos \theta = \frac{R}{Z}$. Thus $\cos \theta = \frac{12}{20} = 0.6$, from which $\theta = 53.1°$.

The relation between apparent power, true power, and reactive power is shown in figure 12-5 (D). The hypotenuse represents the appar-

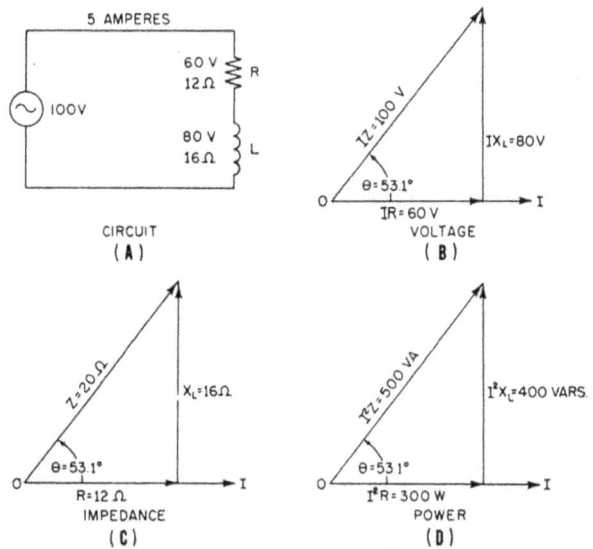

Figure 12-5.—Summary of relation between E, Z, and P in a series LR circuit.

BE.225

ent power and is equal to EI, or 100 x 5 = 500 volt-amperes. The base represents true power and is equal to EI cos θ, or 100 x 5 x 0.6 = 300 watts, where 0.6 = cos 53.1°. The altitude represents reactive power and is equal to EI sin $\theta$, or 100 x 5 x 0.8 = 400 VARS, where 0.8 = sin 53.1°.

In all three vector diagrams the right triangles are similar. The common factor of current, I, makes their corresponding sides proprotional. Thus the voltage triangle of figure 12-5 (B), is obtained by multiplying the corresponding sides of the impedance triangle of figure 12-5 (C), by I. The hypotenuse is equal to the applied voltage, or IZ = 5 x 20 = 100 volts. The base is equal to the voltage across R, or IR = 5 x 12 = 60 volts. The altitude is equal to the voltage across L, or $IX_L$ = 5 x 16 = 80 volts.

Multiplying corresponding sides of the voltage triangle by I gives the power triangle of figure 10-5 (D). The hypotenuse is IZ times I, or $I^2Z$ = $5^2$ x 20 = 500 volt-amperes of apparent power. The base is IR times I, or $I^2R$ = $5^2$ x 12 = 300 watts of true power. The altitude is $IX_L$ times I, or $I^2X_L$ = $5^2$ x 16 = 400 VARS of reactive power.

In the three vector diagrams, the circuit power factor is equal to the following ratios: In the voltage diagram,

$$\cos \theta = \frac{IR}{IZ} = \frac{60}{100} = 0.6$$

In the impedance diagram,

$$\cos \theta = \frac{R}{Z} = \frac{12}{20} = 0.6$$

In the power diagram,

$$\cos \theta = \frac{I^2R}{I^2Z} = \frac{300}{500} = 0.6$$

In all three diagrams $\theta$ = 53.1° and the power factor is 60 percent.

### CAPACITIVE REACTANCE

Capacitance was defined in chapter 11 as that quality of a circuit that enables energy to be stored in an electric field. A capacitor is a device that possesses the quality of capacitance. In simple form it has been shown to consist of two parallel metal plates separated by an insulator, called a DIELECTRIC. The electric field consists of parallel lines of electric force which terminate in a positive charge on one plate and a negative charge on the other. The positive and negative charges on the plates

establish the electric field in the dielectric because the dielectric prevents these charges from neutralizing each other.

### CURRENT FLOW IN A CAPACITIVE CIRCUIT

A capacitor that is initially uncharged tends to draw a large current when a d-c voltage is first applied. During the charging period, the capacitor voltage rises. After the capacitor has received sufficient charge the capacitor voltage equals the applied voltage and the current flow ceases. If a sine-wave a-c voltage is applied to a pure capacitance the current is a maximum when the voltage begins to rise from zero, and the current is zero when the voltage across the capacitor is a maximum (fig. 12-6 (A)). The current leads the applied voltage by 90°, as indicated in the vector diagram of figure 12-6 (B).

### FACTORS THAT CONTROL CHARGING CURRENT

Because a capacitor of large capacitance can store more energy than one of small capacitance, a larger current must flow to charge a large capacitor than to charge a small one, assuming the same time interval in both cases. Also, because the current flow depends on the rate of charge and discharge, the higher the frequency, the greater is the current flow per unit time.

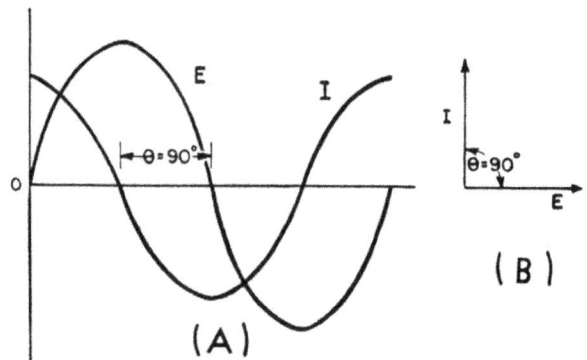

BE.226

Figure 12-6.—Phase relation between E and I in a capacitive circuit.

The charging current in purely capacitive circuit varies directly with the capacitance, voltage, and frequency—

$$I = 2\pi fCE$$

where I is effective current in amperes, f is in hertz, C is the capacitance in farads, and E is in effective volts.

## FORMULA FOR CAPACITIVE REACTANCE

The ratio of the effective voltage across the capacitor to the effective current is called the CAPACITIVE REACTANCE, $X_c$, and represents the opposition to current flow in a capacitive circuit of zero losses—

$$X_c = \frac{1}{2\pi fC}$$

When f is in hertz, and C is in farads, then $X_c$ is in ohms.

Example: What is the capacitive reactance of a capacitor operating at a frequency of 60 hertz and having a capacitance of 133 microfarads, or $1.33 \times 10^{-4}$ farads?

$$X_c = \frac{1}{2\pi fC} = \frac{1}{6.28 \times 69 \times 1.33 \times 10^{-4}} = 20 \text{ ohms}$$

## POWER IN A CAPACITIVE CIRCUIT

With no voltage or charge, the electrons in the dielectric between the capacitor plates rotate around their respective nuclei in normally circular orbits. When the capacitor receives a charge the positive plate repels the positive nuclei and at the same time the electrons in the dielectric are strained toward the positive plate and repelled away from the negative plate. This distorts the orbits of the electrons in the direction of the positive charge. During the time the electron orbits are changing from normal to the strained position there is a movement of electrons in the direction of the positive charge. This movement constitutes the DISPLACEMENT CURRENT in the dielectric. When the polarity of the plates reverses, the electron strain is reversed. If a sine-wave voltage is applied across the capacitor plates the electrons will oscillate back and forth in a direction parallel to the electrostatic lines of force. DISPLACEMENT CURRENT is a result of the movement of bound electrons, whereas CONDUCTION CURRENT represents the movement of free electrons.

Figure 12-7 (A) shows a capacitive circuit and figure 12-7 (B) indicates the sine waveform of charging current, applied voltage, and instantaneous power. The effective voltage is 70.7 volts. The effective current is 7.07 amperes. Because the losses are neglected, the phase angle between current and voltage is assumed to be 90°. The true power is zero, as indicated by the expression

$$P = EI \cos \theta$$
$$= 70.7 \times 7.07 \ (\cos 90° = 0) = 0 \text{ watt}$$

Multiplying instantaneous values of current and voltage over one cycle, or 360°, gives the power curve, P. During the first quarter cycle (from 0° to 90°) the applied voltage rises from zero to a maximum and the capacitor is receiving a charge. The power curve is positive during this period and represents energy stored in the capacitor. From 90° to 180° the applied voltage is falling from maximum to zero and the capacitor is discharging. The corresponding power curve is negative and represents energy returned to the circuit during this interval. The third quarter cycle represents a period of charging the capacitor and the fourth quarter cycle represents a discharge period. Thus, the average power absorbed by the capacitor is zero. The action is like the elasticity of a spring. Storing a charge in the capacitor is like compressing the spring. Discharging the capacitor is like releasing the pressure on the spring, thus allowing it to return the energy that was stored within it on compression.

The apparent power in the capacitor is EI = 70.7 × 7.07 = 500 volt-amperes. The reactive power in the capacitor is

$$EI \sin \theta = 70.7 \times 7.07 \ (\sin 90° = 1)$$
$$= 500 \text{ VARS (leading)}$$

## CAPACITIVE REACTANCE AND RESISTANCE IN SERIES

Losses that appears in capacitive circuits may be lumped in a resistor connected in series with the capacitor, as indicated in figure 12-8. In this example a 39.8-microfarad capacitor is connected in series with a 20-ohm resistor (fig 12-8 (A)). The applied voltage across the RC series circuit is 134 volts and the frequency is 100 hertz.

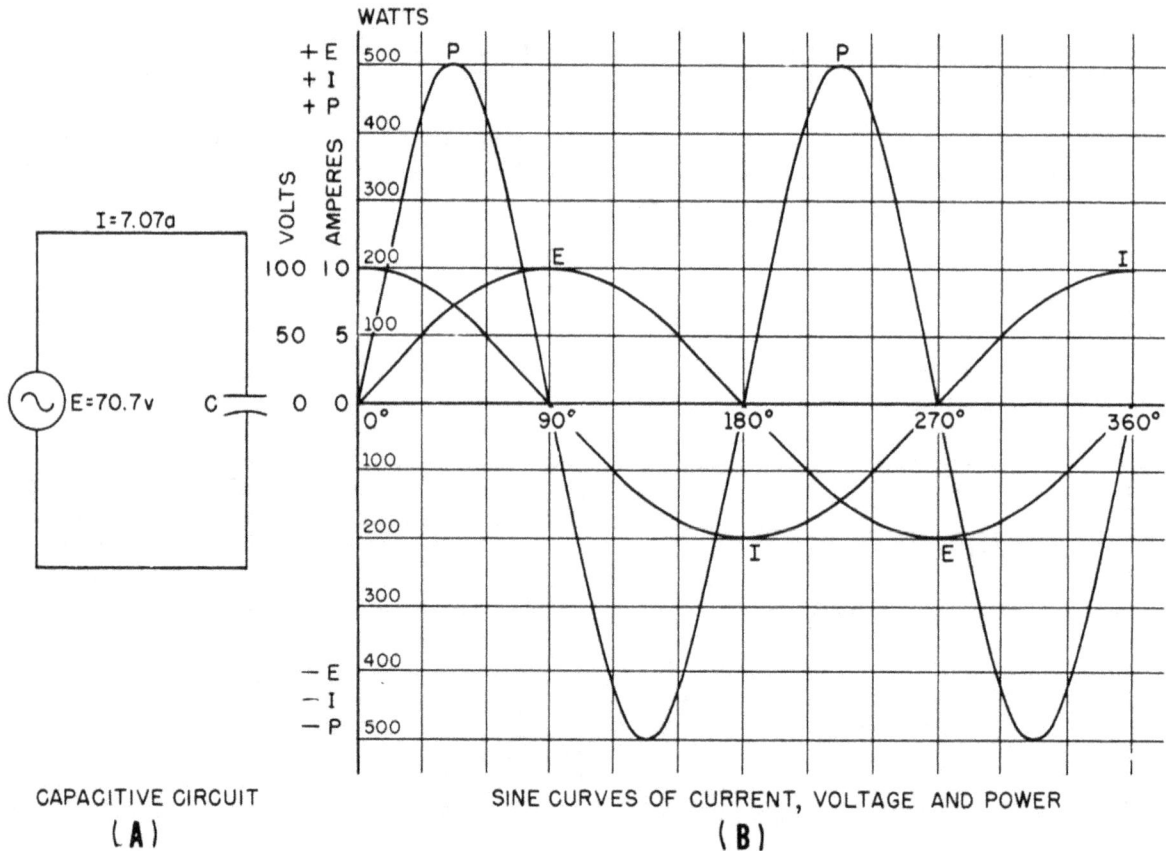

Figure 12-7.—Power in a capacitive circuit.

BE.227

The capacitive reactance at this frequency is

$$X_C = \frac{1}{2\pi fC} = \frac{1}{6.28 \times 100 \times 39.8 \times 10^{-6}} = 40 \text{ ohms}$$

Voltage Relation

The voltage across R and C are 90° out of phase and equal to 60 volts and 120 volts respectively, as shown in the vector diagram of figure 12-8 (B). The voltage across C is represented as $IX_C$ and is plotted vertically downward from the horizontal in order to indicate that the current leads the voltage across the capacitor by 90°. Angle between the voltage across the capacitor and the circuit current is represented as -90° because it is measured clockwise from the horizontal reference vector, OI. The total voltage is equal to the vector sum of IR and $IX_C$ and is represented in the figure as the hypotenuse of a right triangle the base of which represents the voltage across R, having an effective value of 60 volts and the voltage drop across C, having an effective value of 120 volts. The total (applied) voltage is $\sqrt{60^2 + 120^2} = 134$ volts.

Impedance

The total impedance, Z, of the series circuit is

$$Z = \frac{E}{I} = \frac{134}{3} = 44.7 \text{ ohms}$$

The impedance diagram is indicated in figure 12-8 (C). The base of the triangle represents the resistance, R, of the circuit and has a magnitude of 20 ohms. The capacitive reactance, $X_C$ is represented as the altitude of the triangle and has a magnitude of 40 ohms. The total impedance may be represented as the hypotenuse of the triangle and has a magnitude of $\sqrt{20^2 + 40^2} = 44.7$ ohms, which is also the ratio of the applied voltage to the circuit

213

two. The base represents the true power absorbed by the circuit resistance and has a magnitude of $I^2R = 3^2 \times 20 = 180$ watts. The altitude represents the reactive volt-ampere (leading) and has a magnitude of $I^2X_c = 3^2 \times 40 = 360$ VARS. The total apparent power is

$$EI = 134 \times 3 = 402 \text{ volt-amperes}$$

Summary of E, Z, and P
Relations in RC Circuits

As mentioned previously, all three triangles are similar. The common factor is the circuit current; and the phase angle, $\theta$, between E and I is equal to the same value in all three diagrams. Thus, the circuit power factor is equal to

$$\cos \theta = \frac{IR}{IZ} = \frac{60}{134} = 0.446 \text{ (fig. 12-8 (B))}$$

$$\cos \theta = \frac{R}{Z} = \frac{20}{44.7} = 0.446 \text{ (fig. 12-9 (C))}$$

$$\cos \theta = \frac{I^2R}{I^2Z} = \frac{180}{402} = 0.446 \text{ (fig. 12-8 (D))}$$

and angle $\theta = 63.4°$ in all three triangles. The true power is

EI $\cos \theta = 134 \times 3 \times (\cos 63.4° = 0.446) = 180$ watts

The reactive power is

EI $\sin \theta = 134 \times 3 \times (\sin 63.4° = 0.894) = 360$ VARS

The apparent power is

$$EI = 134 \times 3 = 402 \text{ volt-amperes}$$

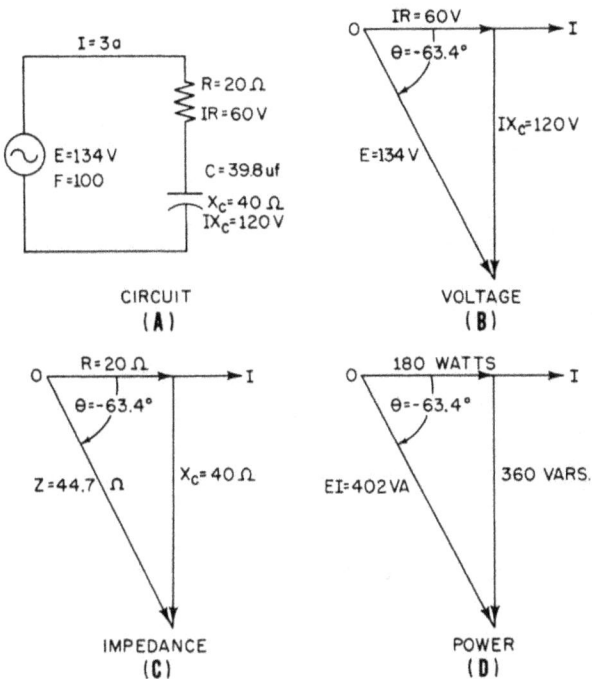

BE.228

Figure 12-8.—Capacitive reactance and resistance connected in series.

current. The impedance diagram is a right triangle that is similar to the triangle representing the voltage relation in figure 12-8 (B).

Power

The power relations are indicated in the vector diagram of figure 12-8 (D). This diagram is also a right triangle similar to the other

# CHAPTER 13

# FUNDAMENTAL ALTERNATING-CURRENT
# CIRCUIT THEORY

In the preceding chapter, terms were clarified, a-c reactance was explained, and the effects of individual inductors and capacitors were described. To do this, only the simplest two-element RL and RC series circuits were employed. In this chapter, the subject coverage is expanded to include more complex reactive circuits.

## RESISTANCE, INDUCTANCE, AND CAPACITANCE IN SERIES

### RELATION OF VOLTAGES AND CURRENT IN AN RLC SERIES CIRCUIT

When resistive, inductive, and capacitive elements are connected in series, their individual characteristics are unchanged. That is, the current through and the voltage drop across the resistor are in phase while the voltage drops across the reactive components (assuming pure reactances) and the current through them are 90° out of phase. However, a new relation must be recognized with the introduction of the three-element circuit. This pertains to the effect on total line voltage and current when connecting reactive elements in series, whose individual characteristics are opposite in nature, such as inductance and capacitance. Such a circuit is shown in figure 13-1.

In the figure, note first that current is the common reference for all three element voltages, because there is only one current in a series circuit, and it is common to all elements. The common series current is represented by the dashed line in figure 13-1 (A). The voltage vector for each element, showing its individual relation to the common current, is drawn above each respective element. The total source voltage E is the vector sum of the individual voltages of IR, $IX_L$, and $IX_C$.

The three element voltages are arranged for summation in figure 13-1 (B). Since $IX_L$ and $IX_C$ are each 90° away from I, they are therefore 180° from each other. Vectors in direct opposition (180° out of phase) may be subtracted directly. The total reactive voltage $E_X$ is the difference of $IX_L$ and $IX_C$. Or, $E_X = IX_L - IX_C$ = 45 - 15 = 30 volts. The final relationship of line voltage and current, as seen from the source, is shown in figure 13-1 (C). Had $X_C$ been larger than $X_L$, the voltage would lag, rather than lead. When $X_C$ and $X_L$ are of equal values, line voltage and current will be in phase.

### IMPEDANCE OF RLC SERIES CIRCUITS

The impedance of an RLC (three-element) series circuit is computed in exactly the same manner described for the two-element circuits in the preceding chapter. However, there is one additional operation to be performed. That is, the difference of $X_L$ and $X_C$ must be determined prior to computing total impedance. When employing the Pythagorean Theorem-based formula for determining series impedance, the net reactance of the circuit is represented by the quantity in parenthesis $(X_L - X_C)$. Applying this formula to the circuit in figure 13-1, the impedance is

$$Z = \sqrt{R^2 + (X_L - X_C)^2}$$
$$= \sqrt{40^2 + (45 - 15)^2}$$
$$= \sqrt{1,600 + 900}$$
$$= \sqrt{2,500}$$
$$Z = 50\Omega$$

Series impedance may also be determined by the use of vectorial layout, or triangulation. In an

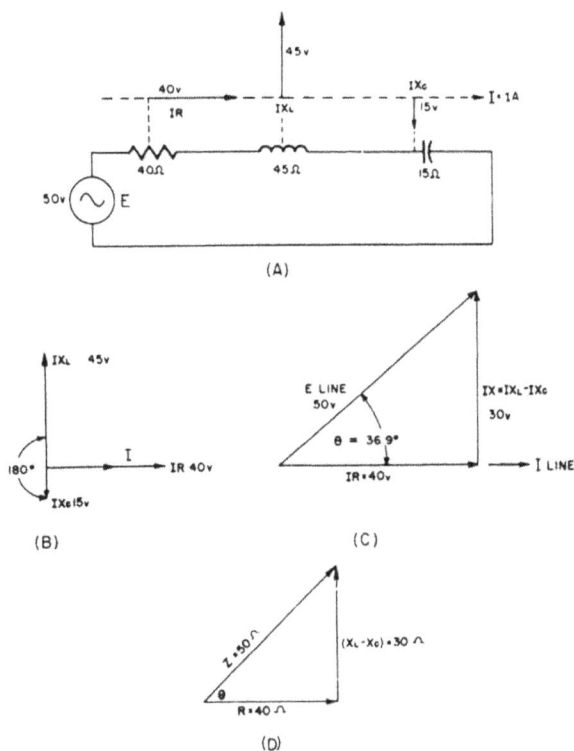

BE. 229

Figure 13-1.—Resistance, inductance,
and capacitance connected in series.

impedance triangle (fig. 13-1 (D)) for a series
circuit, the base always represents the series
resistance, the altitude represents the NET reactance ($X_L - X_C$), and the hypotenuse represents total impedance.

It should be noted that as the difference of $X_L$
and $X_C$ becomes greater, total impedance also
increases. Conversely, when $X_L$ and $X_C$ are
equal, their effects cancel each other, and impedance is minimum, equal only to the series resistance. When $X_L$ and $X_C$ are equal, their individual voltages are 90° out of phase with current,
but their collective effect is zero, because they
are equal and opposite in nature. Therefore,
when $X_L$ and $X_C$ are equal, line voltage and
current are in phase. This condition is the same
as if there were only resistance and no reactances in the circuit. A circuit in this condition
is said to be at resonance.

## POWER IN AN RLC
## SERIES CIRCUIT

Total true power in an RLC series circuit is
the product of line voltage and current, times the

cosine of the angle between them. When $X_L$ and
$X_C$ are equal, total impedance is at a minimum,
and thus current is maximum. When maximum
current flows, the series resistor dissipates
maximum power. When $X_L$ and $X_C$ are made
unequal, total impedance increases, line current
decreases, and moves out of phase with line
voltage. Both the decrease in current and the
creation of a phase difference cause a decrease
in true power.

## INDUCTANCE AND RESISTANCE
## IN PARALLEL

If the voltage applied across a resistor and
inductor in parallel has a sine waveform, the
currents in the branches will also have sine
waveforms. In the parallel circuit of figure
13-2 (A), the applied voltage, E, has a magnitude
of 100 volts (rms). Branch ① contains a 20-ohm
resistor, and the current, $I_1$, is equal to $\frac{100}{20}$, or
5 amperes (rms). This current is in phase with E.
Branch ② contains an inductance of 0.053 henry.
The line frequency is 60 Hz and the inductive
reactance is

$$X_L = 2\pi fL = 6.28 \times 60 \times 0.053 = 20 \text{ ohms}$$

The true power losses associated with the inductor in this example are considered negligible.
Therefore, current $I_2$ is equal to $\frac{100}{20}$, or 5 amperes. This current lags E by an angle of 90°.
(In an inductive circuit the current always lags
the voltage by some angle.)

The sine waveforms of applied voltage and
current are shown in figure 13-2 (B). $I_1$ is in
phase with E. $I_2$ lags E by an angle of 90°. The
total current, $I_t$, lags E by an angle of 45°.

## CURRENT VECTORS

A polar-vector diagram representing the
three currents and the applied voltage is shown
in figure 13-2 (C). Vector OE represents the
effective value of the applied voltage and is the
horizontal reference vector for both branches
because it is common to both branches. Vector
$I_1$ is the effective current of 5 amperes in
branch ① . This vector is in the same line as
vector OE because the current and voltage in
the resistive circuit are in phase. Vector $I_2$ is
the effective current of 5 amperes in branch
② and lags vector OE by an angle of 90°. $I_1$ is
called the energy component of the circuit current. It flows in the resistive branch in which

**(A)**
CIRCUIT

**(B)**
SINE WAVEFORMS OF CURRENT
AND VOLTAGE

**(C)**
POLAR DIAGRAM OF EFFECTIVE
CURRENTS

**(D)**
TOPOGRAPHIC DIAGRAM OF EFFECTIVE
CURRENTS

**(E)**
POWER TRIANGLE

**(F)**
EQUIVALENT SERIES CIRCUIT

**(G)**
IMPEDANCE TRIANGLE

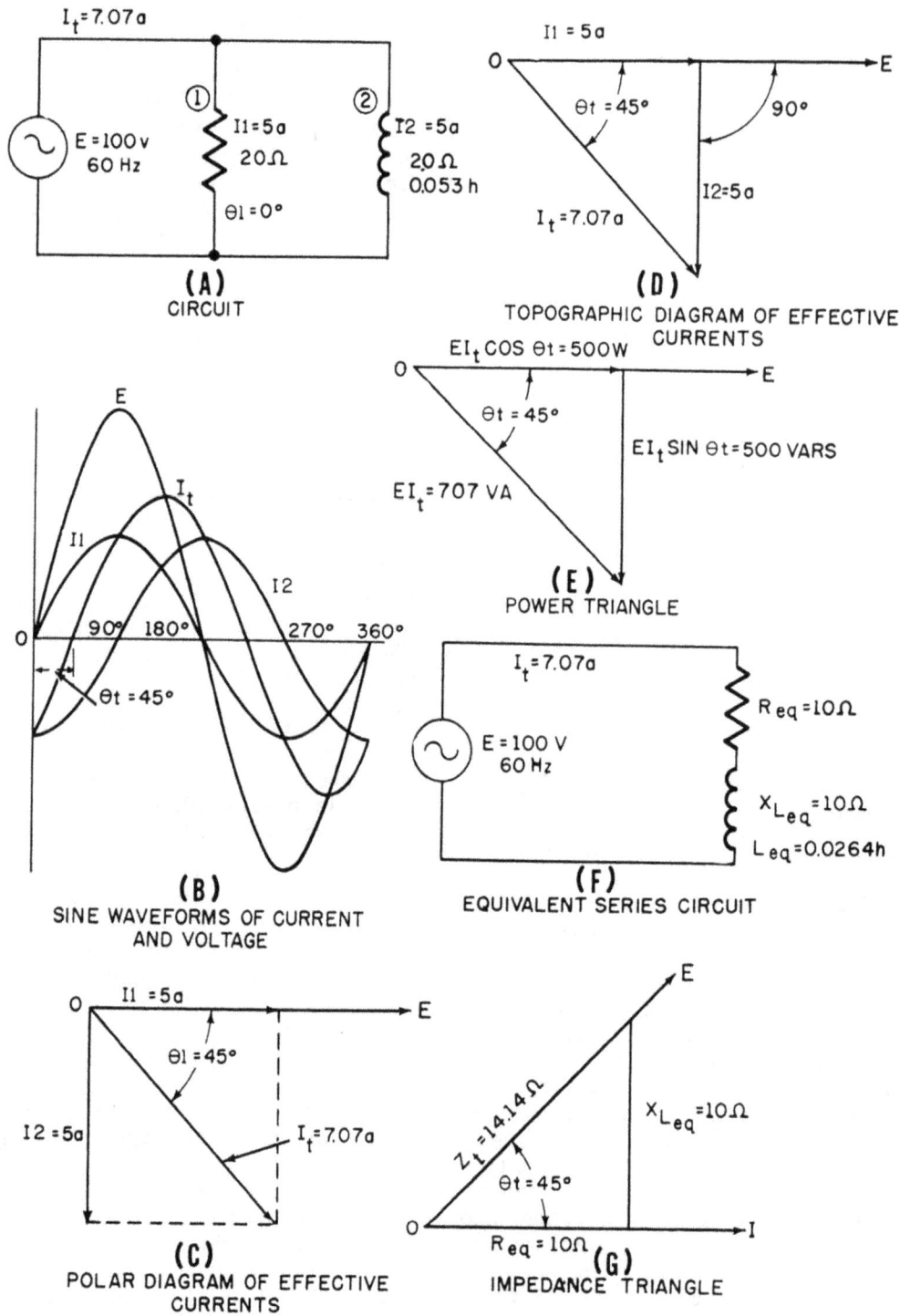

Figure 13-2.—Resistance and inductance, in parallel.

BE.230

true power is absorbed from the source and is dissipated as heat. $I_2$ is called the nonenergy component of the circuit current. This current flows in the inductive branch where the true power is zero, and the reactive power is exchanged with the source twice during each cycle of applied voltage.

The total circuit current, $I_t$, is represented as the diagonal of the parallelogram, the sides of which are $I_1$ and $I_2$ (fig. 13-2 (C)). In this example, the sides are 5 amperes (rms) each and the diagonal is 7.07 amperes (rms).

A topographic-vector diagram representing the three currents and the applied voltage is shown in figure 13-2 (D). As in the polar diagram, OE is the reference vector. $I_1$, in phase with OE, is the base of the triangle. $I_2$, 90° out of phase with OE, is the altitude of the triangle, and is plotted downward to indicate lag. The resultant current, $I_t$, is the hypotenuse of the triangle. The hypotenuse is equal to the square root of the sum of the squares of the other two sides. Thus,

$$I_t = \sqrt{I_1^2 + I_2^2} = \sqrt{5^2 + 5^2} = 7.07 \text{ amperes}$$

the phase angle between $I_t$ and E is the angle whose cosine is

$$\frac{I_1}{I_1} = \frac{5}{7.07} = 0.707$$

This angle is 45°.

## POWER AND POWER FACTOR

The apparent power, true power, and reactive power in the parallel circuit are related as the hypotenuse, base, and altitude, respectively, of a right triangle in a similar manner to that described in the preceding chapter.

The relation between apparent power, true power, and reactive power is shown in figure 13-2 (E). The hypotenuse of the right triangle represents the apparent power and is equal to $EI_t$, or 100 x 7.07 = 707 volt-amperes. The base of the triangle represents the true power and is equal to $EI_t$ cos $\theta_t$, or 100 x 7.07 x 0.707 = 500 watts, where 0.707 = cos 45°. The altitude represents the reactive power and is equal to $EI_t$ sin $\theta_t$, or 100 x 7.07 x 0.707 = 500 vars, where 0.707 = sin 45°. The power triangle is similar to the current triangle and is related to it by the common factor of voltage (the voltage

is the same across both branches of the parallel circuit).

Because branch ① (fig. 13-2 (A)) is purely resistive and branch ② is purely inductive, the true power of the circuit is absorbed in branch ① and the reactive power is developed in branch ②. In branch ① the true power is $EI_1$ cos $\theta_1$, or 100 x 5 x 1 = 500 watts, where 1 = cos 0°. In branch ② the reactive power is $EI_2$ sin $\theta_2$, or 100 x 5 x 1 = 500 vars, where 1 = sin 90°.

The true power in branch ① may also be calculated as $I_1^2 R_1$, or $5^2$ x 20 = 500 watts. The reactive power in branch ② may be calculated as $I_2^2 X_{L_2}$, or $5^2$ x 20 = 500 vars. The total circuit power factor is cos $\theta_t$ = $\frac{\text{true power}}{\text{apparent power}} = \frac{500}{707}$ = 0.707. The power factor of branch ① is cos $\theta_1$ = cos 0° = 1, and the power factor of branch ② is cos $\theta_2$ = cos 90° = 0.

## EQUIVALENT SERIES CIRCUIT IMPEDANCE

The combined impedance of the parallel circuit is

$$Z_t = \frac{E}{I_t} = \frac{100}{7.07} = 14.14 \text{ ohms}$$

The combined impedance is also called the impedance of the equivalent series circuit. The equivalent series circuit (fig. 13-2 (F)) contains a resistor and an inductor in series that combine to give the same impedance as the total impedance, of the given parallel circuit. Thus, the current in the equivalent series circuit is equal to the total circuit current in the parallel circuit when rated voltage and frequency are applied to the circuits.

In figure 13-2 (G), the hypotenuse, $Z_t$, of the impedance triangle is 14.14 ohms and the phase angle, $\theta_t$, between total current and line voltage, is 45°. The equivalent series resistance, $R_{eq}$, is the base of the impedance triangle an is equal to $Z_t$ cos $\theta$, or 14.14 x cos 45° = 10 ohms. The equivalent series reactance, $X_{Leq}$, is the altitude of the impedance triangle and is equal to $Z_t$ sin $\theta_t$, or 14.14 x sin 45 = 10 ohms.

The inductance of the equivalent series circuit is

$$L_{eq} = \frac{X_L}{2\pi f} = \frac{10}{6.28 \times 60} = 0.0264 \text{ henry}$$

Thus, in this example, the 20-ohm resistor in branch ① shunts the 0.053-henry inductor in branch ②, and the source "sees" an equivalent resistor of 10 ohms in series with an equivalent inductor of 0.0264 henry.

## EQUIVALENT CIRCUIT OF A LOW-LOSS INDUCTOR

The losses in air-core inductor occur in the wire with which the inductor is wound. If the losses are small, the resistance of the wire will be small compared with the inductive reactance developed at the operating frequency. The coil resistance acts in series with the coil reactance.

An equivalent parallel circuit for a low-loss coil can be established by substituting an equivalent coil of zero losses, but having the same inductance as the given coil for one branch and a resistor for the other branch. The resistor is of such a magnitude that the same losses will occur in this branch as occur in the given coil when rated voltage and frequency are applied.

In the following example a low-loss circuit is established and then the equivalent parallel circuit is derived. A low-loss coil is indicated in figure 13-3 (A). It has an inductance of 1.59 henry and a d-c resistance of 10 ohms at an operating frequency of 100 Hz. The inductive reactance is

$$X_L = 2\pi fL = 6.28 \times 100 \times 1.59$$
$$= 1,000 \text{ ohms}$$

The equivalent series circuit (fig. 13-3 (B)), is shown with 10 ohms of resistance acting in series with 1,000 ohms of inductive reactance. The impedance triangle is shown in figure 13-3 (C).

A low-loss coil has a low power factor. Thus in the impedance triangle, $\cos \theta = \dfrac{R_{se}}{Z} = \dfrac{10}{1,000} = 0.01$, or 1 percent. As can be seen in this example, Z is approximately equal to $X_L$, and $\cos \theta = \dfrac{R_{se}}{X_L}$. Since $\tan \theta = \dfrac{X_L}{R_{se}}$, it follows that in this case $\cos \theta = \dfrac{1}{\tan \theta}$. Thus $\tan \theta = \dfrac{1,000}{10} = 100$, and the power factor $= \dfrac{10}{1,000}$, or 0.01, or 1 percent.

A useful relationship for finding angle $\theta$ in low-loss circuits is shown in figure 13-3 (D). This concept involves the relation between $\cos \theta$, expressed decimally, and the complementary angle, $90° - \theta$, expressed in radians. The figure indicates a unit-radius circle in which any length of arc BC measures the corresponding angle $(90° - \theta)$ which it subtends. The length of arc is equal to the angle in radians. Since $2\pi$ radians are equal to 360° it follows that 1 radian is approximately equal to $\dfrac{360°}{2\pi}$, or 57.3°. Thus, from a knowledge of the angle in radians, the equivalent angle $\theta$, in degrees, is $\theta = 57.3$ x angle in radians.

For angles greater than 84.3°, $\tan \theta$ is greater than 10 and $\cos \theta$ — which equals OA, in figure 13-3 (D) — is approximately equal to arc BC. Arc BC is a measure of the complementary angle $90 - \theta$ and is expressed in radians. Thus, $\cos \theta$ is approximately equal to the angle in radians by which the current fails to be exactly 90° out of phase with the voltage. In this example, the power factor is 0.01 and therefore the angle by which the current fails to be 90° out of phase with the voltage (complementary angle) is 0.01 radian. This angle, in degrees, becomes 57.3° x 0.01, or 0.573°. Therefore, $\theta$ is equal to 090° - 0.573°, or 89,427°, and $\cos 89.427° = 0.01$.

The relationship described in the preceding example is expressed in general terms as follows: The complementary angle, $(90 - \theta)$ in radians, is equal to the power factor, $\cos \theta$, expressed decimally, where $\tan \theta$ is numerically equal to or greater than 10. From this relationship it is possible to estimate quickly the phase angle between current and voltage in low power-factor (low-loss) circuits.

The equivalent parallel circuit for the 1.59-henry coil discussed in this example is shown in figure 13-3 (E). The equivalent shunt resistor, $R_{sh}$, has a resistance of 100,000 ohms (to be derived later), and this resistor is connected in parallel with a 1.59-henry inductor having zero losses. With rated voltage and frequency applied, the input current to the parallel circuit has the same magnitude and phase with respect to the applied voltage as in the original coil.

The current in the resistive branch is $\dfrac{100}{100,000} = 0.001$ ampere, or 1 milliampere, and is the base of the current triangle (fig. 13-3 (F)).

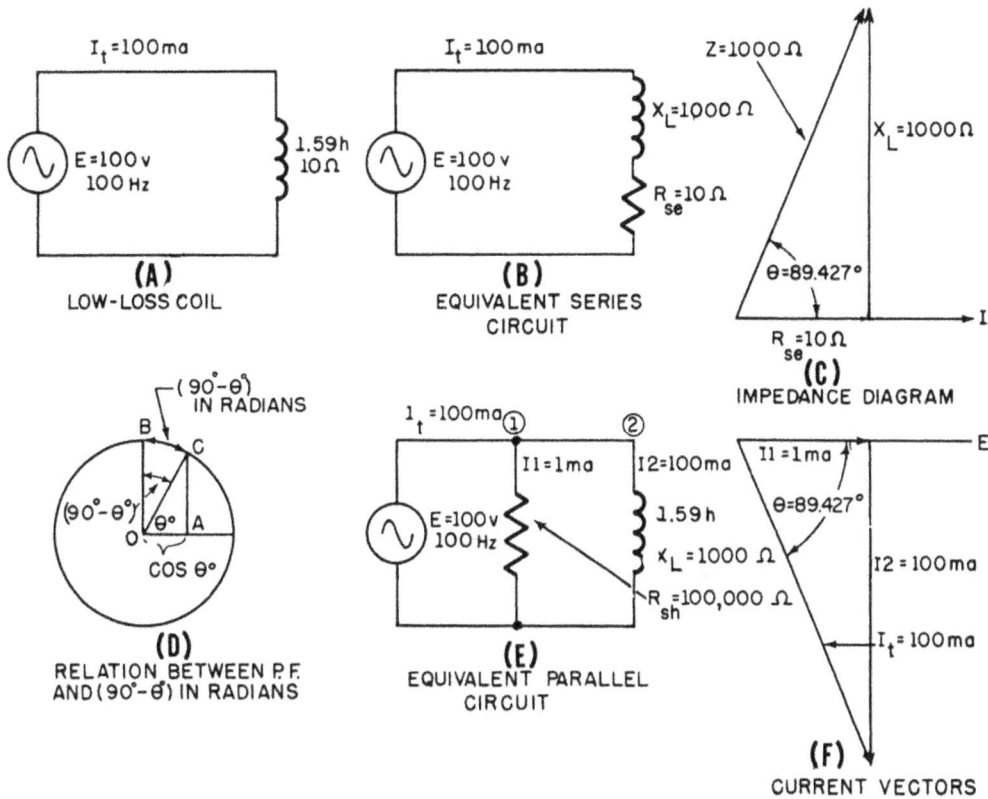

BE.231

Figure 13-3.—Equivalent circuits of a low-loss coil.

The current in the inductive branch is $\frac{100}{1,000} =$ 0.1 ampere, or 100 ma, and is the altitude of the current triangle. The total current in the parallel circuit is equal to $\sqrt{1^2 + 100^2} = 100$ milliamperes (approx.), and is the hypotenuse of the current triangle. This current lags the line voltage by an angle of $90° - (57.3 \times 0.01)$, or $89.427°$.

The equivalent series-circuit impedance triangle (fig. 13-3 (C)) is similar to the current triangle (fig. 13-3 (F)). From the impedance triangle, $\cos \theta \approx \dfrac{R_{se}}{X_L}$; and from the current triangle, $\cos \theta \approx \dfrac{I_1}{I_2}$, where $I_1$ is the energy current and $I_2$ is the nonenergy current.

Therefore,

$$\frac{I_1}{I_2} = \frac{R_{se}}{X_L} \qquad (13\text{-}1)$$

For the resistive branch,

$$I_1 = \frac{E}{R_{sh}} \qquad (13\text{-}2)$$

and for the inductive branch,

$$I_2 = \frac{E}{X_L} \qquad (13\text{-}3)$$

Substituting equations 13-2 and 13-3 in equation 13-1,

$$\frac{\dfrac{E}{R_{sh}}}{\dfrac{E}{X_L}} = \frac{R_{se}}{X_L} \qquad (13\text{-}4)$$

Canceling E and transposing equation 13-4,

$$R_{sh} = \frac{X_L^{\,2}}{R_{se}} \qquad (13\text{-}5)$$

In the example of figure 13-3, $R_{se}$ = 10 ohms, $X_L$ = 1,000 ohms, and $R_{sh}$ = $\frac{(1,000)^2}{10}$ = 100,000 ohms.

The preceding relations are sufficiently accurate for low-loss inductive circuits in which $\theta$ is 84.3° or higher, and tan $\theta$ is 10 or higher.

## COMBINING CURRENTS AT ACUTE ANGLES

Figure 13-4 (A) represents a 2-branch parallel circuit in which branch ① has a power factor of 0.50 and branch ② has a power factor of 0.866. The current in ① lags the applied voltage by an angle of 60° and the current in branch ② lags the applied voltage by an angle of 30°. The series resistance, $R_1$, of branch ① is 10 ohms, the series inductive reactance, $X_{L1}$, is 17.32 ohms, and the impedance, $Z_1$, is $\sqrt{10^2 + (17.32)^2}$ = 20 ohms. The series resistance, $R_2$, of branch ② is 17.32 ohms, the series inductive reactance, $X_L2$, is 10 ohms, and the impedance, $Z_2$, is $\sqrt{(17.32)^2 + 10^2}$ = 20 ohms.

A topographic-vector diagram of the currents in the two branches and the total circuit current is shown in figure 13-4 (B). The current in branch ① is $I_1 = \frac{100}{20}$ = 5 amperes and is the hypotenuse of the right triangle of which the base (energy component) is 5 cos 60° = 2.5 amperes and the altitude (nonenergy component) is 5 sin 60° = 4.33 amperes. The current in branch ② is $I_2 = \frac{100}{20}$ = 5 amperes and is the hypotenuse of the right angle of which the base (energy component) is 5 cos 30° = 4.33 amperes and the altitude (nonenergy component) is 5 sin 30° = 2.5 amperes. The total circuit current is the vector sum of $I_1$ and $I_2$, and is the hypotenuse of the resultant right triangle, the base of which is the sum of the energy components of both branches, 2.5 + 4.33 = 6.83 amperes, and the altitude of which is the sum of the nonenergy components of both branches, 4.33 + 2.5 = 6.83 amperes. Thus,

$$I_t = \sqrt{(2.5 + 4.33)^2 + (4.33 + 2.5)^2}$$

$$= 9.66 \text{ amperes}$$

The circuit power factor is cos $\theta t$ = $\frac{2.5 + 4.33}{9.66}$ = 0.707, and $\theta t$ = 45°.

The power relations for this circuit are shown in figure 13-4 (C). The apparent power in branch ① is $EI_1$ = 100 x 5 = 500 volt-amperes and is the hypotenuse of the right triangle, the base of which is $EI_1 \cos \theta_1$ = 100 x 5 cos 60° = 250 watts of true power, and the altitude of which is $EI_1 \sin \theta_1$ = 100 x 5 sin 60° = 433 VARS of reactive power.

The apparent power in branch ② is $EI_2$ = 100 x 5 = 500 volt-amperes and is the hypotenuse of the right triangle, the base of which is $EI_2 \cos \theta_2$ = 100 x 5 x cos 30° = 433 watts of true power and the altitude of which is $EI_2 \sin \theta_2$ = 100 x 5 x sin 30° = 250 VARS of reactive power.

The total apparent power of the combined parallel circuit is the hypotenuse of the resultant triangle, the base of which is equal to the sum of the true power components in both branches, and the altitude of which is equal to the sum of the reactive power components in both branches. Thus, the total apparent power = $\sqrt{(250 + 433)^2 + (433 + 250)^2}$ = 966 volts amperes.

The equivalent series circuit (fig. 13-4 (D)), representing the combined impedance of the given parallel circuit, has a total impedance, $Z_t$, that is determined by dividing the applied voltage, E, by the total circuit current, $I_t$.

$$Z_t = \frac{E}{I_t} = \frac{100}{9.66} = 10.35 \text{ ohms}$$

The hypotenuse of the impedance triangle (fig. 13-4 (E)) is 10.35 ohms. The impedance triangle is similar to the resultant-current right triangle (fig. 13-4 (B)), and angle $\theta t$, has the same magnitude in both triangles. This angle is equal to 45°, as determined in the current-triangle calculations.

The equivalent series resistance, $R_{eq}$, is the base of the impedance triangle and is equal to $Z_t \cos \theta t$, or 10.35 x cos 45 = 7.33 ohms.

The equivalent series reactance, $X_{Leq}$, is the altitude of the impedance triangle and is equal to $Z_t \sin \theta_t$ or 10.35 x sin 45° = 7.33 ohms.

The impedance triangle is also similar to the resultant power triangle for the combined parallel circuit (fig. 13-4 (C)). The common factor between the two triangles is the square of the total circuit current, $I_t^2$. The base of the resultant power triangle is $I_t^2 R_{eq}$ = $(9.66)^2$ x 7.33 = 683 watts of true power. This product may be checked by adding the true

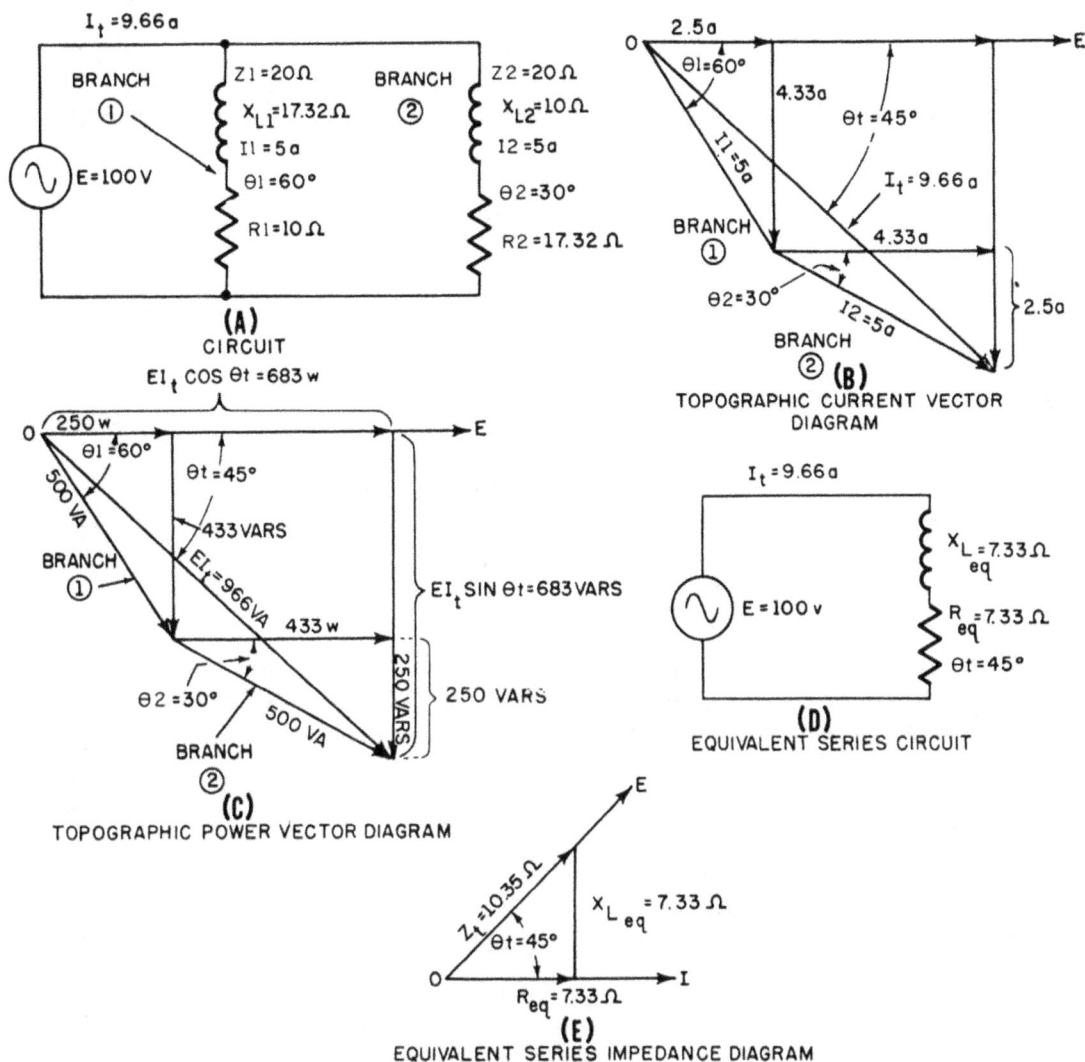

BE.232

Figure 13-4.—Two-branch LR circuit with currents 30° out of phase.

power components in branches ① and ②. Thus, $250 + 433 = 683$ watts of true power.

The altitude of the resultant power triangle is $I_t^2 X_{Leq} = (9.66)^2 \times 7.33 = 683$ VARS of reactive power. This produce may be checked by adding the reactive VAR components in branches ① and ②. Thus, $433 + 250 = 683$ VARS of reactive power.

The hypotenuse of the resultant power triangle is $I_t^2 Z_t = (9.66)^2 \times 10.35 = 966$ volt-amperes of apparent power. This product may be checked by adding vectorially the apparent power in branch ① and the apparent power in branch ②.

This produce is equal to the hypotenuse of the resultant power triangle (fig. 13-4 (C)). This produce cannot be checked accurately by adding arithmetically the apparent power in branch ① to the apparent power in branch ② because these quantities are not in phase with each other.

## CAPACITANCE AND RESISTANCE IN PARALLEL

The action of a capacitor in an a-c series circuit was described in chapter 11 and is amplified further at this time as an introduction to the a-c parallel RC circuit. In figure 13-5 (A), an a-c voltage of sine waveform is applied across a capacitor and a charging current of sine waveform flows around the circuit in a manner something like that of the flow of water in the hydraulic analogy shown in figure 13-5 (B). The pump at the left corresponds to the a-c source and the cylinder at the right corresponds to the capacitor. If the pump piston is driven by means of a crank turning at uniform speed, the resulting motion of the water will be sinusoidal. The motion of the water is transmitted through the flexible diaphragm in the cylinder and the resulting motion, first in one direction and then in the other, corresponds to the electron flow in the wires connecting the capacitor to the a-c source.

The mechanical stress in the diaphragm in the cylinder corresponds to the electric stress in the dielectric between the plates of the capacitor. Electron flow does not occur through the dielectric in the same way that water would flow through the flexible diaphragm if a hole were punctured in it. Instead, the electrons flow around the capacitor circuit on one alternation causing a negative charge to build up on one place, and a corresponding positive charge on the other, and on the next alternation causing a reversal of the polarity of the charges on the plates. Thus, the effective impedance which the capacitor offers to the flow of alternating current can be relatively low at the same time that the insulation resistance which the dielectric offers to the flow of direct current is extremely high.

In the example of figure 13-6, a 2-branch circuit consists of a 100-ohm resistor and a 15.9-microfarad capacitor of negligible losses in parallel with a 100-volt a-c source. The frequency of the source voltage is 100 $H_z$ and the voltage is assumed to have a sine waveform. The current in branch ① is $\frac{100}{100}$ = 1 ampere (rms).

The impedance of branch ② is composed of capacitive reactance; its resistance component is neglected. In branch ② $X_c = \frac{1}{2\pi fC} =$

$$\frac{1}{6.28 \times 100 \times 15.9 \times 10^{-6}} = 100 \text{ ohms.}$$

The current in branch ② is $\frac{100}{100}$ = 1 ampere (rms). The sine waveform of the branch currents and the applied voltage, together with the resultant line current, are shown in figure 13-6 (B).

The current in the resistive branch is in phase with the applied voltage and has a peak value of $\frac{1}{0.707}$ = 1.41 amperes. The current in the capacitor branch leads the applied voltage by an angle of 90° and has a peak value of $\frac{1}{0.707}$ = 1.41 amperes. The resultant line current is the algebraic sum of the instantaneous values of the currents in the two branches and has a peak value of 2 amperes. The resultant line current, $I_t$, leads the applied voltage by an angle of 45°.

## CURRENT VECTORS

A topographic-vector diagram of the effective values of these currents is shown in figure 13-6 (C). The base of the current triangle is 1 ampere and represents the current in branch ①. This current is in phase with the applied voltage and represents the energy component of the total line current.

The altitude of the current triangle is 1 ampere and represents the current in branch ②. This current leads the applied voltage by 90° and represents the nonenergy component of the total line current.

The hypotenuse of the triangle is 1.41 amperes and represents the total line current.

The reference vector, OE, for the current triangle is the line voltage, and in the RC parallel circuit, the altitude extends above the reference vector to indicate the sense of lead. This direction is opposite to that of the altitude of the current triangle for the RL parallel

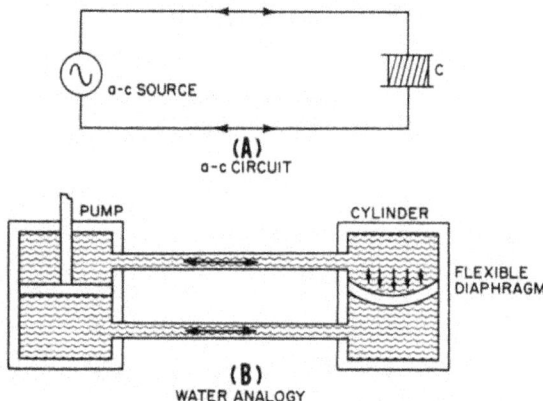

**(A)**
a-c CIRCUIT

**(B)**
WATER ANALOGY

BE. 233

Figure 13-5.—Water analogy of a capacitor in an a-c circuit.

circuit in figure 13-2 (D). In both figures the vectors are assumed to rotate in a counter-clockwise direction to indicate the sense of lead or lag of the currents with respect to the line voltage. In all single-phase circuits such as these, the current vectors are considered in their relative phase by angles that never exceed 90° with respect to their common reference voltage vector.

In figure 13-2 (D), the line current lags the line voltage by an angle of 45°, and in figure 13-6 (C), the line current leads the line voltage by an angle of 45°. If both inductance and capacitance exist in the same parallel circuit as separate branches, these branches will be 180° out of phase with each other, but the current in the inductive branch will never lag the line voltage by an angle in excess of 90°, and the current in the capacitive branch will never lead the line voltage by an angle in excess of 90°.

## POWER AND POWER FACTOR

The power triangle for the parallel RC circuit is shown in figure 13-6 (D). The power in branch ① is

$$EI_1 \cos \theta_1 = 100 \times 1$$
$$x (\cos 0° = 1) = 100 \text{ watts}$$

and forms the base of the triangle in line with the voltage vector, OE. the reactive power in branch ② is

$$EI_2 \sin \theta_2 = 100$$
$$x 1 (\sin 90° = 1) = 100 \text{ VARS}$$

and is the altitude of the power triangle, perpendicular to OE. The reactive power in branch ① is

$$EI_1 \sin \theta_1 = 100$$
$$x 1 (\sin 0° = 0) = 0 \text{ VAR}$$

and the power in branch ② is

$$EI_2 \cos \theta_2 = 100$$
$$x 1 (\cos 90° = 0) = 0 \text{ watt}$$

The apparent power of the parallel RC circuit is

$$EI_t = 100 \times 1.41 = 141 \text{ volt-amperes}$$

and is the hypotenuse of the power triangle.

The power triangle (fig. 13-6 (D)) is similar to the current triangle (fig. 13-6 (C)), since $\theta_t$ has the same magnitude in both triangles. The total circuit power factor, as determined from the current triangle, is

$$\cos \theta_t = \frac{I_1}{I_t} = \frac{1}{1.41} = 0.707$$

The total circuit power factor, as determined from the power triangle, is

$$\cos \theta_t = \frac{\text{true power}}{\text{apparent power}} = \frac{100}{141} = 0.707$$

The power of the total circuit, as determined from the power triangle, is

$$EI_t \cos \theta_t = 100 \times 1.41$$
$$(\cos 45° = 0.707 = 100 \text{ watts}$$

This value is equal to the true power in branch ①.

The reactive power of the total circuit, as determined from the power triangle, is

$$EI_t \sin \theta_t = 100 \times 1.41$$
$$(\sin 45° = 0.707 = 100 \text{ VARS}$$

This value is equal to the reactive power in branch ②.

## EQUIVALENT SERIES IMPEDANCE

The total impedance of the parallel RC circuit is

$$Z_t = \frac{E}{I_t} = \frac{100}{1.41} = 70.7 \text{ ohms}$$

As mentioned previously, the total impedance is also the impedance of the equivalent series circuit (fig. 13-6 (E)). In figure 13-6 (F), the hypotenuse, $Z_t$, of the impedance triangle is 70.7 ohms and the phase angle, $\theta_t$, between total current and line voltage is 45°. The equivalent series resistance, $R_{eq}$, is the base of the impedance triangle and has a magnitude of

$$R_{eq} = Z_t \cos \theta_t = 70.7$$
$$(\cos 45° = 0.707) = 50 \text{ ohms}$$

The equivalent series reactance, $X_{Ceq}$, is the altitude of the impedance triangle and has a magnitude of $Z_t \sin \theta_t = 70.7$ ($\sin 45° = 0.707$) = 50 ohms. The altitude is extended downward, in

**(A)**
CIRCUIT

**(B)**
WAVEFORMS

**(C)**
CURRENT TRIANGLE

**(D)**
POWER TRIANGLE

**(E)**
EQUIVALENT SERIES CIRCUIT DIAGRAM

**(F)**
EQUIVALENT SERIES-IMPEDANCE
VECTOR DIAGRAM

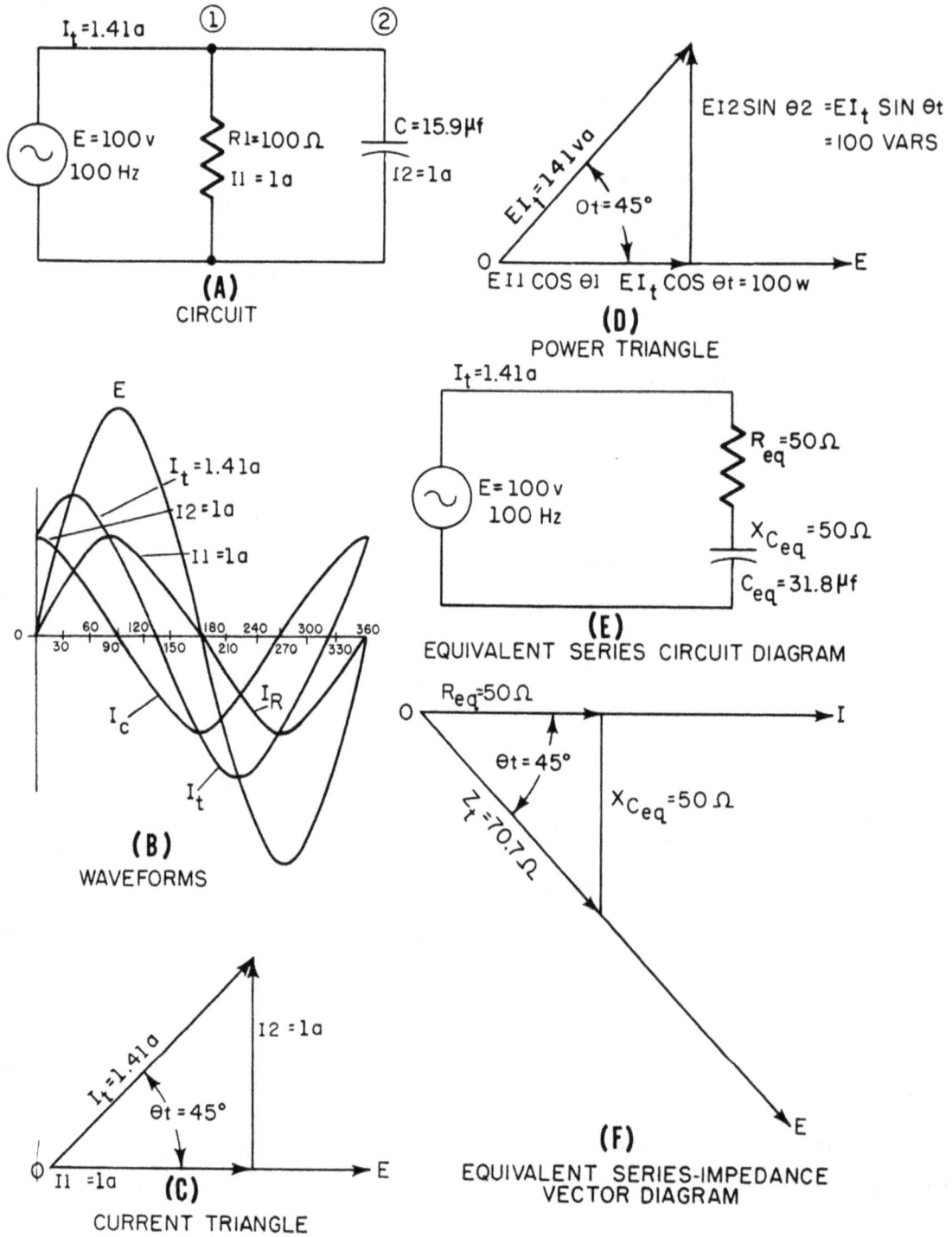

Figure 13-6.—Parallel RC circuit analysis.

BF.234

225

contrast to the upward direction of the altitude of the current triangle in figure 13-6 (C), to maintain the sense of current lead for counter-clockwise vector rotation. The line voltage is the horizontal reference vector, OE (fig. 13-6 (C)), and the line current is the horizontal reference vector, OI (fig. 13-6 (F)).

The capacitance in microfarads of the equivalent series impedance is

$$C_{eq} = \frac{10^6}{2\pi f X_{Ceq}} = \frac{10^6}{6.28 \times 100 \times 50}$$

$$= 31.8 \text{ microfarads.}$$

Thus, in this example, the 100-ohm resistor in branch ① shunts the 15.9-microfarad capacitor in branch ② and the source "sees" and equivalent resistor of 50 ohms in series with an equivalent capacitor of 31.8 microfarads (fig. 13-6 (E)).

## EQUIVALENT CIRCUIT OF A LOW-LOSS CAPACITOR

The action of a capacitor in an a-c circuit was described as being like the flow of water in a cylinder having a flexible diaphragm. The diaphragm corresponds to the dielectric in a capacitor and the mechanical stresses in the diaphragm correspond to the electric stress in the dielectric.

Most of the heating produced in a capacitor is due to the loss in the dielectric. The movement of the electrons in the atoms of a solid dielectric is depicted in figure 13-7. The capacitor is connected to a source of a-c voltage of sine waveform and the conditions are indicated for three successive instants in the cycle of applied a-c voltage.

In figure 13-7 (A), the voltage across the capacitor plates is positive maximum and the capacitor is charged. The atoms in the dielectric are subjected to electric stress that causes their orbital electrons to move in paths that are charged from circular to elliptical patterns. The elliptical pattern forms as a result of the attractive force of the negatively charged plate on the positive nucleus and the simultaneous repulsion of the positively charged plate on the nucleus. At the same time the positive plate attracts the orbital electrons, the negative plate repels them.

In figure 13-7 (B), the charge on the plates is zero and the electric stress is removed from

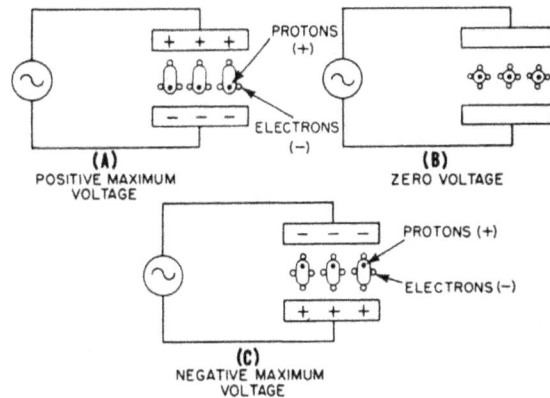

Figure 13-7.—Capacitor dielectric losses.

the dielectric. In this case the orbital electrons travel in circular paths about the nucleus with no external forces applied to them.

In figure 13-7 (C), the charges on the plates are reversed and the orbital electrons are again caused to move in elliptical paths. In this case the protons (entire nucleus) are moved toward the upper plate, the forces being opposite to those developed in figure 13-7 (A).

The rapid reversals of applied voltage accompanied by the change of the orbital electron paths from circular to elliptical and back to circular patterns cause heat to be developed in the dielectric. The change in pattern of the electron orbits constitutes a dielectric displacement current. This current is considered to be made up of two components, one leading the voltage by 90°, the other an energy component in phase with the applied voltage. Factors that determine the dielectric loss are the applied voltage, the dielectric constant, the capacitor power factor, and the frequency of the applied voltage. An air-dielectric capacitor has no appreciable loss and the power factor is 0. A mica capacitor, having a dielectric constant of 7 and a power factor of 0.0001, has relatively low loss and low dielectric heating at high voltages and high frequencies.

The product of the dielectric constant and the power factor is called the loss factor. The loss factor is low for good dielectrics that operate without much dielectric heating. The loss factor is the best indication of the ability of a material to withstand high voltages at high frequencies. The loss factor for the previously mentioned mica capacitor is 7 x

0.0001 = 0.0007. The loss factor for air dielectric is zero because the power factor is zero.

The equivalent circuits of a low power-factor capacitor are represented in figure 13-8. The circuits are derived by assuming a capacitor of the same capacitance as that of the original capacitor, but having no losses (zero power factor), to be connected (1) in series with a resistor that develops the same true power loss as in the original capacitor; and (2) a capacitor of the same capacitance as that of the original capacitor, but having no losses (zero power factor), to be connected in shunt with a resistor that develops the same true power loss as in the original capacitor. The capacitance of the capacitor in this example is 66.4 microfarads. The power factor of the capacitor is $\cos \theta = 0.05$. The effective voltage is 200 volts and the capacitor current is 5 amperes.

The equivalent series circuit is shown in figure 13-8 (B). At the operating frequency of 60 Hz the capacitive reactance of the capacitor is

$$X_C = \frac{1}{2\pi fC} = \frac{1}{6.28 \times 60 \times 66.4 \times 10^{-6}}$$

$$= 40 \text{ ohms approximately}$$

The impedance of the series circuit is

$$Z_t = \frac{E}{It} = \frac{200}{5} = 40 \text{ ohms.}$$

The equivalent series resistance is

$$R_{se} = Z_t \cos \theta_t = 40 \times 0.05 = 2 \text{ ohms}$$

(fig. 13-8 (B)). The base of the impedance triangle is 2 ohms of resistance (fig. 13-8 (C)), the altitude of the impedance triangle is 40 ohms of capacitive reactance, and the hypotenuse is approximately 40 ohms of impedance.

The power factor of this circuit is $\cos \theta_t$ $= \frac{R_{se}}{Z_t} = \frac{2}{40} = 0.05$, or 5 percent. In this example, as in the low-loss inductor of figure 13-3, $Z_t$ is approximately equal to $X_C$, and $\cos \theta_t = \frac{R_{se}}{X_C}$. Since $\tan \theta_t = \frac{X_C}{R_{se}} = \frac{40}{2} = 20$, it follows that $\cos \theta_t = \frac{1}{\tan \theta}$, and the power factor of the capacitor is $\frac{2}{40}$, or 0.05.

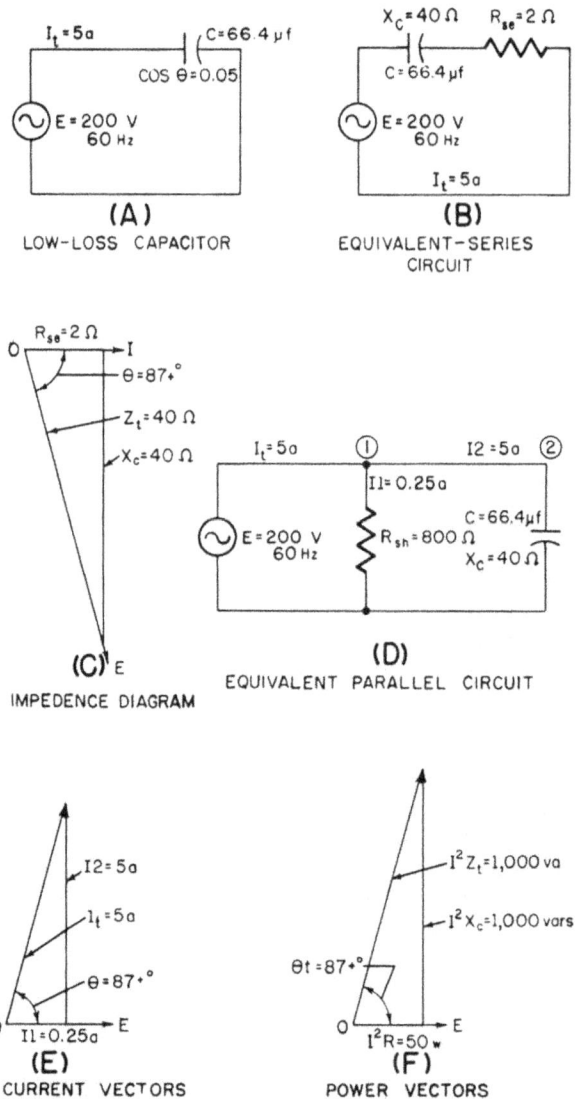

Figure 13-8.—Equivalent circuits of a low-loss capacitor.

The angle whose cosine is 0.05 may be closely approximated by the previously described relation between the power factor and the complementary angle in radians. In this example, the complementary angle by which the current fails to be 90° out of phase with the voltage is 0.05 radian, or 0.05 x 57.3 = 2.865°, and $\theta = 90° - 2.865° = 87.135°$. This angle is indicated in all of the vector diagrams of figure 13-8 as 87+°.

The equivalent parallel circuit for the 66.4-microfarad capacitor discussed in this example is shown in figure 13-8 (D). The equivalent series resistance and the equivalent shunt resistance are related in the equivalent capacitor circuits in the same way that $R_{se}$ and $R_{sh}$ are related in the low-loss inductor circuits of figure 13-3. This relation was stated in equation 13-5. Thus, in the capacitor circuits,

$$R_{sh} = \frac{(X_C)^2}{R_{se}} = \frac{(40)^2}{2} = 800 \text{ ohms}$$

The equivalent shunt resistance of 800 ohms is connected in parallel with a 66.4-microfarad capacitor of zero losses. With rated voltage and frequency applied, the input current to the parallel circuit will have the same magnitude and phase with respect to the applied voltage as in the original capacitor.

The energy current in the resistive branch is

$$I_1 = \frac{E}{R_{sh}} = \frac{200}{800} = 0.25 \text{ ampere and is the base}$$

of the current triangle (figure 13-8 (E)). The nonenergy current in the capacitive branch is

$$I_2 = \frac{E}{X_C} = \frac{200}{40} = 5 \text{ amperes and is the altitude}$$

of the current triangle. The total current in the parallel circuit is equal to

$$\sqrt{(0.35)^2 + (5)^2} = 5$$

amperes (approx.) and is the hypotenuse of the current triangle. This current leads the voltage by an angle of 87.135°.

The power relations are shown in figure 13-8 (F). The true power of the circuit may be found in a number of ways. Three methods are listed as follows:

1. $P = EI_t \cos\theta_t = 200 \times 5 \times 0.05 = 50$ watts (fig. 13-8 (A)).

2. $P = I_t^2 R_{se} = 5^2 \times 2 = 50$ watts (fig. 13-8 (B)).

3. $P = I_1^2 R_{sh} = (0.25)^2 \times 800 = 50$ watts (fig. 13-8 (D)).

The true power of 50 watts is the base of the power triangle (fig. 13-8 (F)).

The reactive power may be calculated in a number of ways. Three methods are indicated as follows:

1. VARS $= E_t I_t \sin\theta_t = 200 \times 5 \times (\sin 87+°$ $= 1$ approx.) $= 1,000$ VARS (fig. 13-8 (A)).

2. VARS $= I_t^2 X_C = 5^2 \times 40 = 1,000$ VARS (fig. 13-8 (B)).

3. VARS $= I_2^2 X_C = 5^2 \times 40 = 1,000$ VARS (fig. 13-8 (D)).

The reactive power of 1,000 VARS is the altitude of the power triangle (fig. 13-8 (F)).

The apparent power of the capacitor may be calculated as:

1. Apparent power $= EI_t = 200 \times 5 = 1,000$ VA (fig. 13-8 (A)).

2. Apparent power $= I_t^2 Z_t = 5^2 \times 40 = 1,000$ VA (fig. 13-8 (B)).

The apparent power of 1,000 volt-amperes is the hypotenuse of the power triange (fig. 13-8 (F)).

## COMBINING CURRENTS AT ACUTE ANGLES

A 2-branch parallel circuit containing a 75-ohm resistor in branch ① and a series combination of a 79.6 microfarad capacitor and a 30-ohm resistor in branch ② is shown in figure 13-9 (A). The capacitor losses are assumed to be included with those of the 30-ohm resistor so that the power factor of that portion of branch ② represented by the capacitor is zero.

The capacitive reactance at the operating frequency of 50 Hz is

$$X_C = \frac{1}{2\pi fC} = \frac{1}{6.28 \times 50 \times 79.6 \times 10^{-6}}$$
$$= 40 \text{ ohms.}$$

The impedance, $Z_2$, of branch ② is the combined opposition of 30 ohms of resistance in series with 40 ohms of capacitive reactance and is the hypotenuse of the right triangle of which 30 ohms is the base and 40 ohms is the altitude (not shown in the figure). Thus,

$$Z_2 = \sqrt{30^2 + 40^2} = 50 \text{ ohms.}$$

The power factor of branch ② $=$ is $\cos\theta_2 = \frac{R_2}{Z_2} = \frac{30}{50} = 0.60$ and the phase angle is 53.1°.

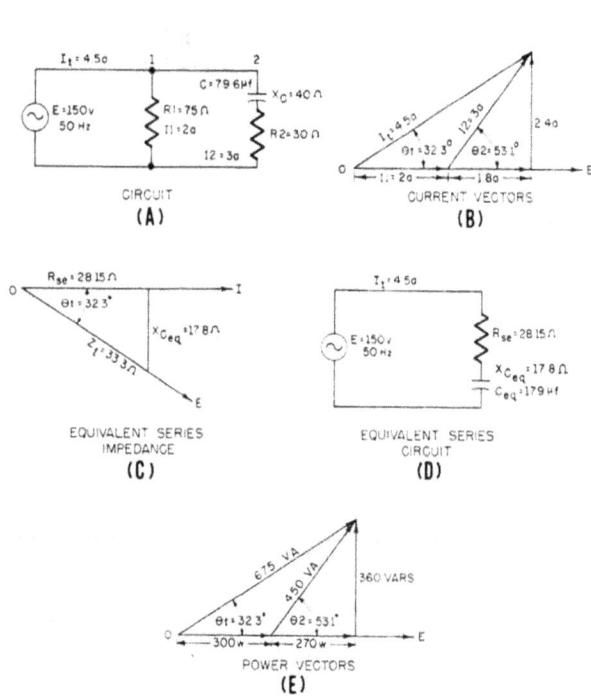

CIRCUIT
**(A)**

CURRENT VECTORS
**(B)**

EQUIVALENT SERIES
IMPEDANCE
**(C)**

EQUIVALENT SERIES
CIRCUIT
**(D)**

POWER VECTORS
**(E)**

BE.237
Figure 13-9.—Two-branch RC circuit with currents out of phase by an angle of 53.1°.

The current vectors for both branches are shown in figure 13-9 (B).

The current in branch ① is $I_1 = \dfrac{E}{Z_1} = \dfrac{150}{75}$ = 2 amperes, and is a portion of the base of the equivalent right triangle in line with the horizontal reference voltage, OE. This current is in phase with the applied voltage because there is no reactance present in branch ①

The current in branch ② is

$$I_2 = \frac{E}{Z_2} = \frac{150}{50} = 3 \text{ amperes}$$

and is the hypotenuse of the right triangle, the base of which is $I_2 \cos \theta_2$ = 3 (cos 53.1° = 0.6) = 1.8 amperes (energy current) and the altitude of which is $I_2 \sin \theta_2$ = 3 (sin 53.1° = 0.8) = 2.4 amperes (nonenergy current). The total current is the vector sum of $I_1$ and $I_2$ and is the hypotenuse of the equivalent right triangle, the base of which is the arithmetic sum of the current in branch ① and the energy component of current in branch ②, or 2 + 1.8 = 3.8 amperes. The altitude of the equivalent right triangle is equal to the nonenergy current in

branch ②, or 2.4 amperes. The total current in the parallel circuit is

$$I_t = \sqrt{(3.8)^2 + (2.4)^2} = 4.5 \text{ amperes}$$

The phase angle, $\theta_t$ between the total circuit current and the applied voltage is the angle whose cosine is $\dfrac{3.8}{4.5}$ = 0.845. This angle is approximately 32.3°.

Thus, the total current leads the applied voltage by an angle of 32.3°.

The total impedance of the parallel circuit is

$$Z_t = \frac{E}{I_t} = \frac{150}{4.5} = 33.3 \text{ ohms}$$

and is the hypotenuse of the impedance triangle (fig. 13-9 (C)). The base of the triangle is

$$R_{se} = Z_t \cos \theta_t = 33.3$$

(cos 32.3° = 0.845) = 28.15 ohms

of resistance and is in line with the horizontal current reference vector for the equivalent series circuit (fig. 13-9 (D)). The altitude of the triangle is

$$X_{Ceq} = Z_t \sin \theta_t = 33.3$$

(sin 32.3° = 0.534) = 17.8 ohms

of capacitive reactance. The capacitance of the equivalent series circuit at the operating frequency of 50 $H_z$ is

$$C_{eq} = \frac{10^6}{2\pi f X_C} = \frac{10^6}{6.28 \times 50 \times 17.8}$$

= 179 microfarads.

The power relations are indicated in the triangles of figure 13-9 (E). The apparent power of the total circuit is $E I_t$ = 150 x 4.5 = 675 volt-amperes and is the hypotenuse of the equivalent right triangle representing the total parallel circuit.

The true power in branch ① is

$$I_1^2 R_1 = 2^2 \times 75 = 300 \text{ watts}$$

The true power in branch ② is

$$I_2^2 R_2 = 3^2 \times 30 = 270 \text{ watts}$$

The total true power is

$$300 + 270 = 570 \text{ watts}$$

and is the base of the equivalent right triangle for the entire circuit. The reactive power in the total circuit—that is, the reactive power of branch ② —is

$$I_2{}^2 X_C = 3^2 \times 40 = 360 \text{ VARS}$$

and is the altitude of the equivalent triangle.

The apparent power in branch ① is equal to the true power because the power factor is unity. The apparent power in branch ② is $EI_2 = 150 \times 3 = 450$ volt-amperes and is the hypotenuse of the power triangle for this branch. The true power of 270 watts in branch ② is the base of this triangle.

The reactive power of 360 VARS in branch ② is the altitude of the triangle. The power factor for branch ② is $\cos\theta_2 = \dfrac{\text{true power}}{\text{apparent power}} = \dfrac{270}{450} = 0.60$, and the phase angle for branch ② is $\theta_2 = 53.1°$.

## PARALLEL CIRCUITS CONTAINING L, R, AND C

Inductance in an a-c circuit causes the current to lag behind the applied voltage. Transformers and induction motors are essentially inductive in nature and the power factor, especially on light loads (as contrasted with full loads), is relatively low.

Most circuits, that supply electric power from the source to the consumer, transmit the power at relatively high voltage and low current in order to keep the $I^2R$ loss in the transmission lines satisfactorily low. Transformers at the point of utilization reduce the voltage to the proper value to operate the equipment.

Capacitance in a-c circuits causes the current to lead the applied voltage and when placed in parallel with inductive components can produce a neutralizing effect so that lagging currents can be brought into phase with the applied voltage or may be made to lead the applied voltage, depending on the relative magnitude of the capacitance and inductance in parallel.

The true power of a circuit is $P = EI \cos\theta$; and for any given amount of power to be transmitted, the current, I, varies inversely with the power factor, $\cos\theta$. Thus, the addition of capacitance in parallel with inductance will, under the proper conditions, improve the power factor (make nearer unity power factor) of the circuit and make possible the transmission of electric power with reduced line loss and improved voltage regulation.

## CURRENT VECTORS

In the example of figure 13-10, the parallel circuit contains three branches. Branch ① contains a 30-ohm resistor. Branch ② consists of an inductor of 0.0612 henry and a resistor of 6.6 ohms. Branch ③ contains a capacitance of 44.3 microfarads having negligible losses. The parallel circuit is shown in figure 13-10 (A). The current vectors are shown in figure 13-10 (B).

The current in branch ① is

$$I_{1en} = \frac{E}{R_1} = \frac{120}{30} = 4 \text{ amperes}$$

The corresponding 4-ampere vector in figure 13-10 (B), is drawn in the same horizontal reference line as the voltage vector because the phase angle between the current and voltage in this branch is zero.

The inductive reactance of branch ② at the operating frequency of 60 Hz is

$$X_L = 2\pi fL + 6.28 \times 60 \times 0.0612 = 23.1 \text{ ohms}$$

The impedance of branch ② is

$$Z2 = \sqrt{R_2{}^2 + X_L{}^2} = \sqrt{(6.6)^2 + (23.1)^2}$$
$$= 24 \text{ ohms}$$

The current in branch ② , at the operating voltage of 120 volts, is $I_2 = \dfrac{E}{Z_2} = \dfrac{120}{24} = 5$ amperes and is the hypotenuse of the current triangle for branch ② . The angle by which the current in branch ② lags the applied voltage is the angle whose cosine is

$$\frac{R_2}{Z_2} = \frac{6.6}{24} = 0.275$$

This angle is $\theta_2 = 74°$. The base of the current triangle for branch ② is

$$I_{2en} = I_2 \cos 74° = 5$$
$$(\cos 74° = 0.275) = 1.38 \text{ amperes}$$

The altitude of the current triangle for branch ② is

$$I_{2 \text{ nonen}} = I_2 \sin 74° = 5$$

$$(\sin 74° = 0.962) = 4.8 \text{ (approx) amperes}$$

and extends below the horizontal voltage reference vector to indicate current lag.

The capacitive reactance in branch ③ is

$$X_C = \frac{1}{2\pi fC} = \frac{1}{6.28 \times 60 \times 44.3 \times 10^{-6}}$$

$$= 60 \text{ ohms}$$

and with negligible losses, the impedance, $Z_3$, is also 60 ohms. The current in branch ③ is

$$I_{3 \text{ nonen}} = \frac{E}{Z_3} = \frac{120}{60} = 2 \text{ amperes}$$

The current in the capacitor branch leads the applied voltage by an angle of 90°. The nonenergy component of the current in the inductive branch lags the applied voltage by an angle of 90°. Therefore, these two currents are 180° out of phase with each other, and the capacity current vector, $I_3$, extends upward from the lower extremity of the vector representing the nonenergy component of current in branch ②.

The total current, $I_t$, in the parallel circuit is the vector sum of the currents in the three branches and is the hypotenuse of the equivalent current triangle, the base of which is the arithmetic sum of the energy components of the currents and the altitude of which is the algebraic sum of the nonenergy components of the currents. Thus, the total current is

$$I_t = \sqrt{(I_{1 \text{en}} + I_{2 \text{en}})^2 + (I_{2 \text{nonen}} - I_{3 \text{nonen}})^2}$$

$$= \sqrt{4 + (1.38)^2 + (4.8 - 2)^2}$$

$$= 6.06 \text{ amperes}$$

The phase angle, $\theta_t$, between the total current and the applied voltage is the angle whose cosine is the ratio of the sum of the energy components of the currents in all the branches to the total current. Thus,

$$\cos \theta_t = \frac{4 + 1.38}{6.06} = 0.888$$

$$\theta_t = 27.5°$$

and the total circuit current lags the applied voltage by an angle of 27.5°.

## EQUIVALENT SERIES IMPEDANCE

The total impedance of the parallel circuit (fig. 13-10 (A)) is equal to the total circuit voltage divided by the total circuit current, or

$$Z_t = \frac{E}{I_t} = \frac{120}{6.06} = 19.8 \text{ ohms}$$

and is the hypotenuse of the impedance triangle (fig. 13-10 (C)). This triangle represents the impedance relations existing in the equivalent series circuit (fig. 13-10 (D)). The phase angle, $\theta_t$, of the equivalent series circuit has the same magnitude as the phase angle between the total circuit current and the applied voltage across the parallel circuit. The impedance of the equivalent series circuit has the same magnitude as the total impedance of the parallel circuit. The base of the triangle is

$$R_{eq} = Z_t \cos \theta_t = 19.8$$

$$(\cos 27.5° = 0.889) = 17.6 \text{ ohms}$$

and represents the resistive component of the equivalent series circuit. The altitude of the triangle is

$$X_{Leq} = Z_t \sin \theta_t = 19.8$$

$$(\sin 27.5° = 0.462) = 9.16 \text{ ohms}$$

and represents the inductive reactance component of the equivalent series circuit.

## POWER AND POWER FACTOR

The true power of the parallel circuit (fig. 13-10 (A)) is the arithmetic sum of the power absorbed in each branch. The true power in branch ① is

$$P_1 = I_1^2 R_1 = 4^2 \times 30 = 480 \text{ watts}$$

The true power in branch ② is

$$P_2 = I_2^2 R_2 = 5^2 \times 6.6 = 165 \text{ watts}$$

The true power in branch ③ is negligible since the power factor of the capacitor is assumed to be zero. The total true power is

$$480 + 165 = 645 \text{ watts}$$

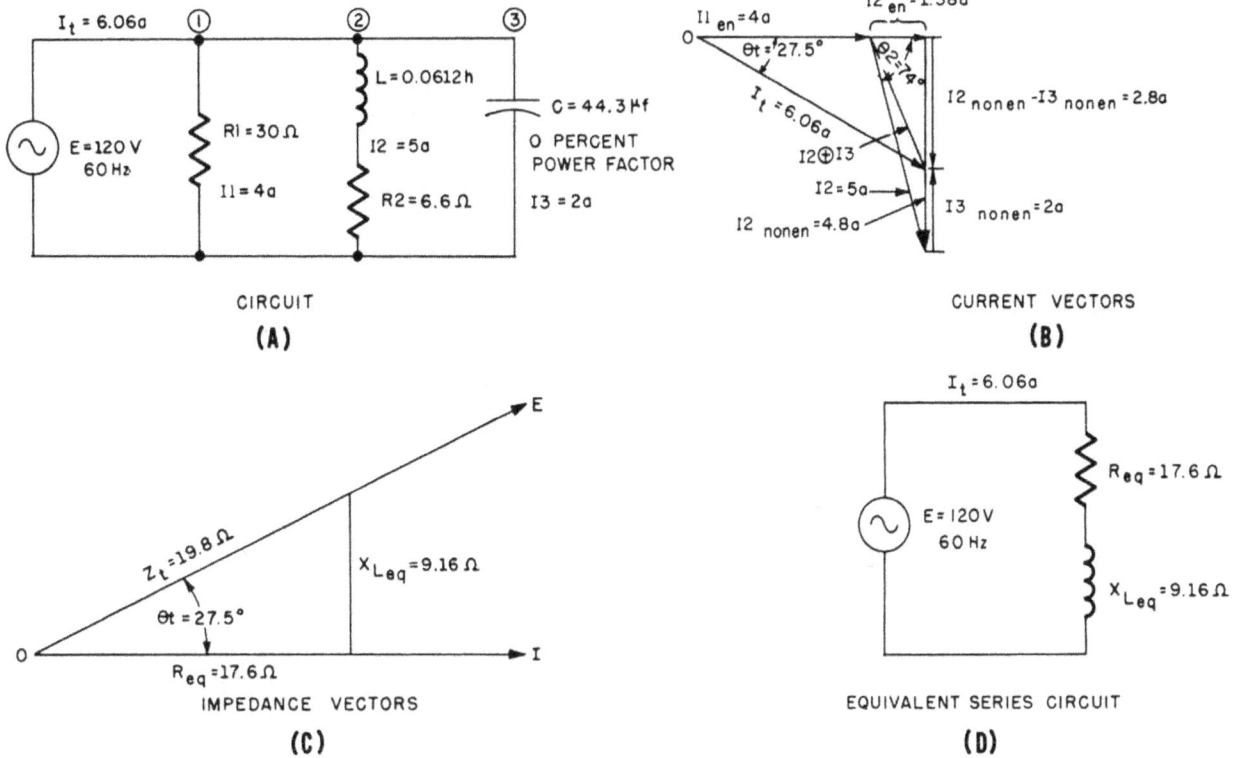

Figure 13-10.—Parallel circuit containing L, R, and C.

The total apparent power of the parallel circuit is the product of the total circuit current and the applied voltage. Thus, the apparent power is

$$EI_t = 120 \times 6.06 = 727.2 \text{ volt-amperes}$$

The parallel circuit power factor is the ratio of the total true power to the total apparent power. Thus, the circuit power factor is

$$\cos \theta_t = \frac{\text{true power}}{\text{apparent power}} = \frac{645}{727.2} = 0.888$$

## POWER-FACTOR CORRECTION

Power-factor correction in parallel circuits is accomplished by placing a capacitor of the proper size in parallel with the circuit at the point where the power-factor correction is to be effected. The leading current of the capacitor branch supplies the lagging component of current in the inductive portion of the parallel circuit and reduces the line current accordingly. As mentioned before, this action improves the efficiency of transmission by reducing the line current and $I^2R$ losses.

In the example of figure 13-11, the load is rated at 10 amperes, 1,000 volts, and a power factor of 50 percent lagging (the current lags the voltage). The true power absorbed by the load (fig. 13-11 (A)) is

$$P = EI \cos \theta = 1,000 \times 10 \times 0.5$$
$$= 5,000 \text{ watts.}$$

The load is supplied through a line having a resistance of 20 ohms and negligible reactance. The power loss in the line is

$$I^2R = 10^2 \times 20 = 2,000 \text{ watts,}$$

and the efficiency of transmission is

$$\frac{\text{output}}{\text{output + losses}} = \frac{5,000}{5,000 + 2,000}$$
$$= 71.4 \text{ percent.}$$

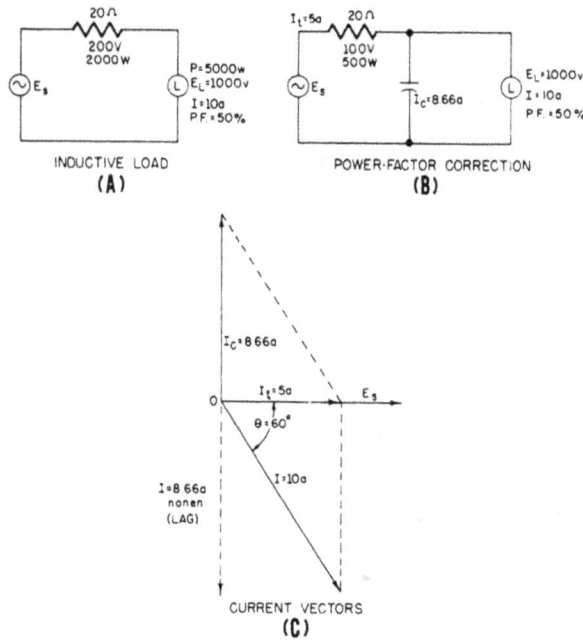

INDUCTIVE LOAD
(A)

POWER-FACTOR CORRECTION
(B)

CURRENT VECTORS
(C)

BE. 239

Figure 13-11.—Power-factor correction.

If a 1,000-volt capacitor of negligible losses supplying a leading current of 8.66 amperes is placed in parallel with the inductive load (fig. 13-11 (B)), the total line current will be reduced from 10 amperes to 5 amperes.

The current vectors are shown in figure 13-11 (C). The nonenergy component of the current in the inductive branch is $I_{nonen} = I \sin \theta = 10$ (sin $60° = 0.866$) $= 8.66$ amperes (lagging) and is $180°$ out of phase with the capacitor current of 8.66 amperes (leading). These currents circulate between the capacitor and the inductive load and do not enter the line. The vector sum of the capacitor current and the total inductive load current is equal to the line current ($I_t = 5$ amperes), and in the vector diagram is represented as the diagonal of the parallelogram of which the 2 branch currents are the sides. The line current vector, $I_t$, is in the same horizontal reference as the load voltage vector, indicating that the line current is in phase with the voltage applied to the parallel combination of the inductive load and the capacitor.

The reduction in line current from 10 to 5 amperes reduces the line loss from 2,000 watts to $5^2 \times 20 = 500$ watts and increases the efficiency of transmission from 71.4 percent to

$$\frac{5,000}{5,000 + 500} = 91 \text{ percent.}$$ Thus, the improvement in efficiency of operation of the line and load is demonstrated. The condition represented involves an interchange of energy between the inductive and capacitive branches known as parallel resonance and is described in the training manual, Basic Electronics, NavPers 10087-B, chapter 4, on the subject of resonance in LRC circuits.

## VOLTAGE REDUCTION WITH RESISTANCE

The example of figure 13-12 illustrates the advantages and disadvantages of controlling the voltage on a load by means of resistance and inductance and also the effect of power-factor correction on the circuit efficiency.

In figure 13-12 (A), the voltage applied to the load is reduced from 100 volts to 50 volts through the action of the dropping resistor. The circuit

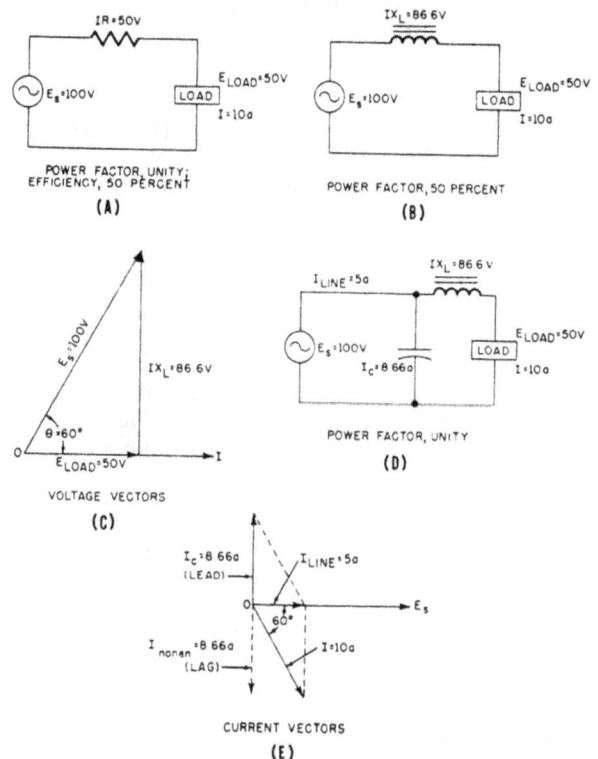

POWER FACTOR, UNITY;
EFFICIENCY, 50 PERCENT
(A)

POWER FACTOR, 50 PERCENT
(B)

VOLTAGE VECTORS
(C)

POWER FACTOR, UNITY
(D)

CURRENT VECTORS
(E)

BE. 240

Figure 13-12.—Voltage reduction and power-factor correction.

current is 10 amperes, and the voltage across the dropping resistor is 50 volts. With this arrangement, the input power to the circuit is divided equally between the resistor and the load, and the circuit efficiency ($\frac{output}{output + losses} \times 100$) is $\frac{500}{500 + 500} \times 100 = 50$ percent. This arrangement represents an inefficient method of voltage reduction.

## VOLTAGE REDUCTION WITH INDUCTANCE

In figure 13-12 (B), the voltage applied to the load is reduced to 50 volts through the action of the series inductor. Neglecting the relatively small losses in the inductor, the circuit efficiency remains high compared to the efficiency of the previous circuit, but the series inductor lowers the power factor from unity (100 percent) to 50 percent. The circuit current is still 10 amperes, and the accompanying line loss between the source and the inductor unnecessarily high.

The voltage vectors for the LR circuit are shown in figure 13-12 (C). The load voltage of 50 volts is the base of the right triangle and is in phase with the load current. The voltage drop across the inductor is 86.6 volts and is the altitude of the voltage triangle. The source voltage, $E_S = 100$ volts, is the vector sum of the load voltage and the inductor voltage and is the hypotenuse of the voltage triangle. The circuit power factor is

$$\cos \theta = \frac{E_{load}}{E_{source}} = \frac{50}{100} = 0.5$$

and

$$\theta = 60°$$

In figure 13-12 (D), the circuit power factor is improved to unity by the addition of a capacitor of negligible losses which supplies a leading current of 8.66 amperes. This current supplies the nonenergy component of the current in the branch containing the inductor and load, and reduces the line current from 10 amperes to 5 amperes.

The current vectors are shown in figure 13-12 (E). The capacitor current of 8.66 amperes is represented by the vector extending above the horizontal voltage reference vector to indicate a lead of 90°. The inductor branch (load) current extends below the horizontal at an angle of 60° to indicate lag. The vector sum of these currents is the diagonal of the parallelogram of which the

branch currents are the sides, and is a horizontal vector of 5 amperes in phase with the source voltage. Thus, the power factor of the parallel circuit is unity and the line current is reduced from 10 amperes to 5 amperes.

## ADVANTAGES OF INDUCTIVE ARRANGEMENT WITH POWER-FACTOR CORRECTION

In the example under consideration, voltage reduction with a series inductor and a shunt capacitor provides a means of supplying a 50-volt 10-ampere load from a 100-volt source with only the small losses associated with the reactive components. The line current is kept to the minimum value required to supply the 500-watt load and the circuit efficiency is high. Inductive control alone provides a means of reducing the load voltage by changing the phase of the applied voltage with respect to the load voltage. The addition of the capacitor reduces the current in the line without altering the load current, and thus reduces the line losses; at the same time the circuit power factor is increased to unity. In most circuits it is not economical to improve the power factor to unity, but to improve it, for example, from 50 percent lagging to 85 percent lagging. The reduction in line losses is most pronounced in this range and any further reduction in losses may not justify the added expense of the capacitance required to further improve the power factor in the range from 85 percent lagging to unity.

## EFFECTIVE RESISTANCE-NONUSEFUL ENERGY LOSSES IN A-C CIRCUITS

### ENERGY CONCEPT OF RESISTANCE

The energy stored in the magnetic field of a pure inductance, as the result of a rise in current through the coil, is returned to the circuit when the current decreases and the field collapses. Similarly, the energy stored in the electric field of a pure capacitance, as a result of the rise in voltage across the capacitor, is returned to the circuit when the voltage falls and the field collapses. Hence, in a pure inductance and a pure capacitance there is no loss or expenditure of energy.

When current flows through a conductor having appreciable resistance the flow is accompanied by the generation of heat. Work is done in moving the electrons through the conductor resistance. The energy converted into heat is not returned to the circuit when the current falls, but is expended rather than stored. Thus, energy is stored periodically in inductance and capacitance but always expended in resistance.

Because resistance is the only circuit component capable of expending electrical energy, all energy expended in any circuit can be identified in electrical terms, one factor of which is effective resistance. The effective resistance, $R_{ac}$, of any circuit may be defined as the ratio of the true power absorbed by the circuit to the square of the effective current flowing in the circuit, or $R_{ac} = \frac{P}{I^2}$ . When the power is expressed in watts and the current is expressed in amperes, the effective resistance will be in ohms. The d-c circuit resistance as measured by an ohmmeter or d-c bridge may be considerably lower than the effective a-c resistance as calculated from the readings on a wattmeter and an ammeter.

For example, assume that a motor draws 1 kilowatt from a 110-volt source. The input current is 10 amperes. The effective resistance between the motor terminals is

$$R_{ac} = \frac{P}{I^2} = \frac{1,000}{(10)^2} = 10 \text{ ohms}$$

The d-c resistance measured between the motor terminals, for example with an ohmmeter, is 0.5 ohm. Thus, in this example the effective a-c resistance is 20 times the d-c resistance. Most of the energy taken from the line is converted into mechanical energy and is not returnable to the electric circuit; hence, it is represented electrically as being expended in an effective a-c resistance of 10 ohms.

The source of power for the motor is unaware of the manner in which the motor expends the electrical energy. To the source, the motor appears as an impedance, $Z = \frac{E}{I} = \frac{110}{10} = 11$ ohms, having a resistive component of 10 ohms. The nature of the various energy conversions taking place inside the motor is important only when the motor itself is being analyzed. From this point of view a motor, electric light, loudspeaker, electron tube, or any other electrical device can be pictured as an equivalent electric circuit containing the fundamental components of inductance, capacitance, and resistance. The energy expended in the circuit is always inter- preted in terms of the effective resistance component.

The energy expended in any electrical device may be divided into two parts: (1) That which is converted into useful form; and (2) that which is not useful. No machine has been built that is capable of perfect conversion—that is, one in which there are no losses. In the motor, for example, there are friction losses in the bearings and heat losses in the windings as a result of the current flow through the resistance.

The number of possible nonuseful losses in a-c circuits is much greater than in d-c circuits. These include: (1) ohmic-resistance loss, (2) skin-effect loss, (3) eddy-current loss, (4) dielectric loss, (5) magnetic-hysteresis loss, (6) corona loss, and (7) radiation loss.

## EFFECTIVE RESISTANCE OF CONDUCTORS

The effective (a-c) resistance of electrical conductors is frequently higher than their d-c resistance especially when they are embedded in iron slots, as in the case of motor and generator armatures; and when they are being used in high-frequency circuits, as in radio transmitters and receivers.

Direct current is distributed uniformly throughout the cross-sectional area of a conductor. For example, if a conductor having a cross-sectional area of 1,000 circular mils is carrying one ampere of direct current then one-thousandth of an ampere (one milliampere) is flowing in each circular mil of cross-sectional area. However, when the current in the conductor varies in amplitude, this uniform distribution throughout the conductor cross section is no longer obtained. The accompanying magnetic field is strongest near the center of the conductor and weaker at the circumference. The varying field induces a voltage in the conductor that opposes the change in current. The voltage induced in that portion of the conductor near the center is greater than the voltage induced in the outer surface of the conductor. The total opposition to the current flow includes the effect of this induced emf and is greater near the center of the conductor than at the surface. Therefore, the current divides inversely with the opposition—more of the current flowing near the circumference, and less near the center of the conductor.

The overall result of this action is a decrease in the available area of cross section to conduct the current and an increase in conductor resistance. The decrease in area and increase in resistance become pronounced at high frequencies, at high current densities, and at high magnetic flux densities. This action is called skin effect. It represents the tendency of a-c conductors to carry the circuit current on the surface, or skin, of the conductors rather than uniformly throughout their cross section. As a result of this tendency, many electrical conductors are made of hollow tubing in order to save the added weight and expense of the unused central portion of the solid conductor. The effective a-c resistance of an isolated circular conductor varies approximately as the product of the square root of the frequency and the length of the conductor, and inversely as the conductor diameter.

## EFFECTIVE RESISTANCE OF INDUCTORS

When a conductor is wound in the form of a coil, the current is concentrated on the inner sides of the turns and into an area much smaller than would be the case in an isolated straight conductor. This action results in a large increase in effective resistance. The area in which the current is concentrated decreases as the frequency increases, hence, effective resistance will increase with frequency. When two or more conductors carrying alternating current are so placed that the magnetic field of one reacts with the field of the other, the resultant field around each conductor is no longer uniform. The change in current distribution in a conductor due to the action of an alternating current in a nearby conductor is called proximity effect.

The proximity effect decreases as the separation between conductors increases. Thus, to lower the effective resistance of radiofrequency inductance coils, it is common practice to space the turns a distance equal to the diameter of the conductor. This decreases the reaction between magnetic fields of adjacent turns and permits the current to distribute itself over a larger area in the cross section of each turn.

The inductance of a hollow-core coil operating at a frequency of 60 Hz is increased many fold when a laminated core of soft silicon steel is inserted in the coil. This increase is due to the high permeability of the transformer-iron laminations. Thus, the skin effect is also increased due to the increased field strength. In addition to the increased skin effect in the coil the effective a-c resistance is further increased because of the magnetic hysteresis loss in the iron. Thus, if the coil is connected to a constant-potential a-c source, the current in the coil will decrease when the iron core is inserted because of a small increase in effective resistance and a large increase in the coil reactance. If the laminated steel core is removed and a piece of steel shafting is inserted in the coil, the effective resistance is further increased due to the eddy-current losses and the larger hysteresis losses in the steel shaft. A wattmeter inserted in the coil circuit will indicate this increase in effective resistance by an increased deflection when the solid steel core is inserted in place of the laminated core.

Powdered iron cores are used in certain types of coils on frequencies as high as 100 megahertz in order to limit the effective resistance of the coil to a satisfactorily low value. The iron particles are separated from each other by an insulated coating and when compressed into cylindrical form and inserted in the coil the induced voltage in each iron particle is so small in relation to the resistance to the path for eddy currents that the accompanying heat loss is negligible. Eddy-current losses are reduced in generator and motor armature conductors of large size by laminating the conductors and insulating the adjacent laminations in a manner similar to that in which the iron of the armature core itself is laminated. Thus, the effective resistance of the armature conductors is reduced.

## EFFECTIVE RESISTANCE OF CAPACITORS

The equivalent circuits of a low-loss capacitor were described earlier in this chapter and the factors affecting the equivalent series resistance noted. The effective a-c resistance of a capacitor is equal to its equivalent series resistance and represents the factor which when multiplied by the square of the effective capacitor charging current will equal the power expended in heat in the capacitor circuit.

As mentioned previously, most of the heating is produced in solid dielectrics, and only a negligible amount is produced in the capacitor plates themselves. The dielectric heating is produced by dielectric displacement currents described in connection with figure 13-7. In most electrical circuits dielectric heating is a nonuseful loss. However, in one commercial application, dielectric heating has been put to good use—that of facilitating the gluing together of stacks of laminated plywood. The plywood laminations are stacked between the plates of a capacitor and moderately high-frequency voltage is applied across the plates. The resulting dielectric displacement currents heat the stack from the inside and the glue is quickly set—much more rapidly than in processes involving the external application of steam heat.

## CORONA LOSS

Corona loss occurs as the result of the emission of electrons from the surface of electrical conductors at high potentials. It is dependent upon the curvature of the conductor surface, with most emission occurring from sharp points and the least emission occurring from surfaces having a large radius of curvature. Corona loss is frequently accompanied by a visual blue glow and an audible hissing sound as the electrons leak off the conductor surface into the atmosphere. Corona loss increases with voltage increase and decreases with increase in atmospheric pressure. This loss is held to a satisfactorily low value by (1) the use of large-diameter conductors, (2) not excessively high voltages, (3) smooth polished surfaces, (4) avoiding sharp points, bends, or turns, and (5) in some devices, for example, high-voltage capacitors, by the use of a compressed gas to retard the electron emission.

## RADIATION LOSS

Radiation loss is not appreciable at power line frequencies, but in the field of communications this loss may become excessive. Power is radiated from transmitting antennas in the form of electric and magnetic fields, and its magnitude varies as the square of the antenna input current and as the so-called radiation resistance of the antenna. The transmission line that connects the transmitter and the antenna may, under certain circumstances, develop a radiation loss. This loss is discussed in chapter 12 of Basic Electronics, NavPers 10087-B, in connection with various types of transmission lines.

# CHAPTER 14

# CIRCUIT PROTECTIVE AND CONTROL DEVICES

Electricity, when properly controlled, is of vital importance to the operation of equipment. When it is not properly controlled, however, it can become dangerous and destructive. It can destroy components or complete units; it can injure personnel and even cause their death.

It is of the greatest importance, then, that all precautions necessary be taken to protect the electrical circuits and units and to keep this force under proper control at all times. In this chapter some of the devices that have been developed to protect and control electrical circuits are discussed.

## PROTECTIVE DEVICES

When an electrical unit is built, the greatest care is taken to insure that each separate electrical circuit is fully insulated from all others so that the current in a circuit will follow its intended individual path. Once the unit is placed into service, however, there are many things that can happen to alter the original circuitry. Some of these changes can cause serious troubles if they are not detected and corrected in time.

Perhaps the most serious trouble we can find in a circuit is a direct short. Recall that this term is used to describe a situation in which some point in the circuit, where full system voltage is present, comes in direct contact with the ground or return side of the circuit. This establishes a path for current flow that contains no resistance other than that present in the wires carrying the current, and these wires have very little resistance.

According to Ohm's Law, if the resistance in a circuit is extremely small, the current will be extremely great. When a direct short occurs, then, there will be an extremely heavy current flowing through the wires. Suppose, for instance, that the two leads from a battery to a motor came in contact with each other. Not only would the

motor stop running, because of the current going through the short, but the battery would become discharged quickly (perhaps ruined), and there would also be danger of fire.

The battery cables in our example would be very large wires, capable of carrying very heavy currents. Most wires used in electrical circuits are considerably smaller, and their current-carrying capacity is quite limited. The size of the wires used in any given circuit is determined by the amount of current the wires are expected to carry under normal operating conditions. Any current flow greatly in excess of normal, such as there would be in case of a direct short, would cause a rapid generation of heat.

If the excessive current flow caused by the short is left unchecked, the heat in the wire will continue to increase until something gives way. Perhaps a portion of the wire will melt and open the circuit so that nothing is damaged other than the wires involved. The probability exists, however, that much greater damage would result. The heat in the wires could char and burn their insulation and that of other wires bundled with them, which could cause more shorts. If a fuel or oil leak is near any of the hot wires, a disastrous fire might be started.

To protect electrical systems from damage and failure caused by excessive current, several kinds of protective devices are installed in the systems. Fuses, circuit breakers, and thermal protectors are used for this purpose.

## DESCRIPTION AND PURPOSE

Circuit protective devices, as the name implies, all have a common purpose: to protect the units and the wires in the circuit. Some are designed primarily to protect the wiring. These open the circuit in such a way as to stop the current flow when the current becomes greater than the wires can safely carry. Other devices are designed to protect a unit in the circuit by stopping current flow to it when the unit becomes excessively warm.

## FUSES

The simplest protective device is a fuse. All fuses are rated according to the amount of current that is safely carried by the fuse element at a rated voltage. Usually, the current rating is in amperes, but some instrument fuses are rated in fractions of an ampere. When a fuse blows, it should be replaced with another of the same rated voltage and current capacity, including the same current-versus-time characteristic.

The most important fuse characteristic is its current-versus-time or "blowing" ability. Three time ranges for existence of overloads can be broadly defined as fast, medium, and delayed. FAST, may range from 5 microseconds through one-half second; MEDIUM, 1/2 to 5 seconds; DELAYED, 5 to 25 seconds.

Normally, when the circuit is overloaded, or a fault develops, the fuse element melts and opens the circuit that it is protecting. However, all fuse openings are not the result of current overload or circuit faults. Abnormal production of heat, aging of the fuse element, poor contact due to loose connections, oxides or corrosion forming within the fuse holder, and the heated condition of the surrounding atmosphere will alter the heating conditions and the time required for the element to melt.

### Delayed-Action Fuses

Some equipment, such as an electric motor, requires more current during starting than for normal running. Thus, a fast-time or medium-time fuse rating that will give running protection might blow during the initial period when high starting current is required. Delayed-action fuses are used to handle these situations.

One type of delayed-action fuse has a heater element connected in parallel with the fuse element in order to get the delayed action. During normal operation, the heat developed in the fuse link is not great enough to melt the link. The melting, or opening, of the fuse link depends on the transfer of heat to the link from the heater. Therefore, more time is needed to melt the link than would be required if the link were directly heated.

Because the heater and fuse element are in parallel, the opening of the fuse element will cause the total circuit current to flow through the heater. The high current will cause the heater to burn out and completely open the circuit.

Another type of delayed-action fuse has the fuse element and heater connected in series. Current above that of the rated value for a short time will have no effect on the fuse or heater. However, prolonged overloads cause the heater section to become hot enough to melt the junction between the elements; this action opens the circuit.

Delayed-action fuses are sometimes called Time-Lag fuses; and three trade names "Slo Blo," "Fusestat," and "Fusetron" are in common use.

### Plug Fuses

The plug fuse is constructed so that it can be screwed into a socket mounted on the control panel or distribution center. The fuse link is enclosed in an insulated housing of porcelain or glass. The construction is so arranged that the fuse link is visible through a window of mica or glass. Therefore, an open element may be located by visual examination. When found to be defective, the fuse is discarded and a new fuse installed in its place. The plug fuse is used primarily to protect low-voltage, low-current circuits. The operating ratings range from 0.5 to 30 amperes up to 150 volts.

### Cartridge Fuses

In operation, the cartridge fuse is exactly the same as the plug fuse. In construction, the fuse link is enclosed in a tube of insulating material with metal ferrules at each end (for contact with the fuse holder). The dimensions of cartridge fuses vary with the current and voltage ratings.

### Blown-Fuse Indicators

It is not always possible to detect a blown fuse by a visual examination. Hence, fuses are often equipped with a device that will provide a visual indication so that a blown-fuse condition can be readily detected (fig. 14-1). These devices consist of the spring-loaded and the neon-lamp types of blown-fuse indicators.

In the spring-loaded type (fig. 14-1), when the link opens, it releases a spring that is held under tension. This action exposes an indicator, which makes the visual location of the blown fuse possible.

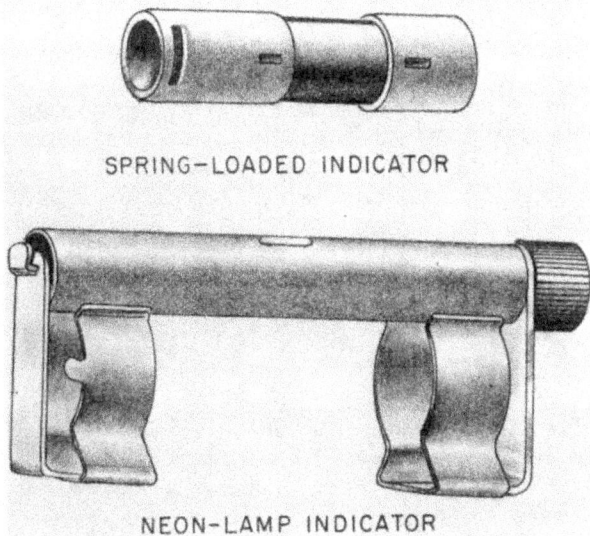

SPRING-LOADED INDICATOR

NEON-LAMP INDICATOR

BE. 241

Figure 14-1.—Blown-fuse indicators.

The neon lamp type (fig. 14-1) is designed to be mounted on the fuse. When the link opens, a neon lamp glows to show a blown fuse.

When no indicator is used, it is necessary to test the fuse continuity with a megger, ohmmeter, or voltmeter. Various methods of testing will be described later in this chapter.

Most fuse panels and switchboards are of the enclosed panel type. The term "dead-front" means that all fuses and bus connections are enclosed in a metal cabinet when the cover is closed. The use of this type of construction reduces the possibility of equipment damage and danger to personnel. Modern switchboards are of the "dead-front" type.

However, the complete enclosure of the equipment makes it less accessible for test purposes. Therefore, most fuses used on "dead-front" switchboards have indicators that show when a fuse is blown. The fuse holder consists of a molded phenolic base, plug, and cap with a built-in indicator lamp (blown-fuse indicator). The lamp is usually a small neon bulb, which normally is shunted by the fuse element. When the fuse opens, the shunt is removed, causing an increase in the voltage across the neon lamp. The lamp then glows, indicating the open fuse.

## TROUBLESHOOTING FUSED CIRCUITS

An electrical system may consist of a comparatively small number of circuits or, in the larger systems, the installation may be equal to that of a fair sized city.

Regardless of the size of the installation, an electrical system consists of a source of power (generators or batteries) and a means of delivering this power from the source to the various loads (lights, motors, and other electrical equipments).

From the main power supply the total electrical load is divided into several feeder circuits and each feeder circuit is further divided into several branch circuits. Each final branch circuit is fused to safely carry only its own load while each feeder is safely fused to carry the total current of its several branches. This reduces the possibility of one circuit failure interrupting the power for the entire system. The feeder distribution boxes and the branch distribution boxes contain fuses to protect the various circuits.

The distribution wiring diagram showing the connections that might be used in a lighting system is illustrated in figure 14-2. An installation might have several feeder distribution boxes, each supplying six or more branch circuits through branch distribution boxes.

Fuses $F_1$, $F_2$, and $F_3$ (fig. 14-2) protect the main feeder supply from heavy surges such as short circuits or overloads on the feeder cable. Fuses A-A1 and B-B1 protect branch No. 1. If trouble develops and work is to be done on branch No. 1, switch $S_1$ may be opened to isolate this branch. Branches 2 and 3 are protected and isolated in the same manner by their respective fuses and switches.

### Branch Circuit Tests

Usually, receptacles for portable equipment and fans are on branch circuits separate from lighting branch circuits. Test procedures are the same for any branch circuit. Therefore, a description will be given of the steps necessary to (1) locate the defective circuit and (2) follow through on that circuit and find the trouble.

Assume that, for some reason, several of the lights are not working in a certain section. Because several lights are out, it will be reasonable to assume that the voltage supply has been interrupted on one of the branch circuits.

To verify this assumption, first locate the distribution box feeding the circuit that is inoperative. Then make sure that the inoperative circuit is not being supplied with voltage. Unless the circuits are identified in the distribution box,

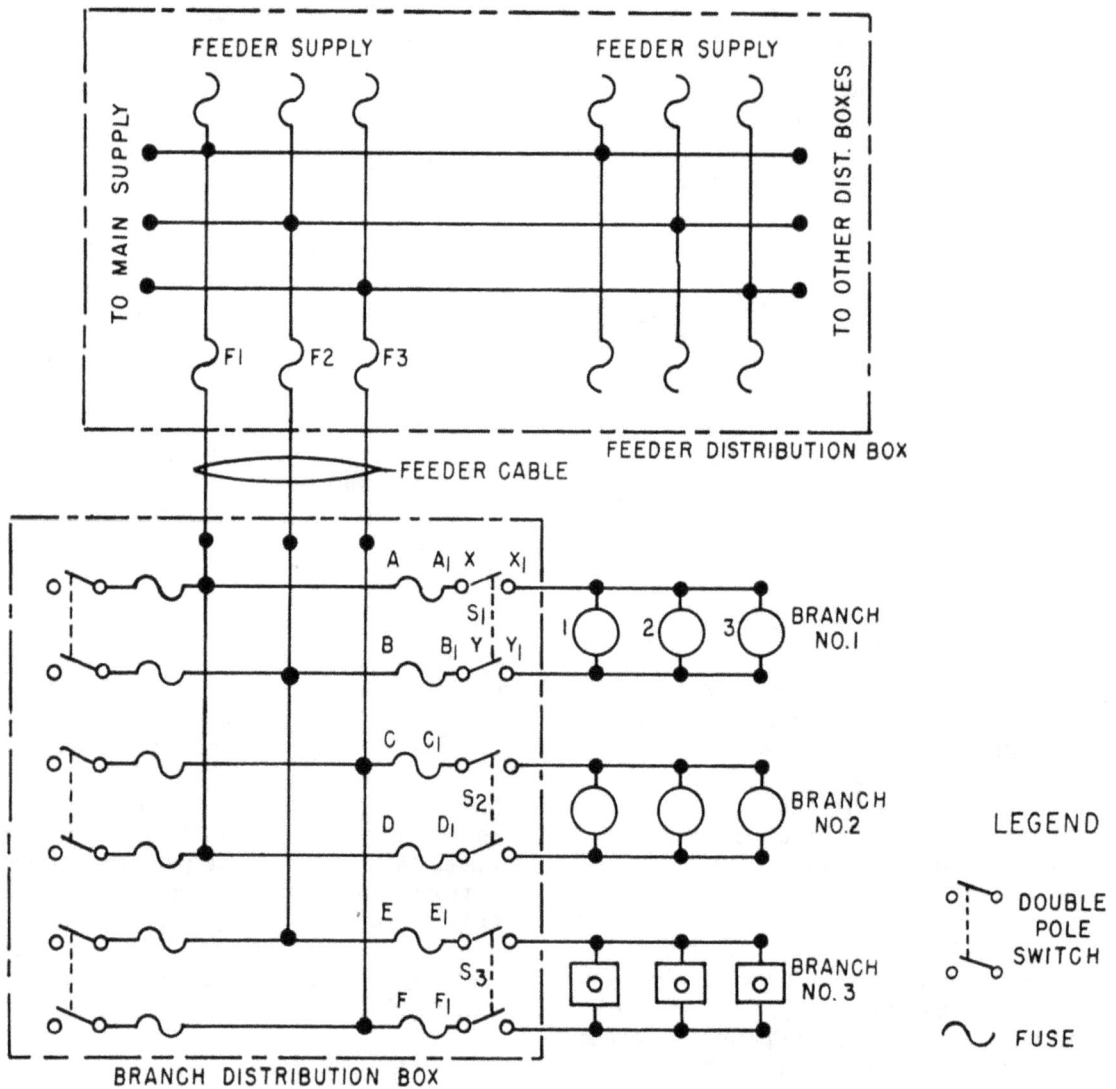

Figure 14-2.—Three-phase distribution wiring diagram.

BE.242

the voltage at the various circuit terminations will have to be measured. For the following procedures, use the circuits shown in figure 14-2 as an example circuit.

To pin down the trouble, connect the voltage tester to the load side of each pair of fuses in the branch distribution box. No voltage between these terminals indicates a blown fuse or a failure in the supply to the distribution box. To find the defective fuse, make certain $S_1$ is closed, then connect the voltage tester across A-A1, and next across B-B1 (fig. 14-2). The full-phase voltage will appear across an open

fuse, provided circuit continuity exists across the branch circuit. However, if there is an open circuit at some other point in the branch circuit, this test is not conclusive. If the load side of a pair of fuses does not have the full-phase voltage across its terminals, place the tester leads on the supply side of the fuses. The full-phase voltage should be present. If the full-phase voltage is not present on the supply side of the fuses, the trouble is in the supply circuit from the feeder distribution box.

Assume that you are testing at terminals A-B (fig. 14-2) and that normal voltage is present. Move the test lead from A to A1. Normal voltage between A1 and B indicates that fuse A-A1 is in good condition. To test fuse B-B1, place the tester leads on A and B, and then move the lead from B to B1. No voltage between these terminals indicates that fuse B-B1 is open. Full-phase voltage between A and B1 indicates that the fuse is good.

This method of locating blown fuses is preferred to the method in which the voltage tester leads are connected across the suspected fuse terminals, because the latter may give a false indication if there is an open circuit at any point between either fuse and the load in the branch circuit.

## CIRCUIT BREAKERS

A circuit breaker is designed to break the circuit and stop the current flow when the current exceeds a predetermined value. It is commonly used in place of a fuse and may sometimes eliminate the need for a switch. A circuit breaker differs from a fuse in that is "trips" to break the circuit and it may be reset, while a fuse melts and must be replaced.

Several types of circuit breakers are commonly used. One is a magnetic type. When excessive current flows in the circuit, it makes an electromagnet strong enough to move a small armature which trips the breaker. Another type is the thermal overload switch or breaker. This consists of a bimetallic strip which, when it becomes overheated from excessive current, bends away from a catch on the switch lever and permits the switch to trip open.

Some circuit breakers must be reset by hand, while others reset themselves automatically. When the circuit breaker is reset, if the overload condition still exists, the circuit breaker will trip again to prevent damage to the circuit.

One common type of circuit breaker now being used is depicted in figure 14-3. This breaker is designed for front or rear connections as required and may be mounted so as to be removable from the front without removing the circuit breaker cover. The voltage ratings of this breaker are 500 volts a.c., 60 Hz, or 250 volts d.c. with a maximum current capacity of 250 amperes. Trip units (fig. 14-4) for this breaker are available with current ratings of 125, 150, 175, 225, and 250 amperes.

The trip unit houses the electrical tripping mechanisms, the thermal element for tripping the circuit breaker on overload conditions, and

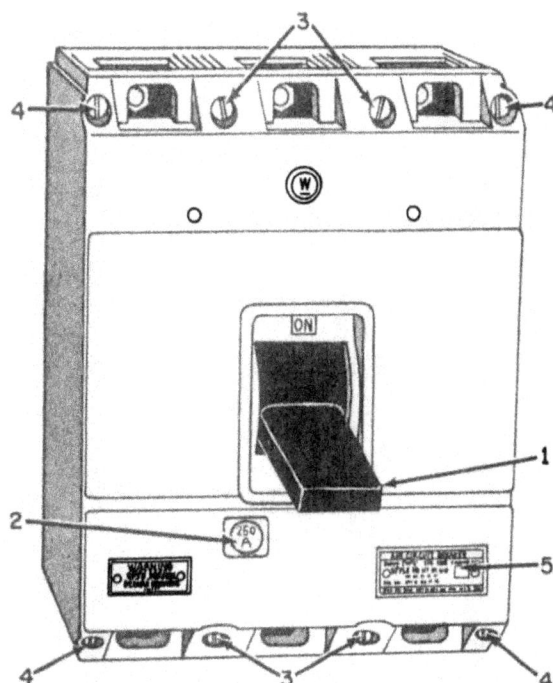

BE. 243

1. Operating handle shown in latched position.
2. Ampere rating maker.
3. Mounting screws.
4. Cover screws.
5. Breaker nameplate.

Figure 14-3.—Circuit breaker, front view.

the instantaneous trip for tripping on short circuit conditions. The automatic trip devices of this circuit breaker are "trip free" of the operating handle; this means the circuit breaker cannot be held closed by the operating handle if an overload exists. When the circuit breaker has tripped due to overload or short circuit, the handle rests in a center position. To reclose after automatic tripping, the handle must be moved to the extreme OFF position which resets the latch in the trip unit; then the handle must be physically moved to the ON position. Metal locking devices are available that can be attached to the handles of circuit breakers to prevent accidental operation.

## THERMAL PROTECTORS

A thermal protector, or switch, is a device used to protect a motor. It is designed to open the circuit automatically whenever the temperature of the motor becomes excessively high. It has two positions, open and closed. The most

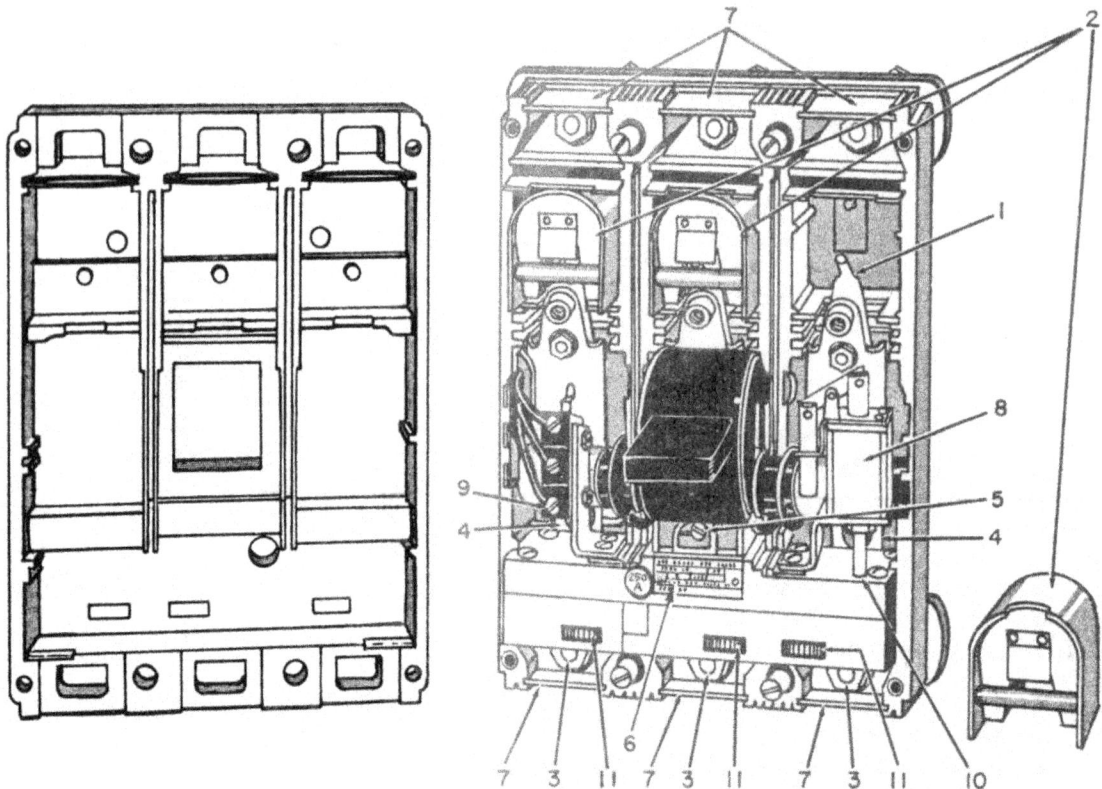

BE. 244

1. Stationary contact.
2. Arc suppressors.
3. Terminal stud nuts and washers.
4. Trip unit line terminal screw-outer poles.
5. Trip unit line terminal screw-center pole.
6. Trip unit nameplate.
7. Terminal barriers.
8. Shunt trip.
9. Auxiliary switch.
10. Hole for shunt trip undervoltage release plunger.
11. Instantaneous trip adjusting wheels.

Figure 14-4.—Circuit breaker, cover and arc suppressor removed.

common use for a thermal switch is to keep a motor from overheating. If some malfunction in the motor causes it to overheat, the thermal switch will break the circuit intermittently. If the trouble is a locked rotor, the intermittent opening and closing of the circuit may release the rotor and allow the motor to resume normal operation.

The thermal switch contains a bimetallic disk, or strip, which bends and breaks the circuit when it is heated. This happens because one of the metals expands more than the other when they

are subjected to the same temperature. When the strip or disk cools, the metals contract and the strip returns to its original position and closes the circuit.

OVERLOAD DEVICE

A prolonged overloaded electrical system can be damaged beyond repair by the resulting heat and flame. Therefore, it is expedient to use a device which can detect an overload before damage occurs and either warns the operator of

243

the hazardous conditon or automatically turns off the power. Relays have been designed which are capable of accomplishing these protective functions. A few of these relays that are commonly used as overload devices are discussed in greater detail later in this manual.

## CONTROL DEVICES

Control devices are those electrical accessories which govern (in some predetermined way) the power delivered to any electrical load.

In its simplest form the control applies voltage to, or removes it from a single load. In more complex control systems, the initial switch may set into action other control devices that govern motor speeds, servomechanisms, temperatures, and numerous other equipments. In fact, all electrical systems and equipment are controlled in some manner by one or more controls. A controller is a device or group of devices which serves to govern, in some predetermined manner, the device to which it is connected.

In large electrical systems, it is necessary to have a variety of controls for operation of the equipment. These controls range from simple pushbottons to heavy duty contactors that are designed to control the operation of large motors. The pushbutton is manually operated while a contactor is electrically operated.

## SWITCHES

A switch may be described as a device used in an electrical circuit for making, breaking, or changing connections under conditions for which the switch is rated. Switches are rated in amperes and volts; the rating refers to the maximum voltage and current of the circuit in which the switch is to be used. Because it is placed in series, all the circuit current will pass through the switch. Because it opens the circuit, the applied voltage will appear across the switch in the open circuit position. Switch contacts should be opened and closed quickly to minimize arcing; therefore, switches normally utilize a snap action.

Many types and classifications of switches have been developed. A common designation is by the number of poles, throws, and positions they have. The number of poles indicates the number of terminals at which current can enter the switch. The throw of a switch signifies the number of circuits each blade or contactor can complete through the switch. The number of positions indicates the number of places at which the operating device (toggle, plunger, etc.) will come to rest. Figure 14-5 presents the schematic diagrams of some often used switches.

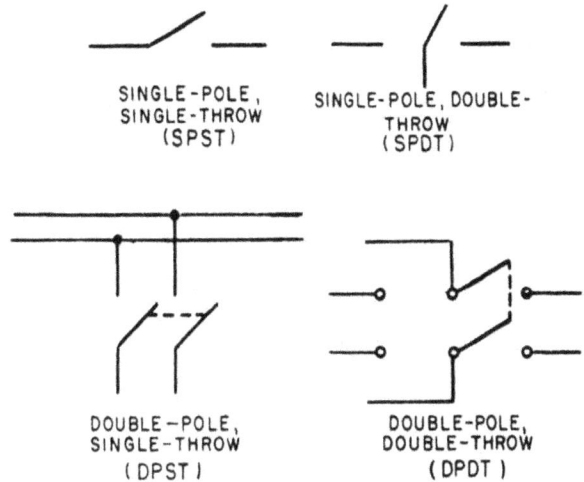

SINGLE-POLE, SINGLE-THROW (SPST)

SINGLE-POLE, DOUBLE-THROW (SPDT)

DOUBLE-POLE, SINGLE-THROW (DPST)

DOUBLE-POLE, DOUBLE-THROW (DPDT)

BE.245

Figure 14-5.—Schematic diagrams of commonly used switches.

An example of the switch position designation is a toggle switch which comes to rest at either of two positions, opening the circuit in one position and completing it in another. This is called a two-position switch. A toggle switch which is spring loaded to the OFF position and must be held in the ON position to complete the circuit is called a momentary contact two-position switch. If the toggle switch will come to rest at any of three positions, it is called a three-position switch.

Another means of classifying switches is the method of actuation; that is, toggle, pushbutton, sensitive, and rotary types. Further classification can be accomplished by a description of switch action such as on-off, momentary on-off, on-momentary off, etc. Momentary contact switches hold a circuit closed or open only as long as the operator deflects the actuating control.

### Manually Operated Switches

One of the most common types of switches is the toggle. Toggle switches have their moving parts enclosed. A double-pole, double-throw, ON-OFF-ON toggle switch is shown in figure 14-6. These switches have many uses and are used especially for applying power to various circuits. They are often provided with a luminous tip on the lever so as to be visible in the dark.

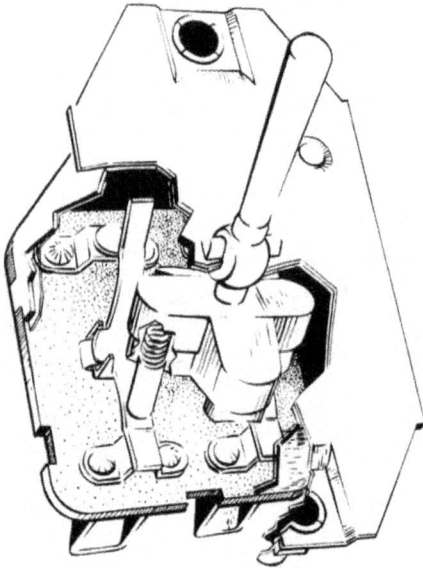

AE. 50

Figure 14-6.—Toggle switch.

Pushbutton switches have one or more stationary contacts and one or more movable contacts. The movable contacts are attached to the pushbutton by an insulator. This switch is usually spring loaded and is of the momentary contact type. These switches have many uses, such as indicator light checks and circuit reset.

A rotary selector switch may perform the functions of a number of switches. As the knob of a rotary selector switch is rotated, it opens one circuit and closes another. This can be seen from an examination of figure 14-7. Some rotary switches have several layers of wafers. By adding wafers, the switch can be made to operate as a large number of switches. Ignition switches and voltmeter selector switches are typical examples of this type.

Mechanically Operated Switches

Mechanically operated switches are used in many applications. They are widely used because of their small size, light weight, and excellent dependability. The term Micro switch, although frequently used in referring to all switches of

this type, is a trade name for the switches made by the Micro Switch Division of the Minneapolis Honeywell Regulator Company.

These switches will open or close a circuit with a very small movement of the tripping device (1/16 inch or less). They are usually of the pushbutton variety and depend upon one or more springs for their snap action. For example, the heart of the Micro switch is a beryllium copper spring, heat-treated for long life and unfailing action. The simplicity of the one-piece spring contributes to the long life and dependability of this switch. The basic Micro switch is shown in figure 14-8.

The versatility of the snap-action switch is shown in figure 14-9, which shows how the basic switch may be used with different type enclosures and actuators. The particular type switching unit used depends upon the function it is to perform and environmental conditions. Figure 14-9 (A) shows a lightweight aluminum enclosure which may contain one or more plastic enclosed basic switches. The plastic enclosure provides electrical insulation between the housing and the energized electrical parts, support for the terminals, and a dusttight box around the electrical contacts.

Figure 14-9 (B) shows one of the various types of switch actuators that may be used with basic switches when complete enclosures are not needed. They provide protection and mounting means. They relay the operating motion from a cam or slide to the basic switch in a way that assures long life and dependability. Figure 14-9 (C) shows a toggle type switch which uses one or more of the basic subminiature switches. These subminiature switches are smaller than the conventional Micro switch and are finding wide use in various switching arrangements.

Pressure-operated switches usually have Bourdon tubes, syphons, or diaphrams against which the fluids or air operate to actuate the switch. Some uses of pressure switches are in connection with fuel, oil, and hydraulic pressure signals and electric heaters.

Thermal switches usually incorporate a bimetallic sheet that bends or snaps at a desired temperature to actuate the switch. They are used extensively as circuit breakers and also find application in controlling the igniter circuit on heater and operating signal lights at critical temperatures.

AE. 51

Figure 14-7.—Rotary selector switch.

Maintenance of Switches

While the switch itself is relatively simple to check, it sometimes offers difficulty in maintenance because of its location in inaccessible places. After a visual inspection of the connections and the switch, a continuity test will indicate any malfunctions. When the switch mechanism is found to be defective, it normally is not repairable and therefore should be replaced.

When enclosed switches are used, failure to seal properly around cable openings may cause difficulty. Altitude changes permit "breathing" of moist air into enclosures with improperly sealed cable openings, and the moisture in the air may condense within the switch enclosure. The condensation can short across the switch terminals and can corrode the switch actuators in a manner that may make them inoperative. This difficulty can be corrected by careful sealing of openings or by using hermetically sealed switches. Hermetically sealed switches will also prevent dust and dirt from reaching the contacts and thereby reduce the possibility of high resistance and open circuits.

Some switches are damaged during installation, particularly those with plastic housings. Proper care in installing or replacing plastic enclosed switches will eliminate this.

Some switch assemblies are equipped with adjustments which enable them to operate at a preset time or pressure. Caution should be exercised in making these adjustments; if they are not accurate, damage can result.

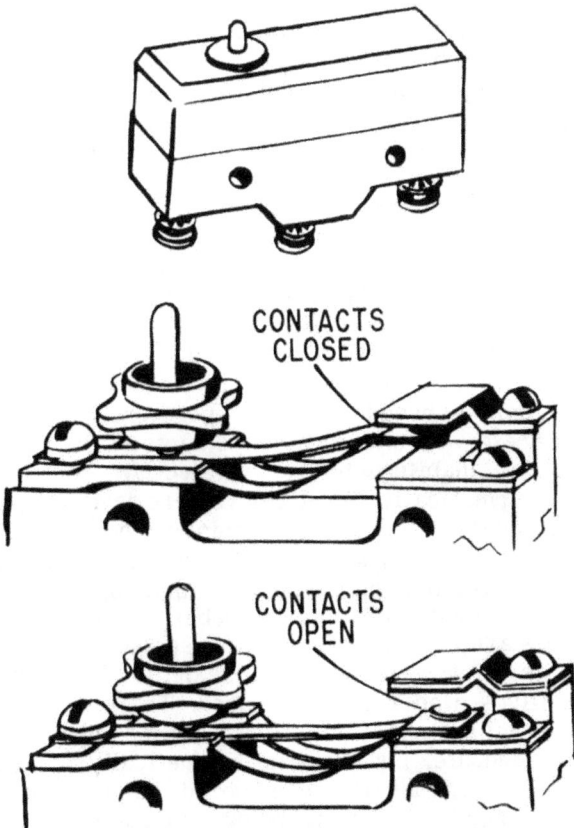

CONTACTS
CLOSED

CONTACTS
OPEN

AE. 52

Figure 14-8.—Micro switch.

switches. Safety is also an important factor in using relays, since high power circuits can be switched remotely without danger to the operator.

Control relays, as their name implies, are frequently used in the control of other relays, although the small control relays find many other uses. With these, small currents can control the larger currents necessary to operate electrical devices. They find frequent use in automatic relaying circuits, where a small electric signal sets off a chain reaction of successively acting relays performing various functions. Control relays can also be used in so-called "lockout" action to prevent certain functions from occurring at the improper time. Various electrical operations in the equipment which must not occur simultaneously can be "interlocked" by control relays. Another important function of control relays in equipment is for "sensing." Control relays are used for sensing undervoltage and overvoltage, reversal of current, and excessive currents.

Another possible classification of relays is open, semisealed, and sealed. Semisealed relays have protextive covers and are gasketed against entrance of salt, dust, and foreign material into the contact or mechanism area. These relays are still considered satisfactory for certain applications in current equipment. Open relays are seldom used outside of black boxes.

For other applications in today's complex equipment, however, it is necessary to go beyond the protection offered by the open type and the semisealed relays. When such relays are used, quick changes in altitude, humidity, or temperature can cause condensation of water vapor within the unit. Subsequent low temperature will then freeze the moisture on the contactor with a resultant inability to carry electric current.

Hermetically sealed relays were developed to answer the demands of complex and delicate equipment. A true hermetic seal is generally considered one that is metal to metal or glass to metal. Plastic or plastic rubber type gasketed seals are not generally considered true hermetic seals. However, both semisealed and hermetically sealed relays are used. There are applications where a gasket type sealed relay may be adequate, but the true hermetically sealed type is generally considered to be more permanent. Besides being independent of environmental changes, the hermetically sealed relay also has the advantage of being protected from improper adjustments.

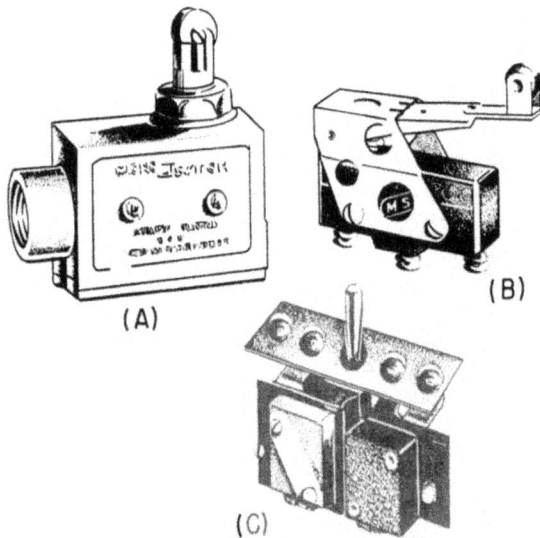

AE. 53

Figure 14-9.—Snap-action switching units.

## RELAYS

Relays are electrically operated switches that are classified according to their use as control relays, power relays, or sensing relays. The power relays are the workhorses of a large electrical system. As such, they control the heavy power circuits.

The function of a control relay is to take a relatively small amount of electrical power and use it either to signal or to control a large amount of power. Where multipole relays are used, several circuits may be controlled simultaneously.

The use of relays saves space and weight by permitting the use of small switches at remote control stations. These switches permit the operator to control large amounts of current at other locations and the heavy power cables need to be run only to the point of use. Only lightweight control wires are connected to the control

In general, the basic components of a relay are as follows: the coil or solenoid, the iron core, the fixed and movable contacts, and the mounting (and if sealed, the can). A manual switch, limit switch, or other small control device starts and stops the flow of current to the magnet coil. The flow of electric current through the coil creates a strong magnetic field around and within the coil. This magnetic field moves a clapper or plunger which completes the magnetic circuit. Figure 14-10 (A) shows a basic single coil clapper type relay. The dashed lines indicate the magnetic lines of flux.

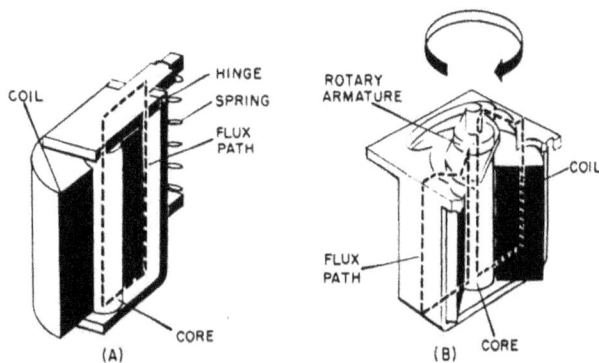

BE. 246

Figure 14-10.—Basic types of relays.

The second basic type of relay is the rotary. (See fig. 14-10 (B).) Although this type of construction is not as common as the clapper type, the rotary type has greater vibration and shock resistance than the others. The disadvantage is that they are somewhat sluggish and require higher operating power for many purposes. The rotary relay operates on the principle of an electric motor, but through only a small arc. The problem of hanging contacts on such a mechanism is a difficult one, and therefore the use of these devices is limited to applications where high shock warrants the larger size and weight. When used with standard wafer switch assemblies, this type of relay provides a means for assembling a switching device of any degree of complexity.

Occasionally relays operating from an a-c supply are encountered. These a-c relays depend upon the same fundamental principles as the d-c relay; that is, magnetic fields. When a.c. is applied to an electromagnet, the current will pass through zero twice every cycle. Since the pull on the armature is proportional to the current through the electromagnet, the armature tends

to open every time the current nears zero, causing chatter. To remedy this, shading coils (sometimes called shaded poles) are used.

A shading coil consists of copper band or stamping which is short-circuited and embedded around part of the electromagnet pole face. By being placed around part of the pole face, it acts as a shorted transformer secondary. The current in the main coil lags the applied voltage by approximately 90°, and the flux is in phase with the current. The voltage of the shading coil is induced voltage and lags the current in the main coil by 90°. Since the shading coil acts like a shorted secondary (resistive), the current in the shading coil is in phase with the induced voltage. Therefore, the magnetic field of the shading coil lags the magnetic field of the main coil by 90°. This means that flux will exist in the electromagnet even when the main coil current becomes zero. Thus, chattering is prevented.

The arrangements of relay contacts are found in many different forms. Usually the number and sequence of switching operations to be performed dictates the contact arrangement.

It is often desirable to introduce time delays by use of relays. One method is to use a thermal relay for a time delay. Due to its simple mechanism, it can be made very small and hermetically sealed, making it ideal for use in aircraft. Because the thermal relay is activated by heat, it can be used on either a.c. or d.c.

Maintenance of Relays

The relay is one of the most dependable electromechanical devices in use, but like any other mechanical or electrical device, relays occasionally wear out or become inoperative for one reason or another. Should relay inspection determine that a relay has exceeded its safe life, the relay should be removed immediately and replaced with another of the same type. Care should be exercised in obtaining the same type replacement because relays are rated in voltage, amperage, type of service, number of contacts, continuous or intermittent duty, and similar characteristics.

For spotting potential relay trouble during preventive maintenance, the following guides are suggested: check for charred or burned insulation on the relay and for darkened or charred terminal leads coming from the relay. Both of these indicate overheating. If there is even a slight indication that the relay has overheated it should be replaced with a new relay of the same type. An occasional cause of relay trouble is not the fault of the relay at all, but is due to overheating caused by the power terminal connectors not being tight enough. This should always be checked during preventive maintenance.

It is recommended that covers not be removed from semisealed relays in the field. Removal of a cover in the field, although it might give useful information to a trained eye, may result in entry of dust or other foreign material which may cause contact discontinuity.

# CHAPTER 15

# ELECTRICAL INDICATING INSTRUMENTS

In the field of electricity, as in all the other physical sciences, accurate quantitative measurements are essential. This involves two important items—numbers and units. Simple arithmetic is used in most cases, and the units are well-defined and easily understood. The standard units of current, voltage, and resistance as well as other units are defined by the National Bureau of Standards. At the factory, various instruments are calibrated by comparing them with established standards.

The technician commonly works with ammeters, voltmeters, ohmmeters, and electron-tube analyzers; but he may also have many occasions to use wattmeters, watt-hour meters, power-factor meters, synchroscopes, frequency meters, and capacitance-resistance-inductance bridges.

Electrical equipments are designed to operate at certain efficiency levels. To aid the technician in maintaining the equipment, technical instruction books and sheets containing optimum performance data, such as voltages and resistances, are prepared for each Navy equipment.

To the technician, a good understanding of the functional design and operation of electrical instruments is important. In electrical service work one or more of the following methods are commonly used to determine if the circuits of an equipment are operating properly.

1. Use an ammeter to measure the amount of current flowing in a circuit.

2. Use a voltmeter to determine the voltage existing between two points in a circuit.

3. Use an ohmmeter or megger (megohm-meter) to measure circuit continuity and total or partial circuit resistance.

The technician may also find it necessary to employ a wattmeter to determine the total POWER being consumed by certain equipments. If he wishes to measure the ENERGY consumed by certain equipments or certain circuits, a watt-hour of kilowatt-hour meter is used.

For measuring other quantities such as power factor and frequency, the technician employs the appropriate instruments. In each case the instrument indicates the value of the quantity measured, and the technician interprets the information in a manner that will help him understand the way the circuit is operating. Occasionally the technician will need to determine the value of a capacitor or an inductor. Inductance or capacitance bridges may be employed for this purpose.

A thorough understanding of the construction, operation, and limitations of electrical measuring instruments, coupled with the theory of circuit operation, is most essential in servicing and maintaining electrical equipment.

## DIRECT-CURRENT INSTRUMENTS

In this chapter a discussion of both the d-c and a-c meters and measuring instruments is presented. The d-c type meters will be discussed first since they are the simpler type. Some of the d-c meters may also be used for the measuring of a-c potentials.

### D'ARSONVAL METER

The stationary permanent-magnet moving-coil meter is the basic movement used in most measuring instruments for servicing electrical equipment. This type of movement is commonly called the D'Arsonval movement because it was first employed by the Frenchman D'Arsonval in making electrical measurements.

The basic D'Arsonval movement consists of a stationary permanent magnet and a movable coil. When current flows through the coil the resulting magnetic field reacts with the magnetic field of the permanent magnet and causes the coil to rotate. The greater the amount of current flow through the coil, the stronger the magnetic field produced; and the stronger this field, the greater the rotation of the coil. In order to determine the amount of current flow a means must be provided to indicate the amount of coil rotation. Either of two methods may be used—(1) the pointer arrangement, and (2) the light and mirror arrangement. In the pointer arrangement, one end of the pointer is fastened to the rotating coil and as the coil turns the pointer also turns. The other end of the pointer moves across a graduated scale and indicates the amount of current flow. A disadvantage of the pointer arrangement is that it introduces the problem of coil balance, especially if the pointer is long. An advantage of this arrangement is that it permits overall simplicity. The use of a mirror and a beam of light simplifies the problem of coil balance. When this arrangement is used to measure the turning of the coil, a small mirror is mounted on the supporting ribbon (fig. 15-1) and turns with the coil. An internal light source is directed to the mirror, and then reflected to the scale of the meter. As the moving coil turns, so does the mirror, causing the light reflection to move over the scale of the meter. The movement of the reflection is proportional to the movement of the coil, thus the amount of current being measured by the meter is indicated.

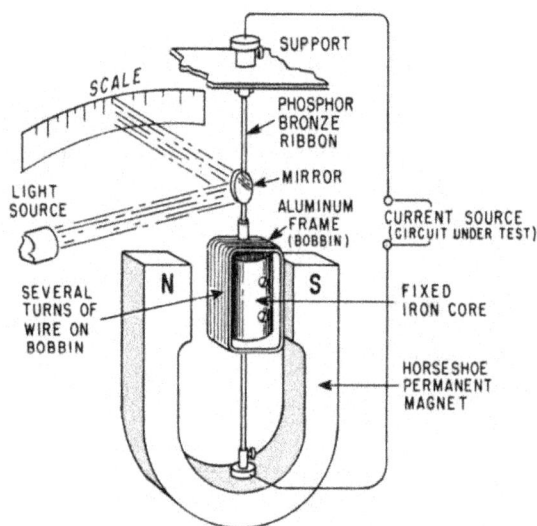

BE.247

Figure 15-1.—Simplified diagram of a galvanometer.

A simplified diagram of one type of stationary permanent-magnet moving-coil instrument is shown in figure 15-1. Such an instrument is commonly called a GALVANOMETER. The galvanometer indicates very small amounts (or the relative amounts) of current or voltage, and is distinguished from other instruments used for the same purpose in that the movable coil is suspended by means of metal ribbons instead of by means of a shaft and jewel bearings.

The movable coil of the galvanometer in figure 15-1 is suspended between the poles of the magnet by means of thin flat ribbons of phosphor bronze. These ribbons provide the conducting path for the current between the circuit under test and the movable coil. They also provide the restoring force for the coil. The restoring force, exerted by the twist in the ribbons, is the force against which the driving force of the coil's magnetic field (to be described later) is balanced in order to obtain a measurement of the current strength. The ribbons thus tend to oppose the motion of the coil, and will twist through an angle that is proportional to the force applied to the coil by the action of the coil's magnetic field against the permanent field. The ribbons thus restrain or provide a counter force, for the magnetic force acting on the coil. When the driving force of the coil current is removed, the restoring force returns the coil to its zero position.

If a beam of light and mirrors are used, the beam of light is swept to the right or left across a central-zero translucent screen (scale) having uniform divisions. If a pointer is used, the pointer is moved in a horizontal plane to the right or left across a central-zero scale having uniform divisions. The direction in which the beam of light or the pointer moves depends on the direction of current through the coil.

This instrument is used to measure minute current as, for example, in bridge circuits. In modified form, the basic D'Arsonval movement has the highest sensitivity of any of the various types of meters in use today.

Operating Principle

In order to understand the operating principle of the D'Arsonval meter it is first necessary to consider the force acting on a current-carrying conductor placed in a magnetic field. The magnitude of the force is proportional to the product of the magnitudes of the current and the field strength. The field is established between the poles of a U-shaped permanent magnet and is concentrated through the conductor by means of a soft-iron stationary member mounted between the poles to complete the magnetic circuit. The conductor is made movable by shaping it in the form of a closed loop and mounting it between fixed pivots so that it is free to swing about the fixed iron member between the poles of the magnets. A convenient method of determining the direction of motion of the conductor is by the use of the RIGHT-HAND MOTOR RULE FOR ELEC-TRON FLOW (fig. 15-2).

To find the direction of motion of a conductor, the thumb, first finger, and second finger of the right hand are extended at right angles to each other, as shown. The first finger is pointed in the direction of the flux (toward the south pole) and the second finger is pointed in the direction of electron flow in the conductor. The thumb then points in the direction of motion of the conductor with respect to the field. The conductor, the field, and the force are mutually perpendicular to each other.

The force acting on a current-carrying conductor in a magnetic field is directly proportional to the field strength of the magnet, the active length of the conductor, and the intensity of the electron flow through it.

In the D'Arsonval type meter, the length of the conductor is fixed and the strength of the field between the poles of the magnet is fixed. Therefore, any change in I causes a proportionate change in the force acting on the coil.

The principle of the D'Arsonval movement may be more clearly shown by the use of the simplified diagram (fig. 15-3) of the D'Arsonval movement commonly used in d-c instruments. In the diagram, only one turn of wire is shown;

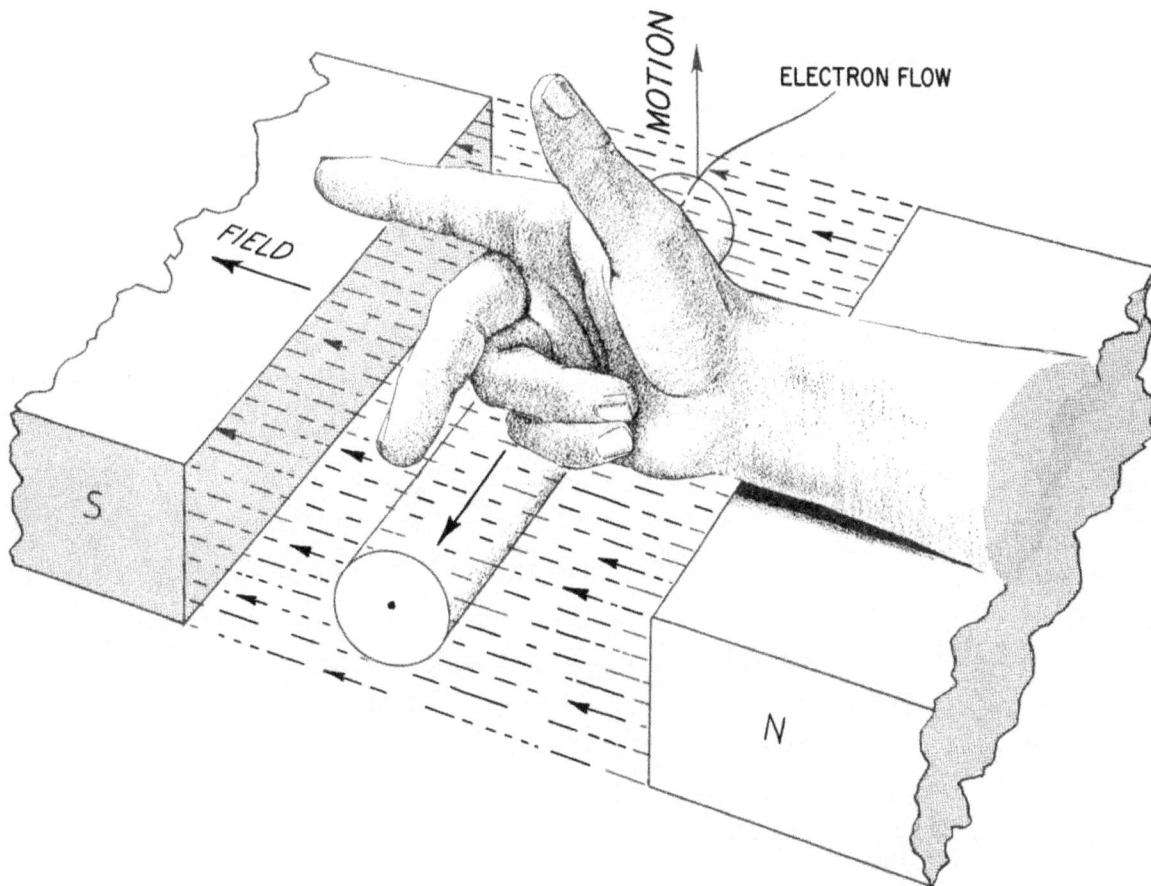

BE.248

Figure 15-2.—Right-hand motor rule.

however, in an actual meter movement many turns of fine wire would be used, each turn adding more effective length to the coil. The coil is wound on an aluminum frame or bobbin, to which the pointer is attached. Oppositely wound hairsprings (one of which is shown in fig. 15-3), are also attached to the bobbin, one at either end. The circuit to the coil is completed through the hairsprings. In addition to serving as conductors, the hairsprings serve as the restoring force that returns the pointer to the zero position when no current flows.

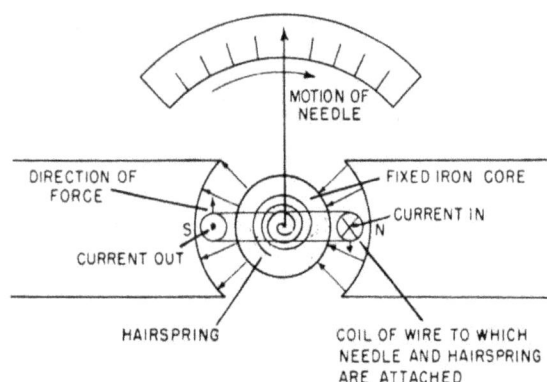

BE.249

Figure 15-3.—D'Arsonval movement.

As has been stated, the deflecting force is proportional to the current flowing in the coil. The deflecting force tends to rotate the coil against the restraining force of the hairsprings. The angle of rotation is proportional to the force that the springs exert against the moving coil (within the elastic limit of the spring). When the deflecting force and the restraining force are equal, the coil and the pointer cease to move. Because the restoring force is proportional to the angle of deflection, it follows that the driving force, and the current in the coil, are proportional to the angle of deflection. When current ceases to flow in the coil, the driving force ceases, and restoring force of the springs returns the pointer to the zero position.

If the current through the single turn of wire is in the direction indicated (away from the observer on the right-hand side and toward the observer on the left-hand side), the direction of force, by the application of the right-hand motor rule, is upward on the left-hand side and downward on the right-hand side. The direction of motion of the coil and pointer is clock-wise. If the current is reversed in the wire, the direction of motion of the coil and pointer is reversed.

A detailed view of the basic D'Arsonval movement, as commonly employed in ammeters and voltmeters, is shown in figure 15-4. This instrument is essentially a MICROAMMETER because the current necessary to activate it is of the order of 1 microampere. The principle of operation is the same as that of the simplified versions discussed previously. The iron core is rigidly supported between the pole pieces and serves to concentrate the flux in the narrow space between the iron core and the pole piece— in other words, in the space through which the coil and the bobbin moves. Current flows into one hairspring, through the coil, and out of the other hairspring. The restoring force of the spiral springs returns the pointer to the normal, or zero, position when the current through the coil is interrupted. Conductors connect the hairsprings with the outside terminals of the meter.

If the instrument is not damped—that is, if friction or some other type of loss is not introduced to absorb the energy of the moving element—the pointer will oscillate for a long time about its final position before coming to rest. This action makes it nearly impossible to obtain a reading and some form of damping is necessary to make the meter practicable. Damping is accomplished in many D'Arsonval movements by means of the motion of the aluminum bobbin upon which the coil is wound. As the bobbin oscillates in the magnetic field, an e.m.f. is induced in it because it cuts through the lines of force. Therefore, according to Lenz's law, induced currents flow in the bobbin in such a direction as to oppose the motion, and the bobbin quickly comes to rest in the final position after going beyond it only once.

In addition to factors such as increasing the flux density in the air gap, the overall sensitivity of the meter can be increased by the use of a lightweight rotating assembly (bobbin, coil, and pointer) and by the use of jewel bearings as shown in figure 15-4 (B).

While D'Arsonval-type galvanometers are useful in the laboratory for measurements of extremely small currents, they are not portable, compact or rugged enough for use in the maintenance of military equipment. The Weston meter movement is used instead.

The Weston meter uses the principles of the D'arsonval galvanometer, but it is portable, compact, rugged and easy to read. In the Weston

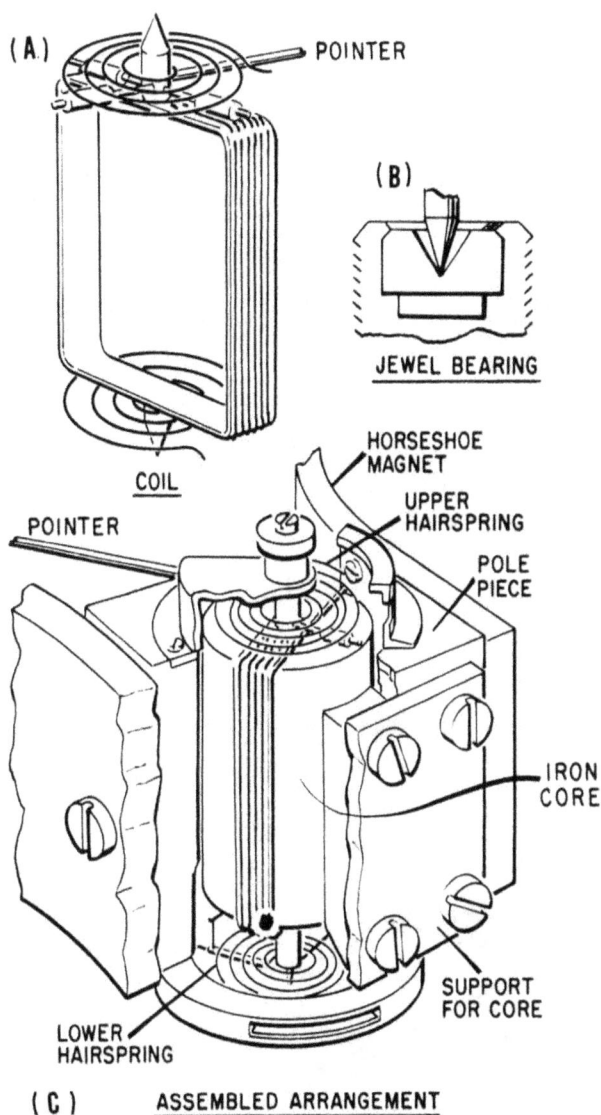

BE. 250

Figure 15-4.—Detailed view of basic
D'Arsonval movement.

meter the coil is mounted on a shaft fitted between two permanently mounted jewel bearings.
A lightweight pointer is attached to and turns with
the coil; the pointer indicates the amount of current flow. Figure 15-5 (A) illustrates this movement.

Balance springs on each end of the shaft exert
opposite turning forces on the coil. By adjusting
the tension of one spring, the meter pointer may
be adjusted to read zero on the meter scale.
Since temperature change affects both coil
springs equally, this factor can be discounted.
As the meter coil turns, one spring tightens to
provide a restoring force; at the same time the
other spring releases its tension. In addition to
providing tension, the springs are also used as
conductors to carry current from the meter
terminals to the moving coil.

In order that the turning force will increase
uniformly as the current increases, the horseshoe magnet poles are shaped to form semi-
circles. The amount of current required to turn
the meter pointer to full-scale deflection depends
upon the magnet's strength and the number of
turns of wire in the moving coil. This amount of
current is the ammeter's MAXIMUM ALLOW-
ABLE AMOUNT. A further increase of meter
current would damage the meter.

## AMMETER

The small size of the wire with which an ammeter's movable coil is wound places severe
limits on the current that may be passed through
the coil. Consequently, the basic D'Arsonval
movement discussed thus far may be used to indicate or measure only very small currents—for
example, microamperes ($10^{-6}$ amperes) or milliamperes ($10^{-3}$ amperes), depending on meter
sensitivity.

To measure a larger current, a shunt must be
used with the meter. A shunt is a heavy low-
resistance conductor connected across the meter
terminals to carry most of the load current. This
shunt has the correct amount of resistance to
cause only a small part of the total circuit current to flow through the meter coil. The meter
current is proportional to the load current. If the
shunt is of such a value that the meter is calibrated in milliamperes, the instrument is called
a MILLIAMMETER. If the shunt is of such a value
that the meter is calibrated in amperes, it is
called an AMMETER.

A single type of standard meter movement is
generally used in all ammeters, no matter what
the range of a particular meter. For example,
meters with working ranges of zero to 10 amperes, zero to 5 amperes, or zero to 1 ampere
all use the same galvanometer movement. The
designer of the ammeter simply calculates the
correct shunt resistance required to extend the
range of the 1-milliampere meter movement to
measure any desired amount of current. This
shunt is then connected across the meter

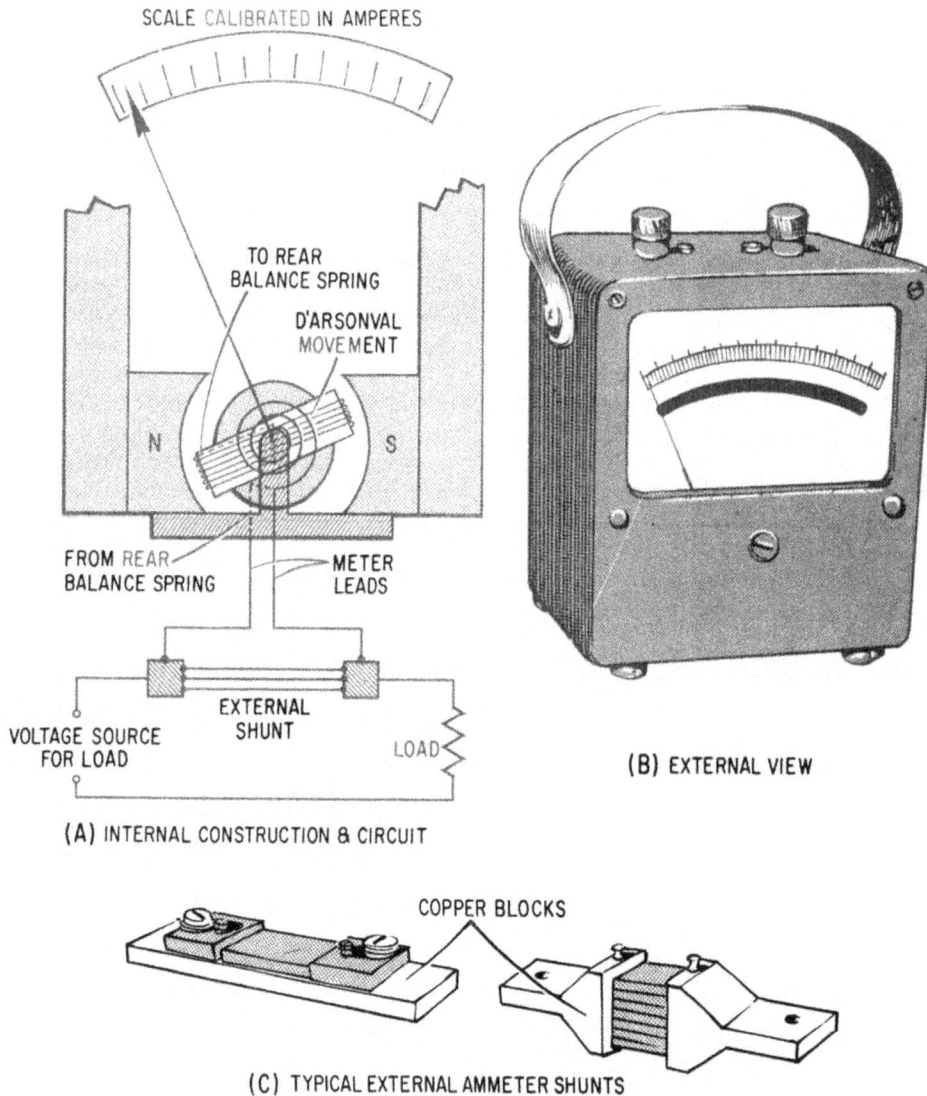

SCALE CALIBRATED IN AMPERES

TO REAR BALANCE SPRING

D'ARSONVAL MOVEMENT

N    S

FROM REAR BALANCE SPRING

METER LEADS

VOLTAGE SOURCE FOR LOAD

EXTERNAL SHUNT

LOAD

(A) INTERNAL CONSTRUCTION & CIRCUIT

(B) EXTERNAL VIEW

COPPER BLOCKS

(C) TYPICAL EXTERNAL AMMETER SHUNTS

BE. 251

Figure 15-5.—Weston ammeter employing D'Arsonval principle in its movement.

terminals. Shunts may be located inside the meter case (internal shunt) or somewhere away from the meter (external shunt), with leads going to the meter. An external shunt arrangement is shown in figure 15-5 (A). Some typical external shunts are shown in part (C).

The shunt strips are usually made of manganin; an alloy having an almost zero temperature coefficient of resistance. The ends of the shunt strips are embedded in heavy copper blocks to which are attached the meter coil leads and the line terminals. To insure accurate readings, the meter leads for a particular ammeter should

not be used interchangeably with those for a meter of a different range. Slight changes in lead length and size may vary the resistance of the meter circuit and thus its current, and may cause an incorrect meter reading. External shunts are generally used where currents greater than 50 amperes must be measured.

It is important to select a suitable shunt when using an external shunt ammeter so that the scale indication is easily read. For example, if the scale has 150 divisions and the load current to be measured is known to be between 50 and 100 amperes, a 150-ampere shunt is suitable.

If the scale deflection is 75 divisions, the load current is 75 amperes, the needle will deflect half-scale when using the same 150-ampere shunt.

A shunt having exactly the same current rating as the estimated normal load current should never be selected because any abnormally high load would drive the pointer off scale and might damage the movement. A good choice would bring the needle somewhere near the mid-scale indication, when the load is normal.

## EXTENDING THE RANGE BY USE OF INTERNAL SHUNTS

For limited current ranges (below 50 amperes), internal shunts are most often employed. In this manner the range of the meter may be easily changed by selecting the correct internal shunt having the necessary current rating. Before the required resistance of the shunt for each range can be calculated, the resistance of the meter movement must be known.

For example, suppose it is desired to use a 100-microampere D'Arsonval meter having a resistance of 100 ohms to measure line currents up to 1 ampere. The meter deflects full scale when the current through the 100-ohm coil is 100 microamperes. Therefore, the voltage drop across the meter coil is IR, or

$$0.0001 \times 100 = 0.01 \text{ volt}$$

Because the shunt and coil are in parallel, the shunt must also have a voltage drop of 0.01 volt. The current that flows through the shunt is the difference between the full-scale meter current and the line current. In this case, the meter current is $100 \times 10^{-6}$, or 0.0001 ampere. This current is negligible compared with the line (shunt) current, so the shunt current is approximately 1 ampere. The resistance, $R_S$, of the shunt is therefore

$$R_S = \frac{E}{I} = \frac{0.01}{1} = 0.01 \text{ ohm (approx.)}$$

and the range of the 100-microampere meter has been increased to 1 ampere by paralleling it with the 0.01-ohm shunt.

The 100-microampere instrument may also be converted to a 10-ampere meter by the use of a proper shunt. For full-scale deflection of the meter the voltage drop, E, across the shunt (and across the meter) is still 0.01 volt. The meter current is again considered negligible, and the shunt current is now approximately 10 amperes. The resistance, $R_S$, of the shunt is therefore

$$R_S = \frac{E}{I} = \frac{0.01}{10} = 0.001 \text{ ohm}$$

The same instrument may likewise be converted to a 50-ampere meter by the use of the proper type of shunt. The current, $I_S$, through the shunt is approximately 50 amperes and the resistance, $R_S$, of the shunt is

$$R_S = \frac{E}{I_S} = \frac{0.01}{50} = 0.0002 \text{ ohm}$$

Various values of shunt resistance may be used, by means of a suitable switching arrangement, to increase the number of current ranges that may be covered by the meter. Two switching arrangements are shown in figure 15-6. Figure 15-6 (A), is the simpler of the two arrangements from the point of view of calculating the value of the shunt resistors when a number of shunts are used. However, it has two disadvantages:

**1.** When the switch is moved from one shunt resistor to another the shunt is momentarily removed from the meter and the line current then flows through the meter coil. Even a momentary surge of current could easily damage the coil.

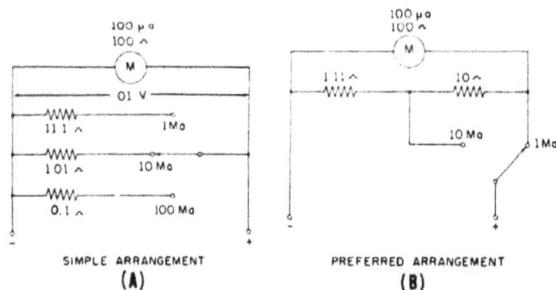

SIMPLE ARRANGEMENT (A)    PREFERRED ARRANGEMENT (B)

BE. 252

Figure 15-6.—Ways of connecting internal shunts.

**2.** The contact resistance—that is, the resistance between the blades of the switch when they are in contact—is in series with the shunt but not with the meter coil. In shunts that must pass high currents the contact resistance becomes an appreciable part of the total shunt resistance. Because the contact resistance is of a variable nature, the ammeter indication may not be accurate.

A more generally accepted method of range switching is shown in figure 15-6 (B). Although only two ranges are shown, as many ranges as needed can be used. In this type of circuit the range selector switch contact resistance is external to the shunt and meter in each range position, and therefore has no effect on the accuracy of the current measurement.

CURRENT - MEASURING INSTRUMENTS MUST ALWAYS BE CONNECTED IN SERIES WITH A CIRCUIT AND NEVER IN PARALLEL WITH IT. If an ammeter were connected across a constant-potential source of appreciable voltage the shunt would become a short circuit, and the meter would burn out.

If the approximate value of current in a circuit is not known, it is best to start with the highest range of the ammeter and switch to progressively lower ranges until a suitable reading is obtained.

Most ammeter needles indicate the magnitude of the current by being deflected from left to right. If the meter is connected with reversed polarity, the needle will be deflected backwards, and this action may damage the movement. Hence the proper polarity should be observed in connecting the meter in the circuit. That is, the meter should always be connected so that the electron flow will be into the negative terminal and out of the positive terminal.

Figure 15-7 shows various circuit arrangements; the ammeter or ammeters are properly connected for measuring current in various portions of the circuits.

### VOLTMETER

The 100-microampere D'Arsonval meter used as the basic meter for the ammeter may also be used to measure voltage if a high resistance is placed in series with the moving coil of the meter. For low-range instruments, this resistance is mounted inside the case with the D'Arsonval movement and typically consists of resistance wire having a low temperature coefficient and wound either on spools or card frames. For higher voltage ranges, the series resistance may be connected externally. When this is done the unit containing the resistance is commonly called a MULTIPLIER.

A simplified diagram of a voltmeter is shown in figure 15-8 (A). The resistance coils are treated in such a way that a minimum amount of moisture will be absorbed by the insulation.

Moisture reduces the insulation resistance and increases leakage currents, which cause incorrect readings. Leakage currents through the insulation increase with length of resistance wire and become a factor that limits the magnitude of voltage that may be measured. An external view of a voltmeter is shown in figure 15-8 (B).

Extending the Range

The value of the necessary series resistance is determined by the current required for full-scale deflection of the meter and by the range of voltage to be measured. Because the current through the meter circuit is directly proportional to the applied voltage, the meter scale can be calibrated directly in volts for a fixed series resistance.

For example, assume that the basic meter (microammeter) is to be made into a voltmeter with a full-scale reading of 1 volt. The coil resistance of the basic meter is 100 ohms, and 0.0001 ampere (100 microamperes) causes a full-scale deflection. The total resistance, R, of the meter coil and the series resistance is

$$R = \frac{E}{I} = \frac{1}{0.0001} = 10,000 \text{ ohms}$$

and the series resistance alone is

$$R_s = 10,000 - 100 = 9,900 \text{ ohms}$$

Multirange voltmeters utilize one meter movement with the required resistances connected in series with the meter by a convenient switching arrangement. A multirange voltmeter with three ranges is shown in figure 15-9. The total circuit resistance for each of the three ranges beginning with the 1-volt range is:

$$R = \frac{E}{I} = \frac{1}{100} = 0.01 \text{ megohm}$$

$$= \frac{100}{100} = 1 \text{ megohm}$$

$$= \frac{1,000}{100} = 10 \text{ megohms}$$

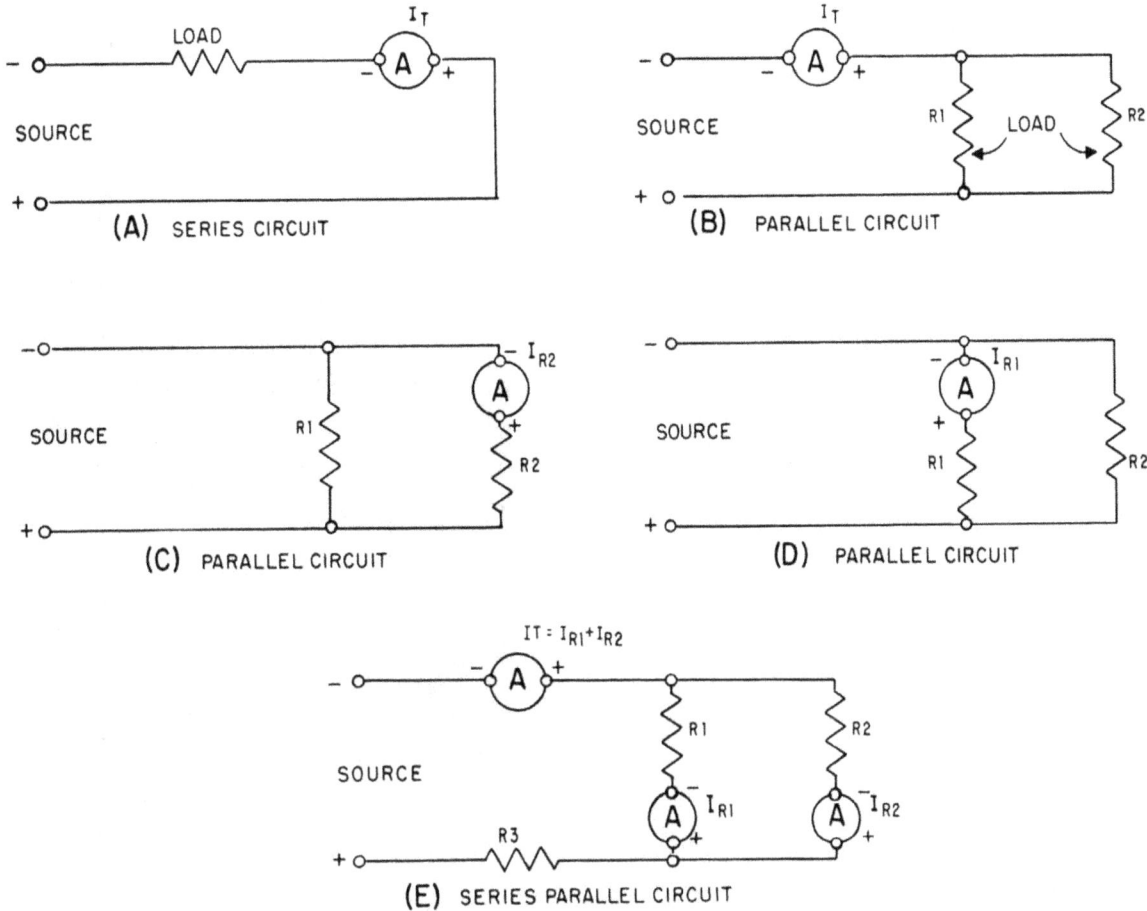

Figure 15-7.—Proper ammeter connections.

VOLTAGE - MEASURING INSTRUMENTS ARE CONNECTED ACROSS (IN PARALLEL WITH) A CIRCUIT. If the approximate value of the voltage to be measured is not known, it is best to start with the highest range of the voltmeter and progressively lower the range until a suitable reading is obtained.

In many cases, the voltmeter is not a central-zero indicating instrument. Thus, it is necessary to observe the proper polarity when connecting the instrument to the circuit, as is the case in connecting the d-c ammeter. The positive terminal of the voltmeter is always connected to the positive terminal of the source, and the negative terminal to the negative terminal of the source when the source voltage is being measured. In any case, the voltmeter is connected so that electrons will flow into the negative terminal and out of the positive terminal of the meter.

Influence in a Circuit

The function of a voltmeter is to indicate the potential difference between two points in a circuit. When the voltmeter is connected across a circuit, it shunts the circuit. If the voltmeter has low resistance it will draw an appreciable amount of current. The effective resistance of the circuit will be lowered and the voltage reading will consequently be lowered.

When voltage measurements are made in high-resistance circuits, it is necessary to use a

SCALE CALIBRATED IN VOLTS

TO REAR
BALANCE SPRING

N          S

FROM REAR
BALANCE SPRING

SERIES
RESISTORS

VOLTAGE TO BE
MEASURED

(A) INTERNAL CONSTRUCTION & CIRCUIT

(B) EXTERNAL VIEW

BE. 254

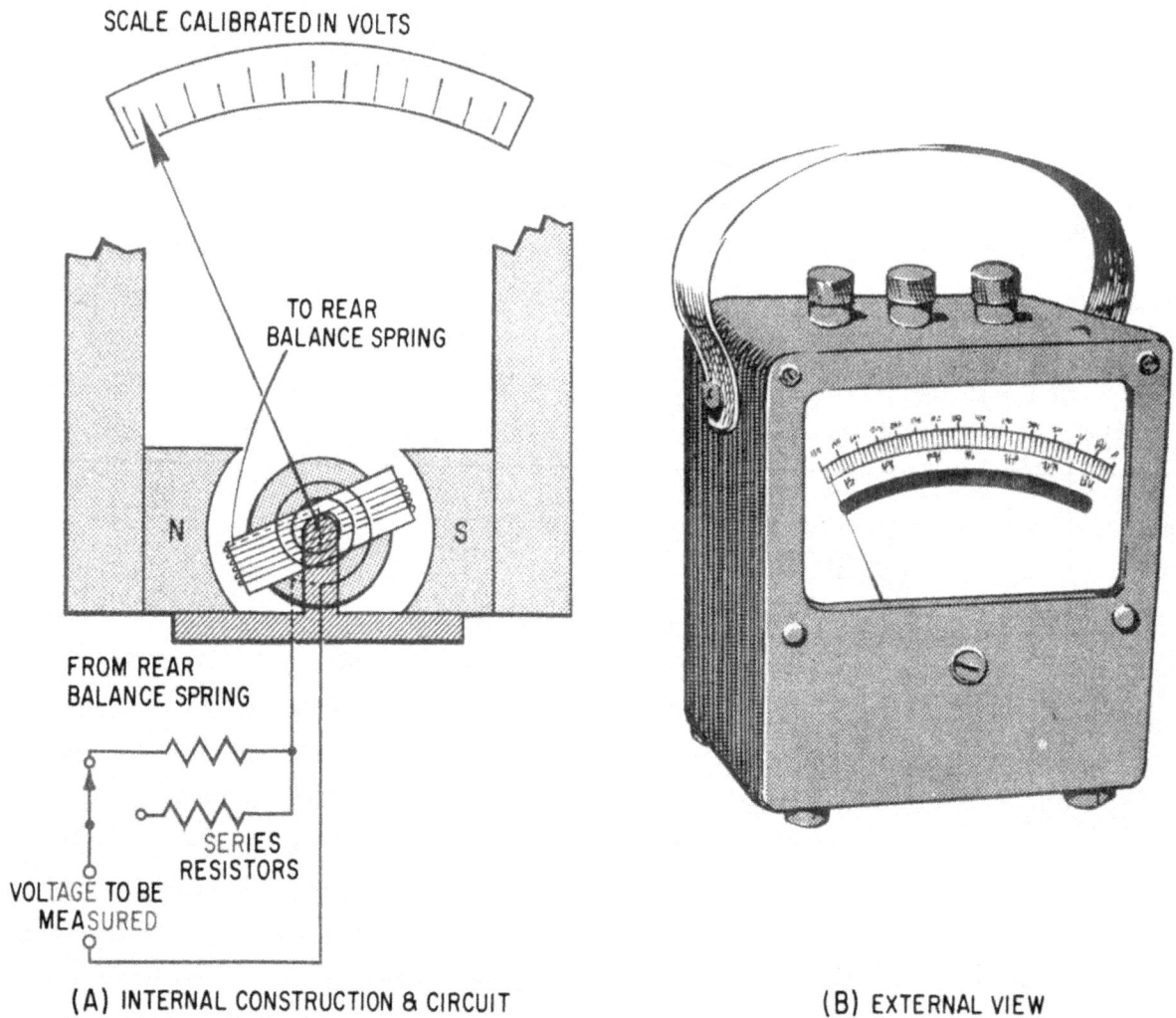

Figure 15-8.—Simplified voltmeter circuit.

high-resistance voltmeter to prevent the shunting action of the meter. The effect is less noticeable in low-resistance circuits because the shunting effect is less.

Sensitivity

The sensitivity of a voltmeter is given in ohms per volt, $(\Omega/E)$, and may be determined by dividing the resistance $R_m$, of the meter plus the series resistance, $R_s$, by the full-scale reading in volts. Thus,

$$\text{sensitivity} = \frac{R_m + R_s}{E}$$

This is the same as saying that the sensitivity is equal to the reciprocal of the current (in amperes)—that is,

$$\text{sensitivity} = \frac{\text{ohms}}{\text{volts}} = \frac{1}{\dfrac{\text{volts}}{\text{ohms}}} = \frac{1}{\text{amperes}}$$

Thus, the sensitivity of a 100-microampere movement is the reciprocal of 0.0001 ampere, or 10,000 ohms per volt.

value that is within 1 percent of the correct value. The statement means that if the correct value is 100 units, the meter indication may be anywhere within the range of 99 to 101 units.

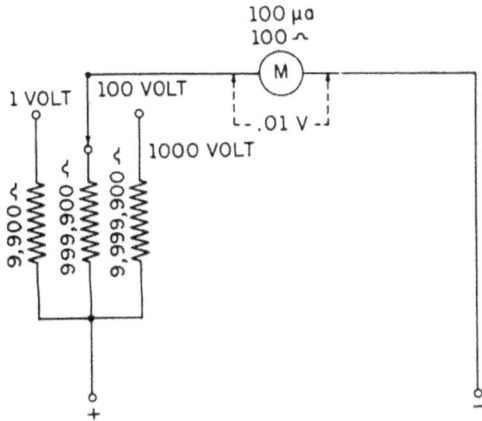

Figure 15-9.—Multirange voltmeter.

BE.255

Table 15-1 shows how the sensitivity of permanent-magnet movable-coil voltmeters has been increased over many years of manufacture. The table indicates the sensitivity of a 0 to 150-volt voltmeter, its resistance, and the current required to produce a full-scale deflection.

## ELECTRODYNAMOMETER TYPE METER

The electrodynamometer-type meter differs from the galvanometer-type meter in that no permanent magnet is used. Instead, two fixed coils are utilized to produce the magnetic field. Two movable coils are also used in this type meter.

The two fixed coils are connected in series and positioned coaxially, with a space between them. The two movable coils are also positioned coaxially, and are connected in series. The two pairs of coils (fixed pair and movable pair) are further connected in series with each other. The movable-coil unit is pivot-mounted between the fixed coils.

This meter arrangement is illustrated in figure 15-10.

The central shaft on which the movable coils are mounted is restrained by spiral springs which hold the pointer at zero when no current is flowing through the coil. These springs also

Table 15-1.—Increase in sensitivity in voltmeters.

| Time of manufacture | Sensitivity | Resistance | Current to deflect full scale |
|---|---|---|---|
| Before 1924........ | 10 ohms per volt | 1.5 k-ohms | 100 ma. |
| After 1924........ | 100 ohms per volt | 15 k-ohms | 10 ma. |
| Today.......... | 1,000 ohms per volt<br>20,000 ohms per volt<br>200,000 ohms per volt | 150 k-ohms<br>3 megohms<br>30 megohms | 1 ma.<br>50 $\mu$a.<br>5 $\mu$a. |

The sensitivity has been increased by increasing the strength of the permanent magnet, by using lighter weight materials for the moving element (consistent with increased number of turns on the coil), and by using sapphire jewel bearings to support the moving coil.

### Accuracy

The accuracy of a meter is generally expressed in percent. For example, a meter that has an accuracy of 1 percent will indicate a

serve as conductors for delivering current to the movable coils. Since these conducting springs are very small, the meter cannot carry a very heavy current.

When used as a voltmeter, no difficulty in construction is encountered, because the current required is not more than 0.1 ampere. This amount of current can be brought in and out of the moving coil through the springs. When the electrodynamometer is used as a voltmeter, its internal connections and construction are as shown in figure 15-11 (A). The fixed coils a and

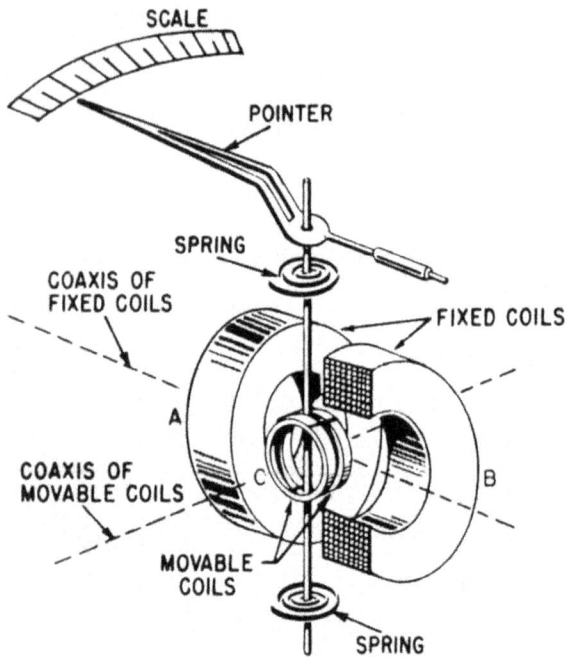

BE.256

Figure 15-10.—Inside construction of an electrodynamometer.

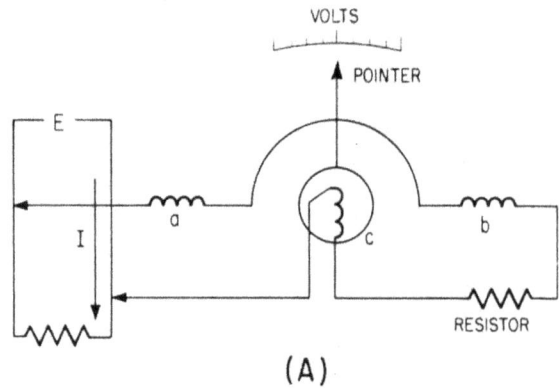

(A)

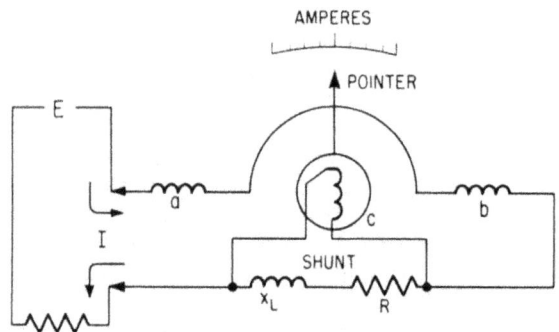

(B)

BE.257

Figure 15-11.—Circuit arrangement of electro-dynamometer type meter. (A) Voltmeter; (B) ammeter.

b are wound with fine wire, since the current through them will be no more than 0.1 ampere. They are connected directly in series with the movable coil c and the series current-limiter resistance. For ammeter applications, however, a special type construction must be used, because the large currents that flow through the meter cannot be carried through the moving coils.

In the ammeter, the stationary coils a and b of figure 15-11 (B), are generally wound of heavier wire, to carry up to 5 amperes. In parallel with the moving coils is an inductive shunt, which permits only a small part of the total current to flow through the moving coil. This current through the moving coil is directly proportional to the total current through the instrument. The shunt has the same ratio of reactance to resistance as has the moving coil, thus the instrument will be reasonably correct at all frequencies with which it is designed to be used.

The meter is mechanically damped by means of aluminum vanes that move in enclosed air chambers. Although electrodynamometer type

meters are very accurate, they do not have the sensitivity of the D'Arsonval type meter. For this reason they are not widely used outside the laboratory.

METERS USED FOR
MEASURING RESISTANCE

The two instruments most commonly used to check the continuity, or to measure the resistance of a circuit or circuit element, are the OHMMETER and the MEGGER (megohmmeter). The ohmmeter is widely used to measure resistance and check the continuity of electrical circuits and devices. Its range usually extends to only a few megohms. The megger is widely used for measuring insulation resistance, such as between a wire and the outer surface of its insulation, and insulation resistance of cables and insulators. The range of a megger may extend to more than 1,000 megohms.

Ohmmeter

The ohmmeter consists of a d-c milliammeter, which was discussed earlier in this chapter, with a few added features. The added features are:

1. A d-c source of potential (usually a 3-volt battery).

2. One or more resistors (one of which is variable). A simple ohmmeter circuit is shown in figure 15-12.

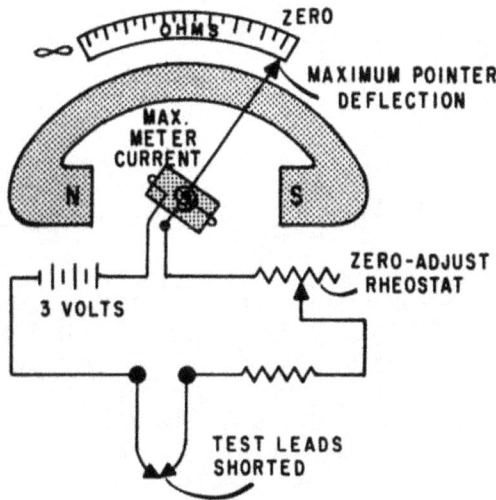

BE.258

Figure 15-12.—Simple ohmmeter circuit.

The ohmmeter's pointer deflection is controlled by the amount of battery current passing through the moving coil. Before measuring the resistance of an unknown resistor or electrical circuit, the test leads of the ohmmeter are first shorted together, as shown in figure 15-12. With the leads shorted the meter is calibrated for proper operation on the selected range. (While the leads are shorted meter current is maximum and the pointer deflects a maximum amount, somewhere near the zero position on the ohms scale.) When the variable resistor (rheostat) is adjusted properly, with the leads shorted, the pointer of the meter will come to rest exactly

on the zero graduation. This indicates ZERO RESISTANCE between the test leads, which in fact are shorted together. The zero readings of series-type ohmmeters are sometimes on the right-hand side of the scale, whereas the zero reading for ammeters and voltmeters is generally to the left-hand side of the scale. When the test leads of an ohmmeter are separated, the pointer of the meter will return to the left side of the scale, due to the interruption of current and the spring tension acting on the movable coil assembly.

After the ohmmeter is adjusted for zero reading, it is ready to be connected in a circuit to measure resistance. A typical circuit and ohmmeter arrangement is shown in figure 15-13.

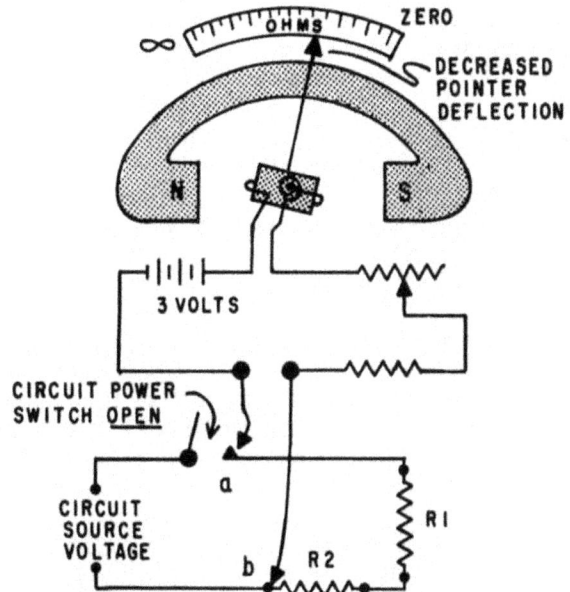

BE.259

Figure 15-13.—Measuring circuit resistance with an ohmmeter.

The power switch of the circuit to be measured should always be in the OFF position. This prevents the circuit's source voltage from being applied across the meter, which could cause damage to the meter movement.

The test leads of the ohmmeter are connected across (in series with) the circuit to be measured. (See fig. 15-13.) This causes the current produced by the meter's 3-volt battery to flow through the circuit being tested. Assume that the meter test leads are connected at points a and b of figure 15-13. The amount of current that flows through the meter coil will depend on the resistance of resistors, $R_1$ and $R_2$, plus the resistance of the meter. Since the meter has been preadjusted (zeroed), the amount of coil movement now depends solely on the resistance of $R_1$ and $R_2$. The inclusion of $R_1$ and $R_2$ raised the total series resistance, decreased the current, and thus decreased the pointer deflection. The pointer will now come to rest at a scale figure indicating the combined resistance of $R_1$ and $R_2$. If $R_1$ or $R_2$, or both, were replaced with a resistor(s) having a larger ohmic value, the current flow in the moving coil of the meter would be decreased still more. The deflection would also be further decreased, and the scale indication would read a still higher circuit resistance. Movement of the moving coil is proportional to the amount of current flow. The scale reading of the meter, in ohms, is inversely proportional to current flow in the moving coil.

The amount of circuit resistance to be measured may vary over a wide range. In some cases it may be only a few ohms, and in others it may be as great as 1,000,000 ohms. To enable the meter to indicate any value being measured, with the least error, scale multiplication features are incorporated in most ohmmeters. For example, a typical meter will have four test lead jacks, marked as follows—COMMON, R x 1, R x 10, and R x 100. The jack marked COMMON is connected internally through the battery to one side of the moving coil of the ohmmeter. The jacks marked R x 1, R x 10, and R x 100 are connected to three different size resistors located within the ohmmeter. This is shown in figure 15-14.

Some ohmmeters are equipped with a selector switch for selecting the multiplication scale desired, so that only two test lead jacks are necessary. Other meters have a separate jack for each range, as shown in figure 15-14. The range to be used in measuring any particular unknown resistance ($R_x$ in fig. 15-14) depends on the approximate ohmic value of the unknown resistance. For instance, assume the ohmmeter scale in figure 15-14 is calibrated in divisions from zero to 1,000. If $R_x$ is greater than 1,000 ohms, and the R x 1 range is being used, the ohmmeter cannot measure it. This occurs because the combined series resistance of resistor R x 1 and $R_x$ is too great to allow sufficient battery current to flow to deflect the pointer away from infinity ($\infty$). The test lead would have to be plugged into the next range, R x 10. With this

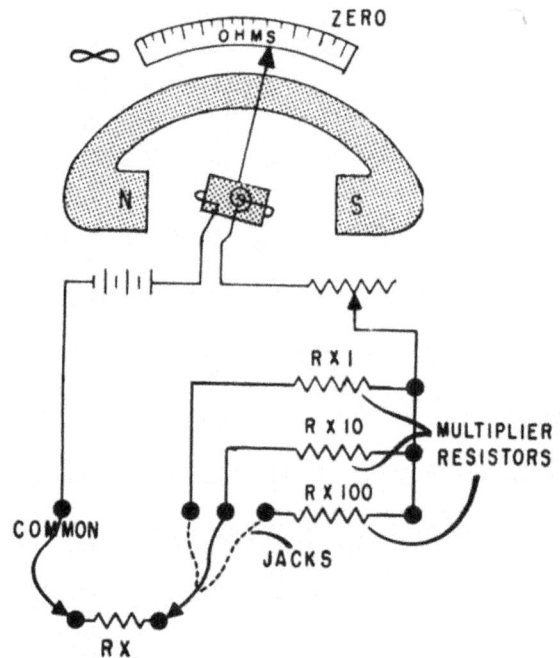

BE.260

Figure 15-14.—Ohmmeter with multiplication jacks.

done, assume the pointer deflects to indicate 375 ohms. This would indicate that $R_x$ has 375 x 10 = 3,750 ohms resistance. The change of range caused the deflection because resistor R x 10 has only 1/10 the resistance of resistor R x 1. Thus, selecting the smaller series resistance permitted a battery current of sufficient amount to cause a useful pointer deflection. If the R x 100 range were used to measure the same 3,750-ohm resistor, the pointer would deflect still further, to the 37.5 ohm position. This increased deflection would occur because resistor R x 100 has only 1/10 the resistance of resistor R x 10.

The foregoing circuit arrangement allows the same amount of current to flow through the meter's moving coil whether the meter measures 10,000 ohms on the R x 1 scale, or 100,000 ohms on the R x 10 scale, or 1,000,000 ohms on the R x 100 scale.

It always takes the same amount of current to deflect the pointer to a certain position on the scale (midscale position for example), regardless of the multiplication factor being used. Since the multiplier resistors are of different values, it is necessary to ALWAYS "zero" adjust the meter for each multiplication factor selected.

The operator of the ohmmeter should select the multiplication factor that will result in the pointer coming to rest as near as possible to the midpoint of the scale. This enables the operator to read the resistance more accurately, because the scale readings are more easily interpreted at or near midpoint.

Megger

An ordinary ohmmeter cannot be used for measuring resistance of multimillions of ohms, such as conductor insulation. To adequately test for insulation breakdown, it is necessary to use a much higher potential than is furnished by an ohmmeter's battery. This potential is placed between the conductor and the outside surface of the insulation.

An instrument called a MEGGER (megohmmeter) is used for these tests. The megger (fig. 15-15 (A)) is a portable instrument consisting of two primary elements—(1) a hand-driven d-c generator, G, which supplies the necessary voltage for making the measurement, and (2) the instrument portion, which indicates the value of the resistance being measured. The instrument portion is of the opposed-coil type as shown in figure 15-15 (A). Coils a and b are mounted on the movable member c with a fixed angular relationship to each other, and are free to turn as a unit in a magnetic field. Coil b tends to move the pointer counterclockwise, and coil a clockwise.

Coil a is connected in series with $R_3$ and the unknown resistance, $R_x$, to be measured. The combination of coil $R_3$, and $R_x$ form a direct series path between the positive (+) and negative (-) brushes of the d-c generator. Coil b is connected in series with $R_2$ and this combination is also connected across the generator. There are no restraining springs on the movable member of the instrument portion of the megger. Therefore, when the generator is not operated, the pointer floats freely and may come to rest at any position on the scale.

The guard ring intercepts leakage current. Any leakage currents intercepted are shunted to the negative side of the generator. They do not flow through coil a; therefore, they do not affect the meter reading.

If the test leads are open-circuited, no current flows in coil a. However, current flows internally through coil b, and deflects the pointer to infinity, which indicates a resistance too large to measure. When a resistance such as $R_x$ is connected between the test leads, current also flows in coil a, tending to move the pointer clockwise. At the same time, coil b still tends to move the pointer counterclockwise. Therefore, the moving element, composed of both coils and the pointer, comes to rest at a position at which the two forces are balanced. This position depends upon the value of the external resistance, which controls the relative magnitude of current in coil a. Because changes in voltage affect both coil a and coil b in the same proportion, the position of the moving system is independent of the voltage. If the test leads are short-circuited, the pointer rests at zero because the current in a is relatively large. The instrument is not injured under the circumstances because the current is limited by $R_3$.

The external view of one type of megger is shown in figure 15-15 (B).

Meggers provided aboard ship usually are rated at 500 volts. To avoid excessive test voltages, most meggers are equipped with friction clutches. When the generator is cranked faster than its rated speed, the clutch slips and the generator speed and output voltage are not permitted to exceed their rated values. For extended ranges, a 1,000-volt generator is available. When extremely high resistances—for example, 10,000 megohms or more—are to be measured, a high voltage is needed to cause sufficient current flow to actuate the meter movement.

MULTIMETER

The MULTIMETER is a multipurpose instrument that can measure resistance, voltage, or current. It contains one milliammeter. The face of the instrument has separate graduated scales to indicate the three values that can be measured. Figure 15-16 shows a multimeter that is widely used in the Navy today.

The front panel of the multimeter is constructed and labeled in such a way that all functions are self explanatory. One pin jack marked -DC ± AC OHMS is common to all functions and ranges. One test lead is always plugged into this common jack and the remaining lead into the jack marked for the particular function or range desired. The internal circuit arrangement of the meter is controlled by means of the FUNCTION switch. This switch selects the proper circuit elements for the type of measurement desired. The desired voltage, current or ohmmeter range is determined by the combined

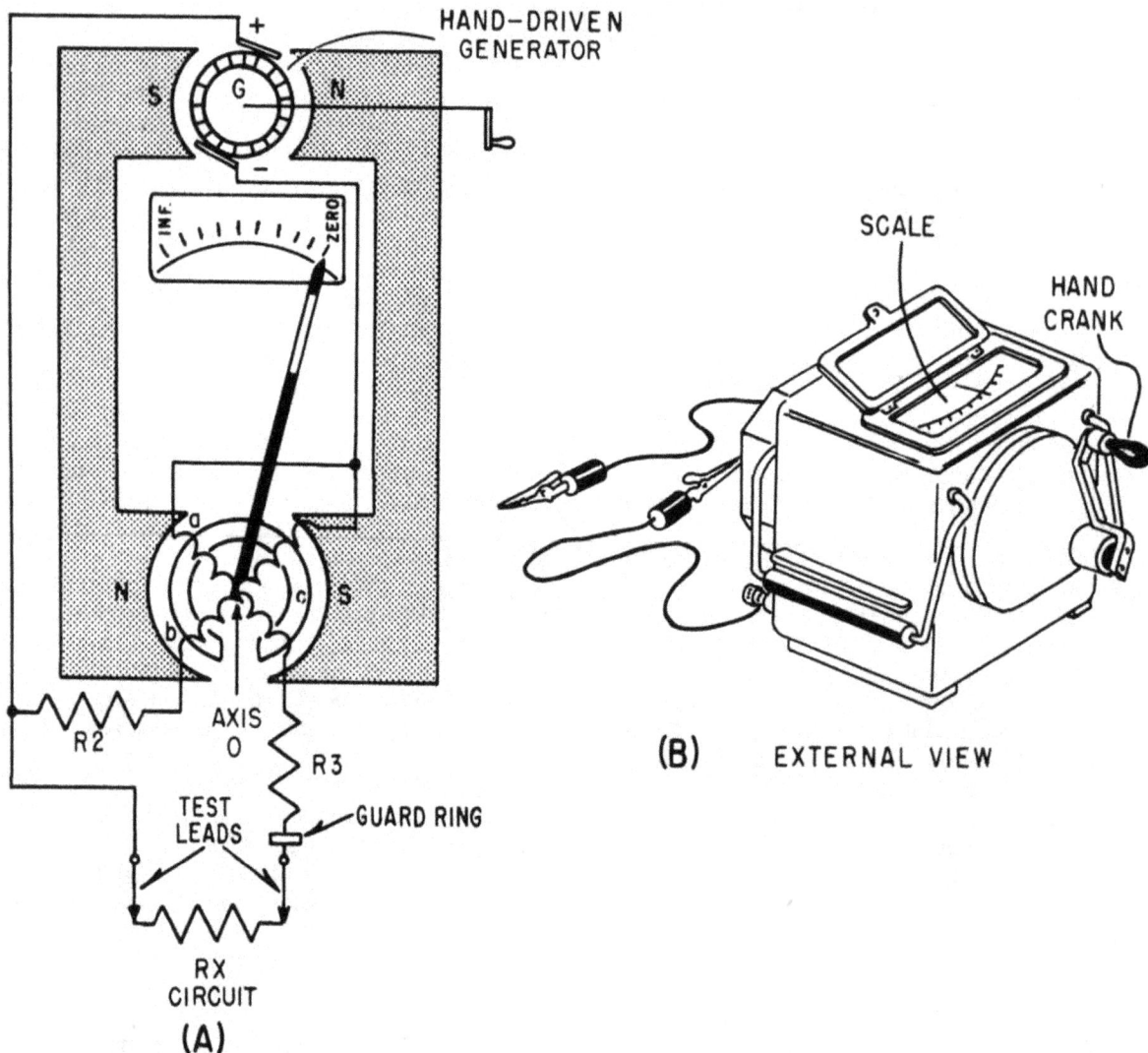

BE.261

Figure 15-15.—(A) Megger internal circuit; (B) external view of megger.

action of positioning the function and range switches and selecting certain pin jacks. The range switch selects the resistance or current range and the pin jacks select the voltage range. The function switch selects the type of operation to be performed. Separate pin jacks are provided for ohms and for the 10-ampere range. A CASE GROUND jack is connected directly to the cast aluminum case.

The meter panel contains the following operating controls:

1. FUNCTION SWITCH. This is a six position rotary switch which is clearly marked with the type of measurement to be taken. It is located

in the lower left-hand corner of the front panel.

2. SELECTOR SWITCH. This is a twelve position switch used in selecting one of five resistance ranges or one of seven current ranges. This switch is located in the lower right-hand corner of the front panel. An additional current range (10 amperes) is brought out to a separate pin jack, because of the magnitude of the current.

3. OHMS ZERO ADJ. This control is a rheostat used to zero the pointer on the ohmmeter. With the test leads shorted the rheostat is turned until the pointer comes to rest at zero. The rheostat also serves to compensate for variations in battery voltage, and is located in the lower center section of the front panel.

BE.262

Figure 15-16.—Multimeter.

The following operational procedures should be used when making measurements with the meter.

1. A-C VOLTS MEASUREMENTS. Turn FUNCTION switch to AC VOLTS. Plug one test lead into the ± AC jack. Plug the other lead into the desired 1,000 OHMS PER VOLT jack.

2. OHMS MEASUREMENTS. Turn FUNCTION switch to OHMS. Plug leads into the two OHMS jacks. Turn the range switch to the desired resistance range. Short the two test leads together and set the indicator pointer to zero ohms (top scale on the indicator) by rotating the OHMS ZERO ADJ.

3. D-C CURRENT MEASUREMENT. Turn FUNCTION switch to DC CURRENT. Plug the black test lead into the -DC jack. For ranges from 250 microamperes to 2.5 amperes full scale, plug the red lead into the jack marked + DC current. Use the jack marked +10 AMPS ONLY for current reading between 2.5 amperes and 10 amperes range.

4. D-C VOLTS MEASUREMENTS at 1,000 OHMS PER VOLT. Turn the FUNCTION switch to 1,000 $\Omega$/VDC. Plug the black lead into the -DC jack. Plug the red lead into the desired range jack.

5. D-C VOLTS MEASUREMENTS at 20,000 OHMS PER VOLT. An indicator reversing position is available on the 20,000 ohms per volt

measurements and is selected by indexing the FUNCTION switch to DIRECT or REVERSE. This permits measurement of voltage, either positive or negative in polarity, while still keeping the instrument case and the common d-c input jack at the same potential as the chassis of the device under check. Accordingly, the following procedure should always be followed when using the 20,000 ohms per volt ranges, particularly where the 1,000 or 5,000 volt ranges are used.

a. Plug the black lead into the -DC jack.

b. Connect the other end of the black lead solidly to the chassis or exposed metal part of the device under check.

c. Turn the FUNCTION switch to "20,000 $\Omega$ /VDC DIRECT" if the voltage to be measured is positive with respect to chassis.

d. Turn the FUNCTION switch to REVERSE if the voltage is expected to be negative with respect to chassis.

NOTE: The CASE GROUND jack is connected directly to the metal case. It is not connected to any part of the multimeter circuit. For protection of operating personnel it should be jumpered to an earth or common ground.

## MOVING IRON-VANE METER

The moving iron-vane meter is another basic type of meter. Unlike the D'Arsonval-type meter, which employs permanent magnets, the moving iron-vane meter depends on induced magnetism for its operation. It employs the principle of repulsion between two concentric iron vanes, one fixed and one movable, placed inside a coil, as shown in figure 15-17 (A). A pointer is attached to the movable vane.

When current flows through the coil, the two iron vanes become magnetized with north poles at their upper ends and south poles at their lower ends for one direction of current through the coil, as shown in the figure. Because like poles repel, the unbalanced component of force tangent to the movable element causes it to turn against the force exerted by the springs.

The movable vane is rectangular in shape, and the fixed vane is tapered. This design permits the use of a relatively uniform scale.

When no current flows through the coil, the movable vane is positioned so that it is opposite the larger portion of the tapered fixed vane, and the scale reading is zero. The amount of magnetization of the vanes depends on the strength of the field, which in turn, depends on the amount

of current flowing through the coil. The force of repulsion is greater opposite the larger end of the fixed vane than it is nearer the smaller end. Therefore, the movable vane moves toward the smaller end through an angle that is proportional to the magnitude of the coil current. The movement ceases when the force of repulsion is balanced by the restoring force of the spring.

Because the repulsion is always in the same direction (toward the smaller end of the fixed vane) regardless of the direction of current flow through the coil, the moving iron-vane instrument operates on either d-c or a-c circuits.

Mechanical damping in this type of instrument is obtained by the use of an aluminum vane attached to the shaft (not shown in the figure) in such a way that, as the shaft moves, the vane moves in a restricted air space.

When the moving iron-vane meter is designed to be used as an ammeter, the coil is wound with relatively few turns of large wire in order to carry the rated current.

When the moving iron-vane meter is designed to be used as a voltmeter the solenoid is wound with many turns of small wire. Portable voltmeters are made with self-contained series resistance for ranges up to 750 volts. Higher ranges are obtained by the use of additional external multipliers. An external view of a moving iron-vane meter is shown in figure 15-17 (B).

The moving iron-vane instrument may be used to measure direct current, but has an error due to residual magnetism in the vanes. The error may be minimized by reversing the meter connections and averaging the readings. When used on a-c circuits the instrument has an accuracy of 0.5 percent. Because of its simplicity, its relatively low cost, and the fact that no current is conducted to the moving element, this type of movement is used extensively to measure current and voltage in a-c power circuits.

However, because the reluctance of the magnetic circuit is high, the moving iron-vane meter requires much more power to produce full-scale deflection than is required by a D'Arsonval meter of the same range. Therefore, the moving iron-vane meter is seldom used in high-resistance low-power circuits.

## INCLINED-COIL
## IRON-VANE METER

The principle of the moving iron-vane mechanism is applied to the inclined-coil type of meter shown in figure 15-18. The inclined-coil

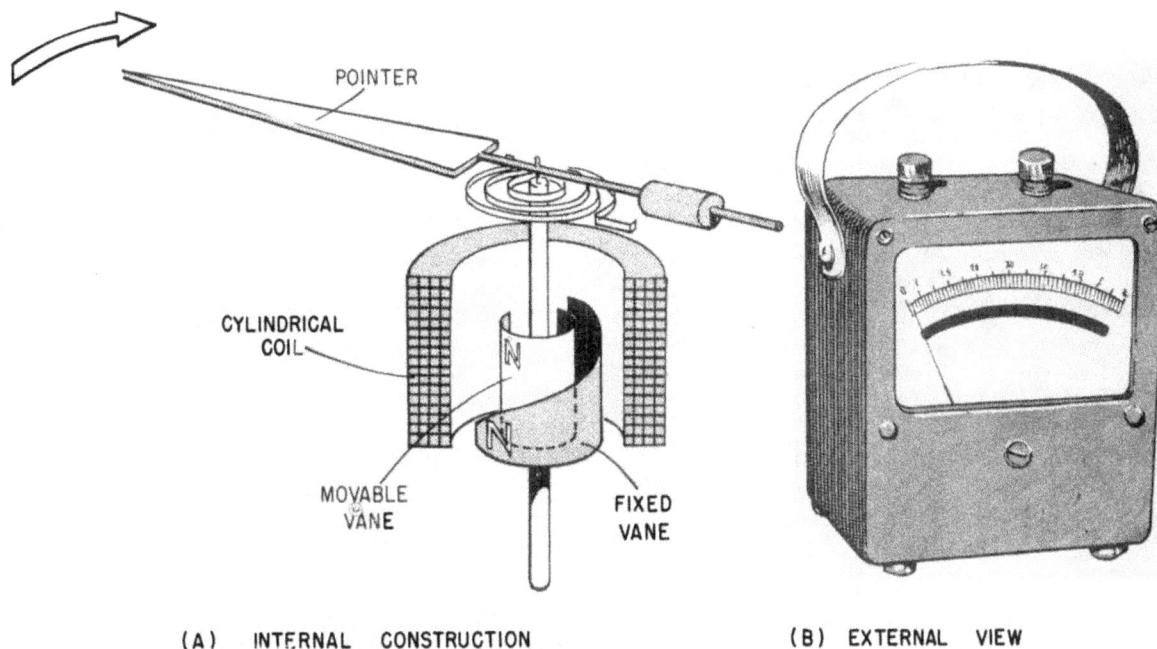

(A) INTERNAL CONSTRUCTION        (B) EXTERNAL VIEW

BE.263

Figure 15-17.—(A) Simplified diagram of a moving iron-vane meter; (B) external view.

iron-vane meter has a coil mounted at an angle to the shaft. Attached obliquely to the shaft, and located inside the coil, are two soft-iron vanes. When no current flows through the coil, a control spring holds the pointer at zero and the iron vanes lie in planes parallel to the plane of the coil. When current flows through the coil, the vanes tend to line up with magnetic lines passing through the center of the coil at right angles to the plane of the coil. Thus the vanes rotate against the spring action to move the pointer over the scale.

The iron vanes tend to line up with the magnetic lines regardless of the direction of current flow through the coil. Therefore, the inclined-coil iron-vane meter can be used to measure either alternating current or direct current. The aluminum disk and the drag magnets provide electromagnetic damping.

Like the moving iron-vane meter, the inclined-coil type requires a relatively large amount of current for full-scale deflection and hence is seldom used in high-resistance low-power circuits.

As in the moving iron-vane instrument, the inclined-coil instrument is wound with few turns of relatively large wire when used as an ammeter and with many turns of small wire when used as a voltmeter.

## THERMOCOUPLE TYPE METER

If the ends of two dissimilar metals are welded together and this junction is heated, a d-c voltage is developed across the two open ends. The voltage developed depends on the material of which the wires are made and on the difference in temperature between the heated junction and the open ends.

In one type of instrument, the junction is heated electrically by the flow of current through a heater element. It does not matter whether the current is alternating or direct because the heating effect is independent of current direction. The maximum current that may be measured depends on the current rating of the heater, the heat that the thermocouple can stand without being damaged, and on the current rating of the meter used with the thermocouple. Voltage may also be measured if a suitable resistor is placed in series with the heater.

268

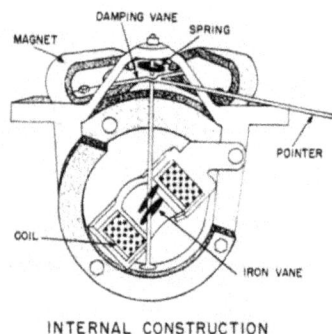

BE.264

Figure 15-18.—Inclined-coil iron-vane meter.

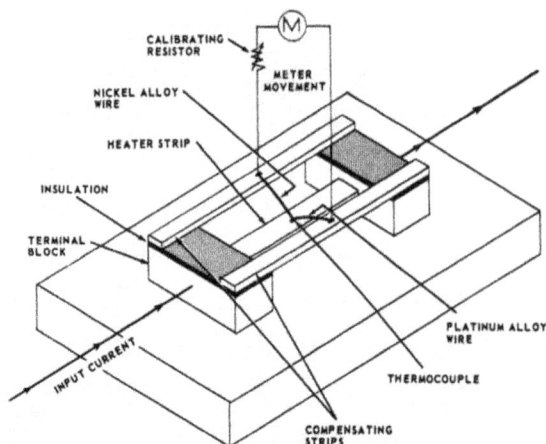

BE.265

Figure 15-19.—Simplified schematic of one type of thermocouple.

A simplified schematic diagram of one type of thermocouple is shown in figure 15-19. The input current flows through the heater strip via the terminal blocks. The function of the heater strip is to heat the thermocouple, which is composed of a junction of two dissimilar wires welded to the heater strip. The open ends of these wires are connected to the center of two copper compensating strips. The function of these strips is to radiate heat so that the open ends of the wires will be much cooler than the junction end of the wires; thus permitting a higher voltage to be developed across the open ends of the thermocouple. The compensating strips are thermally and electrically insulated from the terminal blocks.

The heat produced by the flow of line current through the heater strip is proportional to the square of the heating current ($P = I^2R$). Because the voltage appearing across the two open terminals is proportional to the temperature, the movement of the meter element connected across these terminals is proportional to the square of the current flowing through the heater element. The scale of the meter is crowded near the zero end, and is progressively less crowded near the maximum end of the scale. Because the lower portion of the scale is crowded the reading is necessarily less accurate. For the sake of accuracy in making a given measurement, it is desirable to choose a meter in which the deflection will extend at least to the more open portion of the scale.

The meter used with the thermocouple should have low resistance to match the low resistance of the thermocouple, and it must deflect full scale when rated current flows through the heater. Because the resistance must be low and the sensitivity high, the moving element must be light.

A more nearly uniform meter scale may be obtained if the permanent magnet of the meter is constructed so that as the coil rotates (needle moves up scale), it moves into a magnetic field of less and less density. The torque then increases approximately as the first power of the current instead of as the square of the current and a more linear scale is achieved.

If the thermocouple is burned out by excessive current through the heater strip, it may be replaced and the meter recalibrated by means of the calibrating variable resistor.

## ALTERNATING-CURRENT INSTRUMENTS

So far the simpler electrical indicating instruments were discussed. Only a few could measure a.c. Most were used to measure only direct voltage or current. It will be necessary for the technician to become familiar with additional, more advanced a-c indicating instruments. Those that will be discussed in this chapter are (1) rectifier-type a-c meters; (2) wattmeters and watt-hour meters; (3) instrument transformers; (4) bridge meters for measuring resistance, voltage, current, capacitance, and inductance; (5) frequency meters; and (6) the single-phase power-factor meter.

Many of the instruments discussed in this chapter utilize semiconductor rectifiers. Since these units are common to a number of different instruments, they will be discussed first.

## RECTIFIERS

A rectifier is a device that offers a high opposition to current flow through it in one direction but not in the other. It is therefore effectively a unidirectional conductor and is used mostly for converting alternating current into a unidirectional current (direct current).

A metallic rectifier element, called a cell, consists of a good conductor and a semiconductor (material of high resistivity) separated by a thin insulating barrier layer. The flow of forward current through a cell consists of a flow of electrons from the good conductor, across the barrier layer, and through the semiconductor.

Metallic rectifier cells are usually made in the form of plates, circular or square in shape, with a hole in the center. A number of cells, with the necessary terminals, spacers, and washers, are assembled on an insulated stud passing through their center holes. This assembly is called a rectifier stack. Some assemblies contain fins, which are used to keep the rectifier from overheating; they afford a large surface area for conducting away the heat. Cells and stacks may be connected in series or parallel, with proper polarities, to obtain the required voltage and current ratings and circuit connections for specific applications.

Two types of metallic rectifiers are used in the Navy—(1) a thin film of copper oxide and copper, and (2) selenium and either iron or aluminum. Rectifier units are represented by the symbol shown in figure 15-20 (A). The arrowhead in the symbol points against the direction of electron flow.

Figure 15-20 (B) shows a simple a-c circuit that utilizes a rectifier. It rectifies the a-c voltage and produces a series of d-c voltage pulses as its output. Although the copper-oxide rectifier is shown, the selenium rectifier may be used instead.

In the copper-oxide rectifier shown in figure 15-21 (A), the oxide is formed on the copper disk before the rectifier unit is assembled. In this type of rectifier the electrons flow more readily from the copper to the oxide than from the oxide to the copper. External electrical connections may be made by connecting terminal lugs between the left pressure plate and the copper and between the right pressure plate and the lead washer.

For the rectifier to function properly, the oxide coating must be very thin. Thus, each individual unit can stand only a low inverse voltage. Rectifiers designed for moderate and high-power applications consist of many of these individual units mounted in series on a single support. The lead washer enables uniform pressure to be applied to the units so that the internal resistance may be reduced. When the units are connected in series, they normally present a relatively high resistance to the current flow. The resultant heat developed in the resistance must be removed if the rectifier is to operate satisfactorily. The useful life of the unit is extended by keeping the temperature low (below $140^\circ$ F). The efficiency of this type of rectifier is generally between 60 and 70 percent.

Selenium rectifiers function in much the same manner as copper-oxide rectifiers. A selenium rectifier is shown in figure 15-21 (B). Such a rectifier is made up of an iron disk that is coated with a thin layer of selenium. In this type of rectifier the electrons flow more easily from the selenium to the iron than from the iron to the selenium. This type of rectifier may be operated at a somewhat higher temperature than a copper-oxide rectifier of similar rating. The efficiency is between 65 and 85 percent, depending on the circuit and the loading.

Rectifiers may be used not only as half-wave rectifiers, as shown in figure 15-20, but also in full-wave and bridge circuits. In each of these applications the action of the metallic rectifier is similar to that of a diode.

Many types of semiconductor rectifier diodes are available. Germanium and silicon are used extensively. These rectifiers vary in size from tiny ones (hardly bigger than a pinhead) used in miniature circuitry, to large 500-ampere diode rectifiers used in power supplies. A silicon diode rectifier about an inch long and an inch in diameter will supply a direct current of 50 amperes (peak) and has a peak-inverse voltage rating of 60 volts. They operate in much the same manner as the metallic rectifier and are utilized in similar application.

## RECTIFIER TYPE
## A-C INSTRUMENTS

It is possible to connect a D'Arsonval direct-current type instrument and a rectifier so as to measure a-c quantities. The rectifier is usually of the semiconductor type and is arranged in a

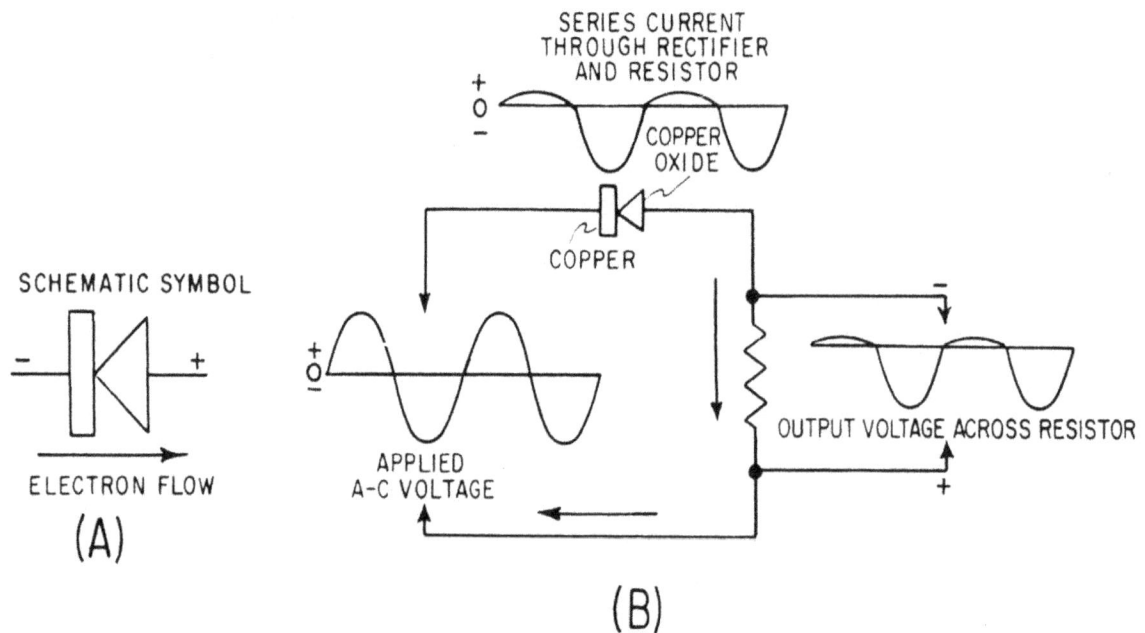

BE.266

Figure 15-20.—(A) Rectifier symbol; (B) waveforms in simple a-c circuit utilizing a rectifier.

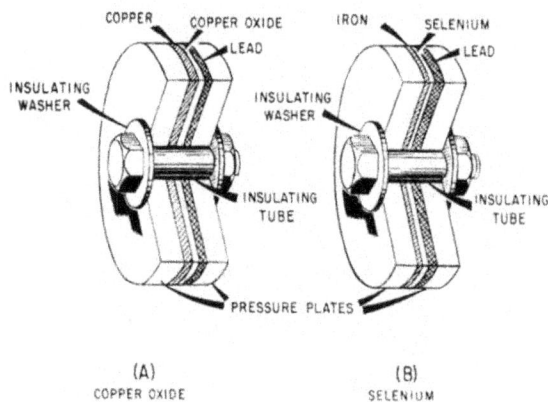

BE.267

Figure 15-21,—Metallic rectifier construction.

bridge circuit, as shown in figure 15-22. By the use of rectifiers in the bridge, it can be seen that current flow through the meter is always in one direction. When the voltage being measured has a waveform as shown in figure 15-22, the path of current flow will be from the lower input terminal through rectifier No. 3 through the instrument, and then through rectifier No. 2, thus

completing its path back to the source's upper terminal. The next half cycle of the input voltage (indicated by dotted sine wave) will cause the current to pass through rectifier No. 1, through the instrument, and through rectifier No. 4, completing its path back to the source.

This type of instrument generally is characterized by errors due to waveform and frequency. Allowances must be made, according to data furnished by the manufacturer. It is possible, however, to add corrective networks to the instrument which will make it practically free from error up to 100 kHz. An instrument of this type requires a current from the line of only about one milliampere for full scale deflection. It is widely used for a-c voltmeters, especially of the lower ranges.

There is some "aging" of the rectifier with a corresponding change in the calibration of the instrument. Because of this aging such instruments must be recalibrated from time to time. A rectifier type instrument reads the average value of the a-c quantity. However, because the EFFECTIVE, or root-meansquare (rms), values are more useful, a-c meters are generally calibrated to read rms values (1.11 times the average of the instantaneous values).

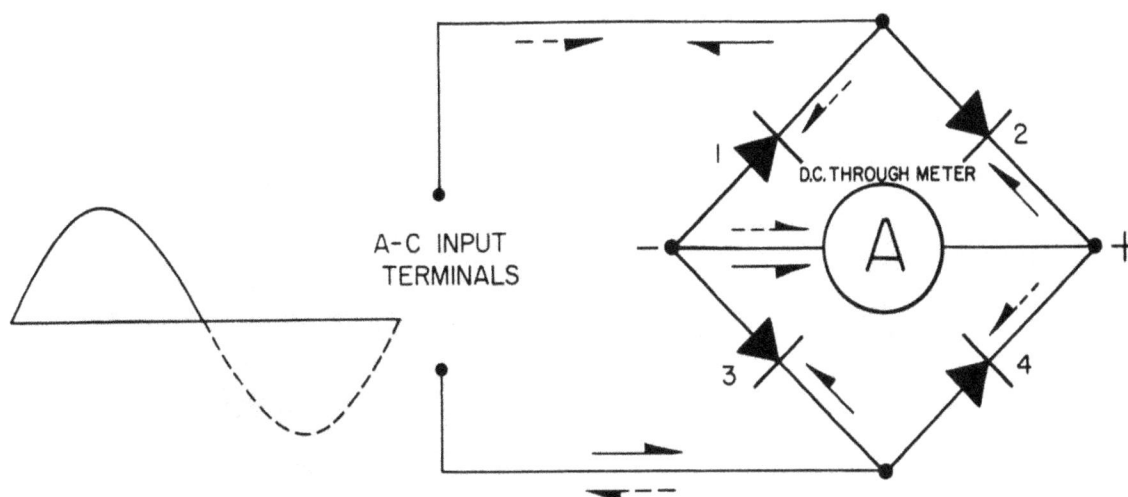

BE.268

Figure 15-22.—Simple connection of a full-wave, rectifier-type, a-c instrument.

## WATTMETER

Electric power is measured by means of a wattmeter. This instrument is of the electrodynamometer type. It consists of a pair of fixed coils, known as current coils, and a movable coil, known as the potential coil. (See fig. 15-23.) The fixed coils are made up of a few turns of comparatively large conductor. The potential coil consists of many turns of fine wire; it is mounted on a shaft, carried in jeweled bearings, so that it may turn inside the stationary coils. The movable coil carries a needle which moves over a suitably graduated scale. Spiral coil springs hold the needle to a zero position, as shown in figure 15-10.

The current coil (stationary coil) of the watt-meter is connected in series with the circuit (load), and the potential coil (movable coil) is connected across the line.

When line current flows through the current coil of a wattmeter, a field is set up around the coil. The strength of this field is proportional to the line current and in phase with it. The potential coil of the wattmeter generally has a high resistance resistor connected in series with it. This is for the purpose of making the potential-coil circuit of the meter as purely resistive as possible. As a result, current in the potential circuit is practically in phase with line voltage.

Therefore, when voltage is impressed on the potential circuit, current is proportional to and in phase with the line voltage.

The actuating force of a wattmeter is derived from the interaction of the field of its current coil and the field of its potential coil. The force acting on the movable coil at any instant (tending to turn it) is proportional to the product of the instantaneous values of line current and voltage.

The wattmeter consists of two circuits, either of which will be damaged if too much current is passed through them. This fact is to be especially emphasized in the case of wattmeters, because the reading of the instrument does not serve to tell the user that the coils are being overheated. If an ammeter or voltmeter is overloaded, the pointer will be indicating beyond the upper limit of its scale. In the wattmeter, both the current and potential circuits may be carrying such an overload that their insulation is burning, and yet the pointer may be only part way up the scale. This is because the position of the pointer depends upon the power factor of the circuit as well as upon the voltage and current. Thus, a low power-factor circuit will give a very low reading on the wattmeter even when the current and potential circuits are loaded to the maximum safe limit. This safe rating is generally given on the face of the instrument.

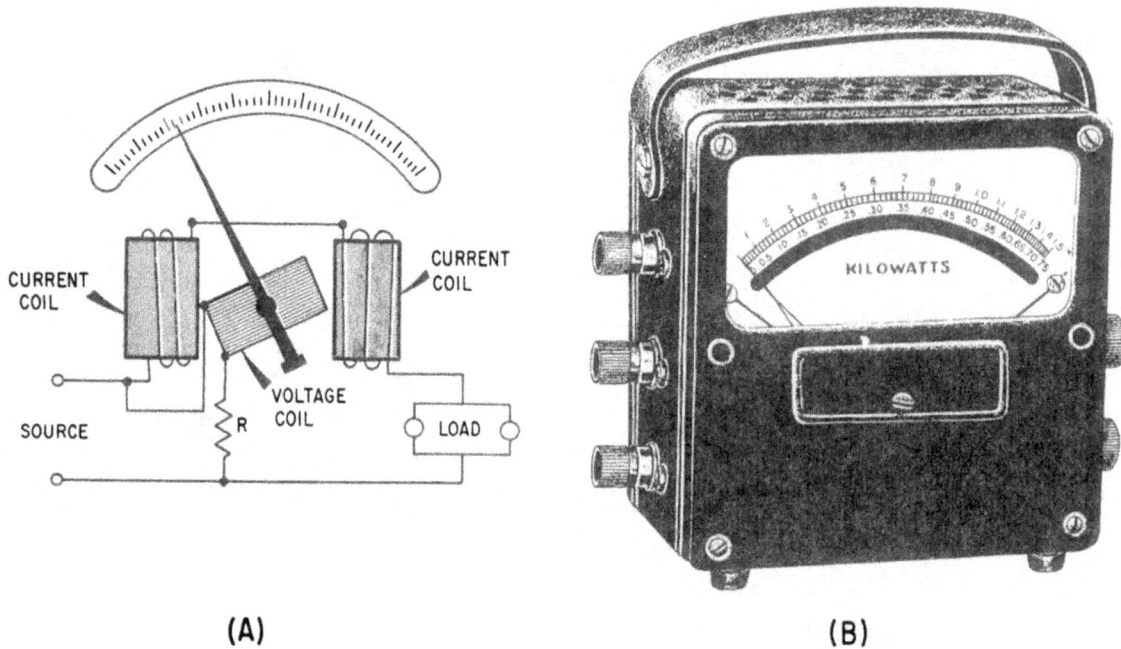

(A)                                    (B)

BE.269

Figure 15-23.—Simplified electrodynamometer wattmeter circuit.

A wattmeter is always distinctly rated, not in watts but in volts and amperes.

Figure 15-24 shows the proper way to connect a wattmeter in various circuits.

## WATT-HOUR METER

The watt-hour meter is an instrument for measuring energy. As energy is the product of power and time; the watt-hour meter must take into consideration both of these factors.

In principle, the watt-hour meter is a small motor whose instantaneous speed is proportional to the POWER passing through it. The total revolutions in a given time are proportional to the total ENERGY or watt-hours consumed during that time.

Referring to figure 15-25, the line is connected to two terminals on the left-hand side of the meter. The upper terminal is connected to two coils FF in series. These coils are wound with wire sufficiently large to carry the maximum current taken by the load. They are connected so that their magnetic fields add. The armature A rotates in the field produced between the coils FF. The other line wire goes directly to the load.

A shunt circuit is connected to the upper line terminal on the left-hand side. This circuit runs through coil F', to the brushes B, commutator C, through the armature A, and the resistor R, to the return line. (The resistor is omitted in some watt-hour meters.)

Since the load current is through coils FF and there is no iron in the circuit, the magnetic field produced by these coils is proportional to the LOAD CURRENT. The armature, in series with a resistance, is connected directly across the line. The current in the meter armature is proportional to the LINE VOLTAGE. Neglecting the small voltage drop in FF, the torque acting on the armature must be proportional to the product of the load current and the load voltage. In other words, it must be proportional to the power passing through the meter to the load.

If the meter is to register correctly, there must be a retarding torque acting on the moving element. This force is proportional to the speed of rotation of the moving element. To meet this condition, an aluminum disk D is mounted on the armature shaft. This disk rotates between the poles of two permanent magnets M and M. In cutting the field produced by these magnets, eddy currents are induced in the disk, retarding its

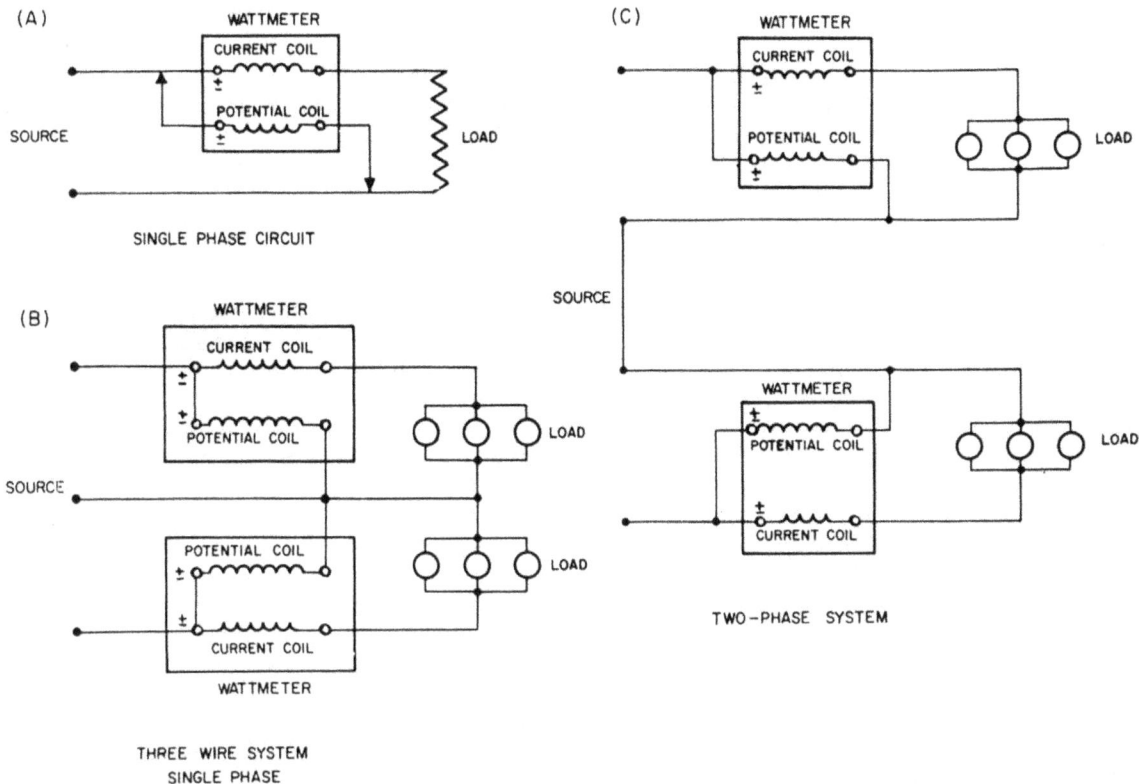

BE.270

Figure 15-24.—Wattmeter connected in various circuits.

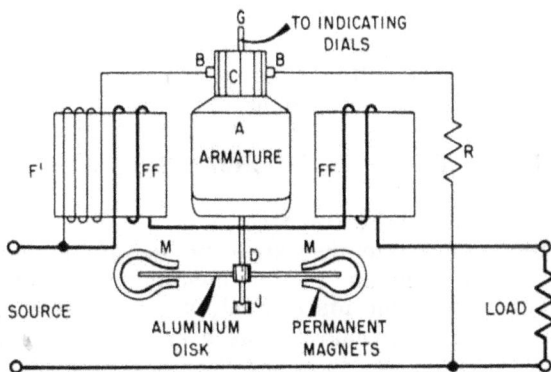

BE.271

Figure 15-25.—Internal connections of watt-hour
meter.

motion. The strength of these currents is proportional to the angular velocity of the disk. Since

they are acting in conjunction with a magnetic
field of constant strength, their retarding effect
is proportional to the speed of rotation.

Friction produced by the rotating element
cannot be entirely eliminated. Near the rated load
of the meter, the effect of the frictional torque is
practically negligible. But, at light loads, the
friction torque, which is nearly constant at all
loads, is a much greater percentage of the load
torque. Since the ordinary meter may operate at
light loads during a considerable portion of the
time, it is desirable that the error due to friction
be eliminated. This is accomplished by means of
coil F', which is connected in series with the
armature. Coil F' is connected so that its field
acts in the same direction as that of coils FF.
Therefore, it assists armature A to rotate. Since
it is connected in the shunt circuit, it acts continuously. The coil is movable and its position
can be adjusted so that the friction error is
eliminated.

To reduce friction and wear, the rotating element of the meter is made as light as practical. The element rests on a jewel bearing J. This bearing is a sapphire in the smaller types, and a diamond in the larger types of meters. The jewel is supported on a spring. A hardened steel pivot rests in the jewel. In time, the pivot becomes dulled and the jewel roughened, which increases friction and causes the meter to register low unless F' is readjusted. The moving element drives the clockwork of the meter through shaft G.

The following directions should be followed when reading the dials of a watt-hour meter. The meter, in this case, is a four-dial type.

The pointer on the right-hand dial (fig. 15-26) registers 1 kw.-hr. or 1,000 watt-hours for each division on the dial. A complete revolution of the hand on this dial will move the hand of the second dial one division and register 10 kw.-hr. or 10,000 watt-hours. A complete revolution of the hand of the second dial will move the third hand one division and register 100 kw.-hr. or 100,000 watt-hours, and so on.

Accordingly, you must read the hands from left to right, and add three zeros to the reading of the lowest dial to obtain the reading of the meter in watt-hours. The dial hands should always be read as indicating the figure which they have LAST PASSED, and not the one they are approaching.

Single-Phase Induction
Watt-Hour Meter

As in the case of series motors, the Thompson watt-hour meter may be used with alternating or direct current because both the armature and the field flux reverse at the same time, and the armature continues to rotate in the same direction. The induction watt-hour meter, however, has certain advantages over the Thompson watt-hour meter and is more commonly used to measure a-c power.

The SINGLE-PHASE INDUCTION WATT-HOUR METER includes a simple induction-drive motor consisting of an aluminum disk, moving magnetic field, drag magnets, current and potential coils, integrating dials, and associated gears. A simplified sketch of an induction watt-hour meter is shown in figure 15-27 (A). The potential coil connected across the load is composed of many turns of relatively small wire. It is wound on one leg of the laminated magnetic circuit. Because of its many turns, the potential

coil has high impedance and high inductance, and therefore, the current through it lags the applied voltage by nearly 90°. The two current coils connected in series with the load are composed of a few turns of heavy wire. They are wound on two legs of the laminated magnetic circuit. Because of the few turns, the current coils have low inductance and low impedance.

The arrangement of the potential coil, the aluminum disk, the current coils, and one of the drag magnets is shown in the phantom view (fig. 15-27 (B)).

The rotating aluminum disk is the moving member that causes the gears to turn and the dials to indicate the amount of energy passed through the meter. This rotation of the aluminum disk is accomplished by eddy currents that are established by the current coils and the potential coil. The speed of rotation of the aluminum disk is proportional to the true power supplied through the line to the load. The total energy supplied to the load is proportional to the number of revolutions of the disk during a given period of time.

A small copper shading disk (not shown in the figure) is placed under a portion of the potential pole face. It is in an adjustable mounting and is used to develop a torque in the disk to counteract static friction. The disk has the effect of a shading pole and provides a light-load adjustment for the meter.

The two drag magnets supply the counter torque against which the aluminum disk acts when it turns. The drag is increased (speed of motor is reduced) by moving the magnets toward the edge of the disk. Conversely, the drag is decreased and the speed is increased by moving the magnets toward the center of the disk. Adjustment of the drag magnets will only be made by an authorized instrument and meter technician.

INSTRUMENT TRANSFORMERS

It is not usually practical to connect instruments and meters directly to high-voltage circuits. Unless the high-voltage circuit is grounded at the instrument, a dangerously high potential-to-ground voltage may exist at the instrument or switch board. Further, instruments become inaccurate when connected directly to a high voltage, because of the electrostatic forces that act on the indicating element. Specially designed instruments may be constructed so that they can be connected directly to high-voltage circuits, but these instruments are usually expensive.

Figure 15-26.—Reading a watt-hour meter.

By means of instrument transformers, instruments may be entirely insulated from the high-voltage circuit and yet indicate accurately the current, voltage, and power in the circuit. Low-voltage instruments having standard current and voltage ranges may be used for all high-voltage circuits, irrespective of the voltage and current ratings of the circuits, if instrument transformers are utilized.

Potential Transformers

Potential transformers do not differ materially from the constant-potential power

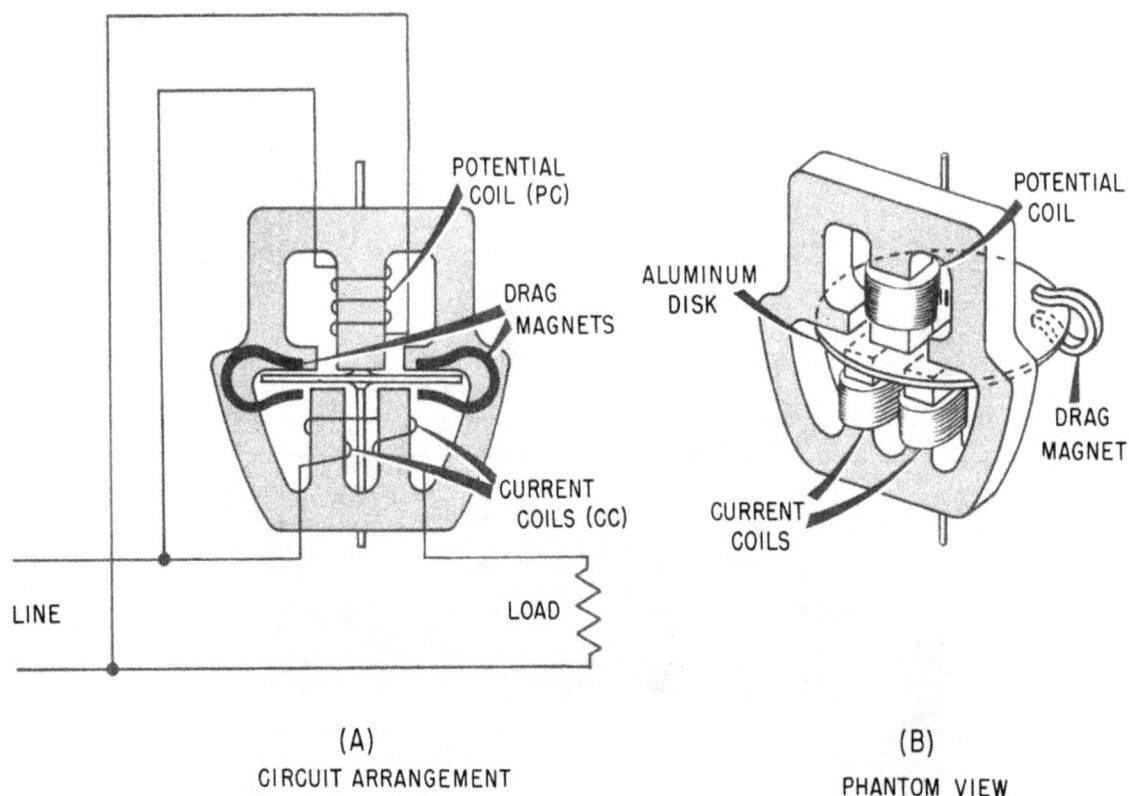

(A)
CIRCUIT ARRANGEMENT

(B)
PHANTOM VIEW

BE.273

Figure 15-27.—Simplified sketch of an induction watt-hour meter.

transformers. Constant-potential power transformers are discussed in chapter 16. An exception is that their power rating is small, and they are designed for minimum ratio and phase angle error. At unity power factor, the impedance drop through the transformer, from no-load to rated-load, should not be greater than 1 percent. For taking measurements below 5,000 volts, potential transformers are usually of the dry type; between 5,000 and 13,800 volts they may be either the dry type of oil immersed, and above 13,800 volts they are oil-immersed.

Since only instruments, meters, and sometimes indicator lights are ordinarily connected to the secondaries of potential transformers, they have ratings from 40 to 500 watts. For primary voltages of 34,500 volts and higher, the secondaries are rated at 115 volts. For primary voltages less than 34,500 volts, the secondaries are rated at 120 volts. For example, a 14,400-volt potential transformer would have a ratio of

$$\frac{14,400}{120} = \frac{120}{1}$$

The ratio of turns may vary about 1 percent from this value to allow for the transformer impedance drop under load. Figure 15-28 shows a simple connection for measuring voltage in a 14,000-volt circuit by means of a potential transformer.

The secondary should always be grounded at one point to eliminate static electricity from the instrument and to insure the safety of the operator.

Current Transformers

To avoid connecting instruments directly into high-voltage lines, current transformers are used. In addition to insulating from high voltage, they step down the line current in a known ratio.

277

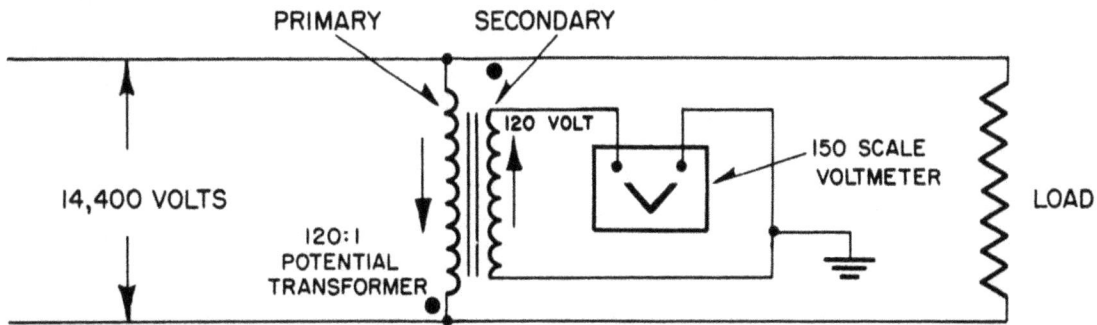

BE.274

Figure 15-28.—Connections of a potential transformer to a 14,400-volt circuit.

This permits the use of a lower-range ammeter than would be required if the instrument were connected directly into the primary line.

The current, or series, transformer has a primary winding, usually of a few turns, wound on a core and connected in series with the line. Figure 15-29 shows a simple connection for measuring current in a 14,400-volt circuit by means of a current transformer.

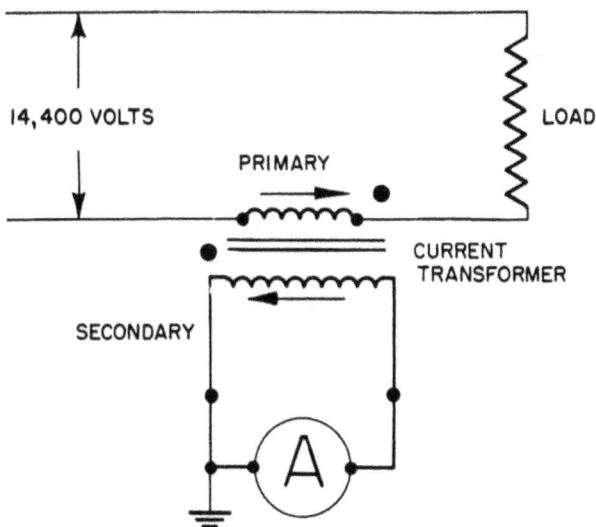

BE.275

Figure 15-29.—Connections of a current transformer in a 14,400-volt circuit to measure current.

The secondary windings of practically all current transformers are rated at 5 amperes regardless of the primary current rating. For example, a 2,000-ampere current transformer has a ratio of 400 to 1, and 50-ampere transformer has a ratio of 10 to 1.

The insulation between the primary and the secondary of a current transformer must be sufficient to withstand full circuit voltage. The current transformer differs from the ordinary constant-potential transformer in that its primary current is determined entirely by the load on the system and not by its own secondary load. If its secondary becomes open-circuited, a high voltage will exist across the secondary, because the large ratio of secondary to primary turns causes the transformer to act as a step-up transformer. Also, since the effects of the counter-ampere-turns of the secondary no longer exist, the flux in the core will depend on the total primary ampere-turns acting alone. This causes a large increase in the flux, producing excessive core loss and heating, as well as a dangerously high voltage across the secondary terminals. THEREFORE, THE SECONDARY OF A CURRENT TRANSFORMER SHOULD NOT BE OPEN-CIRCUITED UNDER ANY CIRCUMSTANCES.

Figure 15-30 shows the method of connecting a complete instrument load, through instrument transformers, to a high-voltage line. The load on the instrument transformers includes an ammeter A, a voltmeter V, a wattmeter, and a watt-hour meter.

Polarity Marking

Instruments, meters, and relays must be connected so that the correct phase relations exist between their potential and current circuits. In instrument transformers it is important that the

278

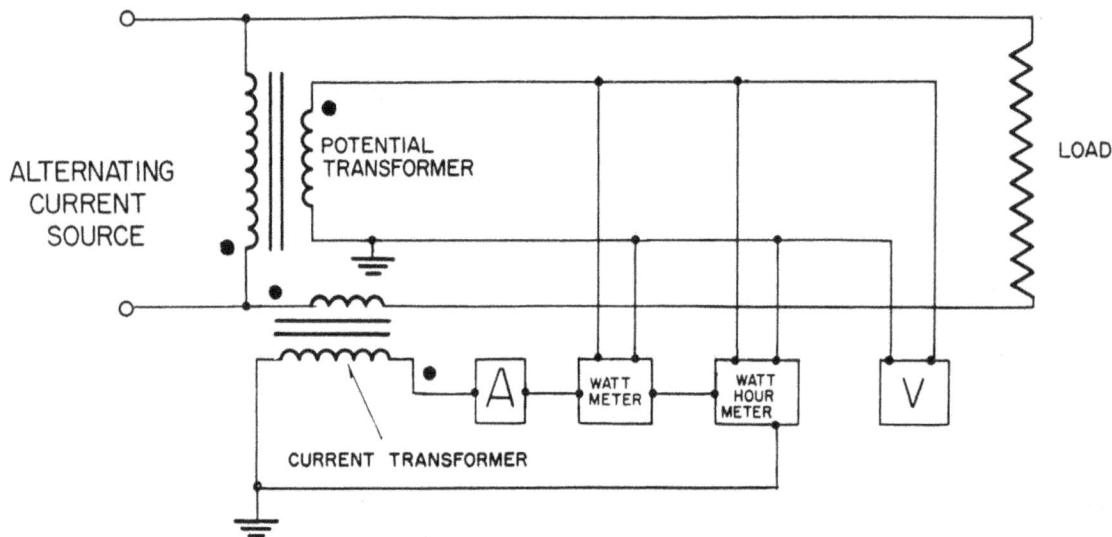

BE. 276

Figure 15-30.—Typical connections of instrument transformers and instruments for single-phase measurements.

relation of the instantaneous polarities of the secondary terminals to the primary terminals be known. It has become standard to designate or mark the primary terminals and the secondary terminals that have the same instantaneous polarity.

In a wiring diagram, primary and secondary terminals having the same instantaneous polarity are marked with a dot. This is shown in figure 15-30.

Hook-On Type Voltammeter

The hook-on a-c ammeter consists essentially of a current transformer with a split core and a rectifier-type instrument connected to the secondary. The primary of the current transformer is the conductor through which the current to be measured flows. The split core permits the instrument to be "hooked on" the conductor without disconnecting it. Therefore the current flowing through the conductor may be measured safely with a minimum of inconvenience, as shown in figure 15-31.

The instrument is usually constructed so that voltages also may be measured. However, in order to read voltage, the meter switch must be set to VOLTS, and leads must be connected from the voltage terminals on the meter to the terminals across which the voltage is to be measured.

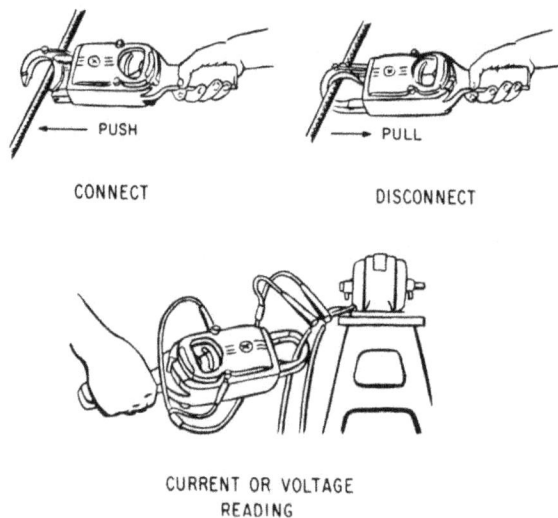

BE. 277

Figure 15-31.—Hook-on type voltammeter.

A-C BRIDGES

D-c bridges, called Wheatstone bridges, are covered in chapter 6. Before continuing this section on a-c bridges you should review bridge principles given in chapter 6. Notice particularly

279

the ratio method of determining, an unknown value that forms one leg of a bridge. Keep in mind that current varies inversely as resistance—that is, as resistance increases, current decreases.

The same circuit may be used with a fixed-frequency a-c voltage applied, in place of a direct voltage. Figure 15-32 shows bridge circuits using alternating current. Circuit (A) is used to determine an unknown capacitance $C_x$. Circuit (B) is used to determine an unknown inductance $L_x$.

## CAPACITANCE BRIDGE

Circuit (A) will be discussed first. You will recall that a capacitor has a certain opposition to current flow. Also, the larger its capacitance the less its opposition becomes. It follows that if an UNKNOWN capacitor is to be measured, there must be a KNOWN capacitor in the bridge with which the unknown capacitor may be compared, as shown in figure 15-32 (A). $C_s$ is the capacitor whose value is known. The resistance of $C_s$ is shown as an equivalent resistance, $R_s$. This capacitor (with its series resistance) forms one leg of the bridge. The unknown capacitor, $C_x$, along with its unknown resistance, $R_x$, forms another leg of the bridge.

The ratio resistors, $R_1$ and $R_2$, are potentiometers that can be varied to bring the circuit into balance—that is, cause equal currents to flow in the two sides of the bridge. When the circuit is in balance, there will be little or no current flow through the indicator (headphones, or an a-c milliammeter). $R_1$ and $R_2$ are adjusted by means of accurately calibrated dials. When they are adjusted so that minimum hum is heard in the headphones, the bridge is in balance, and the unknown capacity may be calculated by the formula:

$$\frac{R_1}{R_2} = \frac{C_x}{C_s} \quad \text{or} \quad C_x = \frac{R_1}{R_2} \times C_s$$

Notice that this is an inverse proportion because current varies inversely with resistance and directly with capacitance. If the ratio of reactance to resistance is the same in the legs containing capacitance, the following direct proportion exists between the resistive components:

$$\frac{R_1}{R_2} = \frac{R_s}{R_x}$$

or

$$R_x = \frac{R_2}{R_1} \times R_s$$

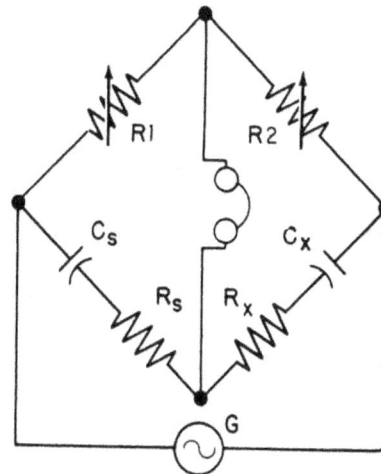

CAPACITANCE BRIDGE

(A)

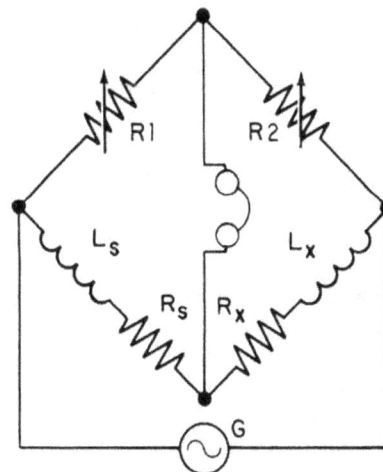

INDUCTANCE BRIDGE

(B)

BE.278

Figure 15-32.—A-c bridge circuits.

Thus, the unknown resistance and capacitance, $R_x$, and $C_x$, can be estimated in terms of the known resistance, $R_1$, $R_2$, and $R_s$, and the known capacitance, $C_s$.

## INDUCTANCE BRIDGE

The value of an unknown inductance may be determined in the same manner as an unknown capacitance, as shown in figure 15-32 (B). You will recall that an inductor also has a certain opposition to current flow. Also, the larger the inductance, the greater becomes its opposition. To measure an UNKNOWN inductance ($L_x$) in a bridge circuit, there must be an inductor of KNOWN VALUE ($L_s$). This known inductor is compared with the unknown inductor.

The ratio resistors, $R_1$ and $R_2$, are potentiometers that can be varied to bring the circuit into balance. When there is minimum hum in the headphones, the bridge is said to be in balance, and the unknown inductance may be calculated by the formula:

$$\frac{R_1}{R_2} = \frac{L_s}{L_x}$$

or

$$L_x = \frac{R_2}{R_1} L_s$$

and

$$\frac{R_1}{R_2} = \frac{R_s}{R_x}$$

or

$$R_x = \frac{R_2}{R_1} R_s$$

## IMPEDANCE BRIDGE

Figure 15-33 is the diagram of an impedance bridge. It consists of four impedances with a source of voltage applied across one diagonal of a square and a detector or meter (M) across the other diagonal. The bridge is balance when the meter reads zero, indicating no current flow through the meter.

thus

$$Z_1 \times Z_3 = Z_2 \times Z_4$$

The impedance bridge is used to measure impedances of a bridge that may contain inductance, capacitance, and resistance.

## FREQUENCY METERS

All alternating voltage sources are generated at a set frequency or range of frequencies. A frequency meter provides a means of measuring this frequency. Two common types of frequency meters are the vibrating-reed frequency meter and the moving-disk frequency meter—both are discussed.

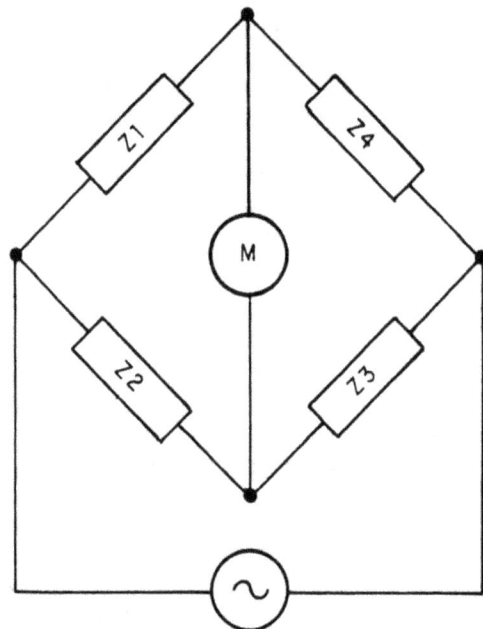

BE. 279

Figure 15-33.—Impedance bridge.

## VIBRATING-REED FREQUENCY METER

The vibrating-reed type of frequency meter is one of the simplest devices for indicating the frequency of an a-c source. A simplified diagram of one type of vibrating-reed frequency meter is shown in figure 15-34.

The current whose frequency is to be measured flows through the coil and exerts maximum attraction on the soft-iron armature TWICE during each cycle (fig. 15-34 (A)). The armature is attached to the bar, which is mounted on a flexible support. Reeds of suitable dimensions to have natural vibration frequencies of 110, 112, 114, and so forth, up to 130 hertz per second are mounted on the bar (fig. 15-34 (B)). The reed having a frequency of 110 hertz is marked "55" hertz, the one having a frequency of 112 hertz is marked "56" hertz, the one having a frequency of 120 hertz is marked "60" hertz, and so forth.

When the coil is energized with a current having a frequency between 55 and 65 hertz, all the reeds are vibrated slightly; but the reed having a natural frequency closest to that of the energizing current (whose frequency is to be measured) vibrates through a larger amplitude.

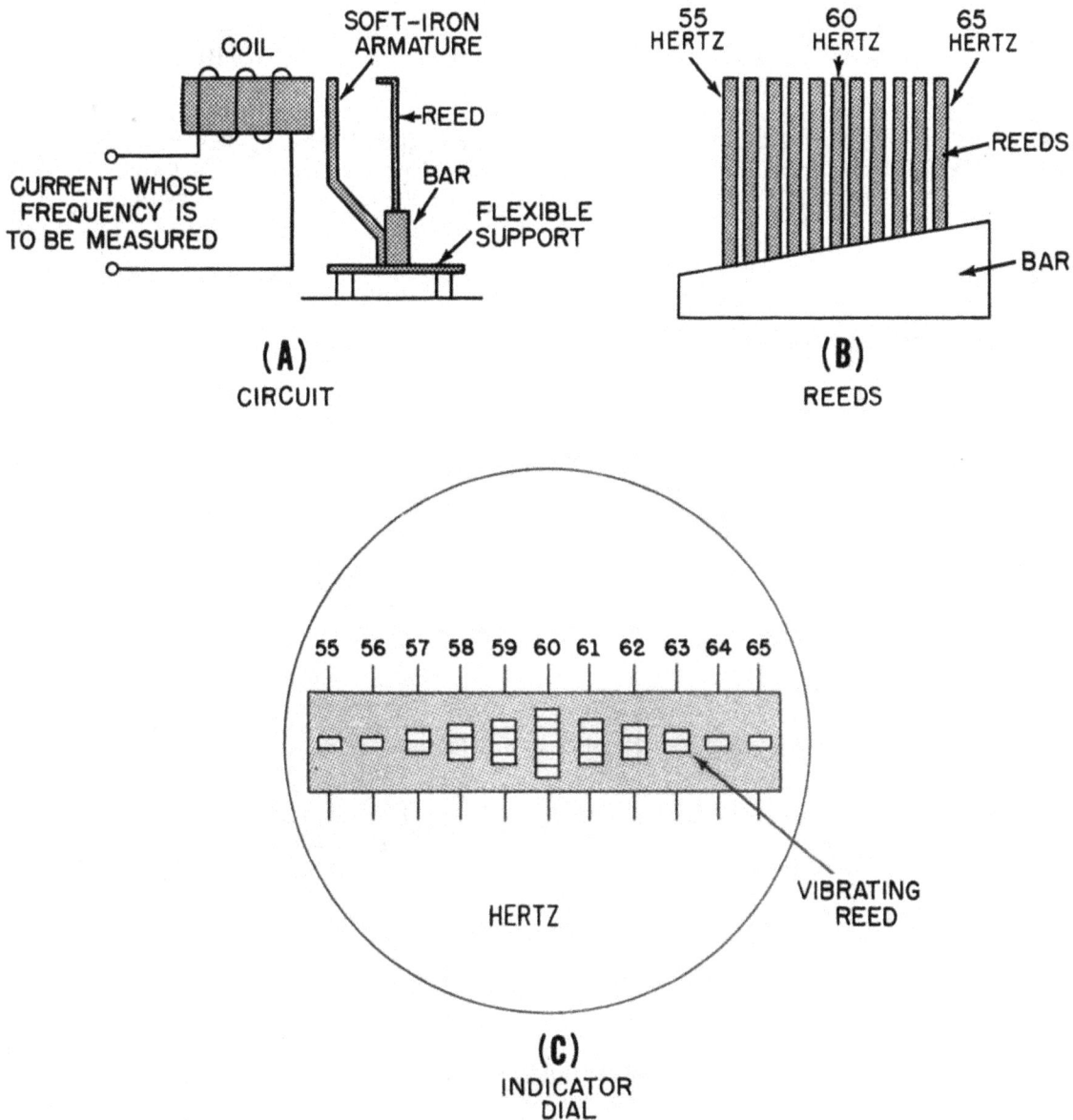

Figure 15-34.—Simplified diagram of a vibrating-reed frequency meter.

BE. 280

The frequency is read from the scale value opposite the reed having the greatest amplitude of vibration.

In some instruments the reeds are the same lengths, but are weighted by different amounts at the top so that they will have different natural rates of vibration.

An end view of the reeds is shown in the indicator dial of figure 15-34 (C). If the energizing current has a frequency of 60 hertz per second, the reed marked "60" hertz will vibrate the greatest amount, as shown.

MOVING-DISK FREQUENCY METER

A moving-disk frequency meter is shown in figure 15-35 (A). Each of the two-pole fields (fig. 15-35 (B)) has a short-circuited turn (shading coil) on one side of the pole face and a magnetizing coil around the associated magnetic circuit. As in a shaded-pole motor, the magnetic field tends to produce rotation toward the shorted turn (A to B). One coil tends to turn the disk clockwise, and the other, counterclockwise. Magnetizing coil A is connected in series with a large value of resistance. Coil B is connected in series with a large inductance and the two circuits are supplied in parallel by the source.

more than the desired amount, the left half of the disk is mounted so that when motion occurs, the same amount of disk area will always be between the poles of coil A. Therefore, the force produced by coil A to rotate the disk is constant for a constant applied voltage. The right half of the disk is offset, as shown in the figure. When the disk rotates clockwise, an increasing area will

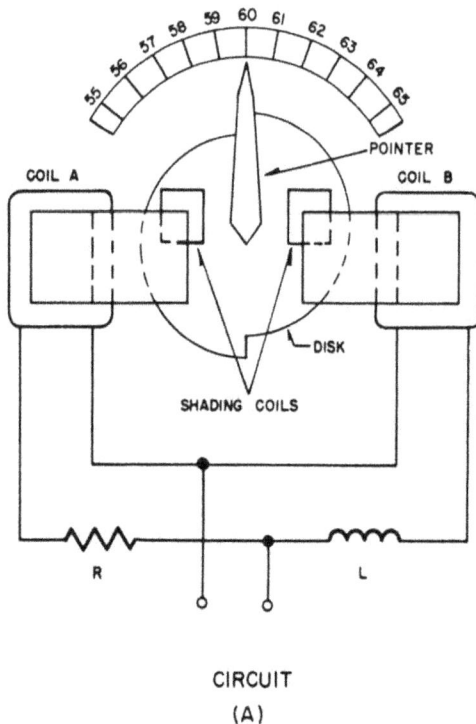

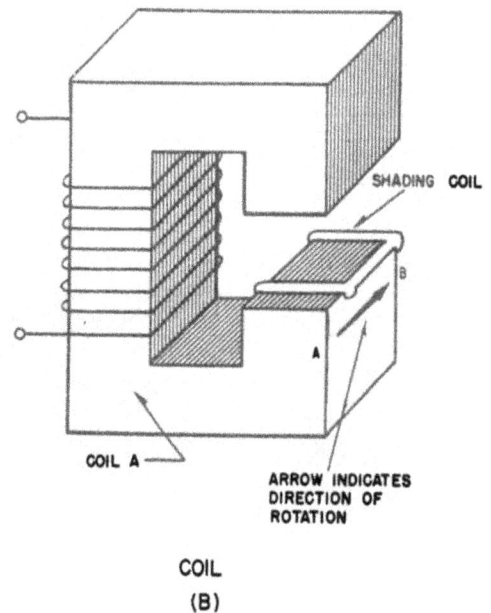

Figure 15-35.—Simplified diagram of a moving-disk frequency meter.

For a given voltage, the current through coil A is practically constant. However, the current through coil B varies inversely with the frequency. At a higher frequency the inductive reactance is greater and the current through coil B is less; the reverse is true at a lower frequency. The disk turns in the direction determined by the stronger coil.

A perfectly circular disk would tend to turn continuously. This is not desirable, and accordingly the disk is constructed so that it will turn only a certain amount clockwise or counterclockwise about the center position, which is commonly marked "60" hertz on commercial equipment. To prevent the disk from turning

come between the poles of coil B; when it rotates counterclockwise, a decreasing area will come between the poles of coil B. The greater the area between the poles, the greater will be the disk current and the force tending to turn the disk.

If the frequency applied to the frequency meter should decrease, the reactance offered by L would decrease and the field produced by coil B would increase; the field produced by coil A would remain the same. Thus, the torque produced by coil B would tend to move the disk and the pointer counterclockwise until the area between the poles was reduced sufficiently to make the two torques equal. The scale is calibrated to indicate the correct frequency.

If the frequency should increase, the reactance offered by L would increase and the field produced by coil B would decrease; the field produced by coil A would remain the same. Thus, the decreasing torque produced by coil B would permit a greater disk area to move under the coil, and therefore the disk and the pointer would move clockwise until the two torques were balanced.

If the frequency is constant and the voltage is changed, the currents in the two coils—and therefore the opposing torques—change by the same amount. Therefore, the indication of the instrument is not affected by a change in voltage.

## SINGLE-PHASE POWER-FACTOR METER

The power factor of a circuit is the ratio of the true power to apparent power. It is also equal to the cosine of the phase angle between the circuit current and voltage. A pure resistance has a power factor of unity (the current and voltage are in phase and the true power and the apparent power are the same); a pure inductance has a power factor of zero (current lags the voltage by $90°$); and a pure capacitance has a power factor of zero (current leads the voltage by $90°$).

The power factor of a circuit may be determined by the use of a wattmeter, a voltmeter, and an ammeter—that is, the power factor may be determined by dividing the wattmeter reading by the product of the voltmeter and ammeter readings. This is inconvenient, however, and instruments have been developed that indicate continuously the power factor and at the same time indicate whether the current is leading or lagging the voltage. An instrument that indicates these values is called a POWER-FACTOR METER.

## CROSSED-COIL POWER-FACTOR METER

One type of power-factor meter is shown schematically in figure 15-36. The instrument consists of movable potential coils A and B, fixed at right angles to each other, and stationary current coil C. Coils A and B are pivoted, and the assembly, together with the attached pointer, is free to move through an angle of approximately $90°$. Coil A is in series with inductor L, and the combination is connected across the line. Coil B is in series with noninductive resistor R, and the combination is also connected across the line.

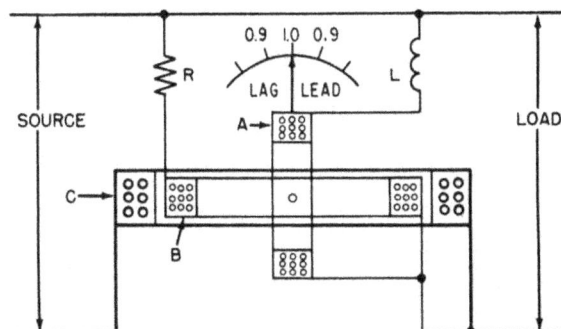

BE. 282

Figure 15-36.—Simplified diagram of a crossed-coil power-factor meter.

Circuit continuity to the coils is provided by three spiral springs (not shown in the figure) that exert negligible restraining force on the coils. Therefore, when no current flows through the coils, the pointer may come to rest at any position on the dial. Coil C (not drawn to scale) is connected in series with the line. In many switchboard installations the coils are energized by instrument transformers, in which case the currents are proportional to line values but not equal to them.

The current in coil B is in phase with the line voltage. The current in coil A lags the line voltage by $90°$. When the line current is in phase with the line voltage, the currents in B and C are in phase and a torque is exerted between them that alines their axes so that the pointer indicates unity power factor. The average torque between A and C is zero because these currents are $90°$ out of phase when the line power factor is unity.

When the current in coil C lags the line voltage—for example, by $45°$ —the currents in coil A and coil B both will be out of phase with the current in C. The current in A lags the current in C by 45° and the current in B leads the current in C by 45°. The flux around coil C will therefore react with the resultant of the fluxes around coils A and B, which is in phase with the current of C, and the pointer will be moved to an intermediate position ($45°$) between zero power factor and unity power factor. In most power-factor meters, lagging current causes the pointer to move to the left of the central position (marked

284

"1" on the scale) and a leading current causes the pointer to move to the right of the central position.

## MOVING IRON-VANE POWER-FACTOR METER

A diagram of a moving iron-vane type of power-factor meter is shown in figure 15-37. In part (A) of the figure, potential coil A in series with resistor R comprises a resistive circuit, which is connected across the line. Potential coil B in series with inductor L comprises an inductive circuit, which is also connected across the line. Current coil C, having a few turns of large wire, is connected in series with the line. All coils are fixed in the positions shown, and only the iron vanes are free to move. The current in B lags the line voltage by 90°. The current in A is in phase with the line voltage. Hence, the current in B lags the current in A by 90° and, since the axes of A and B are displaced 90°, a magnetic revolving field is established by these coils when they are energized. The iron vanes are free to rotate about the axis

of coil C. These vanes are magnetized alternately north and south by the line current flowing in coil C.

None of the moving parts carry current, and therefore springs are not needed. The movement is free to rotate 360°, but is prevented from rotating with the rotating field by an aluminum disk that rotates in a field produced by drag magnets (disk and magnets not shown).

In figure 15-37 (B), a two-pole rotating field is assumed to revolve at synchronous speed in a clockwise direction. The vanes are attracted or repelled by this field depending on the instantaneous direction of the resultant flux produced by A and B and on the instantaneous polarity of the iron vanes as determined by the instantaneous direction of the current through C. The vanes will assume a position out of alinement with the field (position of minimum torque) when the line power factor is unity.

If, at the instant shown in figure 15-37 (B), the current in coil C is maximum and the resultant flux produced by coils A and B passes through the movable element at right angles

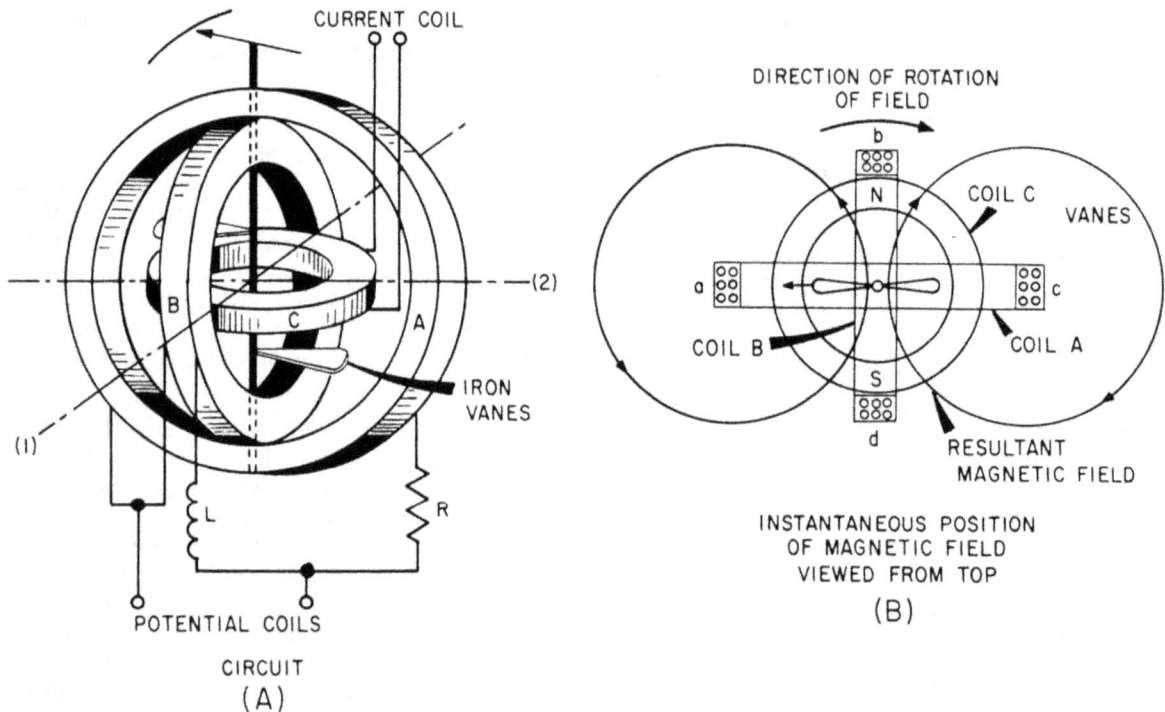

Figure 15-37.—Simplified circuit of a moving iron-vane power-factor meter.

BE.283

to the vanes (and has no effect on them), the pointer will indicate unity power factor, as shown. Ninety degrees later, the resultant north-south field will pass through the iron vanes in alinement with them. At this instant the current in magnetizing coil C is zero and the torque on the vanes is again a minimum. If the current through C leads the line voltage by 45° (line power factor 70% leading), the vanes will move 45° to a new position of minimum torque. The current in C will become maximum earlier in the cycle and the revolving field will occupy a new position, again out of alinement with the vanes at this instant.

If the line current lags the line voltage, the pointer will come to rest on the opposite side of the unity power-factor mark.

# CHAPTER 16

# ALTERNATING-CURRENT GENERATORS
# AND TRANSFORMERS

## A-C GENERATORS

Most electric power utilized today is generated by alternating-current generators. A-c generators are also finding increased use in aircraft and automobiles.

A-c generators are made in many different sizes, depending on their intended use. For example, any one of the generators at Boulder Dam can produce millions of volt-amperes, while generators used on aircraft produce only a few thousand volt-amperes.

Regardless of size, however, all generators operate on the same basic principle—a magnetic field cutting through conductors, or conductors passing through a magnetic field. Thus, all generators have at least two distinct sets of conductors. They are (1) a group of conductors in which the output voltage is generated, and (2) a second group of conductors through which direct current is passed to obtain an electromagnetic field of fixed polarity. The conductors in which the output voltage is generated are always referred to as the armature windings. The conductors in which the electromagnetic field originates are always referred to as the field windings.

In addition to the armature and field, there must also be motion between the two. To provide this, a-c generators are built in two major assemblies, the stator and the rotor. The rotor rotates inside the stator. It may be driven by any one of a number of commonly used power sources, such as gas or hydraulic turbines, electric motors, and steam or internal-combustion engines.

## TYPES OF A-C GENERATORS

There are various types of alternating-current generators utilized today. However, they all perform the same basic function. The types discussed in the following paragraphs are typical of the more predominant ones encountered in electrical equipment.

### Revolving Armature

In the revolving-armature a-c generator, the stator provides a stationary electromagnetic field. The rotor, acting as the armature, revolves in the field, cutting the lines of force, producing the desired output voltage. In this generator, the armature output is taken through sliprings and thus retains its alternating characteristic.

For a number of reasons, the revolving-armature a-c generator is seldom used. Its primary limitation is the fact that its output power is conducted through sliding contacts (sliprings and brushes). These contacts are subject to frictional wear and sparking. In addition, they are exposed, and thus liable to arc-over at high voltages. Consequently, revolving-armature generators are limited to low-power, low-voltage applications.

### Revolving Field

The revolving-field a-c generator (fig. 16-1) is by far the most widely used type. In this type of generator, direct current from a separate source is passed through windings on the rotor by means of sliprings and brushes. This maintains a rotating electromagnetic field of fixed polarity (similar to a rotating bar magnet). The rotating magnetic field, following the rotor, extends outward and cuts through the armature windings imbedded in the surrounding stator. As the rotor turns, alternating voltages are induced in the windings since magnetic fields of first one polarity and then the other cut through them. Since the output power is taken from stationary windings, the output may be connected through fixed terminals T1 and T2 in figure 16-1. This is advantageous, in that there are no sliding contacts, and the whole output circuit is continuously insulated, thus minimizing the danger of arc-over.

Sliprings and brushes are adequate for the d-c field supply because the power level in the field is much smaller than in the armature circuit.

## RATING OF A-C GENERATORS

The rating of an a-c generator pertains to the load it is capable of supplying. The normal-load rating is the load it can carry continuously. Its overload rating is the above-normal load which it can carry for specified lengths of time only. The load rating of a particular generator is determined by the internal heat it can withstand. Since heating is caused mainly by current flow, the generator's rating is identified very closely with its current capacity.

The maximum current that can be supplied by an a-c generator depends upon (1) the maximum heating loss ($I^2R$ power loss) that can be sustained in the armature and (2) the maximum heating loss that can be sustained in the field. The armature current varies with the load. This action is similar to that of d-c generators. In a-c generators, however, lagging power-factor loads tend to demagnetize the field and

terminal voltage is maintained only by increasing the d-c field current. Therefore, a-c generators are rated in terms of armature load current and voltage output, or kilovolt-ampere (kva) output, at a specified frequency and power factor. The specified power factor is usually 80 percent lagging. For example, a single-phase a-c generator designed to deliver 100 amperes at 1,000 volts is rated at 100 kva. This machine would supply a 100-kw load at unit power factor or an 80-kw load at 80 percent power factor. If the a-c generator supplied a 100-kva load at 20 percent power factor, the required increase in d-c field current needed to maintain the desired terminal voltage would cause excessive heating in the field.

Basic Functions of
Generator Parts

A typical rotating-field a-c generator consists of an a-c generator and a smaller d-c generator built into a single unit. The output of the a-c generator section supplies alternating current to the load for which the generator was designed. The d-c generator's only purpose is to supply the direct current required to maintain the a-c generator field. This d-c generator is referred to as the exciter. A typical a-c generator is shown in figure 16-2 (A): figure 16-2 (B) is a simplified schematic of the generator.

Any rotary generator requires a prime moving force to rotate the a-c field and exciter armature. This rotary force is transmitted to the generator through the rotor drive shaft (fig. 16-2 (A)) and is usually furnished by a combustion engine, turbine, or electric motor. The exciter shunt field (2, fig. 16-2 (B)) creates an area of intense magnetic flux between its poles. When the exciter armature (3) is rotated in the exciter field flux, voltage is induced into the exciter armature windings. The exciter output commutator and brushes (4) connects the exciter output directly to the a-c generator field input sliprings and brushes (5). Since these sliprings, rather than a commutator, are used to supply current through the a-c generator field (6), current always flows in one direction through these windings. Thus, a fixed polarity magnetic field is maintained at all times in the a-c generator field windings. When the a-c generator field is rotated, its magnetic flux is passed through and across the a-c generator armature windings (7). Remember, a voltage is

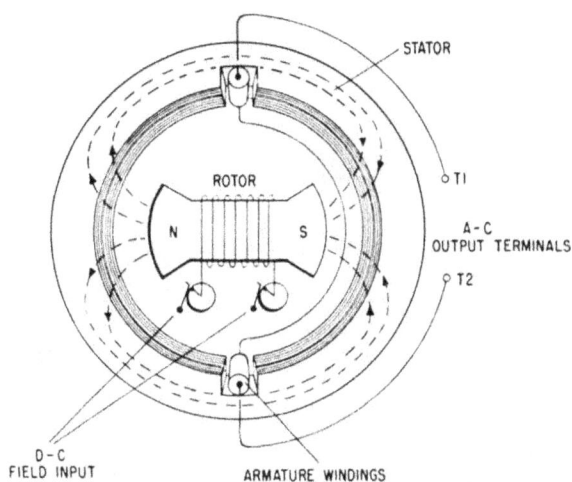

B. 284

Figure 16-1.—Essential parts of a rotating-field
a-c generator.

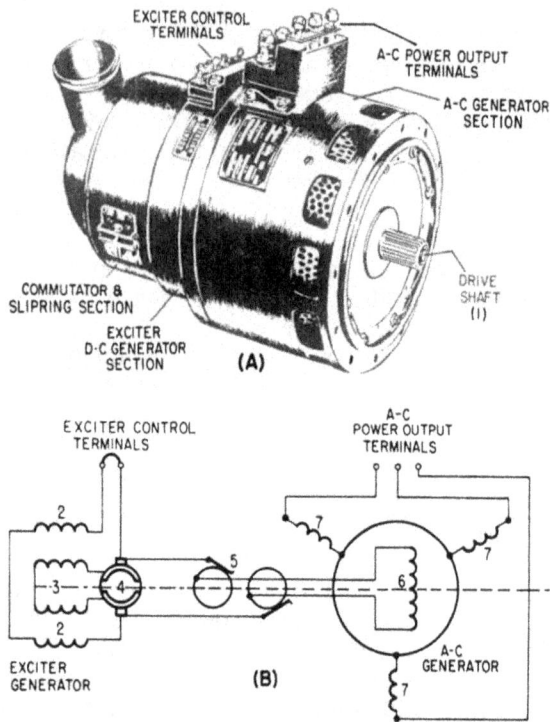

BE.285

Figure 16-2.—A-c generator and schematic.

induced into a conductor if it is stationary and a magnetic field is passed across the conductor, the same as if the field is stationary and the conductor is moved. The alternating voltage induced in the a-c generator armature windings is connected through fixed terminals to the a-c load.

Construction

A-c generators may be divided into three classes according to the type of prime mover. These classes are as follows:

1. Low-speed engine-driven.
2. High-speed turbine-driven.
3. High-speed engine-driven.

The stator, or armature, of the revolving-field a-c generator is built up from steel punchings, or laminations. The laminations of an a-c generator stator form a steel ring that is keyed or bolted to the inside circumference of a steel frame. The inner surface of the laminated ring has slots in which the stator winding is placed.

The low-speed engine-driven a-c generator (fig. 16-3) has a large diameter revolving field with many poles, and a stationary armature

relatively short in axial length. The stator (fig. 16-3 (A)) contains the armature windings, and the rotor (fig. 16-3 (B)) consists of salient poles, on which are mounted the d-c field windings. The exciter armature is the smaller unit shown in the foreground and mounted on an extension of the a-c generator shaft.

The high-speed turbine-driven a-c generator (fig. 16-4) is connected either directly or through gears to a steam turbine. The enclosed metal structure is a part of a forced ventilation system that carries away the heat by circulation of the air through the stator (fig. 16-4 (A)) and rotor (fig. 16-4 (B)). The exciter is a separate unit (not shown). The enclosed stator directs the paths of the circulating air-cooling currents and also reduces windage noise.

The high-speed engine-driven generator (fig. 16-5) may be driven directly by the engine, a hydraulic constant-speed drive, or an air turbine.

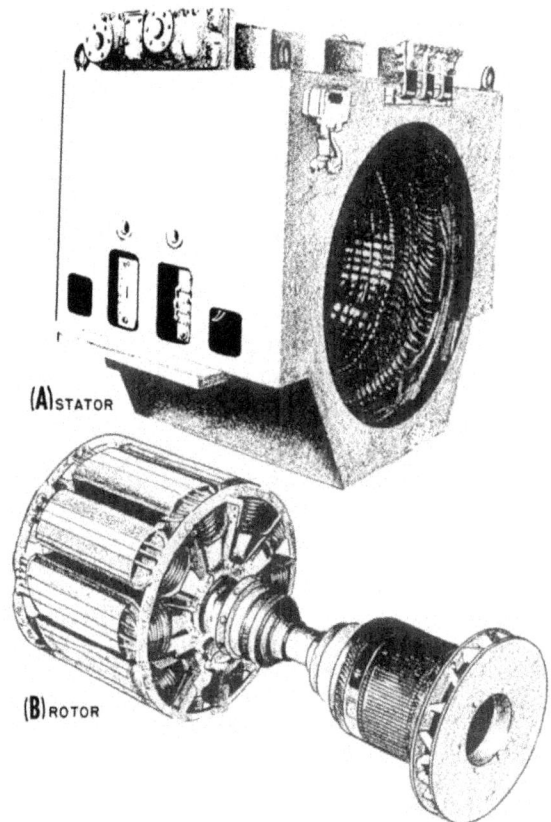

BE.286

Figure 16-3.—Low-speed engine-driven a-c generator.

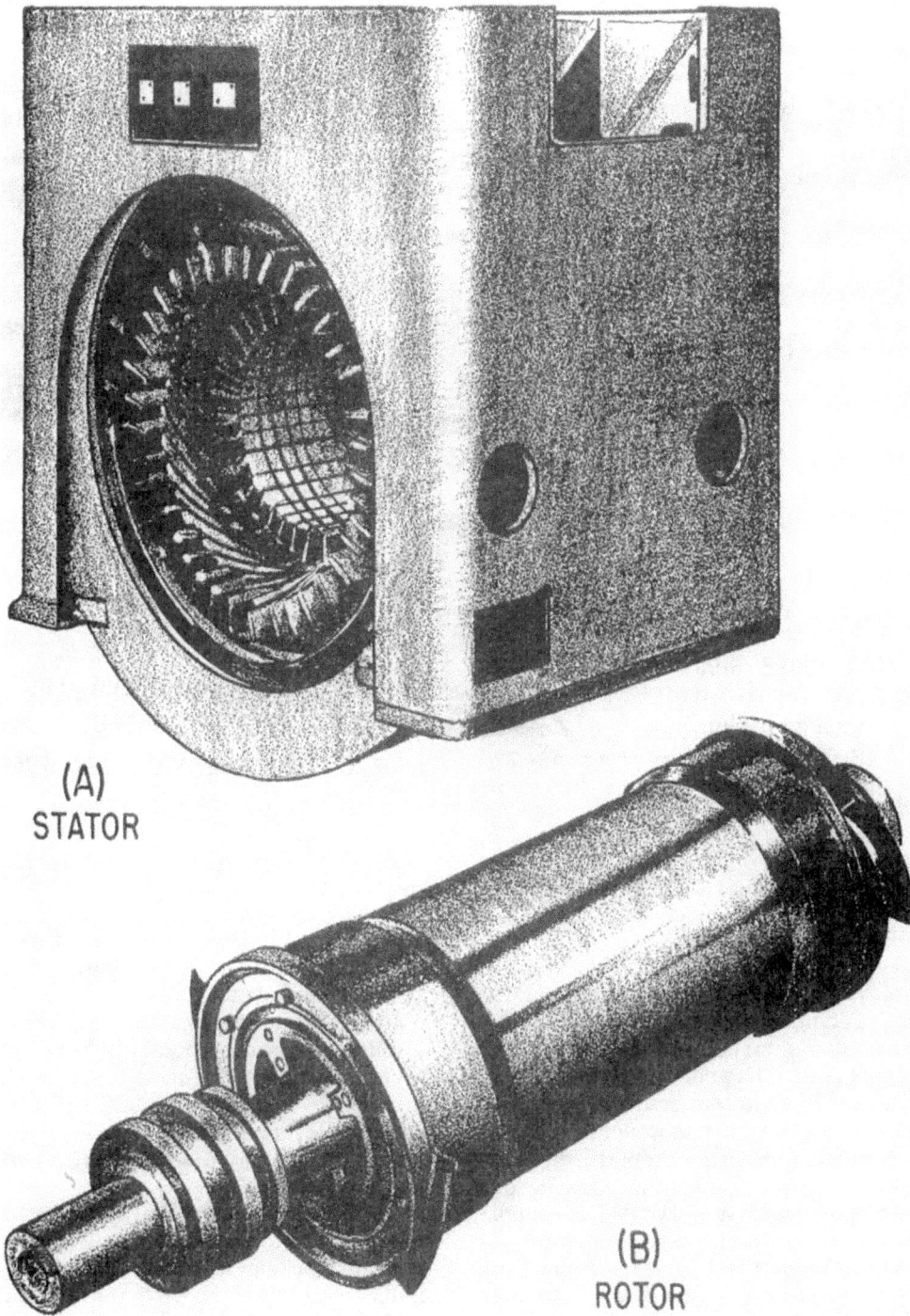

(A)
STATOR

(B)
ROTOR

BE.287

Figure 16-4.—High-speed turbine-driven a-c generator.

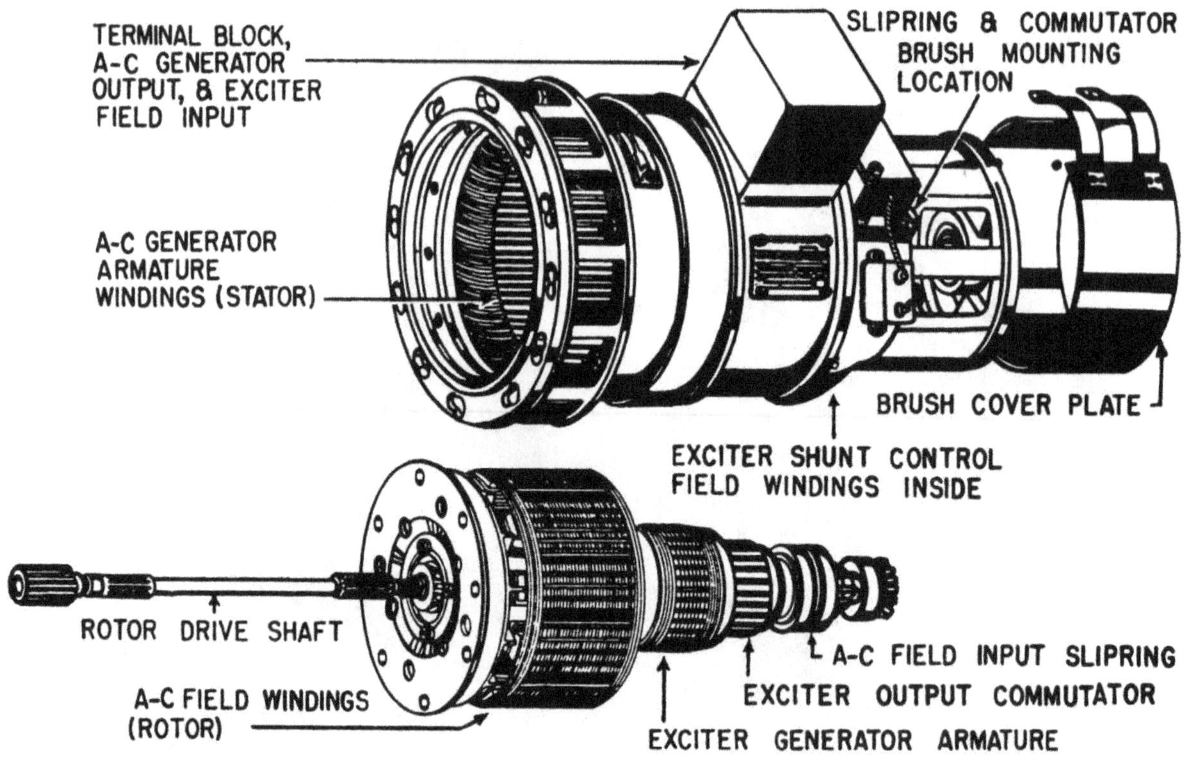

TERMINAL BLOCK,
A-C GENERATOR
OUTPUT, & EXCITER
FIELD INPUT

SLIPRING & COMMUTATOR
BRUSH MOUNTING
LOCATION

A-C GENERATOR
ARMATURE
WINDINGS (STATOR)

BRUSH COVER PLATE

EXCITER SHUNT CONTROL
FIELD WINDINGS INSIDE

ROTOR DRIVE SHAFT

A-C FIELD WINDINGS
(ROTOR)

A-C FIELD INPUT SLIPRING

EXCITER OUTPUT COMMUTATOR

EXCITER GENERATOR ARMATURE

BE.288

Figure 16-5.—High-speed engine-driven a-c generator.

## SINGLE-PHASE GENERATORS

A single-phase a-c generator has a stator made up of a number of windings in series, which form a single circuit in which an output voltage is generated. The principle of the single-phase a-c generator is described first, and the polyphase a-c generator is described later.

Figure 16-6 illustrates a schematic diagram of a single-phase a-c generator having four poles. The stator has four polar groups evenly spaced around the stator frame. The rotor has four poles, with adjacent poles of opposite polarity. As the rotor revolves, a-c voltages are induced in the stator windings. Since one rotor pole is in the same position relative to a stator winding as any other rotor pole, all stator polar groups are cut by equal amounts of magnetic lines of force at any given time. As a result, the voltages induced in all the

windings have the same amplitude or value at any given instant. The four stator windings are connected to each other so that the a-c voltages are in phase, or "series aiding". Assume that rotor pole 1, a south pole, induces a voltage in the direction indicated by the arrow in stator winding 1. Since rotor pole 2 is a north pole, it will induce a voltage in the opposite direction in stator coil 2 with respect to that in coil 1.

In order that the two induced voltages be in series addition, the two coils are connected as shown. Applying the same reasoning, the voltage induced in stator coil 3 (clockwise rotation of the field) is in the same direction (counterclockwise) as the voltage induced in coil 1. Similarly, the direction of the voltage induced in winding 4 is opposite to the direction of the voltage induced in coil 1. All four stator coil groups are connected in series so that the

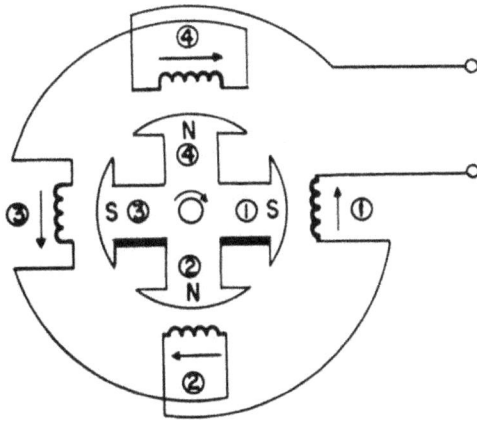

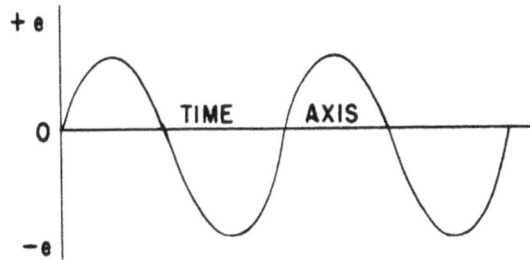

BE.289

Figure 16-6.—Single-phase a-c generator.

voltages induced in each winding add to give a total voltage that is four times the voltage in any one winding.

## TWO-PHASE GENERATORS

Multiphase or polyphase a-c generators have two or more single-phase windings symmetrically spaced around the stator. In a 2-phase a-c generator there are two single-phase windings physically spaced so that the a-c voltage induced in one is 90° out of phase with the voltage induced in the other. The windings are electrically separate from each other. When one winding is being cut by maximum flux, the other is being cut by no flux. This condition establishes the 90° relation between the two phases.

Figure 16-7 is a schematic diagram of a 2-phase 4-pole a-c generator. The stator consists of two single-phase windings (phases) completely separated from each other. Each phase is made up of four windings, which are connected in series so that their voltages add. The rotor is identical with that used in the single-phase a-c generator. In figure 16-7 (A), the rotor poles are opposite all of the coils of phase A. Therefore, the voltage induced in phase A is maximum and the voltages induced in phase B is zero. As the rotor continues rotating in a clockwise direction, it moves away from the windings of phase A and approaches those of phase B. As a result, the voltage in phase A decreases from its maximum value and the voltage in phase B increased

from zero. In figure 16-7 (B), the rotor poles are opposite the windings of phase B. Now the voltage induced in phase B is maximum; whereas the voltage induced in phase A has dropped to zero. Notice that in the 4-pole a-c generator a 45° mechanical rotation of the rotor corresponds electrically to one quarter cycle, or 90 electrical degrees. Figure 16-7 (C) illustrates the waveforms of the voltage generated in each of the two phases. Both are sine curves, and A leads B by 90°. Figure 16-7 (D) illustrates the vectors representing the 2-phase voltages. Vector A leads vector B by 90°.

The two phases of a 2-phase a-c generator can be connected to each other as shown in figure 16-8, so that only three leads are brought out to the load. The a-c generator is then called a 2-phase 3-wire a-c generator. The two phases for a 4-pole a-c generator are shown in figure 16-8 (A). A simplified schematic diagram is shown in figure 16-8 (B), in which the rotor is omitted and each phase is indicated as a single coil. The two coils are drawn at right angles to each other to represent the 90° phase displacement between their respective voltages. The three wires make possible three different load connections—(A) and (B) across phases A and B, respectively, and (C) across both phases in series. The third voltage (C) is displaced 45° from the A and B phase voltages and is their vector sum as shown in figures 16-8 (C) and (D) respectively. If each phase voltage has an effective value of 100 volts, the vector sum of these voltages is the hypotenuse of the right triangle, the base and altitude of

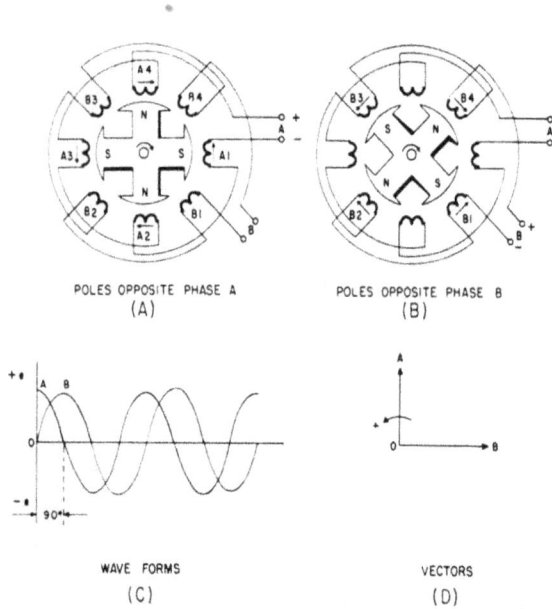

POLES OPPOSITE PHASE A
(A)

POLES OPPOSITE PHASE B
(B)

WAVE FORMS
(C)

VECTORS
(D)

BE.290

Figure 16-7.—Two-phase four-pole a-c generator.

which are each 100 volts. This hypotenuse is equal to

$$\frac{100}{\cos 45^\circ}, \quad \text{or} \quad \frac{100}{0.707} = 141 \text{ volts}$$

THREE-PHASE GENERATORS

The 3-phase a-c generator, as the name implies, has three single-phase windings spaced so that the voltage induced in each winding is 120° out of phase with the voltages in the other two windings. A schematic diagram of a 3-phase stator showing all the coils becomes complex, and it is difficult to see what is actually happening. A simplified schematic diagram, showing all the windings of a single phase lumped together as one winding, is illustrated in figure 16-9 (A). The rotor is omitted for simplicity. The waveforms of voltage are shown to the right of the schematic. The three voltages are 120° apart and are similar to the voltages that would be generated by three single-phase a-c generators whose voltages are out of phase by angles of 120°. The three phases are independent of each other.

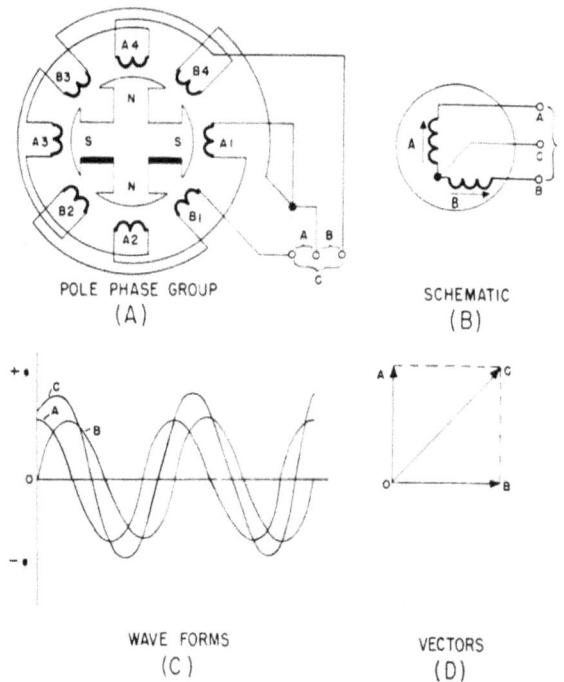

POLE PHASE GROUP
(A)

SCHEMATIC
(B)

WAVE FORMS
(C)

VECTORS
(D)

BE.291

Figure 16-8.—Two-phase three-wire a-c generator.

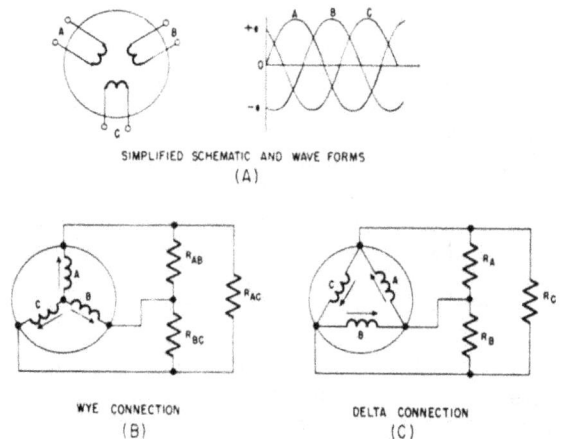

SIMPLIFIED SCHEMATIC AND WAVE FORMS
(A)

WYE CONNECTION
(B)

DELTA CONNECTION
(C)

BE.292

Figure 16-9.—Three-phase a-c generator.

The Wye Connection

Rather than have six leads come out of the 3-phase a-c generator, one of the leads from each phase may be connected to form a common junction. The stator is then called wye, or star, connected. The common lead may or may not be brought out of the machine. If it is brought out, it is called the neutral. The simplified schematic (fig. 16-9 (B)) shows a wye-connected stator with the common lead not brought out. Each load is connected across two phases in series. Thus, $R_{ab}$ is connected across phases A and B in series, $R_{ac}$ is connected across phases A and C in series, and $R_{bc}$ is connected across phases B and C in series. Thus the voltage across each load is larger than the voltage across a single phase. In a wye-connected a-c generator the three start ends of each single-phase winding are connected together to a common neutral point and the opposite, or finish, ends are connected to the line terminals A, B, and C. These letters are always used to designate the three phases of a 3-phase system, or the three line wires to which the a-c generator phases connect. A 3-phase wye-connected a-c generator supplying three separate loads is shown in figure 16-10 (A). When unbalanced loads are used, a neutral may be added as shown in the figure by the broken line between the common neutral point and the loads. The neutral wire serves as a common return circuit for all three phases and maintains a voltage balance across the loads. No current flows in the neutral wire when the loads are balanced. This system is a 3-phase 4-wire circuit and is used to distribute 3-phase power to shore-based installations. The 3-phase 4-wire system is not used aboard ship, but is widely used in industry and for aircraft a-c power systems.

The phase relations in a 3-wire 3-phase wye-connected system are shown in figure 16-10 (B). In constructing vector diagrams of 3-phase circuits, a counterclockwise rotation is assumed in order to maintain the correct phase relation between line voltages and currents. Thus, the a-c generator is assumed to rotate in such a direction as to generate the three phase voltages in the order, $E_a$, $E_b$, $E_c$.

The voltage in phase b, or $E_b$, lags the voltage in phase a, or $E_a$, by 120°. Likewise, $E_c$ lags $E_b$ by 120°, and $E_a$ lags $E_c$ by 120°. In figure 16-10 (A), the arrows $E_a$, $E_b$, and $E_c$, represent the positive direction of generated voltage in the wye-connected a-c generator. The arrows $I_1$, $I_2$, and $I_3$ represent the positive direction of phase and line currents supplied to balanced unity power-factor loads connected in wye. The three voltmeters connected between

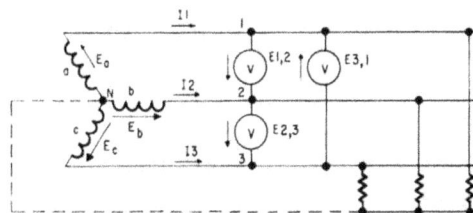

CIRCUIT
(A)

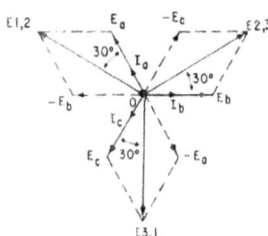

| $E1,2$ | $=$ | $E_a \ominus E_b$ |
| $E2,3$ | $=$ | $E_b \ominus E_c$ |
| $E3,1$ | $=$ | $E_c \ominus E_a$ |
| $I_a$ | $=$ | $I1$ |
| $I_b$ | $=$ | $I2$ |
| $I_c$ | $=$ | $I3$ |
| $E1,2$ | $=$ | $\sqrt{3}E_a$ |
| $E2,3$ | $=$ | $\sqrt{3}E_b$ |
| $E3,1$ | $=$ | $\sqrt{3}E_c$ |

VECTORS
(B)

BE.293

Figure 16-10.—Three-phase wye-connected system.

lines 1-2, 2-3, and 3-1, respectively, indicate effective values of line voltage. The line voltage is greater than the voltage of a phase in the wye-connected circuit because there are two phases connected in series between each pair of line wires, and their voltages combine. Line voltage is not twice the value of phase voltage, however, because the phase voltages are not in phase with each other.

The relation between phase and line voltages is shown in the vector diagram. Effective values of phase voltage are indicated by vectors, $E_a$, $E_b$, and $E_c$. Effective values of line and phase current are indicated by vectors $I_a$, $I_b$, and $I_c$. Because there is only one path for current between any given phase and the line wire to which it is connected, the phase current is equal to the line current. The respective phase currents have equal values because the load is assumed to be balanced. For the same reason, the respective line currents have equal values. When the load has unity power factor, the phase currents are in phase with their respective phase voltages.

In combining a-c voltages, it is necessary to know the direction in which the positive maximum values of the voltages act in the circuit as well as the magnitudes of the voltages. For example, in figure 16-11 (A), the positive maximum voltage generated in coils A and B act in the direction of the arrows, and B leads A by 120°. This arrangement may be obtained by assuming coils A and B to be two armature windings located 120° apart. If each voltage has an effective value of 100 volts, the total voltage is $E_r$ = 100 volts, as shown by the polar vectors in figure 16-11 (B).

If the connections of coil B are reversed (fig. 16-11 (C)) with respect to their original connections, the two voltages are in opposition, as may be seen by tracing around the circuit in the direction of the arrow in coil A. The positive direction of the voltage in coil B is opposite to the direction of the trace; the positive direction of the voltage generated in coil A is the same as that of the trace; hence, the two voltages are in opposition. This effect is the same as though the positive maximum value of $E_b$, were 60° out of phase with that of $E_a$, and $E_b$ acted in the same direction as $E_a$ when the circuit trace was made (fig. 16-11 (D)) to vector $E_a$ is accomplished by reversing the position of $E_b$ from that shown in figure 16-11 (B) to the position shown in figure 16-11 (D) and completing the parallelogram. If $E_a$ and $E_b$ are each 100 volts, $E_r$ = $\sqrt{3}$ x 100, or 173 volts.

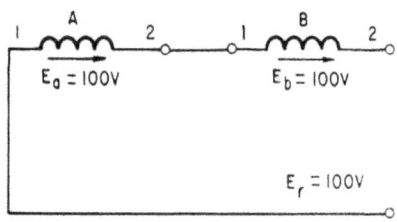

(A)  POSITIVE VOLTAGES
     SAME DIRECTION

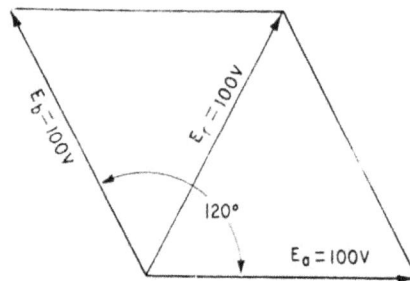

(B)  $E_b$ LEADS $E_a$ BY 120°

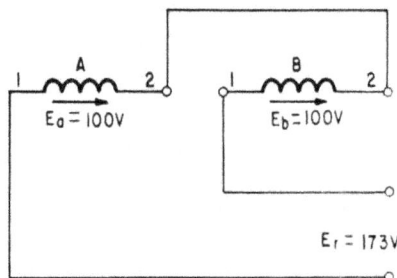

(C)  POSITIVE VOLTAGES
     IN OPPOSITE DIRECTION

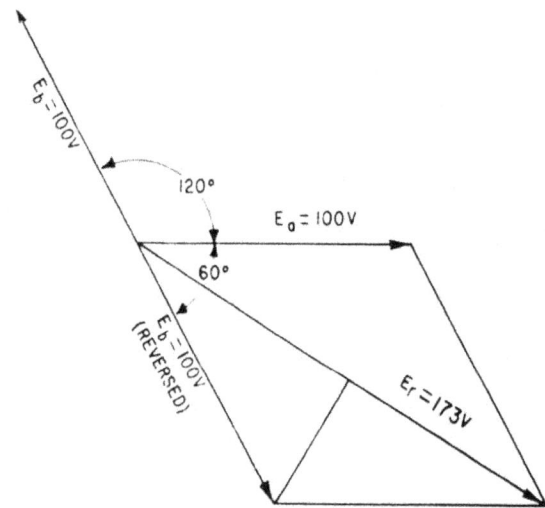

(D)  (REVERSED) $E_b$ LAGS $E_a$ BY 60°

BE.294

Figure 16-11.—Vector analysis of voltages in series aiding and opposing.

The value of $E_r$ may be derived in the following manner: Erecting a perpendicular to $E_r$ divides the isosceles triangle into two equal right triangles each having a hypotenuse of 100 volts and a base of 100 cos 30°, or 86.6 volts. The total length of $E_r$ is 2 x 86.6, or 173.2 volts.

To construct the line voltage vectors $E_{1,2}$, $E_{2,3}$, and $E_{3,1}$, in figure 16-10, it is first necessary to trace a path around the closed circuit which includes the line wires, armature windings, and one of the three voltmeters. For example, in figure 16-10 (A), consider the circuit that includes the upper and middle wires, the voltmeter connected across them, and the a-c generator phases a and b. The circuit trace is started at the center of the wye, proceeds through phase a of the a-c generator, out line 1, down through the voltmeter from line 1 to line 2, and through phase b of the a-c generator back to the starting point. Voltage drops along line wires are disregarded. The voltmeter indicates an effective value equal to the vector sum of the effective value of voltage in phases a and b. This value is the line voltage, $E_{1,2}$. According to Kirchhoff's law, the source voltage between lines 1 and 2 equals the voltage drop across the voltmeter connected to these lines.

If the direction of the path traced through the generator is the same as that of the arrow, the sign of the voltage is plus; if the direction of the trace is opposite to the arrow, the sign of the voltage is minus. If the direction of the path traced through the voltmeter is the same as that of the arrow, the sign of the voltage is minus; if the direction of the trace is opposite to that of the arrow, the sign of the voltage is plus.

The following equations for voltage are based on the preceding rules:

$$E_a \oplus (-E_b) = E_{1,2}, \text{ or } E_{1,2} = E_a \ominus E_b$$

$$E_b \oplus (-E_c) = E_{2,3}, \text{ or } E_{2,3} = E_b \ominus E_c$$

$$E_c \oplus (-E_a) = E_{3,1}, \text{ or } E_{3,1} = E_c \ominus E_a$$

The signs $\oplus$ and $\ominus$ mean vector addition and, vector subtraction, respectively. One vector is subtracted from another by reversing the position of the vector to be subtracted through an angle of 180° and constructing a parallelogram, the sides of which are the reversed vector and the other vector. The diagonal of the parallelogram is the difference vector.

These equations are applied to the vector diagram of figure 16-10 (B), to derive the line voltages $E_{1,2}$, $E_{2,3}$, and $E_{3,1}$, as the diagonals of three parallelograms of which the sides are the phase voltages $E_a$, $E_b$, and $E_c$. From this vector diagram, the following facts are observed: (1) The line voltages are equal and 120° apart; (2) the line currents are equal and 120° apart; (3) the line currents are 30° out of phase with the line voltages when the power factor of the load is 100 percent; and (4) line voltage is the product of the phase voltage and $\sqrt{3}$.

The Delta Connection

A three-phase stator may also be connected as shown in figure 16-9 (C). This is called the delta connection. In a delta-connected a-c generator, the start end of one phase winding is connected to the finish end of the third; and the start of the third phase winding is connected to the finish of the second phase winding; and the start of the second phase winding is connected to the finish of the first phase winding. The three junction points are connected to the line wires leading to the load. A 3-phase delta-connected a-c generator is depicted in figure 16-12 (A). The generator is connected to a 3-phase 3-wire circuit which supplies a 3-phase delta-connected load at the right-hand end of the 3-phase line. Because the phases are connected directly across the line wires, phase voltage is equal to line voltage. When the generator phases are properly connected in delta, no appreciable current flows within the delta loop when there is no external load connected to the generator. If any one of the phases is reversed with respect to its correct connection, a short-circuit current flows within the windings of no load, causing damage to the windings.

To avoid connecting a phase in reverse it is necessary to test the circuit before closing the delta. This may be accomplished by connecting a voltmeter or fuse wire between the two ends of the delta loop before closing the delta. The two ends of the delta loop should never be connected if there is an indication of any appreciable current or voltage between them when no load is connected to the generator.

The three phase currents, $I_a$, $I_b$, and $I_c$, are indicated by accompanying arrows in the generator phases in figure 16-12 (A). These arrows point in the direction of the positive current and voltage of each phase. The three

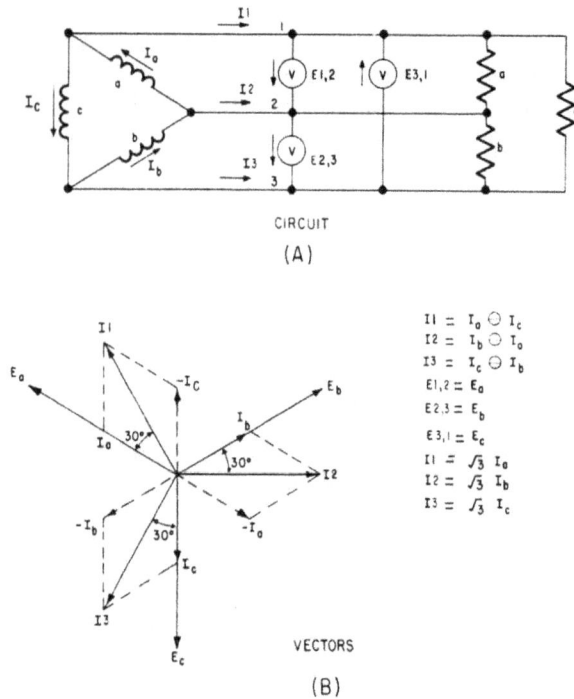

CIRCUIT

(A)

$I_1 = I_a \ominus I_c$
$I_2 = I_b \ominus I_a$
$I_3 = I_c \ominus I_b$
$E_{1,2} = E_a$
$E_{2,3} = E_b$
$E_{3,1} = E_c$
$I_1 = \sqrt{3} \, I_a$
$I_2 = \sqrt{3} \, I_b$
$I_3 = \sqrt{3} \, I_c$

VECTORS

(B)

BE.295

Figure 16-12.—Three-phase delta-connected system.

voltmeters connected across lines 1-2, 2-3, and 3-1, respectively, indicate effective values of line and phase voltage. Line current $I_1$ is supplied by phase a and c, which are connected to line 1. Line current is greater than phase current, but not twice as great because the phase currents are not in phase with each other. The relation between line currents and phase currents is shown in figure 16-12 (B).

Effective values of line and phase voltages are indicated by vectors $E_a$, $E_b$, and $E_c$. Note that the vector sum of $E_a$, $E_b$, and $E_c$ is zero. The phase currents are equal to each other because the loads are balanced. The line currents are equal to each other for the same reason. At unity-power-factor loads, the phase current and phase voltage have a $0°$ angle between them.

To construct the line current vectors it is first necessary to consider the direction of currents at the junctions of the line wires and the generator phases. Consider the junction formed by phase a, phase c, and line 1. According to Kirchhoff's law, line current $I_1$ is equal

to the vector difference between $I_a$ and $I_c$. In figure 16-12 (B), current vector $I_a$ is in the same general direction as current vector $I_1$ (less than $90°$ displaced from $I_1$) and current vector $I_c$ is in opposition to vector $I_1$ (more than $90°$ displaced from it). $I_1$ is the vector sum of $I_a$ and $-I_c$. These are two equal values of phase current and they are $60°$ out of phase with each other. The vector sum of all currents entering and leaving the junction is zero. Current arrows that point toward the junction are considered to represent positive currents. Current arrows that point away from the junction are considered to represent negative currents. The following equations for current are based on the preceding rules:

$$I_a \oplus (-I_c) = I_1, \text{ or } I_1 = I_a \ominus I_c.$$

$$I_b \oplus (-I_a) = I_2, \text{ or } I_2 = I_b \ominus I_a.$$

$$I_c \oplus (-I_b) = I_3, \text{ or } I_3 = I_c \ominus I_b.$$

Here also, the signs $\oplus$ and $\ominus$ mean vector addition and vector subtraction, respectively. Vector subtraction is accomplished again by reversing the vector being subtracted through an angle of $180°$ and combining it with the other vector by the parallelogram method to form the diagonal which is the vector difference.

If the current equations are applied to the vector diagram of figure 16-12 (B), the line current $I_1$, $I_2$, and $I_3$ are derived as diagonals of the three parallelograms in which the sides are the phase currents $I_a$, $I_b$, and $I_c$. From this diagram the following facts are observed: (1) The line currents are equal and $120°$ apart; (2) the line voltages are equal and $120°$ apart; (3) the line currents are $30°$ out of phase with the line voltages when the power factor of the load is unity; and (4) the line current is equal to the product of the phase current and $\sqrt{3}$.

The power delivered by a balanced 3-phase wye-connected system is equal to three times the power delivered by each phase. The total true power is

$$P_t = 3E_{phase}I_{phase} \cos \theta$$

Because

$$E_{phase} = \frac{E_{line}}{\sqrt{3}} \text{ and } I_{phase} = I_{line}$$

297

the total true power is

$$P_t = 3 \frac{E_{line}}{\sqrt{3}} I_{line} \cos \theta$$

$$= \sqrt{3} E_{line} I_{line} \cos \theta$$

The power delivered by a balanced 3-phase delta-connected system is also three times the power delivered by each phase.

Because

$$E_{phase} = E_{line} \text{ and } I_{phase} = \frac{I_{line}}{\sqrt{3}}$$

the total true power is

$$P_t = 3E_{line} \frac{I_{line}}{\sqrt{3}} \cos \theta$$

$$= \sqrt{3} E_{line} I_{line} \cos \theta$$

Thus, the expression for 3-phase power delivered by a balanced delta-connected system is the same as the expression for 3-phase power delivered by a balanced wye-connected system. Two examples are given to illustrate the phase relations between current, voltage, and power in (1) a 3-phase wye-connected system and (2) a 3-phase delta-connected system.

Example 1: A 3-phase wye-connected a-c generator has a terminal voltage of 450 volts and delivers a full-load current of 300 amperes per terminal at a power factor of 80 percent. Find (a) the phase voltage, (b) the full-load current per phase, (c) the kilovolt-ampere, or apparent power, rating, and (d) the true power output.

(a) $E_{phase} = \dfrac{E_{line}}{\sqrt{3}} = \dfrac{450}{\sqrt{3}} = 260$ volts

(b) $I_{phase} = I_{line} = 300$ amperes

(c) Apparent power $= \sqrt{3} E_{line} I_{line}$

$= \sqrt{3} \times 450 \times 300 = 233,600$ va, or 233.6 kva

(d) True power $= \sqrt{3} E_{line} I_{line} \cos \theta$

$= \sqrt{3} \times 450 \times 300 \times 0.8$

$= 186,800$ watts, or 186.8 kw

Example 2: A 3-phase delta-connected a-c generator has a terminal voltage of 450 volts and the current in each phase is 200 amperes. The power factor of the load is 75 percent. Find (a) the line voltage, (b) the line current, (c) the apparent power, and (d) the true power.

(a) $E_{phase} = E_{line} = 450$ volts

(b) $I_{line} = \sqrt{3} I_{phase} = 1.732 \times 200 = 346$ amperes

(c) Apparent power $= \sqrt{3} E_{line} I_{line}$

$= 1.732 \times 450 \times 346 = 269,000$ va, or 269 kva

(d) True power $= \sqrt{3} E_{line} I_{line} \cos \theta$

$= 1.73 \times 450 \times 346 \times 0.75$

$= 202,020$, or 202.02 kw

## MEASUREMENT OF POWER

The wattmeter connections for measuring the true power in a 3-phase system are shown in figure 16-13. The method shown in figure 16-13 (A) uses three wattmeters with their current coils inserted in series with the line wires and their potential coils connected between line and neutral wires. The total true power is equal to the arithmetic sum of the three wattmeter readings.

The method shown in figure 16-13 (B) uses two wattmeters with their current coils connected in series with two line wires and their potential coils connected between these line wires and the common, or third, wire that does not contain the current coils. The total true power is equal to the algebraic sum of the two wattmeter readings. If one meter reads backward, its potential coil connections are first reversed to make the meter read up-scale and the total true power is then equal to the difference in the two wattmeter readings. If the load power factor is less than 0.5 and the

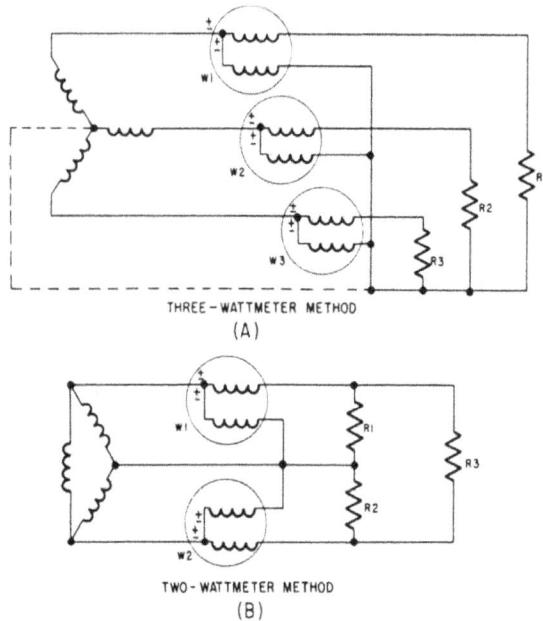

THREE-WATTMETER METHOD
(A)

TWO-WATTMETER METHOD
(B)

BE.296

Figure 16-13.—Wattmeter connection for measurement of power in a 3-phase system.

loads are balanced, the total true power is equal to the difference in the two wattmeter readings. If the load power factor is 0.5, one meter indicates the total true power and the other indicates zero. If the load power factor is above 0.5, the total true power is equal to the sum of the two wattmeter readings.

## FREQUENCY

The frequency of the a-c generator voltage depends upon the speed of rotation of the rotor and the number of poles. The faster the speed, the higher the frequency will be; and the lower the speed, the lower the frequency becomes. The more poles there are on the rotor, the higher the frequency is for a given speed. When a rotor has rotated through an angle such that two adjacent rotor poles (a north and a south pole) has passed one winding, the voltage induced in that winding will have varied through one complete cycle. For a given frequency, the more pairs of poles there are, the lower is the speed of rotation. A 2-pole generator rotates at twice the speed of a 4-pole generator for the same frequency of generated voltage. The frequency of the generator in Hz (cycles per second) is

related to the number of poles and the speed, as expressed by the equation

$$f = \frac{P}{2} \times \frac{N}{60} = \frac{PN}{120}$$

where P is the number of poles and N the speed in rpm. For example, a 2-pole 3,600-rpm generator has a frequency of $\frac{2 \times 3,600}{120}$ = 60 Hz; a 4-pole 1,800-rpm generator has the same frequency; and a 6-pole 500-rpm generator has a frequency of $\frac{6 \times 500}{120}$ = 25 Hz, and a 12-pole 4,000-rpm generator has a frequency of $\frac{12 \times 4,000}{120}$ = 400 Hz.

## GENERATED VOLTAGE

Once the machine has been designed and built, the generated voltage of an a-c generator is controlled, in practice, by varying the d-c excitation voltage applied to the field winding.

In the design, however, many factors must be taken into account, as the following text illustrates.

As mentioned before, the conductors in a generator armature are arranged in one or more groups called phases, according to whether the machine is designed to deliver single-phase, 2-phase, or 3-phase voltages. The effective voltage, E, per phase is

$$E = 2.22 \Phi \, Zf \, 10^{-8} \, K_b K_p$$

where $\Phi$ is the number of magnetic lines of flux per pole, Z the number of conductors in series per phase, f the frequency in Hz, $K_b$ a breadth (or belt) factor, and $K_p$ a pitch factor.

Generally, the coils of each phase are distributed uniformly around the stator. The voltages generated in the various coils are not in time phase with each other because of the displacement of the coils. A breadth, or belt, factor $K_b$ takes the displacement into account and reduces the total generated voltage per phase below the voltage that would be generated if all the active conductors were in the form of a concentrated winding. Also, if the two halves of a coil are less than 180 electrical degrees apart, the voltage generated in the coil is less than it would be if the coil pitch were a full 180 electrical degrees. The pitch factor, $K_p$, accounts for this reduction in voltage.

For example, a certain 3-phase, 60-Hz generator has 96 conductors in series per phase and a field pole flux of $2.54 \times 10^6$ lines per pole. The breadth factor is 0.958, and the pitch factor is 0.966. The effective voltage per phase is

$$E = \frac{2.22 \times 2.54 \times 10^6 \times 96 \times 60 \times 0.958 \times 0.966}{10^8}$$

$$= 300 \text{ volts}$$

## CHARACTERISTICS

When the load on a generator is changed, the terminal voltage varies with the load. The amount of variation depends on the design of the generator and the power factor of the load. With a load having a lagging power factor, the drop in terminal voltage with increased load is greater than for unity power factor. With a load having a leading power factor, the terminal voltage tends to rise. The causes of a change in terminal voltage with load change are (1) armature resistance (2) armature reactance, and (3) armature reaction.

### Armature Resistance

When current flows through a generator armature winding, there is an IR drop due to the resistance of the winding. This drop increases with load, and the terminal voltage is reduced. The armature resistance drop is small because the resistance is low.

### Armature Reactance

The armature current of an a-c generator varies approximately as a sine wave. The continuously varying current in the generator armature is accompanied by an $IX_L$ voltage drop in addition to the IR drop. Armature reactance in an a-c generator may be from 30 to 50 times the value of armature resistance because of the relatively large inductance of the coils compared with their resistance.

A simplified series equivalent circuit of one phase of an a-c generator is shown in figure 16-14. The voltage generated in the phase winding is equal to the vector sum of the terminal voltage for the phase and the internal voltage loss in the armature resistance, R, and the armature reactance, $X_L$, associated with that phase. The voltage vectors for a unity power-factor load are shown in figure 16-14 (A). The armature IR drop is in phase with the current, I, and the terminal voltage $E_t$. Because the armature $IX_L$ drop is 90° out of phase with the current, the terminal voltage is approximately equal to the generated voltage, less the IR drop in the armature.

The voltage vectors for a lagging power-factor load are shown in figure 16-14 (B). The load current and IR drop lag the terminal voltage by angle $\theta$. In this example, the armature IZ drop is more nearly in phase with the terminal voltage and the generated voltage. Hence, the terminal voltage is approximately equal to the generated voltage, less the armature IZ drop. Because the IZ drop is much greater than the IR drop, the terminal voltage is reduced that much more. The voltage vectors for a leading power-factor load are shown in figure 16-14 (C). The load current and IR drop lead the terminal voltage by angle $\theta$. This condition results in an increase in terminal voltage above the value of $E_g$. The total available voltage of the a-c generator phase is the combined effect of $E_g$ (rotationally induced) and the self-induced voltage (not shown in the vectors). The self-induced voltage, as in any a-c circuit, is caused by the varying field (accompanying the varying armature current) linking the armature conductors. The self-induced voltage always lags the current by 90°; hence, when I leads $E_t$, the self-induced voltage aids $E_g$, and $E_t$ increases.

### Armature Reaction

When an a-c generator supplies no load, the d-c field flux is distributed uniformly across the airgap. When an a-c generator supplies a reactive load, however, the current flowing through the armature conductors produces an armature magnetomotive force (mmf) that influences the terminal voltage by changing the magnitude of the field flux across the airgap. When the load is inductive, the armature mmf opposes the d-c field and weakens it, thus lowering the terminal voltage. When a leading current flows in the armature the d-c field is aided by the armature mmf and the flux across the airgap is increased, thus increasing the terminal voltage.

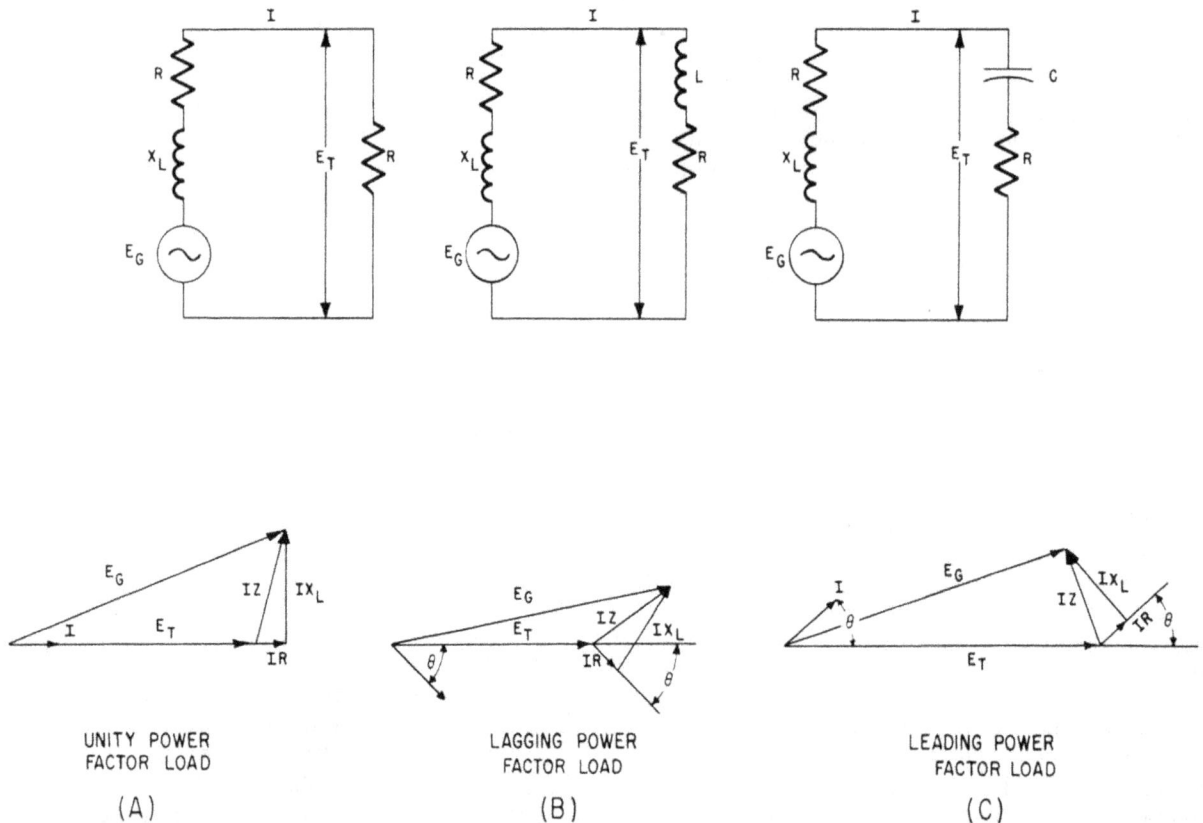

Figure 16-14.—A-c generator voltage characteristics.

## VOLTAGE REGULATION

The voltage regulation of an a-c generator is the change of voltage from full load to no load, expressed in percentage of full-load volts, when the speed and d-c field current are held constant.

$$\text{Percent regulation} = \frac{E_{NL} - E_{fL}}{E_{fL}} \times 100$$

For example, the no-load voltage of a certain generator is 250 volts and the full-load voltage is 220 volts. The percent regulation is

$$\frac{250 - 220}{220} \times 100 = 13.6 \text{ percent}$$

## PRINCIPLES OF A-C VOLTAGE CONTROL

In an a-c generator, an alternating voltage is induced into the armature windings when magnetic fields of alternating polarity are passed across these windings. The amount of voltage induced into the a-c generator windings depends mainly on three things: the number of conductors in series per winding, the speed at which the magnetic field passes across the winding (generator rpm), and the strength of the magnetic field. Any of these three could conceivably be used to control the amount of voltage induced into the a-c generator windings.

The number of windings, of course, is fixed when the generator is manufactured. Also, if the frequency of a generator's output is required to be of a constant value, then the speed of the rotating field must be held constant. This prevents the use of the generator rpm as a means of controlling the voltage output. Thus, the only

301

practical remaining method for obtaining voltage control is to control the strength of the rotating magnetic field. In some specialized applications, the field is furnished by a permanent magnet. The a-c generator in figure 16-2 uses an electromagnetic field rather than a permanent magnetic type field. The strength of this electromagnetic field may be varied by changing the amount of current flowing through the coil. This is accomplished by varying the amount of voltage applied across the coil. It can be seen that by varying the exciter armature d-c output voltage, the a-c generator field strength is also varied. Thus, the magnitude of the generated a-c voltage depends directly on the value of the exciter output voltage. This relationship allows a relatively large a-c voltage to be controlled by a much smaller d-c voltage.

The next function of the a-c generator that must be understood is how the d-c exciter output voltage is controlled. Voltage control in a d-c generator is obtained by varying the strength of the d-c generator shunt field. This is accomplished by using any of a number of different types of voltage regulators. A device which will vary the exciter shunt field current in accordance with changes in the a-c generator voltage is called an a-c generator voltage regulator. This regulator must also maintain the correct value of exciter shunt field current when no a-c voltage corrective action is required (steady state output). In figure 16-15, note that a pair of connections labeled a-c sensing input feeds a voltage proportional to the a-c generator output voltage to the a-c voltage regulator. Note also that a portion of the exciter's armature output is connected through the exciter's field rheostat, RX, then through the exciter shunt field windings, and finally back to the exciter armature. Obviously, the exciter supplies direct current to its own control field, in addition to the a-c generator field, as determined by the setting of RX. The setting of RX is controlled by the magnetic strength of the control coil L. The magnetic strength of L is in turn controlled by the voltage across resistor R. The voltage across R is rectified d.c., and is proportional to the a-c line voltage. (Rectifiers are devices that change a.c. to d.c.)

Thus, the essential function of the voltage regulator is to use the a-c output voltage, which it is designed to control, as a sensing influence to control the amount of current the exciter supplies to its own control field. A drop in the output a-c voltage will change the setting of RX in one direction and cause a rise in the exciter control field current. A rise in the output a-c voltage will change the setting of RX in the opposite direction and cause a drop in the exciter control field current. These latter two characteristics are caused by actions within the voltage regulator. These characteristics are common to both the resistive and magnetic (magnetic amplifier) types of a-c voltage regulators. Both types of regulators perform the same functions, but accomplish them through different operating principles.

## PARALLEL OPERATION OF A-C GENERATORS

A-c generators are connected in parallel to (1) increase the plant capacity beyond that of a single unit, (2) serve as additional reserve power for expected demands, or (3) permit shutting down one machine and cutting in a standby machine without interrupting the power supply. When a-c generators are of sufficient size, and are operating at unequal frequencies and terminal voltages, severe damage may result if they are suddenly connected to each other through a common bus. To avoid this, the generators must be synchronized as closely as possible before connecting them together. This may be accomplished by connecting one generator to the bus (referred to as bus generator), and then synchronizing the other (incoming generator) to it before closing the incoming generator's main power contactor. The generators are synchronized when the following conditions are set:

1. Equal terminal voltages. This is obtained by adjustment of the incoming generator's field strength.

2. Equal frequency. This is obtained by adjustment of the incoming generator's prime mover speed.

3. Phase voltages in proper phase relation. (Connecting phase voltages must reach peak values of the same polarity at the same instant.) The generators could have the same frequency, and still not be in phase. That is, if the two generators have the same frequency, but one is lagging the other, the lagging generator will remain a fixed number of degrees behind the leading generator, until it is accelerated slightly to catch up.

4. Phases connected in proper sequence. One pair of phases may be properly connected, while the other two pairs may be crossed. In

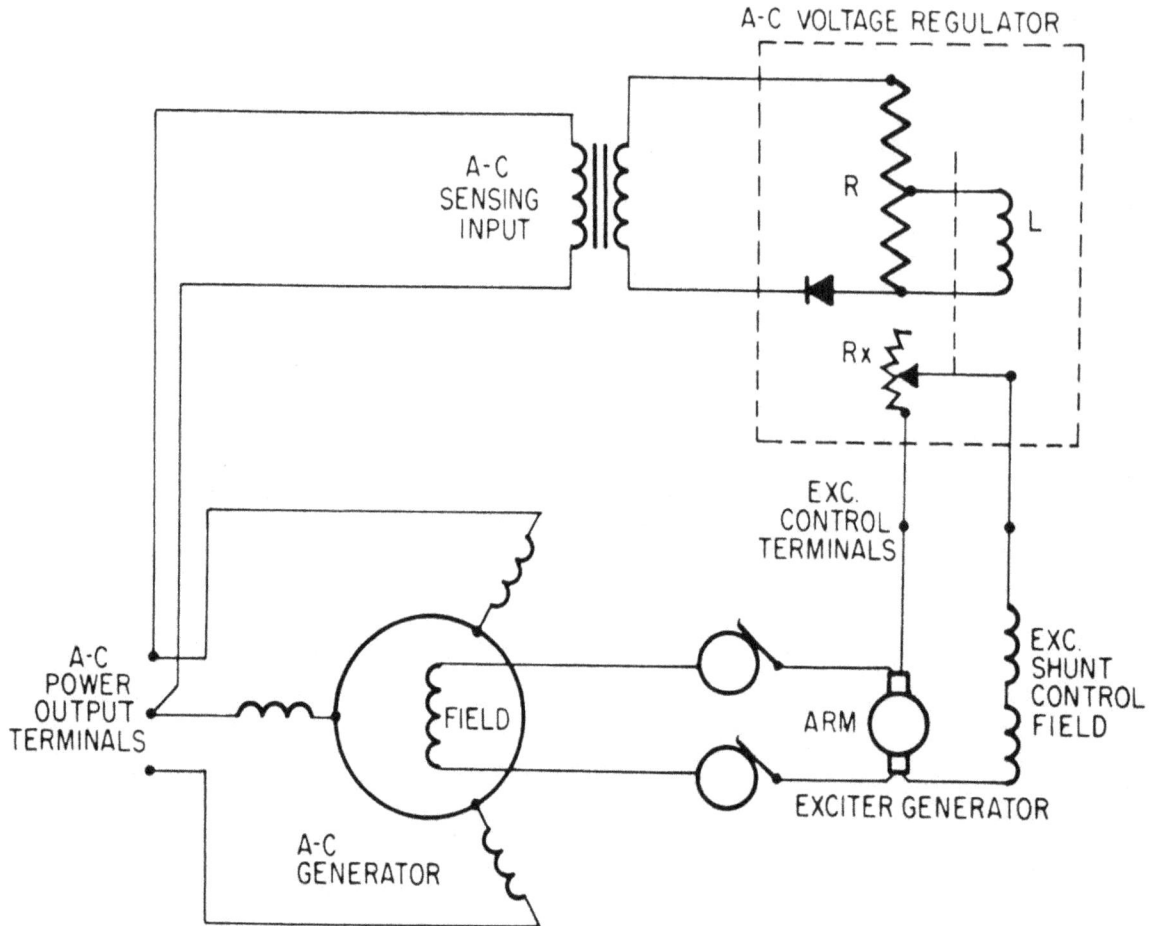

BE.298

Figure 16-15.—Simplified voltage regulator circuit.

this case, the phase sequence of one generator may be ACB, while the other is ABC.

Synchronizing A-C Generators

All of the foregoing conditions may be set by the following methods:

Equal voltages may be checked by using a voltmeter. The remaining conditions, equal frequency, phase relations, and phase sequence, are checked by using either synchronizing lamps or a synchroscope. There are a number of ways in which synchronizing lamps may be connected, but the most satisfactory arrangement is the two-bright one-dark method. This connective arrangement is shown in figure 16-16. Note that the lamps are connected directly between

the incoming generator's output and the buses. In this way, the two a-c sources may be synchronized before the incoming generator's main power contactor is closed. At the instant the two generators are to be paralleled $L_2$ and $L_3$ will glow with maximum brilliance and $L_1$ will be dark.

Assume that the incoming generator is far out of synchronism—lagging. All three lamps will appear to glow steadily, because the frequency of the voltage across them is the difference in the frequencies of the two generators, and thus is too high for individual alterations to be observed. As the lagging generator is accelerated, however, the lamp (differential) frequency decreases until their light flickers visibly. Their flickering will have a rotating

303

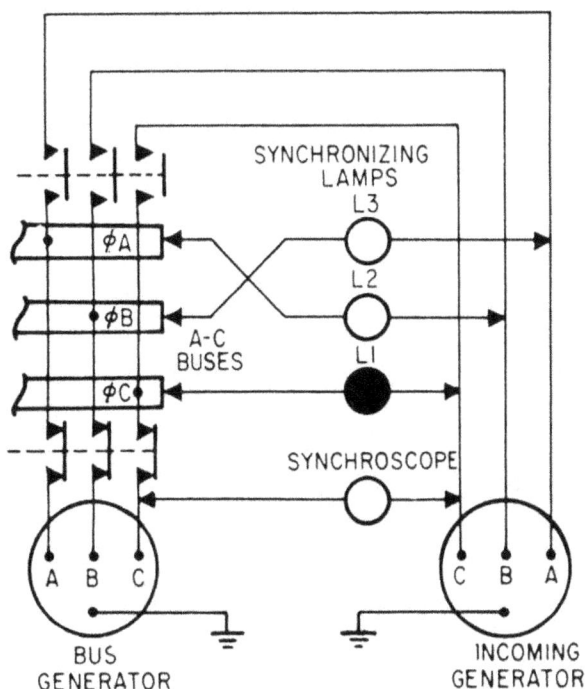

BE.299

Figure 16-16.—Synchronizing lamp connection
for two-bright one-dark method.

sequence, if connections are correct, and will
indicate which generator is faster.

At a point approaching synchronism, lamp
$L_1$ will be dark because it is connected between
like phases. That is, the two phase C voltages
will be so nearly synchronized that their differential voltage across $L_1$ will be insufficient
to make it glow visibly. However, this differential may still be of sufficient magnitude to damage
the generators should they be connected at this
time. The reason for cross-connecting $L_2$ and
$L_3$ is now indicated. Under perfectly synchronized conditions, the phase voltages across $L_2$
and $L_3$ are both $120°$ apart, because of their
cross-connection, and both glow with equal
brilliance. However, if the generators are not
in complete synchronism, but only very near it,
one of the bright lamps would be increasing in
voltage as the other was decreasing. This
action would cause a visible difference in the
brilliance of $L_2$ and $L_3$. Thus, by adjusting the
incoming generator's frequency so that no visible difference of brilliance exists between $L_2$
and $L_3$, the exact point of synchronism can be
approached very closely before paralleling the
incoming generator with the generator on the
bus.

When the lamps have been used to bring the
generators as close to synchronization as is
visibly possible, the synchroscope is used to
make the final fine adjustment to the incoming
generator's frequency. A synchroscope is a
highly sensitive instrument used to detect a difference in frequency between two a-c sources.

TRANSFORMERS

A transformer is a device that has no moving
parts and that transfers energy from one circuit
to another by electromagnetic induction. The
energy is always transferred without a change
in frequency, but usually with changes in voltage
and current. A step-up transformer receives
electrical energy at one voltage and delivers it
at a higher voltage. Conversely, a stepdown
transformer receives energy at one voltage and
delivers it at a lower voltage. Transformers
require little care and maintenance because of
their simple, rugged, and durable construction.
The efficiency of transformers is high. Because
of this, transformers are responsible for the
more extensive use of alternating current than
direct current. The conventional constant-
potential transformer is designed to operate with
the primary connected across a constant-
potential source and to provide a secondary
voltage that is substantially constant from no
load to full load.

Various types of small single-phase transformers are used in electrical equipment. In
many installations, transformers are used on
switchboards to stepdown the voltage for indicating lights. Low-voltage transformers are included in some motor control panels to supply
control circuits or to operate overload relays.

Instrument transformers include potential,
or voltage, transformers and current transformers. Instrument transformers are commonly used with a-c instruments when high voltages
or large currents are to be measured.

Electronic circuits and devices employ many
types of transformers to provide necessary
voltages for proper electron-tube operation,
interstage coupling, signal amplification, and so
forth. The physical construction of these transformers differs widely.

The power-supply transformer used in electronic circuits is a single-phase constant-
potential transformer with one or more secondary windings, or a single secondary with several
tap connections. These transformers have a low

volt-ampere capacity and are less efficient than large constant-potential power transformers. Most power-supply transformers for electronic equipment are designed to operate at a frequency of 50 to 60 Hz. Aircraft power-supply transformers are designed for a frequency of 400 Hz. The higher frequencies permit a saving in size and weight of transformers and associated equipment.

## CONSTRUCTION

The typical transformer has two windings insulated electrically from each other. These windings are wound on a common magnetic core made of laminated sheet steel. The principal parts are: (1) The core, which provides a circuit of low reluctance for the magnetic flux; (2) the primary windings, which receives the energy from the a-c source; (3) the secondary winding, which receives the energy by mutual induction from the primary and delivers it to the load, and (4) the enclosure.

When a transformer is used to step up the voltage, the low-voltage winding is the primary. Conversely, when a transformer is used to step down the voltage, the high-voltage winding is the primary. The primary is always connected to the source of the power; the secondary is always connected to the load. It is common practice to refer to the windings as the primary and secondary rather than the high-voltage and low-voltage windings.

The principal types of transformer construction are the core type and the shell type, as illustrated respectively in figure 16-17 (A) and (B). The cores are built of thin stampings of silicon steel. Eddy currents, generated in the core by the alternating flux as it cuts through the iron, are minimized by using thin laminations and by insulating adjacent laminations with insulating varnish. Hysteresis losses, caused by the friction developed between magnetic particles as they are rotated through each cycle of magnetization, are minimized by using a special grade of heat-treated grain-oriented silicon-steel laminations.

In the core type transformer, the copper windings surround the laminated iron core. In the shell type transformer the iron core surrounds the copper windings. Distribution transformers are generally of the core type; whereas some of the largest power transformers are of the shell type.

If the windings of a core-type transformer were placed on separate legs of the core, a relatively large amount of the flux produced by the primary winding would fail to link the secondary winding and a large leakage flux would result. The effect of the leakage flux would be to increase the leakage reactance drop, $IX_L$, in both windings. To reduce the leakage flux and reactance drop, the windings are subdivided and half of each winding is placed on each leg of the core. The windings may be cylindrical in form and placed one inside the other with the necessary insulation, as shown in figure 16-17 (A). The low-voltage winding is placed with a large part of its surface area next to the core, and the high-voltage winding is placed outside the low-voltage winding in order to reduce the insulation requirements of the two windings. If the high-voltage winding were placed next to the core, two layers of high-voltage insulation would be required, one next to the core and the other between the two windings.

In another method, the windings are built up in thin flat sections called pancake coils. These pancake coils are sandwiched together, with the required insulation between them, as shown in figure 16-17 (B).

The complete core and coil assembly (fig. 16-18 (A)) is placed in a steel tank. In some transformers the complete assembly is immersed in a special mineral oil to provide a means of insulation and cooling, while in other transformers they are mounted in drip-proof enclosures as shown in figure 16-18 (B).

Transformers are built in both single-phase and polyphase units. A 3-phase transformer consists of separate insulated windings for the different phases, wound on a 3-legged core capable of establishing three magnetic fluxes displaced $120°$ in time phase.

## VOLTAGE AND CURRENT RELATIONS

The operation of the transformer is based on the principle that electrical energy can be transferred efficiently by mutual induction from one winding to another. When the primary winding is energized from an a-c source, an alternating magnetic flux is established in the transformer core. This flux links the turns of both primary and secondary, thereby inducing voltages in them. Because the same flux cuts both windings, the same voltage is induced in each turn of both windings. Hence, the total

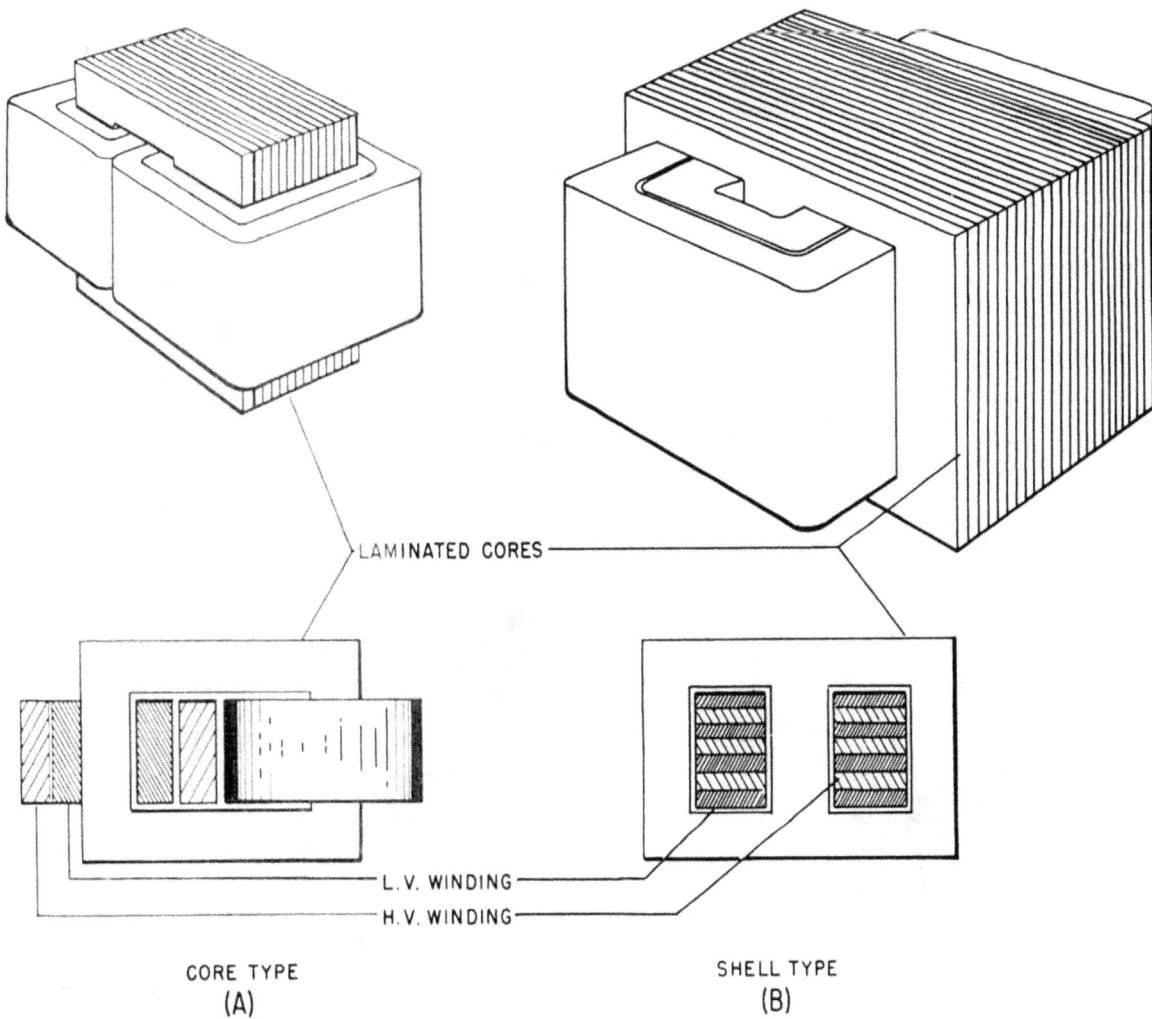

Figure 16-17.—Types of transformer construction.

induced voltage in each winding is proportional to the number of turns in that winding; that is,

$$\frac{E_1}{E_2} = \frac{N_1}{N_2}$$

where $E_1$ and $E_2$ are the induced voltages in the primary and secondary windings, respectively, and $N_1$ and $N_2$ are the number of turns in the primary and secondary windings, respectively. In ordinary transformers the induced primary voltage is almost equal to the applied primary voltage; hence, the applied primary voltage and the secondary induced voltage are approximately proportional to the respective number of turns in the two windings.

A constant-potential single-phase transformer is represented by the schematic diagram in figure 16-19 (A). For simplicity, the primary winding is shown as being on one leg of the core and the secondary winding on the other leg. The equation for the voltage induced in one winding of the transformer is

$$E = \frac{4.44 \, BSfN}{10^8}$$

306

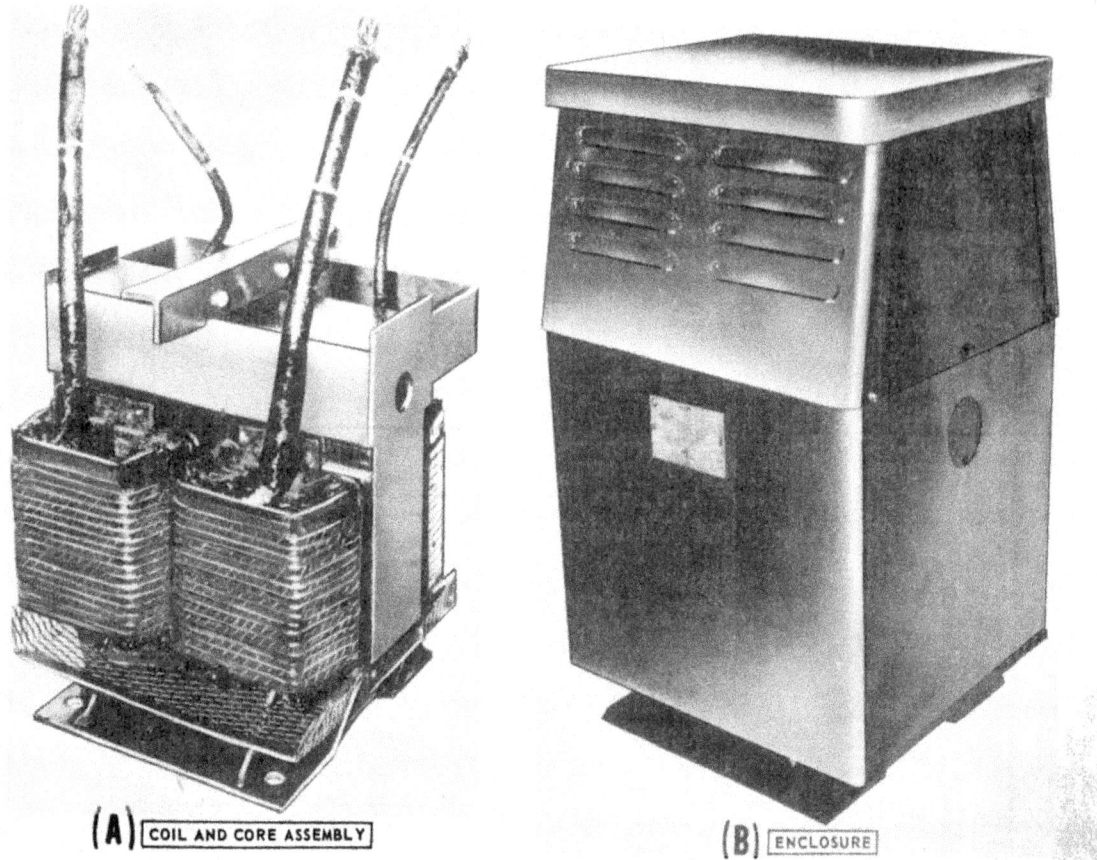

(A) COIL AND CORE ASSEMBLY    (B) ENCLOSURE

BE.301

Figure 16-18.—Single-phase transformer.

where E is the rms voltage, B the maximum value of the magnetic flux density in lines per square inch in the core, S the cross-sectional area of the core in square inches, f the frequency in hertz, and N the number of complete turns in the winding.

For example, if the maximum flux density is 90,000 lines per square inch, the cross-sectional area of the core is 4.18 square inches, the frequency is 60 Hz, and the number of turns in the high-voltage winding is 1,200, the voltage rating of this winding is

$$E_1 = \frac{4.44 \times 90,000 \times 4.18 \times 60 \times 1,200}{10^8}$$

$$= 1,200 \text{ volts}$$

If the primary-to-secondary turns ratio of this transformer is 10 to 1, the number of turns in the low-voltage winding will be

$$\frac{1,200}{10} = 120 \text{ turns}$$

and the voltage induced in the secondary will be

$$E_2 = \frac{1,200}{10} = 120 \text{ volts}$$

The waveforms of the ideal transformer with no load are shown in figure 16-19 (B). When $E_1$ is applied to the primary winding, $N_1$, with the

(A)

(B)                    (C)                    (D)

BE.302

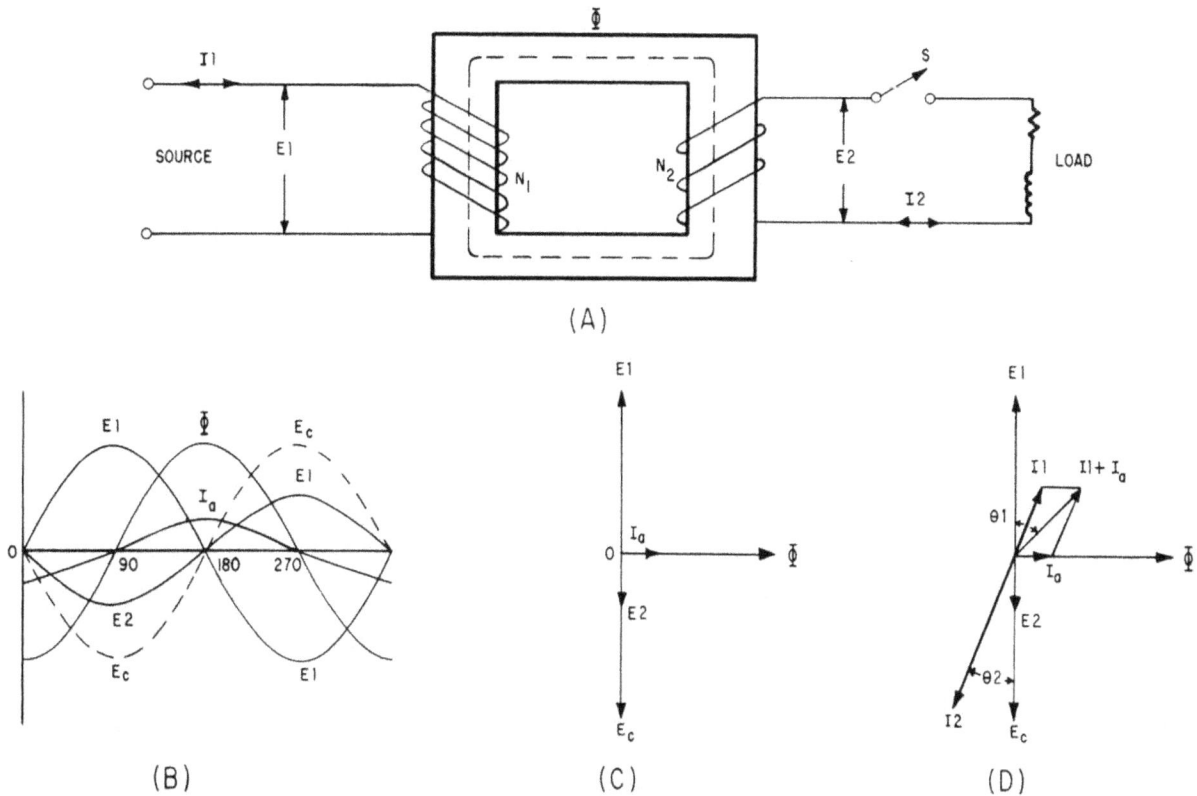

Figure 16-19.—Constant-potential transformer.

switch, S, open, the resulting current $I_a$, is small and lags $E_1$ by almost $90°$ because the circuit is highly inductive. This no-load current is called the exciting, or magnetizing, current because it supplies the magnetomotive force that produces the transformer core flux $\Phi$. The flux produced by $I_a$ cuts the primary winding, $N_1$, and induces a counter voltage $E_c$ $180°$ out of phase with $E_1$, in this winding. The voltage, $E_2$, induced in the secondary winding is in phase with the induced (counter) voltage, $E_c$, in the primary winding, and both lag the exciting current and flux, whose variations produce them, by an angle of $90°$. These relations are shown in vector form in figure 16-19 (C). The values are only approximate and are not drawn exactly to scale.

When a load is connected to the secondary by closing switch S (fig. 16-19 (A)), the secondary current $I_2$, depends upon the magnitude of the secondary voltage, $E_2$, and the load impedance, Z. For example, if $E_2$ is equal to 120 volts and the load impedance is 20 ohms, the secondary current will be

$$I_2 = \frac{E_2}{Z_2} = \frac{120}{20} = 6 \text{ amperes}$$

If the secondary power factor is 86.6 percent, the phase angle, $\theta_2$, between secondary current and voltage will be the angle whose cosine is 0.866, or $30°$.

The secondary load current flowing through the secondary turns comprises a load component of magnetomotive force, which according to Lenz's law is in such a direction as to oppose the flux which is producing it. This opposition tends to reduce the transformer flux a slight amount. The reduction in flux is accompanied by a reduction in the counter voltage induced in the primary winding of the transformer. Because the internal impedance of the primary winding is low and the primary current is limited principally by the counter emf in the winding, the transformer primary current increases when the counter emf in the primary is reduced.

The increase in primary current continues until the primary ampere-turns are equal to the secondary ampere-turns, neglecting losses. For example, in the transformer being considered, the magnetizing current, $I_a$, is assumed to be negligible in comparison with the total primary current, $I_1 + I_a$, under load conditions because $I_a$ is small in relation to $I_1$ and lags it by an angle of $60°$. Hence, the primary and secondary ampere-turns are equal and opposite; that is,

$$N_1 I_1 = E_2 I_2$$

In this example,

$$I_1 = \frac{N_2}{N_1} I_2 = \frac{120}{1200} \times 6 = 0.6 \text{ ampere}$$

Neglecting losses, the power delivered to the primary is equal to the power supplied by the secondary to the load. If the load power is $P_2 = E_2 I_2 \cos \theta_2$, or $120 \times 6$ (cos $30° = 0.866$) $= 624$ watts, the power supplied to the primary is approximately $P_1 = E_1 I_1 \cos \theta$, or $1200 \times 0.6$ (cos $30° = 0.866$) $= 624$ watts.

The load component of primary current, $I_1$, increases with secondary load and maintains the transformer core flux at nearly its initial value. This action enables the transformer primary to take power from the source in proportion to the load demand, and to maintain the terminal voltage approximately constant. The lagging-power-factor load vectors are shown in figure 16-19 (D). Note that the load power factor is transferred through the transformer to the primary and that $\theta_2$ is approximately equal to $\theta_1$, the only difference being that $\theta_1$ is slightly larger than $\theta_2$ because of the presence of the exciting current which flows in the primary winding but not in the secondary.

The copper loss of a transformer varies as the square of the load current; whereas the core loss depends on the terminal voltage applied to the primary and on the frequency of operation. The core loss of a constant-potential transformer is constant from no load to full load because the frequency is constant and the effective values of the applied voltage, exciting current, and flux density are constant.

If the load supplied by a transformer has unity power factor, the kilowatt (true power) output is the same as the kilovolt-ampere (apparent power) output. If the load has a lagging power factor, the kilowatt output is proportionally less than the kilovolt-ampere output. For example, a transformer having a

full-load rating of 100 kva can supply a 100-kw load at unity power factor, but only an 80-kw load at a lagging power factor of 80 percent.

Many transformers are rated in terms of the kva load that they can safely carry continuously without exceeding a temperature rise of $80°$ C when maintaining rated secondary voltage at rated frequency and when operating with an ambient (surrounding atmosphere) temperature of $40°$ C. The actual temperature rise of any part of the transformer is the difference between the total temperature of that part and the temperature of the surrounding air.

It is possible to operate transformers on a higher frequency than that for which they are designed, but it is not permissable to operate them at more than 10 percent below their rated frequency, because of the resulting overheating. The exciting current in the primary varies directly with the applied voltage and, like any impedance containing inductive reactance, the exciting current varies inversely with the frequency. Thus, at reduced frequency, the exciting current becomes excessively large and the accompanying heating may damage the insulation and the windings.

EFFICIENCY

The efficiency of a transformer is the ratio of the output power at the secondary terminals to the input power at the primary terminals. It is also equal to the ratio of the output to the output plus losses. That is,

$$\text{efficiency} = \frac{\text{output}}{\text{input}}$$

$$= \frac{\text{output}}{\text{output} + \text{copper loss} + \text{core loss}}$$

The ordinary power transformer has an efficiency of 97 to 99 percent. The losses are due to the copper losses in both windings and the hysteresis and eddy-current losses in the iron core.

The copper losses vary as the square of the current in the windings and as the winding resistance. In the transformer being considered, if the primary has 1,200 turns of number 23 copper wire, having a length of 1,320 feet, the resistance of the primary winding is 26.9 ohms. If the load current in the primary is 0.5 ampere, the primary copper loss in $(0.5)^2 \times 26.9 = 6.725$ watts. Similarly, if the secondary winding contains 120 turns of number 13 copper wire, having a length of approximately 132 feet, the secondary resistance will be 0.269 ohm. The secondary copper loss is $I_2^2 R_2$, or $(5)^2$ x 0.269 = 6.725 watts, and the total copper loss is 6.725 x 2 = 13.45 watts.

The core losses, consisting of the hysteresis and eddy-current losses, caused by the alternating magnetic flux in the core are approximately constant from no load to full load, with rated voltage applied to the primary.

In the transformer of figure 16-19 (A), if the core loss is 10.6 watts and the copper loss is 13.4 watts, the efficiency is

$$\frac{output}{output + copper\ loss + core\ loss} =$$

$$\frac{624}{624 + 13.4 + 10.6} = \frac{624}{648} = 0.963$$

or 96.3 percent. The rating of the transformer is

$$\frac{E_1 I_1}{1,000} = \frac{1,200 \times 0.5}{1,000} = 0.60\ kva$$

The efficiency of this transformer is relatively low because it is a small transformer and the losses are disproportionately large.

## SINGLE-PHASE CONNECTIONS

Single-phase distribution transformers usually have their windings divided into two or more sections, as shown in figure 16-20. When the two secondary windings are connected in series (fig. 16-20 (A)), their voltages add. In figure 16-20 (B), the two secondary windings are connected in parallel, and their currents add. For example, if each secondary winding is rated at 120 volts and 100 amperes, the series-connection output rating will be 240 volts at 100 amperes, or 24 kva; the parallel-connection output rating will be 120 volts at 200 amperes, or 24 kva.

In the series connection, care must be taken to connect the coils so that their voltages add. The proper arrangement is indicated in the figure. A trace made through the secondary circuits from $X_1$ to $X_4$ is in the same direction as that of the arrows representing the maximum positive voltages.

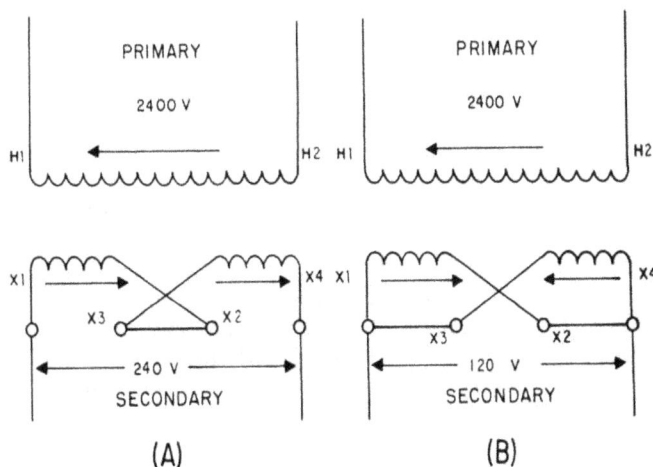

BE.303

Figure 16-20.—Single-phase transformer secondary connections.

In the parallel connection, care must be taken to connect the coils so that their voltages are in opposition. The correct connection is indicated in the figure. The direction of a trace made through the secondary windings from $X_1$ to $X_2$ to $X_4$ to $X_3$ and returning to $X_1$ is the same as that of the arrow in the right-hand winding. This condition indicates that the secondary voltages have their positive maximum values in directions opposite to each other in the closed circuit, which is formed by paralleling the two secondary windings. Thus, no circulating current will flow in these windings on no load. If either winding were reversed, a short-circuit current would flow in the secondary, and this would cause the primary to draw a short-circuit current from the source. This action would, of course, damage the transformer as well as the source.

## THREE-PHASE CONNECTIONS

Power may be supplied through 3-phase circuits containing transformers in which the primaries and secondaries are connected in various wye and delta combinations. For example, three single-phase transformers may supply 3-phase power with four possible combinations of their primaries and secondaries. These connections are: (1) primaries in delta and secondaries in delta, (2) primaries in wye and secondaries in wye, (3) primaries in wye and secondaries in delta, (4) primaries in delta and secondaries in wye.

Delta and wye connections were described earlier in this chapter under the subject "Three-Phase Generators," and the relations between line and phase voltages and currents were also described. These relations apply to transformers as well as to a-c generators.

If the primaries of three single-phase transformers are properly connected (either in wye or delta) to a 3-phase source, the secondaries may be connected in delta, as shown in figure 16-21. A topographic vector diagram of the 3-phase secondary voltages is shown in figure 16-21 (A). The vector sum of these three voltages is zero. This may be seen by combining

any two vectors, for example, $E_A$ and $E_B$, and noting that their sum is equal and opposite to the 3rd vector, $E_C$. A voltmeter inserted within the delta will indicate zero voltage, as shown in figure 16-21 (B), when the windings are connected properly.

Assuming all three transformers have the same polarity, the delta connection consists of connecting the $X_2$ lead of winding A to the $X_1$ lead of B, the $X_2$ lead of B to $X_1$ of C, and the $X_2$ lead of C to $X_1$ of A. If any one of the three windings is reversed with respect to the other two windings, the total voltage within the delta will equal twice the value of one phase; and if

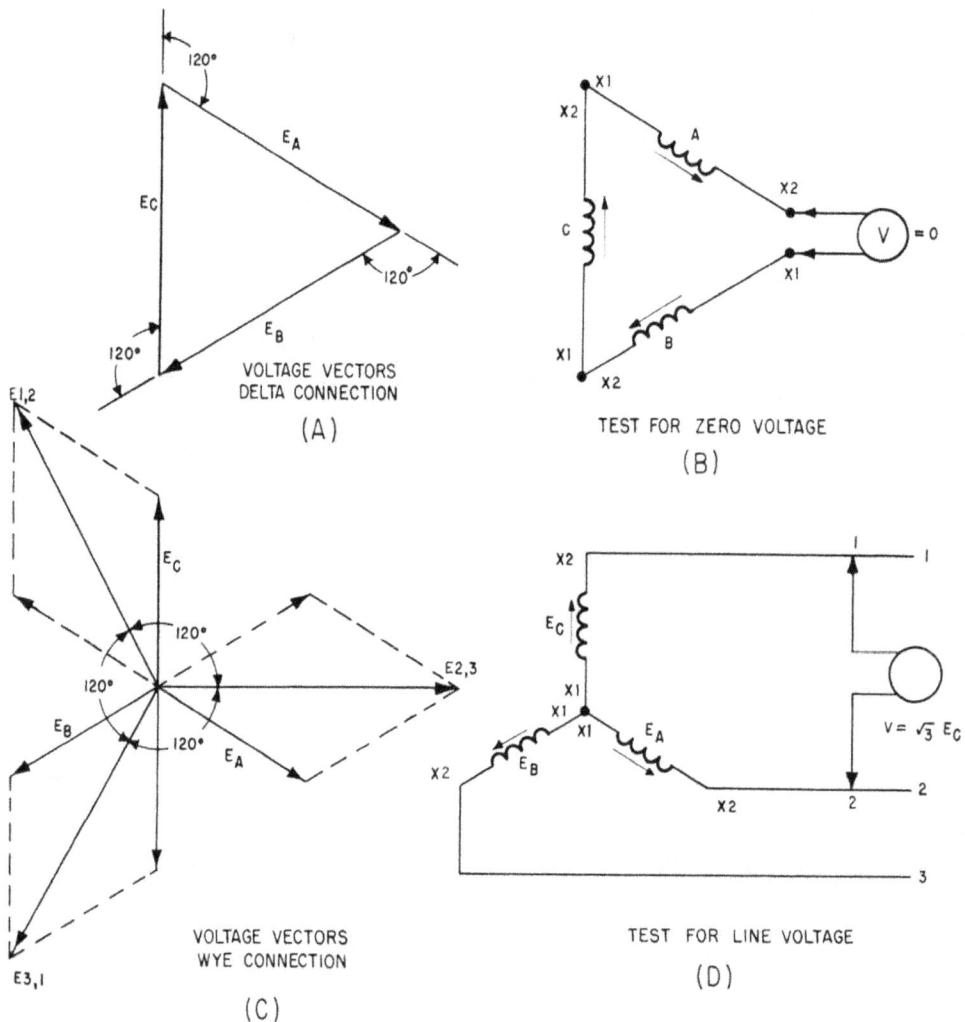

Figure 16-21.—Delta-connected transformer secondaries.

BE.304

the delta is closed on itself, the resulting current will be of short-circuit magnitude, with resulting damage to the transformer windings and cores. The delta should never be closed until a test is first made to determine that the voltage within the delta is zero or nearly zero. This may be accomplished by using a voltmeter, fuse wire, or test lamp. In the figure, when the voltmeter is inserted between the $X_2$ lead of A and the $X_1$ lead of B, the delta circuit is completed through the voltmeter, and the indication should be approximately zero. Then the delta is completed by connecting the $X_2$ lead of A to the $X_1$ lead of B.

If the three secondaries of an energized transformer bank are properly connected in delta and are supplying a balanced 3-phase load, the line current will be equal to 1.73 times the phase current. If the rated current of a phase (winding) is 100 amperes, the rated line current will be 173 amperes. If the rated voltage of a phase is 120 volts, the voltage between any two line wires will be 120 volts.

The three secondaries of the transformer bank may be reconnected in wye in order to increase the output voltage. The voltage vectors are shown in figure 16-21 (C). If the phase voltage is 120 volts, the line voltage will be 1.73 x 120 = 208 volts. The line voltages are represented by vectors, $E_{1,2}$, $E_{2,3}$, and $E_{3,1}$. A voltmeter test for the line voltage is represented in figure 16-21 (D). If the three transformers have the same polarity, the proper connections for a wye-connected secondary bank are indicated in the figure. The $X_1$ leads are connected to form a common or neutral connection and the $X_2$ leads of the three secondaries

are brought out to the line leads. If the connections of any one winding are reversed, the voltages between the 3 line wires will become unbalanced, and the loads will not receive their proper magnitude of load current. Also the phase angle between the line currents will be changed, and they will no longer be 120° out of phase with each other. Therefore, it is important to properly connect the transformer secondaries in order to preserve the symmetry of the line voltages and currents.

Three single-phase transformers with both primary and secondary windings delta connected are shown in figure 16-22. The $H_1$ lead of one phase is always connected to the $H_2$ lead of an adjacent phase, the $X_1$ lead is connected to the $X_2$ terminal of the corresponding adjacent phase, and so on; and the line connections are made at these junctions. This arrangement is based on the assumption that the three transformers have the same polarity.

An open-delta connection results when any one of the three transformers is removed from the delta-connected transformer bank without disturbing the 3-wire 3-phase connections to the remaining two transformers. These transformers will maintain the correct voltage and phase relations on the secondary to supply a balanced 3-phase load. An open-delta connection is shown in figure 16-23. The 3-phase source supplies the primaries of the two transformers, and the secondaries supply a 3-phase voltage to the load. The line current is equal to the transformer phase current in the open-delta connection. In the closed-delta connection, the transformer phase current, $I_{phase} = \frac{I_{line}}{\sqrt{3}}$.

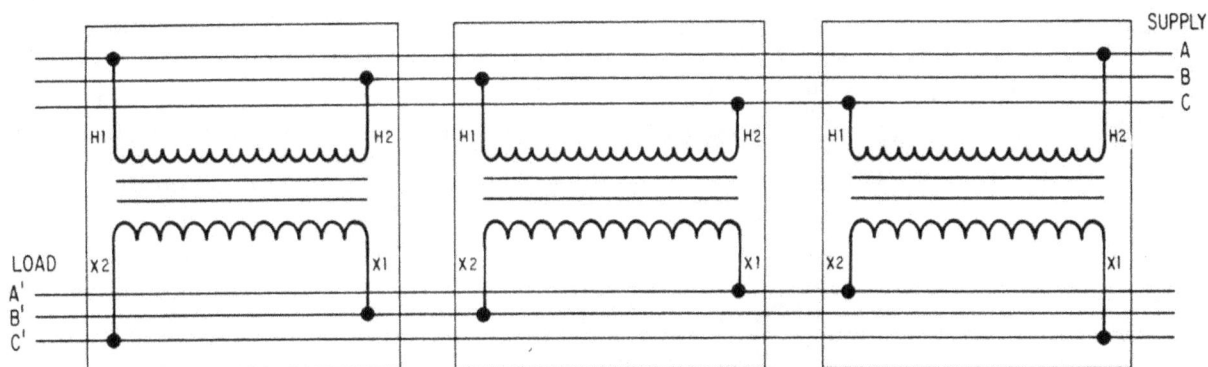

BE.305

Figure 16-22.—Delta-delta transformer connections.

312

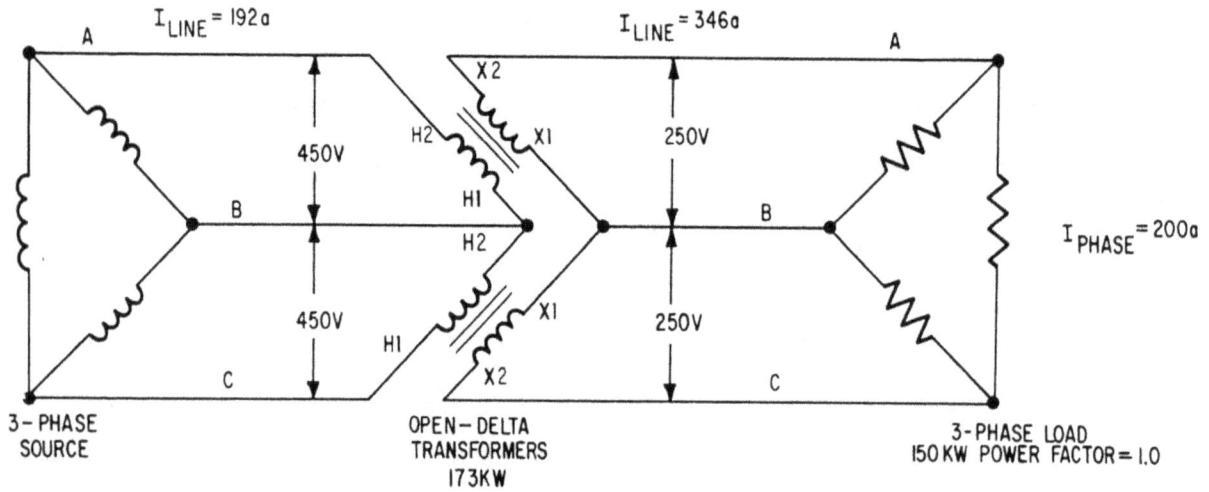

Figure 16-23.—Open-delta transformer connection.

BE.306

Thus, when one transformer is removed from a delta-connected bank of three transformers, the remaining two transformers will carry a current equal to $\sqrt{3}$ $I_{phase}$. This value amounts to an overload current on each transformer of 1.73 times the rated current, or an overload of 73.2 percent.

Thus, in an open-delta connection, the line current must be reduced so as not to exceed the rated current of the individual transformers if they are not to be overloaded. The open-delta connection therefore results in a reduction in system capacity. The full-load capacity in a delta connection at unity power factor is

$$P_\Delta = 3I_{phase}E_{phase} = \sqrt{3}E_{line}I_{line}$$

In an open-delta connection, the line current is limited to the rated phase current of $\dfrac{I_{line}}{\sqrt{3}}$, and the full-load capacity of the open-delta, or V-connected, system is

$$P_v = \sqrt{3}E_{line} \frac{I_{line}}{\sqrt{3}} = E_{line}I_{line}$$

The ratio of the load that can be carried by two transformers connected in open delta to the load that can be carried by three transformers in closed delta is

$$\frac{P_v}{P_\Delta} = \frac{E_{line}I_{line}}{\sqrt{3}E_{line}I_{line}} = \frac{1}{\sqrt{3}}$$

$$= 0.577, \text{ or } 57.7 \text{ percent}$$

of the closed-delta rating.

For example, a 150-kw 3-phase balanced load operating at unity power factor is supplied at 250 volts. The rating of each of three transformers in closed delta is $\dfrac{150}{3}$ = 50 kw, and the phase current is $\dfrac{50,000}{250}$ = 200 amperes. The line current is $200\sqrt{3}$ = 346 amperes. If one transformer is removed from the bank, the remaining two transformers would be overloaded 346 - 200 = 146 amperes, or $\dfrac{146}{200}$ x 100 = 73 percent. To prevent overload on the remaining two transformers, the line current must be reduced from 346 amperes to 200 amperes and the total load reduced to

$$\frac{\sqrt{3} \times 250 \times 200}{1,000} = 86.6 \text{ kw}$$

or

$$\frac{86.6}{150} \times 100 = 57.7 \text{ percent}$$

of the original load.

313

The rating of each transformer in open delta necessary to supply the original 150-kw load is $\frac{E_{phase}I_{phase}}{1,000}$, or $\frac{250 \times 346}{1,000} = 86.6$ kw, and two transformers require a total rating of $2 \times 86.6 = 173.2$ kw, compared with 150 kw for three transformers in closed delta. The required increase in transformer capacity is $173.2 - 150 = 23.2$ kw, or $\frac{23.2}{150} \times 100 = 15.5$ percent, when two transformers are used in open delta to supply the same load as three 50-kw transformers in closed delta.

Three single-phase transformers with both primary and secondary windings wye connected are shown in figure 16-24. Only 57.7 percent of the line voltage $(\frac{E_{line}}{\sqrt{3}})$ is impressed across each winding, but full-line current flows in each transformer winding.

Three single-phase transformers delta connected to the primary circuit and wye connected to the secondary circuit are shown in figure 16-25. This connection provides 4-wire, 3-phase service with 208 volts between line wires A'B'C', and $\frac{208}{\sqrt{3}}$, or 120 volts, between each line wire and neutral N. The wye-connected secondary is desirable in installations when a large number of single-phase loads are to be supplied from a 3-phase transformer bank. The neutral, or grounded, wire is brought out from the midpoint of the wye connection, permitting the single-phase loads to be distributed evenly across the three phases. At the same time, 3-phase loads can be connected directly across the line wires. The single-phase loads have a voltage rating of 120 volts, and the 3-phase loads are rated at 208 volts. This connection is often used in high-voltage plate-supply transformers. The phase voltage is $\frac{1}{1.73}$, or 0.577 of the line voltage.

Three single-phase transformers with wye-connected primaries and delta-connected secondaries are shown in figure 16-26. This arrangement is used for stepping down the voltage from approximately 4,000 volts between line wires on the primary side to either 120 volts or 240 volts, depending upon whether the secondary windings of each transformer are

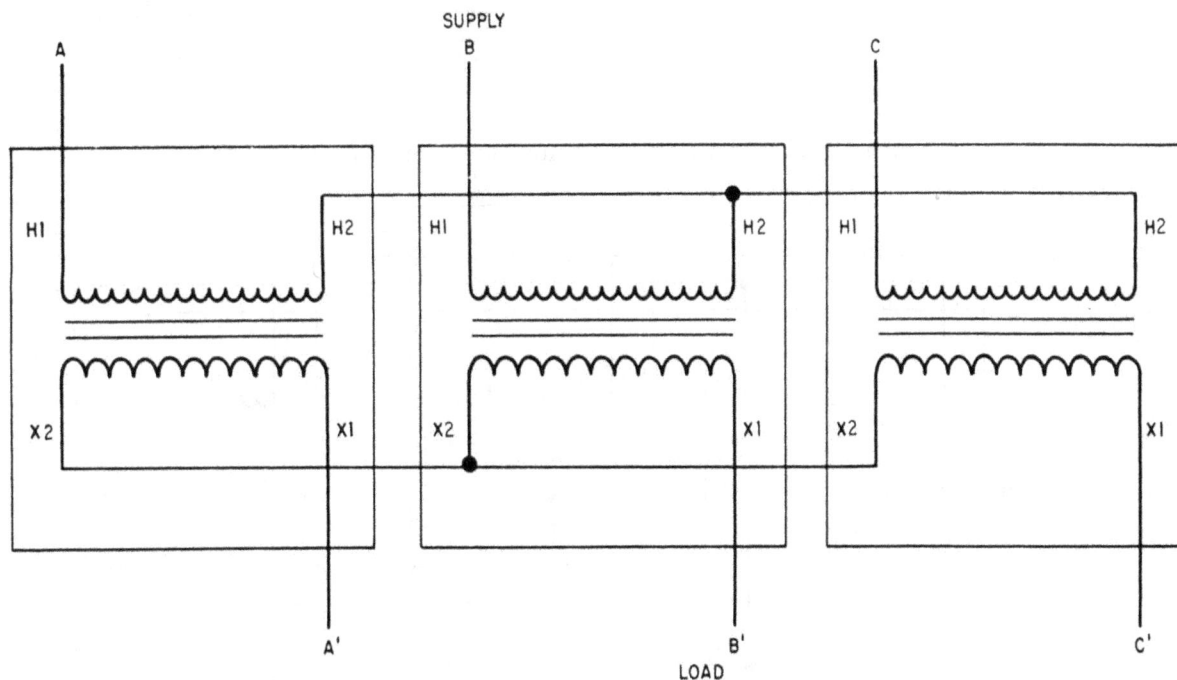

Figure 16-24.—Wye-wye transformer connections.

BE.307

314

connected in parallel or in series. In the figure, the two secondaries of each transformer are connected in parallel, and the secondary output voltage is 120 volts. There is an economy in transmission with the primaries in wye because the line voltage is 73 percent higher than the phase voltage, and the line current is accordingly less. Thus, the line losses are reduced and the efficiency of transmission is improved.

## POLARITY MARKING OF TRANSFORMERS

It is essential that all transformer windings be properly connected; therefore, it behooves the technician to have a basic understanding of the coding and the marking of transformer leads.

Small power transformers, of the size used in electronic equipment, are generally color

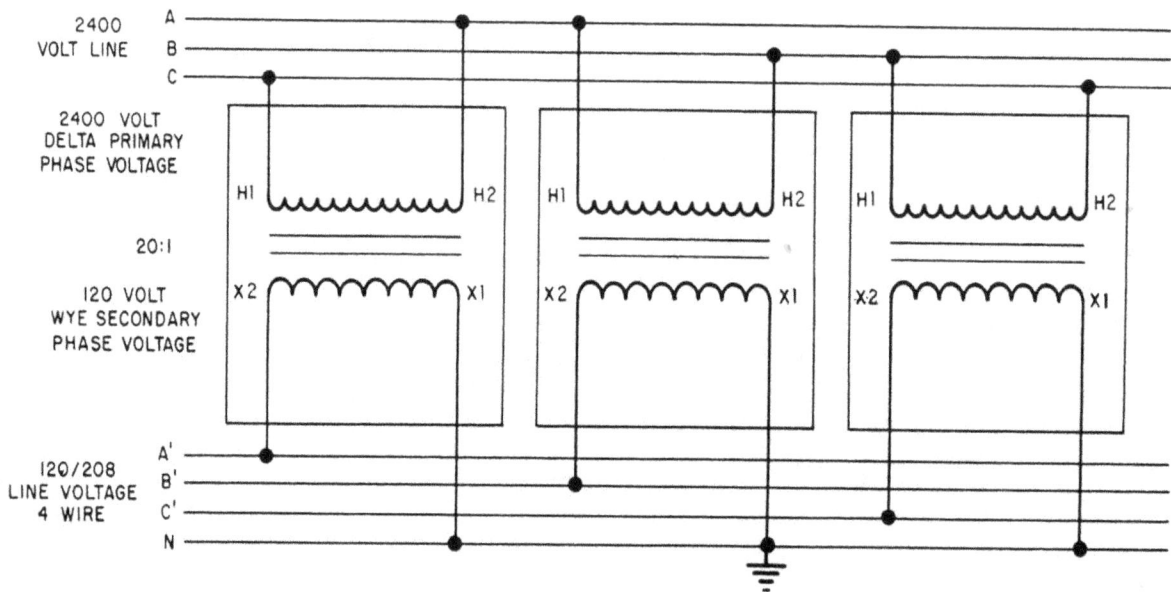

Figure 16-25.—Delta-wye transformer connections.

BE.308

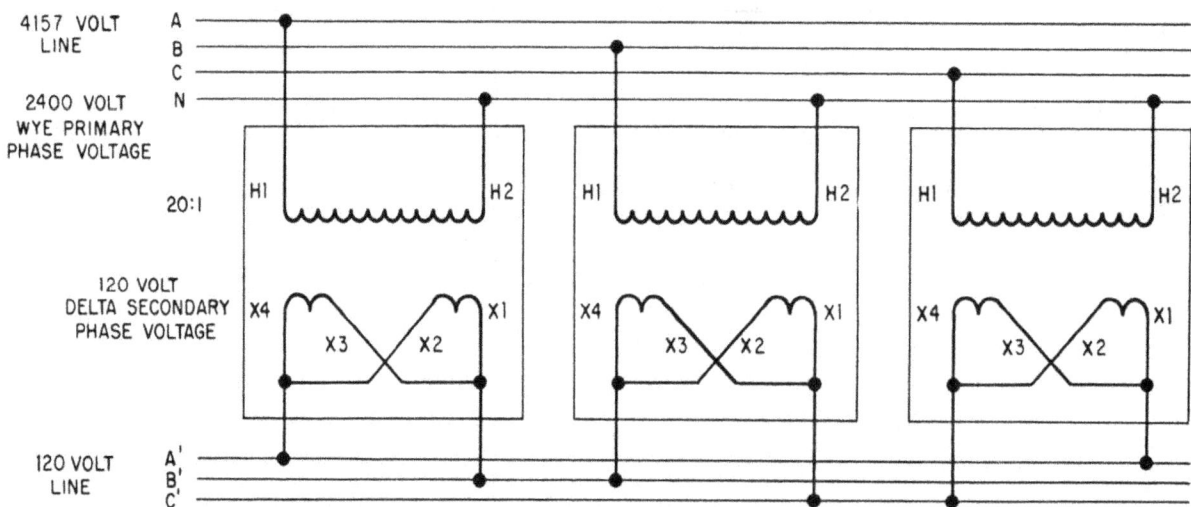

Figure 16-26.—Wye-delta transformer connections.

BE.309

coded as shown in figure 16-27. In an untapped primary, both leads are black. If the primary is tapped, one lead is common and is colored black, the tap lead is black and yellow, and the other lead is black and red.

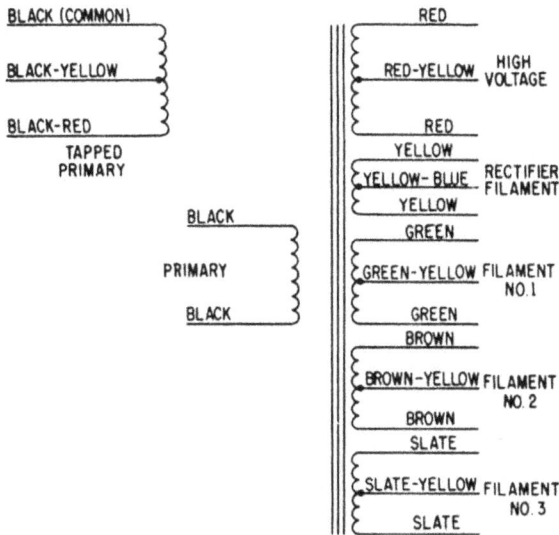

SUBTRACTIVE POLARITY (A)　　ADDITIVE POLARITY (B)

BE.311

Figure 16-28.—Polarity markings for large transformers.

BE.310

Figure 16-27.—Color coding of small power transformer leads.

On the transformer secondary, the high-voltage winding has two red leads if untapped, or two red leads and a yellow tap lead if tapped. On the rectifier filament windings, yellow leads are used across the whole winding, and the tap lead is yellow and blue. If there are other filament windings, they will be either green, brown, or slate. The tapped wire will be yellow in combination with one of the colors just named, that is green and yellow, brown and yellow, or slate and yellow.

The leads of large power transformers, such as those used for lighting and public utilities, are marked with numbers, letters, or a combination of both. This type of marking is shown in figure 16-28. Terminals for the high-voltage windings are marked $H_1$, $H_2$, $H_3$, and so forth. The increasing numerical subscript designates an increasing voltage. Thus, the voltage between $H_1$ and $H_3$ is higher than the voltage between $H_1$ and $H_2$.

The secondary terminals are marked $X_1$, $X_2$, $X_3$, and so forth. There are two types of markings that may be employed on the secondaries. These are shown in figure 16-28. When the $H_1$ and $X_1$ leads are brought out on the same side of the transformer (fig. 16-28 (A)), the polarity is called subtractive. The reason this arrangement is called subtractive is as follows: If the $H_1$ and $X_1$ leads are connected and a reduced voltage is applied across the $H_1$ and $H_2$ leads, the resultant voltage which appears across the $H_2$ and $X_2$ leads in the series circuit formed by this connection will equal the difference in the voltages of the two windings. The voltage of the low-voltage winding opposes that of the high-voltage winding and subtracts from it, hence the term "subtractive polarity."

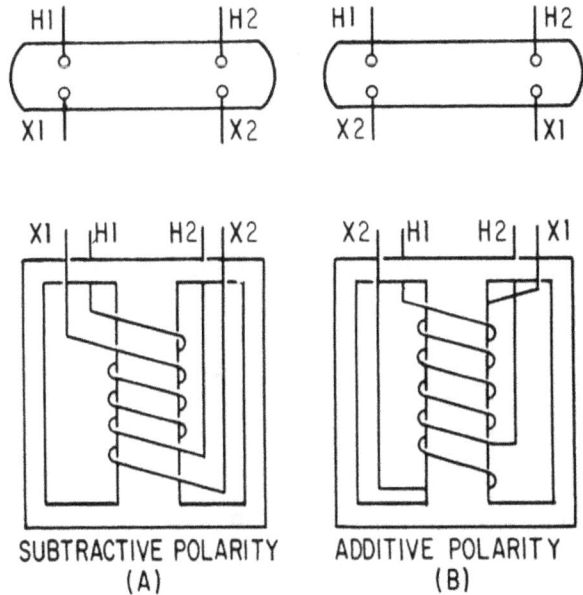

When the $H_1$ and $X_1$ leads are brought out on opposite corners of the transformer (fig. 16-28 (B)), the polarity is additive. If the $H_1$ and $X_2$ leads are connected and a reduced voltage is applied across the $H_1$ and $H_2$ leads, the resultant voltage across the $H_2$ and $X_1$ leads in the series circuit formed by this connection will equal the sum of the voltages of the two windings. The voltage of the low-voltage winding aids the voltage of the high-voltage winding and adds to it, hence the term "additive polarity."

Polarity markings do not indicate the internal voltage stress in the windings but are useful only in making external connections between transformers, as previously mentioned.

# CHAPTER 17

# ALTERNATING-CURRENT MOTORS

Since alternating voltage can be easily transformed from low voltages to high voltages or vice versa, it can be moved over a much greater distance without too much loss in efficiency. Most of the power generating systems today, therefore, produce alternating current. Thus, it follows that a great majority of the electrical motors utilized today are designed to operate on alternating current. However, there are other advantages in the use of a-c motors besides the wide availability of a-c power. In general, a-c motors are less expensive than d-c motors. Most types of a-c motors do not employ brushes and commutators. This eliminates many problems of maintenance and wear; in addition, it eliminates the problem of dangerous sparking.

D-c motors are best suited for some uses, such as applications that require variable-speed motors. However, in the great majority of applications, the a-c motor is best and the design and increasing use, in recent years, of variable-speed controllers for induction motors, which offer wide speed control have greatly increased the use of a-c motors where speed control is required.

A-c motors are manufactured in many different sizes, shapes, and ratings, for use on an even greater number of jobs. They are designed for use with either polyphase or single-phase power systems.

This chapter cannot possibly cover all aspects of the subject of a-c motors. Consequently, it will deal mainly with the operating principles of only the two most common types—the rotating-field induction motor and the synchronous motor.

## ROTATING FIELD

The rotating field is set up by out-of-phase currents in the stator windings. Figure 17-1 illustrates the manner in which a rotating field is produced by stationary coils, or windings, when they are supplied by a 3-phase current source. For purpose of explanation, rotation of the field is developed in the figure by "stopping" it at six selected positions, or instants. These instants are marked off at 60° intervals on the sine waves representing currents in the three phases A, B, and C.

At instant 1 the current in phase B is maximum positive. (Assume plus 10 amperes in this example.) Current is considered to be positive when it is flowing out from a motor terminal, and negative when it flows into a motor terminal. At the same time (instant 1) current flows into the A and C terminals at half value (minus 5 amperes each in this case). These currents combine at the neutral (common connection) to supply plus 10 amperes out through the B phase.

The resulting field at instant 1 is established downward and to the right as shown by the arrow NS. The major portion of this field is produced by the B phase (full strength at this time) and is aided by the adjacent phases A and C (half strength). The weaker portions of the field are indicated by the letters "n" and "s." The field is a two-pole field extending across the space that would normally contain the rotor.

At instant 2 the current in phase B is reduced to half value (plus 5 amperes in this example). The current in phase C has reversed its flow from minus 5 amperes to plus 5 amperes, and the current in phase A has increased from minus 5 to minus 10 amperes.

The resulting field at instant 2 is now established upward and to the right as shown by the arrow NS. The major portion of the field is produced by phase A (full strength) and the weaker portions by phases B and C (half strength).

At instant 3 the current in phase C is plus 10 amperes and the field extends vertically upward;

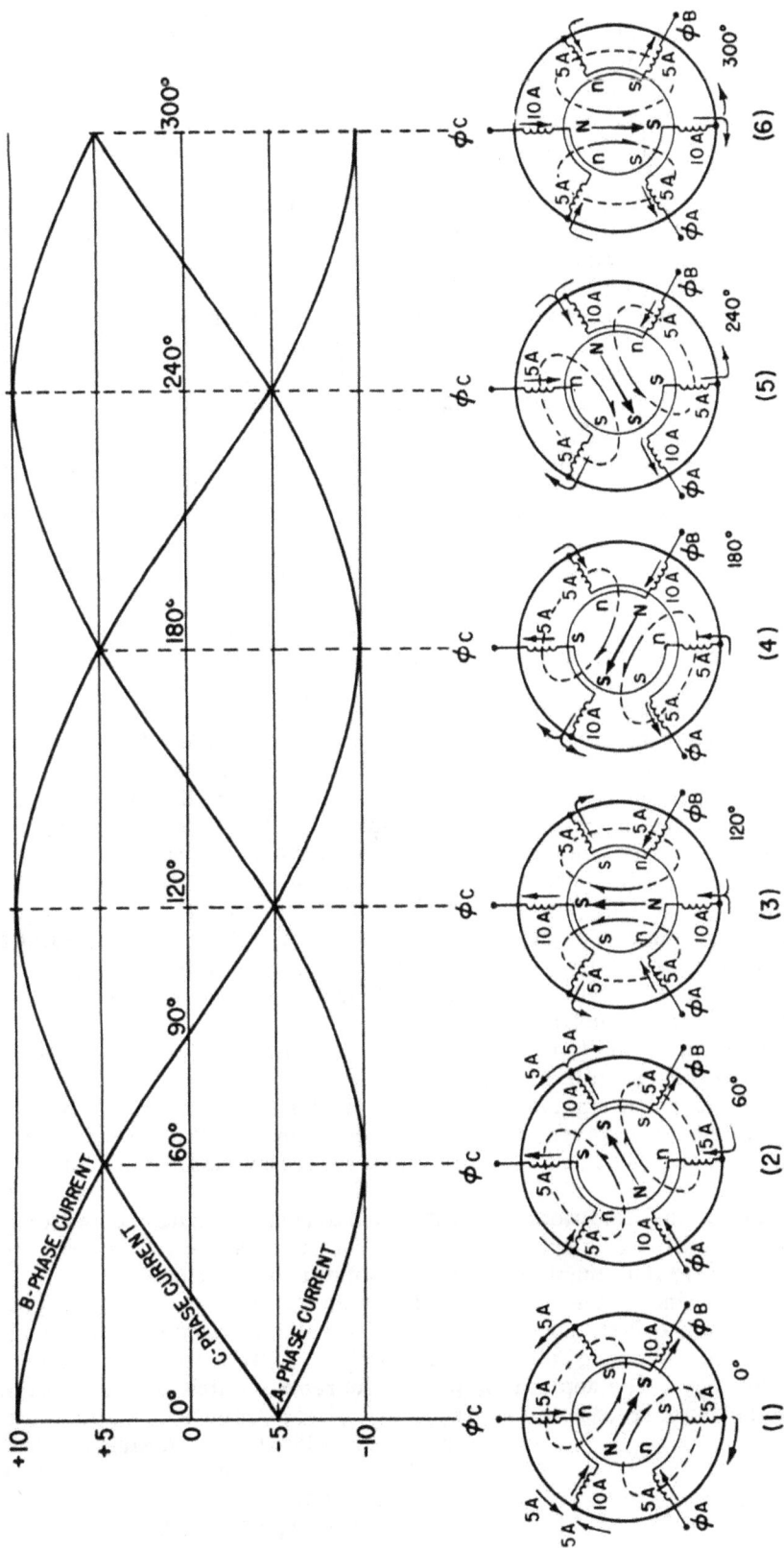

Figure 17-1.—Development of a rotating field.

BE.312

at instant 4 the current in phase B becomes minus 10 amperes and the field extends upward and to the left; at instant 5 the current in phase A becomes plus 10 amperes and the field extends downward and to the left; at instant 6 the current in phase C is minus 10 amperes and the field extends vertically downward. Instant 7 (not shown) corresponds to instant 1 when the field again extends downward and to the right.

Thus a full rotation of the two-pole field has been accomplished through one full cycle of 360 electrical degrees of the 3-phase currents flowing in the windings.

The direction of rotation of the magnetic revolving field may be changed by interchanging any two line leads to the three motor terminals. For example, in figure 17-2 (A), if line 1 connects to phase A, line 2 to phase B, and line 3 to phase C, and the line currents reach their positive maximum values in the sequence 1, 2, 3, the phase sequence is A, B, C and the rotation is arbitrarily clockwise. If lines 1 and 2 are interchanged, the phase sequence becomes B, A, C, and the revolving field turns counterclockwise.

Most induction motors utilized today are designed to operate on single-phase power supplies, or the 3-phase supply used in the preceding discussion. The out-of-phase currents necessary to produce a rotating field are inherent in the 2-phase power supply, since 2-phase voltages are generated $90°$ apart. When a single-phase supply is used, it is necessary to split the power supply into two separate coil groups. The required phase difference is then generally obtained by inserting capacitance in series with one of the groups (other methods are sometimes used). This causes the single-phase or "split-phase" motor to have characteristics similar in many respects to the true 2-phase motor.

In figure 17-1, note that the sine waves of current traversed $300°$ through the six positions shown. Accordingly, the field rotated $300°$. If the supplied current were 60 hertz, the field would rotate at 60 revolutions per second ($60 \times 60$ 3,600 revolutions per minute). However, if the number of stator coils were doubled, producing a 4-pole field, the field will rotate only half as fast. Thus, it can be seen that the speed of the revolving field varies directly as the frequency of the applied voltage and inversely as the number of poles. Thus,

$$N = \frac{120 \, f}{P}$$

where N is the number of revolutions that the field makes per minute, f the frequency of the applied voltage in hertz ($H_z$), and P the number of poles produced by the 3-phase winding.

The speed at which an induction motor field rotates is referred to as its SYNCHRONOUS speed, because it is synchronized to the frequency of the power supply at all times. A motor having a 2-pole 3-phase stator winding connected to a 60-$H_z$ source has a synchronous speed (magnetic revolving field speed) of 3,600 rpm. A 2-pole 25-$H_z$ motor has a synchronous speed of 1,500 rpm. Increasing the number of poles lowers the speed. Thus, a 4-pole 25-$H_z$ motor has a synchronous speed of 750 rpm. A 12-pole 60-$H_z$ motor has a synchronous speed of 600 rpm. Increasing the frequency of the line supply increases the speed with which the field revolves. Thus, if the frequency is increased from 50 to 60 $H_z$, and the motor has 4 poles, the speed of the field is increased from 1,500 rpm to 1,800 rpm.

The speed of the rotating field is always independent of load changes on the motor, provided the line frequency is maintained constant. The magnetic revolving field always runs at the same speed, pole for pole, as the a-c generator supplying it. If a 2-pole 60-$H_z$ a-c generator supplies a 2-pole motor, the motor has a synchronous speed of 3,600 rpm. which is the same as the speed of the a-c generator. If a 4-pole 60-$H_z$ a-c generator runs at 1,800 rpm and supplies a 4-pole 60-$H_z$ motor, the motor has a synchronous speed of 1,800 rpm. If this same a-c generator supplies an 8-pole 60-$H_z$ motor, the motor has a synchronous speed of 900 rpm.

## POLYPHASE INDUCTION MOTORS

The driving torque of both d-c and a-c motors is derived from the reaction of current-carrying conductors in a magnetic field. In the d-c motor, the magnetic field is stationary and the armature, with its current-carrying conductors, rotates. The current is supplied to the armature through a commutator and brushes.

In induction motors, the rotor currents are supplied by electromagnetic induction. The stator windings contain two or more out-of-time-phase currents, which produce corresponding mmf's. These mmf's establish a rotating magnetic field across the airgap. This magnetic field rotates continuously at constant speed regardless of the load on the motor. The stator winding corresponds to the armature winding of

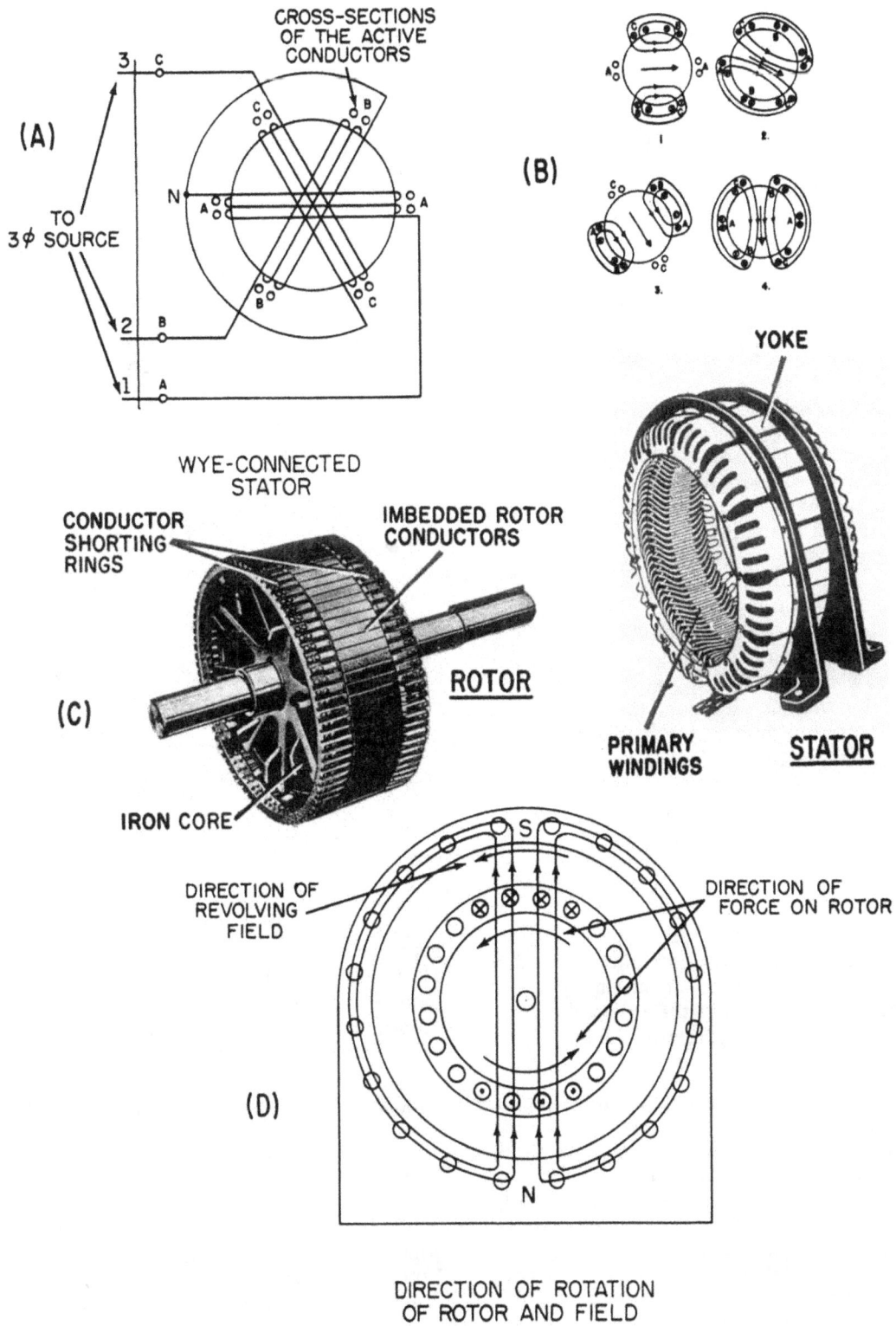

Figure 17-2.—Three-phase induction motor.

BE.313

a d-c motor or to the primary winding of a transformer. The rotor is not connected electrically to the power supply. The induction motor derives its name from the fact that mutual induction (transformer action) takes place between the stator and the rotor under operating conditions. The magnetic revolving field produced by the stator cuts across the rotor conductors, inducing a voltage in the conductors. This induced voltage causes rotor current to flow. Hence, motor torque is developed by the interaction of the rotor current and the magnetic revolving field.

Figure 17-2 (A) represents the winding of a 3-phase induction motor stator. Part (B) shows how the active stator conductors produce a rotating field, as previously described. Part (C) shows the essential parts of both stator and rotor. The purpose of the iron rotor core is to reduce airgap reluctance and to concentrate the magnetic flux through the rotor conductors. Induced current flows in one direction in half of the rotor conductors, and in the opposite direction in the remainder. The shorting rings on the ends of the rotor complete the path for rotor current. In part (D) a 2-pole field is assumed to be rotating in a counterclockwise direction at synchronous speed. At the instant pictured, the south pole field cuts across the upper rotor conductors from right to left, and the lines of force extend upward. Applying the left-hand rule for generator action to determine the direction of the voltage induced in the rotor conductors, the thumb is pointed in the direction of motion of the conductors with respect to the field. Since the field sweeps across the conductors from right to left, their relative motion with respect to the field is to the right. Hence the thumb points to the right. The index finger points upward and the second finger points into the page, indicating that the rotationally induced voltage in the upper rotor conductors is away from the observer.

Applying the left-hand rule to the lower rotor conductors and the north-pole field, the thumb points to the left, the index finger points upward, and the second finger points toward the observer, indicating that the direction of the rotationally induced voltage is out of the page. The rotor bars are connected to end rings that complete their circuits, and the rotationally induced voltages act in series addition to cause rotor currents to flow in the rotor conductors in the directions indicated. For simplification, the rotor currents are assumed to be in phase with the rotor voltages.

Motor action is analyzed by applying the right-hand rule for motors to the rotor conductors in figure 17-2 (D), to determine the direction of the force acting on the rotor conductors. For the upper rotor conductors, the index finger points upward, the second finger points into the page and the thumb points to the left, indicating that the force on the rotor tends to turn the rotor counterclockwise. This direction is the same as that of the rotating field. For the lower rotor conductors the index finger points upward, the second finger points toward the observer, and the thumb points toward the right, indicating that the force tends to turn the rotor counterclockwise—the same direction as that of the field.

The STATOR of a polyphase induction motor consists of a laminated steel ring with slots on the inside circumference. The motor stator winding is similar to the a-c generator stator winding and is generally of the two-layer distributed preformed type. Stator phase windings are symmetrically placed on the stator and may be either wye or delta connected.

There are two types of ROTORS—the CAGE ROTOR and the FORM-WOUND ROTOR. Both types have a laminated cylindrical core with parallel slots in the outside circumference to hold the windings in place. The cage rotor has an uninsulated bar winding; whereas the form-wound rotor has a two-layer distributed winding with preformed coils like those on a d-c motor armature.

CAGE ROTORS

A cage rotor is shown in figure 17-3 (A). The rotor bars are of copper, aluminum, or a suitable alloy placed in the slots of the rotor core. These bars are connected together at each end by rings of similar material. The conductor bars carry relatively large currents at low voltage. Hence, it is not necessary to insulate these bars from the core because the currents follow the path of least resistance and are confined to the cage winding.

FORM-WOUND ROTOR

A form-wound rotor (fig. 17-3 (B)) has a winding similar to 3-phase stator windings. Rotor windings are usually wye connected with the free ends of the winding connected to three slip rings mounted on the rotor shaft. An external variable wye-connected resistance (fig. 17-3 (C)) is connected to the rotor circuit through the

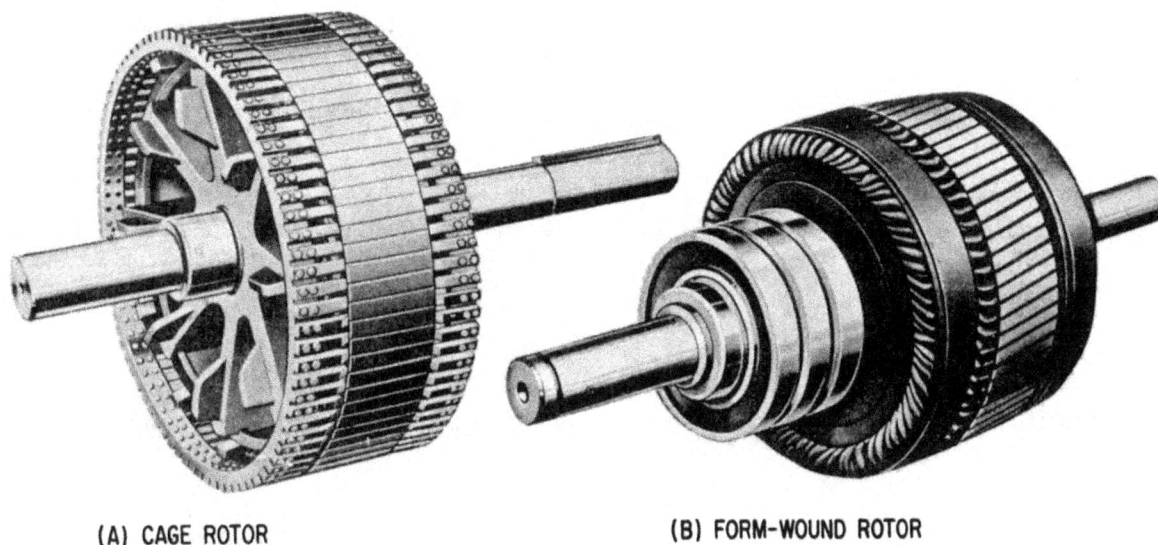

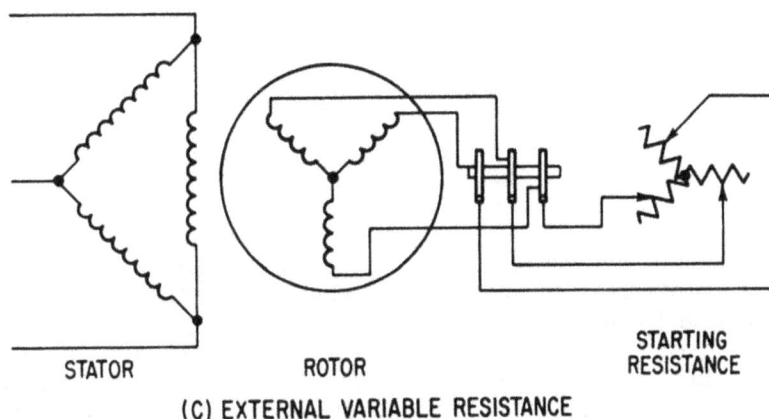

(A) CAGE ROTOR

(B) FORM-WOUND ROTOR

(C) EXTERNAL VARIABLE RESISTANCE

STATOR    ROTOR    STARTING RESISTANCE

BE.314

Figure 17-3.—Induction motor rotors.

sliprings. The variable resistance provides a means of increasing the rotor-circuit resistance during the starting period to produce a high starting torque. As the motor accelerates, the rheostat is cut out. When the motor reaches full speed, the sliprings are short-circuited and the operation is similar to that of the cage motor.

TORQUE

As previously described, the revolving field produced by the stator windings cuts the rotor conductors and induces voltages in the conductors. Rotor currents flow because the rotor end-rings provide continuous metallic circuits. The resulting torque tends to turn the rotor in the direction of the rotating field. This torque is proportional to the product of the rotor current, the field strength, and the rotor power factor.

The simplified cross section of a 2-pole cage-rotor motor is shown in figure 17-4. The magnetic field is rotating in a clockwise direction. Applying the left-hand rule for generator action, note that in figure 17-4 (A), the induced currents flow outward in the upper half and inward in the

322

lower half of the rotor conductors. Applying the right-hand rule for motor action, note that the force acting on the rotor conductors is to the right on the upper group and to the left on the lower group.

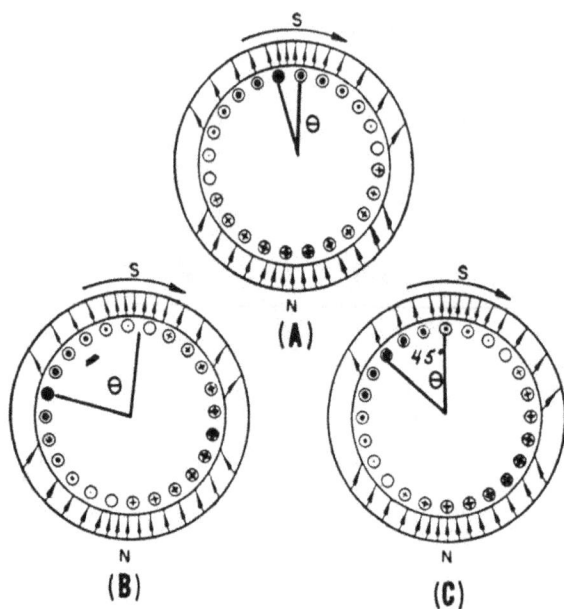

BE.315

Figure 17-4.—Development of torque.

As previously stated, a d-c motor receives its armature current by means of conduction through the commutator and brushes; whereas an induction motor receives its rotor current by means of induction. In this respect the induction motor is like a transformer with a rotating secondary. The primary is the stator which produces the revolving field; the secondary is the rotor. At start, the frequency of the rotor current is that of the primary stator winding. The reactance of the rotor is relatively large compared with its resistance, and the power factor is low and lagging by almost 90°. The rotor current therefore lags the rotor voltage by approximately 90°, as shown in figure 17-4 (B). Because almost half of the conductors under the south pole carry current outward and the remainder of the conductors carry current inward, the net torque on the rotor as a result of the interaction between the rotor and the rotating field is small.

As the rotor comes up to speed in the same direction as the revolving field, the rate at which the revolving field cuts the rotor conductors is

reduced and the rotor voltage and frequency of rotor currents are correspondingly reduced. Hence, at almost synchronous speed the voltage induced in the rotor is very small. The rotor reactance, $X_L$, also approaches zero, as may be seen from the relationship

$$X_L = 2\pi f_O \, LS,$$

where $f_O$ is the frequency of the stator current, L the rotor inductance, and S the ratio of the difference in speed (between the stator field and the rotor) to the synchronous speed—that is, the slip. Slip is expressed mathematically as

$$S = \frac{N_S - N_r}{N_S},$$

where $N_S$ is the number of revolutions per minute of the stator field, and $N_r$ the number of revolutions per minute of the rotor. The frequency of the induced rotor current is $f_O S$.

In figure 17-4 (A), the rotor current is nearly in phase with the rotor voltage, and the direction of flow under the south pole is the same in all conductors. The torque would be ideally high except for the very small rotor current, which is produced by the low rotor voltage. The rotor voltage is proportional to the difference in speed between the rotor and the rotating field—that is, proportional to the slip.

The frequency of the rotor current varies directly with the slip. Thus, when the slip and the frequency of rotor current are almost zero, the rotor reactance and angle of lag are very small. When the rotor is starting, the difference in speed between the rotor and the rotating field is maximum; hence, the rotor reactance is maximum because the frequency of rotor current is maximum and approaches that of the line supplying the stator primary.

Normal operation is between these two extremes of rotor slip—that is, when the rotor is not turning at all or when it is turning almost at synchronous speed. The motor speed under normal load conditions is rarely more than 10 percent below synchronous speed. At the extreme of 100 percent slip, the rotor reactance is so high that the torque is low because of low power factor. At the other extreme of zero rotor slip, the torque is low because of low rotor current. The equation for torque, T, is

$$T = K\Phi I_r \, \cos\theta_r,$$

where K is a constant, $\Phi$ the strength of the magnetic revolving field, $I_r$ the rotor current, and cos $\theta_r$ the power factor of the rotor current.

In the previous formula the product of rotor current and rotor power factor for a given strength of magnetic revolving field is a maximum value when the phase angle between rotor current and rotor induced voltage is 45° lagging, as indicated in figure 17-4, (C). In this case the reactance of the rotor equals the resistance of the rotor circuit, and the rotor power factor is 70.7 percent. This condition of operation is called the PULL-OUT POINT. Beyond this point, the motor speed falls off rapidly with added load, and the motor stalls.

Variations in applied stator voltage affect the motor torque. This applied voltage establishes mmf's which create the rotating field. The rotating field, in turn, establishes the rotor current. Because rotor torque varies as the product of these factors, the torque of an induction motor varies as the square of the voltage applied to the stator primary winding.

The rotating field also sweeps across the stator winding which produces it. This action induces a counter emf in the winding. The counter emf opposes the applied voltage and limits the stator currents. If the applied voltage is increased to magnetic saturation, the counter emf is limited, and the primary current becomes dangerously high.

### SYNCHRONOUS SPEED AND SLIP

The speed, N, of the rotating field is called the SYNCHRONOUS SPEED of the motor. As previously stated, the torque on the rotor tends to turn the rotor in the same direction as the revolving field. If the motor is not driving a load, it will accelerate to nearly the same speed as the revolving field. During the starting period, the increase in rotor speed is accompanied by a decrease in induced rotor voltage because the relative motion between the rotating field and rotor conductors is less. If it were possible for the rotor to attain synchronous speed there would be no relative motion between the rotor and the rotating field. There would then be no induced emf in the rotor, no rotor current, and thus no torque.

It is obvious that an induction motor cannot run at exactly synchronous speed. Instead, the rotor always runs just enough below synchronous speed at no load to establish sufficient rotor current to produce a torque equal to the resisting torque that is caused by the rotor losses.

The frequency of the alternating voltage induced in the rotor depends on the speed of the revolving field with respect to the rotor. One cycle of alternating voltage is induced in the rotor when the stator field sweeps completely around the rotor once. The rotor frequency, $f_r$, is directly proportional to the percent slip—where S

$$f_r = \frac{Sf_s}{100}$$

is the percent slip and $f_s$ is the frequency of the supply. For example, if the frequency of the supply is 60 hertz and the slip is 5 percent, the rotor frequency will be $\frac{5 \times 60}{100} = 3$ hertz. The frequency and magnitude of the induced rotor voltage decrease as the rotor speed increases. Both the rotor voltage and frequency would become zero if the rotor could attain synchronous speed.

### LOSSES AND EFFICIENCY

The losses of an induction motor include (1) stator copper loss, $I_s^2 R_s$, and rotor copper loss, $I_r^2 R_r$; (2) stator and rotor core loss; and (3) friction and windage loss. For all practical purposes, the core, bearing friction, and windage losses are considered to be constant for all loads of an induction motor having a small slip. The power output may be measured on a mechanical brake or calculated from a knowledge of the input and the losses. The efficiency is equal to the ratio of the output power to input power; and at full load, it varies from about 85 percent for small motors to more than 90 percent for large motors.

### Characteristics of the Cage-Rotor Motor

As stated previously, the cage-rotor induction motor is comparable to a transformer with a rotating secondary. At no load, the magnetic revolving field produced by the primary stator winding cuts the turns of the stator winding. This action generates a counter emf in the stator winding, which limits the line current to a small value. This no-load value is called EXCITING CURRENT. Its function is to maintain the revolving field. Because the circuit is highly inductive, the power factor of the motor with no load is very poor. It may be as much as 30 percent lagging. Because there is no drag on the rotor, it runs at almost synchronous speed and the rotor current is quite small. Hence, the reaction of the rotor mmf on the primary revolving field is small.

When load is added to the motor, the rotor slows down slightly; but the rotating field continues at synchronous speed. Therefore, the rotor current and slip increase. The motor torque increases more than the decrease in speed and the power output increases. The increased rotor mmf opposes the primary field flux and lowers it slightly. The primary counter emf therefore decreases slightly and primary current increases. The load component of primary current maintains the rotating field and prevents its further weakening because of the rotor-current opposition. Because of the relatively low internal impedance of the motor windings, a small reduction in speed and counter emf in the primary may be accompanied by large increases in motor current, torque, and power output. Thus, the cage-rotor motor has essentially constant-speed variable torque characteristics.

If the induction motor is stalled by overload, the resulting increased rotor current lowers the primary counter emf and causes excessive primary current. This excessive current may damage the motor winding. When the rotor of an induction motor is locked, the voltage applied to the primary winding should not exceed 50 percent of its rated voltage.

When the motor is operating at full load, the load component of stator current is more nearly in phase with the voltage across each stator phase because of the mechanical output (true power component) of the motor. The power factor is considerably improved over the no-load condition.

The torque and current curves for a 3-phase induction motor with cage rotor are shown in figure 17-5 (A). Rotor reactance increases with slip and increasingly affects the rotor current and power factor as the motor load is increased. The pull-out point on the torque curve occurs at about 25 percent slip. Maximum torque at this condition is about 3.5 times the normal full-load value, and, as previously mentioned, corresponds to a rotor power factor of 70.7 percent. Thus, for the pull-out condition, the rotor resistance equals the rotor reactance and the rotor power factor angle equals 45°. Any additional load on the motor beyond this point causes the rotor to pull out of its normal speed range and to stall quickly. At standstill, stator current is nearly 5 times normal; hence, constant-potential motor circuits like the one supplying this motor

are equipped with time-delay automatic-overload protective devices. Sustained overload causes a circuit breaker to open and thus protect both the motor and the circuit from damage.

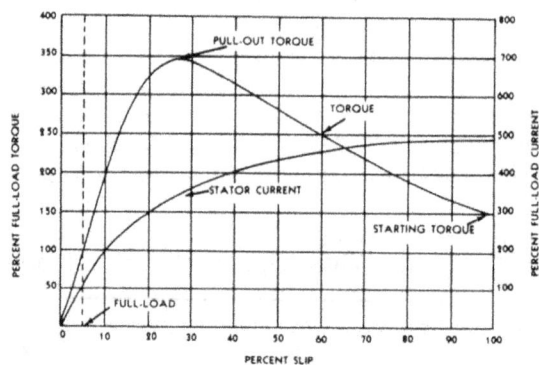

TORQUE AND CURRENT CURVES

(A)

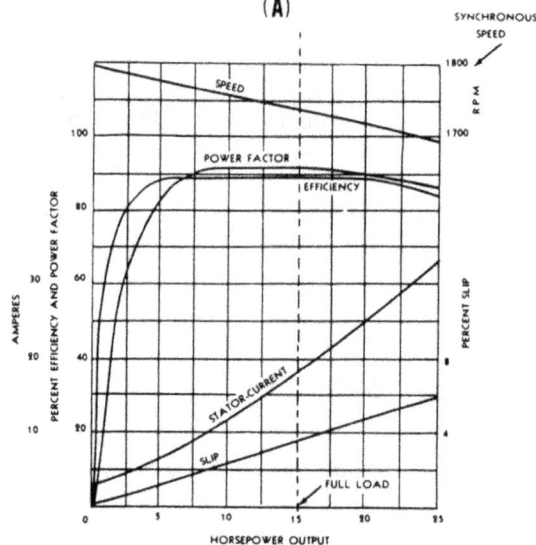

PERFORMANCE CURVES FOR 4-POLE 3-PHASE 440-VOLT 15-HP CAGE-ROTOR MOTOR

(B)

BE.316

Figure 17-5.—Characteristic curves of a cage-rotor motor.

The performance curves of a 4-pole 3-phase 440-volt 15-horsepower cage-rotor motor are shown in figure 17-5 (B). The full-load slip is only about 3.5 percent. At standstill, the rotor reactance of this type of motor is nearly 5 times as great as the rotor resistance. At full load, however, the rotor reactance is much less than the rotor resistance. When the motor is running, the rotor current is determined principally by the rotor resistance. The torque increases up to the pull-out point as the slip increases. Beyond this point the torque decreases and the motor stalls. Because the change in speed from no load to full load is relatively small, the motor torque and the horsepower output are considered to be directly proportional.

The cage-rotor induction motor has a fixed rotor circuit. The resistance and inductance of the windings are determined when the motor is designed and cannot be changed after it is built. The standard cage-rotor motor is a general-purpose motor. It is used to drive loads that require a variable torque at approximately constant speed with high full-load efficiency—such as blowers, centrifugal pumps, motor-generator sets, and various machine tools.

If the load requires special operating characteristics, such as high-starting torque, the cage rotor is designed to have high resistance. The starting current of a motor with a high-resistance rotor is less than that of a motor with a low-resistance rotor. The high-resistance rotor motor, like the cumulative compounded d-c motor, has wider speed variations than the low-resistance rotor motor. The high rotor resistace also increases the rotor copper losses, resulting in a lower efficiency than that of the low resistance type. These motors are used to drive cranes and elevators when high-starting torque and moderate-starting current are required and when it is desired to slow down the motor without drawing excessive currents.

## CHARACTERISTICS OF WOUND-ROTOR MOTOR

The wound-rotor, or slipring, induction motor is used when it is necessary to vary the rotor resistance in order to limit the starting current or to vary the motor speed. As previously explained, the starting torque may be made equal to the pull-out torque by increasing the rotor resistance to the point where it equals the rotor reactance at standstill. Maximum torque at start can be obtained with a wound-rotor motor with about 1.15 times full-load current; whereas a cage-rotor motor may require 5 times full-load current to produce maximum torque at start. Because the rotor-circuit copper losses are largely external to the rotor winding, the wound-rotor motor is desirable for an application which requires frequent starts.

The advantages of the wound-rotor induction motor over the cage-rotor induction motor are: (1) high-starting torque with moderate starting current, (2) smooth acceleration under heavy loads, (3) no excessive heating during starting, (4) good running characteristics, and (5) adjustable speed. The chief disadvantage of the wound-rotor is that the initial and maintenance costs are greater than those of the cage-rotor motor.

## SYNCHRONOUS MOTORS

The synchronous motor differs from the induction motor in several ways. The synchronous motor requires a separate source of d-c for the field. It also requires special starting components. These include a salient-pole field with starting grid winding. The rotor of the conventional type synchronous motor is essentially the same as that of the salient-pole a-c generator. The stator windings of induction and synchronous motors are essentially the same. The stator of a synchronous motor is illustrated in figure 17-6 (A).

If supplied with proper voltage, a d-c generator operates satisfactorily as a d-c motor, and there is practically no difference in construction and rating between the two. Similarly, an a-c generator becomes a synchronous motor if electric power is supplied to its terminals from an external source. Synchronous-motor rotating fields are generally of the salient-pole type, as shown in figure 17-6 (B).

Assume, for example, that two a-c generators of the salient-pole type are operating in parallel and feeding the same bus. If the prime mover is disconnected from one of the generators, it will become a synchronous motor and continue to run at the same speed, drawing its power from the other a-c generator.

With the exception of certain modifications to make its operation more efficient and also to make it self-starting, the synchronous motor is very similar to the salient-pole rotating-field a-c generator. The rotor fields of both are separately excited from a d-c source, and both run at synchronous speeds under varying load conditions. An 8-pole generator rotor revolving at 900 rpm generates 60 Hz; likewise, and 8-pole synchronous motor supplied with 60-Hz current will rotate at 900 rpm.

CAGE
WINDING

ROTOR
(B)

STATOR
(A)

BE.317

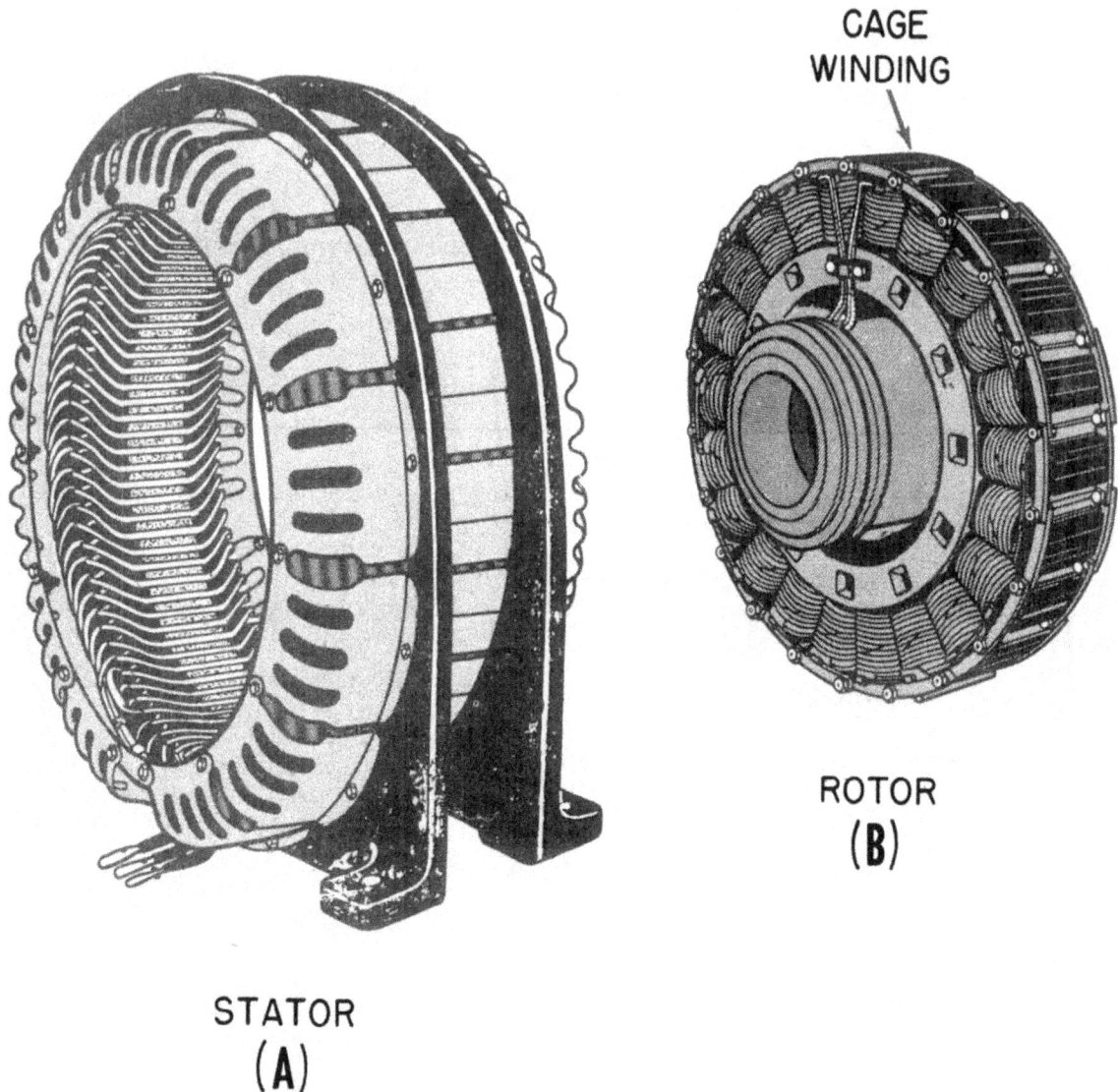

Figure 17-6.—Synchronous motor.

PRINCIPLE OF OPERATION

A polyphase current is supplied to the stator winding of a synchronous motor and produces a rotating magnetic field the same as in an induction motor. A direct current is supplied to the rotor winding, thus producing a fixed polarity at each pole. If it could be assumed that the rotor had no inertia and that no load of any kind were applied, then the rotor would revolve in step with the revolving field as soon as power was applied to both of the windings. This, however, is not the case. The rotor has inertia, and in addition there is a load.

The reason a synchronous motor has to be brought up to synchronous speed by special means, may be understood from a consideration of figure 17-7.

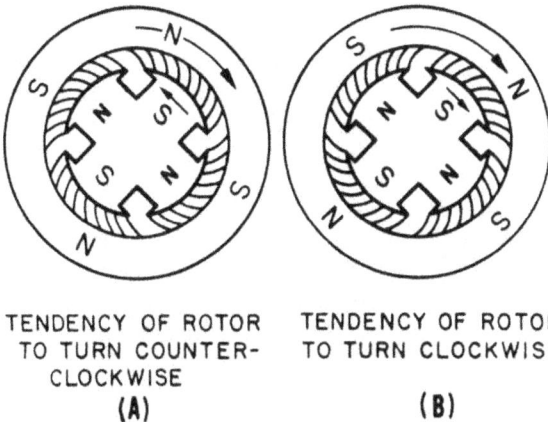

TENDENCY OF ROTOR
TO TURN COUNTER-
CLOCKWISE
**(A)**

TENDENCY OF ROTOR
TO TURN CLOCKWISE
**(B)**

BE.318
Figure 17-7.—Operating principles
of a synchronous motor.

If the stator and rotor windings are energized, then as the poles of the rotating magnetic field approach rotor poles of opposite polarity (fig. 17-7 (A)), the attracting force tends to turn the rotor in the direction opposite to that of the rotating field. As the rotor starts in this direction, the rotating-field poles are leaving the rotor poles (fig. 17-7 (B)), and this tends to pull the rotor poles in the same direction as the rotating field. Thus, the rotating field tends to pull the rotor poles first in one direction and then in the other, with the result that the starting torque is zero.

STARTING

As has been explained, some type of starter must be used with the synchronous motor to bring the rotor up to synchronous speed. Although a small induction motor may be used to bring the rotor up to speed, this is not generally done. Sometimes, if direct current is available, a d-c motor coupled to the rotor shaft may be used to bring the rotor up to synchronous speed. After synchronous speed is attained, the d-c motor is converted to operate as a generator to supply the necessary direct current to the rotor of the synchronous motor.

In general, however, another method is used to start the synchronous motor. A cage-rotor winding is placed on the rotor of the synchronous motor to make the machine self-starting as an

induction motor. At start, the d-c rotor field is deenergized and a reduced polyphase voltage is applied to the stator windings. Thus, the motor starts as an induction motor and comes up to a speed which is slightly less than synchronous speed. The rotor is then excited from the d-c supply (generally a d-c generator mounted on the shaft) and the field rheostat adjusted for minimum line current.

If the armature has the correct polarity at the instant synchronization is reached, the stator current will decrease when the excitation voltage is applied. If the armature has the incorrect polarity, the stator current will increase when the excitation voltage is applied. This is a transient condition, and if the excitation voltage is increased further the motor will slip a pole and then come into step with the revolving field of the stator.

If the rotor d-c field winding of the synchronous machine is open when the stator is energized, a high a-c voltage will be induced in it because the rotating field sweeps through the large number of turns at synchronous speed.

It is therefore necessary to connect a resistor of low resistance across the rotor d-c field winding during the starting period. During the starting period, the d-c field winding is disconnected from the source and the resistor is connected across the field terminals. This permits alternating current to flow in the d-c field winding. Because the impedance of this winding is high compared with the inserted external resistance, the internal voltage drop limits the terminal voltage to a safe value.

STARTING TORQUE

Both the alternating currents induced in the rotor field winding and the cage-rotor winding during starting are effective in producing the starting torque. The torques produced by the rotor d-c field winding and the cage-rotor winding at different speeds are shown by the curves $T_r$ and $T_s$, respectively, of figure 17-8. Curve T is the sum of $T_r$ and $T_s$ and indicates the total torque at different speeds during the starting period. Note that $T_r$ is very effective in producing torque as the rotor approaches synchronous speed, but that both windings contribute no torque at synchronous speed because the induced voltage is zero and no d-c excitation is yet applied to the d-c winding.

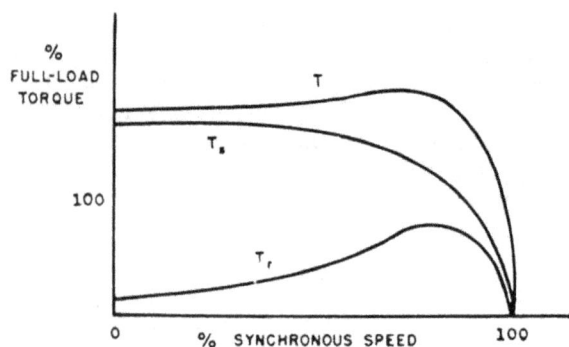

BE.319

Figure 17-8.—Starting torque of a
synchronous motor.

## EFFECT OF VARYING LOAD
## AND FIELD STRENGTH

The power factor of an induction motor depends on the load and varies with it. The power factor of a synchronous motor carrying a definite load may be unity or less than unity, either lagging or leading, depending on the d-c field strength.

The induced (counter) emf in armature coil C, shown in figure 17-9 (A), is maximum when its sides are opposite the pole centers and minimum when its sides are midway between the pole tips; and it varies as indicated by the solid-line curve, $E'_c$. When the motor carries no load, the counter voltage $E'_c$ is practically 180° out of phase with the applied voltage $E_a$ (fig. 17-9 (A) and (B)). If the field is adjusted so that $E_c$ almost equals $E_a$, the stator current $I_s$ is small and corresponds to the exciting current in a transformer.

When load is applied to the motor it causes the poles to be pulled α degrees behind their no-load position, as indicated by the broken curve in figure 17-9 (A), and the counter emf occurs α degrees later. This is indicated by curve $E_c$ in figure 17-9 (A), and by vector $E_c$ in figure 17-9 (C). The resultant voltage, E, causes the stator current $I_s$, to lag behind $E_a$ by angle $\theta$.

In a d-c motor the armature current is determined from the equation

$$I_a = \frac{E_a - E_c}{R_a}$$

Similarly, in a synchronous motor the stator current is determined as

$$I_S = \frac{\text{vector sum of } E_a \text{ and } E_c}{Z_S}$$

where $I_S$ is the stator current, $E_a$ the applied voltage, $E_c$ the counter voltage, and $Z_S$ the stator impedance. This vector sum is indicated by E in figure 17-9 (C). The stator reactance is large compared to its resistance, and therefore the current $I_S$ lags the resultant voltage E by nearly 90°. Hence, $I_S$, lags, $E_a$, by angle $\theta$. In this condition the synchronous motor operates with a lagging power factor. If the load is increased, the rotor poles are pulled further behind the stator poles, which causes $E_c$ to lag further, and angle α to increase. This action causes the resultant voltage, E, and stator current, $I_S$, to increase.

Because the speed is constant, if the field excitation is decreased, the counter voltage, $E_c$, decreases; and the resultant voltage, E, becomes greater. The stator current becomes greater and lags the applied voltage, $E_a$, a greater amount. On the other hand, if the field excitation is increased until the current, $I_S$, is in phase with the applied voltage, $E_a$, the power factor of the motor becomes unity for a given load (fig. 17-9 (D)). For a definite load at unity power factor, E and $I_S$ are both at their minimum. If the field excitation is further increased, $I_S$ increases and leads $E_a$ (fig. 17-9 (E)). Thus, for a definite load, the power factor is governed by the field excitation—that is, a weak field produces lagging current and a strong field produces a leading current. Normal field excitation for a given load occurs when $I_S$ is in phase with $E_a$.

The so-called synchronous motor V curves, which indicate the variations of current for a constant load and varied field excitation, are shown in figure 17-10. The corresponding variations of power factor are also depicted.

A synchronous motor is frequently used in certain systems to change the power factor of the system to which it is paralleled by adjusting its field excitation. If operated without a load, its power factor may be adjusted to a value as low as 10 percent leading. When operated under this condition, the motor is generally referred to as a synchronous condenser because it takes a leading current in the same manner as capacitors. In this case the synchronous condenser takes only enough true power from the line to supply its losses. At the same time, it supplies

329

PRIMARY RESISTOR starters insert a resistor in the PRIMARY circuit (the stator circuit) of the motor for starting, or for starting and speed control. This starter is used when it is necessary to limit the starting current of a large a-c motor so as not to place too great a load on the system. If the resistor is used only during starting, its rating is based on intermittent operation, in which case it is relatively small and operates at a higher temperature. If the resistor is used also for speed control of small motors, such as for ventilating fans, its rating is based on continous operation. In this case the resistor is relatively large and operates at a lower temperature.

SECONDARY RESISTOR starters insert a resistor in the secondary circuit (the rotor circuit of the form-wound type of induction motor) for starting and speed control. This starter may be used to limit starting currents, but is usually found where speed control of a large a-c motor is required, in which case the resistors are rated for continuous duty. Examples are some elevators and hoists equipped with direct a-c electric drives.

COMPENSATOR, or AUTOTRANSFORMER, starters start the motor at reduced voltage through an autotransformer, and subsequently connect the motor to full voltage after acceleration. The compensator may be either of two types.

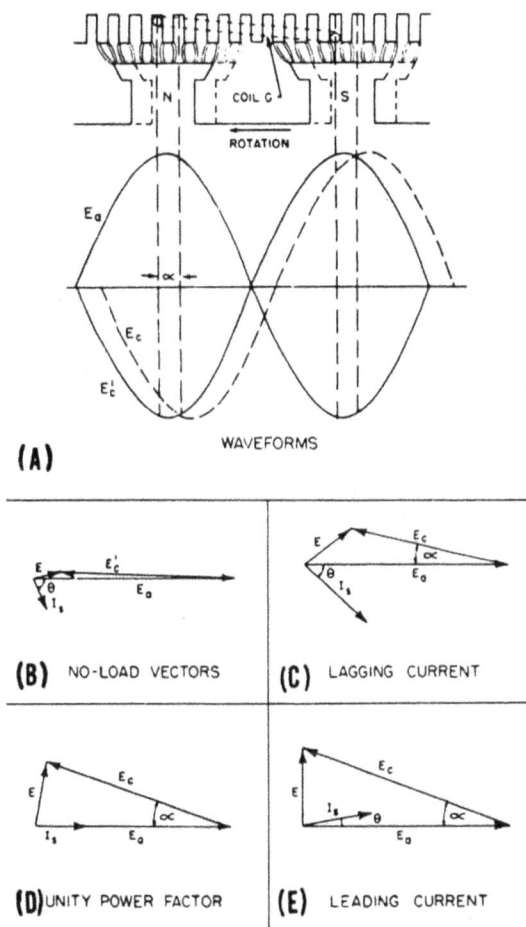

(A) WAVEFORMS

(B) NO-LOAD VECTORS

(C) LAGGING CURRENT

(D) UNITY POWER FACTOR

(E) LEADING CURRENT

BE. 320

Figure 17-9.—Effect of varying the load and field strength of a synchronous motor.

a high leading reactive power, which cancels the lagging reactive power taken by the parallel inductive loads, and the system power factor is thereby improved.

## ALTERNATING-CURRENT MOTOR STARTERS

As in the case of d-c motors, some type of starter (controller) may also be employed on a-c motors to limit the initial inrush of current.

ACROSS-THE-LINE starters are the most common because of their simplicity. This type of starter throws the stator winding of the motor directly across the main supply line. This may be feasible if the motor is not too large (5 horsepower or less) and if the generating capacity of the a-c generator can take care of the added load.

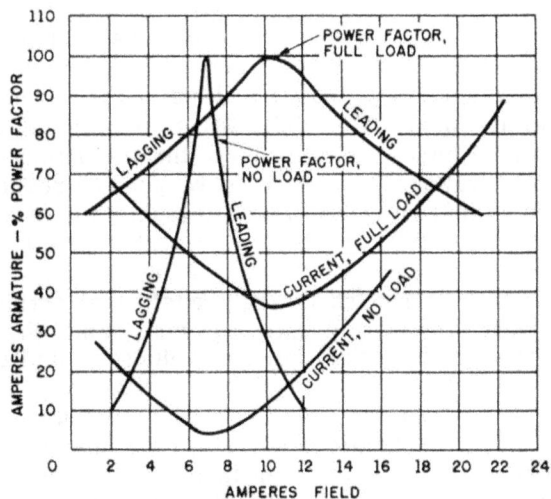

BE. 321

Figure 17-10.—V curves for a synchronous motor of 15 kva.

1. The OPEN-TRANSITION type, during the transition period of shifting the motor from the autotransformer to direct connection with the supply lines, disconnects the motor from all power for a short period of time during which, if the motor is of the synchronous type, it may coast and slip out of phase with the power supply. When the motor is then connected directly to the power lines, a high transition current may result.

2. The CLOSED-TRANSITION type keeps the motor connected to the power supply at all times during the transition period, thus not permitting the motor to decelerate. Accordingly, no high transition current is developed.

The AUTOTRANSFORMER starter is the most common form of the reduced-voltage type used for limiting the starting current of a motor. The open-transition form has the disadvantage of allowing a high transition current to develop, which can cause circuit breakers to open. The closed-transition form is preferable because no high transition current is developed.

REACTOR STARTERS insert a reactor in the primary circuit of an a-c motor during starting, and subsequently short-circuit the reactor to apply full voltage to the motor. This type of starter is not very widely used at present, but is becoming more common for starting large motors because it does not have the high transition current problem of the open-transition compensator and is smaller than the closed-transition compensator.

A simplified schematic diagram of a compensator, or autotransformer starter is shown in figure 17-11. Assume that the line voltage is 100 volts and that when the taps on the auto-transformer are positioned as shown (in the starting position) 40 volts will be applied to the 3-phase motor stator. With the reduction in motor voltage, there is a corresponding reduction in starting current drawn from the line. At the same time, the motor current supplied by the secondary low-voltage windings is proportionately increased by the transformer action.

After the proper time interval, during which acceleration occurs, full-line voltage is applied to the motor.

A resistance starter could be used in place of autotransformer to lower the voltage applied to the stator. The power factor would be improved, but the line current would be greater. The autotransformer has the advantage in permitting the motor to draw a relatively large starting current from the secondary with a relatively low line current.

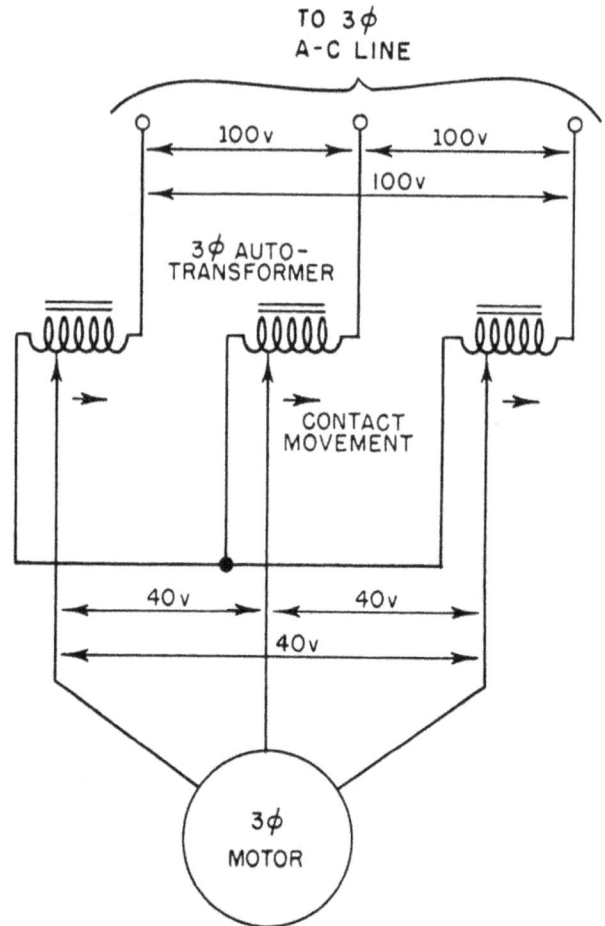

BE. 322

Figure 17-11.—Simplified schematic diagram of an autotransformer starter.

## SINGLE-PHASE MOTORS

Single-phase motors, as their name implies, operate on a single-phase power supply. These motors are used extensively in fractional horse-power sizes in commercial and domestic applications. The advantages of using single-phase motors in small sizes are that they are less expensive to manufacture than other types, and they eliminate the need for 3-phase a-c lines. Single-phase motors are used in interior communications equipment, fans, refrigerators, portable drills, grinders, and so forth.

A single-phase induction motor with only one stator winding and a cage rotor is like a 3-phase induction motor with a cage rotor except that the single-phase motor has no magnetic revolving field at start and hence no starting torque. However, if the rotor is brought up to speed by external means, the induced currents in the rotor will cooperate with the stator currents to produce a revolving field, which causes the rotor to continue to run in the direction in which it was started.

Several methods are used to provide the single-phase induction motor with starting torque. These methods identify the motor as split phase, capacitor, shaded pole, repulsion, and so forth.

Another class of single-phase motors is the a-c series (universal) type. Only the more commonly used types of single-phase motors are described. These include the (1) split-phase motor, (2) capacitor motor, (3) shaded-pole motor, (4) repulsion-start motor, and (5) a-c series motor.

## SPLIT-PHASE MOTOR

The split-phase motor (fig. 17-12 (A)), has a stator composed of slotted laminations that contain an auxiliary (starting) winding and a running (main) winding. The axes of these two windings are displaced by an angle of 90 electrical degrees. The starting winding has fewer turns and smaller wire than the running winding, hence has higher resistance and less reactance. The main winding occupies the lower half of the slots and the starting winding occupies the upper half. The two windings are connected in parallel across the single-phase line supplying the motor. The motor derives its name from the action of the stator during the starting period. The single-phase stator is split into two windings (phases), which are displaced in space by 90°, and which contain currents displaced in time phase by an angle of approximately 15° (fig. 17-12 (B)). The current, $I_S$, in the starting winding lags the line voltage by about 30° and is less than the current in the main winding because of the higher impedance of the starting winding. The current, $I_m$, in the main winding lags the applied voltage by about 45°. The total current, $I_{line}$, during the starting period is the vector sum of $I_S$ and $I_m$.

At start, these two windings produce a magnetic revolving field that rotates around the stator airgap at synchronous speed. As the rotating field moves around the airgap, it cuts across the rotor conductors and induces a voltage in them, which is maximum in the area of highest field intensity and therefore is in phase with the stator field. The rotor current lags the rotor voltage at start by an angle that approaches 90° because of the high rotor reactance. The

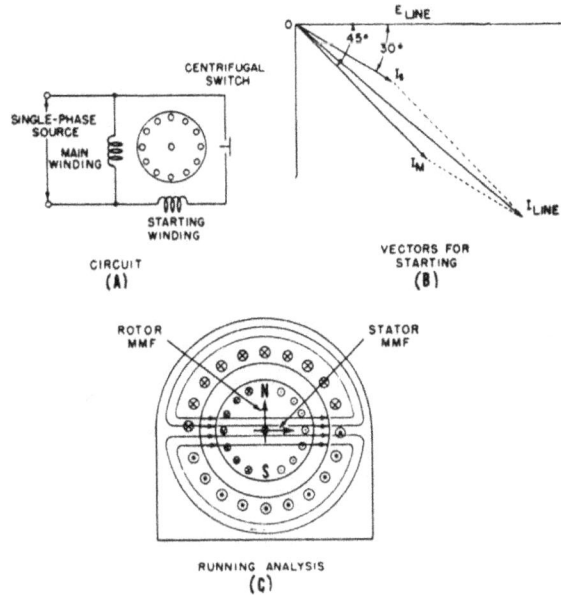

BE. 323

Figure 17-12.—Split-phase motor.

interaction of the rotor currents and the stator field cause the rotor to accelerate in the direction in which the stator field is rotating. During acceleration, the rotor voltage, current, and reactance are reduced and the rotor currents come closer to an in-phase relation with the stator field.

When the rotor has come up to about 75 percent of synchronous speed, a centrifugally operated switch disconnects the starting winding from the line supply, and the motor continues to run on the main winding alone. Thereafter, the rotating field is maintained by the interaction of the rotor magnetomotive force and the stator magnetomotive force. These two mmf's are pictured as the vertical and horizontal vectors respectively in the schematic diagram of figure 17-12 (C).

The stator field is assumed to be rotating at synchronous speed in a clockwise direction, and the stator currents correspond to the instant that the field is horizontal and extending from left to right across the airgap. The left-hand rule for magnetic polarity of the stator indicates that the stator currents will produce an N pole on the left side of the stator and an S pole on the right side. The motor indicated in the figure is wound for two poles.

Applying the left-hand rule for induced voltage in the rotor (the thumb points in the direction of motion of the conductor with respect to the field), the direction of induced voltage is back on the left side of the rotor and forward on the right side. The rotor voltage causes a rotor current to flow, which lags the rotor voltage by an angle whose tangent is the ratio of rotor reactance to rotor resistance. This is a relatively small angle because the slip is small. Applying the left-hand rule for magnetic polarity to the rotor winding, the vertical vector pointing upward represents the direction and magnitude of the rotor mmf. This direction indicates the tendency to establish an N pole on the upper side of the rotor and an S pole on the lower side, as indicated in the figure. Thus, the rotor and stator mmf's are displaced in space by 90° and in time by an angle that is considerably less than 90°, but sufficient to maintain the magnetic revolving field and the rotor speed.

The motor has the constant-speed variable-torque characteristics of the shunt motor. Many of these motors are designed to operate on either 120 volts or 240 volts. For the lower voltage the stator coils are divided into two equal groups and these are connected in parallel. For the higher voltage the groups are connected in series. The starting torque is 150 to 200 percent of the full-load torque and the starting current is 6 to 8 times the full-load current. Fractional-horsepower split-phase motors are used in a variety of equipments such as washers, oil burners, and ventilating fans. The direction of rotation of the split-phase motor can be reversed by interchanging the starting winding leads.

CAPACITOR MOTOR

The capacitor motor is a modified form of split-phase motor, having a capacitor in series with the starting winding. An external view is shown in figure 17-13, with the capacitor located on top of the motor. The capacitor produces a greater phase displacement of currents in the starting and running windings than is produced in the split-phase motor. The starting winding is made of many more turns of larger wire and is connected in series with the capacitor. The starting winding current is displaced approximately 90° from the running winding current. Since the axes of the two windings are also displaced by an angle of 90°, these conditions produce a higher starting torque than that of the split-phase motor. The starting torque of the capacitor motor may be as much as 350 percent of the full-load torque.

If the starting winding is cut out after the motor has increased in speed, the motor is called a CAPACITOR-START MOTOR. If the starting winding and capacitor are designed to be left in the circuit continuously, the motor is called a CAPACITOR-RUN MOTOR. Electrolytic capacitors for capacitor-start motors vary in size from about 80 microfarads for 1/8-horsepower motors to 400 microfarads for one-horsepower motors. Capacitor motors of both types are made in sizes ranging from small fractional horsepower motors up to about 10 horsepower. They are used to drive grinders, drill presses, refrigerator compressors, and other loads that require relatively high starting torque. The direction of rotation of the capacitor motor may be reversed by interchanging the starting winding leads.

SHADED-POLE MOTOR

The shaded-pole motor employs a salient-pole stator and a cage rotor. The projecting poles on the stator resemble those of d-c machines except that the entire magnetic circuit is laminated and a portion of each pole is split to accommodate a short-circuited copper strap called a SHADING COIL (fig. 17-14). This motor is generally manufactured in very small sizes, up to 1/20 horsepower. A 4-pole motor of this type is illustrated in figure 17-14 (A). The shading coils are placed around the leading pole tip and the main pole winding is concentrated and wound around the entire pole. The 4 coils comprising the main winding are connected in series across the motor terminals. An inexpensive type of 2-pole motor employing shading coils is illustrated in figure 17-14 (B).

During that part of the cycle when the main pole flux is increasing, the shading coil is cut by the flux, and the resulting induced emf and current in the shading coil tend to prevent the flux from rising readily through it. Thus, the greater portion of the flux rises in that portion of the pole that is not in the vicinity of the shading coil. When the flux reaches its maximum value, the rate of change of flux is zero, and the voltage and current in the shading coil also are zero. At this time the flux is distributed more uniformly over the entire pole face. Then as the main flux decreases toward zero, the induced voltage and current in the shading coil reverse their polarity, and the resulting magnetomotive force tends to prevent the flux from collapsing through the iron

BE. 324

Figure 17-13.—Capacitor motor.

in the region of the shading coil. The result is that the main flux first rises in the unshaded portion of the pole and later in the shaded portion. This action is equivalent to a sweeping movement of the field across the pole face in the direction of the shaded pole. The cage rotor conductors are cut by this moving field and the force exerted on them causes the rotor to turn in the direction of the sweeping field.

Most shaded-pole motors have only one edge of the pole split, and therefore the direction of rotation is not reversible. However, some shaded-pole motors have both leading and trailing pole tips split to accommodate shading coils. The leading pole tip shading coils form one series group, and the trailing pole tip shading coils form another series group. Only the shading coils in one group are simultaneously active, while those in the other group are on open circuit.

The shaded-pole motor is similar in operating characteristics to the split-phase motor. It has the advantages of simple construction and low cost. It has no sliding electrical contacts and is reliable in operation. However, it has low starting torque, low efficiency, and high

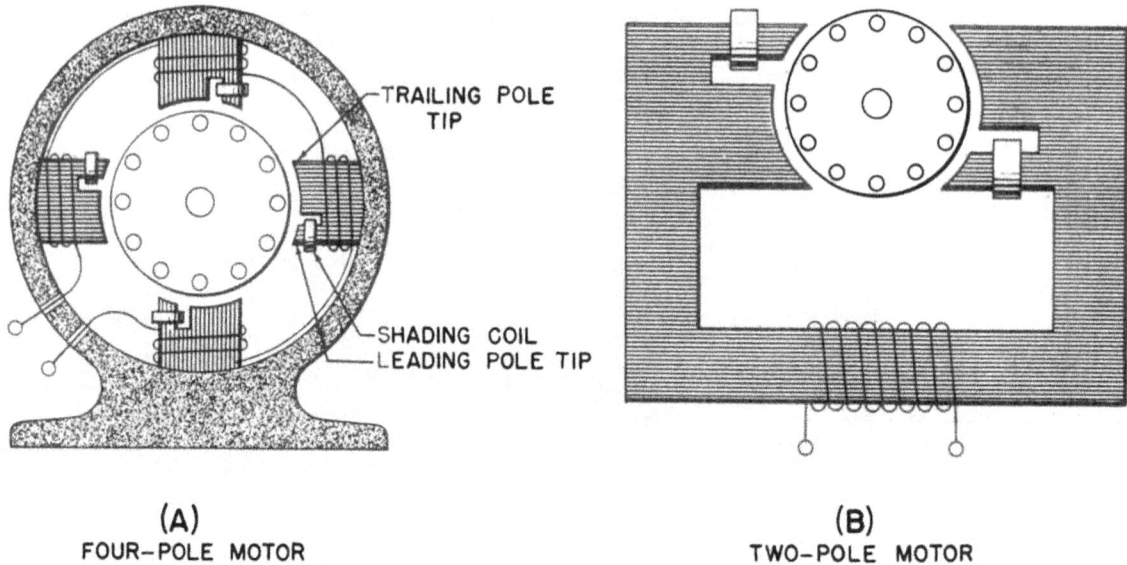

**(A)**
FOUR-POLE MOTOR

**(B)**
TWO-POLE MOTOR

BE.325

Figure 17-14.—Shaded-pole motor.

noise level. It is used to operate small fans. The shading coil and split pole are used in clock motors to make them self-starting.

REPULSION-START MOTOR

The repulsion-start motor has a form-wound rotor with commutator and brushes. The stator is laminated and contains a distributed single-phase winding. In its simplest form, the stator resembles that of the single-phase motor. In addition, the motor has a centrifugal device which removes the brushes from the commutator and places a short-circuiting ring around the commutator. This action occurs at about 75 percent of synchronous speed. Thereafter, the motor operates with the characteristics of the single-phase induction motor.

The starting torque of the repulsion-start induction motor is developed through the interaction of the rotor currents and the single-phase stator field. Unlike the split-phase motor, the stator field does not rotate at start, but alternates instead. The rotor currents are induced through transformer action. For example, in the 2-pole motor of figure 17-15 (A), the stator currents are shown for the instant when a north pole is established on the upper side of the stator,

and a south pole on the lower side. The induced voltage in the rotor causes the rotor currents to flow in such a direction as to oppose the stator field. These currents flow in opposite directions under the left and right portions of the north pole and in similar manner under the south pole. Thus, the net force to turn the rotor is zero when the brushes are located in the positions shown.

In figure 17-15 (B), the brushes are moved 90° from their original positions, and again there is no rotor turning effort because in this case the rotor current is zero. There can be no rotor current in this position because the transformer induced voltages are equal and opposite to each other in the two halves (upper and lower) of the rotor winding.

In figure 17-15 (C), the brush axis is displaced from the stator polar axis by an angle of about 25°, and in this position maximum torque is developed.

The direction of the induced currents in the rotor under the north pole of the stator is toward the observer, and under the south pole, away from the observer. Applying the right-hand rule for motors, the force acting on the conductors under the north pole is toward the left, and under the south pole, toward the right, thus tending to turn the rotor in a counterclockwise direction. When

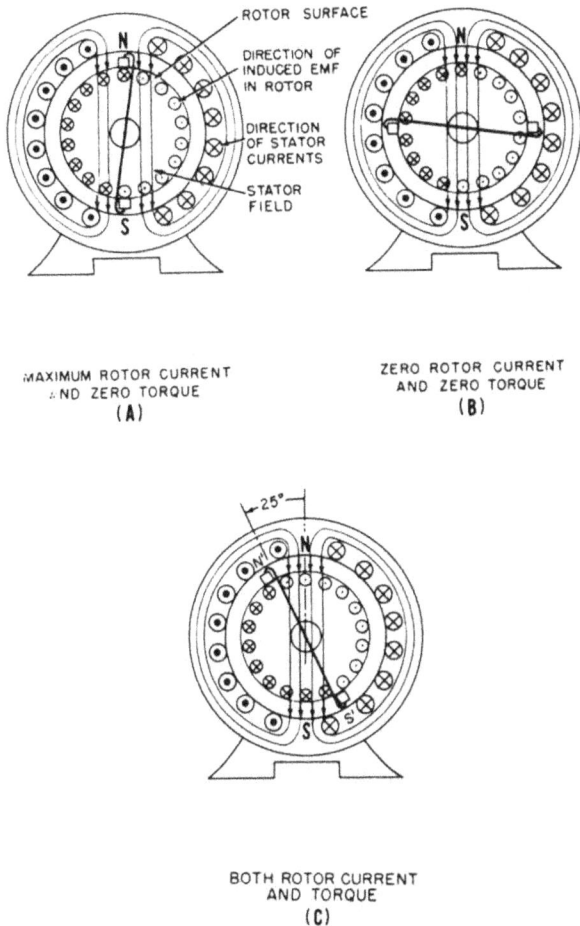

MAXIMUM ROTOR CURRENT
AND ZERO TORQUE
(A)

ZERO ROTOR CURRENT
AND ZERO TORQUE
(B)

BOTH ROTOR CURRENT
AND TORQUE
(C)

BE.326

Figure 17-15.—Repulsion-start induction motor.

the stator polarity reverses, the direction of the rotor current also reverses, thereby maintaining the same direction of rotation. The function of the commutator and brushes is to divide the rotor currents along an axis that is displaced from the axis of the stator field in a counterclockwise direction. The motor derives its name from the repulsion of like poles between the rotor and stator. Thus, the rotor currents establish the rotor poles N'—S', which are repelled by the stator poles N—S.

The starting torque is 250 to 450 percent of the full-load torque, and the starting current is 375 percent of the full-load current. This motor is made in fractional horsepower sizes and in

larger sizes up to 15 horsepower, but has been replaced in large part by the cheaper and more rugged capacitor motor. The repulsion-start motor has higher pull-out torque (torque at which the motor stalls) than the capacitor-start motor, but the capacitor-start motor can bring up to full speed loads that the repulsion motor can start but cannot accelerate.

## A-C SERIES MOTOR

The a-c series motor will operate on either a-c or d-c circuits. The direction of rotation of a d-c series motor is independent of the polarity of the applied voltage, provided the field and armature connections remain unchanged. Hence, if a d-c series motor is connected to an a-c source, a torque will be developed which tends to rotate the armature in one direction. However, a d-c series motor does not operate satisfactorily from an a-c supply for the following reasons:

1. The alternating flux sets up large eddy-current and hysteresis losses in the unlaminated portions of the magnetic circuit and causes excessive heating and reduced efficiency.

2. The self-induction of the field and armature windings causes a low power factor.

3. The alternating field flux establishes large currents in the coils that are short-circuited by the brushes; this action causes excessive sparking at the commutator.

To design a series motor for satisfactory operation on alternating current the following changes are made:

1. The eddy-current losses are reduced by laminating the field poles, yoke, and armature.

2. Hysteresis losses are minimized by using high-permeability transformer-type silicon-steel laminations.

3. The reactance of the field windings is kept satisfactorily low by using shallow pole pieces, few turns of wire, low frequency (usually 25 $H_z$ for large motors), low flux density, and low reluctance (a short airgap).

4. The reactance of the armature is reduced by using a compensating winding embedded in the pole pieces. If the compensating winding is connected in series with the armature (fig. 17-16 (A)), the armature is CONDUCTIVELY compensated. If the compensating winding is short-circuited on itself (fig. 17-16 (B)), the armature is INDUCTIVELY compensated. If the motor is designed for operation on both d-c and a-c circuits, the compensating winding is connected in series with the armature. The axis of the compensating

winding is displaced from the main field axis by an angle of 90°. This arrangement is similar to the compensating winding used in some d-c motors and generators to overcome armature reaction. The compensating winding establishes a' counter magnetomotive force which neutralizes the effect of the armature magnetomotive force, thereby preventing distortion of the main field flux and reducing the armature reactance. The inductively compensated armature acts like the primary of a transformer, the secondary of which is the shorted compensating winding. The shorted secondary receives an induced voltage by the action of the alternating armature flux, and the resulting current flowing through the turns of the compensating winding establish the opposing magnetomotive force that neutralizes the armature reactance.

5. Sparking at the commutator is reduced by the use of PREVENTIVE LEADS $P_1$, $P_2$, $P_3$, and so forth (fig. 17-16 (C)). A ring armature is shown for simplicity. When coils at A and B are shorted by the brushes, the induced current is limited by the relatively high resistance of the leads. Preventive leads are used in some of the older type a-c series motors on electric locomotives. Sparking at the brushes is also reduced by using armature coils having only a single turn and multipolar fields. High torque is obtained by having a large number of armature conductors and a large-diameter armature. Thus, the commutator has a large number of commutator bars that are very thin, so that the armature voltage is limited to about 250 volts.

The operating characteristics of the a-c series motor are similar to those of the d-c series motor. The speed will increase to a dan-

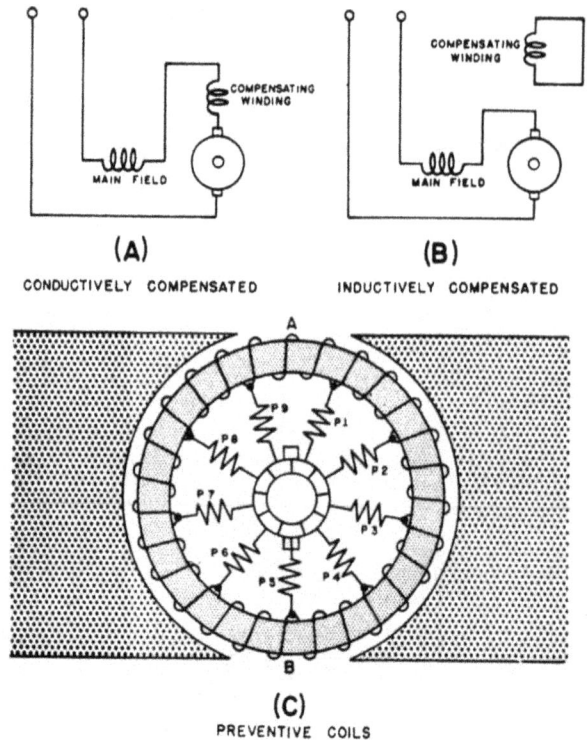

(A) CONDUCTIVELY COMPENSATED

(B) INDUCTIVELY COMPENSATED

(C) PREVENTIVE COILS

BE.327

Figure 17-16.—A-c series motor.

gerously high value if the load is removed from the motor.

Fractional horsepower a-c series motors are called UNIVERSAL MOTORS. They do not have compensating windings or preventive leads. They are used extensively to operate fans and portable tools, such as drills, grinders, and saws.

# CHAPTER 18

# DIRECT-CURRENT GENERATORS

Generators can be designed to supply small amounts of power or they can be designed to supply many thousands of kilowatts of power. Also, generators may be designed to supply either direct current or alternating current. Direct-current generators are described in this chapter. Alternating-current generators were described in chapter 16.

A d-c generator is a rotating machine that converts mechanical energy into electrical energy. This conversion is accomplished by rotating an armature, which carries conductors, in a magnetic field, thus inducing an emf in the conductors. As stated before, in order for an emf to be induced in the conductors, a relative motion must always exist between the conductors and the magnetic field in such a manner that the conductors cut through the field. In most d-c generators the armature is the rotating member and the field is the stationary member. A mechanical force is applied to the shaft of the rotating member to cause the relative motion. Thus, when mechanical energy is put into the machine in the form of a mechanical force or twist on the shaft, causing the shaft to turn at a certain speed, electrical energy in the form of voltage and current is delivered to the external load circuit.

It should be understood that mechanical power must be applied to the shaft constantly so long as the generator is supplying electrical energy to the external load circuit.

The power source used to turn the armature is commonly called a PRIME MOVER. Many forms of prime movers are in use, such as steam turbines, diesel engines, gasoline engines, and steam engines.

A d-c generator consists essentially of the following components:

1. A steel frame or yoke containing the pole pieces and field windings.

2. An armature consisting of a group of copper conductors mounted in a slotted cylindrical core, made up of thin steel disks called laminations.

3. A commutator for maintaining the current in one direction through the external circuit.

4. Brushes with brush holders to carry the current from the commutator to the external load circuit.

D-c generators come in various sizes and appearances. A typical large generator as depicted in figure 18-1 (A) may weigh hundreds of pounds, while a smaller aircraft generator as depicted in (B) of figure 18-1 rarely exceeds 100 pounds.

## CONSTRUCTION

The basic components of a typical generator consist of a frame, field windings, pole pieces, an armature, a commutator, and brushes and brush holders. These components are discussed in the following paragraphs.

## FRAME

The d-c generator field frame, or yoke, is usually made of annealed steel. The use of steel reduces the reluctance of the magnetic circuit to a low value and therefore reduces the necessary size of the field windings. The frame provides a mechanical support for the pole pieces and serves as a portion of the magnetic circuit to provide the necessary flux across the air gap.

The end bells are bolted to the frame structure and support the armature shaft bearings. One end bell also supports the brush rigging and extends over the commutator.

## FIELD WINDINGS

The field windings are connected so that they produce alternate north and south poles (fig. 18-2)

FIELD POLE
&
FIELD CORE

YOKE

BRUSHES

COMMUTATOR

ARMATURE

(A)

CAP SCREWS FOR
MOUNTING
FIELD POLES

TERMINAL
BLOCK

COOLING-AIR
INLET

COOLING-AIR
OUTLET

DRIVE
SHAFT
EXTENSION

MOUNTING
FLANGE

YOKE

BRUSH AND
COMMUTATOR
COVER

(B)

BE.328

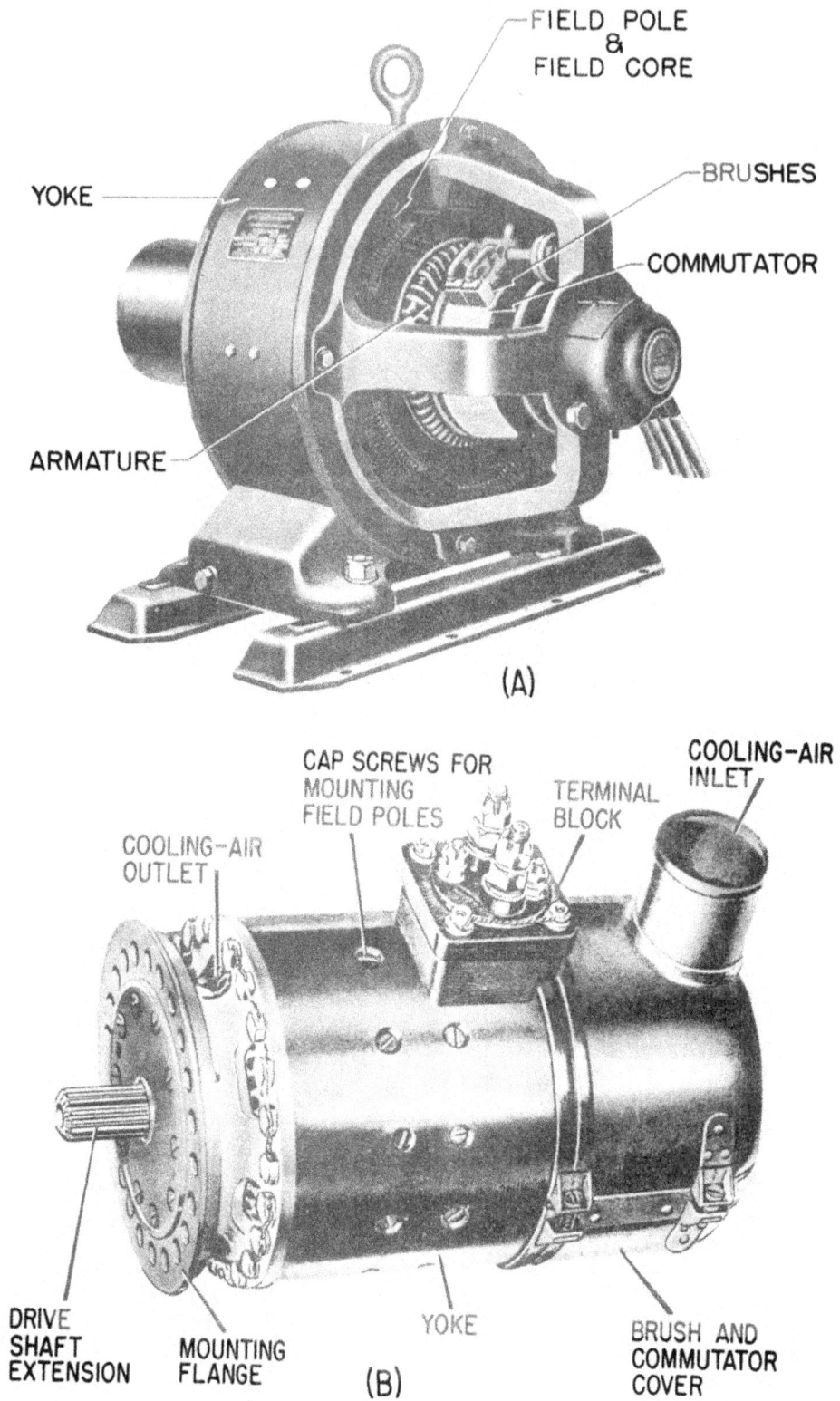

Figure 18-1.—D-c generators.

(A) and (F)) to obtain the correct direction of emf in the armature conductors. The field windings form an electromagnet which establishes the generator field flux. These field windings may receive current from an external d-c source or they may be connected directly across the armature, which then becomes the source of voltage. When they are so energized, they establish magnetic flux in the field yoke, pole pieces, airgap, and armature core, as shown in figure 18-2 (A).

## POLE PIECES

The pole pieces, which support the field windings, are mounted on the inside circumference of the yoke, with capscrews that extend through the frame (fig. 18-1). These pole pieces are usually built of sheet steel laminations riveted together. The pole faces are shaped to fit the curvature of the armature, as shown in figure 18-2 (A). The preformed field coil shown in figure 18-2 (B) is mounted on the laminated core from the back. The coil is held securely in place between the frame and the flanged end of the pole.

## ARMATURE

The armature (fig. 18-2 (C)) is mounted on a shaft and rotates through the field. If the output of the armature is connected across the field windings, the voltage and the field current at start will be small because of the small residual flux in the field poles. However, as the generator continues to run, the small voltage across the armature will circulate a small current through the field coils and the field will become stronger. This action causes the generator voltage to rise quickly to the proper value and the machine is said to "build up" its voltage.

The armature core is made of sheet steel laminations. In small machines these laminations are keyed dirctly to the shaft. In large machines the laminations are assembled on a spider which is keyed to the shaft. The outer surface of this cylindrical core is slotted to provide a means of securing the armature coils. Radial ventilating ducts are provided in the core by inserting spacers between the laminations at definite intervals. These ducts permit air to circulate through the core and to carry off heat produced in the armature winding and core. The armature coils on most generators are form-wound to the correct size and shape. Additional insulation between the core and windings is obtained by placing sheet insulation in the slots. The windings are secured by fiber wedges driven into the tops of the slots. The free ends of the armature coils are connected to the commutator riser, as indicated in figure 18-2 (C). The entire armature winding forms a closed circuit.

## COMMUTATOR

The commutator consists of a number of wedge-shaped segments, or bars, of hard drawn copper that are assembled into a cylinder and held together by V-rings, as shown in figure 18-2 (D). The commutator segments are insulated from each other by sheet mica and the entire commutator is insulated from the supporting rings on the shaft by mica collars. Because the brushes bear on the outside surface of the commutator, better brush contact, less sparking, and less noise, are obtained by undercutting the mica to about 1/64 inch below the level of the commutator surface. The voltage and the number of poles in the generator determine the number of commutator segments. To prevent flashover between segments, generators are designed so that a voltage not to exceed 15 volts exists between adjacent segments. Therefore, a high-voltage generator requires more commutator segments than a low-voltage generator.

## BRUSHES AND BRUSH HOLDERS

The brushes carry the current from the commutator to the external circuit. They are usually made of a mixture of carbon and graphite. For low-voltage machines the brushes are made of a mixture of graphite and a metallic powder. The brushes must be free to slide in their holders (fig. 18-2 (E)) so that they may follow any small irregularities in the curvature of the commutator. In addition, the brushes must be able to "feed in" as they wear. However, excessive play not only would encourage brush vibration, but also might cause misalinement of the brush with the axis of commutation. This would result in excessive sparking.

The proper pressure of the brushes against the commutator is maintained by means of springs and should be from 1-1/2 to 2 pounds per square inch of brush contact area. A low resistance connection between the brushes and brush holders is maintained by means of braided copper wires, or pigtails, that are attached between each brush holder and brush.

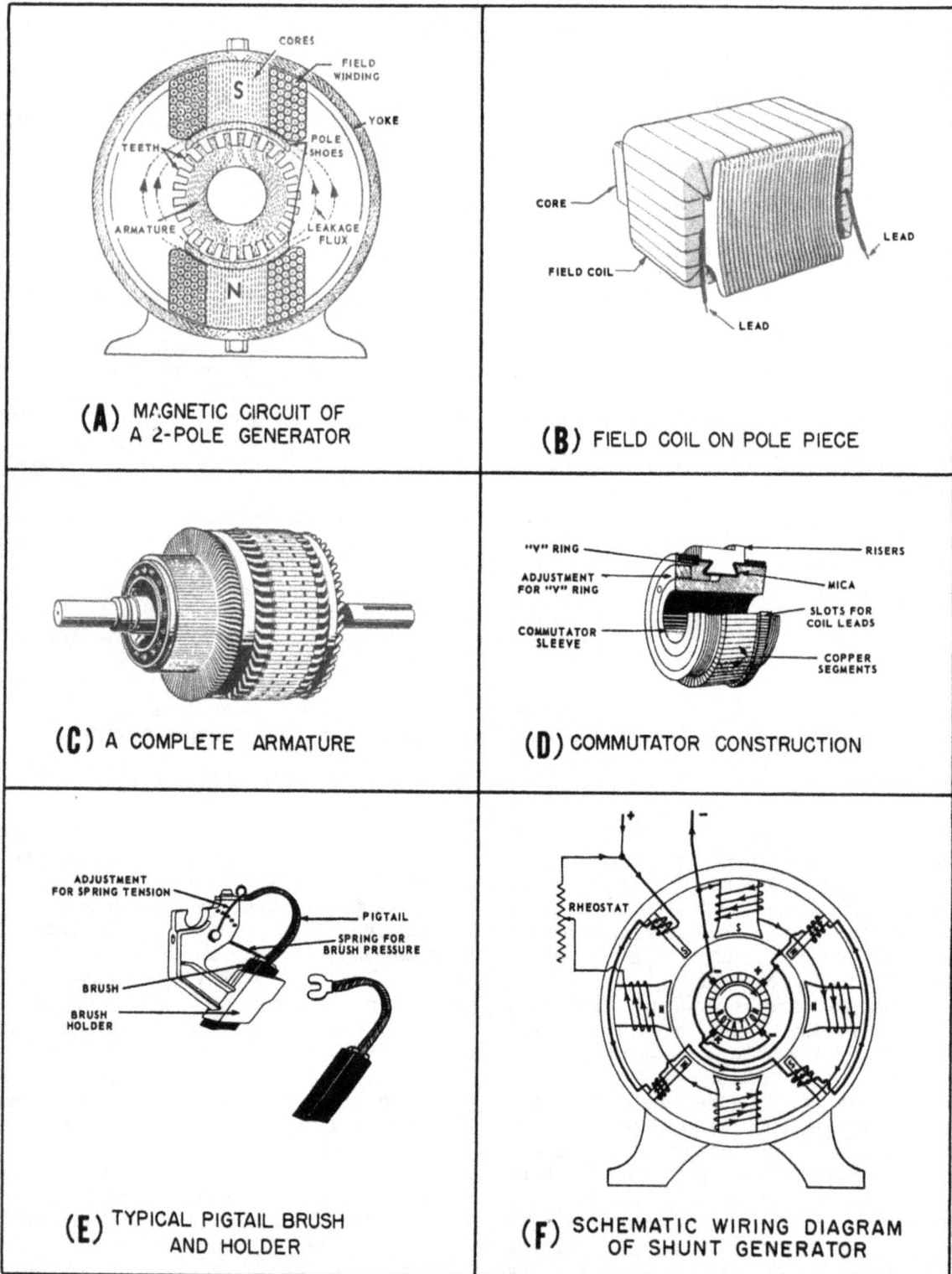

(A) MAGNETIC CIRCUIT OF A 2-POLE GENERATOR

(B) FIELD COIL ON POLE PIECE

(C) A COMPLETE ARMATURE

(D) COMMUTATOR CONSTRUCTION

(E) TYPICAL PIGTAIL BRUSH AND HOLDER

(F) SCHEMATIC WIRING DIAGRAM OF SHUNT GENERATOR

BE.329

Figure 18-2.—D-c generator parts.

The brush holders, which are attached to brush studs, hold the brushes in their proper positions on the commutator. The brush studs are fastened to a rocker arm, or brush holder yoke, that is attached to the frame. Multipolar generators usually have as many brush studs as there are main poles. The brush studs are of alternate positive and negative polarity and those of like polarity are connected together as indicated in figure 18-2 (F).

## ARMATURE WINDINGS

Armature windings for generators range from the simple to the complex. The material presented in this chapter will discuss the following types of windings:
1. Simple coil armature.
2. Gramme-ring winding.
3. Drum armature.
4. Simplex lap winding.
5. Simplex wave winding.

## SIMPLE COIL ARMATURE

The simplest generator armature winding is a loop or single coil. Rotating this loop in a magnetic field will induce an emf whose strength is dependent upon the strength of the magnetic field and the speed of rotation of the conductor.

A single-coil generator with each coil terminal connected to a bar of a 2-segment metal ring is shown in figure 18-3. The two segments of the split ring are insulated from each other and the shaft, thus forming a simple commutator. The commutator mechanically reverses the armature coil connections to the external circuit at the same instant that the direction of the generated voltage reverses in the armature coil. This action involves the process known as commutation, which is described later in the chapter.

When the coil rotates clockwise from the position shown in figure 18-3 (A), to the position shown in figure 18-3 (B), an emf is generated in the coil in the direction (indicated by the heavy arrows) that deflects the galvanometer to the right. Current flows out of the negative brush, through the galvanometer, and back to the positive brush to complete the circuit through the armature coil. If the coil is rotated to the position shown in figure 18-3 (C), the generated voltage and the current fall to zero, as in figure 18-3 (A). At this instant the brushes make contact with both bars of the commutator and short-circuit the coil. As the coil moves to the position

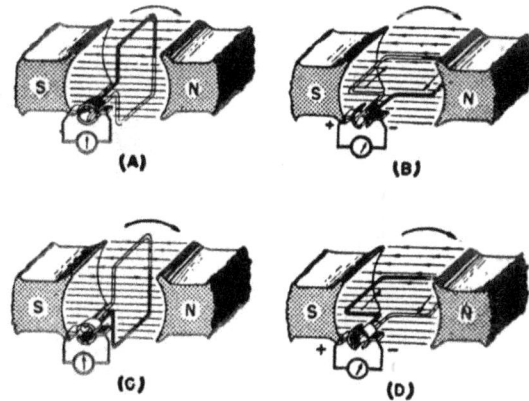

BE.330

Figure 18-3.—Single-coil generator with commutator.

shown in figure 18-3 (D), an emf is generated again in the coil but of opposite polarity.

The emf's generated in the two sides of the coil shown in figure 18-3 (D), are in the reverse direction to that of the emf's shown in figures 18-3 (B). Since the bars of the commutator have rotated with the coil and are connected to opposite brushes, the direction of the flow of current through the galvanometer remains the same. The emf developed across the brushes is pulsating and unidirectional (in one direction only), varying twice during each revolution between zero and maximum.

Figure 18-4 is graph of the pulsating direct emf for one revolution of a single-loop 2-pole armature. A pulsating direct voltage of this characteristic (called RIPPLE) is unsuitable for most applications. Therefore, in practical generators more coils and more commutator bars are used to produce an output voltage waveform with less ripple.

## EFFECTS OF MORE COILS

Figure 18-5 shows the reduction in ripple component of the voltage obtained by the use of two coils instead of one. Since there are now four commutator segments in the commutator and only two brushes, the voltage cannot fall any lower than point A; Therefore, the ripple is limited to the rise and fall between points A and B. By adding still more armature coils, the ripple voltage can be reduced still more.

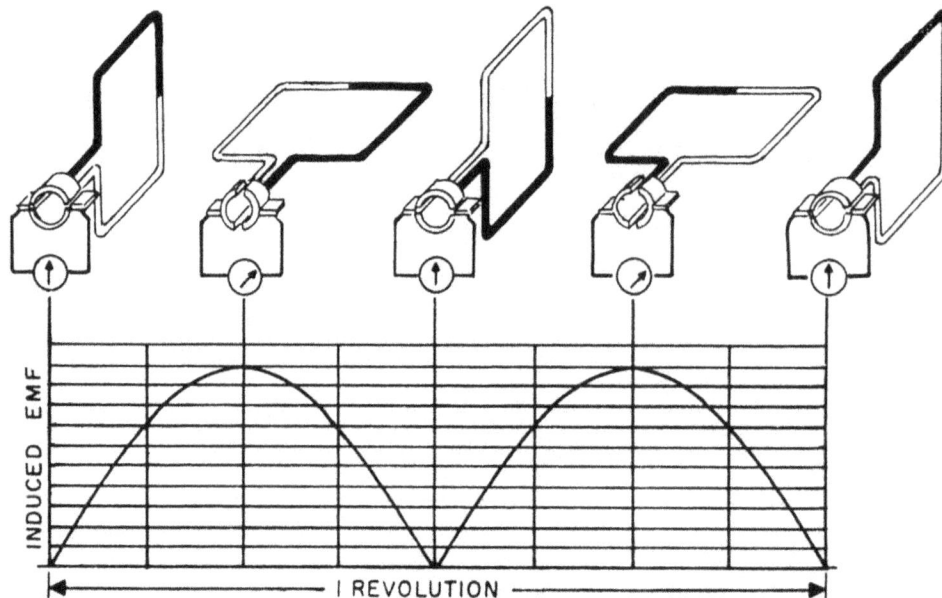

BE.331

Figure 18-4.—Voltage from a single coil armature.

## GRAMME-RING WINDING

A Gramme-ring winding (fig. 18-6) is formed by winding insulated wire around a hollow iron ring and tapping it at regular intervals to the commutator bars. The portions of the windings on the inside of the ring cut practically no flux and act as connectors for the active portions of the conductors, which lie on the outer surface of the ring. Because only a small fraction of the conductor is used to generate voltage in such a winding, a relatively large amount of copper is required to produce a given voltage. The schematic wiring diagram of a Gramme-ring winding is frequently used as a simplified equivalent circuit for the practical but more complex drum armature wiring diagrams. The Gramme-ring winding is used in this discussion only to simplify the drum winding circuits.

In the 2-pole winding of figure 18-6 (A), there are two parallel paths for armature current between the brushes. One path is through the coils on the left side and the other is through the coils on the right side. The voltage between the brushes is the vector sum of the voltages generated in all the coils of each path. No circulating current flows between the two paths because the generated voltages in the two paths are equal and in opposition. The polarity of the brushes may be determined by the left-hand rule

for generator action. The negative terminal is the one from which electrons flow out to the load; the positive terminal is the one to which the electrons return from the load.

## DRUM ARMATURE

In the basic drum winding shown in figure 18-6 (B), all the conductors lie in slots near the surface of the armature. The armature conductors are indicated in the figure as circles. Those on the left contain a dot to indicate that the direction of generated voltage is toward the observer; those on the right contain an x to show that the direction of the generated voltage is away from the observer. The dotted lines connecting the circles represent end connections on the back of the armature between the two halves of each coil. The solid lines between the commutator and the armature conductors represent the end connections on the front of the armature between the coils and commutator segments.

There are 8 coils, 8 slots, and 8 commutator segments in this simplified basic drum winding. One half of a coil lies in the upper portion of a slot and the other half of the coil lies in the lower portion of a slot that is approximately halfway around the armature. This arrangement permits the generated voltages in the two halves of a coil to be a maximum at almost the same

343

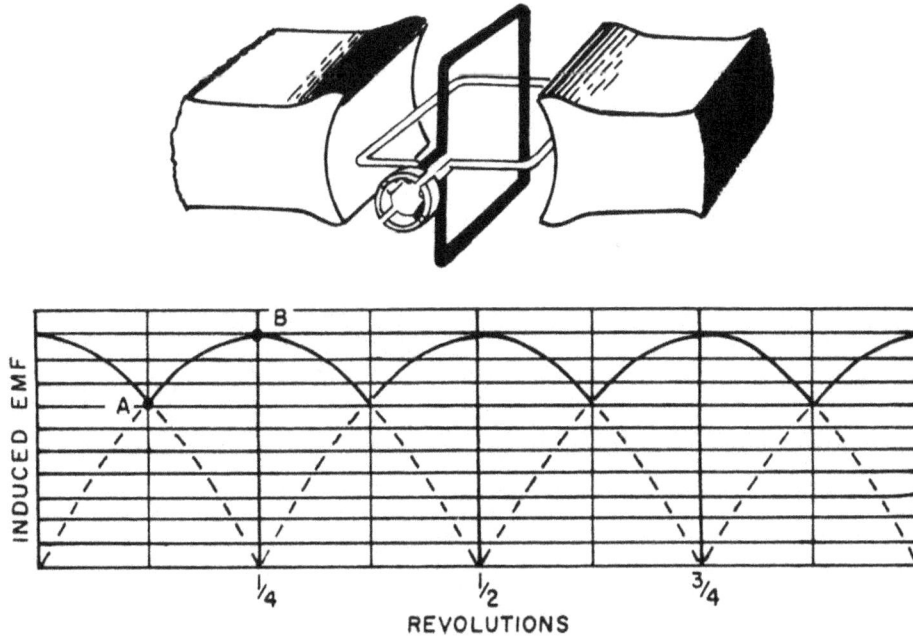

Figure 18-5.—Voltage from a two coil armature.

instant, because the armature is rotating in a 2-pole field. The polarity of the brushes may be determined by the left-hand rule for generator action. The brushes are shown on the inside of the commutator surface to simplify the drawing. Normally, the brushes contact the outside surface of the commutator.

In drum armatures, with the exception of the coil-end connections, all of the copper is used to generate the emf. The distance between the two sides of a coil is known as the COIL PITCH. This distance should be about the same as the distance between the centers of adjacent poles as mentioned previously.

The following analysis is made for a no-external-load condition. There are two paths for armature current through the drum winding—one through conductors a, b, c, d, e, f, g, and h; the other through conductors i, j, k, l, m, n, o, and p. The generated voltages in all conductors in the first path are in the same direction—from the positive brush to the negative brush. The generated voltages in all conductors in the second path are also all in the same direction—from the positive brush to the negative brush. The two paths form a closed circuit and the voltages of each path are equal and in opposition with respect to each other. Therefore no circulating current will flow.

Certain similarities exist in the Gramme-ring winding (fig. 18-6 (A)) and the drum winding (fig. 18-6 (B)). The generated voltage distribution is the same. The number of paths for load current is the same. In both types, when a brush contacts two segments, a coil is short-circuited. In both types, the brushes are positioned so that only coils that are moving approximately parallel to the field are short-circuited. That is, a coil is short-circuited only at the exact instant that it has no voltage induced in it. Thus, sparking is minimized. In both types, there are 8 commutator bars and 2 brushes and the armatures are wound for 2-pole fields. Also, in both types there are single-turn coils and 16 active conductors in which the voltage of the machine is generated.

The differences that exist in the two types of armatures are in the position of the brush axis, the method of mounting the coils, and the path for the magnetic field flux. In the Gramme-ring winding the brush axis is perpendicular to the field axis. In the drum winding the brush axis coincides with the field axis. In the Gramme-ring armature the coils are wound on a ring. In the drum type they are preformed and inserted in slots. In the Gramme-ring type the magnetic field crosses the airgap and is confined to the ring. In the drum type the field crosses the

GRAMME-RING WINDING
**(A)**

TWO-LAYER DRUM WINDING
**(B)**

BE.333

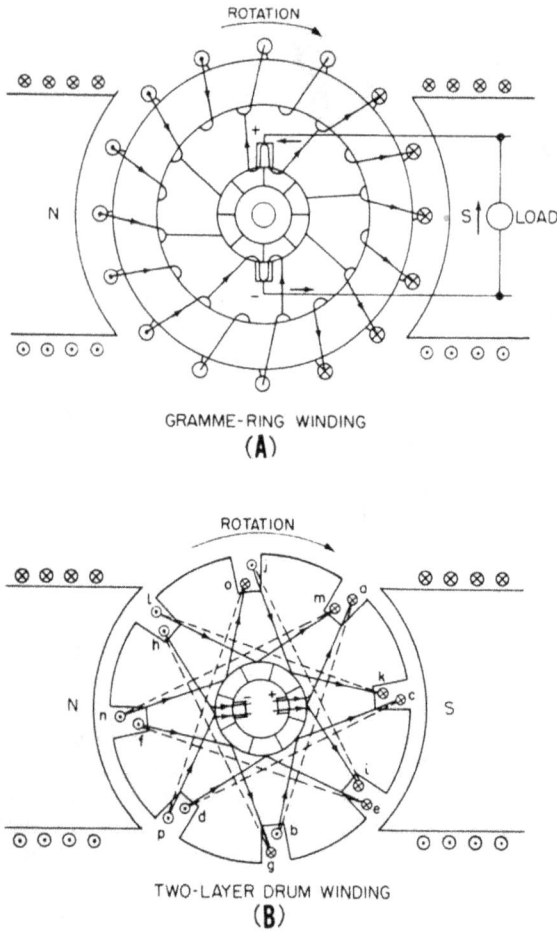

Figure 18-6.—Basic d-c generator armature windings.

airgap and permeates the entire cross section of the armature including the outer slotted portion as well as the inner section. Finally, the end connections form a much higher percentage of the total winding in the Gramme-ring type than in the drum type. It is principally for this reason that Gramme-ring armatures have been replaced by the more efficient drum-type armature.

## SIMPLEX LAP WINDING

As mentioned previously, direct-current armatures are generally wound with preformed coils, as shown in figure 18-7. The term "span" is the distance from one winding element of a coil to the other winding element of the same coil and is usually given in terms of the number of slots included between them. A winding element is a coil side consisting of one or more active (face) conductors in series and taped together to form that part of the coil that is inserted in the armature slot. The span of the armature coil should be about equal to the peripheral distance between the centers of adjacent field poles, so that the voltages generated in the two sides of the coil will be in series addition. This distance is called POLE PITCH. When the span of a coil is less than the pole pitch the winding is called a FRACTIONAL-PITCH WINDING. A fractional pitch coil can be as low as 0.8 pole pitch. Fractional pitch coils have reduced emf because the coil side voltages do not reach their maximum values at the same time. Ordinarily, the reduction in emf is small. The savings in copper, in shorter end-connections, warrants the loss.

Direct-current armature windings are generally 2-layer windings. In this arrangement, the coils are placed on the armature with one side of coil occupying the top of one slot and its other side occupying the bottom of another slot. The distance between the slots is approximately 1 pole span. Such an arrangement allows the windings to fit readily on the armature with an equal number of coils and slots. Thus each slot contains two layers of conductors in which voltages are generated as the armature rotates through the field. This arrangement is shown in the 2-layer drum winding of figure 18-6 (B).

A 4-pole simplex-lap winding is shown in figure 18-8. Starting with commutator bar 1, the circuit may be traced through the heavy black coil to the adjacent commutator bar 2. The trace may be continued through successive coils until

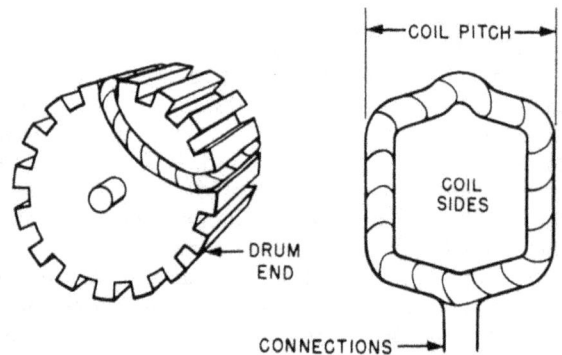

BE.334

Figure 18-7.—Formed armature coil for 4-pole armature.

the entire armature circuit has been traced from one end to the other. Upon reentering the starting commutator bar 1, the trace is completed. Thus, the circuit is seen to be a closed-circuit winding.

There are four groups of coils generating the same voltage between brushes of opposite polarity in the example shown in figure 18-8. One group consists of coils occupying winding spaces 1, 12, 3, 14, 5, 16, 7, 18, 9, and 20. A second group consists of coils occupying winding spaces 11, 22, 13, 24, 15, 26, 17, 28, 19, and 30. A third group consists of coils occupying winding spaces 21, 32, 23, 34, 25, 36, 27, 38, 29, and 40. The fourth group consists of coils occupying winding spaces 31, 42, 33, 2, 35, 4, 37, 6, 39, 8 41, and 10. These four groups of coils are placed in parallel by connecting the two positive brushes together and the two negative brushes together as indicated in the schematic diagram of figure 18-8 (B). No circulating current flows between these four parallel paths when no load is connected between the positive and negative brushes, because the voltage of each of the groups is equal and of opposite polarity to the voltages of the other groups.

A simplified schematic of the 4-pole simplex-lap armature winding is shown in figure 18-9. Connecting a load between the positive and negative brushes causes current to flow through the armature in these four paths. The current is distributed equally in the four paths. Thus if the total output current of the armature is 400 amperes, each path through the armature will carry 100 amperes. In this winding there are as many parallel paths for current through the armature as there are poles in the field.

The simplex-lap winding described in this chapter is the simplest type of lap winding. It is identified as a single winding to distinguish it from more complex double and triple lap windings. It is single-reentrant—that is, the winding closes on itself at the end of one complete turn around the armature. This arrangement distinguishes the winding from more complex types that reenter upon themselves after one, two, or three complete turns around the armature.

The characteristics of the (single) simplex-lap winding may be summarized as follows:

1. There are as many paths for current as there are field poles.

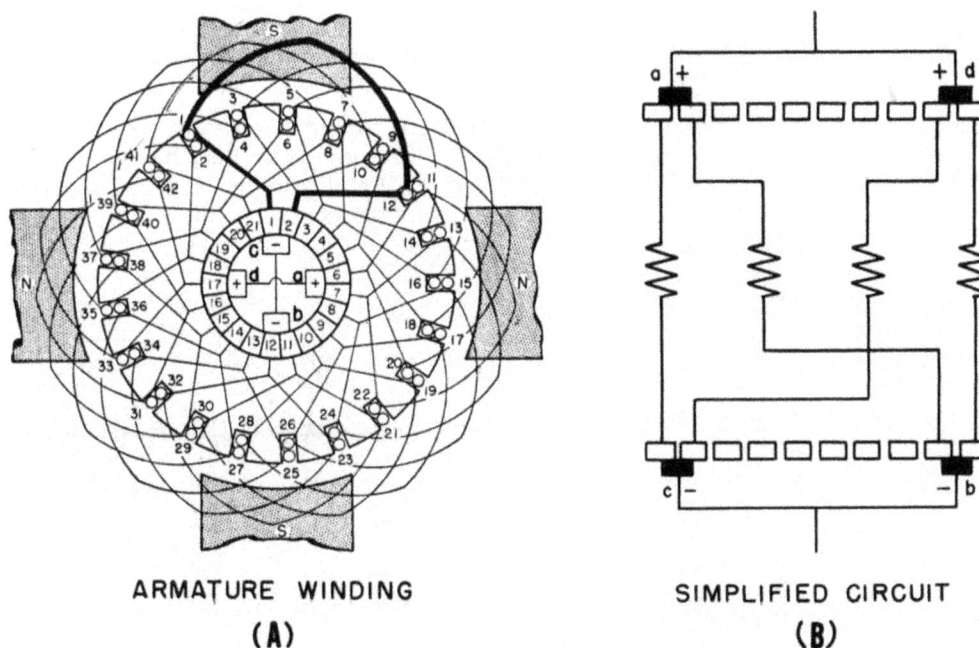

ARMATURE WINDING

(A)

SIMPLIFIED CIRCUIT

(B)

BE.335

Figure 18-8.—Simplex-lap 4-pole armature winding.

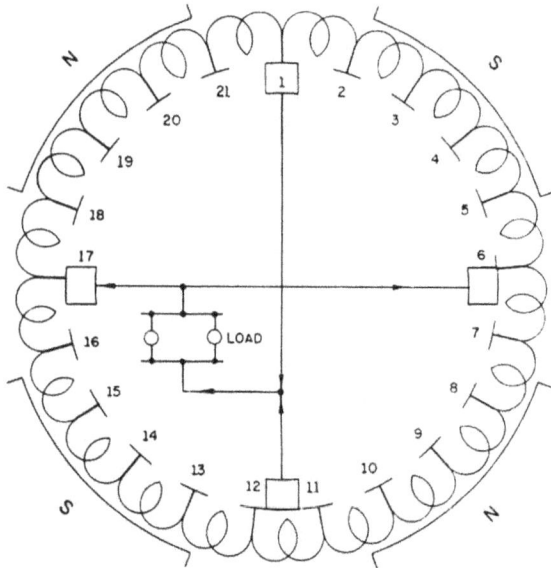

BE.336

Figure 18-9.—Simplified schematic of simplex-lap 4-pole armature winding.

2. There are as many brush positions on the commutator as there are field poles.

3. The two ends of an armature coil connect to adjacent commutator segments.

4. The armature winding forms a continuous closed circuit.

5. The lap armature winding is used for relatively high-current low-voltage loads.

6. The voltage generated between positive and negative brushes is

$$E = \frac{\Phi ZN}{10^8}$$

where E is the generated emf in volts, Z the number of armature face conductors, $\Phi$ the lines of magnetic flux per pole, and N the armature speed in revolutions per second.

A face conductor lies in a slot and generates a voltage as the armature rotates. An armature coil includes one or more turns, each of which consists of two face conductors and their respective end connections. For example, a 4-pole d-c generator with a simplex-lap armature winding having $10^6$ lines of magnetic flux per pole, 440 armature conductors, and a speed of 50 revolutions per second (3,000 rpm) has a generated voltage of

$$E = \frac{10^6 \times 440 \times 50}{10^8} = 220 \text{ volts.}$$

If the load current is 1,000 amperes, each armature coil will carry $\frac{1,000}{4}$, or 250 amperes.

SIMPLEX WAVE WINDING

In the single simplex-wave winding (fig. 18-10) groups of coils under similar pairs of poles at any instant are generating equal voltages and are connected in series. This winding is also called a SERIES WINDING. Each coil has its ends connected to the commutator bars that are two pole spans apart. This distance is measured around the armature circumference from the center of one field pole to the center of the next pole of like sign. In this example, the distance covered by two pole spans is halfway around the commutator; in a 6-pole machine, two pole spans are one-third the way around; and in an 8-pole machine, one-fourth the way around. In the single simplex-wave winding there are as many coils connected in series between adjacent commutator bars as there are pairs of poles in the field. In this example there are two pairs of poles, so there are two coils in series between adjacent commutator bars. This arrangement may be seen by starting at segment 1 and tracing through the circuit of the two coils occupying winding spaces 1, 10, 17, and 26, and ending at commutator segment 2.

The single simplex-wave winding is a closed-circuit winding as may be seen by starting with one coil and tracing through the entire armature circuit to complete the trace by ending at the

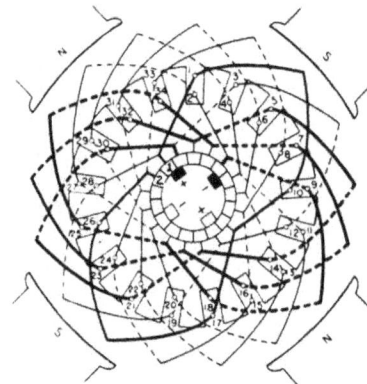

BE.337

Figure 18-10.—Single simplex-wave winding.

347

starting position. In the single simplex wave-wound armature there are two paths for current regardless of the number of poles for which it is wound and regardless of the number of brushes. Starting with the upper positive brush, one path includes the armature conductors that occupy winding spaces 1, 10, 17, 26, 33, 8, 15, 24, 31, 6, 13, 22, 29, 4, 11, 20, 27, and 2, returning to the upper negative brush. The other path from the upper positive brush includes the armature conductors that occupy winding spaces 28, 19, 12, 3, 30, 21, 14, 5, 32, 23, 16, 7, 34, 25, 18, and 9, returning to the upper negative brush.

Although four brushes are shown, only two are required. The full voltage of the winding is developed across the upper brushes, and the full voltage is also developed across the lower brushes. The load current divides equally between the two pairs of brushes when they are connected in parallel because the resistance of the two paths through the armature is the same. For example, if the armature has a 20-ampere load at 220 volts the upper brushes alone will carry 20 amperes at 220 volts. Adding the lower brushes will reduce the current to 10 amperes per brush. If the commutator is designed for four brushes, the removal of one pair will overload the remaining brushes unless the load is halved. However, where commutators are inaccessible, wave windings permit the use of two brush positions irrespective of the number of poles, so that servicing the brushes and commutator is facilitated.

The characteristics of the simplex-wave winding may be summarized as follows:

1. There are two parallel paths irrespective of the number of field poles.

2. There is a minimum of two brush positions irrespective of the number of field poles.

3. The two ends of a coil connect to commutator segments that are two pole spans apart (fig. 18-10).

4. The winding has relatively high voltage and low current.

5. The winding forms a continuous closed circuit.

6. There are as many coils in series between adjacent commutator segments as there are pairs of poles in the field.

7. The voltage generated between positive and negative brushes is

$$E = \frac{\Phi ZNP}{10^8},$$

where P is the number of pairs of field poles and the other symbols are the same as those used in the equation for generated voltage in a simplex-lap winding.

For example, a 6-pole d-c generator having a simplex-wave armature winding, $1.05 \times 10^6$ lines of flux per pole, 3,500 armature conductors, and a speed of 10 revolutions per second (600 rpm) has a generated voltage of

$$E = \frac{\Phi ZNP}{10^8} = \frac{1.05 \times 10^6 \times 3,500 \times 10 \times 3}{10^8} = $$

1,100 volts.

If the armature supplies 200 amperes to a load, each armature coil will carry $\frac{200}{2} = 100$ amperes.

## ARMATURE LOSSES

There are three losses in every d-c generator armature. These are as follows:

1. The $I^2R$ or copper loss in the winding.

2. The eddy current loss in the core.

3. The hysteresis loss due to the friction of the revolving magnetic particles in the core.

## COPPER LOSSES

The copper loss is the power lost in heat in the windings due to the flow of current through the copper coils. This loss varies directly with the armature resistance and as the square of the armature current. The armature resistance varies with the length of the armature conductors and inversely with their cross-sectional area.

Armature conductor size is based on an allowance of from 300 to 1,200 circular mils per ampere. For example, a 2-pole armature that is required to supply 100 amperes may use a wire size based on 800 circular mils per ampere, or $\frac{100}{2} \times 800 = 40,000$ circular mils. This value corresponds to a No. 4 wire. Very small armature windings may use only 300 circular mils per ampere with a resulting high current density. Large generators (5,000 kw) require an allowance of 1,200 circular mils per ampere with a resulting low current density in the windings. These variations are the result of the variable nature of the heat-radiating ability of the armature conductors.

Very small round conductors have a much higher ratio of surface to volume than do large round conductors. For example, a 0.1-inch diameter round conductor of a given length has a surface-to-volume ratio of

$$\frac{4\pi D}{\pi D^2}, \text{ or } \frac{4}{0.1} = 40.$$

A 1.0-inch diameter conductor of the same length has a surface-to-volume ratio of $\frac{4}{1}$, or 4. Since the heat-radiating ability of a round conductor varies as the ratio of its surface to volume, the 0.1-inch diameter conductor has $\frac{40}{4}$, or 10 times the heat radiating ability of the 1-inch diameter conductor, other factors being equal.

High-speed generators use a lower circular-mil-per-ampere allowance than low-speed generators because of better cooling. The temperature rise is limited by ventilating ducts, and in some cases by the use of forced ventilation, as in aircraft d-c generators.

The hot resistance of an armature winding is higher that its cold resistance. A 2.5° centigrade increase in temperature of a copper conductor corresponds to an increase in resistance of approximately 1 percent. For example, if the no-load temperature of an armature winding is 20°C and its full-load temperature is 70°C, the increase in resistance is $\frac{70-20}{2.5}$, or 20 percent. Thus, if the no-load resistance is 0.05 ohm between brushes, the hot resistance will be 1.2 x 0.05, or 0.06 ohm. If the full-load armature current is 100 amperes, the full-load armature copper loss will be $(100)^2$ x 0.06, or 600 watts. The armature copper loss varies more widely with the variation of electrical load on the generator than any other loss occurring in the machine. This is because most generators are constant-potential machines supplying a current output that varies with the electrical load across the brushes. The limiting factor in load on a generator is the allowable current rating of the generator armature.

The armature circuit resistance includes the resistance of the windings between brushes of opposite polarity, the brush contact resistance, and the brush resistance.

## EDDY-CURRENT LOSSES

If a d-c generator armature core were made of solid iron (fig. 18-11 (A)) and rotated rapidly in the field, excessive heating would develop even with no-load current in the armature windings. This action would be the result of a generated voltage in the core itself. As the core rotates, it cuts the lines of magnetic field flux at the same time the copper conductors of the armature cut them. Thus, induced currents alternate through the core, first in one direction and then in the other, with accompanying generation of heat.

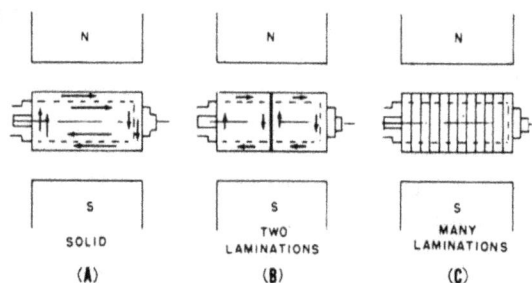

BE.338

Figure 18-11.—Eddy currents in d-c generator armature cores.

These induced currents are called EDDY CURRENTS. They are kept to a low value by sectionalizing (laminating) the armature core. For example, if the core is split into two equal parts (fig. 18-11 (B)) and these parts are insulated from each other, the voltage induced in each section of iron is halved and the resistance of the eddy-current paths is doubled (resistance varies inversely with cross-sectional area). If 10 volts is induced in the core (fig. 18-11 (A)) and the resistance of the path for eddy currents is 1 ohm, the eddy-current loss is $\frac{E^2}{R}$, or $\frac{10^2}{1}$ = 100 watts. The voltage is each section (fig. 18-11 (B)) will be $\frac{10}{2}$, or 5 volts, because the length is halved and the resistance of the eddy-current path for each section is 2 ohms. The loss in each section is $\frac{E^2}{R}$, or $\frac{5^2}{2}$ = 12.5 watts, and the total loss in both sections is 12.5 x 2, or 25 watts. This value represents one-fourth of the power loss in figure 18-11 (A).

Reducing the thickness of the core to one-half its original value reduces the loss to one-fourth of the original loss. Thus, eddy-current losses vary as the square of the thickness of the core laminations.

If the armature core is sufficiently subdivided into multiple sections or laminations (fig. 18-11 (C)), the eddy-current loss can be reduced to a negligible value. Reducing the thickness of the laminations reduces the magnitude of the induced emf in each section and increased the resistance of the eddy-current paths. Laminations in small generator armatures are 1/64 inch thick. The laminations are insulated from each other by a thin coat of lacquer or in some instances simply by the oxidation of the surfaces due to contact with the air while the laminations are being annealed. The insulation need not be high because the voltages induced are very small.

All electrical rotating machines are laminated to reduce eddy-current losses. Transformer cores are laminated for the same reason.

The eddy-current loss is also influenced by speed and flux density. Because the induced voltage, which causes the eddy-currents to flow, varies with the speed and flux density, the power loss, $\frac{E^2}{R}$, varies as the square of the speed and the square of the flux density.

### HYSTERESIS LOSSES

When an armature revolves in a stationary magnetic field, the magnetic particles of the armature are held in alinment with the field in varying numbers depending upon the strength of the field. If the field is that of a 2-pole generator, these magnetic particles will rotate, with respect to the particles not held in alinement, one complete turn for each revolution of the armature. The rotation of magnetic particles in the mass of iron produces friction and heat.

Heat produced in this manner is identified as magnetic HYSTERESIS LOSS. The hysteresis loss varies with the speed of the armature and the volume of iron. The flux density varies from approximately 50,000 lines per square inch in the armature core to 130,000 lines per square inch in the iron between the bottom of adjacent armature slots (called the TOOTH ROOT). Heat-treated sillicon steel having a low hysteresis loss is used in most d-c generator armatures. After the steel has been formed to the proper shape, the laminations are heated to a dull red

heat and allowed to cool. This annealing process reduces the hysteresis loss to a low value.

### ARMATURE REACTION

Armature reaction in a generator is the effect on the main field of the armature acting as an electromagnet. With no armature current, the field is undistorted, as shown in figure 18-12 (A). This flux is produced entirely by the ampere-turns of the main field windings. The neutral plane AB is perpendicular to the direction of the main field flux. When an armature conductor moves through this plane its path is parallel to the undistorted lines of force and the conductor does not cut through any flux. Hence no voltage is induced in the conductor. The brushes are placed on the commutator so that they short-circuit coils passing through the neutral plane. With no voltage generated in the coils, no current will flow through the local path formed momentarily between the coils and segments spanned by the brush. Therefore, no sparking at the brushes will result.

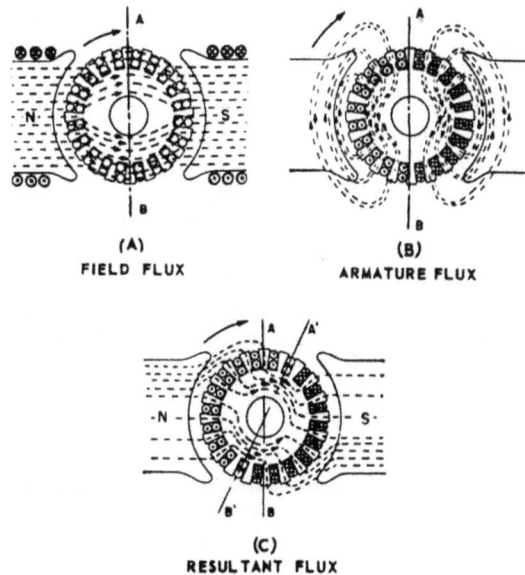

(A) FIELD FLUX

(B) ARMATURE FLUX

(C) RESULTANT FLUX

BE.339

Figure 18-12.—Flux distribution in a d-c generator.

When a load is connected across the brushes, armature current flows through the armature conductors, and the armature itself becomes a source of magnetomotive force. The effect of the armature acting as an electromagnet is considered in figure 18-12 (B), with the assumption that the main field coils are deenergized and full-load current is introduced to the armature circuit from an external source. The currents in the conductors on the left of the neutral plane all carry current toward the observer, and those on the right carry current away from the observer. These directions are the same as those in which the current would flow if it were under the influence of the normal emf generated in the armature with normal field excitation.

These armature current-carrying conductors establish a magnetomotive force that is perpendicular to the axis of the main field, and in the figure the force acts downward. This magnetizing action of the armature current is called CROSS MAGNETIZATION and is present only when current flows through the armature circuit. The amount of cross magnetization produced is proportional to the armature current.

When current flows in both the field and armature circuits, the two resulting magnetomotive forces distort each other. They twist in the direction of rotation of the armature. The mechanical (no load) neutral plane, AB (fig. 18-12 (C)), is now advanced to the electrical (load) neutral plane, A'B'. When armature conductors move through plane A'B' their paths are parallel to the distorted field and the conductors cut no flux, hence no voltage is induced in them. The brushes must therefore be moved on the commutator to the new neutral plane. They are moved in the direction of armature rotation. The absence of sparking at the commutator indicates correct placement. The amount that the neutral plane shifts is proportional to the load on the generator because the amount of cross-magnetizing magnetomotive force is directly proportional to the armature current.

When the brushes are shifted into the electrical neutral plane A'B' the direction of the armature magnetomotive force is downward and to the left, as shown in figure 18-13 (A), instead of vertically downward. The armature magnetomotive force may now be resolved into two components, as shown in figure 18-13 (B).

The conductors included at the top and bottom of the armature within sectors BB produce a magnetomotive force that is directly in opposition to the main field and weakens it. This component is called the ARMATURE DEMAGNETIZING MMF. The conductors included on the right and left sides of the armature within sector AA produce a cross-magnetizing mmf at right

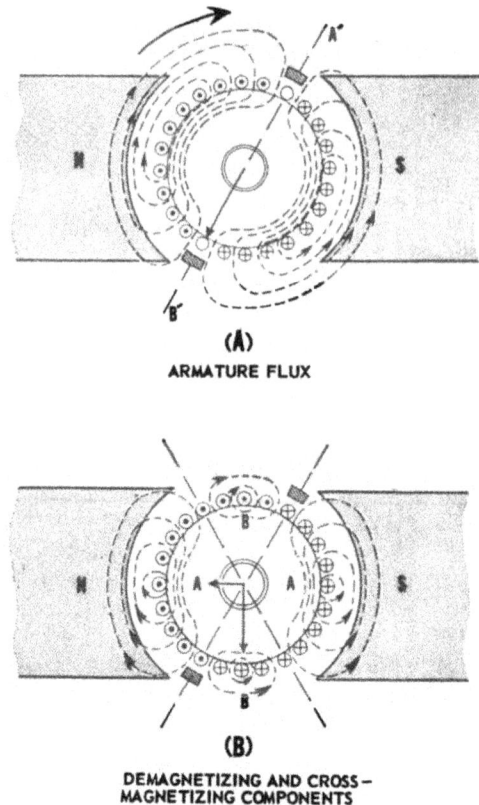

**(A)**
ARMATURE FLUX

**(B)**
DEMAGNETIZING AND CROSS—
MAGNETIZING COMPONENTS

BE.340

Figure 18-13.—Effect of brush shift on armature reaction.

angles to the main field axis. This cross-magnetizing force tends to distort the field in the direction of rotation. As mentioned previously, the distortion of the main field of the generator is the result of ARMATURE REACTION. Armature reaction occurs in the same manner in multipolar machines.

## COMPENSATING FOR ARMATURE REACTION

The effects of armature reaction are reduced in d-c machines by the use of (1) high flux density in the pole tips, (2) a compensating winding, and (3) commutation poles.

The cross-sectional area of the pole tips is reduced by building the field poles with laminations having only one tip. These laminations are alternately reversed when the pole core is stacked so that a space is left between alternate laminations at the pole tips. The reduced cross section of iron at the pole tips increases the flux density so that they become saturated and the cross-magnetizing and demagnetizing forces of the armature will not affect the flux distribution in the pole face to as great an extent as they would at reduced flux densities.

The compensating winding consists of conductors imbedded in the pole faces parallel to the armature conductors. The winding is connected in series with the armature and is arranged so that the ampere-turns are equal in magnitude and opposite in direction to those of the armature. The magnetomotive force of the compensating winding therefore neutralizes the armature magnetomotive force, and armature reaction is practically eliminated. Because of the relatively high cost, compensating windings are ordinarily used only on high-speed and high-voltage generators of large capacity.

Commutating poles are discussed after the description of the process of commutation.

## COMMUTATION

Commutation is the process of reversing the current in the individual armature coils and conducting the direct current to the external circuit during the brief interval of time required for each commutator segment to pass under a brush. In figure 18-14, commutation occurs simultaneously in the two coils that are undergoing momentary short circuit by the brushes—coil B by the negative brush, and the diametrically opposite coil by the positive brush. As mentioned previously, the brushes are placed on the commutator in a position that short-circuits the coils that are moving through the electrical neutral plane because there is no voltage generated in the coils at the time and no sparking occurs between commutator and brush.

There are two paths for current through the armature winding. If the load current is 100 amperes, each path will contain a current of 50 amperes. Thus each coil on the left side carries 50 amperes in a given direction and each coil on the right side carries a current of 50 amperes in the opposite direction. The reversal of the current in a given coil occurs during the time that particular coil is being short-circuited by a brush. For example, as coil A approaches the negative brush it is carrying the full value of 50 amperes which flows through commutator segment 1 and the left half of the negative brush where it joints 50 amperes from coil C.

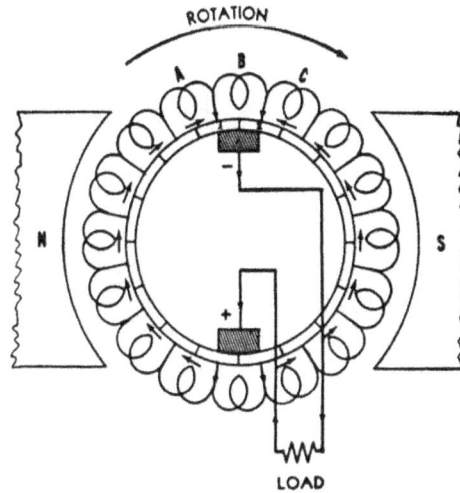

BE.341

Figure 18-14.—Commutation in a d-c generator.

At the instant shown, the negative brush spans half of segment 1 and half of segment 2. Coil B is on short circuit and is moving parallel to the field so that its generated voltage is zero, and no current flows through it. As rotation continues in a clockwise direction the negative brush spans more of segment 1 and less of segment 2. Consider, for example, the interval included between the instant shown in the figure and the instant that the negative brush spans only segment 1. During this time the current in segment 1 increases from 50 to 100 amperes and the current in segment 2 decreases from 50 to 0 amperes. When segment 2 leaves the brush, no current flows from segment 2 to the brush and commutation is complete.

As coil A continues into the position of coil B the current in A decreases to zero. Thus, the current in the coils approaching the brush is reducing to zero during the brief interval of time that it takes for coil A to move to the position of coil B. During this time the flux collapses around the coil and induces an emf of self-induction which opposes the decrease of current. Thus, if the emf of self-induction is not neutralized, the current will not decrease in coil A and the current in the coil lead to segment 1 will not be zero when segment 1 leaves the brush. This delay causes a spark to form between the toe of the brush and the trailing edge of the segment. As the segment breaks contact with the brush, this action burns and pits the commutator.

The reversal of current in the coils takes place very rapidly. For example, in an ordinary 4-pole generator, each coil passes through the process of commutation several thousand times per minute. It is important that commutation be accomplished without sparking to avoid excessive commutator wear.

## ADVANCING THE BRUSHES

The emf of self-induction in the armature coils is caused by the inductance of the coils and the changing current in them and cannot be eliminated. The effects of this emf can be neutralized however, by introducing into the coil during the process of commutation and emf that is equal and in opposition to the induced emf. This neutralization can be accomplished by shifting the brushes in the direction of rotation or by using interpoles (commutating poles).

If the brushes are shifted in the direction of rotation until the coils undergoing commutation cut a small amount of flux from the on-coming main pole, sufficient emf is induced to neutralize the effect of the self-induced emf. This action allows the coil current to decrease to zero and increase to the desired values in the opposite direction without sparking at the brushes.

The flux necessary to generate the emf that neutralizes the emf of self-induction is called the COMMUTATING FLUX. This method of reducing sparking at the brushes is satisfactory only under steady-load conditions because the amount of commutating flux required varies with the load and, therefore, the brushes must be shifted with each change in load.

## COMMUTATING POLES

Commutating poles, or interpoles, provide the required amount of commutating flux without shifting the brushes from mechanical neutral. They are narrow auxiliary poles located midway between the main poles, as shown in figure 18-2 (F). They establish a flux in the proper direction and of sufficient magnitude to produce satisfactory commutation. They do not contribute to the generated emf of the armature as a whole because the voltages generated by their fields cancel each other between brushes of opposite polarity.

The interpole magnetomotive force neutralizes that portion of the armature reaction within the zones of commutation and produces the

proper flux to generate an emf in the short-circuited coil that is equal and opposite to the emf of self-induction. Thus, no sparking occurs at the brushes and "black" commutation indicates the ideal condition. Because both the armature reaction and the self-induced emf in the commutated coils vary with armature current, the interpole flux varies with the armature current. This condition is obtained by connecting the interpole windings in series with the armature and operating the interpole iron at flux densities well below saturation. The magnetic polarities are such that an interpole always has the same polarity as the adjacent main field pole in the DIRECTION OF ROTATION. This relation always exists for d-c generators.

## MOTOR REACTION IN A GENERATOR

Whenever a generator delivers current to a load, the load current creates an opposition force that opposes the rotation of the generator armature. An armature conductor is represented in figure 18-15. When the conductor is stationary, no voltage is generated and no current flows; hence, no force acts on the conductor. When the conductor is moved downward and the circuit is completed through an external load, current flows through the conductor in the direction indicated, setting up lines of force around it that have a clockwise direction.

The interaction of the conductor field and the main field of the generator weakens the field above the conductor and strengthens it below the conductor. The field consists of lines that act like stretched rubber bands. Thus, an upward reaction force is produced that acts in opposition

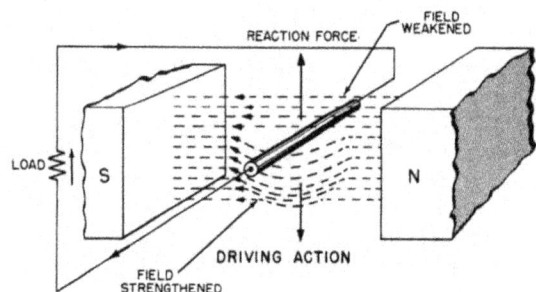

BE.342

Figure 18-15.—Motor reaction in a generator.

to the downward driving force applied to the generator armature. If the current in the conductor increases, the reaction force increases, and more force must be applied to the conductor to keep it from slowing down.

With no armature current, no magnetic reaction exists and the generator input power is low. As the armature current increases, the reaction of each armature conductor against rotation increases and the driving power to maintain the generator armature speed must be increased. If the prime mover driving the generator is a gasoline engine, this effect is accomplished by opening the throttle on the carburetor. If the prime mover is a steam turbine, the main steam-admission valve is opened wider, thus permitting more steam to flow through the turbine.

### D-C GENERATOR CHARACTERISTICS

#### METHODS OF CONNECTING THE FIELD WINDINGS

Usually d-c generators are classified according to the manner in which the field windings are connected to the armature circuit (fig. 18-16).

A SEPARATELY EXCITED D-C GENERATOR is indicated in the simplified schematic diagram of figure 18-16 (A). In this machine the field windings are energized from a separate d-c source other than its own armature.

A SHUNT GENERATOR has its field windings connected across the armature in shunt with the load, as shown in figure 18-16 (B). The shunt generator is widely used in industry.

A SERIES GENERATOR has its field windings connected in series with the armature and load, as shown in figure 18-16 (B). Series generators are seldom used.

COMPOUND GENERATORS contain both series and shunt field windings, as shown in figure 18-16 (B). Compound generators are widely used in industry.

#### FIELD SATURATION CURVES

The strength of the field of a d-c generator depends on the number of ampere-turns in the field windings and the reluctance of the magnetic circuit. The number of turns is generally fixed. Hence, the ampere-turns vary directly with the field current. The generated voltage is directly proportional to the product of the field strength and the speed. Thus, if the field strength is zero or the speed is zero, the generated voltage will be zero. As the current through the field windings increases, the field flux and voltage output will increase. The field strength, however, is not directly proportional to the field current because the reluctance of the magnetic circuit varies with the degree of magnetization. With increasing flux density in the field and armature iron the permeability decreases, thereby increasing the reluctance of the magnetic circuit

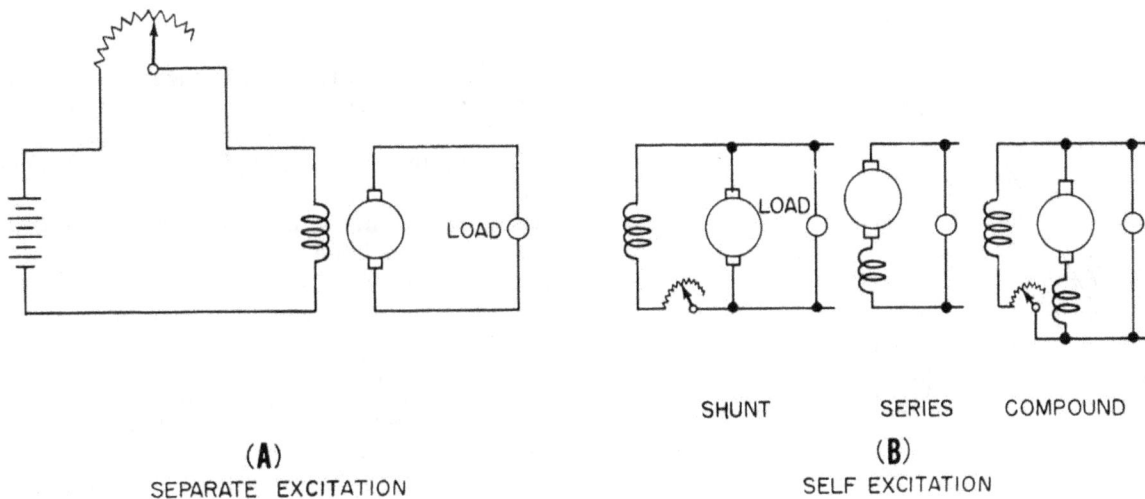

(A)
SEPARATE EXCITATION

SHUNT      SERIES      COMPOUND
(B)
SELF EXCITATION

BE.343

Figure 18-16.—Types of d-c generators.

and it becomes more difficult to increase the voltage.

Certain operating characteristics of the d-c generator are very closely related to the no-load and the full-load field saturation curves of the machine. The no-load saturation curve is determined with no armature current and the full-load saturation is determined with full-load armature current. The speed is held constant for both curves and the field current is increased in equal steps. The terminal voltage corresponding to each value of field current is plotted and forms the field saturation curves as shown in figure 18-17. The no-load saturation curve is preferably taken with the field separately excited. However, for a shunt or a compound generator these curves can be taken with the machine self-excited because the armature voltage drop due to the shunt field current is negligible.

With a certain field current, OA (fig. 18-17), a no-load emf of AD volts is generated, but the terminal voltage at full load is AF. The difference, DF, is caused by athe armature IR drop and armature reaction. The saturation curves bend to the right at high values of field current and thereby show the tendency of the field iron to saturate. The no-load saturation curve is the magnetization curve of the machine. Because this curve is obtained with no armature current, the observed voltage is the generated emf of the machine. At zero field current the emf is not zero because of the residual flux of the machine. This fact is important because a self-excited generator depends on residual flux to build up its voltage. At low values of field current the voltage varies proportionally, but as the field current increases, the steel portion of the magnetic circuit becomes partly saturated and the voltage increases more slowly.

The abrupt bend in the curve is called the KNEE of the curve. Shunt generators are designed to operate at a point slightly above the knee so that a slight change in speed does not cause a great change in voltage. Compound generators operate at a lower point in order to avoid the use of a large series field winding.

## SHUNT GENERATOR

The shunt generator (fig. 18-16 (B)) has field coils of many turns of small size wire connected in shunt with the armature and load. The armature current is equal to the sum of the field current and the load current. The field current is small compared with the load current and is approximately constant for normal variations in load. Thus, the armature current varies directly with the load. The field flux produced by the field current is normally constant so that the terminal voltage does not vary widely with load change. Hence, the generator is essentially a constant-potential machine that delivers current to the load in accordance with the load demand.

### Buildup of Voltage

After the generator is brought up to normal speed, and before any load is connected across the armature, the generator must "build up" its voltage to the rated value (fig. 18-18). A schematic wiring diagram (fig. 18-18 (A)) is shown without load, and with only a few turns of the field coil around one pole indicated for simplicity in tracing the direction of current flow and the polarity of the voltage generated in the armature winding.

Figure 18-18 (B) indicates the field saturation curve for the generator. Line OA represents the relation between voltage and current in the field-coil circuit for one value of field-circuit resistance. Line OA is called an IR-drop curve for the field circuit and is assumed to be a straight line on the basis of constant temperature

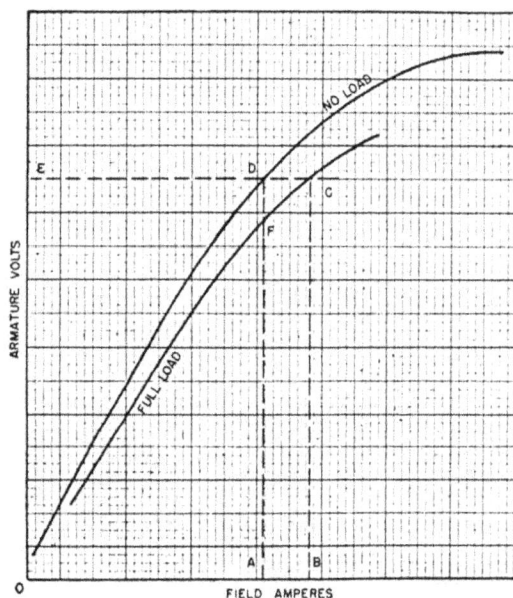

BE. 344

Figure 18-17.—Field saturation curves for d-c generator.

355

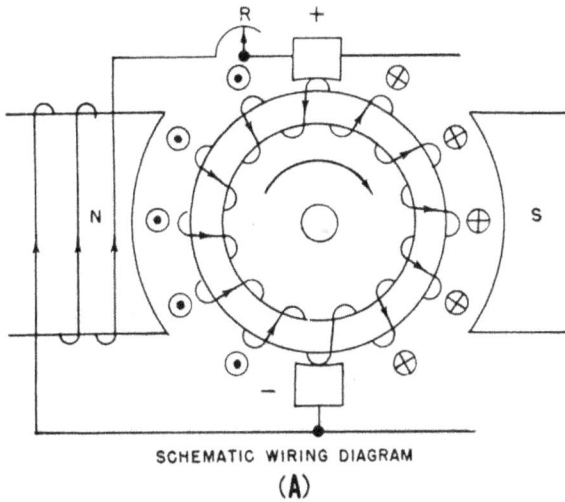

SCHEMATIC WIRING DIAGRAM

(A)

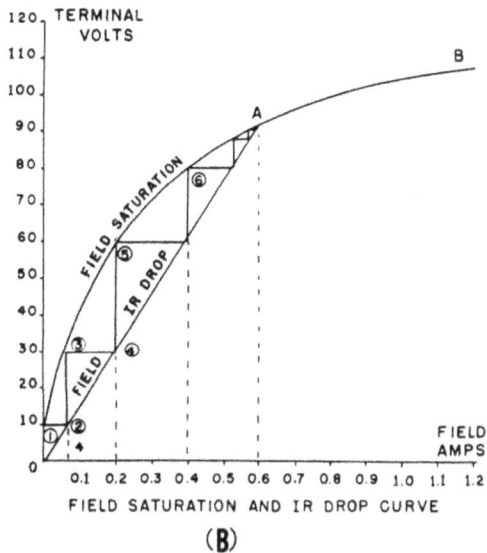

FIELD SATURATION AND IR DROP CURVE

(B)

BE. 345

Figure 18-18.—Buildup of voltage of
a shunt generator.

operation. The field rheostat permits adjustment of the field current through a relatively wide range and is designed so that its resistance is at least equal to that of the field winding.

At start, the generator is brought up to rated speed. The load is not connected. The armature conductors cut the small residual field and generate about 10 volts across the brushes (point ① on the saturation curve). The left-hand rule for generator action indicates that the generated voltage is applied to the field winding in a direction to supplement the residual field. Ten volts

applied across the field circuit will cause approximately 0.07 ampere to flow through the field coils (point ② on the IR-drop curve).

This current strengthens the field, and the armature voltage increases to 30 volts, as shown on the field saturation curve (point ③). During the process of voltage buildup, the effect of the armature circuit resistance on the terminal voltage is neglected since the field current is only a small fraction of an ampere. This current flows through the armature and the accompanying IR drop is negligible. A generated voltage of 30 volts applied to the field circuit causes a field current of 0.2 ampere to flow, as shown by the IR-drop curve (point ④). This current in turn increases the field strength and generated voltage in the armature to 60 volts, as indicated on the field saturation curve (point ⑤). This voltage applied to the field causes 0.4 ampere to flow and this action increases the terminal voltage to 80 volts (point ⑥). The rise in voltage continues until the field current has increased to 0.6 ampere and the terminal voltage levels off at 90 volts (point A). No further increase in generated voltage occurs for the given value of field circuit resistance because the amount of field saturation flux produced is enough to generate the voltage required (90 volts) to circulate this current (0.6 ampere) through the field coils. For this condition the resistance of the field

circuit is $\frac{90}{0.6}$, or 150 ohms. Field saturation limits the generated voltage to this value as determined by the setting of the field rheostat.

To increase the shunt generator terminal voltage further, it is necessary to decrease the field circuit resistance. For example, if the terminal voltage is to be increased to 110 volts, the corresponding field current from the saturation curve is 1.2 amperes (point B). The field circuit resistance is now represented by the

slope of line OB, or $\frac{110}{1.2}$ = 91.8 ohms. Thus, if the field resistance is decreased from 150 ohms to 91.8 ohms, the terminal voltage will increase from 90 volts to 110 volts.

INHERENT REGULATION OF
SHUNT GENERATOR

Internal changes, both electrical and magnetic, that occur in a generator automatically with load change, give the generator certain typical characteristics by which it may be identified. These internal changes are referred to as the inherent regulation of a generator. At no load, the armature current is equal to the field current. With low armature resistance and low field current there is little armature IR drop, and the generated voltage is equal to the terminal voltage. With load applied, the armature IR drop increases, but is relatively small compared with the generated voltage. Also the armature reaction voltage loss is small. Therefore, the terminal voltage decreases only slightly provided the speed is maintained at the rated value.

Load is added to a shunt generator by increasing the number of parallel paths across the generator terminals. This action reduces the total load circuit resistance with increased load. Since the terminal voltage is approximately constant, armature current increases directly with the load. Since the shunt field is in a separate circuit it receives only a slightly reduced voltage and its current does not change to any great extent.

Thus, with low armature resistance and a relatively strong field there is only a small variation in terminal voltage between no load and full load.

## EXTERNAL VOLTAGE CHARACTERISTICS

A graph of the variation in terminal voltage with load on a shunt generator is shown in curve A of figure 18-19. This curve shows that the terminal voltage of a shunt generator falls slightly with increase in load from the no-load condition to the full-load condition. It also shows that with heavy overload the terminal voltage falls more rapidly. The shunt field current is reduced and the magnetization of the field falls to a low value. The dotted portion of curve A indicates the way the terminal voltage falls beyond the breakdown point. In large generators the breakdown point occurs at several times rated load current. Generators are not designed to be operated at these large values of load current and will overheat dangerously even at twice the value of full-load current.

Curve B represents the external voltage characteristic of a shunt generator with constant field current for variations in load between zero and approximately 25 percent over the rated load condition. Curve C represents the external voltage characteristic for the same range of load with simulated conditions of zero armature reaction and constant field current. The diver-

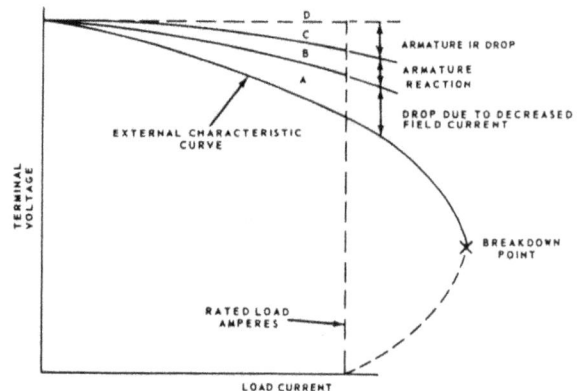

BE. 346

Figure 18-19.—Shunt generator external characteristics.

gence of curve C from curve D represents the voltage variations resulting from the IR drop in the armature circuit. As mentioned previously, the terminal voltage of the shunt generator is prevented from varying widely with load change by providing it with a (1) low armature resistance, (2) strong main field with low armature reaction, and (3) separate field circuit in shunt with the armature and load.

## COMPOUND GENERATOR

Compound generators employ a series field winding in addition to the shunt field winding, as shown in figure 18-16 (B). The series field coils are made of a relatively small number of turns of large copper conductor, either circular or rectangular in cross section, and are connected in series with the armature circuit. These coils are mounted on the same poles on which the shunt field coils are mounted and therefore contribute a magnetomotive force that influences the main field flux of the generator.

### Effect of Series Field

If the ampere-turns of the series field act in the same direction as those of the shunt field the combined magnetomotive force is equal to the sum of the series and shunt field components. Load is added to a compound generator in the same manner in which load is added to a shunt generator—by increasing the number of parallel paths across the generator terminals. Thus, the decrease in total load resistance with added load is accompanied by an increase in armature-circuit and series-field circuit current.

The effect of the additive series field is that of increased field flux with increased load. The extent of the increased field flux depends on the degree of saturation of the field iron as determined by the shunt field current. Thus, the terminal voltage of the generator may increase with load or it may decrease depending upon the influence of the series field coils. This influence is referred to as the degree of compounding.

For example, a FLAT-COMPOUND generator is one in which the no-load and full-load voltages have the same value. An UNDER-COMPOUND generator is one in which the full-load voltage is less than the no-load value. An OVER-COMPOUND generator is one in which the full-load voltage is higher than the no-load value. The way the terminal voltage changes with increasing load depends upon the degree of compounding.

A variable shunt is connected across the series field coils to permit adjustment of the degree of compounding. This shunt is called a DIVERTER. Decreasing the diverter resistance increases the amount of armature circuit current that is bypassed around the series field coils, thereby reducing the degree of compounding.

A field rheostat in the shunt field coil circuit permits adjustment of the no-load voltage of the compound generator. The diverter across the series field coils permits adjustment of the full-load voltage.

External Voltage Characteristics

The variation of terminal voltage with load is indicated in the external characteristic curves of figure 18-20. Curve A is the graph of terminal voltage versus armature current for a flat-compound generator. The no-load and full-load voltages are the same. Neither the diverter nor the field rheostat are altered in this test. The speed is maintained constant at the rated value. The hump in the curve is caused by the increased influence of the series ampere-turns on the field iron at half-load when the degree of saturation is reduced and the armature reaction and armature IR drop are approximately half their normal values.

Curve B is the external characteristic of an over-compound generator. As load is added to this machine its terminal voltage increases so

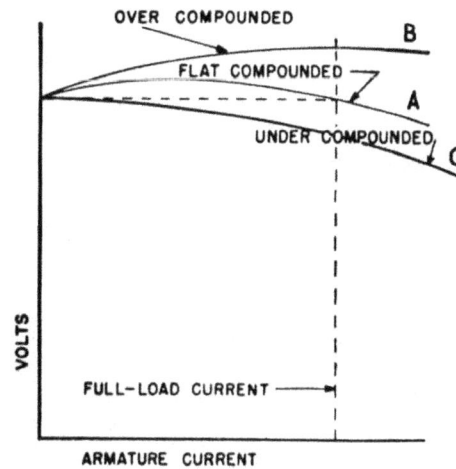

BE. 347

Figure 18-20.—Compound generator external characteristics.

that the field load voltage is higher than the no-load value. Such a characteristic might be desirable where the generator is located some distance from the load and the rise in voltage compensates for the voltage loss in the feeder. This action holds the load voltage approximately constant from no-load to full-load by increasing the generator terminal voltage an amount that is just equal to the voltage drop in the feeder at full load.

Curve C represents the external characteristic for an under-compound generator. A stabilized shunt generator has a small series field winding of a few turns in the coils to partially compensate for the voltage loss due to armature IR drop and armature reaction. Thus, the external characteristic is almost the same as that of a shunt generator except that the terminal voltage does not fall off quite as rapidly with increased load.

## VOLTAGE REGULATION

The external characteristic of a generator is sometimes called the VOLTAGE-REGULATION CURVE of the machine. The regulation of a generator refers to the VOLTAGE CHANGE that takes place when the load is changed. It is usually expressed as the change in voltage from no-load to full-load voltage in percent of full-load voltage. Expressed as a formula,

$$\frac{E_{nL} - E_{fL}}{E_{fL}} \times 100 = \text{percent regulation,}$$

where $E_{nL}$ is the no-load terminal voltage and $E_{fL}$ is the full-load terminal voltage of the generator. For example, the percent regulation of a generator having a no-load voltage of 237 volts and a full-load voltage of 230 volts is

$$\frac{237 - 230}{230} \times 100, \text{ or } 3 \text{ percent.}$$

## VOLTAGE CONTROL

Flat-compound generators were formerly used to supply d-c electric power because they provided a more constant voltage under varying load conditions. Shunt generators are simpler in design and have greater reliability when operating in parallel. The stabilized shunt generator represents a compromise between the two types. As mentioned previously, it has a very light series winding and a slightly drooping voltage characteristic and is generally used to supply d-c electric power requirements.

Voltage control is either (1) manual or (2) automatic. In most cases the process involves changing the resistance of the field circuit. By changing field circuit resistance, field current is controlled. Controlling field current permits control of the terminal voltage. Thus, the major difference between various voltage regulator systems is merely the method by which field circuit resistance and current are controlled. When the load changes are infrequent and small, manual control is sufficient to hold the d-c system voltage to the desired value. When the load changes are frequent and large, it may be desirable to employ some form of automatic control.

Inherent voltage regulation should not be confused with voltage control. As described previously, voltage regulation is an internal action occurring within the generator whenever the load changes. Voltage control is a superimposed action usually by external adjustment. Certain electromechanical devices are used extensively in automatic voltage regulators. These include solenoids, relays, carbon piles, transistors, rocking disks, and tilted plates.

## MANUAL OPERATION

A hand-operated field rheostat (fig. 18-21) connected in series with the shunt field circuit provides the simplest method of controlling the terminal voltage of a d-c generator.

One form of field rheostat contains tapped resistors with leads to a multiterminal switch.

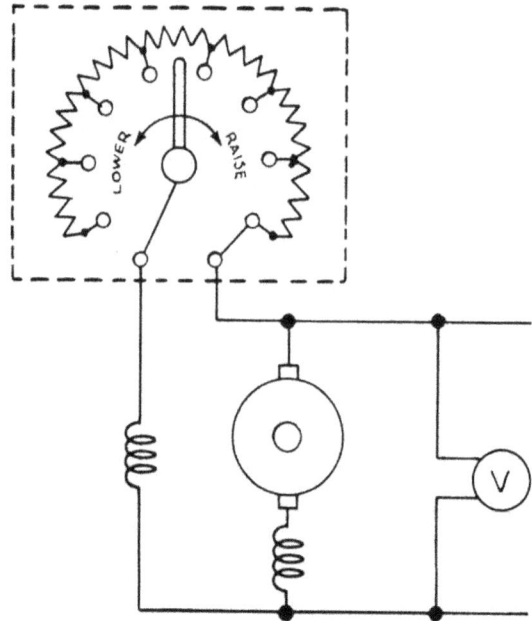

BE. 348

Figure 18-21.—Hand-operated field rheostat.

The arm of the switch may be rotated through an arc to make contact with the various resistor taps, thereby varying the amount of resistance in the field circuit. Rotating the arm in the direction of the LOWER arrow increases the resistance and lowers the terminal voltage. Rotating the arm in the direction of the RAISE arrow, decreases the resistance and increases the terminal voltage.

Field rheostats for generators of moderate size employ resistors of alloy wire having a high specific resistance and a low temperature coefficient. These alloys include copper, nickel, manganese, and chromium and are marketed under trade names such as nichrome, advance, manganin, and so forth. Large generator field rheostats use cast-iron grids and a motor-operated switching mechanism.

### Automatic Operation

Several types of automatic voltage regulators are used to control the terminal voltage of d-c generators. One of the earliest types was the vibrating regulator. More recent types include the tilted plate regulator and the rocking disk regulator.

The VIBRATING REGULATOR operates on the principle that an intermittent short circuit applied across the generator field rheostat will cause the terminal voltage of the generator to pulsate within narrow voltage limits and thus it will maintain an average steady value of voltage that is independent of load change. A simplified circuit is shown in figure 18-22.

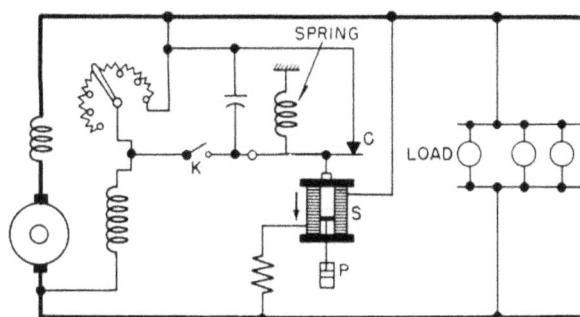

BE.349

Figure 18-22.—Vibrating type of voltage regulator.

With the generator running at normal speed and switch K open, the field rheostat is adjusted so that the terminal voltage is about 60 percent of normal. Solenoid S is weak and contact C is held closed by the spring. When K is closed, a short circuit is placed across the field rheostat. This action causes the field current to increase and the terminal voltage to rise.

When the terminal voltage rises above a certain critical value, for example 111 volts, the solenoid downward pull exceeds the spring tension and contact C opens, thus reinserting the field rheostat in the field circuit and reducing the field current and terminal voltage.

When the terminal voltage falls below a certain critical voltage, for example 109 volts, the spring tension overcomes the solenoid pull and contact C is closed. The field rheostat is now shorted and the terminal voltage starts to rise. The cycle repeats and the action is rapid and continuous. The average voltage of 110 volts is maintained with or without load change.

The dashpot, P, provides smoother operation by acting as a damper to prevent hunting. The capacitor across contact C eliminates sparking. Added load causes the field rheostat to be shorted for a longer period of time and thus the solenoid armature vibrates more slowly. If the load is reduced and the terminal voltage tends to rise, the armature vibrates more rapidly. Thus the regulator holds the terminal voltage to a steady value for any change in load from no load to full load on the generator.

The TILTED PLATE REGULATOR (fig. 18-23) consists of two or more stacks of graphite plates mounted on metallic supports at the center of each stack. Each plate is balanced on its fulcrum (the metallic support) at the center. On one end of the stack the plates are separated by mica spacers. On the other end they are separated by silver contacts. Before the silver contacts are closed, the path for field circuit current is from graphite plate to graphite plate through the connecting fulcrums at the center of the stack. Between the graphite plates and their supporting fulcrums there exists a certain amount of contact resistance. A weight acting on a lever tilts the plate in a manner that brings the contacts together. A solenoid, which responds to terminal voltage change, tilts the plates in the opposite direction. Bringing the silver contacts together shorts out the resistance of the stack. Opening the silver contacts increases the field circuit resistance.

The regulator responds to load change and is not designed to overshoot in the manner of the vibrating regulator previously described. For example, at rated voltage the solenoid is in balance with the lever weight and the arm does not move. An increase in load decreases the terminal voltage only slightly because the lever weight immediately overcomes the solenoid and tilts the plates in a direction to lower the resistance, thereby strengthening the field and the generated voltage. This action checks the fall of terminal voltage with load.

The ROCKING DISK REGULATOR (fig. 18-24) is another device that is quick acting and sensitive to slight changes in terminal voltage. It may be designed to be in a constant-motion condition of overshooting the voltage a slight amount through each cycle of operation or it may be operated in a static condition in which load change alone causes it to operate.

A spring presses the rocking disk against the flat surface of a silver-plated commutator, the segments of which connect to taps on the field circuit resistors. The disk has a carbon shoe which makes contact with the segments one at a time from top to bottom as the disk is rocked downward. A small range of motion of the spring is converted into a long range of motion of the contact point of the carbon shoe with the various

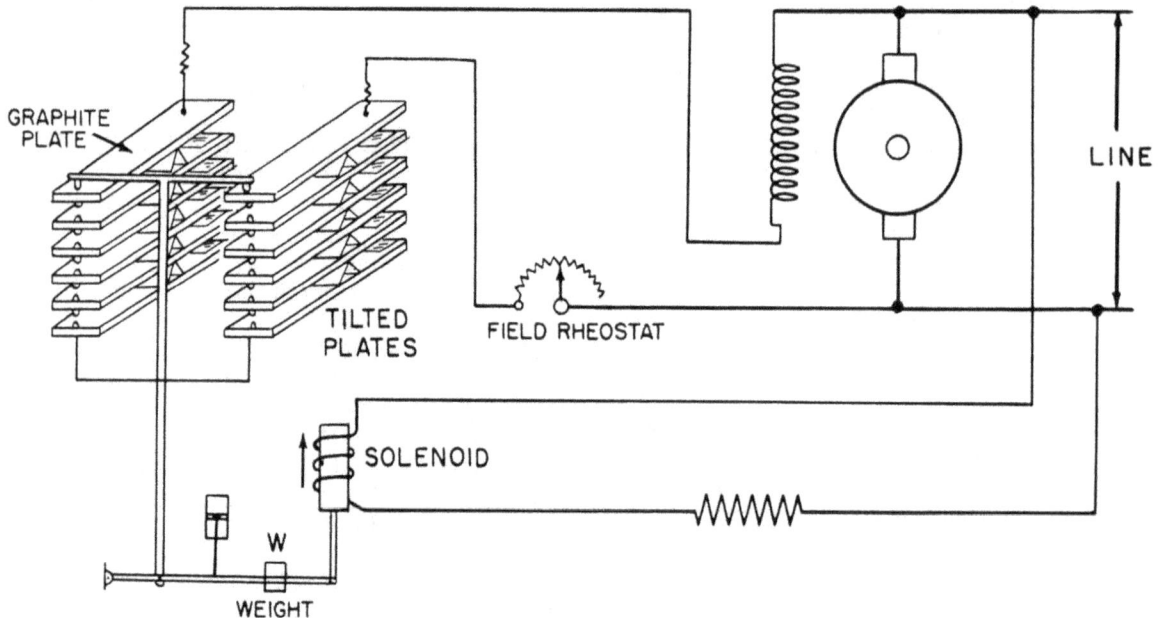

BE.350

Figure 18-23.—Tilted plate voltage regulator.

commutator segments. The lever weight tends to rock the disk downward. The solenoid tends to rock the disk upward. For a constant load on the generator, the lever weight and the solenoid are in a state of balance. Thus, the shoe is in contact with one commutator segment, and a portion of the field resistors is short-circuited.

When the load on the generator increases, the terminal voltage of the generator starts to fall. The solenoid is weakened and the weight rocks the disk downward, thus short-circuiting more of the field resistors and increasing the generated voltage. The decrease in terminal voltage is almost instantly checked by the increase in field strength and generated voltage, and the weight is now in balance with the solenoid at a new position where the disk contacts a commutator segment nearer the lower end of its travel.

A decrease in load causes the terminal voltage to start to rise and this strengthens the solenoid. Thus the disk rocks in an upward direction and more resistance is inserted in the field circuit, with accompanying decrease in field strength and generated voltage.

The decrease in generated voltage checks the rise in terminal voltage. For example, at one load the generated voltage is 120 volts, the internal voltage loss is 10 volts, and the terminal voltage is 120 - 10, or 110 volts. With decrease in load the internal voltage loss might decrease to 5 volts, and without automatic regulation the terminal voltage would increase to 120 - 5, or 115 volts. However, with automatic regulation the terminal voltage momentarily rises to 111 volts and the regulator causes the generated voltage to fall from 120 volts to 111 + 5, or 116 volts, and to become stable at 115 volts when

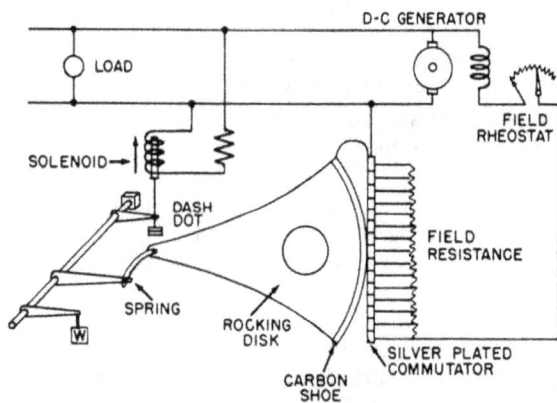

BE.351

Figure 18-24.—Rocking disk voltage regulator.

361

the terminal voltage is again 115 - 5 or 110 volts.

## REGULATORS FOR VARIABLE-SPEED GENERATORS

### VIBRATOR-TYPE REGULATOR

If the speed of a shunt generator varies the output voltage will vary. In cars, trucks, small boats, and some aircraft, the generator is usually driven from the main source of power, which may be a variable-speed internal-combustion engine. If the generator is of the shunt type, a 3-unit regulator is often employed. One unit consists of a vibrating voltage regulator that places an intermittent short-circuit across a resistor in series with the field. This action is similar to the vibrating type voltage regulator described earlier in this chapter. The second unit is a current limiter that limits the output current to a value determined by the rating of the generator. The three-unit regulator (fig. 18-25) is designed for use in a power system employing a battery as an auxiliary power supply. The third unit in the regulator is a reverse-current cutout that prevents the battery from motorizing the generator at low speeds by disconnecting the battery from the generator. If

the voltage of the generator falls below that of the battery, the battery will discharge through the generator armature; thereby tending to drive the generator as a motor. This action is called "motorizing" the generator, and unless it is prevented, will discharge the battery in a short time.

The action of vibrating contact $C_1$ in the voltage-regulator unit (fig. 18-25) places an intermittent short circuit across $R_1$ and $L_2$. When the genrator is not operating, spring $S_1$ holds $C_1$ closed. $C_2$ is also closed by $S_2$, and the shunt field is connected directly across the armature.

When the generator is started, its terminal voltage will rise as the generator comes up to speed, and the armature will apply the field with current through closed contacts $C_2$ and $C_1$.

As the terminal voltage rises, the current flow through $L_1$ increases and the iron becomes more strongly magnetized. At a certain speed and voltage, when the magnetic attraction on the movable arm becomes strong enough to overcome the tension of spring $S_1$, contact points $C_1$ are separated. The field current now flows through $R_1$ and $L_2$. Because resistance is added to the field circuit, the field is momentarily weakened and the rise in terminal voltage is checked. Additionally, because the $L_2$ winding

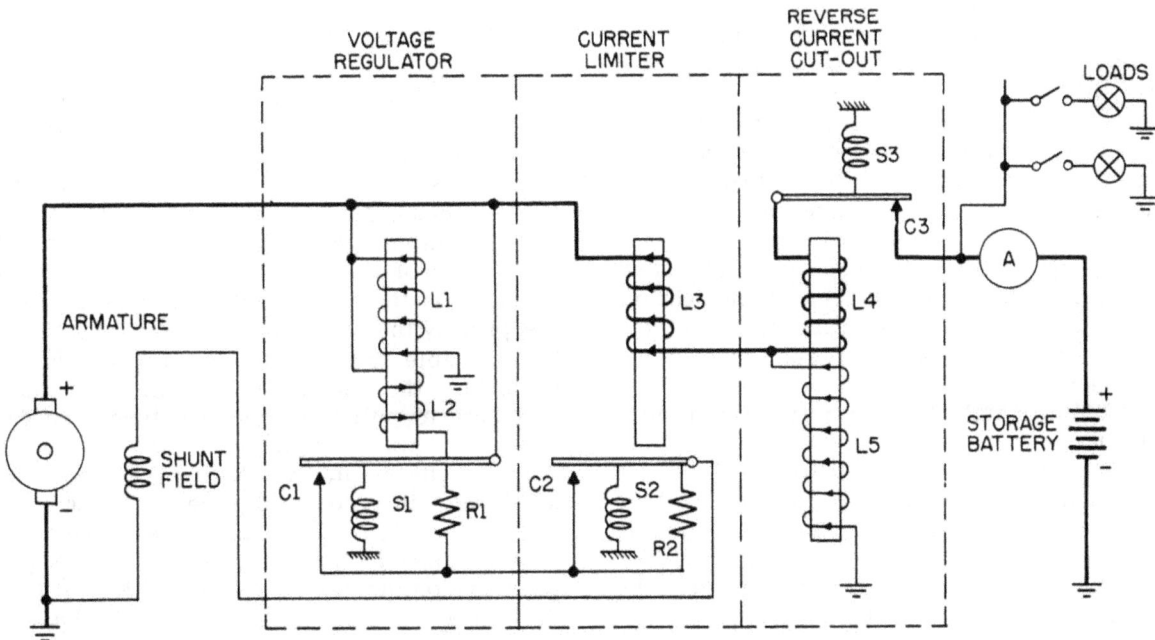

Figure 18-25. —Regulator for variable-speed generator.

BE.352

is opposed to the $L_1$ winding, the magnetic pull of $L_1$ against $S_1$ is partially neutralized, and spring $S_1$ closes contact $C_1$. $R_1$ and $L_2$ are again shorted out of the circuit, the field current again increases, the output voltage increases, and $C_1$ is opened because of the action of $L_1$. The cycle of events occurs very rapidly, many times per second. The terminal voltage of the generator thus varies slightly but rapidly above and below an average value determined by the tension of spring $S_1$, which may be adjusted.

The purpose of the vibrator-type current limiter is to limit the output current of the generator automatically to its maximum rated value in order to protect the generator. As shown in figure 18-25, $L_3$ is in series with the main line and load. Thus, the amount of current flowing in the line determines when $C_2$ will be opened and $R_2$ placed in series with the generator field. By contrast, the voltage regulator is actuated by line voltage, whereas the current limiter is actuated by line current. Spring $S_2$ holds contact $C_2$ closed until the current through the main line and $L_3$ exceeds a certain value, as determined by the tension of spring $S_2$, and causes $C_2$ to be opened. The increase in current is due to an increase in load. This action inserts $R_2$ into the field circuit of the generator and decreases the field current and the generated voltage. When the generated voltage is decreased, the generator current is reduced. The core of $L_3$ is partly demagnetized, and the spring closes the contact points. This causes the generator voltage and current to rise until the current reaches a value sufficient to start the cycle again. A certain minimum value of load current is necessary to cause the current limiter to vibrate.

The purpose of the reverse-current cutout relay (fig. 18-25) is to disconnect the battery automatically from the generator whenever the generator voltage is less than the battery voltage. If this device were not used in the generator circuit, an especially harmful action would occur if the engine were shut off. The battery would discharge through the generator. This would tend to make the generator operate as a motor, but because the generator is coupled to the engine it could not rotate such a heavy load. Under this condition, the generator winding may be severely damaged by excessive current.

There are two windings, $L_4$ and $L_5$, on the soft-iron core. The current winding, $L_4$, consisting of a few turns of heavy wire, is in series with the line and carries the entire line current.

The voltage winding, $L_5$, consisting of a large number of turns of fine wire, is shunted across the generator terminals.

When the generator is not operating, the contacts, $C_3$, are held open by spring $S_3$. As the generator voltage builds up, $L_5$ magnetizes the iron core. When the current (as a result of the generated voltage) produces sufficient magnetism in the iron core, contact $C_3$ is closed, as shown. The battery then receives a charging current. The coil spring, $S_3$, is so adjusted that the voltage winding will not close the contact points until the voltage of the generator is in excess of the normal voltage of the battery. The charging current passing through $L_4$ aids the current in $L_5$ in holding the contacts tightly closed. Unlike $C_1$ and $C_2$, contacts $C_3$ do not vibrate. When the generator slows down, or for any other cause the generator voltage decreases to a certain value below that of the battery, the current reverses through $L_4$ and the ampere-turns of $L_4$ oppose those of $L_5$. Thus, a momentary discharge current from the battery reduces the magnetism of the core and $C_3$ is opened, thereby preventing the battery from discharging into the generator and motorizing it. $C_3$ will not close again until the generator terminal voltage exceeds that of the battery by a predetermined value.

## CARBON-PILE TYPE REGULATOR

The carbon-pile voltage regulator is not used too extensively, but the principle upon which it operates is basically common to most regulators. Its essential parts are shown in figure 18-26. A stack of carbon washers, the "pile," is in series with shunt field. The resistance of the pile, and thus field current, depends on the mechanical pressure applied to the pile by the wafer spring which acts through the movable iron armature. In a steady-state condition, the magnetic force of the potential coil acts to pull the iron armature away from the pile. This force is balanced by the spring. When a change in line voltage occurs, due to a change in generator speed or load, the potential coil current also changes. Its magnetic strength must also change. As a result, the spring pressure on the carbon pile is INCREASED by a drop in line voltage, and DECREASED by a rise in line voltage. The resultant change in shunt field current will raise or lower the generator output voltage as needed. Variations in voltage characteristics from bench adjustment and service installation may be adjusted by the rheostat, which controls the potential coil current.

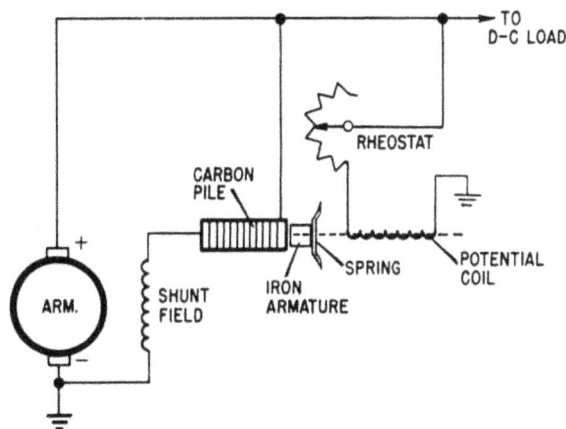

BE.353

Figure 18-26.—Carbon-pile regulator.

## PARALLEL OPERATION

Electric power may be supplied by more than one generator. When two or more generators are sharing a common load, they are said to be operating in parallel. The generators may be physically located quite a distance apart; however, they are connected to the common load through the power distribution system.

There are a number of reasons for operating generators in parallel. First, the number of generators used may be selected in accordance with the load demand. Thus, by operating each generator as near as possible at its rated capacity, maximum efficiency is achieved. Also, disabled or faulty generators may be taken off the line without interrupting normal operations.

In most d-c electric power systems, the lighting circuits are designed to operate on 115 volts while most motors and other relatively large power-consuming devices are designed to operate on 230 volts. To avoid the use of separate generators for lighting and for power, a 3-wire generator is used that supplies both types of circuits simultaneously. The 3-wire generator is similar to a 2-wire generator except that the armature winding is tapped at points 180 electrical degrees (1-pole span) apart and these are brought out to slip rings on the back of the armature. These rings are connected by means of brushes to a reactance coil. The center tap of the coil connects with the neutral, or third, wire of the 3-wire system.

The reactance coil (or balance coil) provides a method of establishing the potential of the neutral wire, which must be 115 volts positive with respect to one of the outer wires and 115 volts negative with respect to the other. Alternating current is produced in the armature of the d-c generator; and this is taken from the machine through the sliprings to which the balance coil is attached. The action of the alternating current in the balance coil is similar to that occurring in the primary winding of a single-phase transformer. That is, the potential of the center point on the balance coil is at the electrical center of the total voltage generated by the machine. Hence, this point can serve as the neutral, or center, between the positive and negative brushes. The neutral current passes through a portion of the balance coil, which has low resistance to direct current.

The connections for operating stabilized shunt generators in parallel are shown in figure 18-27. Each generator develops 230 volts across the two outside (positive and negative) leads, and 115 volts across either outside leads and neutral. Usually one ammeter and one voltmeter are provided on each generator panel.

Assume that generator No. 1 is supplying the load and it is desired to put generator No. 2 in service. The procedure for paralleling the d-c generators is as follows:

1. The brush rigging and armature of the incoming generator should be inspected for loose gear, and the field rheostat should be adjusted to the lowest voltage position.

2. The prime mover of the incoming generator (No. 2) is brought up to rated speed.

3. The voltage of the incoming machine is adjusted to about 2 volts higher than the bus voltage.

4. Closing the circuit breaker and switch places the generator in parallel with generator No. 1. The ammeter will indicate that generator No. 2 is now carrying a small portion of the load.

5. The field of generator No. 2 is strengthened until the load which it supplies is the proper value. At the same time the field of generator No. 1 should be weakened to maintain the normal voltage.

The procedure for removing a generator from its parallel connection with another is as follows:

1. The field of the generator being secured is weakened at the same time the field of the remaining generator is strengthened until the

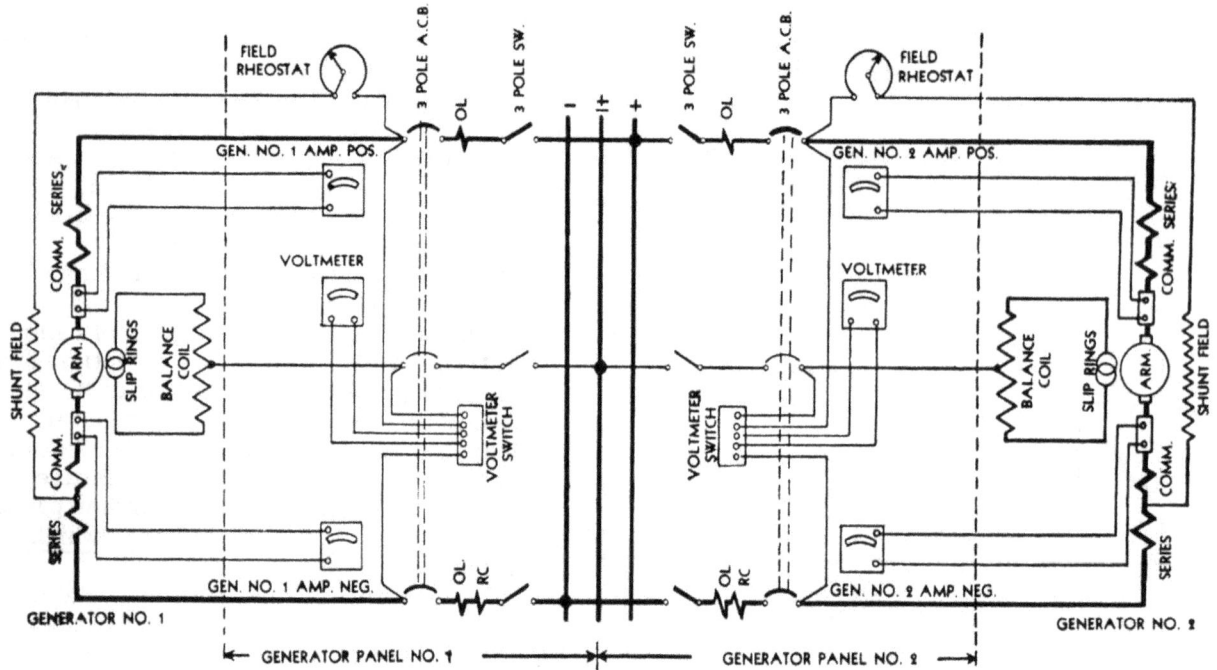

BE.354

Figure 18-27.—Parallel operation of d-c generators.

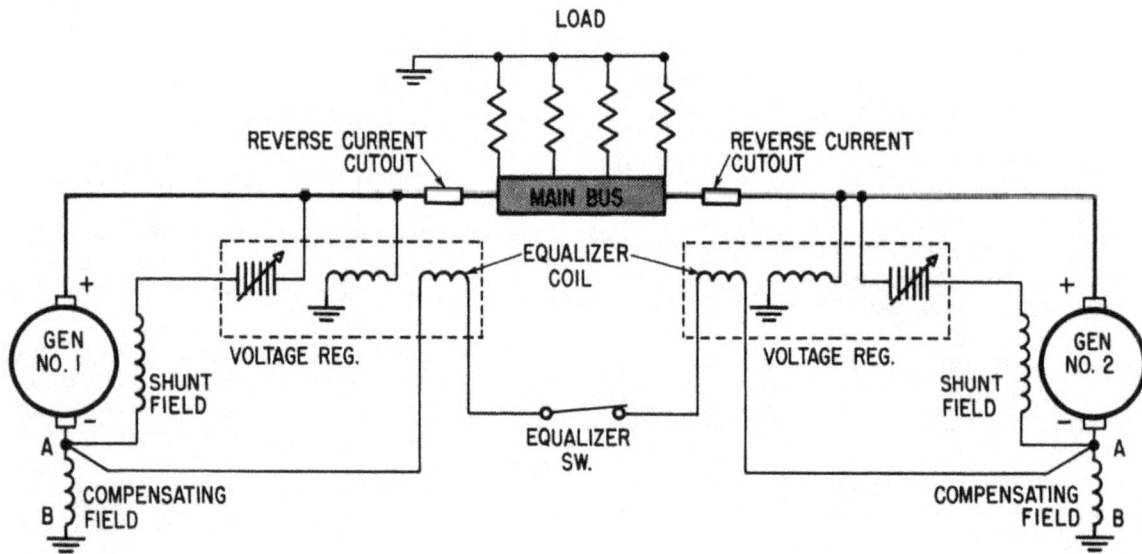

BE.355

Figure 18-28.—Aircraft d-c power system.

load on the outgoing machine is about 5 percent of the rated load current.

2. The circuit breaker is tripped and the switch is opened.

The two-wire d-c system is used for d-c power supply on some installations such as aircraft. On multiengine aircraft, the d-c generators are normally operated in parallel. Figure 18-28 is a schematic of the main parts of a typical aircraft d-c power system. Both generators are connected to the main bus, or common load, through their respective reverse-current cutout relays.

The voltage regulators for aircraft d-c generators are designed and adjusted to permit a slight drooping characteristic. That is, the voltage regulators will permit the terminal voltage of their respective generators to decrease slightly as the bus load is increased. The amount of droop of two or more generators may not be equal, due to internal differences in the generators or adjustment of their regulators. In such cases, one generator will assume more load than the other, as loads are added to the bus.

To assure that the generators will carry equal loads, an automatic load balancing circuit is built into the regulator system. It is referred to as the equalizer. As shown in figure 18-28, the equalizer circuit consists essentially of two equalizer coils and an equalizer switch, all in series, connected between point A in generator No. 1 and point A on generator No. 2. Current through the equalizer circuit flows only when the voltage drops across the compensating fields A-B are unequal. This would be the case when the generators are supplying unequal shares of the total load. The equalizer coils either aid or oppose the regulator potential coils, depending on the DIRECTION of equalizer current. When one generator takes a greater share of the load than the other, the equalizer coil in its regulator will AID the potential coil. This will cause carbon pile decompression, more resistance and less current in the shunt field, less generated voltage, and consequently a decrease in load. The other generator's equalizer coil will OPPOSE its potential coil, increase its shunt field current, raise its generated voltage, and thus cause the generator to assume more of the load.

When only one generator is being operated, the equalizer switch is used to open the equalizer circuit. Otherwise, current would flow from the active generator through the dead generator's compensating field and equalizer coil.

# CHAPTER 19

# DIRECT-CURRENT MOTORS

The construction of a d-c motor is essentially the same as that of a d-c generator. The d-c generator converts mechanical energy into electrical energy, and the d-c motor converts the electrical energy back into mechanical energy. A d-c generator may be made to function as a motor by applying a suitable source of direct voltage across the normal output electrical terminals.

There are various types of d-c motors, (fig. 19-1), depending on the way the field coils are connected. Each has characteristics that are advantageous under given load conditions.

SHUNT MOTORS (fig. 19-1 (A)) have the field coils connected in parallel with the armature circuit. This type of motor, with constant potential applied, develops variable torque at an essentially constant speed, even under changing load conditions. Such loads are found in machine-shop drives. They include lathes, milling machines, drills, planers, shapers, and so forth.

SERIES MOTORS (fig. 19-1 (B)) have the field coils connected in series with the armature circuit. This type of motor, with constant potential applied, develops variable torque but its speed varies widely under changing load conditions. That is, the speed is low under heavy loads, but becomes excessively high under light loads. Series motors are commonly used to drive electric cranes, hoists, winches, and certain types of vehicles (for example, electric trucks). Series motors are used extensively to start internal combustion engines.

COMPOUND MOTORS (Fig. 19-1(C)) have one set of field coils in parallel with the armature circuit, and another set of field coils in series with the armature circuit. This type of motor is a compromise between shunt and series motors. It develops an increased starting torque over that of the shunt motor, and has less variation in speed than the series motor. Shunt, series, and compound motors are all d-c motors

designed to operate from constant-potential variable-current d-c sources.

STABILIZED SHUNT MOTORS (fig. 19-1 (D)) have a light series winding in addition to the shunt field. The action is similar to ordinary shunt motors except that stabilized shunt motors have less field iron and are lighter in weight. Without the stabilizing series field, they have the characteristics of a shunt motor with a strong field and low armature resistance.

Direct-current motors may also be classified in other ways. For example, they may be classified according to the dgree of enclosure, such as OPEN (fig. 19-2 (A)), DRIP-PROOF (fig. 19-2 (B)), ENCLOSED (fig. 19-2 (C)), and so forth.

The open-type motor shown in figure 19-2 (A) has end bells which offer little or no restriction to ventilation. The drip-proof motor shown in figure 19-2 (B) is protected from falling moisture and dirt from any direction up to a 45° angle with respect to the vertical. A motor of this type has all ventilation openings protected with wire screens or perforated covers, as shown in the figure. An enclosed motor like the one in figure 19-2 (C) is totally enclosed except for openings provided for the admission and discharge of air. These openings are connected to inlet and outlet ducts or pipes.

Other motor types include a spray-tight motor so constructed that a stream of water from a hose may be played upon it from any direction without leakage into the motor.

A submersible motor is one that will operate while submerged in liquid.

With reference to cooling, the following types of motors are found aboard ship:

1. NATURAL VENTILATED MOTORS, cooled by the natural circulation of the air caused by the rotation of the armature.

2. SELF-VENTILATED MOTORS, cooled by a fan attached to the armature of the motor.

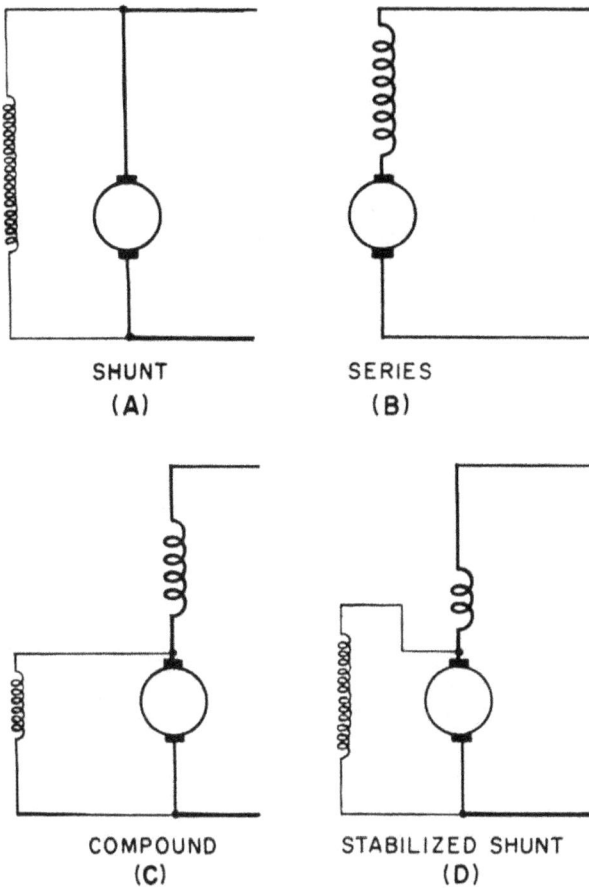

SHUNT
(A)

SERIES
(B)

COMPOUND
(C)

STABILIZED SHUNT
(D)

BE.356

Figure 19-1.—Schematic diagrams of four types
of d-c motors.

3. SEPARATELY VENTILATED MOTORS,
cooled by an independent fan or blower apart from
the motor.

With reference to speed, motors are classified as:

1. CONSTANT-SPEED MOTORS in which the
speed varies only slightly between no load and
full load. A shunt motor is a constant-speed
motor.

2. MULTISPEED MOTORS which can be
operated at any one of several definite speeds,
each speed being nearly constant with load
change. Such motors cannot be operated at intermediate speeds. A motor with two windings on
the armature is an example of a multispeed
motor.

3. ADJUSTABLE-SPEED MOTORS whose
speed can be varied gradually over a wide range,
but when once adjusted remains at nearly constant speed with load change.

4. VARYING SPEED MOTORS in which the
speed varies with the load. Ordinarily the speed
decreases as the load increases. The series
motor is an example of this type of motor.

5. ADJUSTABLE VARYING SPEED MO-
TORS in which the speed may be adjusted over
a wide range for any given load; but if the speed
is adjusted at a given load, the speed will vary
with any change in load.

Motors are also classified according to the
type of duty they are to perform. For example,
a CONTINUOUS-DUTY MOTOR is capable of
operating continuously at its rated output without exceeding specified temperature limits. An
INTERMITTENT-DUTY MOTOR is capable of
being operated at its rated output for a limited
period without exceeding its specified temperature limit.

Most d-c motors on Navy vessels are designed to operate on constant-potential 2-wire
circuits of either 115 volts or 230 volts. These
circuits may be combined in a 3-wire d-c supply
with 230 volts between each outside wire, and
115 volts between each outside wire and middle or
neutral wire.

D-c motors find extensive use on aircraft.
The motors used on aircraft operate on the same
principles as the larger motors. However, the
airborne motors differ considerably in size, rating, and appearance, as illustrated in figure 19-3.
Also, they are designed to operate from a 24-28
volt d-c source.

Figure 19-3 (A) is an aircraft engine starter
motor. Starters are series motors, selected to
supply the typically high torque required for
engine starting. They are usually the largest
motors on the aircraft, with power outputs up to
twenty hp, but are designed only for intermittent
use. Figure 19-3 (B) is one of many types of d-c
motors used to actuate mechanical loads, in this
case cowl flaps. Such loads may include wing
flaps, cowl flaps, landing gear, cockpit canopies,
and bomb bay doors. Actuator motors are also
of the series type, since they must start under
full mechanical load. Many airborne actuator
motors are of the series split-field type, and are
reversible. Most are also designed only for intermittent use. Figure 19-3(C) is a continuous-
duty d-c motor, in this case one that is used to
drive windshield wipers. Continuous-duty motors
on aircraft are usually of the compound type. The

OPEN-TYPE MOTOR
(A)

DRIP-PROOF MOTOR
(B)

ENCLOSED MOTOR
(C)

BE.357

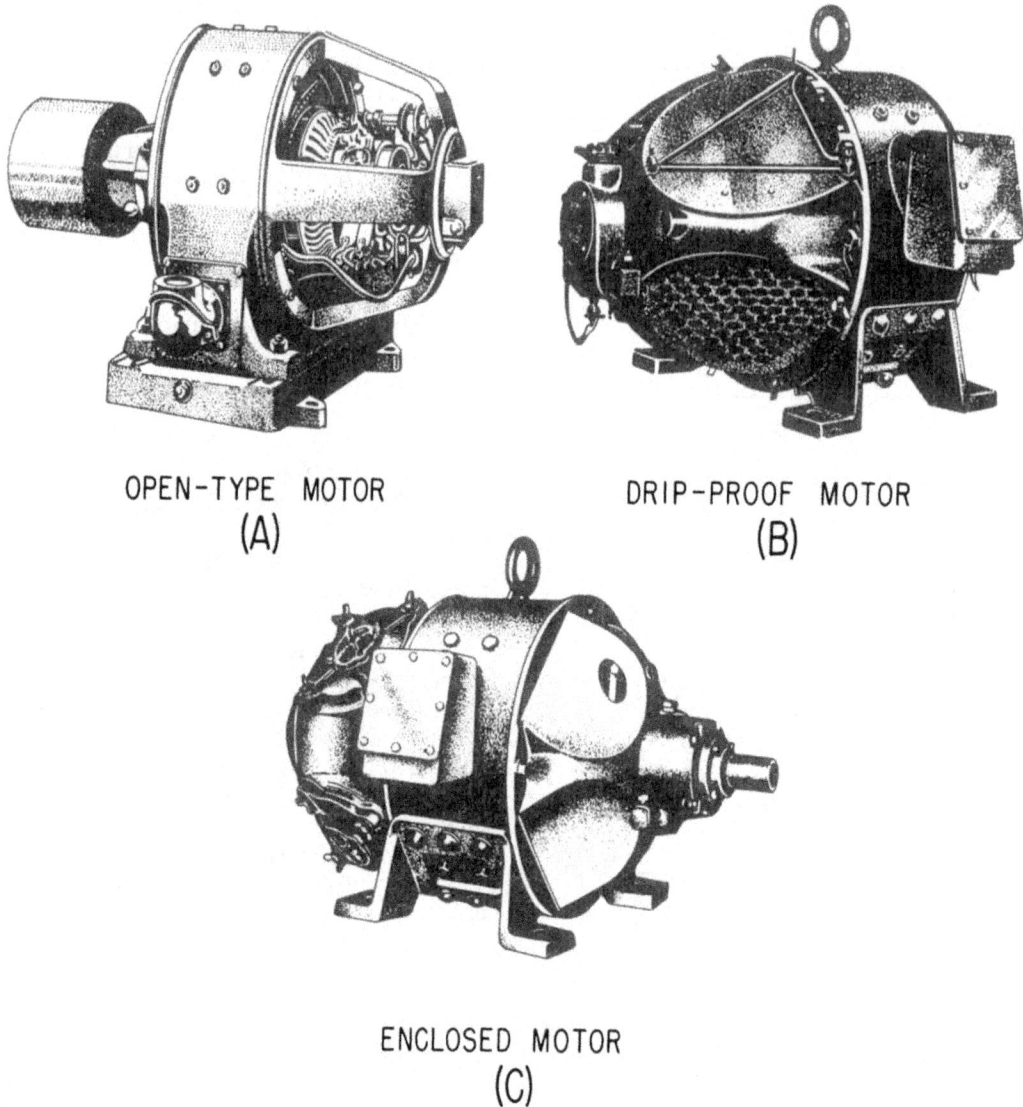

Figure 19-2.—Types of d-c motors classified according to enclosure.

most recent trend in aircraft design is to use d-c motors for intermittent duty, and a-c motors for continuous duty.

## PRINCIPLES OF D-C MOTORS

### FORCE ACTING ON A CONDUCTOR

The operation of a d-c motor depends on the principle that a current-carrying conductor placed in, and at right angles to, a magnetic field tends to move at right angles to the direction of the field, as shown in figure 19-4. This action was previously described under the operating principle of the D'Arsonval meter movement.

The magnetic field between a north and a south pole of a magnet is shown in figure 19-4 (A). The lines of force, comprising the field extend from the north pole to the south pole. A cross section of a current-carrying conductor is shown in figure 19-4 (B). The plus sign in the wire indicates that the electron flow is away from the observer. The direction of the flux loops around the wire is counterclockwise, as shown. This follows

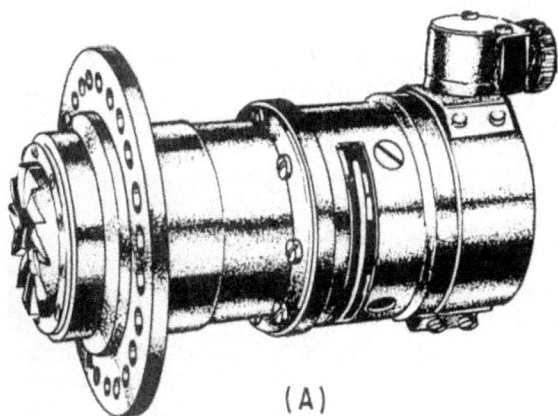

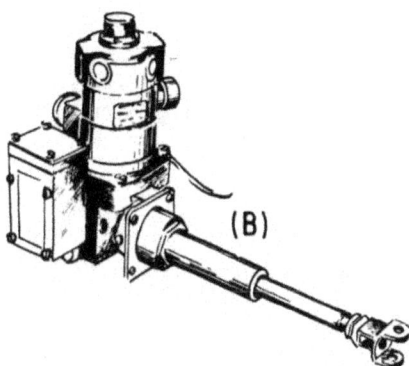

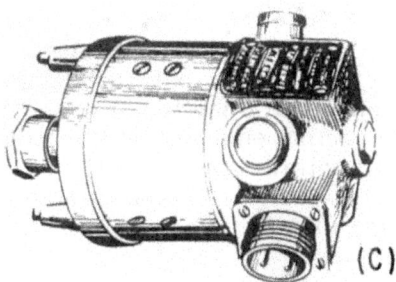

BE. 358

Figure 19-3.—Aircraft d-c motors.

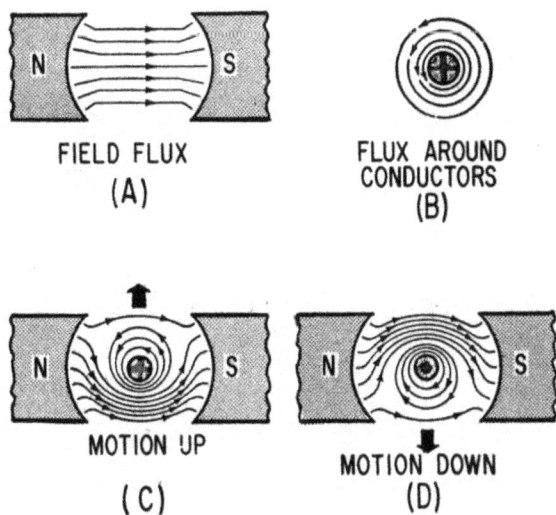

FIELD FLUX
(A)

FLUX AROUND
CONDUCTORS
(B)

MOTION UP
(C)

MOTION DOWN
(D)

BE.359

Figure 19-4.—Force acting on a current-carrying conductor in a magnetic field.

from the left-hand flux rule which states that if the conductor is grasped in the left hand with the thumb extended in the direction of the current flow, the fingers will curve around the conductor in the direction of the magnetic flux.

If the conductor (carrying the electron flow away from the observer) is placed between the poles of the magnet, as in figure 19-4 (C), both fields will be distorted. Above the wire the field is weakened, and the conductor tends to move upward. The force exerted upward depends on the strength of the field between the poles and on the strength of the current flowing through the wire.

If the current through the conductor is reversed, as in figure 19-4 (D), the direction of the flux around the wire is reversed. The field below the conductor is now weakened, and the conductor tends to move downward.

A convenient method of determining the direction of motion of a current-carrying conductor in a magnetic field is by the use of the right-hand motor rule. Practical d-c motors depend for their operation on the interaction between the field flux and a large number of current-carrying conductors. As in d-c generators, the conductors are wound in slots in the armature; and the armature is mounted in bearings and is free to rotate in the magnetic field. An armature with two slots and two conductors (that is, a single conductor wound in the two slots) is shown in figure 19-5.

Figure 19-5 (A) shows the uniform distribution of the main field flux when the field magnets are energized, and no current flows through the armature. The flux is concentrated in the airgap between the field and armature because of the relatively high permeability of the soft-iron armature core and the low reluctance of the magnetic circuit.

the field polarity and the armature current are reversed, rotation continues in the same direction.

A practical d-c motor has many coils in the armature winding. The armature has many slots into which are inserted many turns of wire. This increases the number of armature conductors and thus produces a greater, more constant torque. Power output is also increased by increasing the number of poles in the field.

As in the case of the d-c generator, the d-c motor is equipped with a commutator and brushes.

The force, F, acting on a current-carrying conductor in a magnetic field is directly proportional to the field strength, the active length of the conductor (that part of the conductor contained in the armature slot and lying under a pole face), and the current flowing through it. The force, the conductor, and the field are assumed to be mutually perpendicular. This relationship is expressed algebraically as

$$F = \frac{8.85 \times BLI}{10^8}$$

where F is the force in pounds, B the flux density in lines per square inch, L the active length of the conductor in inches, I the current in amperes flowing through the conductor, and 8.85 is a constant that must be used when the above units are employed.

In a given motor, the length of the conductor is a fixed quantity; therefore, the only variables are the current and the flux. If the field flux is constant, the force acting on the conductor varies directly as the armature current. In other words, F is proportional to I. The following example will illustrate how the force acting on a conductor is calculated.

If a conductor having an active length of 10 inches and a current of 30 amperes is placed in a uniform field of 37,700 lines per square inch, what force will be exerted on the conductor?

$$F = \frac{8.85 \ BLI}{10^8} = \frac{8.85 \times 37,700 \times 10 \times 30}{10^8} = 1 \text{ pound}$$

If the conductor is wound around an armature (fig. 19-5 (C)) and the armature current flows in the directions shown, there will be an upward force of 1 pound on the left of the armature and a downward force of 1 pound on the right of the armature. The net force acting to turn the armature is the sum of these two forces, or 2 pounds.

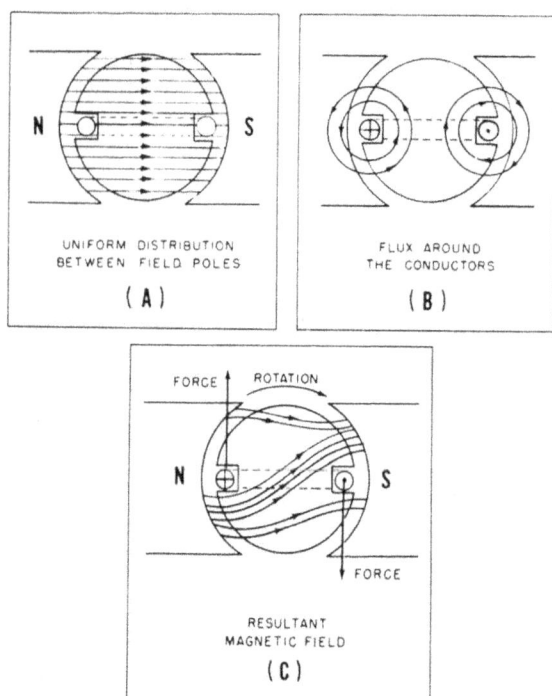

BE. 360

Figure 19-5.—Two-conductor armature.

Figure 19-5 (B) shows the magnetic fields surrounding the two active conductors when current flows through the armature coil in the direction indicated, and the field coils are not energized. The direction of the magnetic fields surrounding the two conductors is determined by means of the left-hand flux rule.

Figure 19-5 (C) shows the resultant magnetic field produced by the interaction of the main field magnetomotive force and the armature magnetomotive force. Note that the flux is strengthened below the conductor at the north pole and above the conductor at the south pole because the lines are in the same directions at these points. Conversely, the flux is weakened above the conductor at the north pole and below the conductor at the south pole because the lines are opposite in direction and tend to cancel each other. The lines of force act like stretched rubber bands that tend to contract, with the result that the armature rotates in a clockwise direction. If either the direction of current through the armature coil or the polarity of the field is reversed (but not both), the direction of the force on the armature conductors reverses. If both

## TORQUE

The TORQUE (or twist) on the armature is the product of the force acting at the surface of the armature times the perpendicular distance to the center of rotation of the armature. Assume the radius of the armature being considered is 1.5 feet, the torque exerted by each conductor is $1 \times 1.5 = 1.5$ pound-feet; and the total torque exerted by both conductors is $1.5 + 1.5 = 3$ pound-feet.

The armature conductors for a motor are assembled in coils and connected to the commutator riser exactly as in a generator. In the 2-pole motor (fig. 19-6 (A)), the current divides equally in the two paths through the armature. Current flows in one direction in the conductors under the north pole and in the opposite direction in the conductors under the south pole. To develop a continuous motor torque the current in a coil must reverse when the coil passes the dead center position (top and bottom). The function of the commutator is to reverse the current at the proper time to maintain the current flow in the same direction in all conductors under a given pole. The total torque is the sum of the individual torques contributed by all the armature conductors. If there are 200 active conductors each developing a force of 1 pound the total torque will be $200 \times 1 \times$ the average moment arm.

The AVERAGE MOMENT ARM is the average of all the perpendicular distances from the armature conductors to the center of rotation. If the field is assumed to consist of uniformly spaced horizontal lines of flux, the average moment arm will be 0.637 of the maximum moment arm, or radius, of the armature.

If the radius of the armature is 1.57 feet, the average moment arm is $0.637 \times 1.57 = 1$ foot. The total torque exerted by 200 armature conductors developing a force of 1 pound each is $200 \times 1 \times 1 = 200$ pound-feet.

In a 4-pole parallel-wound armature, the current divides into four paths as shown in figure 19-6 (B). The current flows in opposite directions under alternate poles, as in the 2-pole armature, and develops a unidirectional force on all of the armature conductors. The total torque produced by the armature current is equal to the sum of the individual torques developed by each conductor. This torque, T, is proportional to the product of armature current and field strength.

$$T = K_t \, \Phi \, I_a \qquad (19-1)$$

where $K_t$ is a constant that includes the number of armature conductors, number of paths, and other factors which are constant for a particular machine; $\Phi$ the flux per pole and $I_a$ the total armature current. When the speed of a motor is constant, the generated torque due to the armature current is just equal to the retarding torque caused by the combined effect of the friction losses in the motor and the mechanical load. The torque developed by the motor armature is a steady one and does not pulsate as in the case of reciprocating steam engines or internal combustion engines.

## HORSEPOWER OF A MOTOR

If the number of revolutions that the armature of a motor makes per minute (or per second), the effective armature radius at which the force acts, and the total force acting at and tangent to this effective radius are known, the horsepower may be determined. Briefly, the horsepower may be determined in the following manner.

Work is accomplished when force acts through distance. For example, when a force of 1 pound acts through a distance of 1 foot, 1 foot-pound of work is accomplished. If 33,000 foot-pounds of work is done in 1 minute, 1 horsepower of work is accomplished.

Assume that an armature makes 100 revolutions per minute and that the effective radius is 1.59 feet. The circumference (the distance through which the force moves in one revolution of the armature) is

$$\text{circumference} = 2\pi r$$

$$= 2 \times 3.14 \times 1.59 = 10 \text{ feet}$$

Assuming that a total effective force of 200 pounds acts on the armature tangent to the 10-foot circumference, the work done in 1 revolution is

$$\text{work} = \text{force} \times \text{distance} = 200 \times 10 = 2,000 \text{ foot-pounds}$$

The work done in 100 revolutions (1 minute) is

$$2,000 \times 100 = 200,000 \text{ foot-pounds}$$

The horsepower is therefore,

$$hp = \frac{\text{foot-pounds per minute}}{33,000} = \frac{200,000}{33,000}$$

$$= 6.06 \text{ horsepower}$$

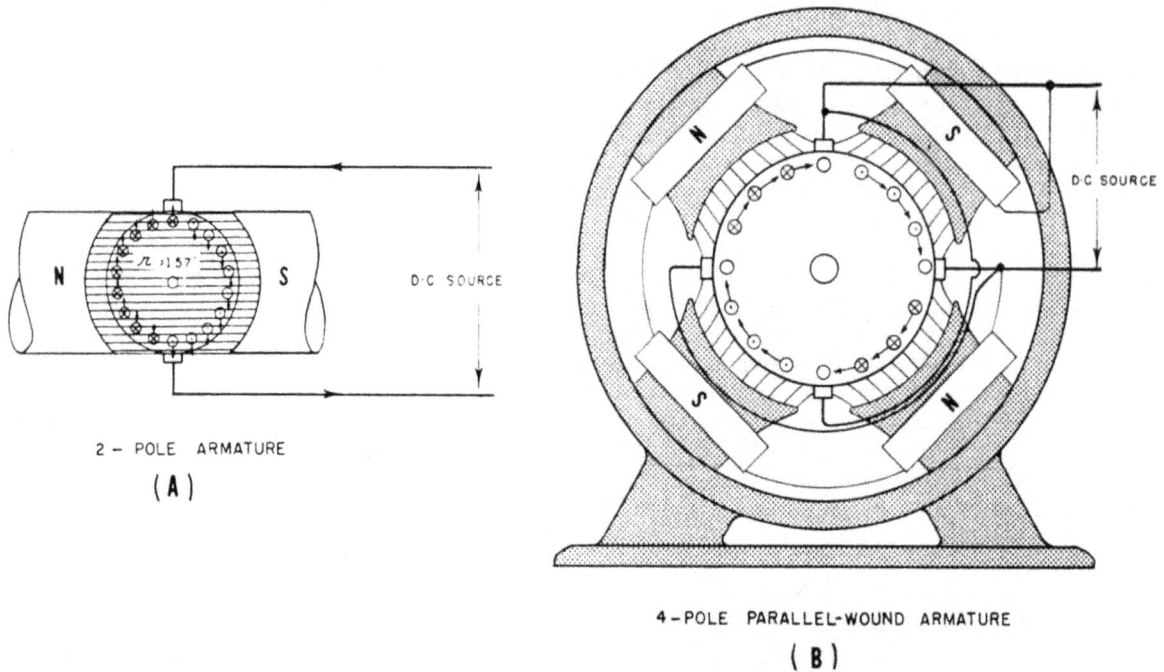

BE.361

Figure 19-6.—Torque developed in a motor armature.

The horsepower developed by a motor armature may be derived from the general expression.

$$hp = \frac{FV}{33,000} \qquad (19-2)$$

where F is the total force in pounds tangent to the effective circumference of the armature, and V the velocity of a point on this circumference in feet per minute. Velocity is determined as

$$V = 2\pi rN \qquad (19-3)$$

where r is the effective radius of the armature in feet, and N is the armature speed in revolutions per minute. The effective circumference is equal to $2\pi r$.

Substituting equation (19-3) in equation (19-2),

$$hp = \frac{2\pi rNF}{33,000} \qquad (19-4)$$

The torque in pound-feet produced by the motor armature is

$$T = rF \qquad (19-5)$$

Substituting equation (19-5) in equation (19-4) and dividing numerator and denominator by $2\pi$,

$$hp = \frac{NT}{5,252} \qquad (19-6)$$

Substituting the values N = 100 r.p.m. and T = 1.59 x 200 = 318 pound-feet of the preceding example in equation (19-6), the result is

$$hp = \frac{100 \times 318}{5,252} = 6.06$$

Thus, the horsepower developed by a motor depends on its speed and torque. Large, slow-speed motors develop a large torque; whereas other much smaller motors of the same horsepower rating operate at reduced torque and increased speed.

COUNTER EMF

The MOTOR ACTION of a generator was treated in chapter 18: the GENERATOR ACTION of a motor is now considered. By applying the right-hand motor rule to an armature conductor

373

carrying current in the direction indicated in figure 19-7, it will be found that the conductor is forced upward. As the conductor is moved up through the field, it cuts lines of flux and has a voltage induced in it. Applying the left-hand generator rule, it is found that this generated voltage is in opposition to the impressed emf. This counter voltage is induced in the windings of any rotating motor armature and always opposes the impressed voltage. It is called COUNTER ELECTROMOTIVE FORCE and is directly proportional to the speed of the armature and the field strength. That is, the counter emf is increased or decreased if the speed is increased or decreased respectively; the same is true if the field strength is increased or decreased.

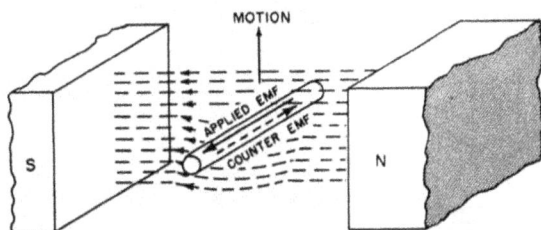

BE.362

Figure 19-7.—Generator action in a motor.

The EFFECTIVE VOLTAGE (IR drop) in the armature is equal to the impressed voltage minus the counter emf. The armature IR drop varies directly with the current flowing in the armature and the resistance of the armature.

To produce an armature current, $I_a$, in an armature of resistance, $R_a$, requires an effective voltage of $I_aR_a$.

The current flowing through the armature can be found by the equation

$$I_a = \frac{E_a - E_c}{R_a}$$

where $I_a$ is the current flowing through the armature, $E_a$ the impressed (or applied) voltage across the armature, $E_c$ the counter emf, and $R_a$ the armature resistance. This equation can be transposed and written as

$$E_c = E_a - I_aR_a \qquad (19-7)$$

In the case of the generator, the generated emf is equal to the terminal voltage plus the armature resistance drop; and in the case of the motor the generated or counter emf is equal to the terminal voltage minus the voltage drop in the armature resistance.

Expressing $E_a$ in terms of $E_c$ and $I_aR_a$

$$E_a = E_c + I_aR_a$$

This formula is valid for any d-c motor regardless of speed or load.

The counter emf, $E_c$, that is induced in a motor armature may be determined by an equation similar to that used to determine the generator voltage in a generator in chapter 18. This equation is

$$E_c = \frac{\Phi ZNP}{10^8 p} \qquad (19-8)$$

where $\Phi$ is the number of lines of flux per pole, Z the number of face conductors, N the armature speed in revolutions per second, P the number of field poles, and p the number of parallel paths through the armature circuit.

ARMATURE REACTION

The armature current in a generator flows in the same direction as the generated emf, but the armature current in a motor is forced to flow in the opposite direction to that of the counter emf. Assume that the field of the motor (fig. 19-8 (A)) is of the same polarity as the field of the generator (fig. 18-12 (A)). For the same direction of armature rotation, the armature flux of the motor (fig. 19-8 (B)) is in the opposite direction to that of the generator (fig. 18-12 (B)). In a motor the main field flux is always distorted in the opposite directon to rotation (fig. 19-8 (C)), whereas in a generator the main field flux is distorted (fig. 18-12 (C)) in the same direction as that of rotation. Note that the resultant field in the motor (fig. 19-8 (C)) is strengthened at the leading pole tips and weakened at the trailing pole tips. This action causes the electrical neutral plane to be shifted back to A'B'. Thus, to obtain good commutation in a motor without interpoles, it is necessary to shift the brushes from the mechanical neutral, AB, in a direction opposite to that of the armature rotation.

The armature reaction is overcome in a motor by the same methods as for the generator —that is, by the use of laminated pole tips with slotted pole pieces, and compensating windings.

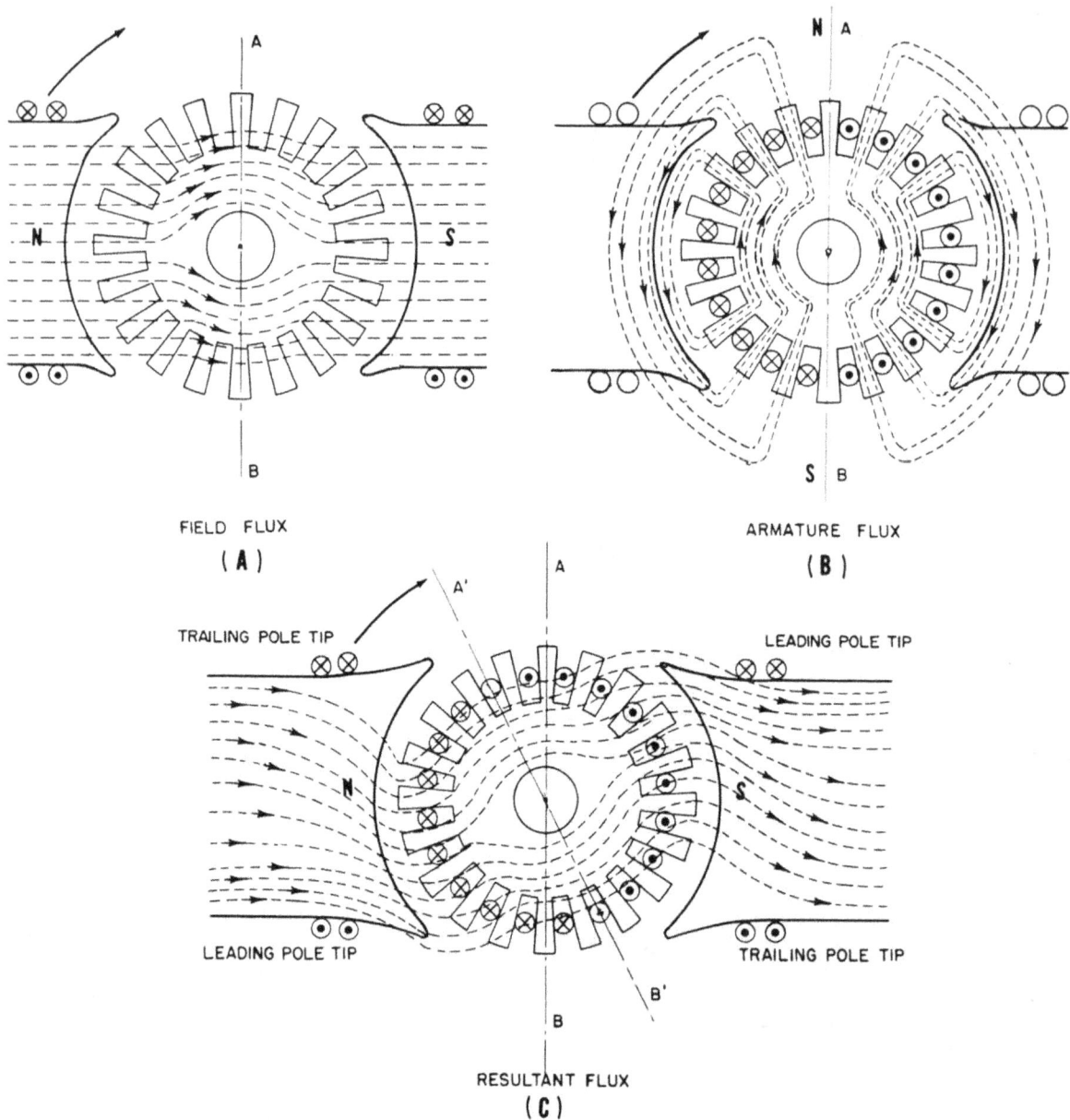

FIELD FLUX

(A)

ARMATURE FLUX

(B)

RESULTANT FLUX

(C)

Figure 19-8.—Armature reaction in a motor.

BE. 363

In each case the effect produced is the same as that produced in the generator, but it is in the opposite direction.

COMMUTATION

The brush axis, A'B' (fig. 19-8 (C)), could be made to coincide exactly with the neutral plane of the combined field. This would eliminate sparking at the brushes if it were not for the self-induced emf in the commutated coils. Because of the necessity for neutralizing this emf to eliminate sparking, the brushes must be set slightly behind the neutral plane in a motor. Thus, in both the generator and the motor, the brushes must be moved from the mechanical

375

neutral slightly beyond the electrical neutral plane to neutralize the effect of self-inductance.

The current flowing in the armature coils of a motor (fig. 19-9) must be periodically reversed when the coils are being short-circuited by the brushes in order to maintain a unidirectional torque as the coils move under alternate poles. When coil A is short circuited by a brush, its current immediately starts to decrease to zero; from zero, the current then builds up to a maximum in the reversed direction by the time the coil moves from position (1) to position (2), where the brush no longer short circuits it. As a result of the changing current, a self-induced voltage is set up in the shorted coil, which tends to keep the current from changing. To obtain sparkless commutation it is necessary to overcome this self-induced voltage.

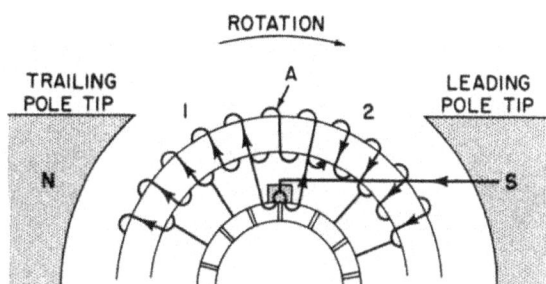

BE.364

Figure 19-9.—Commutation in a motor.

In any motor the current flows as a result of the applied voltage, and the counter emf opposes the flow of current. As the current in the commutated coil decreases to zero, the emf of self-induction tends to keep the current flowing in the same direction, as indicated by the arrow in position (1). Therefore, it aids the applied voltage during this portion of the cycle. As the coil moves into position (2) the current increases in the opposite direction, and the self-induced voltage again opposes this current increase. To overcome the self-induced voltage another voltage is introduced into the coil, which opposes the self-induced voltage. The counter emf is in the proper direction to accomplish this, but the coil being commutated is in the neutral plane and consequently has no counter emf generated in it.

When the brushes are shifted against the direction of rotation, the coils that are being shorted by the brushes still cut lines of flux from the previous north pole and have a small amount of counter emf generated in them. Because this counter emf opposes the applied voltage it also opposes the self-induced voltage, thus resulting in rapid reversal of coil current and in sparkless commutation.

When the load on the motor increases, armature reaction increases and the electrical neutral plane is shifted further in the direction opposite to that of the armature rotation. To maintain sparkless commutation, the plane of the brushes will have to be shifted slightly beyond the electrical neutral plane. When the load is reduced, the brushes are shifted in the opposite direction. Thus, for sparkless commutation, it is necessary to manually shift the brushes when the load varies.

COMMUTATING POLES

Commutating poles are as important in motors as in generators. Nearly all motors of more than one horsepower depend on COMMUTATING POLES, sometimes called INTER-POLES, rather than on the shifting of the brushes to obtain sparkless commutation. Essentially the only difference between the interpole generator and the interpole motor is that in the generator the interpole has the same polarity as the main pole AHEAD of it in the direction of rotation, but in the motor the interpole has the same polarity as the main pole BACK of it.

Without interpoles the emf of self-induction is overcome by commutating the coil while it is cutting the flux of the main pole it is just leaving. Therefore, if the interpole is to overcome the emf of self-induction, it must have the same polarity as the main pole it is leaving. This is a north pole in the example of figure 19-9.

The interpole series field coil in a motor is connected to carry the armature current, as in the generator. As the load varies, the interpole flux varies, and commutation is automatic with load change. It is not necessary to shift the brushes where there is an increase or decrease in load. The brushes are located on the no-load neutral and remain in that position for all conditions of load.

The motor may be reversed by reversing the direction of the current in the armature. When the armature current is reversed, the current through the interpole is also reversed, and therefore the interpole still has the proper polarity to provide automatic commutation.

The direction of the counter emf in the interpole motor with clockwise rotation (fig 19-10) illustrates that coils under the interpoles do not contribute to the armature counter voltage, but that only those coils under the main poles contribute to it. The total counter voltage generated in the conductors on each half of the armature is equal to the algebraic sum of the voltages generated in each conductor between the upper and lower brush. Because those under the upper north interpole generate an equal and opposite voltage to those under the lower south interpole, the remaining conductors under the main field poles generate the armature counter emf that controls the armature load current and the speed-torque characteristics of the motor. It is important that the brushes be set in the correct neutral plane, otherwise the commutation will be impaired and the motor characteristics will be altered.

## SPEED REGULATION

SPEED REGULATION is the ability of a motor to maintain its speed when a load is applied. It is an inherent characteristic of a motor and remains the same as long as the applied voltage does not vary—that is, unless a physical, or mechanical, change is made in the machine. The speed regulation of a motor is a comparison of its no-load speed to its full-load speed and is expressed as a percentage of full-load speed. Thus,

$$\text{percent speed regulation} = \frac{\text{no-load speed - full-load speed}}{\text{full-load speed}} \times 100$$

For example, if the no-load speed of a shunt motor is 1,600 rpm and the full-load speed is 1,500 r.p.m the speed regulation is

$$\frac{1,600 - 1,500}{1,500} \times 100 = 6.6 \text{ percent}$$

The LOWER the speed-regulation percentage figure of a motor, the more constant the speed will be under varying load conditions and the BETTER will be the speed regulation. The HIGHER the speed-regulation percentage figure, the POORER is the speed regulation.

SPEED CONTROL refers to the external means of varying the speed of a motor under any load.

## SHUNT MOTORS

### SPEED REGULATION

The field circuit of a shunt motor is connected across the line and is thus in parallel with the armature the same as it is in the shunt generator (fig. 18-16 (B)). If the supply voltage is constant, the current through the field coils, and consequently the field flux, will be constant.

When there is no load on the shunt motor, the only torque necessary is that required to overcome bearing friction and windage. The rotation of the armature coils through the field flux establishes a counter emf that limits the armature current to the relatively small value required to establish the necessary torque to run the motor on no load.

When an external load is applied to the shunt motor it tends to slow down slightly. The slight decrease in speed causes a corresponding decrease in counter emf. If the armature resistance is low, the resulting increase in armature current and torque will be relatively large. Therefore, the torque is increased until it matches the resisting torque of the load. The speed of the motor will then remain constant at the new value as long as the load is constant.

Conversely, if the load on the shunt motor is reduced, the motor tends to speed up slightly.

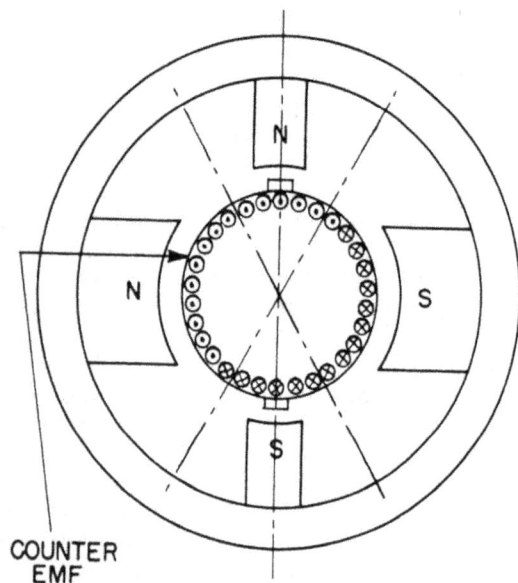

BE. 365

Figure 19-10.—Effect of interpoles on armature counter emf.

The increased speed causes a corresponding increase in counter emf and a relatively large decrease in armature current and torque.

Thus, it may be seen that the amount of current flowing through the armature of a shunt motor depends on the load on the motor. The larger the load, the larger the current; and conversely, the smaller the load, the smaller the current. The change in speed causes a change in counter emf and armature current in each case.

Figure 19-11 indicates graphically what happens when the load is applied or removed from a shunt motor. In this figure it is assumed that the field strength remains constant.

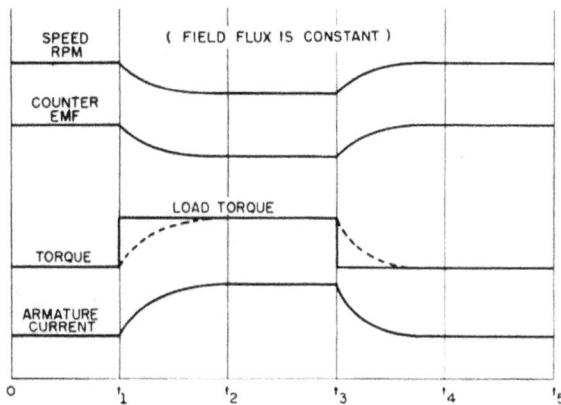

BE. 366

Figure 19-11.—Shunt motor load-speed-torque-emf relationships with respect to time.

During the interval between 0 and $t_1$ the motor operates under conditions of equilibrium —that is, the speed, counter emf, torque, and armature current have steady values. At $t_1$ a load is applied to the motor. Instantly, the load-torque curve rises, but the inertia of the armature prevents the generated torque from rising instantly with it. Because the load torque exceeds the generated torque in the motor armature, the motor speed is reduced until a balance is again obtained. When the generated torque is again equal to the load torque deceleration ceases and the motor operates at a constant reduced speed.

At time $t_2$ this state of equilibrium, corresponding to the new load, is reached. At time $t_3$ the load is suddenly removed. The load torque drops instantly, but the armature inertia

prevents the generated torque from falling instantly with it. Likewise, the armature inertia prevents the armature speed from increasing suddenly. Because the generated torque exceeds the load torque. the motor accelerates. As the speed increases, the counter emf increases proportionately. This opposes the applied emf, and the armature current is reduced until the generated torque and load torque are again in a state of equilibrium. The speed then levels off to a constant value.

The operational features of a shunt motor are indicated in table 19-1. The variation of line current, armature current, counter emf, speed, torque, and efficiency, with varying load conditions are indicated. The values are based on a 1-horsepower 100-volt shunt motor having a field current of 1 ampere and an armature resistance of 1 ohm.

When the armature current is 1 ampere, the armature IR drop is 1 volt, and the counter emf is the difference between the applied voltage and the IR drop, or 100 - 1 = 99 volts. The speed at this load is assumed to be 990 rpm. At this load the motor torque is assumed to be 0.7 pound-foot. The output power, $P_0$, in watts is

$$P_0 = \frac{NT}{5,252} \times 746 = 0.142NT = 0.142 \times 990 \times 0.7$$
$$= 98.4 \text{ watts.}$$

The constant 746 is used to convert horsepower output into an equivalent number of watts output (1 horsepower = 746 watts).

The total input current to the motor is equal to the sum of the field current and the armature current, or 1 + 1 = 2 amperes. The input power is the product of the applied voltage and the total input current, or 100 x 2 = 200 watts. The efficiency is

$$\frac{\text{output}}{\text{input}} \times 100 = \frac{98.4}{200} = 49.2 \text{ percent at this load}$$

The torque calculations for other values of load are based on the proportionality between armature current and torque. For example, if the torque is 0.7 pound-foot when the armature current is 1 ampere, the torque for an armature current of 2 amperes will be 2 x 0.7 = 1.4 pound-feet.

The speed calculations are based on the proportionality between speed and counter voltage. For example, when the armature current is 2 amperes the armature IR drop is 2 x 1 = 2 volts, and the counter emf is 100 - 2 = 98 volts. If the speed is 990 rpm when the counter emf is 99 volts, the speed at 98 volts counter emf will be 980 rpm.

The characteristic curves of a 5-horsepower 230-volt shunt motor are shown in figure 19-12. These curves include speed, torque, efficiency, and input current. The speed curve is almost a horizontal line, and the regulation is good. The torque varies directly with the output load and armature current because the field is approximately constant over the load range.

The speed regulation of a shunt motor is generally better than the voltage regulation of the same machine when operating as a shunt generator, this action is due to the fact that armature reaction in a generator weakens the field and lowers the terminal voltage, whereas armature reaction in a motor weakens the field and tends to increase the motor speed. Armature resistance is low, and a decrease in field strength and counter emf (for example, of 5 percent) might result in an increase in armature current and accelerating force on the armature conductors of 50 percent with resulting increase in armature speed. In general, however, armature reaction in a motor weakens the field only slightly and does not cause the speed to increase with load, but to decrease less than it would if the armature reaction were not present.

UTILIZATION

From the preceding considerations it is evident that a shunt motor is essentially a constant-speed machine. Although the speed may be varied by varying the current through the field winding (for example, by means of a field rheostat), the speed remains nearly constant for a given field current.

The constant-speed characteristic makes the use of shunt motors desirable for driving machine tools or any other device that requires a constant-speed driving source.

Where there is a wide variation in load, or where the motor must start under a heavy load, series motors have desirable features not found in shunt motors.

SERIES MOTORS

SPEED REGULATION

The field coils of a series motor are connected in series with the armature like those of the series generator in figure 18-16 (B). With relatively low flux density in the field iron, the series field strength is proportional to armature current $I_a$. The torque developed by the motor armature is proportional to $I_a \Phi$; and because $\Phi$ is also proportional to $I_a$, the torque is proportional to $I_a^2$.

If the supply voltage is constant, the armature current and the field flux will be constant only if the load is constant. If there were no load (this is never done) on the motor, the armature would speed up to such an extent that the windings might be thrown from the slots and the commutator destroyed by the excessive centrifugal forces. The reason why the speed becomes excessive on no load is explained in the following manner.

The speed of a series motor may be expressed mathematically as

$$rpm = \frac{K [E - I_a(R_a + R_s)]}{\Phi} \qquad (19\text{-}9)$$

Table 19-1.—Operational features of a shunt motor.

| I Line (amperes) * | I Armature (amperes) * | Counter emf (volts) * | Speed (rpm) * | Torque (lb-ft) * | Output 0.142NT (watts) | Input (watts) | Efficiency (percent) | Load |
|---|---|---|---|---|---|---|---|---|
| 2 | 1 | 99 | 990 | 0.7 | 98.4 | 200 | 49.2 | Light. |
| 4 | 3 | 97 | 970 | 2.1 | 289 | 400 | 72.2 | Do. |
| 6 | 5 | 95 | 950 | 3.5 | 472 | 600 | 78.5 | Do. |
| 8 | 7 | 93 | 930 | 4.9 | 646 | 800 | 80.8 | Do. |
| 10 | 9 | 91 | 910 | 6.3 | 815 | 1,000 | 81.5 | Normal |
| 21 | 20 | 80 | 800 | 14 | 1,590 | 2,100 | 76 | Over. |

* Assumed.

Where K is a constant that depends on such factors as the number of poles, the number of current paths, and the total number of armature conductors; E is the voltage applied across the entire motor; $I_a$ is the armature current; $R_a$ is the armature resistance; $R_S$ is the resistance of the series field; and $\Phi$ is the field strength. The factor $E - I_a (R_a + R_S)$ represents mathematically the counter emf, $E_c$. The applied voltage must then be equal to the counter voltage plus the voltage drop in the armature, or

$$E = E_c = I_a(R_a + R_S) \qquad (19\text{-}10)$$

The armature always tends to rotate at such a speed that the sum of the counter voltage, $E_c$, and the $I_aR_a$ drop in the armature will equal the applied voltage, E. If the load is removed from the motor, the armature will speed up and a higher counter emf will be induced in the armature. This reduces the current through the armature and the field. The weakened field causes the armature to turn faster. The counter emf is thus increased, and the speed is further increased until the machine is destroyed. For this reason, series motors are never belt-connected to their loads. The belt might come off and the motor would then overspeed and destroy itself. Series motors are connected to their loads directly, or through gears.

As the load increases, the armature slows down and the counter emf is reduced. The current through the armature is increased and likewise the field strength is increased. This reduces the speed to a very low value. The armature current, however, is not excessive because the torque developed depends on BOTH the field flux and the armature current.

The operational features of a series motor are indicated in table 19-2. The variation of line current, counter emf, speed and torque under varying load conditions are indicated. The values are based on a 0.5-horsepower 100-volt series motor having a field resistance of 1 ohm and an armature resistance of 1 ohm.

When the armature current is 5 amperes, the armature and field IR drop is 10 volts, and the counter emf is the difference between the applied voltage and the IR drop, or 100 - 10 = 90 volts. At this load the speed is assumed to be 900 rpm and the motor torque is assumed to be 3 pound-feet. The output power, $P_O$, in watts is

$$P_O = 0.142NT = 0.142 \times 900 \times 3 = 383.4 \text{ watts}$$

The input power is the product of the applied voltage and the current through the field and the armature, or $100 \times 5 = 500$ watts. The efficiency (the output divided by the input) is $\frac{383.4}{500}$, or 76.8 percent.

The speed calculations are based on the following considerations. As has been explained (see equation 19-8), the counter emf, $E_c$ is determined as

$$E_c = \frac{\Phi ZNP}{10^8 p}$$

It is also determined as

$$E_c = E - I_a (R_a + R_f)$$

These equations may be equated as

$$\frac{\Phi ZNP}{10^8 p} = E - I_a(R_a + R_f)$$

Assume that when 5 amperes flow through the field coil (and the armature) $\Phi = 10^6$ lines of force. Assume also that $Z = 600$ conductors. There are 2 poles (P = 2) and 2 paths (p = 2) through the armature. If N is divided by 60, the number of revolutions will be determined for an interval of 1 minute. Thus,

$$\frac{10^6 \times 600 \times N \times 2}{10^8 \times 60 \times 2} = 100 - 5(1 + 1)$$

and,

$$N = 900 \text{ rpm}$$

If the current is increased to 10 amperes (doubled), the flux is also assumed to double. Thus,

$$\frac{2 \times 10^6 \times 600 \times N}{10^8 \times 60} = 100 - 10 (1 + 1)$$

The number of revolutions per minute becomes

$$N = 400 \text{ rpm}$$

The torque varies as the square of the current. Thus, when the current is doubled, the torque is increased four times.

The characteristic curves of a series motor are shown in figure 19-13. As in the case of the shunt motor, these curves include speed, torque,

Table 19-2.—Operational features of a series motor.

| I Armature * (amperes) | Counter * emf (volts) | Speed * ( rpm ) | Torque * (lb-ft) | Output 0.142 NT (watts) | Input (watts) | Efficiency (percent) |
|---|---|---|---|---|---|---|
| 5 | 90 | 900 | 3 | 383.4 | 500 | 76 |
| 10 | 80 | 400 | 12 | 681.6 | 1,000 | 68 |
| 15 | 70 | 233 | 27 | 893.3 | 1,500 | 59 |
| 20 | 60 | 150 | 48 | 1,022.4 | 2,000 | 51 |
| 25 | 50 | 100 | 75 | 1,065.0 | 2,500 | 42 |

* Assumed.

efficiency, and input current. As stated previously, the torque varies as the square of the armature current (below saturation) and the speed decreases rapidly as the load is increased.

UTILIZATION

As indicated in figure 19-13, the torque increases with load. When the load on a series motor is increased, the speed and the counter emf decrease and the armature current and field strength increase. There is therefore an increase in torque with decrease in speed with the result that the increased load on the motor is limited by the decrease in speed.

When a heavy load is suddenly thrown on a shunt motor, it attempts to take on the load at only slightly reduced speed and counter emf. The flux remains essentially constant and therefore the increased torque is proportional to the increase in armature current. With heavy overload the armature current becomes excessive and the temperature increases to a very high

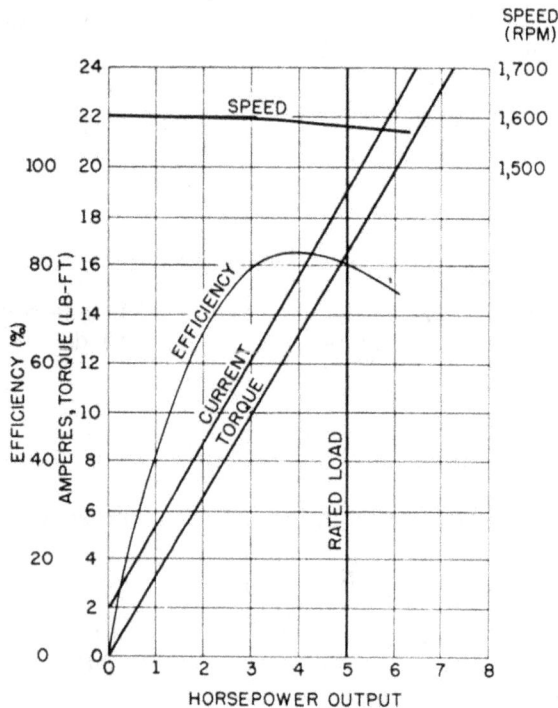

BE.367

Figure 19-12.—Characteristic curves of a shunt motor.

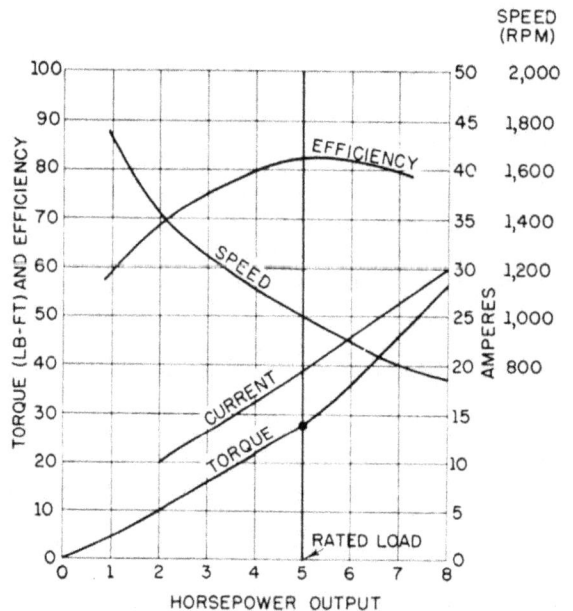

BE.368

Figure 19-13.—Characteristic curves of a series motor.

value. The shunt motor cannot slow down appreciably on heavy load, as can the series motor; hence the shunt motor is more susceptible to overload.

For example, assume that a d-c motor is used to move an electric truck up an incline. If it is a shunt motor it will attempt to maintain almost the same speed up the incline as on the level and consequently excessive current may be drawn through the armature, since the field flux remains almost constant. If it is a series motor, the speed must decrease more than the flux increases in order to have a reduction in counter emf and an increase in armature current ($E_c - \dfrac{\Phi ZNP}{10^8 p}$). The decrease in speed protects the series motor from excessive overload, and the armature current is limited by the counter voltage and the combined resistance of the armature and series field.

The series motor is therefore used where there is a wide variation in both torque and speed such as in traction equipment, blower equipment, hoists, cranes, etc.

### COMPOUND MOTORS

Compound motors, like compound generators, have both a shunt and a series field. In most cases the series winding is connected so that its field aids that of the shunt winding, as shown in figure 19-14 (A). Motors of this type are called CUMULATIVE COMPOUND motors. If the series winding is connected so that its field opposes that of the shunt winding, as shown in figure 19-14 (B), the motor is called a DIFFERENTIAL COMPOUND motor. Under full load, the ampere-turns of the shunt coil are greater than the ampere-turns of the series coil.

In the CUMULATIVE COMPOUND motor the speed decreases (when a load is added) more rapidly than it does in a SHUNT motor, but less rapidly than in a series motor. Series field strength increases as in a series motor. Transposing equation (19-10) for armature current,

$$I_a = \frac{E - E_c}{R_a + R_s}$$

To increase $I_a$, $E_c$ must decrease. As in the series motor, the decrease in speed is necessary to allow the counter emf to decrease at the same time the field increases. Because the torque varies directly as the product of the armature current and the field flux (equation 19-1), it is evident that the cumulative compound motor has greater starting torque than does the shunt motor for equal values of armature current and shunt field strength. The performance of this type of motor approaches that of a shunt motor as the ratio of the ampere-turns of the shunt winding to the ampere-turns of the series winding becomes greater. The performance approaches that of a series motor as this ratio becomes smaller.

If the load is removed from this type of motor it tends to speed up, and the counter emf increases. The current in the series field is reduced, and the greater portion of the field flux is produced by the shunt field coils. The compound motor then has characteristics similar to a shunt motor; and unlike the series motor, there is a definite no-load speed.

Because the total flux increases when there is an increase in load, there is a greater proportional increase in torque than in armature current. Therefore, for a given torque increase, this type of motor requires less increase in armature current than the shunt motor, but more than the series motor.

In some operations it is desirable to use the cumulative series winding to obtain good starting torque; and when the motor comes up to speed, the series winding is shorted out. The motor then has the improved speed regulation of a shunt motor.

In a differential compound motor, because of the opposition of the series field to the shunt field, the flux decreases as the load and armature current increase. From equation (19-9), it is seen that a decrease in flux causes an increase in speed. However, because the speed is proportional to $\dfrac{E_c}{\Phi}$, if both factors vary in the same proportion, the speed will remain constant. This action may occur in the differential compound motor. If more turns are added to the series coil, it is possible to cause the motor to run faster as the load is increased.

More armature current is required of the differential compound motor than of the shunt motor for the same increase in torque. This results from the fact that an increase in armature and series-field current reduces the field flux. Since the torque acting on the armature is proportional to the product of armature current and field strength, a decrease in field strength must be

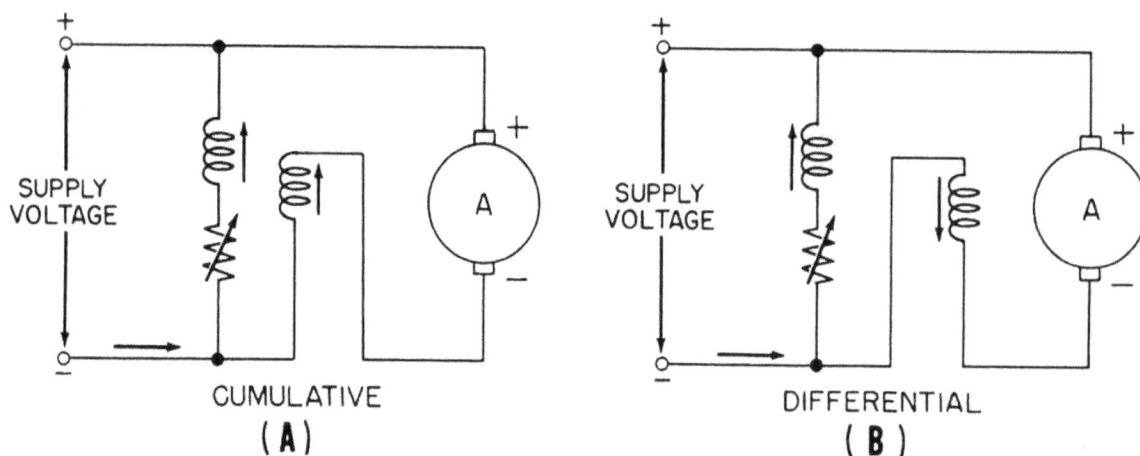

BE.369

Figure 19-14.—Types of compound motors.

accompanied by a disproportionate increase in armature current in order that the product will increase.

Under heavy load the speed of the differential compound motor is unstable; and if the overload current is very heavy, the direction of rotation may be reversed. Thereafter, the motor will run as a series motor with the danger of overspeed on no load that is the inherent characteristic of all series motors.

The characteristics of the differential compound motor are somewhat similar to those of the shunt motor, only exaggerated in scope. Thus, the starting torque is very low, and the speed regulation is very good if the load is not excessive. However, because of some of the undesirable features, this type of motor does not have wide use.

Figure 19-15 (A) shows the relative speed-load characteristics of shunt motors, cumulative compound motors, and differential compound motors. Although the speed of each of the motors is reduced when the current through the armature is increased (load is increased), the cumulative motor has the greatest decrease in speed. Likewise, in figure 19-15 (B) the cumulative motor has the greatest increase in torque with increase in load.

## MANUAL STARTERS

Because the armature resistance of many motors is low (0.05 to 0.5 ohm) and because the

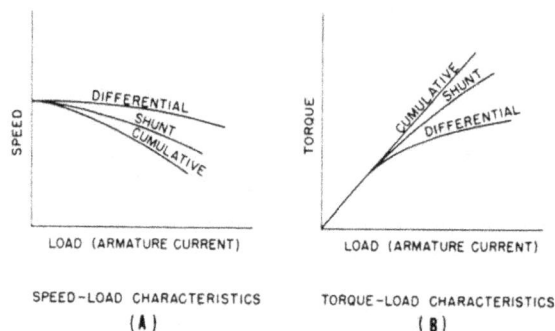

BE.370

Figure 19-15.—Speed-load and torque-load characteristics of shunt and compound motors.

counter emf does not exist until the armature begins to turn, it is necessary to employ an external resistance in series with the armature to keep the initial armature current to a safe value. As the armature begins to turn, the counter emf increases. Because the counter emf opposes the applied voltage, the armature current is reduced. The resistance in series with the armature is then reduced either manually or automatically as the motor comes up to normal speed and full voltage is applied across the armature.

The relation between armature current, $I_a$, applied voltage $E_a$, counter emf, $E_c$, armature

383

resistance, $R_a$, and starting resistance, $R_S$, for a shunt motor is

$$I_a = \frac{E_a - E_c}{R_a + R_S}$$

Assume that for the shunt motor shown in figure 19-16, $E_a$ = 100 volts, $R_a$ = 0.1 ohm, $R_S$ = 0.9 ohm, and the normal full-load armature current is 10 amperes. Without the starting resistance, the armature current would be

$$I_a = \frac{E_a}{R_a} = \frac{100}{0.1} = 1{,}000 \text{ amperes}$$

This is 100 times the normal current for the motor and would obviously burn up the armature insulation and cause excessive acceleration forces.

In order to limit the initial surge of current to a maximum of 100 amperes (arbitrarily, 10 times the full-load current), the starting resistance is

$$R_S = \frac{E_a - I_a R_a}{I_a} = \frac{100 - 100 \times 0.1}{100} = 0.9 \text{ ohm}$$

The relation of $E_a$; $I_a$; the starting resistance, $R_S$, motor speed, and counter emf during the accelerating period is shown in the curves of figure 19-17 and in table 19-3.

Initially, the full starting resistance of 0.9 ohm is inserted in series with the armature resistance of 0.1. The speed and counter emf are zero, and the armature current is limited to 100 amperes by the armature resistance of 0.1 ohm, acting in series with the starting resistance of

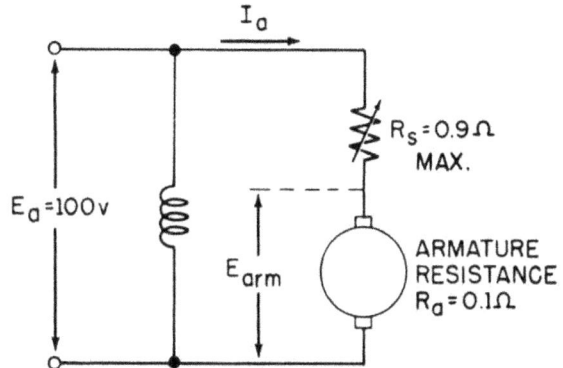

BE.371

Figure 19-16.—Shunt motor with manual starting resistance.

0.9 ohm. As the speed and counter emf increase, the counter emf subtracts a greater magnitude of voltage from the applied emf, and the armature current is reduced from 100 amperes to some lower value. When the speed is 500 rpm the counter emf is 50 volts and the current is

$$I_a = \frac{100 - 50}{0.1 + 0.9} = 50 \text{ amperes}$$

Suppose that at the instant $t_1$, when the motor speed is 500 rpm, the starting switch is moved to the 0.4-ohm tap. The counter emf of 50 volts subtracts from the 100 volts applied, leaving 50 volts to cause the armature current to again increase to $\frac{100 - 50}{0.4 + 0.1} = 100$ amperes at time

Table 19-3.—Starting characteristics of a shunt motor.

| Armature current (amperes) | Starting resistance (ohms) | Armature volts ($E_a - I_a R_S$) | Speed ( rpm ) | Counter e. m. f. volts $[E_a - I_a(R_a + R_S)]$ |
|---|---|---|---|---|
| 100 | 0.9 | 10 | 0 | 0 |
| 50 | 0.9 | 55 | 500 | 50 |
| 100 | 0.4 | 60 | 500 | 50 |
| 50 | 0.4 | 80 | 750 | 75 |
| 100 | 0.15 | 85 | 750 | 75 |
| 50 | 0.15 | 92.5 | 875 | 87.5 |
| 100 | 0.025 | 97.5 | 875 | 87.5 |
| 50 | 0.025 | 98.75 | 937.5 | 93.75 |
| 62.5 | 0 | 100 | 937.5 | 93.75 |
| 30 | 0 | 100 | 970 | 97 |

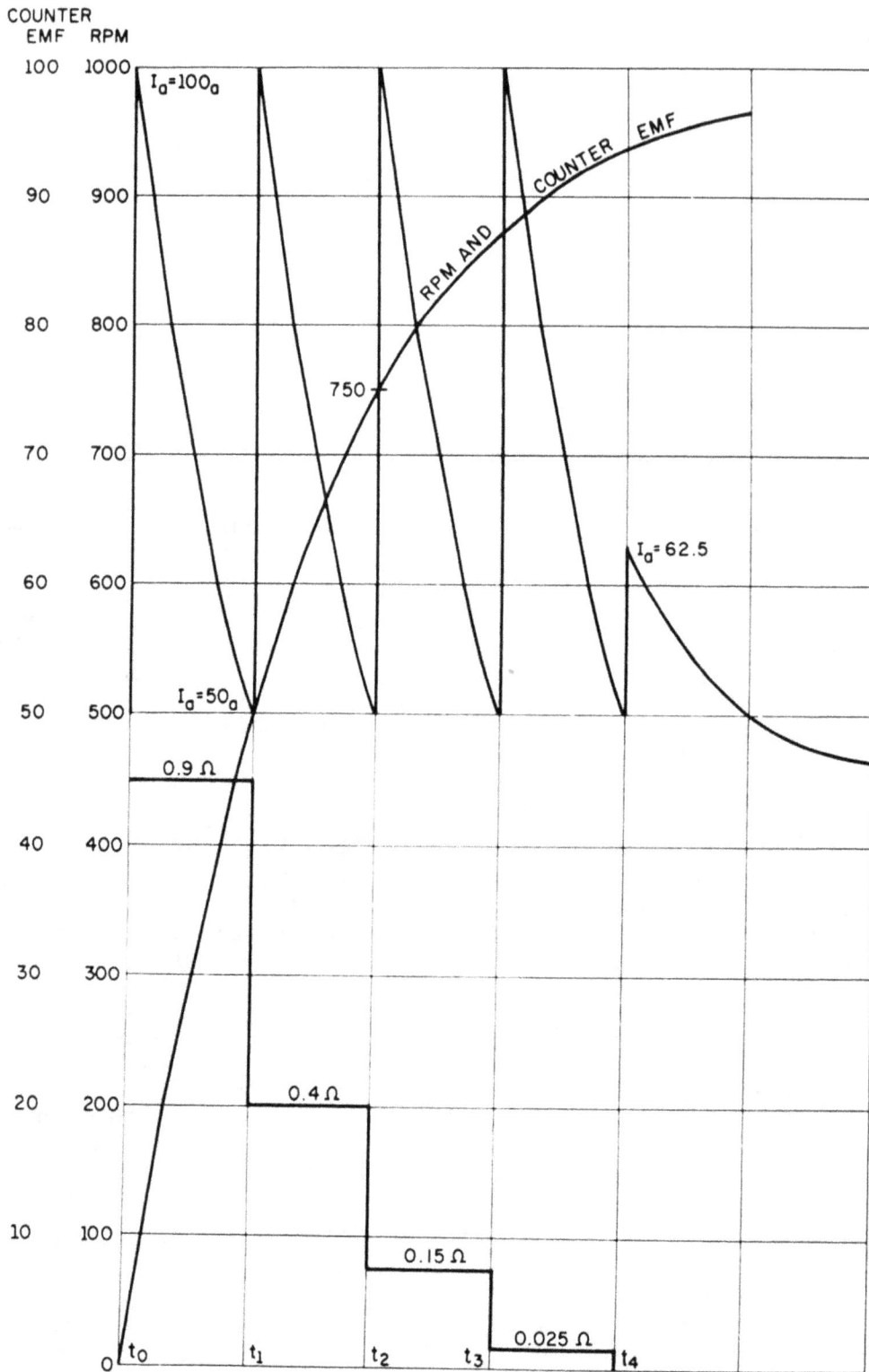

Figure 19-17.—Shunt motor starting curves.

$t_1$. The armature speed and counter emf cannot increase suddenly, but rise gradually during the interval from $t_1$ to $t_2$. Therefore, the armature current is decreased gradually, and at $t_2$, when the counter emf is 75 volts and the speed is 750 rpm, the current is

$$I_a = \frac{100 - 75}{0.4 + 0.1} = 50 \text{ amperes}$$

Moving the starting switch to the 0.15-ohm tap, the 0.025-ohm tap, and the 0-ohm tap at instants $t_2$, $t_3$, and $t_4$ causes the motor to accelerate to a speed of 970 rpm with variations in the armature current as indicated in the figure and in the table.

## AUTOMATIC STARTERS

If a motor is at a remote location, or if a large motor is to be started, it is generally desirable to use an automatic starter to bring the motor up to operating speed. Although small motors are started by means of manually operated starters, difficulties are encountered when starting large motors. In the manually-operated starter, the skill with which the operator regulates the magnitude of the starting current is dependent upon his experience. Too much time may be taken by an overcautious operator or the maximum permissible current for the motor may be exceeded by a careless one. In order that time shall not be wasted and the motor receive maximum safe current during the acceleration period, an automatic device should be used that will interpret the conditions of the load and act accordingly.

The following automatic starters will be discussed: (1) The time-element type, (2) the counter emf type, (3) the shunt current-limit type, and (4) the series current-limit type.

### TIME-ELEMENT STARTER

A time-element starter is one in which the resistance in series with the armature is reduced a certain amount during each succeeding unit of time regardless of the load on the motor. One type of time-element starter is shown in figure 19-18. Initially, the full value of the starting resistance is inserted in series with the armature, and its current is therefore limited to a safe value. When the switch is closed, the full-line voltage is impressed across the accelerating solenoid and across the shunt field. The solenoid immediately begins to draw the iron core up-

wards, and the starting resistance is gradually cut out of the armature circuit. The speed with which the resistance is cut out depends on the size of the orifice (the hole through which the oil flows) in the dashpot. The smaller the orifice, the greater the time delay. The time delay should be such that as the speed and the counter emf builds up, the starting current is maintained at the maximum safe value during the accelerating period. During the time required to reach full speed, the resistance is completely removed from the circuit.

The advantages of the time-element type of automatic starter are that its cost is low, the wiring is simple, and starting is generally sure. The disadvantages are that the amount of load cannot be sensed—that is, the resistance is removed at the same rate whether the load is heavy or light, the motor is not protected on overload, and the dashpot is a source of trouble.

### COUNTER EMF STARTER

The counter emf starter depends for its operation on the counter emf developed across the armature. As may be seen from figure 19-19 (A), the solenoid that activates the contacts is connected directly across the armature. It therefore responds to armature voltage, which is only slightly higher than the counter emf.

At start (fig. 19-19 (B)), time $t_0$, the armature current flowing through the starting resistance causes a reduced voltage to be applied to the armature and the accelerating contactor will not immediately close its contacts due to low voltage across its operating coil. As the motor accelerates, the rising counter emf causes the armature current and voltage drop across the starting resistance to reduce. At the same time, the voltage across the armature increases. When this voltage rises to the proper value, time $t_1$, the accelerating contactor closes and shorts out the starting resistance, thus applying full voltage across the armature. The accompanying increase in starting current causes continued acceleration until the motor comes up to rated speed, and the armature current is again reduced to the normal value. For simplicity, only one resistor and starting contactor are shown, although large motors usually employ several contactors and several steps of starting resistance. If the load is heavy, the acceleration is slower and the rise in voltage on the accelerating contactor

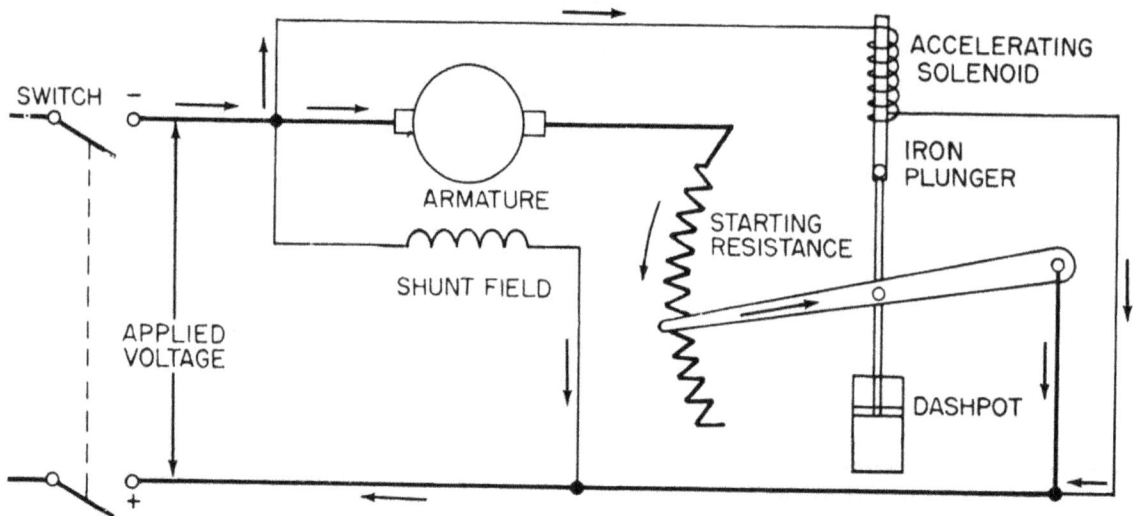

BE.373

Figure 19-18.—Time-element starter.

operating coil is delayed so that the starting resistance is not cut out until the rise in speed and counter emf permits the contactor to close.

The advantages of the counter emf starter are that the load is interpreted and the resistance cut out accordingly, the cost is low, the wiring is simple, and the dashpot is not used.

The disadvantage of the counter emf starter is that if the voltage varies the acceleration may become erratic. For example, if the line voltage rises, the starting resistance may be removed too soon; and if it falls, the starting contactor may not be activated at all. The starting contactor coils are designed to close the contactor on one value of voltage and to operate continuously on an increased voltage. Thus, they are sensitive to voltage change and will not operate properly if line voltage fluctuations occur.

SHUNT CURRENT-LIMIT STARTER

The disadvantages of the counter emf starter are overcome in the shunt current-limit starter. This motor starter employs accelerating contactors with shunt-type operating coils wound for full-line voltage. Each contactor includes an interlocking series relay, the operation of which is limited by the motor starting current. When the series relay closes its contacts, the shunt operating coil of the accelerating contactor is ener-

gized and a section of starting resistance is cut out of the armature circuit. This type of motor starter thus derives its name from the fact that the accelerating contactors are shunt operated and the closure of the series relays is limited by the magnitude of the armature current.

Figure 19-20 is a simplified schematic diagram of a shunt current-limit starter operating in connection with a pushbutton starter with no-voltage release holding-contacts. When the ON button is pressed down momentarily, the no-voltage release coil is energized. The holding relay contacts and the line contacts are closed. The series relay is mechanically released, and its contacts would close were it not for the fact that the armature current flows through the series coil and holds its core up, thus keeping its contacts open. Thus, it may be seen that current flows from the negative terminal of the voltage source through the armature, through the starting resistor, through the series relay coil, and back to the positive terminal of the voltage source.

As the motor speeds up, the counter emf increases, and the armature current is reduced accordingly. The field of the series relay is thereby weakened and the relay contacts are closed. This action energizes the coil of the accelerating contactor and its contacts close, thus shorting out the starting resistor and the series

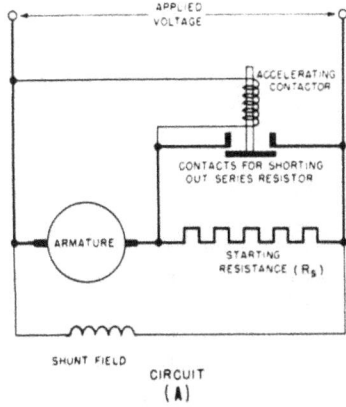

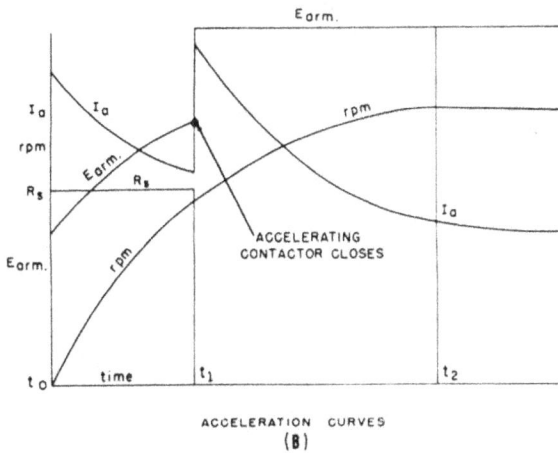

safe value irrespective of line voltage fluctuations. If the load is too great for the motor to accelerate normally, the counter emf cannot build up sufficiently to reduce the current through the series relay to cause its contacts to fall closed. Therefore, the starting resistance remains in the circuit and the armature is protected against excessive current.

The disadvantage is that the accelerating units are complicated and expensive.

## SERIES CURRENT-LIMIT STARTER

The series current-limit type of motor starter was designed to accelerate motors and to provide the same protection afforded by the shunt current-limit type of starter, but to do the job with less complicated equipment. The interlocking series relays are omitted and the accelerating contactors are designed to lock open on the initial in-rush of starting current. When the current falls to a predetermined value the contactor closes and shorts out the starting resistance.

In the series current-limit starter, the relay coil which operates the accelerating contactor is series wound—that is, it is connected in series with the armature circuit. The simplified circuit,

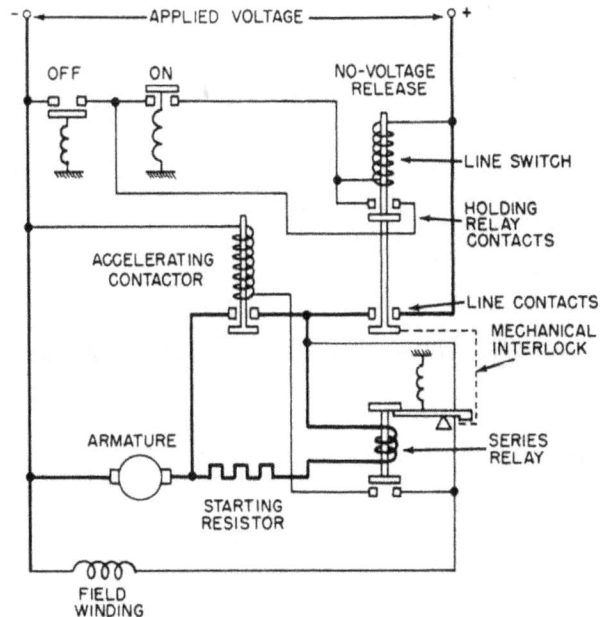

BE.374

Figure 19-19.—Counter emf starter.

relay. The full-line voltage is then safely applied across the armature. The series relay and accelerating contactor contacts remain closed until the line contacts are opened.

When the OFF button is pressed, current through the no-voltage release coil is interrupted and the holding relay contacts, the line contacts, and the series relay contacts are opened. Line voltage is thus removed from the motor.

Although for simplicity only one starting resistor is shown in a practical system, two or three are commonly used. Additional accelerating contactors and interlocking series relays are then used to remove the starting resistance in steps as the motor comes up to speed.

The advantage of the shunt current-limit starter is that starting resistance is cut out of the armature circuit only after the speed has built up and the motor current has been reduced to a

BE.375

Figure 19-20.—Shunt current-limit starter.

showing only one starting resistor, is shown in figure 19-21 (A).

One type of magnetically operated switch that makes this type of starter possible is shown in figure 19-21 (B). The initial current through the coil causes the accelerating contactor to lock open. When the current falls to a certain predetermined value, the switch will close.

The coil is wound with a few turns of heavy wire. When current flows through the coil, the movable armature is acted upon by two forces, one across the main airgap, the other across the auxiliary airgap. The force across the main airgap tends to close the contactor; that across the auxiliary airgap tends to open it. The flux paths are first 1, 2, 3, 4, 5, 6, 1; and second 1, 2, 3, 7, 8, 9, 6, 1. The first path includes only the main airgap. The second includes both gaps. When a small amount of current flows through the coil, the flux path is principally through 1, 2, 3, 4, 5, and 6 and the force exerted across the main airgap tends to close the switch. When a large amount of current flows through the coil, the flux path is through 1, 2, 3, 7, 8, and 9. This path includes both the main and auxiliary airgap.

Because the auxiliary gap is smaller, the force exerted across this gap to hold the contactor open exceeds the force across the main gap to close it and the contactor locks open.

To prevent the flux from rushing through the shorter path, 4, 5, 6, and closing the contactor before it has time to lock open, a short-circuited coil (damper) is placed around this flux path, as shown in figure 19-21 (B). As the flux links this coil, the induced current in the copper damper sets up a counter magnetomotive force that opposes the original flux and causes it to take the path through 7, 8, and 9 during the time that the current is rising. When the series operating coil current levels off at a high value, the flux path 4, 5, and 6 becomes saturated and sufficient flux is established through 7, 8, and 9 to maintain the contactor in the locked-out condition.

This type of contactor is especially satisfactory when the motor is always under load. The principal disadvantage is that the switch may drop open on light load. However, a shunt holding coil may be used to prevent this.

## MOTOR EFFICIENCY

The efficiency of any type of machine is the ratio of the output power to the input power. This ratio is commonly expressed as a percent. Because all machines have some losses, the efficiency will never be 100 percent. Expressed as a percentage, efficiency becomes

$$\text{efficiency} = \frac{\text{output}}{\text{input}} \times 100$$

$$\text{efficiency} = \frac{\text{output}}{\text{output} + \text{losses}} \times 100$$

or

$$\text{efficiency} = \frac{\text{input} - \text{losses}}{\text{input}} \times 100$$

The first equation is general. The second is advantageous when applied to electric generators and transformers. The third is most useful when applied to electric motors. The output must be expressed in the same units as the input. For example, if the output is expressed in horse-power and the input is expressed in watts, they can be changed to common units by means of the relation,

746 watts = 1 horsepower

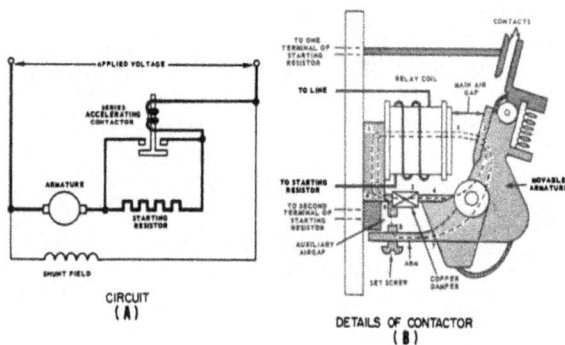

BE.376

Figure 19-21.—Series current-limit starter.

There are various types of mechanical and electrical losses in d-c motors, some of which are common to other types of motors.

One loss is in BEARING FRICTION. This type of friction is reduced by proper oiling. Roller bearings have less friction loss than sleeve bearings. However, at excessive speed or under excessive load the loss due to bearing friction may be appreciable. A loss is also introduced by BRUSH FRICTION. This may be reduced by the use of a well-polished commutator and properly fitted brushes. WINDAGE LOSS is that due to the resistance of the air to a rapidly revolving armature. IRON LOSSES include eddy-current loss and hysteresis in the armature iron. COPPER LOSSES include the $I^2R$ loss in the armature and field windings.

As an example, assume that the motor shown in figure 19-22 is supplied with 10 amperes at 100 volts.

The COPPER LOSS in the field is

$$I_f^2 \times R_f = I^2 \times 100 = 100 \text{ watts}$$

The armature current is 10 - 1, or 9 amperes and the COPPER LOSS in the armature is

$$I_a^2 \times R_a = 9^2 \times 1 = 81 \text{ watts}$$

Assume that the total MECHANICAL LOSS (friction, windage, and so forth) is 90 watts. The TOTAL LOSS is therefore 271 watts. The input is

$$E_a \times I_L = 100 \times 10 = 1,000 \text{ watts}$$

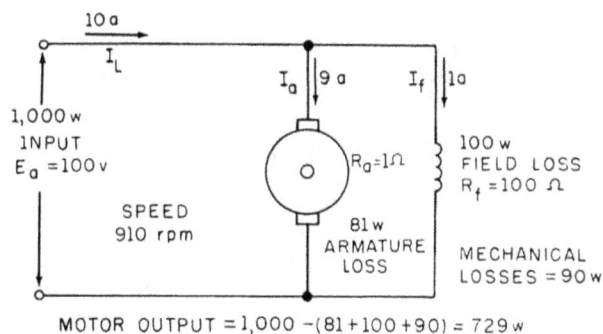

MOTOR OUTPUT = 1,000 −(81+100+90) = 729w

BE.377

Figure 19-22.—Distribution of losses in a motor.

The output is

input - losses = 1,000 - 271 = 729 watts

The efficiency is

$$\frac{\text{output}}{\text{input}} \times 100 = \frac{729}{1,000} \times 100 = 72.9 \text{ percent}$$

If it is determined (by means of a prony brake test) that the power developed by the motor is 0.976 horsepower, the efficiency is also determined as

$$\text{efficiency} = \frac{\text{output}}{\text{input}} \times 100$$

$$= \frac{0.976 \times 746}{1,000} \times 100 = 72.9 \text{ percent}$$

## SPEED CONTROL

The speed, N, of a d-c motor is directly proportional to the armature counter emf, $E_c$, and inversely proportional to the field strength, $\Phi$.

Transposing equation (19-8) for N, $N = \frac{E_c 10^8 p}{\Phi ZP}$.

Such factors as the numbers of poles, P, the number of current paths, p, and the total number of armature conductors, Z, can be lumped into a single constant, K. Simplified, the expression becomes

$$\text{rpm} = \frac{KE_c}{\Phi}$$

In a shunt motor

$$E_c = E_a - I_a R_a$$

and

$$\text{rpm} = \frac{K(E_a - I_a R_a)}{\Phi} \quad (19-11)$$

where $E_a$ is the voltage applied across the armature. This is similar to, but not identical with, equation (19-9) for a series motor.

From equation 19-11, it is seen that the speed may be INCREASED by DECREASING the field strength, and vice versa, or the speed may be increased by increasing $E_a$, and vice versa. Practically, these variations are most simply accomplished by inserting a field rheostat in series with the field circuit, or a variable resistor in series with the armature circuit.

## BY ARMATURE SERIES RESISTANCE

Figure 19-23 shows how speed control may be accomplished by means of a variable resistor in

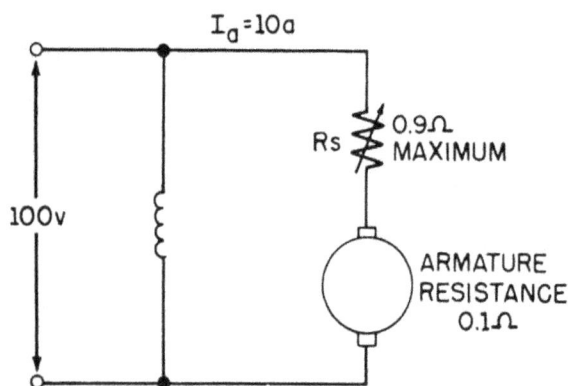

Figure 19-23.—Speed control by means of armature series resistor.

series with the armature. The circuit is the same as that used in the motor starter circuit in figure 19-16. However, the wattage rating is different. The starting resistance should only be inserted for a short interval, whereas the speed control resistance can remain in the circuit indefinitely.

In the example of the motor shown in figure 19-23, assume that $\frac{K}{\Phi}$, as indicated in the previous equation, is replaced by the number 10. The equation for speed in rpm then becomes

$$rpm = 10\left[E_a - I_a(R_s - R_a)\right]$$

When $R_s$ is set at zero, the resistance, $R_a$, offered by the armature is 0.1 ohm, and $I_a$ is assumed to be 10 amperes. Therefore,

$$rpm = 10\left[100 - 10(0 + 0.1)\right] = 990$$

When $R_s$ is increased to 0.1 ohm, $I_a$ is lowered. The motor speed is reduced, the counter emf is reduced, and $I_a$ is assumed to come back to the original value of 10 amperes. At the stable condition, the new speed is

$$rpm = 10\left[100 - 10(0.1 + 0.1)\right] = 980$$

The rheostatic losses involved in this method of speed control are appreciable at low speeds and the resultant reduction in efficiency makes this type of control undesirable if the motor is to be operated at greatly reduced speed for prolonged intervals.

## BY ADJUSTING THE FIELD STRENGTH

A more economical method of speed control is by rheostatic adjustment of the field current. If the field strength is weakened, the speed of the motor is increased; and if the strength of the field is increased, the speed of the motor is decreased.

Figure 19-24 indicates a method of varying the strength of the field by means of a field rheostat. When the field rheostat is cut out, the field current is 1 ampere. The counter emf is 90 volts, and the armature current is 10 amperes. The MEASURED speed is 900 rpm. When the resistance of the field rheostat is increased to 11.1 ohms, the field current is reduced to 0.9 amperes and the field flux is reduced. This causes a reduction in counter emf and a sharp increase in armature current, which causes an increase in force on the armature and an increase in armature speed. As the speed builds up, the counter emf builds up to 90 volts again, and the armature current is again reduced to 10 amperes.

The speed corresponding to a field current of 0.9 ampere may be computed if it is recalled that the speed varies inversely with the flux and that the flux (below saturation) varies directly with the current. Thus, if the speed is 900 rpm when the field current is 1 ampere it will be

$$900 \times \frac{1}{0.9} = 1,000 \text{ rpm, when the field current is}$$

0.9 ampere.

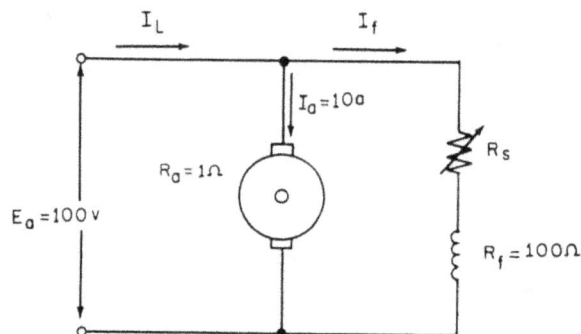

Figure 19-24.—Speed control by varying the strength of the field.

391

## BY WARD-LEONARD SYSTEM

As has been explained in the previous paragraphs, the speed of a d-c motor can be controlled by at least two methods—either the armature voltage can be varied, as in figure 19-23 or the field voltage can be varied as in figure 19-24. The first method gives a range of speeds below the rated full-load speed. The second, gives a range of speed above the rated full-load speed. The first method takes a way the constant speed characteristics from the shunt motor when the load varies. The second method permits control with a physically smaller resistor and greatly reduced power and gives essentially constant speed characteristics at any speed setting of the field rheostat provided the field is not weakened excessively. With a very weak field the motor may stall. The inherent difficulties of both methods are removed by the Ward-Leonard method shown in figure 19-25.

The d-c motor armature whose speed is to be controlled, is fed directly from a d-c generator armature which is driven by a constant-speed prime mover. The d-c field supply to the generator is variable in both magnitude and polarity by means of a rheostat and reversing switch, as shown. Therefore, the motor armature is supplied by a generator having smoothly varying voltage output from zero to full-load value. The motor field is supplied with a constant voltage from the same source as that supplying the generator fields. The generator drive power could be from a single-phase or three-phase a-c motor, from an engine, or other constant-speed source. In the same way, the d-c supply can be supplied from a rectifier, from an exciter on the end of the generator shaft, or from any other suitable d-c source. In the figure the field exciter is a direct-connected unit on the shaft of the motor-generator set. The drive motor for the M-G (MOTOR-GENERATOR) set is shown as a 3-phase motor. Both the main generator and the speed-control motor have commutating poles, and in some instances, compensating windings.

The advantages of this type of speed control are that it does away with the armature rheostatic losses and instability of speed with variable loads.

The disadvantage of this method of speed control is the initial expense of the added equipment—that is, the M-G set with its control equipment.

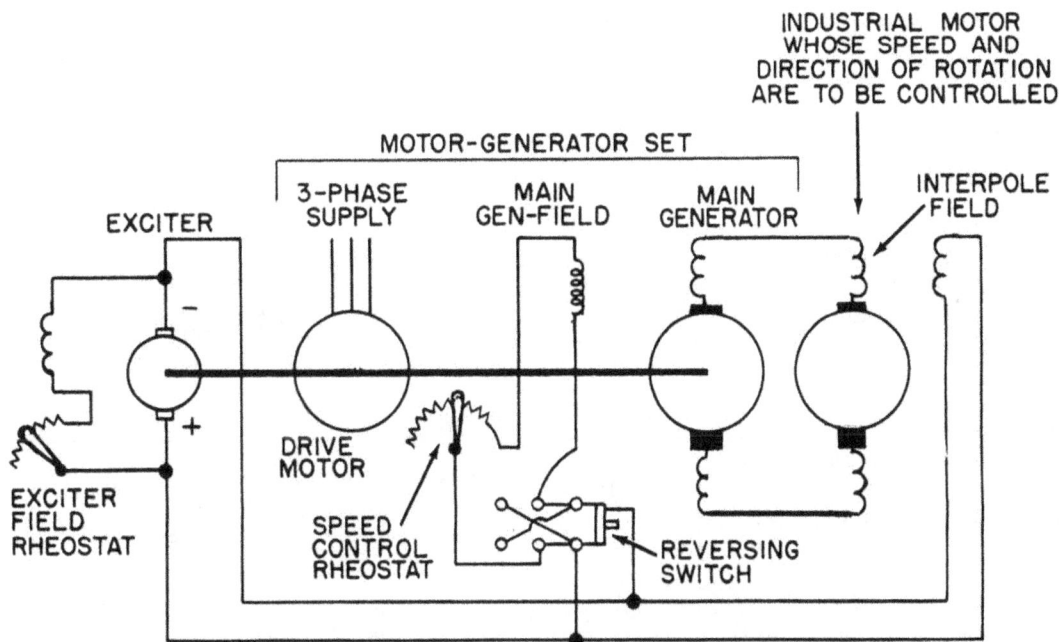

BE.380

Figure 19-25.—Ward-Leonard speed control system.

# CHAPTER 20

# MAGNETIC AMPLIFIERS

Amplification of voltage and power can be accomplished by means of the magnetic amplifier, which employs as its controllable element an iron-core saturable reactor. The magnitude of the impedance of the saturable reactor depends upon the range of flux change that occurs in its core, and this action, in turn depends upon the magnitude of the control current. If the control current is varied a small amount, the power delivered to a load is varied through a much wider range. Herein lies the action of amplification.

The term "amplification," in general, refers to the process of increasing the amplitude of voltage, current, or power. The term "amplification factor" refers to the ratio of output to input. The amplitude of the input signal and the amplification factor determine the level of the output signal.

Magnetic amplifiers are not new devices by any means. The principles of magnetic amplification were discovered toward the end of the 19th century, but the magnetic amplifier did not reach its present state of development and application until the 1950's.

The first types were what are now commonly referred to as saturable reactors. Their use was limited, for the most part, to the control of large industrial power devices and lighting loads, because vacuum and gaseous tubes, partly due to their novelty and attractiveness to engineers, offered serious competition as low-powered amplifiers during the first quarter of the 20th century. It was not until after the development of the "self-saturating" principle in the 1930's that the device began to find wide application as a low-powered amplifier. During World War II, the Germans used magnetic amplifiers in a wide variety of equipments, and the United States conceived or produced servomechanisms and other electronic devices in which they were used. At the end of the war the magnetic amplifier was a "rising star" in electronics.

For a time in the 1950's. it appeared that the transistor would doom the magnetic amplifier. However, it was soon realized that all types of amplifiers have advantages and disadvantages for different applications. Today, the magnetic amplifier finds its proper place alongside, and in many cases in conjunction with, transistors and tubes in hybrid composite circuits.

The magnetic amplifier combines many of the advantages of both the vacuum tube and the thyratron (a gas tube with a control grid) and, in addition, contributes several of its own characteristics to advantageous uses in handling large amounts of power. Although their ability to control large amounts of power is an important feature, they are also well suited and frequently used in devices involving signal or voltage amplification. It is usually necessary or desirable to add control currents from several different sources. This is easily accomplished by putting additional windings on the magnetic amplifier core.

The magnetic amplifier has certain advantages over other types of amplifiers. These include (1) high efficiency (90 percent): (2) reliability (long life, freedom from maintenance, reduction of spare parts inventory); (3) ruggedness (shock and vibration resistance, overload capability, freedom from effects of moisture); and (4) no warmup time. The magnetic amplifier has no moving parts and can be hermetically sealed within a case similar to the conventional dry type transformer.

Also, the magnetic amplifier has a few disadvantages. For example, it cannot handle low-level signals; it is not useful at high frequencies; it has a time delay associated with magnetic effects; and the output waveform is not an exact reproduction of the input waveform.

The magnetic amplifier is important, however, to many phases of naval engineering because it provides a rugged, trouble-free device that has many applications aboard ship and in aircraft. These applications include throttle controls on the main engines of ships, speed, frequency, voltage, current, and temperature controls on auxiliary equipment; fire control, servomechanisms, and stabilizers for guns, radar, and sonar equipment.

## CHARACTERISTICS

The basic component in any magnetic amplifier circuit is a saturable reactor. Any magnetic-cored reactor can be saturated if sufficient current flows in a winding on the core. Thus, in a sense, all reactors with magnetic cores could be classified as saturable reactors. However, such a unit designed for magnetic amplifiers has a very special core, compared to a unit designed for use as a linear reactor or transformer.

As an amplifier, the nonlinear reactor is used alone or with additional electronic components to control the power that is delivered to a load. In its simplest form, it consists of a magnetic core and two windings—a load or gate winding, and a control or signal winding. The load winding is generally in series with the a-c power source and the load. The control winding is generally connected in series with a d-c control power source. In more useful and complex circuits, the unit has multiple cores, multiple load and control windings and, possibly, additional windings for bias and feed-back. Rectifiers are also a part of self-saturating circuits, and of saturable reactor circuits if a d-c load is involved. Such circuits are classified as amplifiers because they are used to control the amount of power delivered to a load in the load circuit by amplification of a lesser amount of power in the control circuit. This is the general definition of any amplifier.

Many magnetic amplifiers look like transformers. Some small units appear to be small, hermetically sealed plug-in relays. Some have windings on an uncased magnetic core and look like a doughnut. Many high grade, two-winding transformers could be used as the basic element to form a poor magnetic amplifier, but this is not usually done.

## REVIEW OF MAGNETICS

Every magnetic amplifier reactor contains a core of magnetic material. The shape of the core and the characteristics of the core material are similar to the cores used in iron-core transformers and linear reactors; but the core usually has special magnetic characteristics which give the reactor the extreme nonlinearity that high gain magnetic amplifiers require. Because of this special core material, complete understanding of magnetic amplifiers requires more than a knowledge of linear transformer and reactor theory; however, knowledge of those fundamental principles is necessary before certain deviations can be appropriately applied.

Information on magnetism and magnetic circuits was discussed in detail in chapter 8 of this manual.

It is the intent of the discussion in this chapter to build upon these basic considerations and develop the subject of magnetic amplification in greater detail.

Magnetic Variables

If a good understanding of magnetic circuits is to be achieved, the magnetic variables involved and their units of measurement must be considered. In the following paragraphs some of these are described.

INDUCTANCE.—Basically, the magnetic amplifier is nothing more than a controlled variable inductance in series with an a-c power source and a load. It may be recalled that any coil possesses a property of electrical inertia called inductance. The inductance of a coil may be increased by providing it with an iron-core. The inductance of coils with magnetic cores may be determined with fair accuracy by the following formula:

$$L = \frac{1.256N^2A\mu \times 10^{-8}}{1}$$

L = Inductance in henrys

N = Number of turns

A = Area of core in square centimeters

$\mu$ = Permeability of magnetic material

1 = Length of core in centimeters

Permeability is a measure of the relative ease with which flux flows in a material. The permeability of air is one (1). The permeability of the core of a coil can be increased if the air is replaced by some type of ferromagnetic material such as iron or steel. By observation of the preceding formula, it can be seen that if the permeability ($\mu$) is increased by 1,000 when the iron core is inserted, the inductance will be increased 1,000 times over the inductance of the same coil with an air core.

Figure 20-1 shows a simple magnetic amplifier in which the iron core is physically moved in and out of the coil to vary the inductance and hence the inductive reactance of the coil. When the iron core is completely within the coil, the inductance of the coil is maximum. Therefore, the voltage drop across the inductance is maximum, reducing the voltage across the load resistor $R_L$ to a low value. However, removing the core from the coil lowers the inductance of the coil (hence the impedance of the circuit), decreasing the voltage across the inductance and increasing the voltage across the load to nearly 115 volts. In effect, this is the operating principle of a magnetic amplifier. Since it requires relatively few watts of power to move the core within the coil, which in turn can control perhaps several horsepower, the device is an amplifier.

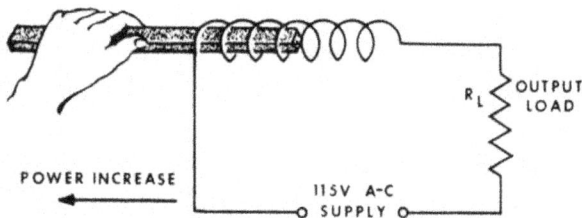

Figure 20-1.—Varying coil inductance.

MAGNETIZATION AND PERMEABILITY CURVES.—Although the arrangement just described is effective, control of the permeability of the coil core is normally obtained by saturating the core with a relatively small amount of d-c or properly phased a-c voltage. An unsaturated core provides a relatively high impedance to the circuit. A saturated core, however, acts similar to an air core, offering practically no impedance except for the low d-c resistance of the coil.

When current flows through a coil, a magnetic force is set up across the coil, the intensity of which depends upon the amount of current flowing through the coil. This force in a magnetic

circuit may be compared to the voltage of an ordinary electric circuit. The magnetomotive force sets up a flux about the coil, which is comparable to the current of an ordinary circuit and has a density which depends on the reluctance of the core of the coil. The core reluctance, which may be compared to the resistance of an electric circuit, has a value which depends on the material forming the core. The reluctance of an air core remains constant regardless of current. This results in an increase of flux density which is proportional to the increase in both current and magnetomotive force.

When a magnetic substance makes up the core of a coil, the reluctance is no longer constant, regardless of current. Instead, as a current begins to flow, the reluctance is very low and the flux is very high compared to that existing in an air-core coil under similar conditions. With an increase in current, the reluctance gradually increases and the rate-of-flux increase is reduced. After the current reaches a certain value, which depends on the core material used, the reluctance increases very rapidly until its value approaches that of air. In this region, around point b in figure 20-2, any further increase in current will produce no appreciable increase in flux. This condition is known as saturation.

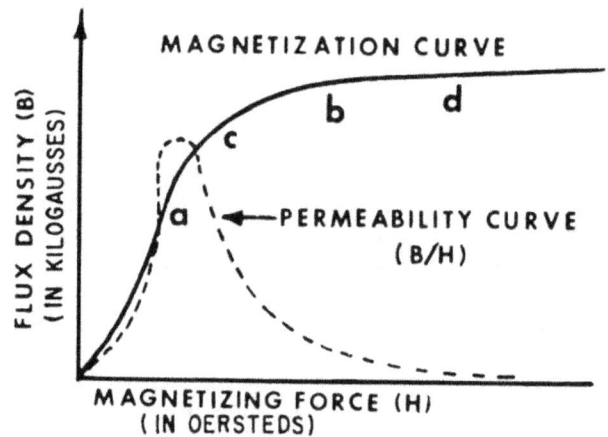

Figure 20-2.—Magnetization and permeability curves.

As previously stated, an iron-core coil that is saturated acts similar to an air-core coil for it offers practically no impedance to changes in current except for the low d-c resistance of the coil. This stems from the fact that the inductance of a coil is proportional to its incremental permeability, which is the ratio of a small change of flux density. $\Delta B$, over a correspondingly small change in magnetizing force or ampere-turns, $\Delta H$. (Incremental $\mu$ = $\Delta B / \Delta H$.)

This can be graphically shown by considering a typical magnetization curve, such as that shown in figure 20-2. The flux density, B, is shown in kilogausses and the magnetizing force, H, is shown in oersteds which are proportional to ampere turns. The inductance of the coil is proportional to the rate of change of flux density, B, with respect to magnetizing force (ampere-turns), H. It will be seen that in the region of the curve around point a, a small change of ampere turns, $\Delta H$, will produce a large change of flux density, $\Delta B$. This represents the unsaturated condition of the coil, and therefore its inductance is large.

In the region of the curve around point b, the same small change of ampere turns, H, will now produce only a very small change in flux density, B. This represents the saturated condition of the coil and its inductance in this region is very low.

An application of the basic saturable reactor may be found in a special type of filter choke that is often used in power supplies where the current drain varies considerably. They are called swinging chokes and are designed to operate on the portion of the curve between point a and b. For low values of current, their inductance values are quite high. For high values of current, their inductance values are low since the core has become saturated.

The airgap in the swinging choke is much smaller than in the ordinary filter choke and, consequently, the total reluctance or opposition to the magnetic lines of force through the core is small. This permits the choke to become saturated for high values of current. This type of choke has special applications in circuits where the direct-current variations are quite large, since it has been found that they give better direct voltage regulation with varying amounts of direct current. That is, the voltage drop across the choke remains nearly constant even though the amount of current varies over a wide range.

The magnetic amplifier works on the principle that varying the average magnetizing force will vary the inductance. Referring again to figure 20-2, the curve from the origin to the

region below point c is quite steep and approximately straight. The inductance in this region is therefore high but constant, that is, given changes of ampere-turns will produce high but constant changes in flux density. Operating the coil in this region will therefore produce little or no change in inductance for changes in ampere turns and no amplifier-action will result.

The region of saturation around point b is relatively flat, so that the inductance around this region is very small and changes very little with changes in ampere-turns. Operation of the coil in this region will also be unsuitable for amplifier action.

In the curved or nonlinear portion of the curve around point c, the inductance changes quite rapidly with small changes of ampere-turns; that is, the rate of change of flux density with respect to ampere-turns is large. This region is the desired operating region of the coil for magnetic amplifier action because small changes of average ampere-turns produce large and significant changes of inductance, therefore producing high gain.

## D-C EXCITED REACTOR

The electrical arrangement of a reactor is shown in figure 20-3 (A). The reactor consists of a laminated magnetic core, a magnetic circuit length 1, a cross-section area A (width w of a leg times the height h of the stack of laminations), and a coil of N turns that is excited by a d-c power source. Core length, width, and height are usually measured in centimeters or inches. Figure 20-3 (B) shows a d-c magnetization curve, the static magnetic characteristics of a core material.

Assuming no previous magnetization and E set so H is at point O, B will then be at 0. Increasing E will apply a magnetizing force H to the core due to current flow in the coil and this will create a magnetic flux density B within the core material. Thus, as H increases to $+ H_m$, B increases to $+ B_m$. From point 0 to (1) (fig. 20-3 (B)) B increases rapidly as H increases, and from point (1) to (2) an increase in H to $+ H_s$ causes only a small increase in B. At point (2) any further increase in H causes no appreciable increase in B, this being the point of saturation $+ B_s$. Then as E is lowered so that H decreases to 0, B will also decrease; however, the retentivity of the core does not allow B to decrease to 0 as H does. As H decreases to 0,

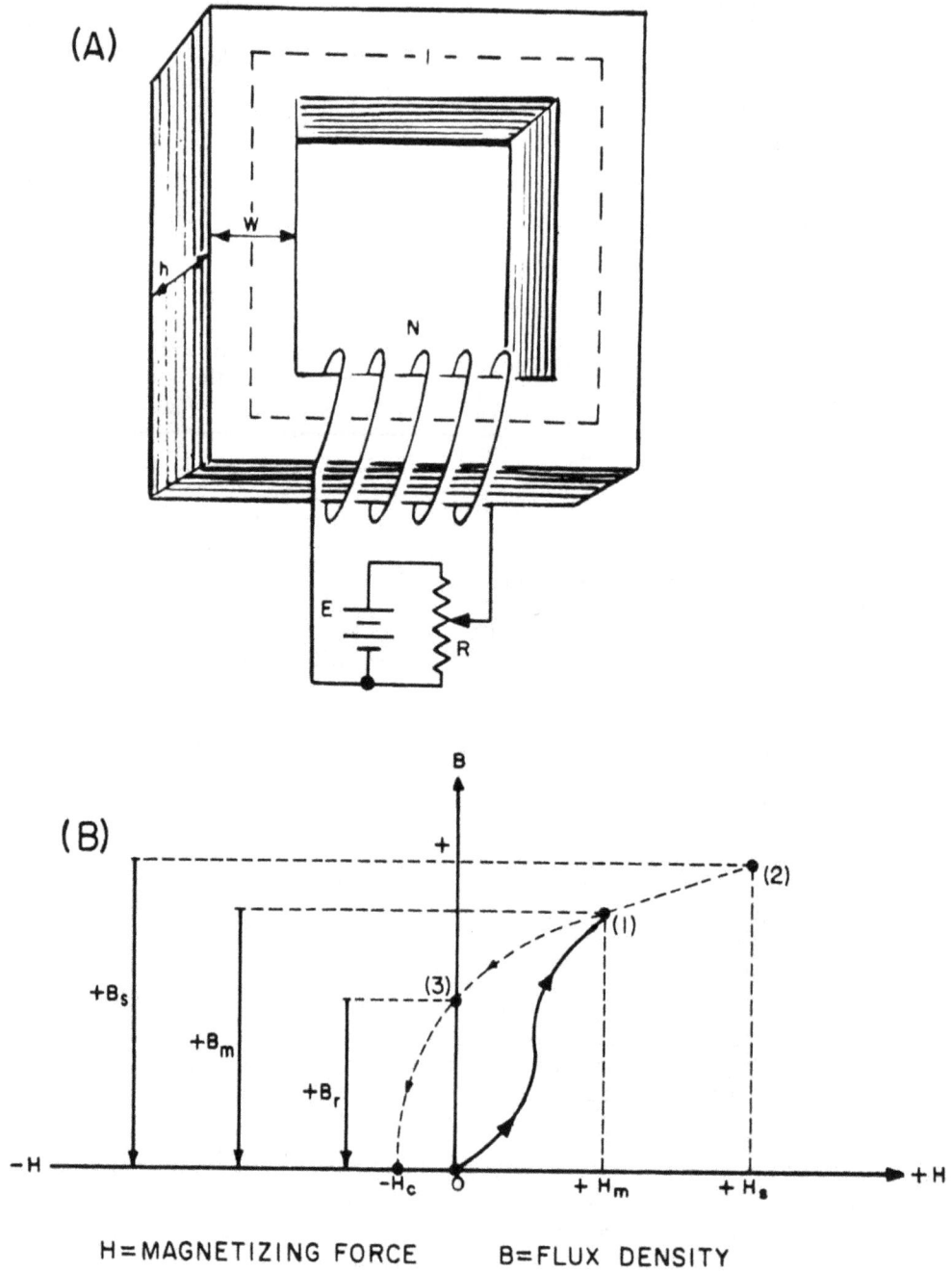

H=MAGNETIZING FORCE          B=FLUX DENSITY

BE. 383

Figure 20-3.—D-c excited reactor and the magnetization curve.

B decreases to point (3) (the remanence level + $B_r$). To reduce B to 0 a coercive (reversed) magnetizing force -$H_c$ would have to be applied.

The 0 to (2) portion of the curve in figure 20-3 (B) is known as the d-c magnetization curve. It shows the static B-H relationships for a unit length and unit cross sectional area of the core. It does not show the total flux $\phi$ established, nor the magnetomotive force F applied.

## A-C EXCITED REACTOR

The reactor of figure 20-3 is d-c excited; and unless the polarity of E can be reversed, H can be applied in only one direction and flux established in only one direction. Both are shown as positive in figure 20-3.

A curve in both the positive and negative directions will be generated when the reactor is excited by a.c., as in figure 20-4 (A); and a B-H or hysteresis loop for the material will be needed to show the B-H relationship. Figure 20-4 (B) shows a plot of E, B, H, and I for the circuit, with proper phase relationships if the reactor were operated as a linear reactor having only inductance and no resistance. This would mean that the B-H loop would have to be symmetrically located along the dashed line between points 1-4 and that no degree of core saturation was reached—an impossible condition because of material nonlinearity and a coil with resistance.

If the core flux density is zero and E is applied at the 0° point of its cycle, the first quarter cycle of E will cause a current to flow in N and apply an H, establishing B along the dotted line between points 0-1. This portion, which is the same as for the d-c excited reactor, is the magnetization curve. The next quarter cycle will reduce B to point 2. During the next half cycle, B will move from point 2 to points 3, 4, and 5. The flux density will be at point 5 at the beginning of the next cycle of H. After a few cycles, the dynamic B-H loop will be established. It will be somewhat wider than if established with d-c excitation.

Since the assumed circuit is entirely inductive in nature, examine its operation during a later cycle of E, beginning at point 4 of the loop. I, H, and B will be found to lag E by 90°. The flux in the core will always assume a shape, dependent on the core's characteristics, that will tend to cause the self-induced voltage of the coil to be of the same shape as E. H and I are not always the same shape; but in this case they are, because it is assumed that the B-H loop is not as shown, but lies along points 1-4, that it is symmetrical, and that the same degree of core saturation exists during the entire cycle of H.

Actually, for this B-H loop and operation, somewhat less than saturation is shown in figure 20-4 (B). I and H will be a peaked sort of sine wave; but B will remain a good sine wave. B, H, and I will lag E by less than 90° because of resistive, eddy current, and hysteresis losses.

## EFFECTS OF SPECIAL CORE

It has already been mentioned that cores for magnetic amplifiers should have special characteristics. These special characteristics are best identified in terms of the B-H loop of the material. It is desirable to use a material having a rectangular B-H loop, as shown in figures 20-4 (C) and 20-5 (A). However, the degree of rectangularity illustrated cannot actually be achieved; even with the best materials the loop will be tilted slightly clockwise. If such a core is used in figure 20-4 (A), the important thing to note from figure 20-4 (C) is that H and I are square waves so long as core saturation is not reached; B and E are sine waves; and if the load presented to E is purely inductive, B, H, and I lag E by 90°. But even though H and I are nearly square for high quality cores until saturation is reached, the magnetic amplifier circuits are mostly resistive after saturation; and the variables are nearly in phase with each other.

Various types of nickel-iron alloys that have more suitable magnetic properties than previous ones for use as core materials for saturable reactors have been developed and are commercially available. These materials are the (1) high permeability alloys and (2) grain-oriented alloys.

High-permeability materials, such as Permalloy A, Mumetal, 1040 alloy and equivalents have low and intermediate values of saturation flux density but relatively narrow and steep hysteresis loops. These materials are used extensively as the cores in low-level input amplifier stages.

Grain-oriented materials, such as Orthonol, Deltamax, Hypernik V, Orthonik, Permeron, and equivalents, have higher values of saturation flux density and more rectangular-shaped hysteresis loops (fig. 20-5 (A)) than the high-permeability materials. Grain-oriented materials are referred to as square-loop materials because of the flat top and bottom of the hysteresis loop. A conventional loop is shown in figure 20-5 (B). These grain-oriented materials are used as the cores in high-level output amplifier stages in which maximum permeability occurs close to

(A)

(B)

(C)

BE. 384

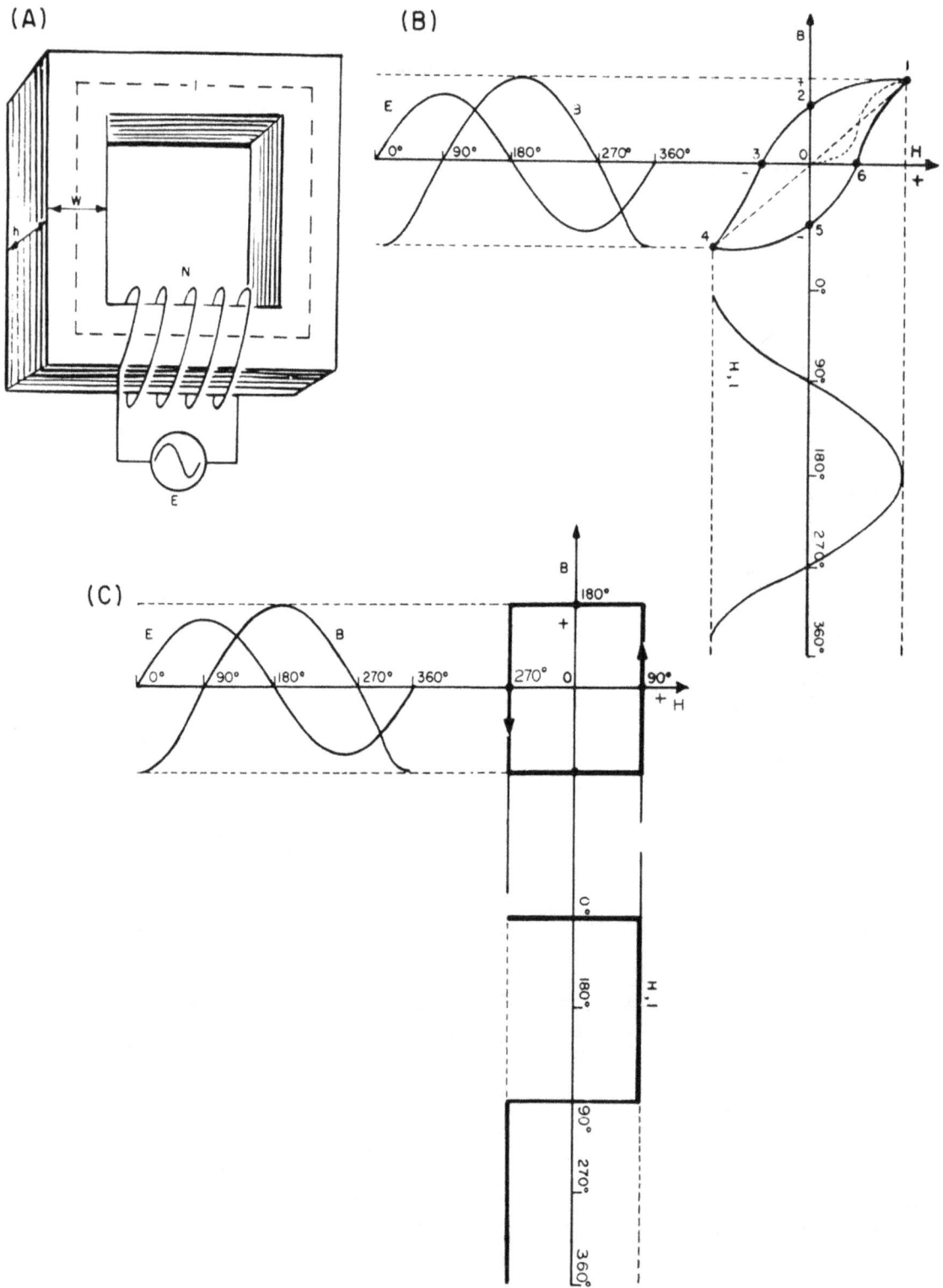

Figure 20-4.—A-c excited reactor, B-H loop, and waveforms.

saturation flux density, resulting in a substantial increase in the power-handling capacity for a given weight of core material.

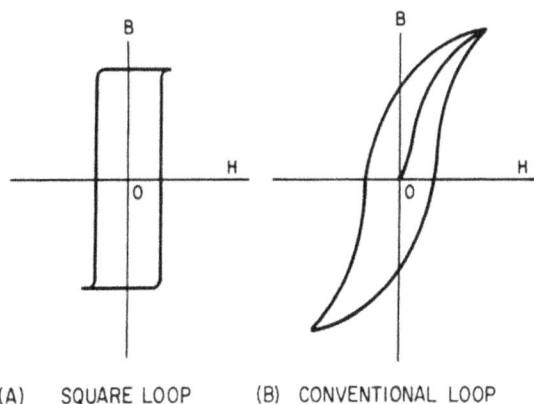

(A)  SQUARE LOOP      (B)  CONVENTIONAL LOOP

BE.385

Figure 20-5.—Hysteresis loops.

In the manufacture of saturable cores, the characteristics of the grain structure of the material can be altered considerably by rolling and annealing processes.

A great improvement in the magnetic properties of some materials is obtained by cold-rolling the material before it is annealed. The cold-rolling process develops an orientation of the grain in the direction of rolling. If a magnetizing force is applied to the material so that the flux is in the direction of the grain, a rectangular hysteresis loop is obtained. Thus, in some materials cold-rolling produces almost infinite permeability up to the knee and almost complete saturation beyond the knee.

## TRANSFORMER PRINCIPLES

There is a wide difference between the operation of a transformer and the operation of a magnetic amplifier. The transformer, except in special purpose applications, is operated as nearly linearly as possible; it is required to reproduce the voltage waveshapes faithfully. Normally, it uses a magnetic core at the low frequencies to provide a good magnetic flux path and produce a high inductance with few turns, and to couple the windings closely for high efficiency. The magnetic amplifier also has a magnetic core to provide a good magnetic flux path and allow the core to be saturated easily, but it does not faithfully reproduce the wave-

forms of the power source. Until saturation sets in, however, magnetic amplifiers, transformers, and linear reactors have much in common, and a refresher in the fundamental principles of transformers should make the operation of magnetic amplifiers easier to understand.

The transformer shown in figure 20-6 (A) has a common type construction for small, two-winding power and audio transformers. However, assume that this transformer has a core of high grade material—a rectangular B-H loop which is slightly tilted—and that E is not sufficient in amplitude to cause core saturation. With switch S1 open, apply E to the primary winding terminals 1 and 2. This will cause a small current, known as the magnetizing current $I_m$, to flow in the primary. (See fig. 20-6 (B).) The current $I_m$ establishes a flux, labeled $\phi_p$, in the core. It divides and circulates throughout the legs uniformly, creating the same flux density in all parts of the core. The flux induces voltages in the same direction in both windings.

The windings are marked with dots to indicate relative polarities between the two windings. The primary and secondary windings are wound on the same leg, in the same direction; and the induced voltages are positive at the upper ends of both windings during the positive half cycle of $E_p$. The primary induced voltage opposes rather than aids the supply voltage E. It is usually stated that the induced voltage is equal and opposite to the applied voltage. This is a true statement in a sense, but it would be more accurate to state that it is equal and in opposition to the applied voltage. To oppose completely the applied terminal voltage, it would have to be equal in amplitude and phase at every point of the applied voltage. If that were the case, no magnetizing current would flow.

The primary induced voltage can never quite equal the terminal voltage because of the resistive drop and the necessity of keeping the flux established. Therefore, the exciting current is automatically regulated to a very low value, usually less than 10 percent of the secondary load current. The induced voltage of the secondary is also positive for the reasons mentioned above (wound in the same direction, located on the same core leg, and cut in the same direction by the flux), and for other reasons. If the primary and secondary are wound in the same direction but located on different core legs, the relative polarity of the induced

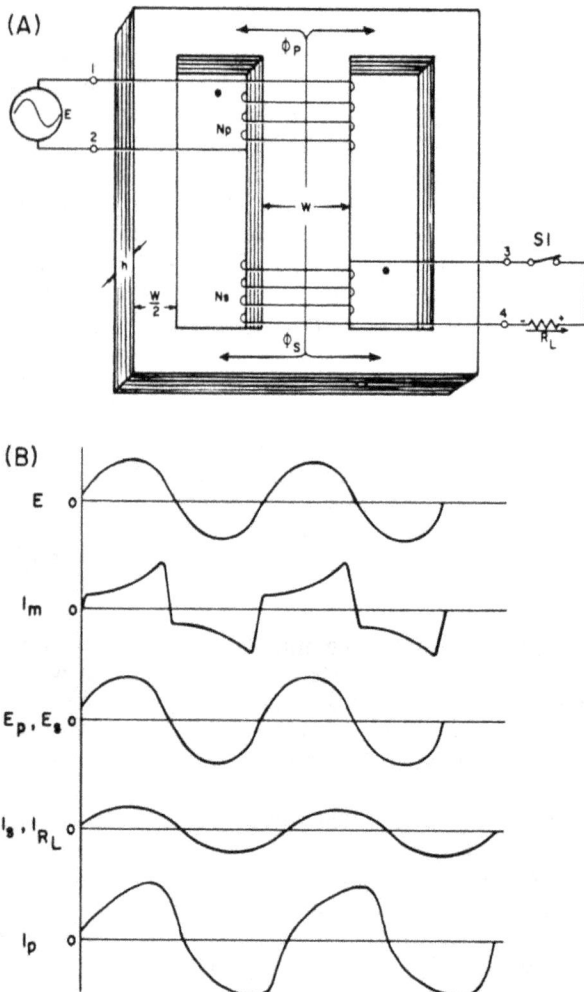

BE.386

Figure 20-6.—An ordinary transformer and waveforms.

voltages will be opposite. The ratio of the primary to secondary voltage amplitudes is dependent on the turns-ratio $N_p/N_s$:

$$\frac{E_p}{E_s} = \frac{N_p}{N_s}$$

If S1 is closed, (fig. 20-6 (A)) allowing a current to flow in the secondary winding and $R_L$, the action when power is transferred from the primary to secondary circuits is as follows: Since the secondary voltage is positive at terminal 3 during the positive half cycle of E and is the source voltage for the load, current will flow

upward through $R_L$, as shown in figure 20-6 (A), during the positive half cycle of the source voltage E. This load current through the secondary will set up a flux $\phi_s$, which is in the opposite direction to the flux created by the primary exciting current $I_m$. This tends to cancel the induced voltages of the primary and secondary, but in doing so it causes an increase in the primary winding current. Thus, the load current is reflected back to the primary winding due to the coupling of the windings and the opposition that the secondary flux provides. This increase in $I_p$ is dependent on the value of the secondary current $I_s$ and the turns-ratio between the secondary and primary:

$$I_p N_p = I_s N_s$$

This is known as the law of equal ampere-turns; it will always hold true. Figure 20-6 (B) shows the waveforms. Note that $I_p$ is the algebraic sum of $I_m$ and $I_{R_L}$.

While a great deal more could be said on transformer principles, the preceding points provide some information needed for understanding the operation of magnetic amplifiers.

## FLUX ACTION IN CORE MATERIAL

The real key to understanding magnetic amplifier operation lies in a knowledge of the flux action in the core materials used in the magnetic amplifiers. Just as an understanding of vacuum tube amplifiers requires analysis using tube curves, understanding magnetic amplifiers requires reference to the B-H curves for the core material. Since the core materials normally used for magnetic amplifiers differ considerably from those normally used in linear reactors and transformers, a closer examination of these materials and an introduction to the phenomenon of saturation should be valuable.

Assuming that the reactor of figure 20-7 (A) has a core material possessing a B-H loop characteristic as in figure 20-7 (B), the supply voltage will cause the flux to vary over certain limits which depend on the amplitude and frequency of E, the cross section area A and the characteristics of the material, and the number of turns $N_L$. The fundamental equation of transformers relates these variables and states

$$E_{rms} = 4.44 \ NAfB_s \ \times \ 10^{-8}$$

where $B_s$ is the flux density, $+B_s$ or $-B_s$, required for saturation of a given material grade or type.

401

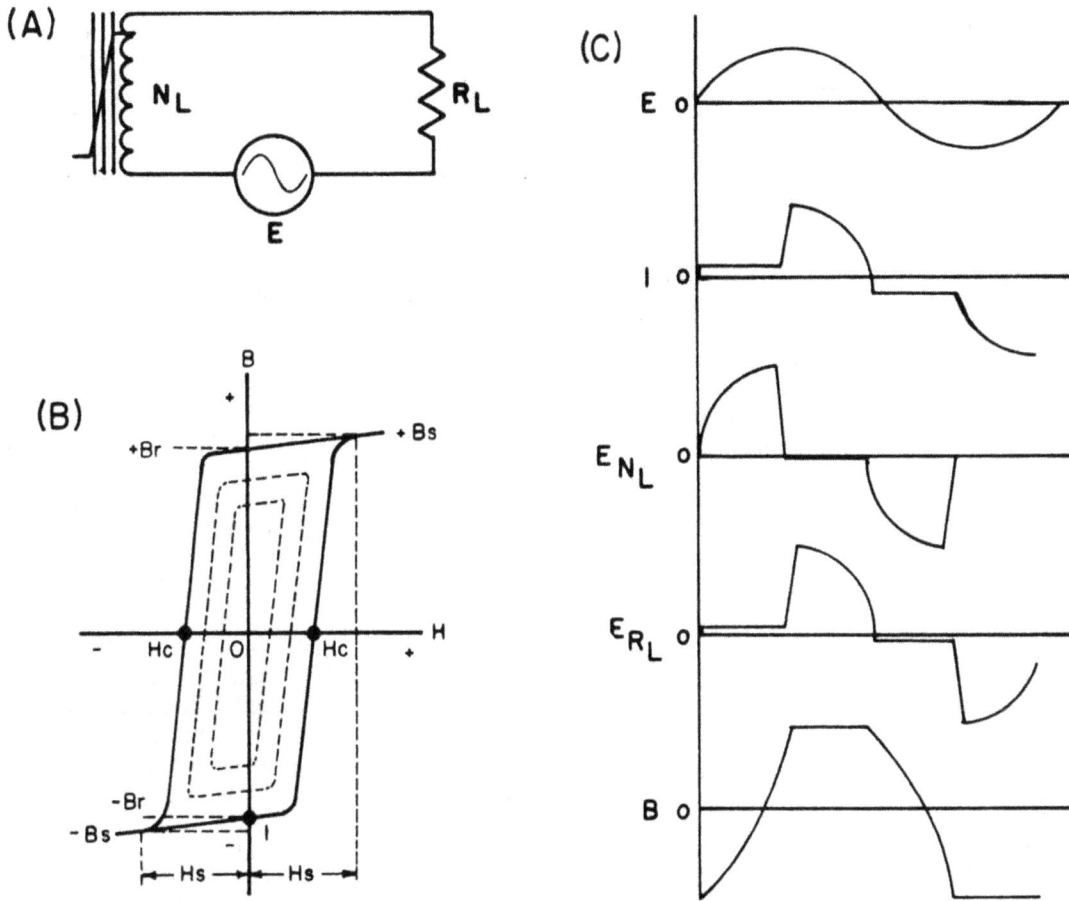

Figure 20-7.—A saturating reactor, B-H loop, and waveforms.

BE. 387

The flux density can be varied by varying E. If E is a smaller amount than that required to satisfy the previous equation, the flux density during a cycle of E will vary over smaller limits. The two inner dotted B-H loops (fig. 20-7 (B)) depict the operation for two reduced values of E. The material does not reach saturation, and a very low value sine wave of voltage is developed across $R_L$ because only a small value of magnetizing current $I_m$ flows in the circuit. The proper value of E will just barely cause core saturation.

Increasing E above the rms value calculated for core saturation will materially change the operation of the circuit from that of a reactor where a reduced E is applied to a very nonlinear reactor.

Assume that several cycles of E have elapsed and that the flux density is at $-B_r$ or point 1 when E begins a new cycle. When sufficient magnetizing force is applied, flux density begins to change, first decreasing toward 0 and then increasing toward $+B_s$. As the flux changes between the two levels of $B_r$, the impedance of $N_L$ is very high because the permeability of the material in this region is very great—as high as a million—and the impedance of $N_L$ is increased by this value of $\mu$ over what it would be for the same coil with an air core, which has a $\mu$ value of unity. Since only a small value of magnetizing current flows in the circuit during this time, the high impedance of $N_L$ will drop nearly all of the circuit voltage. At a flux density level of $+B_s$, the core saturates; effective permeability becomes very low; the impedance of

402

$N_L$ is very low; and most of the circuit voltage is dropped across $R_L$. Since the levels of $B_S$ and $B_r$ are almost the same for the material depicted, the core is essentially saturated for the rest of this half cycle of E. Note the waveshape of the voltages across $N_L$ and $R_L$ (fig. 20-7 (C)). As sufficient negative magnetizing force is applied, the flux density moves between + $B_r$ and - $B_r$; and $N_L$ regains its high impedance properties. At some point in the negative half cycle, saturation is again reached at -$B_S$; and $R_L$ drops nearly all of the circuit voltage. The flux density level is at -$B_r$ at the beginning of the next half cycle.

Several important points should be apparent from the circuit of figure 20-7 and its operation. A core material with a rectangular B-H loop is highly desirable for high gain magnetic amplifiers because little magnetizing force is required to produce saturation abruptly. The core, when fully saturated, has a very low permeability; but its permeability is extremely high when it is not saturated. The permeability of a material is the slope of a line tangent to a point on the B-H loop and varies from a very high value along the vertical portion to a low value at saturation. Also, $N_L$ is a nonlinear impedance which, for all practical purposes, can be neglected except when the flux density is changing levels; a stationary coil on a stationary core must be cut by changing flux level to possess the property of inductance.

Desirable Core Properties

The properties desirable in a core material are a rectangular B-H loop, minimum width, and a high level of maximum flux density. The high degree of rectangularity is needed to provide abrupt changes from unsaturated to saturated states, and vice versa, while steep sides represent high permeability; a narrow width for the B-H loop represents high control sensitivity, and low hysteresis and eddy current losses; and high maximum flux density represents good power handling capabilities. In other words, the B-H loop of the ideal material would be a long vertical line which changes abruptly to a horizontal line at saturation.

Core Shapes

Cores for high-quality, low-powered magnetic amplifiers are usually toroidal, ring, or U shaped. The E-I and C cores are limited to high-powered amplifiers, but may find application if economy is a more important factor than high gain. (See fig. 20-8.)

Because of reduced airgaps, the toroidal and ring cores have the best magnetic properties of the group. The ring core stack is made up of individual washers punched from sheets of core material. The toroidal, a strip of tape from a sheet, is wound into a doughnut-shaped core. A C-shaped core is wound from a strip of tape into an oblong loop form and then sawed into two pieces. The joint faces are precision ground to reduce the amount of the airgap when they are held together with a high tension core band. The U and E-I cores are made by stacking and interleaving lamination pieces. The U cores in small magnetic amplifiers are generally of a nondirectional material. The effective airgap is small because the end legs are twice the width of the side legs and the flux flows from lamination to lamination at the corners; however, the cross sectional areas of the end legs and side legs are the same. For this reason, and because of economy of coil construction, the U core finds considerable low-level application. The E-I core has the largest airgap. Generally it is used only in ordinary transformers and in large low gain amplifiers. Any small airgap has the effect of materially reducing the permeability—tilting the B-H loop.

All the cores depend upon deposited or residual oxides to insulate the washers, turns or tape, or laminations from one another in order to reduce eddy currents. Eddy currents effectively increase the width of the B-H loop. Thin tape strips or laminations also reduce eddy currents. Since materials thinner than about 0.002 inch have strains introduced during cutting and punching, cores are generally made from 0.002- to 0.014-inch material. Because of eddy current losses, thinner material is needed for 400- and 800-Hz amplifiers than for 60- Hz amplifiers.

BASIC MAGNETIC AMPLIFIERS

The basic circuit for controlling the load power by means of a control winding is illustrated in figure 20-9. Both windings are on a common core; the primary is usually referred to as the control winding.

The primary winding is supplied with a d-c source for control purposes. A potentiometer is connected in series with this circuit and is used to control the amount of current flow. This

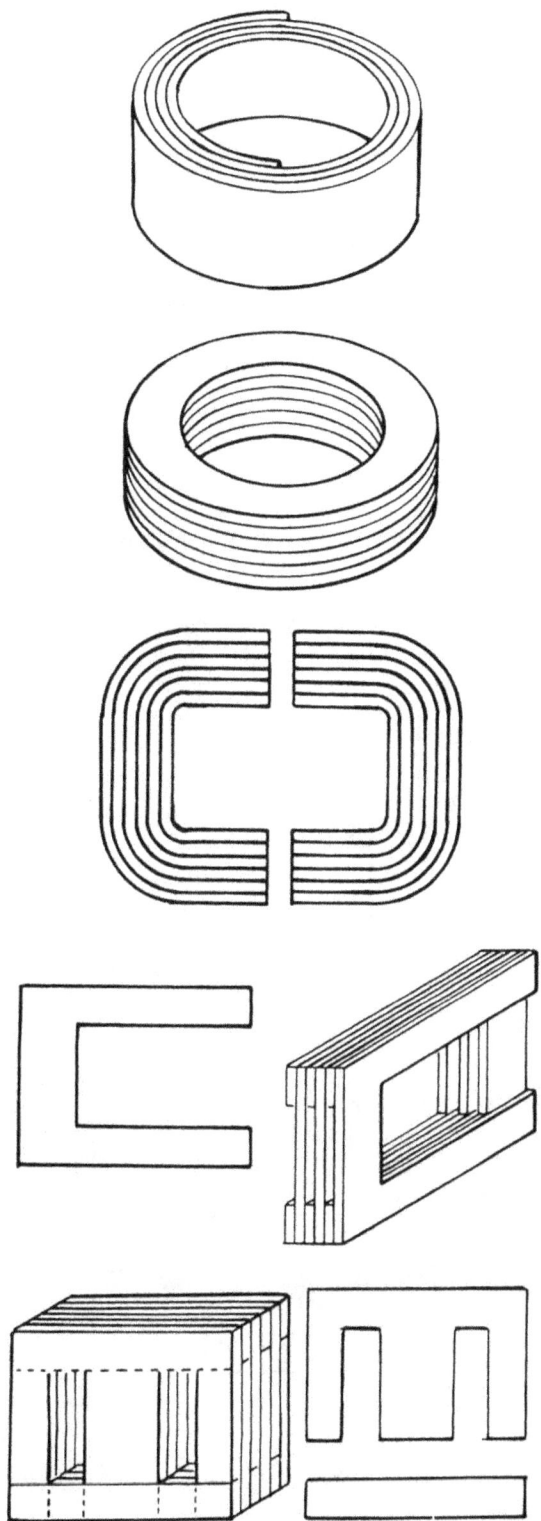

Figure 20-8.—Some typical core shapes.

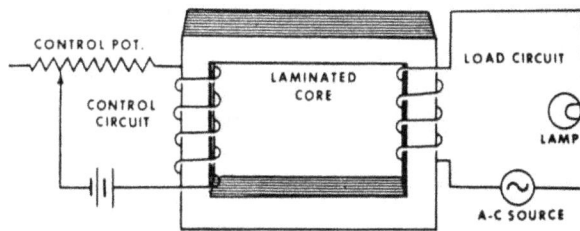

BE. 389

Figure 20-9.—A basic magnetic amplifier.

current flow will establish a unidirectional flux in the core at a level depending upon the amount of d-c magnetizing force. The second winding is connected in series with an a-c source and the load, which is shown as a lamp.

Changes of current through the control winding correspond to variations in magnetic field intensity (ampere-turns). Referring to figure 20-2, these variations over the portion of the curve between the origin and point a will produce little or no change in inductance or permeability due to the nearly straight characteristic of the curve. In order for these changes in current to produce changes in inductance, they must take place on the curved portion, or knee of the curve. This means that sufficient current must flow through the control winding to place the operating point at the knee, such as at point c. This is very similar to setting the operating point of a vacuum tube by means of bias.

With the coil operating on the knee of the magnetization curve, a small increase of current through the control winding will lower the inductance of the coil, or permeability of the core, because the magnetization curve's direction or slope is moving toward a horizontal direction. As a result, the a-c reactance of the load winding is decreased. The decreased impedance of the load circuit will then allow a larger current to flow, thereby increasing the power developed across the load. With the core completely saturated, the load winding reactance drops to nearly zero, leaving only the resistance of the copper wire to impede the flow of current. Under this condition, the maximum supply voltage will be applied to the load; thus the lamp will be brightest. Conversely, decreasing the current through the control winding will increase the reactance of the load coil, thereby decreasing the power developed across the load.

In the region between the origin and point a on the curve in figure 20-2, the core permeability (coil inductance) is maximum, resulting in minimum load current and, minimum output power. Since the reactance of the load coil winding has been changed by varying the degree of core saturation, the device is referred to as a saturable reactor. The following rule will apply to any saturable reactor or magnetic amplifier: a control current which tends to saturate the core will increase the output of the amplifier.

## IMPROVED MAGNETIC AMPLIFIER

The basic magnetic amplifier circuit shown in figure 20-9 is very inefficient because of transformer action between the two windings. The alternating flux from the secondary will induce a voltage into the primary control winding. If the control winding has a large number of turns, the voltage may become excessive and may even break down the insulation.

Another reason why this form of magnetic amplifier is seldom used is the fact that the control winding would act as a short-circuited secondary. This would dissipate a considerable amount of a-c energy that is applied to the control winding. It is possible to prevent this loss by inserting an isolating impedance, in the form of an inductance, in series with the control winding. Another method of preventing this loss is by utilizing a three-legged core. This method is described in detail.

## THREE-LEGGED CORE

The transformer action previously mentioned would also cause an oscillating flux in the core and the usual B-H hysteresis loop of the core material would be traced out during one full cycle of the a-c source voltage as in transformer operation. In order for this circuit to function as an amplifier, sufficient d-c magnetizing force is required to counteract the a-c flux, and an additional amount of d-c magnetizing force is needed to set the operating point.

A much more satisfactory arrangement for a saturable reactor is shown in figure 20-10. This figure shows a three-legged core having an a-c winding on each outer leg connected in series and the d-c control winding on the center leg. In the center leg of a saturable reactor of this type, any alternating flux produced by one of the a-c coils is balanced by the flux produced by the other a-c coil. This is true only when both a-c coils have an equal number of turns and are connected either in series or parallel such that

the flux fields produced by the two coils oppose each other through the center leg. This is shown by the dotted lines in figure 20-10.

Since these forces are opposite and equal, they do not pass through the center leg but join in a common path through the outer legs of the core. This is shown by the broken lines in figure 20-10. The figure also shows that the current through the d-c winding on the center leg of the core produces a magnetic flux as indicated by the arrows on the solid lines. The flux lines are caused by the direct current flow in two paths, as shown, and magnetize the entire core. Although the d-c coil is used to magnetize the cores of the a-c coils, no a-c voltage is induced into the control coil when a three-legged reactor is used.

During operation, variation of current in the control coil will vary in magnetization of the core. As has been shown previously, when the magnetizing force is altered, the permeability of the core is varied and, in turn, the inductance of the a-c coils is also varied. In other respects the three-legged magnetic amplifier functions similar to the basic circuit.

As mentioned previously, the a-c coils can be connected either in series or parallel. The series-connected a-c coils offer the advantages of higher voltage gain and faster response

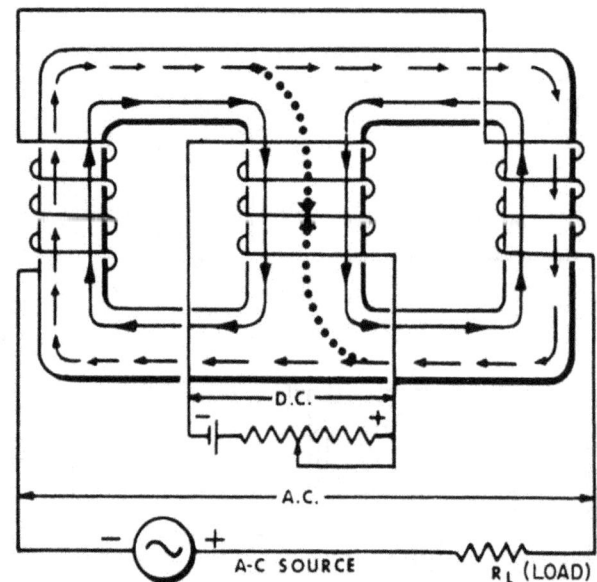

BE. 390

Figure 20-10.—A magnetic amplifier with a three-legged core.

characteristics. The parallel-connected a-c coils are capable of handling loads with medium to high current consumption as may be required by large servomotors. The time of response is longer for the parallel windings; however, this is discussed later in the chapter.

## HALF-WAVE RECTIFIER

Another improvement to the basic magnetic amplifier is the use of half-wave rectifiers. A description of a simple half-wave circuit (fig. 20-11) will be given as an example of the operating principles of the magnetic amplifier.

In the example of figure 20-11 (A), the solid arrow at the source indicates the direction of current flow through the circuit during the positive half cycles of applied voltage, $e_{ac}$. The polarity markings (dots at one end of the windings) are indicative of the way the turns are wound on the core. The dotted ends of the windings on a core are assumed to always have a particular instantaneous polarity with respect to the undotted ends of the windings. Also, the dotted ends of two or more windings on a common core are considered to have the same instantaneous polarity with respect to each other. For example, in figure 20-11 (A), if the voltage applied to the control winding is of a polarity at some instant to cause current to flow INTO the dot-marked end of that winding, the induced voltage of the other winding will be of a polarity (at the same instant) such as to cause current to flow OUT of the dot-marked end of that winding. The control voltage, $e_c$, is assumed to be a direct voltage. The rectifier arrow-heads are pointed against the direction of electron flow.

## FUNCTION OF RECTIFIERS

Rectifiers are placed in the load and control circuits to prohibit current flow in the control circuit during the gating half cycle and in the load, or gating circuit during the reset half cycle. The magnetic amplifier is not an amplifier in the sense of a step-up transformer. Voltages generated by mutual induction (transformer action) between the control and load windings exist in these windings, but they have only a small effect on the amplifier operation under the established conditions.

During the first half cycle (solid arrow of source) of the applied voltage, the direction of the INDUCED voltages in the load windings is negative at the polarity-marked terminals.

This action is in the forward direction of the rectifier and against $e_{ac}$. Thus, the rectifier in the load circuit prevents current flow through the load and is subjected to an inverse voltage equal to the difference between $e_{ac}$ and the mutually induced voltage in the load winding. The time interval, corresponding to the first half cycle, is called the "reset" half cycle. The reset action is described later.

Analysis With Zero D-C
Control Voltage

As mentioned previously, figure 20-11 is used in the analysis of the action of the basic half-wave magnetic amplifier. Figure 20-11 (A) represents the basic circuit. Figure 20-11 (B) represents the square-type hysteresis loop for the core material used in this circuit. Figure 20-11 (C), (D), and (E), represents the waveforms of current and voltage for three conditions to be considered. The symbol representing a quantity is common to all parts of the figure. For example, the magnetizing current, $I_m$, is represented in figure 20-11 (B), (C), and (E). The hysteresis loop is enlarged for clarity and is not drawn to the same scale as parts (C), (D), and (E).

RESET HALF CYCLE.—The first condition to be described is with the control voltage, $e_c$, at zero. At the beginning of the reset half cycle, the core is assumed to possess a residual or negative saturation remanent flux level, $\Phi_1$ (fig. 20-11 (B)). The direction of this flux is indicated by the arrow, $\Phi_1$, in figure 20-11 (A). As $e_{ac}$ increases from 0 in a positive direction (indicated by the solid arrow at the source, figure 20-11 (A) and by the part of the sine curve, point 1 to point 2 (fig. 20-11 (C)), the current in the control winding establishes an m.m.f. represented by half the width of the hysteresis loop, $+I_m$ (fig. 20-11 (B)). The applied voltage establishes an m.m.f. that acts in a direction to oppose the residual core flux, $\Phi_1$, and therefore to demagnetize the core. The amount of change of flux will depend upon the MAGNITUDE of the applied voltage across the control winding and the TIME INTERVAL during which this voltage is applied.

In this example, the first half cycle of applied voltage is assumed to reverse the core magnetism and to establish its flux density at essentially the positive saturation level, $\Phi_2$ (fig. 20-11 (B)). This action is called reset.

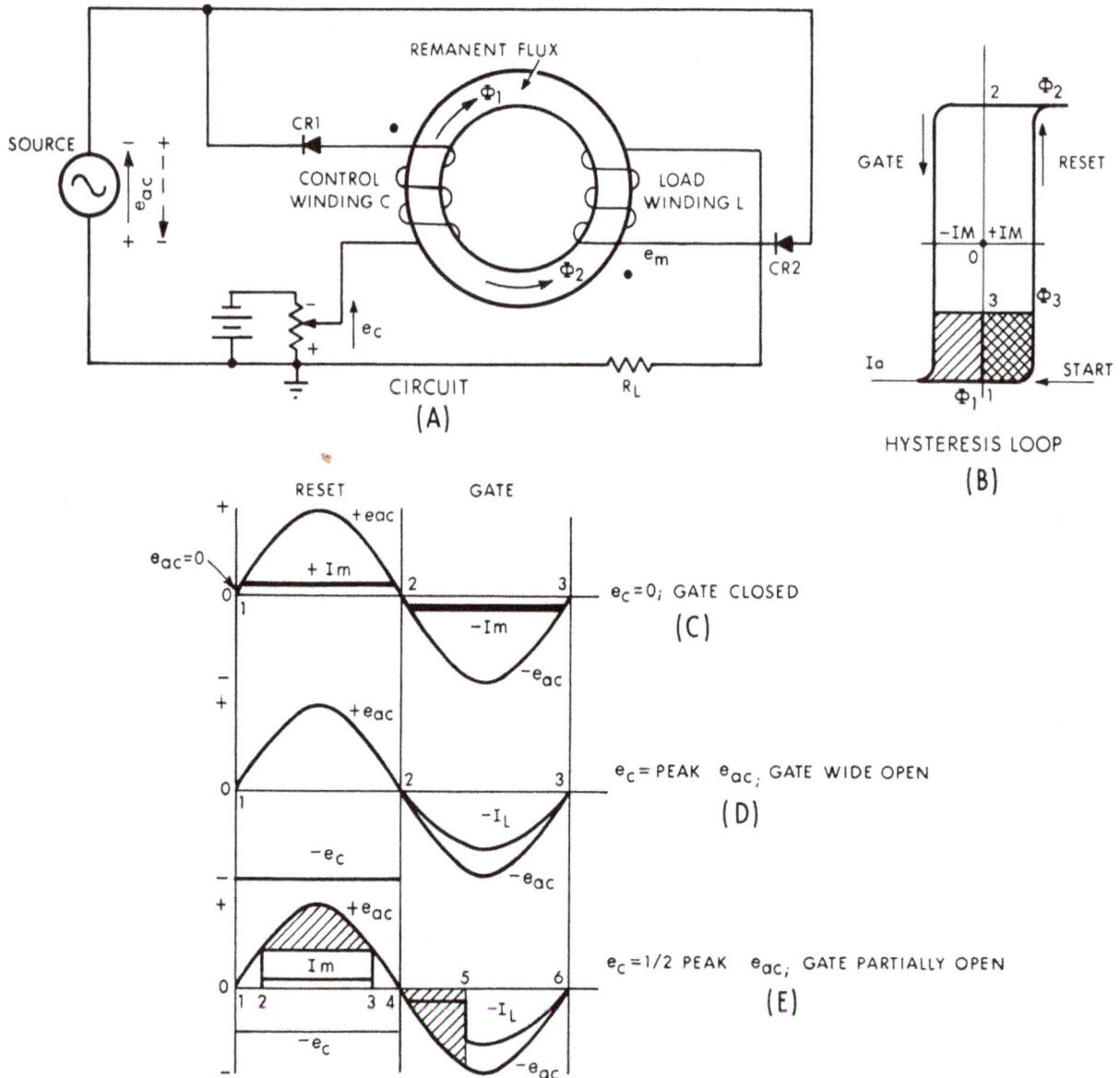

Figure 20-11.—Basic half-wave magnetic amplifier.

BE. 391

As $e_{ac}$ increases from 0 in a positive direction in the vicinity of point 1 (fig. 20-11 (C)), there is no change in core flux until the current increases to the value of $+I_m$, corresponding to one-half of the width of the hysteresis loop. Thus, with no flux change, the current rises abruptly and is limited only by the resistance of the circuit and the low value of $e_{ac}$. When the current reaches the value, $+I_m$, the core flux starts to change from the $\Phi_1$ level toward the $\Phi_2$ level (fig. 20-11 (B)). The accompanying self-induced voltage opposes $e_{ac}$ and limits the current to a constant small value during the flux variation from the level $\Phi_1$ to $\Phi_2$.

This flux change continues during the time interval between point 1 and point 2 (fig. 20-11 (C)). As it continues, the induced voltage continues to vary in magnitude with $e_{ac}$ and to

407

oppose $e_{ac}$ in such a manner that $I_m$ remains constant over the half-cycle interval.

GATING HALF CYCLE.—The next half cycle is called the "gating" half cycle. It starts at point 2 (fig. 20-11 (C)), at which time the polarity of the applied voltage reverses. The direction of current for this half cycle is indicated by the dotted arrow (fig. 20-11 (A)). During this time interval, the rectifier in the control circuit blocks the flow of control circuit current. However, the rectifier in the load circuit permits current from the source to flow in that circuit. This current will magnetize the core in a negative direction—that is, in a direction to change the flux from the $\Phi_2$ level to the $\Phi_1$ level. The applied voltage is assumed to be of the correct magnitude to cause the core to be magnetized to the $\Phi_1$ level (fig. 20-11 (B)). A condition of equilibrium is indicated.

The large flux change from the $\Phi_2$ level to the $\Phi_1$ level causes a self-induced voltage in the load winding, L, and a mutually induced voltage in the control winding, C. The self-induced voltage in the load winding opposes $e_{ac}$. The mutually induced voltage in the control winding also opposes $e_{ac}$. The rectifier in the control circuit is subject to an inverse voltage equal to the difference between $e_{ac}$ and the mutually induced voltage in the control winding. Because of the maximum flux change in the core, maximum impedance is presented by the load winding to the circuit containing $R_L$ throughout the gating half cycle, and therefore $e_{ac}$ will appear across the load winding and not across $R_L$. For this condition, the current through the load is limited to a very small magnetizing component that is negligible compared to normal values of load current. The gate is closed.

Analysis With Maximum
D-C Control Voltage

The second condition described is for the condition that $e_c$ is equal to the peak value of $e_{ac}$. At point 1 (fig. 20-11 (D)) the remanent magnetism is again at the $\Phi_1$ level (fig. 20-11 (B)).

FIRST HALF CYCLE.—The applied voltage, $e_{ac}$, rises from 0 to maximum during the first 90° of the cycle, but has no effect on the core flux because the control voltage, $e_c$, has a magnitude equal to the maximum value of $e_{ac}$, and the polarity opposite to that of $e_{ac}$. Thus, the

rectifier prevents the flow of battery control current during the time that $e_c$ is greater than $e_{ac}$. Because there is no voltage across the control winding (the rectifier is essentially an open circuit), no flux change can occur from point 1 to point 2 (fig. 20-11 (D)). Figure 20-11 (B), does not apply. Thus, no change in flux occurs during the reset half cycle for this assumed condition.

SECOND HALF CYCLE.—When $e_{ac}$ reverses its polarity (solid to dotted arrow), the rectifier in the control circuit continues to block the flow of current in that circuit. In the load circuit the polarity of $e_{ac}$ during the gating interval, points 2 to 3 (fig. 20-11 (D)), is such as to tend to drive the core further into negative saturation (point Ia, fig. 20-11 (B)). Because the core is already saturated, no further flux change occurs, and $e_{ac}$ appears across $R_L$ because the load winding offers no impedance to $I_L$.

The full value of load current flows, and its magnitude is $e_{ac}/R_L$. The gate is wide open. The waveform of this current. —$I_L$ is illustrated in figure 20-11 (D).

Analysis With Partial
D-C Control Voltage

The third condition assumes that $e_c$ is approximately half peak value of $e_{ac}$. During the reset half cycle, voltage is applied to the control winding during the time interval from point 2 to point 3 (fig. 20-11 (E)). The magnitude of this voltage is $e_{ac}-e_c$. This voltage will be less than the peak value of $e_{ac}$, but greater than zero; and a new set of conditions will be established.

FIRST HALF CYCLE.—The reset cycle is just beginning. During the interval 1 to 2 (fig. 20-11 (E)), $e_c$ is greater than $e_{ac}$, and the rectifier opposes any current flow in the control winding. During the interval 2 to 3, $e_{ac}$ exceeds $e_c$, and magnetizing current flows in the control winding. As mentioned previously, the extent of the change in core flux will depend on the time interval and magnitude of the voltage applied across the control winding within the half cycle. Because the time interval is very short, and the net voltage applied to the control winding is much less than $e_{ac}$, the core flux level is assumed to change from $\Phi_1$ to the level along the line through $\Phi_3$ (fig. 20-11 (B)). During the interval 3 to 4 (fig. 20-11 (E)), $e_{ac}$ is again less than $e_c$, and the rectifier prevents any further flow of control current. As in the previous examples, the rectifier in the load circuit prevents any current flow in that circuit during the reset half cycle.

SECOND HALF CYCLE.—When $e_{ac}$ reverses, magnetizing current flowing through the load winding changes the core flux from the level through point 3 to the level through point 1 (fig. 20-11 (B)). This change is assumed to take place during the interval 4 to 5 (fig. 20-11 (E)). The impedance of the load winding is high during this interval, and current flow through the load is restricted to the magnetizing current. However, at point 5 the core becomes saturated, and no further flux change occurs. The impedance of the core drops to zero, and current $e_{ac}/R_L$ flows through the load during the interval 5 to 6. The load voltage is in phase with $e_{ac}$ and has the same waveform as that of $e_{ac}$ for this part of the cycle.

## EFFECT OF HYSTERESIS ON OPERATION

Up to this point, in order to simplify the introduction to magnetic amplifiers, the effect of hysteresis has been neglected. In order to fully understand the operating characteristics of magnetic amplifiers, it is important to know how these characteristics are affected by the shape of the hysteresis loop.

The curve in figure 20-12 (A) represents a typical rectangular shaped B-H curve; the curves in (B), (C), and (D) show the operation for various amounts of control action.

Further use of figure 20-11 is made in explaining the effects of hysteresis. This circuit actually has two distinct periods of operation. The period during which the rectifier CR-2 conducts is known as the "operating period"; the other half-cycle of the alternating current is known as the "control period."

In the explanation of the effect of the hysteresis loop of figure 20-12 (A), assume that control current is zero and that the load winding current is of such amplitude that it will saturate the core during peak conduction. During the first operating period, conduction through the load winding will cause the core flux to build up along the dotted line from the point of origin, o, to the point of saturation, a. At the end of this period, the flux level in the coil returns to point b rather than to zero because of the retentivity of the core. This is called the remanence point in magnetics and leaves the core in a magnetic state, similar

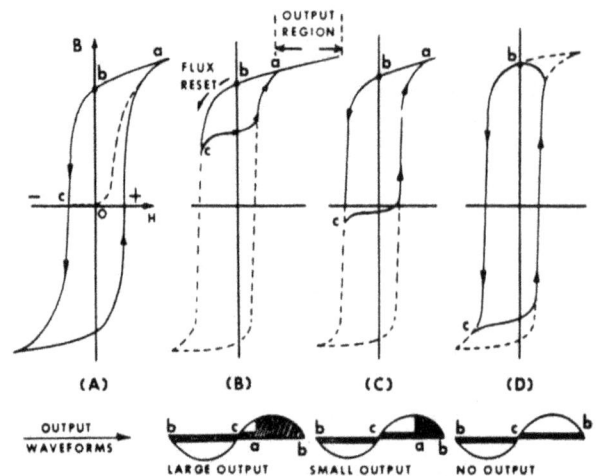

Figure 20-12.—Operation of a magnetic amplifier showing effects of hysteresis.

to a permanent magnet. Thus, with no control current the core flux will remain at point b and during each positive half-cycle the core will saturate immediately. Under this condition the load current will be maximum, for no control is exercised.

Figure 20-12 (B) shows the action when a small control current is added. The operation during the first half-cycle is similar to that just described in connection with figure 20-12 (A), but during the second half-cycle (control period), the flux will reset from point b to point c. This occurs because the direction of the control flux in the core is opposite to that of the operating flux.

This positioning of the residual flux is called "resetting." During the next half-cycle, the flux starts increasing from point c in the figure and a small period of this half-cycle is used before the flux reaches saturation at point a. At this point the core is saturated and the rectifier is in maximum conduction for the remaining part of this half-cycle. At the end of the operating period the flux returns to point b where the control period starts and resets the flux to point c where the cycle starts repeating itself. The output for this condition is shown as the shaded portion of the waveform of figure 20-12 (B). Figure 20-12 (C) and (D) shows the operation for increasing amounts of control action. The operation in each case is basically the same as for figure 20-12 (B).

It may be seen that the control signal sets the point of amplitude at which the source voltage causes saturation of the core and reduces the impedance in series with the load winding to a very low value. The term "firing" is frequently used in connection with magnetic amplifiers interchangeably with the term saturation.

Using Bias for Flux Reset

When the core material possesses hysteresis properties that produce a rectangular-shaped B-H curve, it may be necessary to bias the core in order to retain control. This is accomplished by allowing a bias current to either flow through the control winding or through a separate bias winding. This bias current provides a means of resetting the flux to the initial operating point during the control period. Thus, no control current is necessary to reset the flux. Figure 20-13 shows a separate bias winding wound on the center leg of a three-legged core.

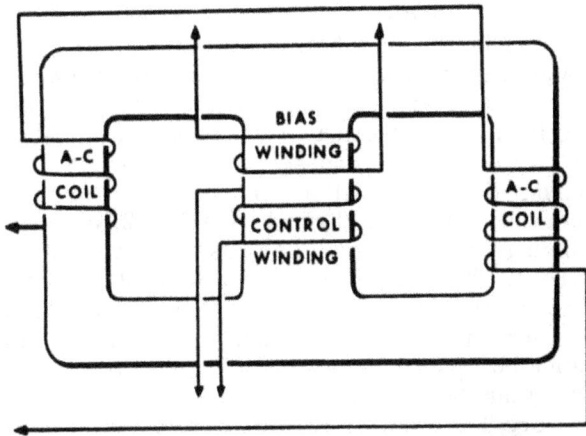

BE. 393

Figure 20-13.—Bias winding.

The core is usually biased so that with zero control current the load winding flux saturates the core midway in the operation cycle. This bias setting enables the control current to either advance or delay the point of saturation. In other words, if the polarity of the control current is such that the flux produced by it adds to the bias flux, the point of saturation is delayed. (Control and bias flux are in opposition to load winding flux.) A control current of opposite polarity produces a flux that opposes the bias flux and this advances the point of saturation.

The bias may be either alternating or direct current. In most cases the polarity of d-c bias is such that it opposes the load current flux. The magnitude of the bias might be such that it would reset the flux to a point between a and c of figure 20-2. There are some applications where the opposite is true; in these applications the purpose of the bias is to provide an initial d-c saturation in the core in order to obtain greater amplification of weak signals.

FULL-WAVE RECTIFIER

The magnetic amplifier that has already been discussed produces a pulsating or half-wave output. In most applications of magnetic amplifiers, full-wave operation is more desirable for it is more efficient since the load is energized during both halves of the a-c cycle. Full-wave operation may be obtained by using a pair of simple half-wave magnetic amplifiers operating in parallel. A typical circuit arrangement is shown in (A) of figure 20-14.

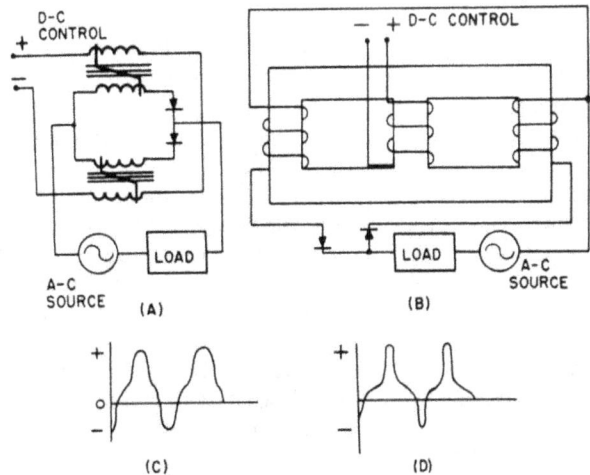

BE. 394

Figure 20-14.—Full-wave magnetic amplifier and output waveforms.

The load current is controlled by means of two control windings connected in series. Each amplifier unit contains a rectifier so that load current flows alternately in the two amplifiers. As one conducts, the core of the series coil in the other is being set by the action of the control current. And with each amplifier conducting during approximately one-half of each cycle of

410

load current, the output variations are full-wave in nature.

Output waveforms for two different control currents are shown in (C) and (D) of figure 20-14. The waveform in (C) is produced when the control current biases the core near the point of saturation. In this condition, saturation of the magnetic cores of the amplifiers is reached early in each half-cycle, and the resulting output current variations are almost sinusoidal in shape.

When the current in the control winding biases the core in such a way as to produce the waveform of figure 20-14 (D), the average value of load current is considerably less than when the control current biased the core to near saturation. Thus, as in the half-wave circuit, the output power developed in the load can be varied by controlling the load current. The output that is developed is an alternating current and is in phase with the source voltage. Figure 20-14 (B) shows a magnetic amplifier that functions similar to the one shown in (A). The major difference is that the two cores shown in (A) have been combined to form a three-legged core.

## CROSSOVER WINDINGS

In discussing the theory of operation of the full-wave magnetic amplifier circuits shown in figure 20-14, it was assumed that the rectifiers have zero forward resistance and infinite back resistance. In actual practice this assumption is not valid since rectifiers do not have infinite back resistance. It should be apparent that any back current flowing through the rectifiers shown in figure 20-14 (A) during their non-conducting half cycle, will produce a magnetic flux which will tend to drive the core out of saturation. During the conducting half-cycle a part of the load current would be absorbed in bringing the core back to the point of saturation, thus reducing the gain of the amplifier.

One method of counteracting this effect is by the addition of a crossover winding to the two magnetic amplifier cores. Figure 20-15 illustrates a magnetic amplifier utilizing full-wave rectification and crossover windings. It should be noted that this circuit is similar to the circuit shown in figure 20-14 (A) with the addition of the crossover windings.

The crossover winding consists of a few turns which produce a magnetic field in a direction opposite to that produced by the backward current through the load winding. Referring to figure 20-15, consider the action of the load and

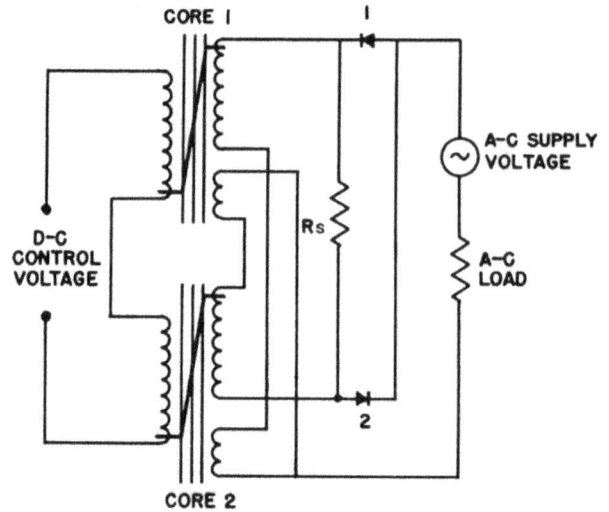

BE. 395

Figure 20-15.—Full-wave magnetic amplifier utilizing crossover windings.

crossover windings on the two cores. Assume that during the first half cycle of the a-c supply voltage, rectifier 1 is conducting in the forward direction. The heavy current, flowing through rectifier 1 and the load winding on core 1, also flows through the crossover winding of core 2. During the same half-cycle a small backward current is flowing through rectifier 2, the load winding on core 2, and the crossover winding on core 1. This small backward current, flowing through the large number of turns in the load winding, produces a magnetic field that is canceled by the heavy current flowing through the small number of turns of the crossover winding on core 2. A similar cancellation is achieved on the second half-cycle of the a-c supply voltage by the other crossover winding.

## PRINCIPLES OF FEEDBACK CIRCUITS

It is possible to change some of the characteristics of a magnetic amplifier by the use of external feedback. (NOTE: The action of rectifiers in series with the load current discussed earlier in this chapter constitutes internal feedback.)

Feedback is the returning of a portion of the output of an amplifier stage to the input of that or a preceding stage and may be either positive or negative depending on the effect desired. Regenerative or positive feedback makes the circuit more sensitive to change in control current; thus, the power gain of a single stage may reach 40 to 50 db. However, this type of feedback has two disadvantages for it increases the time-lag characteristics of the amplifier and may also cause instability. Degenerative or negative feedback reduces the gain of the amplifier; however, it increases the linearity of the magnetization curve and reduces the time-lag characteristics of the amplifier.

The curves in figure 20-16 illustrate the effect of both positive and negative feedback on load current as a function of input control current. It should be noted that the same amount of direct current is needed to saturate the core regardless of the amount of feedback provided. However, when using positive feedback, less input control current is required to saturate the core since the feedback current aids the control current toward this end. The gain characteristics of magnetic amplifiers with positive and negative feedback may be compared by studying the curves shown in figure 20-16. The following should be concluded:

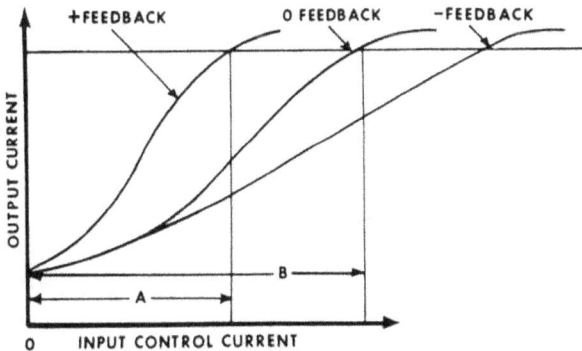

BE. 396

Figure 20-16.—Effects of external feedback.

1. Output current is the same when saturation is attained regardless of feedback.

2. Linearity and stability improve with degenerative (negative) feedback.

3. Gain improves with regenerative (positive) feedback.

A schematic diagram of two magnetic amplifier circuits utilizing external feedback is show in figure 20-17. Feedback is provided by an additional winding on the center leg of each core. The direct current flowing through the feedback winding creates a magnetic field which either aids or opposes the magnetic field created by the control current.

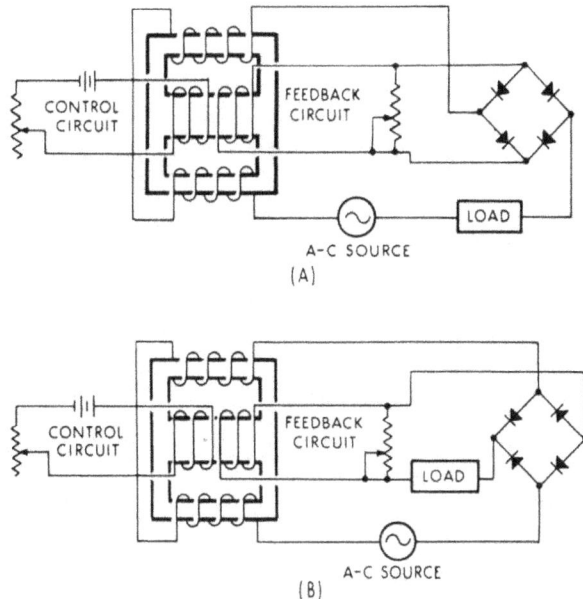

BE. 397

Figure 20-17.—Magnetic amplifiers using external feedback.

In (A) of figure 20-17 the a-c load current is rectified prior to being fed through the feedback winding. It should be noted that the feedback current consists of the total load current and it will always flow in the same direction regardless of the input signal. To change the feedback from regenerative to degenerative or vice versa, the feedback winding connections must be reversed.

Figure 20-17 (B) illustrates the use of external feedback in a magnetic amplifier supplying a d-c load. The amount of feedback is controlled in both circuits by the setting of the potentiometer paralleling the feedback winding.

External feedback, as applied to magnetic amplifiers, may be divided into two basic categories, these being magnetic and electric. These types may be used individually or in combination, depending on the results desired, and also that distinctly different effects can be had from each type. Further, in most cases, the magnetic feedback is a built-in feature of the amplifier, whereas the electric feedback may be applied to any amplifier.

MAGNETIC FEEDBACK

Magnetic feedback is accomplished by feeding back a portion or all of the output current in such a fashion as to directly affect the core flux. This is not, however, referred to as current feedback. Magnetic feedback is achieved by forcing feedback current through a tertiary winding placed on the core. If the output of the amplifier is d.c., then this current may be fed to the feedback winding directly without the need of rectification. If, however, the output of the amplifier is a.c., then the current must first be rectified, and then applied to the feedback winding. The feedback current will then be a direct current, proportional to the load current. This current may be applied in such a fashion that it is either regenerative or degenerative, depending on the action of the generated flux in respect to the circuits action.

Magnetic feedback is considered to be external feedback, where the self-saturating magnetic amplifier is considered as employing a comparable inherent feedback. It should be realized at this point that both internal and external feedback may be applied to the same amplifier. As was previously mentioned, magnetic feedback is one of the basic types of feedback, the other type being electric. It is not uncommon to encounter an amplifier employing both internal, magnetic, and electric feedback.

ELECTRIC FEEDBACK

This type of feedback may be further divided into two classes, these being voltage and current. Both types are accomplished by forcing all or a portion of the output to flow in the control circuit in such a manner as to influence the signal applied to the control winding. The feedback circuitry is completely external to the amplifier proper; as a consequence, it does not affect the gain of the amplifier proper, but does affect the stage gain.

The characteristics of the voltage feedback circuit are as follows:

1. Output appears as a constant voltage source.
2. Output voltage is insensitive to load fluctuations.
3. Output voltage remains relatively constant.
4. Output appears as a low impedance source.

Current feedback has the following characteristics:
1. Output appears as a constant current source.
2. Output current is insensitive to load fluctuations.

3. Output current remains relatively constant.
4. Output appears as a high impedance source.

In summation, the basic amplifier employing no feedback is seldom encountered in the modern circuitry. The amplifiers in use today will almost invariably employ some form of feedback. The type used will be determined by the application of the amplifier.

APPLICATIONS OF
MAGNETIC AMPLIFIERS

The magnetic amplifier has found application in almost all circuits which use vacuum tubes and many which do not. They may be found in such circuits as audio amplifiers, voltage regulators, servoamplifiers, computer counting and dividing circuits, and many high frequency circuits.

POWER SUPPLY REGULATORS

Many electrical and electronic equipments contain voltage regulated power supplies. The accuracy of the voltage produced by these supplies is important insofar as proper operation and maximum life of the equipment are concerned. Present day electronic equipments contain many timing, waveshaping, and other critical circuits that require a highly regulated voltage source. The magnetic amplifier can be used to obtain this regulated voltage by being used as the regulating element for the output of the power supply. The magnetic amplifier type regulated power supply is very reliable under varying load conditions.

SERVOSYSTEMS

One common use of magnetic amplifiers in electrical systems is in connection with servosystems. Figure 20-18 illustrates the use of a magnetic amplifier as the output stage of a servoamplifier. The function of the servoamplifier is to amplify the servo error signal sufficiently to drive the servomotor. Servosystems are discussed later in this manual.

The output of the servoamplifier is connected to one of the motor windings (controlled winding W1), while the other winding (uncontrolled winding W2) is connected across the a-c source in series with a capacitor. The capacitor provides the required 90° phase shift necessary

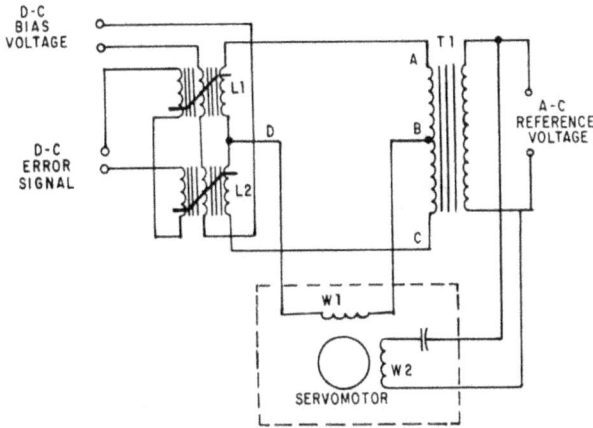

BE.398

Figure 20-18.—Magnetic amplifier used as an output stage of a servoamplifier.

to cause motor rotation. The phase relationship of the current through the two windings determines the direction of rotation of the servomotor.

The magnetic amplifier consists of a transformer (T1) and two saturable reactors, (L1 and L2), each having three windings. Note that the d-c bias current flows through a winding of each reactor and the windings are connected in series-aiding. This bias current is supplied by a d-c bias power source. A d-c error current also flows through a winding in each reactor; however, these windings are connected in series-opposing.

The reactors L1 and L2 are equally and partially saturated by the d-c bias current when no d-c error signal is applied. The reactance of L1 and L2 are now equal, resulting in points B and D being at equal potential. There is no current flow through the controlled phase winding.

If an error signal is applied causing the current to further saturate L2, the reactance of its a-c winding is decreased. This current through L1 tends to cancel the effect of the d-c bias current and increase the reactance of its a-c winding. Within the operating limits of the circuit, the change in reactance is proportional to the amplitude of the error signal. Hence, point D is now effectively connected to point C causing motor rotation. Reversing the polarity of the error signal causes the direction of rotation to reverse since point D is effectively connected to point A.

## COMPUTER SYSTEMS

The magnetic amplifier is found in computer circuits. When used in summing circuits, they are capable of adding and subtracting currents with no interaction between source currents. They can also be used to multiply, divide, compute powers, and extract roots when used in conjunction with other circuit components.

# CHAPTER 21

# SYNCHROS AND SERVOMECHANISMS

## SYNCHROS

Synchros are electromagnetic devices which are used primarily for the transfer of angular-position data. Synchros are, in effect, single-phase transformers whose primary-to-secondary coupling may be varied by physically changing the relative orientation of these two windings.

Synchro systems are used throughout the Navy to provide a means of transmitting the position of a remotely located device to one or more indicators located away from the transmitting area.

Figure 21-1 (A) shows a simple synchro system. The synchro transmitter sends an electrical signal to the synchro receivers, via interconnecting leads. The synchro transmitter signal is generated when the handwheel turns the transmitter's shaft, and the indicator's shaft will rotate to aline with the transmitter's shaft.

Figure 21-1 (B) shows a simple mechanical system that uses gears and shafts for transmitting position data. Mechanical systems are usually impractical because of associated belts, pulleys, and flexible rotating shafts.

In appearance, synchros resemble small electric motors. They consist of a rotor (R) and a stator (S). The letters R and S are used to identify rotor and stator connections in wiring diagrams and blueprints. Synchros are represented schematically by the symbols shown in figure 21-2. The symbols shown in (A) and (B) are used when it is necessary to show only the external connections to a synchro, while those shown in (C), (D), and (E) are used when it is important to see the positional relationship between the rotor and stator. The small arrow on the rotor indicates the angular displacement of the rotor; in these illustrations the displacement is zero degrees.

## SYNCHRO CONSTRUCTION

Knowledge of a synchro's construction and its characteristics will enable you to better understand synchro operation. As stated previously, synchros are, in effect, transformers whose primary-to-secondary coupling may be varied by physically changing the relative orientation of the two windings. This is accomplished by mounting one of the windings so that it is free to rotate inside the other. The inner, usually movable, winding is called the rotor, and the outer, usually stationary, winding is called the stator. The rotor consists of either one or three coils wound on sheet-steel laminations. The stator normally consists of three coils wound in internally slotted laminations. In some units the rotor is the primary and the stator is the secondary, in other units the reverse is true.

### Rotor Construction

The laminations of the rotor core are stacked together and rigidly mounted on a shaft. Sliprings are mounted on, but insulated from, the shaft. The ends of the coil or coils are connected to these sliprings. Brushes riding on the sliprings provide continuous electrical contact during rotation, and low-friction ball bearings permit the shaft to turn easily. In standard synchros, the bearings must permit rotation from very low speeds to speeds as high as 1,200 rpm.

There are two common types of synchro rotors now in use, (1) the salient-pole rotor, and (2) the drum or wound rotor.

The salient-pole rotor, which frequently is called a "dumb-bell" or "H" rotor because of the shape of its core laminations, is shown in figure 21-3.

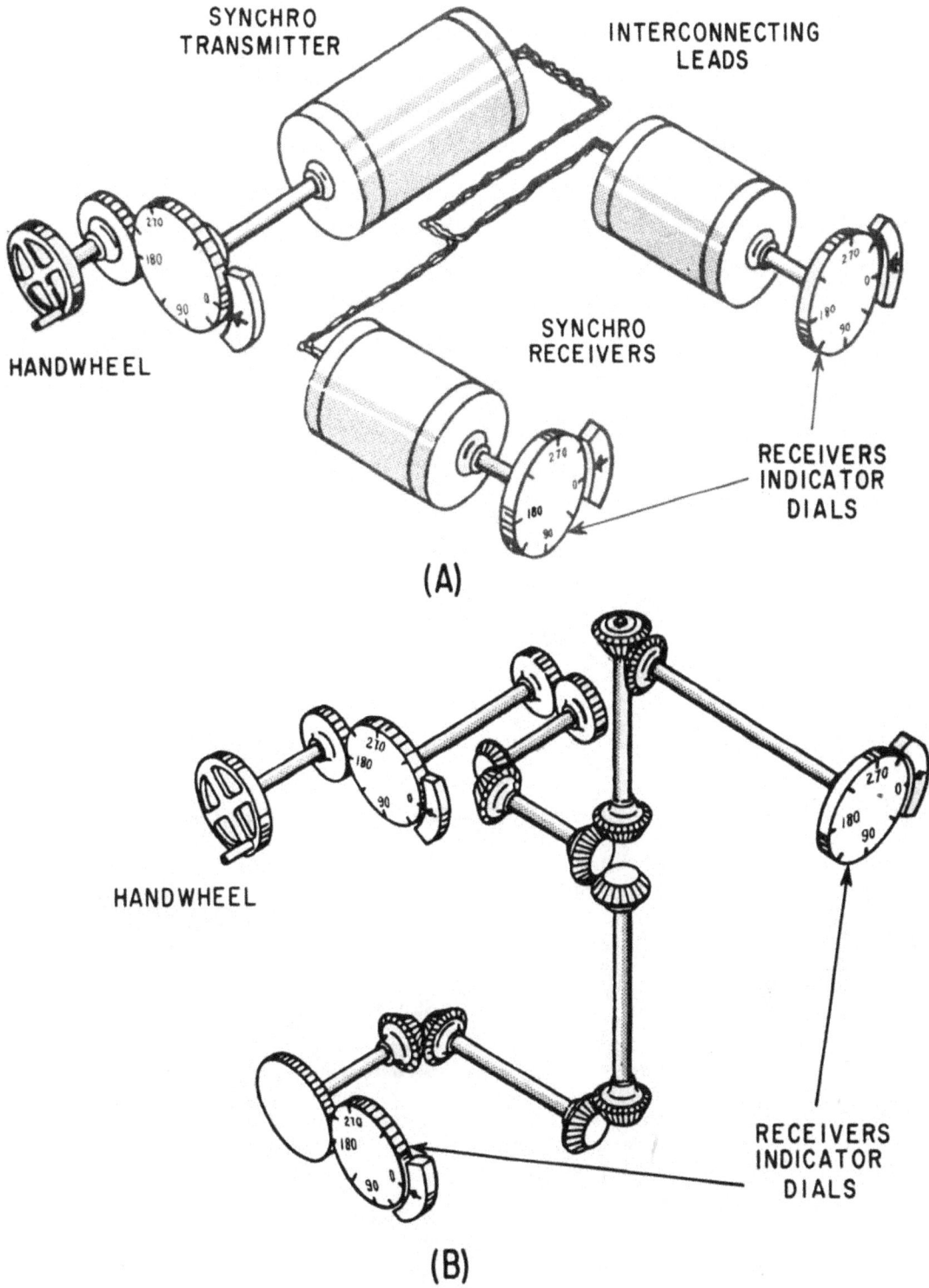

Figure 21-1.—(A) Data transfer with synchros; (B) data transfer with gears.

BE.399

TRANSMITTERS, RECEIVERS
CONTROL TRANSFORMERS

DIFFERENTIALS

(A)

(B)

TRANSMITTERS
AND RECEIVERS

(C)

CONTROL
TRANSFORMERS

(D)

DIFFERENTIALS

(E)

BE.400

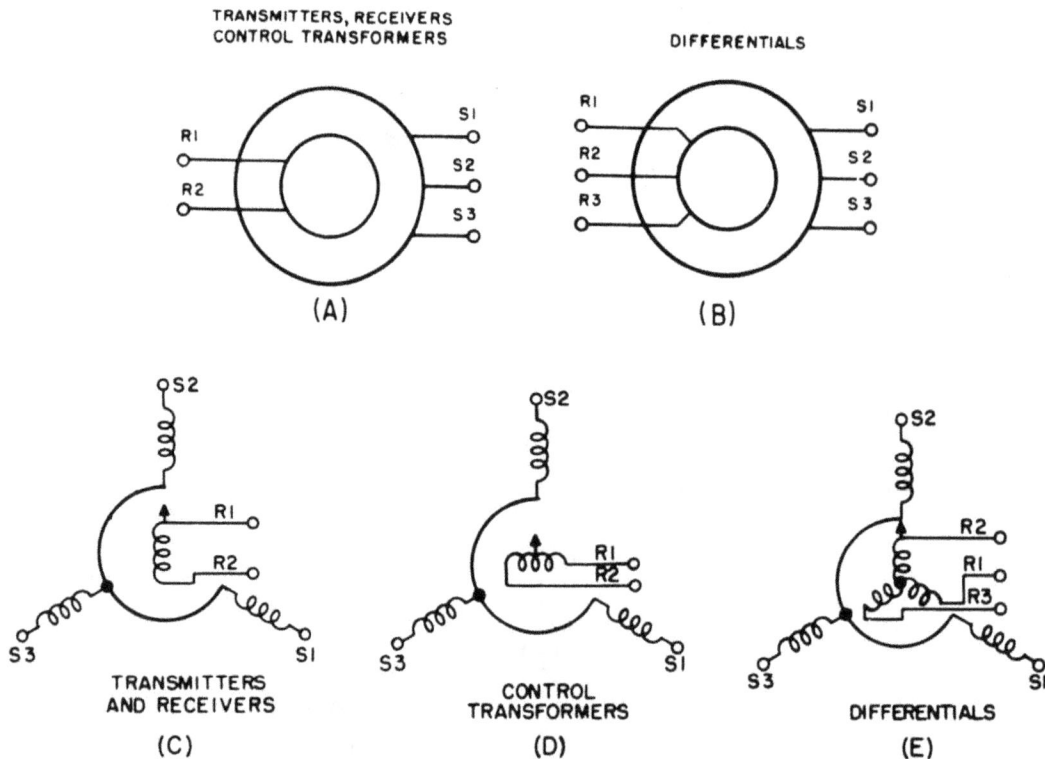

Figure 21-2.—Schematic symbols used to show external connections and relative positions of windings for Navy synchros.

The winding consists of a single machine-wound coil whose turns lie in planes perpendicular to a line through the center called the axis of the coil and along which the mmf of the coil is developed. The coil axis coincides with that of the salient poles. When used in transmitters and receivers, the rotor functions as the excitation or primary winding of the synchro. When energized, it becomes an a-c electromagnet with the poles assuming opposite magnetic polarities. During one complete excitation cycle, the magnetic polarity changes from zero to maximum in one direction, to zero and reversing to maximum in the opposite direction, returning to zero.

The drum or wound rotor is shown in figure 21-4. This type of rotor is used in most synchro control transformers. The winding of the wound rotor may consist of three coils, so wound that their axes are displaced from each other by 120°.

One end of each coil terminates at one of three sliprings on the shaft, while the other end are connected together. Synchro windings of this type are called Y-connected. In other wound rotors, a single coil of wire or group of coils is connected in series to produce either a concentrated winding effect, or the same distributed winding effect as that of the salient-pole rotor.

Stator Construction

The stator of a synchro is a cylindrical structure of slotted laminations on which three Y-connected coils are wound with their axes 120° apart. Figure 21-5 (A) shows a typical stator assembly and figure 21-5 (B) shows a stator lamination.

Stator windings function as the secondary windings in synchro transmitters and receivers. Normally, stators are not connected directly to an a-c source. Their excitation is supplied by the a-c magnetic field of the rotor.

Some synchros are constructed so that both the stator and rotor may be turned. Connections

417

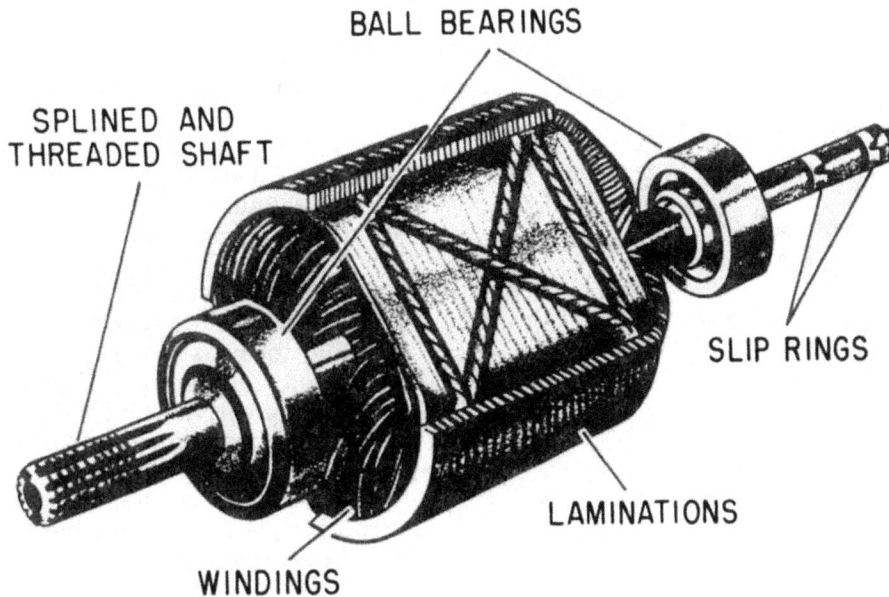

BE. 401

Figure 21-3.—Salient-pole rotor.

to the stator of this type are made via sliprings and brushes. In some units, the sliprings are secured to the housing and the brushes turn with the stator. In other units, the brushes are fixed and the sliprings are mounted on a flat insulated plate secured to the stator.

Unit Assembly

The rotor is mounted so that it may turn within the stator. A cylindrical frame houses the assembled synchro. Standard synchros have an insulated terminal block secured to one end of the housing at which the internal connections to the rotor and stator terminate, and to which external connections are made. Special type synchros often have pigtail leads brought out from inside the unit, rather than terminals.

THEORY OF OPERATION

Transmitters

The conventional synchro transmitter, (fig. 21-6), uses a salient-pole rotor and a stator with skewed slots (fig 21-5 (A)). When an a-c excitation voltage is applied to the rotor, the resultant current produces an a-c magnetic field.

The lines of force, or flux, vary continually in amplitude and direction, and, by transformer action, induce voltages into the stator coils. The effective voltage induced in any stator coil depends upon the angular position of that coil's axis with respect to the rotor axis.

When the maximum coil effective (rms) voltage is known, the effective voltage induced at any angular displacement can be determined. Figure 21-7 is a cross section of a synchro transmitter and shows the effective voltages induced in one stator coil as the rotor is turned to different positions.

The turns ratio between the rotor and stator is such that when single-phase 115-volt power is applied to the rotor, the highest value of effective voltage that will be induced in any one stator coil will be 52 volts. For example, the highest effective voltage that can be induced in stator coil A will occur when the rotor is turned to angle $\theta$ (fig. 21-7 (A)). In this position the rotor axis is in alinement with the axis of winding A and the magnetic coupling between the primary and secondary coil A is maximum.

The effective voltage, $E_s$, induced in any one of the secondary windings A, B, or C is approximately equal to the product of the effective voltage, $E_p$, on the primary; the

BE. 402

Figure 21-4.—Drum of wound rotor.

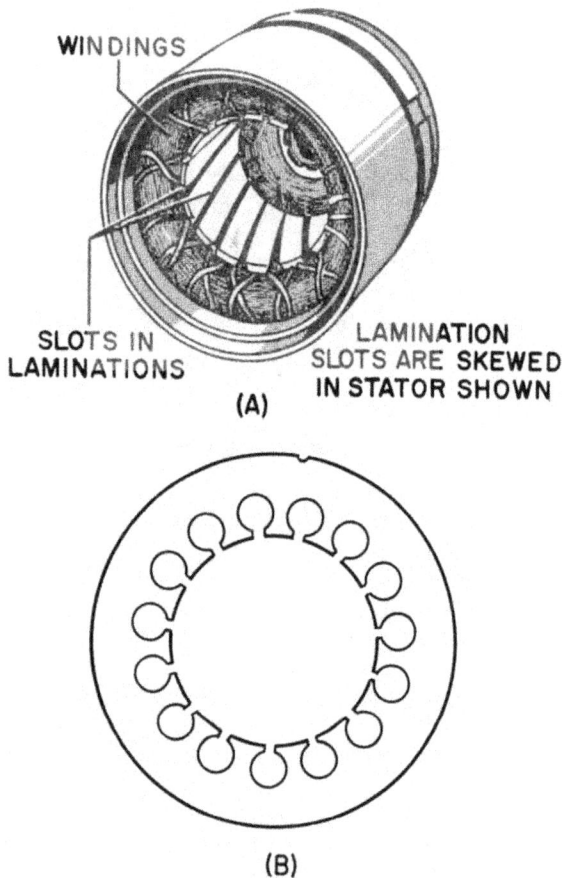

BE. 403

Figure 21-5.—(A) Typical stator;
(B) stator lamination.

secondary-to-primary turns ratio, N; and the magnetic coupling between primary and secondary which depends upon the cosine of the angle, $\theta$, between the rotor axis and the axis of the corresponding secondary winding. Expressed mathematically,

$$E_S = E_p \times N \times \cos \theta$$

Therefore, because the effective primary voltage and the turns ratio are constant, the effective secondary voltage varies with angle $\theta$.

TERMINAL - TO - TERMINAL STATOR VOLTAGES.—Because the common connection between the stator coils is not accessible, it is possible to measure only the terminal-to-terminal voltages. When the maximum terminal-to-terminal effective voltage is known, the terminal-to-terminal effective voltage for any rotor displacement can be determined. Figure 21-8 shows how these voltages vary as the rotor is turned. Values are above the line when the terminal-to-terminal voltage is in phase with the $R_1$ to $R_2$ voltage and below the line when the voltage is 180° out of phase with the $R_1$ to $R_2$ voltage thus negative values indicate a phase reversal. As an example, when the rotor is turned 50 degrees from the reference (zero degree) position, the $S_3$ to $S_1$ voltage will be about 70 volts and in phase with the $R_1$ to $R_2$ voltage, the $S_2$ to $S_3$ voltage will be about 16 volts and also in phase with the $R_1$ to $R_2$ voltage, and the $S_1$ to $S_2$ voltage will be about 85 volts, 180° out of phase with the $R_1$ to $R_2$ voltage. Although the curves of figure 21-8 resemble timegraphs of a-c voltages, they show only the variations in effective voltage amplitude and phase as a function of the mechanical rotor position.

Receivers are electrically identical to transmitters of the same size. In some sizes of the 400 hertz standard synchros, the units may be used as transmitters or receivers.

Normally the receiver rotor is unrestrained in movement except for brush and bearing friction. When power is first applied to a system, the transmitter position quickly changes, or if the receiver is switched into the system, the receiver rotor turns to correspond to the position of the transmitter rotor. This sudden motion can cause the rotor to oscillate (swing back and forth) around the synchronous position; should the movement of the rotor be great enouth it may even spin. Some method of preventing excessive oscillations or spinning must be used.

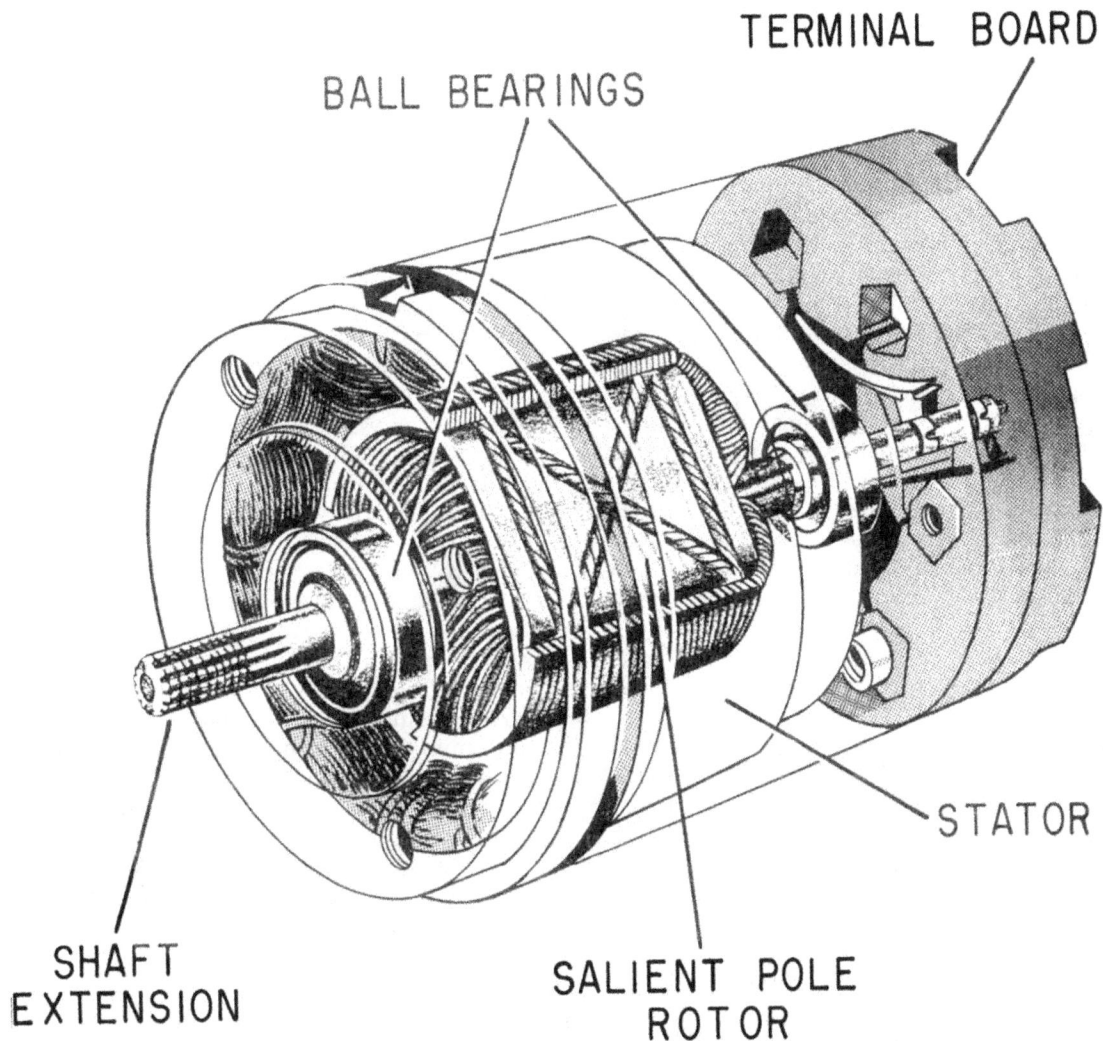

Figure 21-6.—Synchro transmitter or receiver.

In small synchro units, a retarding action may be produced by a shorted winding on the quadrature axis, at right angles to the direct axis. In larger units, a mechanical device known as an inertia damper is more effective. Several variations of the inertia damper are in use. One of the more common types consists of a heavy brass flywheel which is free to rotate around a bushing which is attached to the rotor shaft. A tension spring on the bushing rubs against the flywheel so that they both turn during normal operation. If the rotor shaft turns or tends to change its speed or direction of rotation suddenly, the inertia of the damper opposes the changing condition.

## SIMPLE SYNCHRO SYSTEM

A simple synchro transmission system consists of a transmitter connected to a receiver, as shown in figure 21-9. The $R_1$ transmitter and $R_1$ receiver leads are connected to one side of the a-c supply line and the $R_2$ transmitter and $R_2$ receiver leads are connected to the other side of the supply line. The stators of both the transmitter and the receiver are connected $S_1$ to $S_1$, $S_2$ to $S_2$, and $S_3$ to $S_3$ so that the voltage in each of the transmitter stator coils opposes the voltage in the corresponding coils of the receiver stator. The voltage directions are indicated by arrows for the instant of time shown by the dot on the sine wave of the rotor supply voltage. The a-c generator symbol inserted in series with each of the synchro windings indicates the presence of a transformer-induced voltage.

The field of the transmitter is alined with the axis of $S_2$ (fig. 21-9 (A)) because the transmitter rotor axis is alined with $S_2$. Assume for the moment that the receiver rotor is not connected to the single-phase supply. Under this condition the voltage induced in the transmitter stator windings will be impressed on the receiver stator windings through the three leads connecting the $S_1$, $S_2$, and $S_3$ terminals. Exciting currents that are proportional to the transmitter stator voltages will flow in the receiver stator windings, and the magnetomotive forces produced by these currents will establish a 2-pole field that orients itself in the receiver stator in exactly the same manner that the transmitter field is oriented in its own stator. Thus, if the transmitter field is turned by turning the transmitter rotor, the receiver field will also turn in exact synchronism with the transmitter rotor.

A soft-iron bar placed in a magnetic field will always tend to aline itself so that its longest axis is parallel to the axis of the field. Thus, the salient pole receiver rotor will also aline itself with the receiver field even though the receiver rotor winding is open-circuited. Operation with an open-circuited rotor is not desir-

able, however, because for a given position of the transmitter rotor and field there are two positions, 180° apart, in which the salient pole receiver rotor may come into alinement with the receiver field. This difficulty is eliminated by energizing the receiver rotor with alternating current of the same frequency and phase as that supplied to the transmitter rotor. Now the receiver rotor electromagnet comes into alinement with the receiver field in such a position as to always aid the magnetomotive forces of the three windings of the receiver stator. A rotor position in which the magnetomotive force of the rotor opposes the magnetomotive forces of the stator would not be stable. For example, a compass needle always alines itself with the earth's field so that the north pole of the needle points toward the south magnetic pole (north geographic).

When both rotors are displaced from zero by the same angle, they are properly alined with their respective stator fields. Under this condition they are in correspondence and the induced voltages in each of the three pairs of corresponding stator coils are equal and in opposition. Hence, there is no resultant voltage between corresponding stator terminals and no current flows in the stator coils.

The angle through which a transmitter rotor is mechanically rotated is called a SIGNAL. For example, the angle that the transmitter rotor is displaced (fig. 21-9 (B)) is 60°. As soon

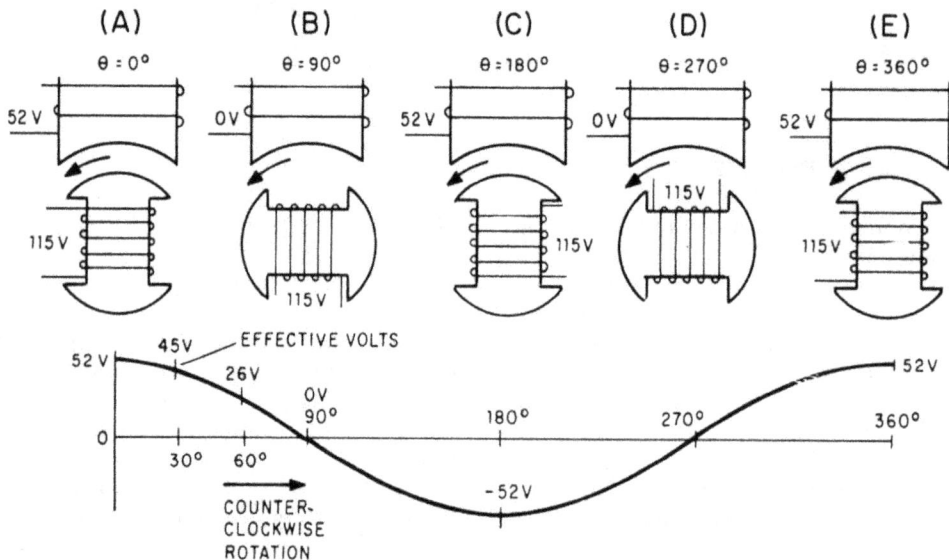

Figure 21-7.—Stator voltage vs rotor position.

BE. 405

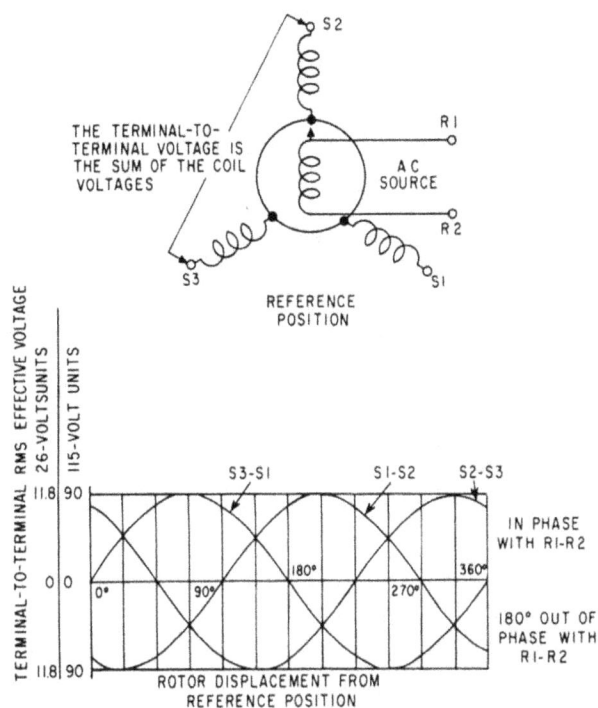

BE. 406

Figure 21-8.—Terminal-to-terminal
voltages vs rotor position.

as the transmitter rotor is turned, the transmitter $S_2$ coil voltage decreases, the $S_1$ coil
voltage reverses direction, and the $S_3$ coil
voltage increases. Current immediately flows
between the transmitter stator and the receiver
stator in the direction of the stronger voltages.
The unbalanced voltages are absorbed in the
line drop and in the internal impedances of the
windings.

When the transmitter (fig. 21-9(A)) is turned
closkwise 60° (fig. 21-9 (B)), a torque is developed in the receiver rotor causing it to attempt to follow through the same angle. When
the receiver and the transmitter are again in
alinement, the torque is reduced to zero, as
shown in 21-9 (C).

Receiver alinement is pictured in figure
21-9 (D). Here the rotor axis coincides with the
axis of the stator field. Notice that currents on
the right side of the vertical axis are back and
those on the left side are forward. This action
produces a magnetic field as shown by the letters

S at the top and N at the bottom of the salient-
pole rotor. Notice that the rotor current also
produces the same polarity (left-hand rule.)

For the balanced condition shown, the currents in the conductors opposite the salient rotor
poles are equal in magnitude and flow in opposite
directions on either side of the centerline
through the poles. The forces developed are
pictured in the detail drawing above figure 21-9
(D). They are developed as a result of the crowding of the magnetic lines of force at the right of
conductor A and at the left of conductor B. Since
they are equally spaced from the centerline, no
net unbalance exists and the rotor torque is zero.

In figure 21-9 (E), the receiver stator field
is rotated clockwise 60° so that its axis coincides with that of phase A. The rotor is deliber-
ately held in the vertical position to enable showing the development of the torque on the rotor
when the rotor and stator field axes are not
alined. Notice that currents to the right of the
A-phase axis are back and those to the left of
the axis are forward. This action produces the
stator field across the rotor and extends from
the upper right to the lower left portions of the
stator.

Now the currents in the stator conductors
opposite the rotor salient poles are all in the
same direction on both sides of the centerline
through the poles. The forces are unbalanced
as shown by the crowding of the magnetic lines
on the right portion of the conductors. (See
detail above fig. 21-9 (E).) This action indicates
a tendency to force the stator conductors to the
left. The equal and opposite reaction force on
the rotor tends to turn it in a clockwise direction.

The amount of torque developed varies with
the sine of the angle of displacement and is
maximum at 90°.

In figure 21-9 (F), the rotor has turned
clockwise 60° and its axis is again in alinement with that of the stator field. The torque is
again reduced to zero. Notice that the current
in conductors B is forward and in conductors C
is back. The flux lines crowd to the right of
conductors B (see detail drawing at right) and to
the left of conductors C. Since the currents in B
and C are equal and opposite in direction and are
symmetrically spaced about the rotor axis, the
net force on the rotor is zero.

As a receiver approaches correspondence,
the stator voltages of the transmitter and receiver approach equality. This action decreases
the stator currents and produces a decreasing
torque on the receiver that reduces as the

422

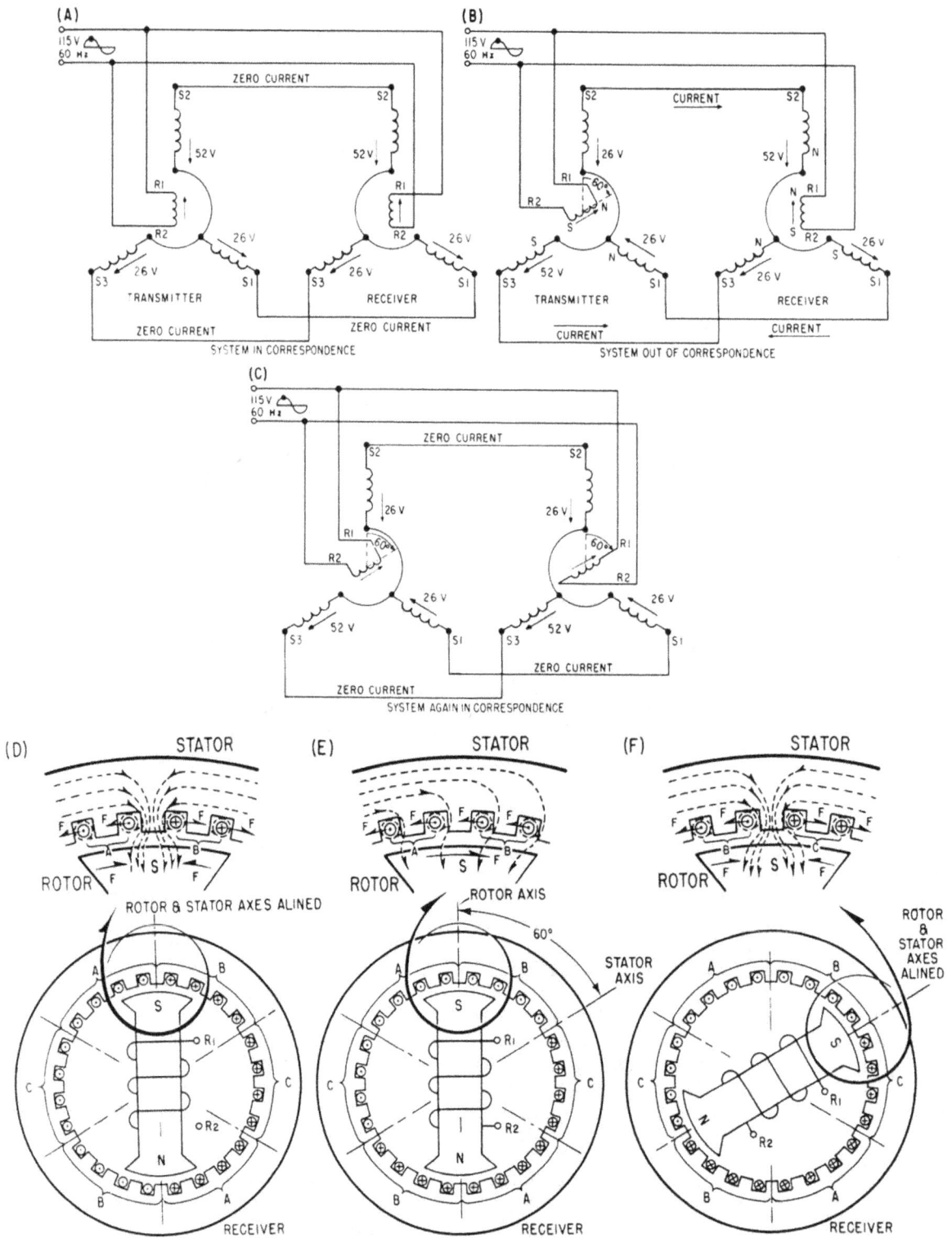

BE.407

Figure 21-9.—Development of torque in a synchro.

position of correspondence is reached. At correspondence, the torque on the receiver rotor is only the amount that is caused by the tendency of the salient-pole rotor to aline itself in the receiver field. Hence, a receiver can position only a very light load.

Synchro transmission systems are used where torque requirements are small. However, when drive power heavier than that required to drive a dial or pointer is necessary, some method of amplifying the synchro's weak torque must be employed. Two common methods used to provide this amplification are the SERVO SYSTEM and the AMPLIDYNE. These two systems are discussed in detail later in this chapter.

Many installations connect one transmitter to several receivers in parallel. This procedure requires a transmitter large enough to carry the total load current of all the receivers. All the $R_1$ rotor leads are connected to one side of the a-c power supply, and all the $R_2$ rotor leads are connected to the other side of the supply. The receiver stator leads are connected lead for lead to the transmitter stator leads.

The sizes of Navy synchro transmitters are designated by numbers. Synchro transmitters in general use are sizes 1, 5, 6, and 7. Synchro receivers in general use are sizes 1 and 5. A size-1 transmitter can control one or two size-1 receivers; a size-5 transmitter can control two size-5 receivers; a size-6 transmitter can control as many as nine size-5 receivers; and a size-7 transmitter can control as many as eighteen size-5 receivers.

Synchros employed in airborne equipments are usually designed for 400-hertz operation, and as a result, are considerably lighter in weight and smaller in size than those that operate with 60-hertz line voltages.

REVERSING DIRECTION OF RECEIVER ROTATION.—When the teeth of two mechanical gears are meshed and a turning force is applied, the gears turn in opposite directions. If a third gear is added, it turns in the same direction as the first. This is important here because synchro receivers are often connected through a train of mechanical gears to the device which they operate, and whether or not force is applied to the device in the same direction as that in which the receiver rotor turns depends on whether the number of gears in the train is odd or even.

The important thing, of course, is to move the dial or other device in the proper direction,

and even when there are no gears involved, this may be opposite to the direction in which the receiver rotor of a normally connnected system would turn.

Either of these two factors, and sometimes a combination of both, may make it necessary to have the transmitter turn the receiver rotor in a direction opposite to that of its own rotor. This is accomplished by reversing the $S_1$ and $S_3$ connections of the transmitter-receiver system, so that $S_1$ of the transmitter is connected to $S^3$ of the receiver and vice versa. This is shown in figure 21-10.

With both rotors at $0°$, conditions within the system remain the same as during normal stator connections, since the rotor coupling to $S_1$ and $S_3$ is equal. But suppose that the transmitter rotor is turned counterclockwise to $60°$, as shown in figure 21-10. In the transmitter, maximum rotor coupling induces maximum voltage across $S_1$, which causes maximum current to flow through $S_3$ in the receiver. The magnetic forces produces turn the receiver rotor clockwise into line with $S_3$, the rotor's $300°$ position, at which point the rotor again induces voltages in its stator coils which equal those of the transmitter coils to which they are connected. Notice that only the direction of rotation changes, not the amount; the $300°$ position is the same as the minus $60°$ position.

It is important to emphasize that the $S_1$ and $S_3$ connections are the only ones ever interchanged in a standard synchro system. Since $S_2$ represents electrical zero, changing the $S_2$ lead would introduce $120°$ errors in indication, and also reverse the direction of rotation.

STATOR CURRENTS.—Whenever the rotors of two interconnected synchros are in different positions, current flows in the stator windings. The amount of current flowing in each stator lead depends on the difference between the voltages induced in the two coils to which that lead connects. This voltage unbalance, in turn, depends on two things—(1) the actual positions of the rotors, and (2) the difference between the two positions.

To observe the effect of the latter, suppose that an ammeter is inserted in any one of the stator leads, the $S_2$ lead for example. Also, suppose that the two rotors are held so that there is a constant difference between their positions while they are rotated together until a point is found at which the ammeter indicates maximum stator current. If the difference between rotor positions is then increased and the

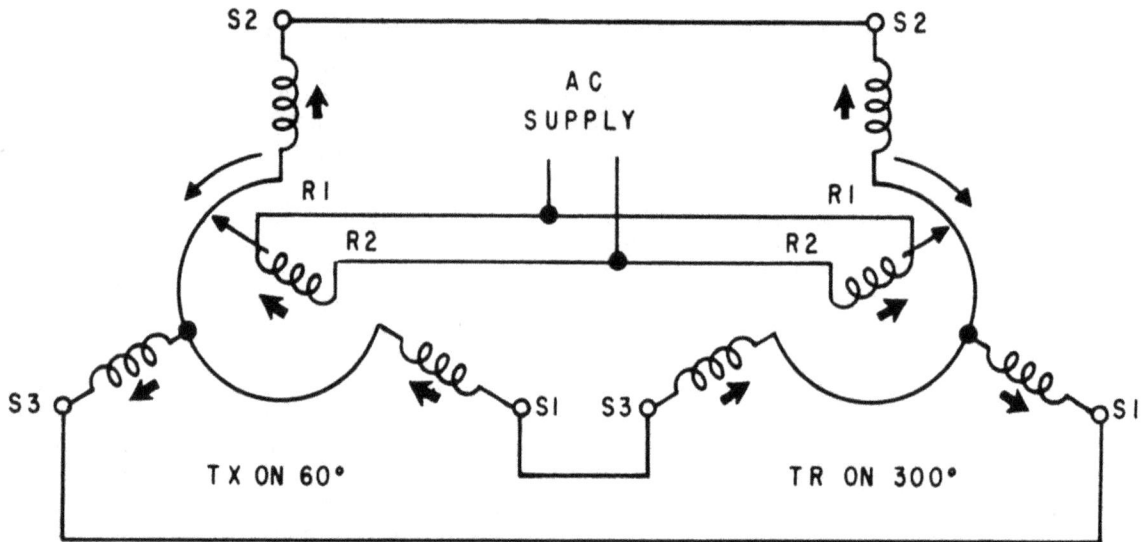

BE.408

Figure 21-10.—Effect of reversing S1 and S3 connections between the transmitter and receiver.

rotation repeated, a different maximum reading is obtained. Each time the difference between rotor positions is changed, the maximum stator current that can be obtained by varying actual rotor positions is changed. Figure 21-11 is a graph showing how the value of this maximum stator current depends on the difference between rotor positions in a typical case.

Note that, in practical operation, the position of the receiver's shaft would never be more than a degree or so away from the transmitter's shaft, so that maximum stator current under normal conditions would be less than one-tenth of an ampere for the synchros used in this example.

To see how the actual rotor positions, as well as the difference between them, affect stator currents, it is only necessary to make a comparison. To do this, compare the maximum stator current in each lead with the strength of the current in that lead as the two rotors are turned through 360°, maintaining the difference in rotor position with which the maximum stator current was established.

The graphs in figure 21-12 show how the maximum current in each of the three stator leads depends on the mean (average) shaft position; the position halfway between the transmitter rotor position and the receiver rotor position.

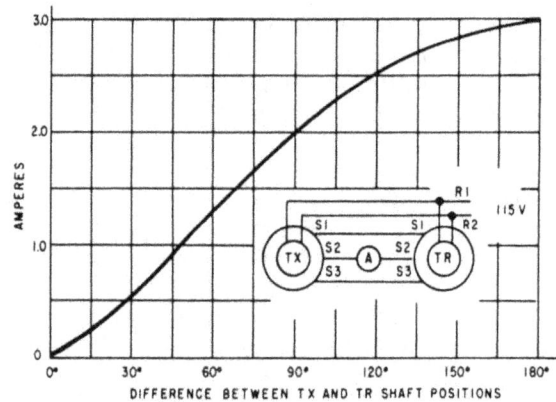

BE.409

Figure 21-11.—Effect of rotor position difference on maximum stator current.

ROTOR CURRENTS.—A synchro transmitter or receiver acts like a transformer, and an increase in the stator, or secondary current, results in a corresponding increase in the rotor (primary) current. When the rotor current of either of the units is plotted on a graph under the same conditions as those used for maximum stator currents, it appears as shown in figure 21-13.

425

ALL CURRENTS ARE RMS VALUE.
CURRENTS SHOWN ABOVE THE 0 LINE ARE IN PHASE
  WITH SI CURRENT AT 0°
CURRENTS SHOWN BELOW THE 0 LINE ARE 180° OUT
  OF PHASE WITH SI CURRENT AT 0°

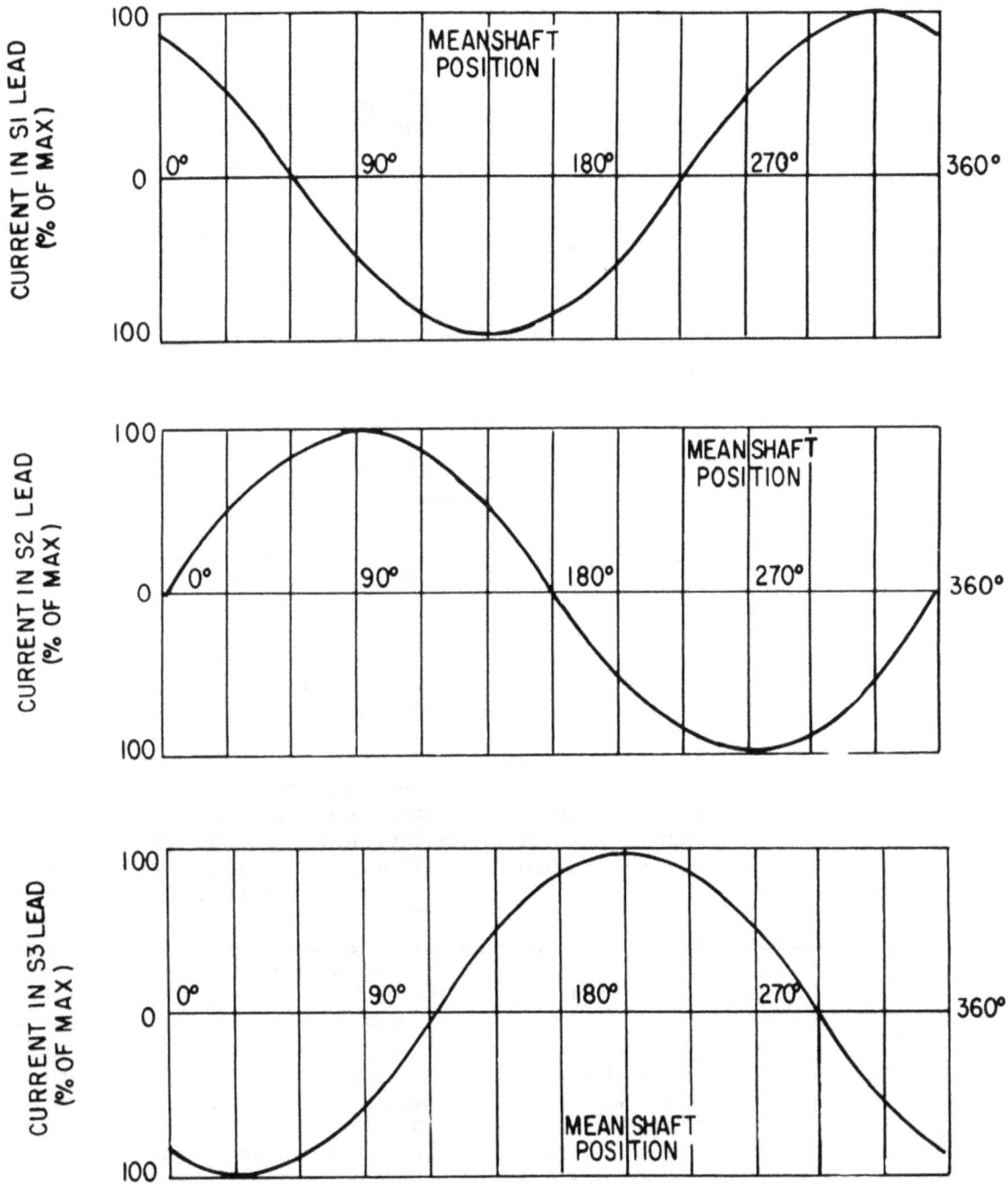

Figure 21-12.—Effect of actual rotor position on individual stator current.

BE. 410

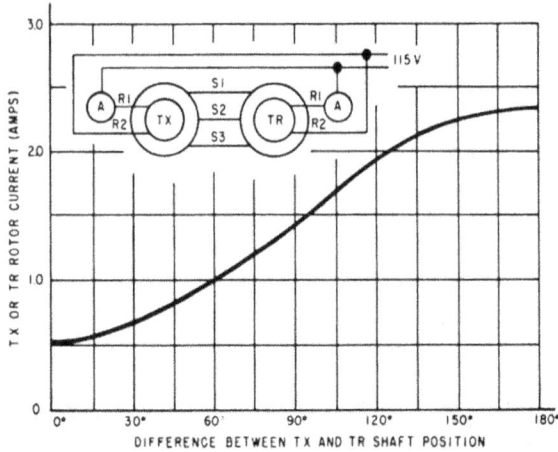

BE.411

Figure 21-13.—Effect of rotor position difference on rotor current.

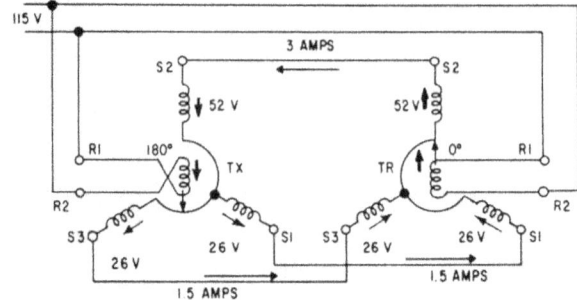

BE.412

Figure 21-14.—Internal conditions in TX-TR system when TX and TR rotors are 180° apart.

Although all three stator currents are zero when the shafts are in the same positions, the rotor current is not. As in any transformer, the primary draws some current with no load on the secondary. This current produces magnetization of the rotor and is equivalent to the transformer exciting current.

WHY THERE IS NO TORQUE AT 180°.—Since the stator current increases with a change in rotor position, and since the magnetic field of each stator increases in strength with the stator current, it appears the torque or turning force exerted by the receiver's shaft should be greatest when the transmitter and receiver rotors are 180° apart. However, exactly the opposite is true. Under these circumstances the torque is zero. To understand this, first consider the current conditions in the synchro system shown in figure 21-14.

Since all the voltages aid each other, strong currents flow in all three stator leads. If the two units are the same size, the currents are the same as those which would flow if the three stator leads were shorted together.

Obviously, these strong currents produce powerful magnetic forces. As powerful as they are, the forces exerted on the rotor work against each other in such a way that their effects are equal and opposite. The resulting torque is zero. In actual situations, the two shafts do not stay in these positions unless held there. The slightest displacement of either

one destroys the balance, and they are rapidly brought into corresponding positions.

Torque Differential Synchros

The demands on a synchro system are not always as simple as the positioning of an indicating device in response to the information received from a single source (transmitter). For example, an error detector used in checking fire-control equipment employs a synchro system to determine the error in a gun turret's position with respect to the training order supplied by a dummy director. To do this, the synchro system must accept two signals, one containing the training order and the other corresponding to the turret's actual position. The system must then compare the two and position an indicator to show the difference between them, which is the error.

Obviously, the simple synchro transmitter-receiver system considered up to now could not handle a job of this sort. A different type of synchro is needed, one which can accept two position-data signals simultaneously, add or subtract the data, and furnish an output proportional to the sum or difference of the two. This is where the synchro differential enters the picture. A differential can perform all three of these functions.

In a differential synchro transmitter, both the rotor and stator windings consist of three Y-connected coils. The stator is normally the primary, and receives its excitation from a synchro transmitter. The voltages appearing across

427

the differential's rotor terminals are determined by the magnetic field produced by the stator currents AND the physical position of the rotor. The magnetic field created by the stator currents assumes an angle corresponding to that of the magnetic field in the transmitter supplying the excitation. If the rotor position changes, the voltages induced into its windings also change, so that the voltages present at the rotor terminals change.

DIFFERENTIAL RECEIVERS.—As torque receivers were previously compared to torque transmitters, so may torque differential receivers be compared to torque differential transmitters. Both rotor and stator receive energizing currents from torque transmitters. The two resultant magnetic fields interact and the rotor turns.

In a torque differential synchro system, the system will consist either of a torque transmitter (TX), a torque differential transmitter (TDX), and a torque receiver (TR); or the system may consist of two torque transmitters (TX) and one torque differential receiver (TDR).

TX-TDX-TR SYNCHRO SYSTEM.—Assume that the stator leads of a torque transmitter are connected to the corresponding stator leads of a torque differential transmitter, as shown in figure 21-15. The resultant stator mmf, shown by the open arrow, produced in the TX directly opposes the TX rotor mmf, shown by the solid arrow. Corresponding stator coils of the two units are in series; for example, $S_2$ of TX is in series with $S_2$ of the TDX, and the current flow produces a resultant stator mmf

of equal strength in the TDX. However, currents in corresponding stator coils of the TDX are opposite in direction. The direction of the stator mmf in the TDX is therefore opposite to the direction of the TX stator mmf, but identical to the direction of its rotor mmf.

The TDX rotor coils are angularly spaced $120°$ apart, in the same manner as the TX stator coils. The TDX stator mmf is identical to the TX rotor mmf, neglecting small circuit losses.

Before considering such an arrangement, however, it must be made clear that the controlling relationship in the TDX is the position of the TDX stator field with respect to the rotor, not with respect to the TDX stator. Suppose that the TX rotor in the previous example is turned to $75°$, and the TDX rotor to $30°$, as shown in figure 21-16. The TDX stator field is now positioned at $75°$ with respect to $S_2$, but the angle at which it cuts the TDX rotor is $45°$, using the $R_2$ axis as a reference. This is the angle which determines the signal which the TDX transmits. Notice that turning the TDX rotor $30°$ counterclockwise decreased the angle between the TDX stator field and $R_2$ by that amount.

HOW THE DIFFERENTIAL TRANSMITTER SUBTRACTS.—The manner in which the torque system containing a TDX subtracts or adds two inputs can be figuratively described as the positioning of magnetomotive forces. Figure 21-17 shows such a system connected for subtraction. A mechanical input of $75°$ is applied to the TX and converted to an electrical signal

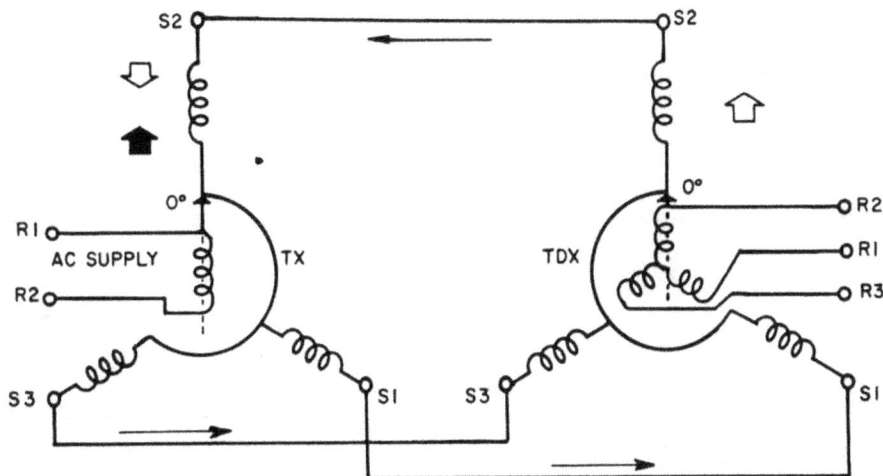

BE.413

Figure 21-15.—Position of TDX stator mmf when TX rotor is at zero degree.

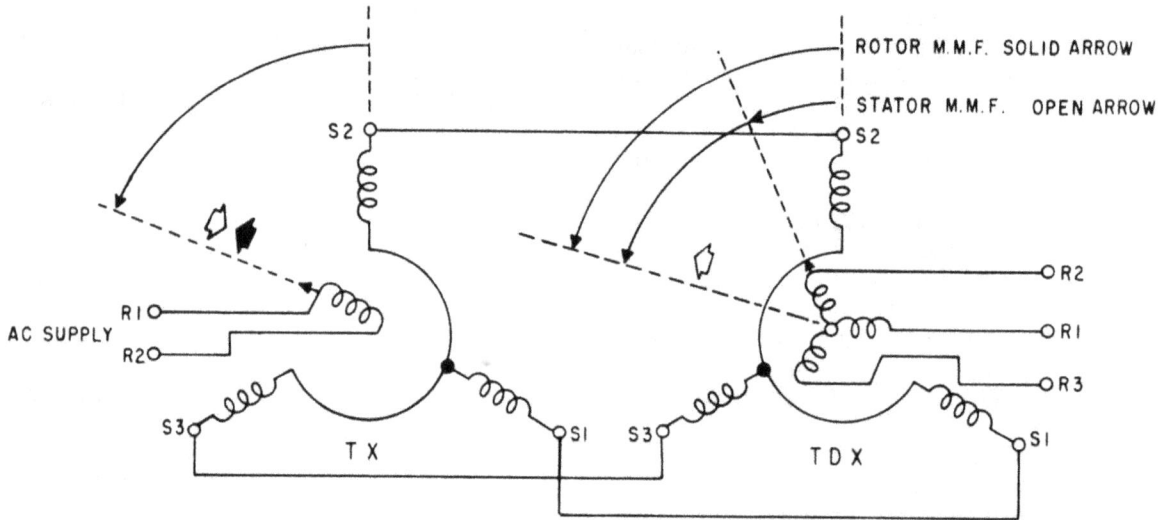

BE.414

Figure 21-16.—Rotating TDX stator field by turning the TX rotor.

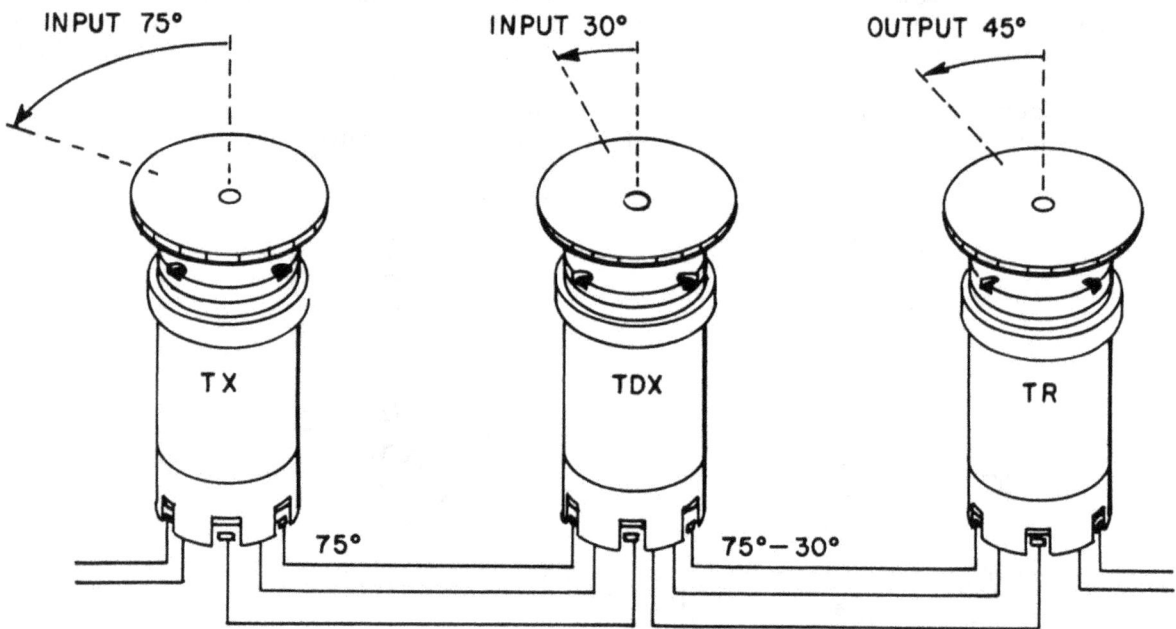

BE.415

Figure 21-17.—Subtraction with TDX.

429

which the TX transmits to the TDX stator. The TDX subtracts its own mechanical input from this signal, and transmits the result to the TR, which indicates the torque system's mechanical output by the position of its rotor.

To understand how this result is accomplished, first consider the conditions in a TX-TDX-TR system when the TX and TDX rotors are turned to $0°$, as in figure 21-18.

It has been shown how torque is developed in a synchro receiver to bring its rotor into a position which corresponds to that of an associated transmitter. The rotor voltages of the TDX depend upon the position of the magnetic field in relation to the rotor windings in the same way that the stator voltages of a TR depend on the position of the magnetic field in relation to the TR stator windings. The TDX stator field axis orients itself in the TDX stator in the same relative position that the TX rotor axis is oriented in its stator. The TR rotor therefore follows the angular position of the TDX stator field with respect to $R_2$ of the TDX. Since this is $0°$, the TR rotor turns to that position, and indicates $0°$.

HOW THE DIFFERENTIAL TRANSMITTER ADDS.—Frequently, it is necessary to set up a TX-TDX-TR system for addition. This is done by reversing the $S_1$ and $S_3$ leads between the TX and TDX stators, and the $R_1$ and $R_3$ leads between the TDX rotor and the TR. With these connections the system behaves as illustrated in figure 21-19. The $75°$ and $30°$ mechanical inputs, applied respectively to the rotor of the

TX and the rotor of the TDX, are added and transmitted to the TR, whose rotor provides an output equal to their sum by turning to $105°$.

Differential Receiver

As previously explained, the differential receiver differs chiefly from the differential transmitter in its application. The TDX in each of the previous examples combined its own input with the signal from a synchro transmitter and transmitted the sum or difference to a synchro receiver, which provided the system's mechanical output. In the case of the differential receiver in a torque system, the differential unit itself provides the system's mechanical output, usually as the sum or difference of the electrical signals received from two synchro transmitters. Figure 21-20 illustrates this operation in a system consisting of two TX's and a TDR (torque differential receiver) interconnected for subtraction.

Control Synchro Systems

There is wide usage of synchros as followup links in automatic control systems. Synchros alone do not possess sufficient torque (turning power), to rotate such loads as radar antennas or gun turrets; however, they can control power amplifying devices, which in turn can move these heavy loads. For such applications, servomotors are used. They are placed in a closed-loop servomechanism employing a special type of

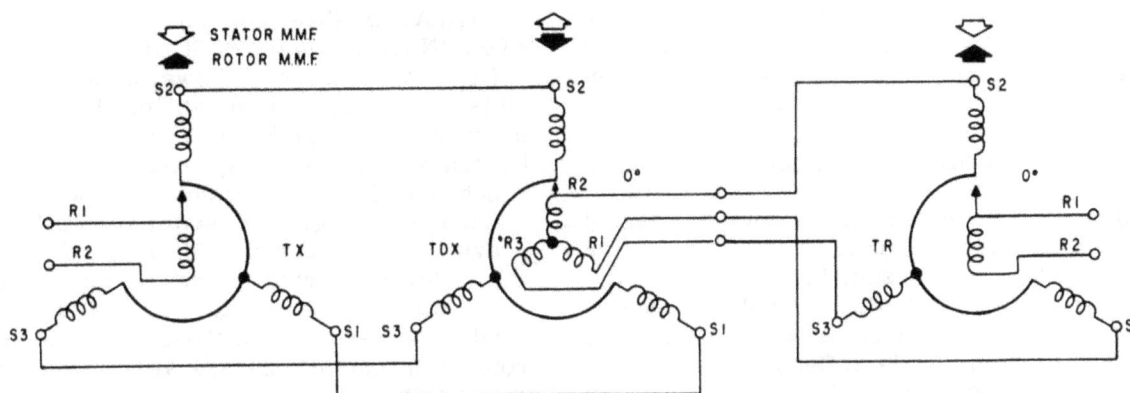

BE.416

Figure 21-18.—Magnetomotive force positions in TX-TDX-TR system with all rotors at zero degree.

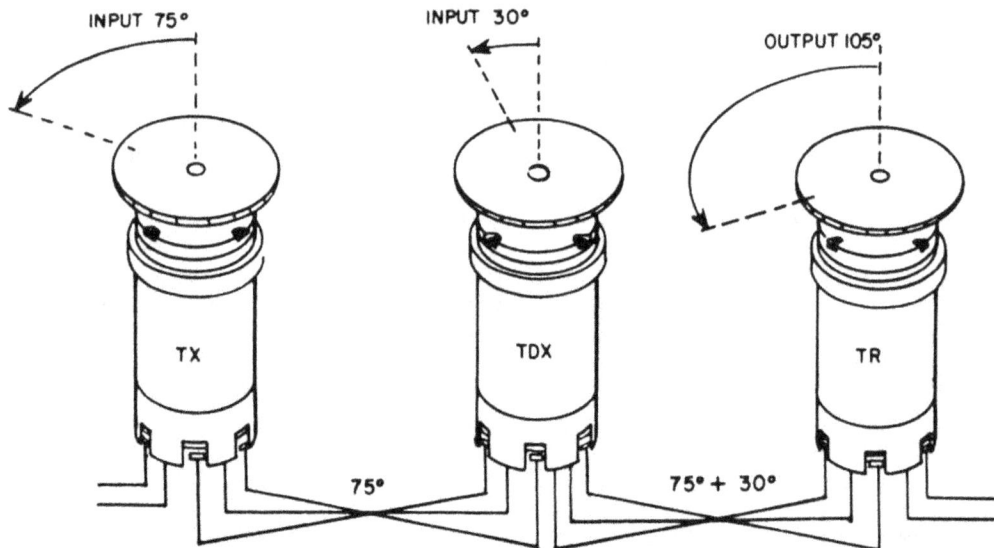

BE.417

Figure 21-19.—Addition with TDX.

synchro called a synchro control transformer (CT) to detect the difference (error) signal between the input and output of the loop. A block diagram of such a transformer is shown in figure 21-21 (A). Figure 21-21 (B) is a phantom view of a typical transformer, CT, with a drum rotor.

CONTROL TRANSFORMER.—The distinguishing unit of any synchro control system is the control transformer, CT. The CT is a synchro designed to supply, from its rotor terminals, an a-c voltage whose magnitude and phase is dependent on the rotor position, and on the signal applied to the three stator windings. The behavior of the CT in a system differs from that of the synchro units previously considered in several important respects.

Since the rotor winding is never connected to the a-c supply, it induces no voltage in the stator coils. As a result, the CT stator currents are determined only by the voltages applied to them. The rotor itself is wound so that its position has very little reflected effect on the stator currents. Also, there is never any appreciable current flowing in the rotor, because its output voltage is always applied to a high-impedance load, 10,000 ohms or more.

Therefore, the rotor does not turn to any particular position when voltages are applied to the stators.

The rotor shaft of a CT is always turned by an external force, and produces varying output voltages from its rotor winding. Like synchro transmitters, the CT requires no inertia damper, but unlike either transmitter or receivers, rotor coupling to $S_2$ is minimum when the CT is at electrical zero. (See fig. 21-22.)

RELATIONSHIP OF THE STATOR VOLTAGES IN A CT TO THE RESULTANT MAGNETIC FIELD.—When current flows in the stator circuits of a CT, a resultant magnetic field is produced. This resultant field can be rotated by the signal from a synchro transmitter, or synchro differential transmitter, in the same manner as the resultant stator field of the TDX previously considered. When the field of the CT stator is at right angles to the axis of the rotor winding, the voltage induced in the rotor winding is zero. When the stator field and the rotor's magnetic axis are alined, the induced rotor voltage is maximum. Since the CT's output is expressed in volts, it is convenient to consider its operation in terms of stator voltages as well as in terms of the position of the

431

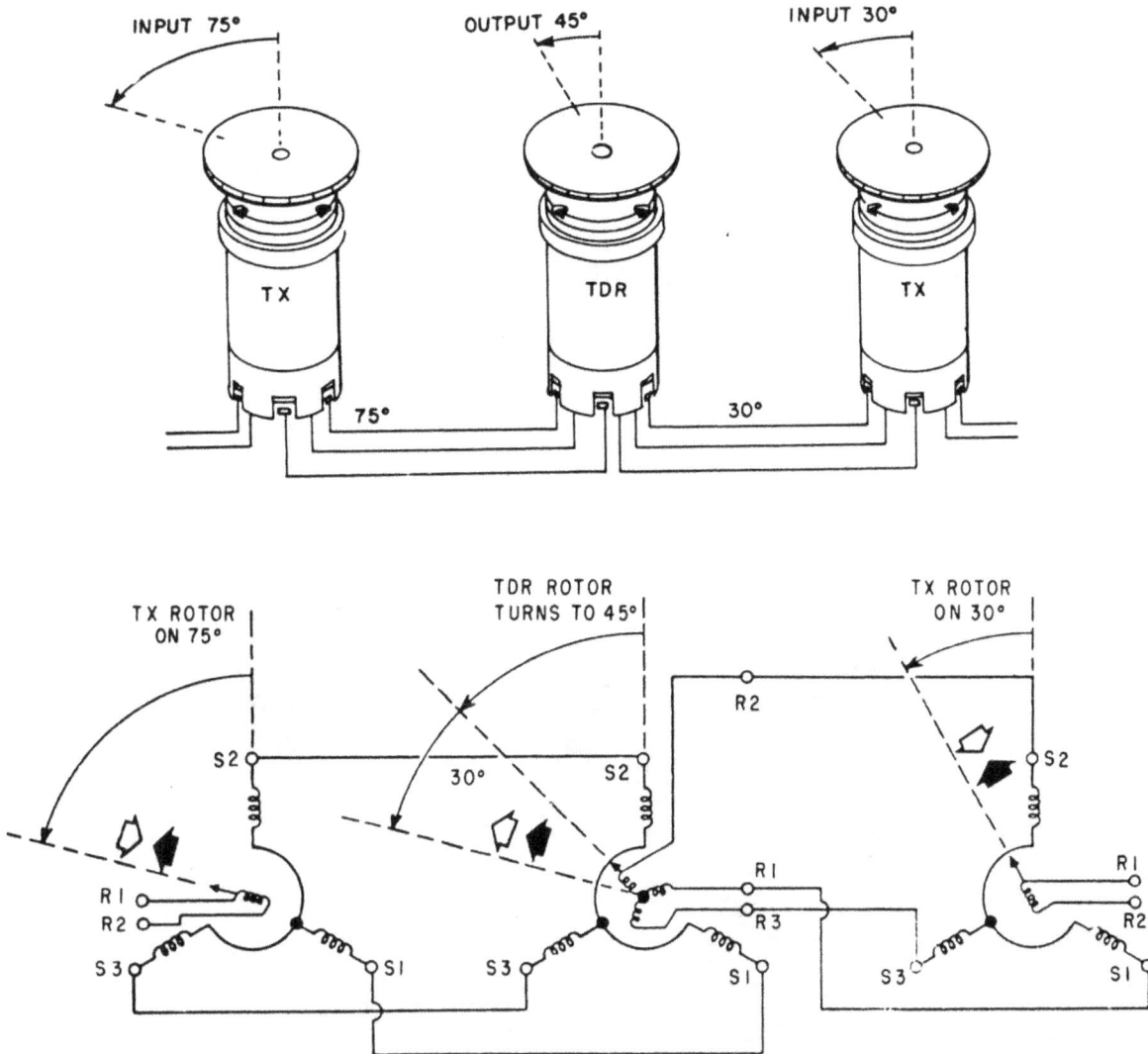

BE.418

Figure 21-20.—Subtraction with TDR.

resultant magnetic field, but it should be remembered that it is the ANGULAR POSITION, with respect to the rotor axis, that determines the output of the CT.

OPERATION OF THE CONTROL TRANSFORMER WITH A SYNCHRO TRANSMITTER.—Consider the conditions existing in the system shown in figure 21-22, where a CT is connected for operation with a TX and the rotors of both units are positioned at $0°$. The relative phases of the individual stator voltages with respect to the $R_1$ to $R_2$ voltage of the transmitter are

indicated by the small arrows. The resultant stator field of the CT is shown in the same manner as for the TDX. With both rotors in the same position, the CT stator field is at right angles to the axis of the rotor coil. Since no voltage is induced in a coil by an alternating magnetic field perpendicular to its axis, the output voltage appearing across the rotor terminals of the CT is zero.

Now assume that the CT rotor is turned to $90°$, as in figure 21-23, while the TX rotor remains at $9°$. Since the CT's rotor position

432

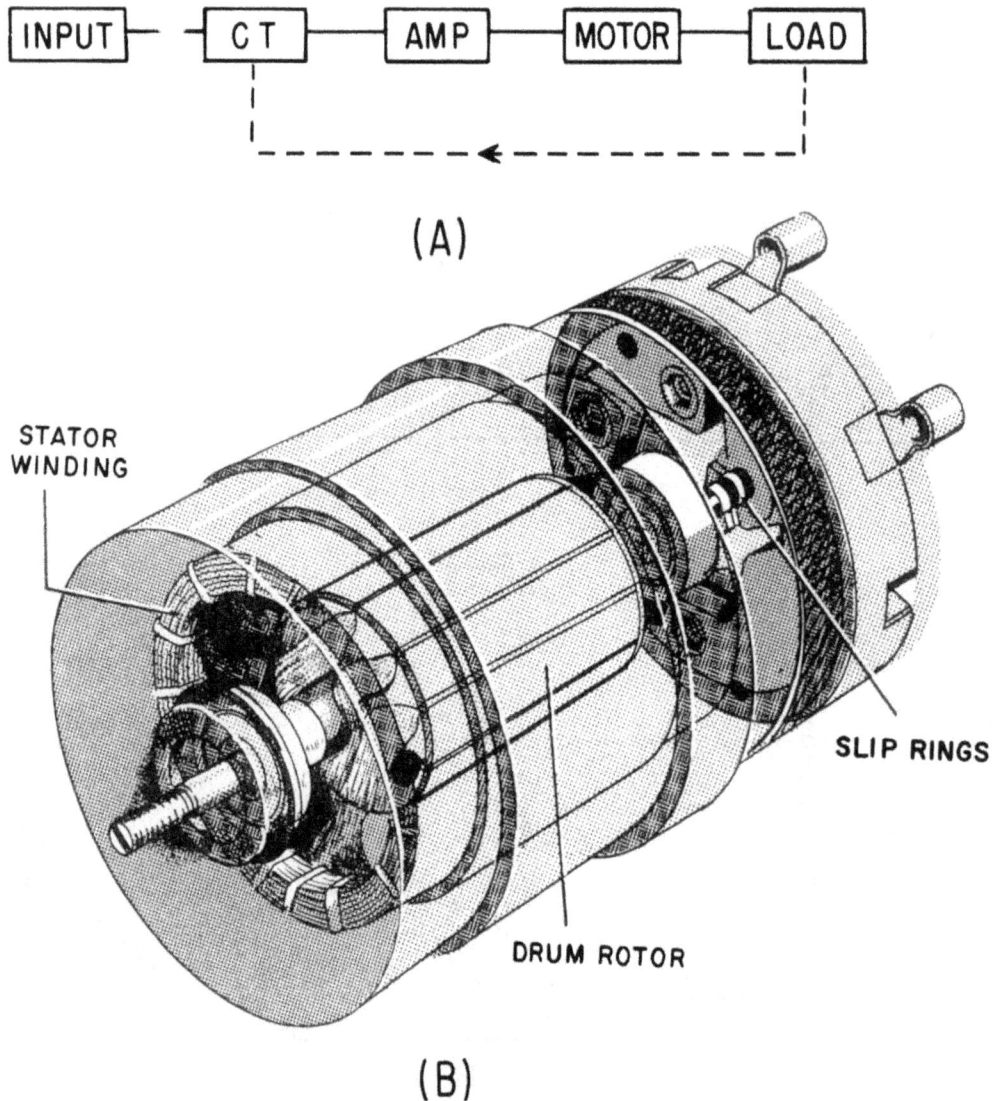

(A)

(B)

BE.419

Figure 21-21.—(A) Block diagram with control transforming; (B) typical control transformer.

does not affect stator voltages or currents, the resultant stator field of the CT remains alined with $S_2$. The axis of the rotor coil is now in alinement with the stator field. Maximum voltage, approximately 55 volts, is induced in the coil and appears across the rotor terminals as the output of the CT.

Next, assume the TX rotor is turned to 180°, as in figure 21-24. The electrical positions of the TX and CT are 90° apart, the CT

stator field and rotor axis are alined, and the CT's output is maximum again, but the direction of the rotor's winding is now reversed with respect to the direction of the stator field. The phase of the output voltage is therefore opposite to that of the CT in the preceding example. This means that the phase of the CT's output voltage indicates the direction in which the CT rotor is displaced with respect to the position-data signal applied to its stators.

433

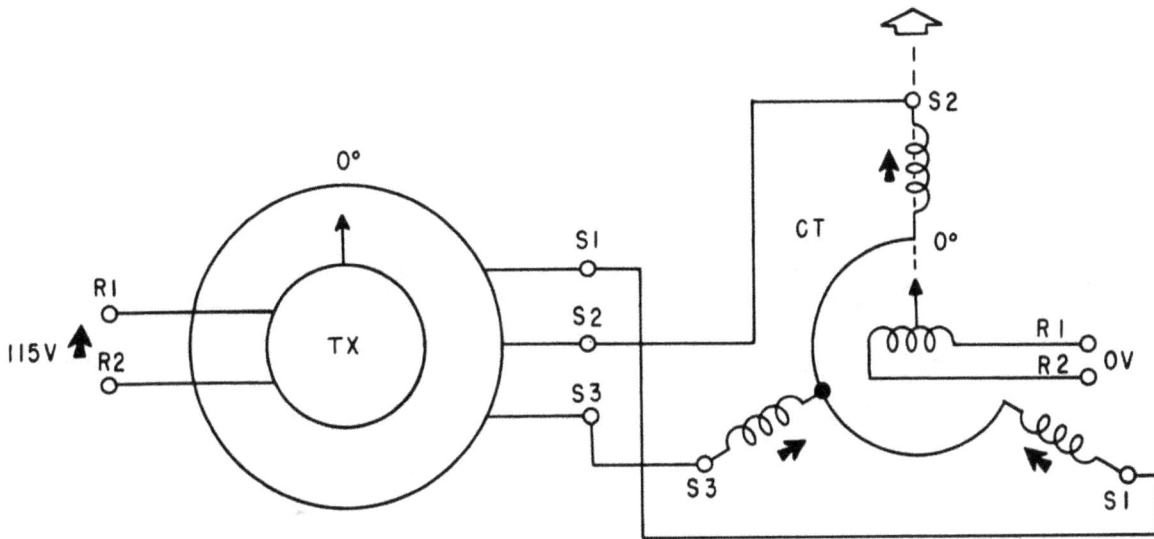

BE.420

Figure 21-22.—Conditions in CX-CT system with rotors in correspondence.

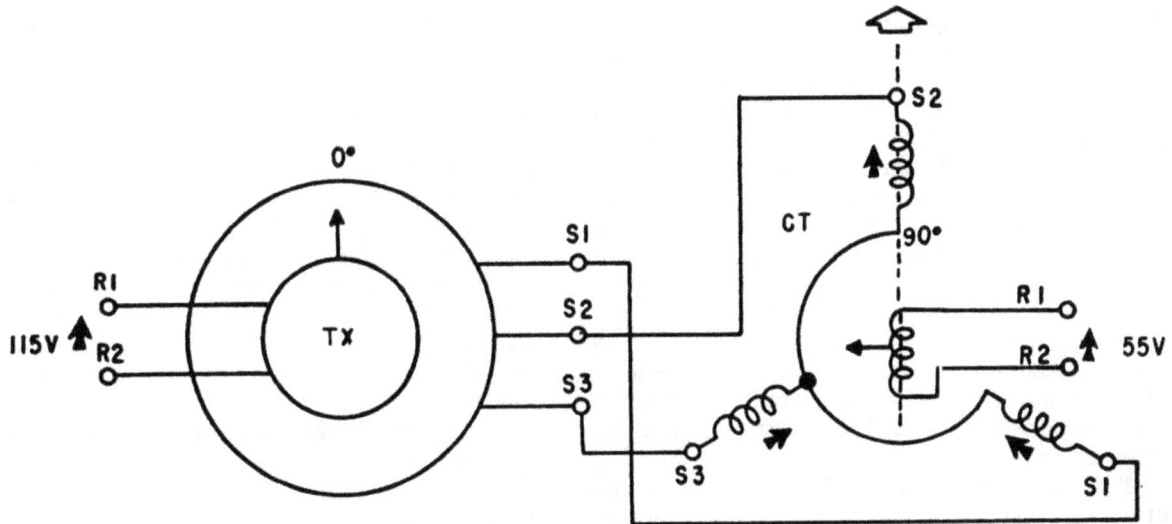

BE.421

Figure 21-23.—Conditions in TX-CT system with TX rotor at $0°$ and CT rotor at $90°$.

It is evident that the CT's output can be varied by rotating either its rotor or the position-data signal applied to its stators. It can also be seen that the magnitude and phase of the output depend on the relationship between signal and rotor rather than on the actual position of either.

### Synchro Resolvers

The synchro resolver is an electromechanical device used to generate trigonometric functions. A resolver, like the synchro, works on the principle of a variable transformer capable of unlimited rotation. In its most common form

434

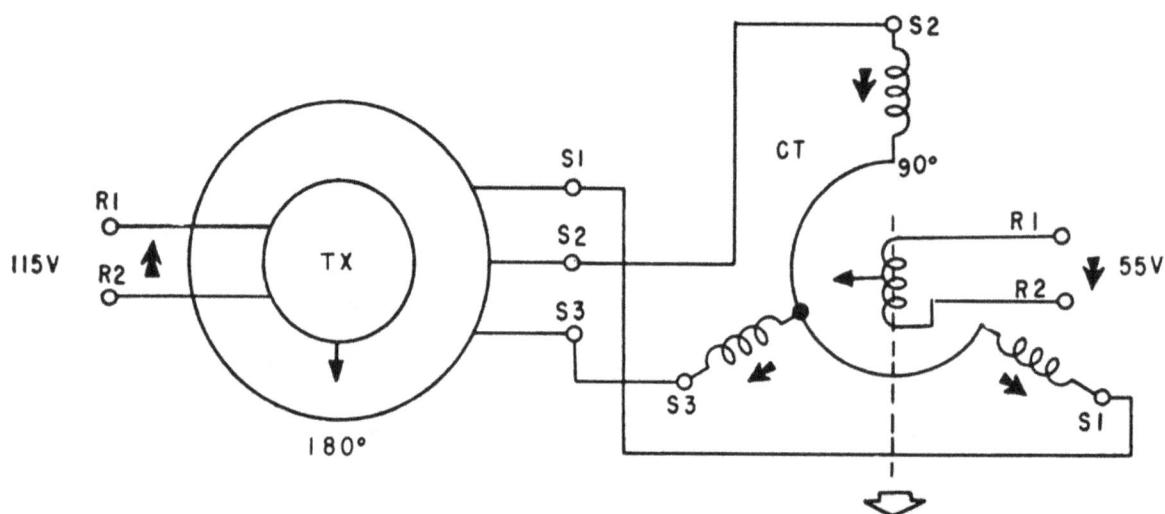

BE.422

Figure 21-24.—Conditions in TX-CT system with TX rotor at 180° and CT rotor at 90°.

it consists of a stator and a rotor, each wound with two separate coils placed precisely at right angles (90°) to each other (fig. 21-25). The two stator windings (S-1 and S-2) are normally stationary, and serve in most applications as the primaries. However, they may be used as the secondary windings in some applications of resolvers. The rotor windings (R-1 and R-2) serve as secondaries and primaries, depending on the application of the resolver. The rotor assembly can be rotated with respect to the stator windings. Thus, the rotor can be set at any angle with respect to the stator. Each winding of the stator and rotor has the same number of turns so there is a 1:1 turns ratio. The 90° displacement of the windings of the stator prevents any magnetic coupling between its windings. This also applies to the rotor.

The inputs to the resolver are (1) the magnitude of the value expressed as a voltage, and (2) its direction expressed as a mechanical rotation. Since the value has both magnitude and direction, it can be represented as a vector. The resolver is basically a right triangle solver, using windings to represent the sides and magnetic flux to represent the hypotenuse. The shaft rotation corresponds to one of the acute angles of the right triangle.

Resolvers are used extensively in computers, coordinate converters, in some radar sets, in weapon direction equipment, target designation equipment, and for data trans-

mission. These devices perform electrical computations involving (1) resolution, (2) composition, and (3) combination. Figure 21-26 shows the rotor in three different positions with respect to the stator. Since similar action takes place in all windings, only one set is shown for simplicity.

In figure 21-26 (A), the rotor and stator windings have an angle of 0° between them. With the rotor in this position, all the flux established by the stator winding voltage cuts the rotor winding and the voltage induced in the rotor is maximum. For example, with a turns ratio between the stator and rotor of 1:1 and an input voltage of 1 volt, ignoring the small transformer losses, the output voltage will be 1 volt.

In figure 21-26 (B), the rotor is turned so that the two windings are displaced by 30°. Now only a part of the stator flux cuts the rotor winding, and the voltage induced in the rotor is 0.866 volt.

In figure 21-26 (C), the rotor is turned so that the two windings are displaced by 90°. At this angle there is no magnetic coupling between the windings, and the output voltage is zero.

The output voltage corresponds numerically to the cosine function of the angular displacement between the rotor and stator. This rotor winding is therefore called the cosine winding. With the second rotor winding displaced 90° from the cosine winding, the voltage induced in this

435

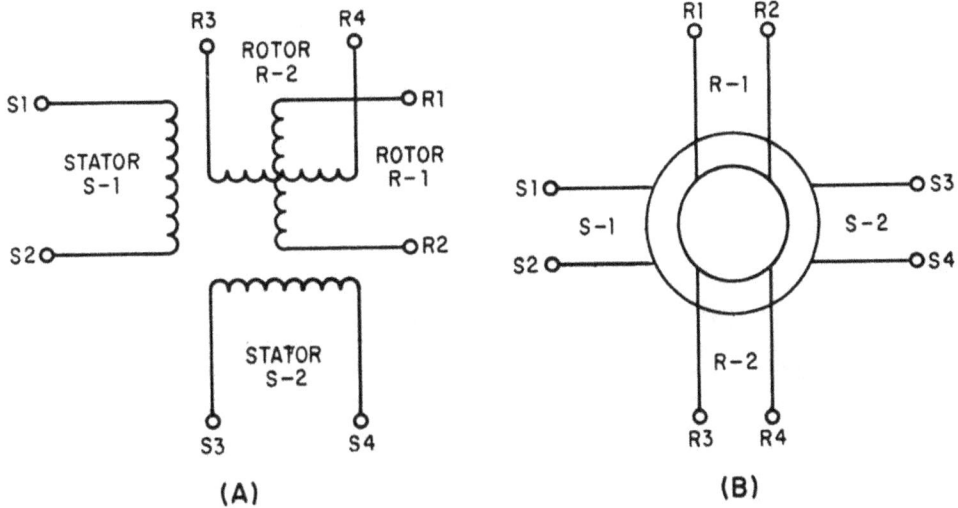

Figure 21-25.—Typical schematic symbols for resolvers.

BE.423

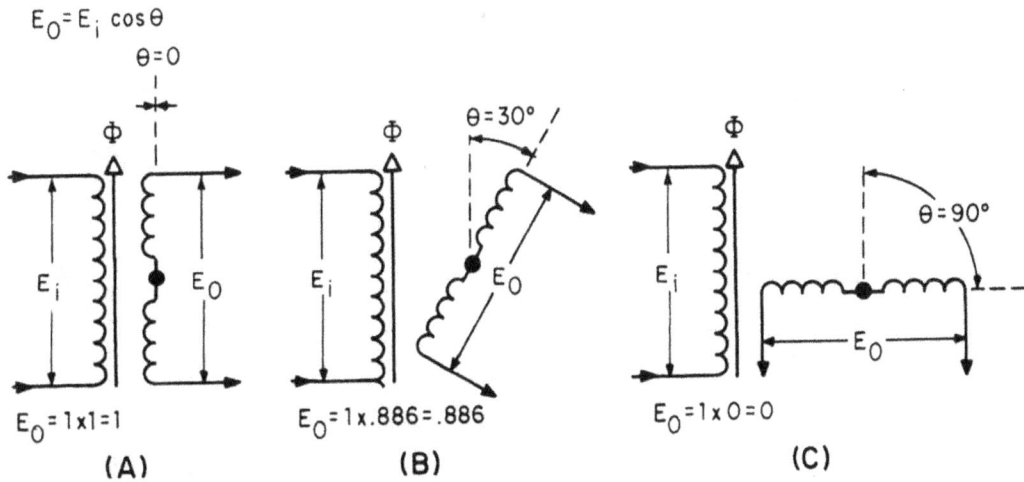

Figure 21-26.—Resolver principle, rotor and stator.

BE.424

winding would correspond to the sine function of the angle of displacement between the rotor and the stator. This naturally would be called the sine winding.

Here are some of the types of problems a resolver is capable of solving. First, the resolver can separate a vector into its two right-angle components; this is called resolution. The hypotenuse of the triangle in figure 21-27 is the vector to be resolved. The angle A represents the direction of the vector; and the voltage applied to the stator winding, given a value of unity, represents the magnitude of the vector. The output voltages, $E_{01}$ and $E_{02}$, are the answers to the vector problem. The magnitude of $E_{01}$ represents the horizontal component and $E_{02}$ the vertical component of the vector.

In figure 21-28, values have been assigned to the inputs. The angle of 30° represents the elevation of a target. The hypotenuse represents the range to the target in the slant plane. In this case unity voltage represents the maximum range of the instrument. The voltage applied to the stator is directly proportional to the target's range. The voltage induced in the cosine winding would represent the horizontal range to the target, while the voltage induced in the sine winding would represent the vertical component of the target's range—that is, its height.

Combining the two components of a vector to produce that vector is the second type of problem a resolver can solve. This is called composition. Here two sides of a right triangle are known and the resolver solves for the hypotenuse. The voltages A sin $\theta$ and A cos $\theta$ are applied to the two stator windings, as shown in figure 21-29. Each winding produces a flux field proportional to the strength of its voltage. These two flux fields are at right angles because the stator windings are at right angles. Right angle flux fields combine according to the formula

$$(\phi_1)^2 + (\phi_2)^2 = (\phi_t)^2$$

where $\phi_1$ is the field strength of stator winding 1 and $\phi_2$ is the field strength of stator winding 2.

(This formula is similar to the formula used to solve for the hypotenuse of a right triangle—the square of the hypotenuse is equal to the sum of the squares of the other two sides.)

The rotor is turned through the angle $\theta$ so that the axis of the rotor winding is alined with the axis of the resultant stator flux field. The voltage induced in the rotor winding is then proportional to the hypotenuse, which is the vector A.

In a computer, the second rotor winding of the resolver is used to furnish a signal to a servomechanism to position the rotor. The second rotor winding would have zero voltage induced when it is exactly 90° from the resultant flux field. Therefore the rotor would remain in this position and the first rotor winding would have minimum voltage induced.

By the use of more than one resolver in a computing circuit, a combination of the two basic problems can be solved simultaneously.

## TROUBLESHOOTING SYNCHRO SYSTEMS

Many times synchro troubleshooting is limited to determining whether the trouble is in the synchro, or in the system connections. Repairs can be made to the system connections; however, defective synchro units must normally be replaced.

In a newly installed system, or unit, the trouble probably is the result of improper zeroing or wrong connections. Make certain

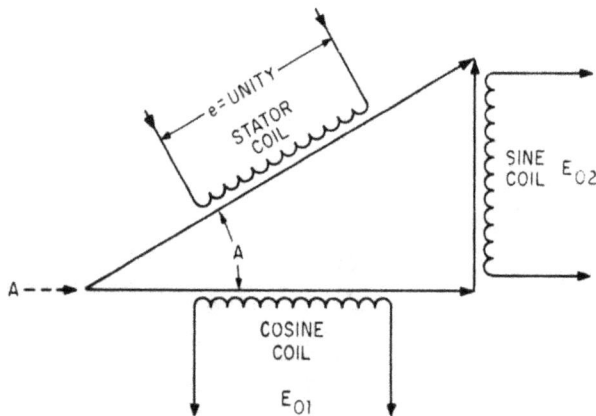

BE.425

Figure 21-27.—Resolution No. 1.

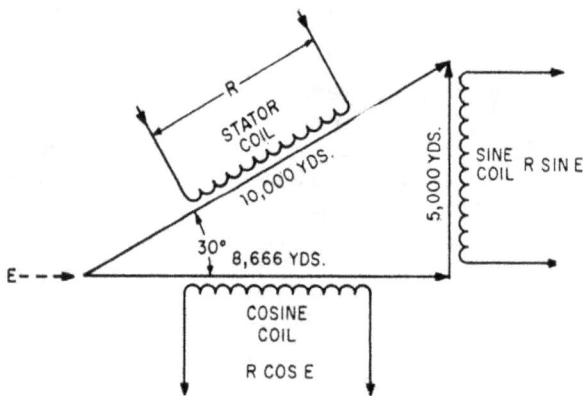

BE.426

Figure 21-28.—Resolution No. 2.

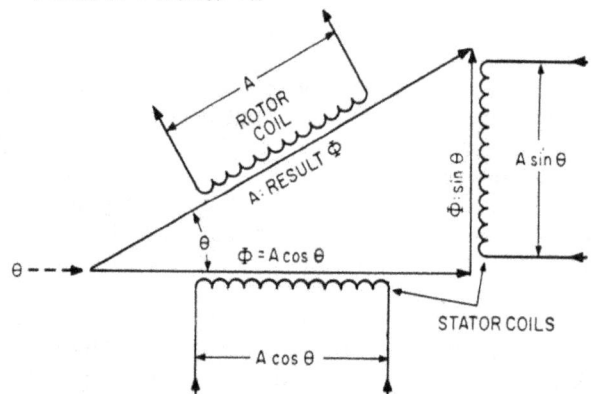

BE.427

Figure 21-29.—Composition.

all units are zeroed correctly; then check the wiring. Do not trust the color coding of the wires; check them with an ohmmeter.

In systems which have been working, the most common trouble source are as follows:

Switches . . . . . . . . . . . . Shorts, opens, grounds, or corrosion.
Nearby equipment . . . . Water or oil leaking into synchro from other devices. If this is the trouble, correct it before installing a new synchro.
Terminal boards . . . . . Loose lugs, frayed wires, or corrosion.
Zeroing . . . . . . . . . . . . Units improperly zeroed.

Wrong connections and improper zeroing in any system are usually the result of careless work or inadequate information. Do not rely on memory when removing or installing units. Refer to the applicable instruction book.

Most synchro systems involve units which are in widely separated locations. If trouble occurs in such a system, it must be localized as quickly as possible. To save time, use overload indicators and blown fuse indicators located on a control board to locate a faulty unit.

Tables 21-1 and 21-2 summarize, for a simple torque transmitter-torque receiver system, some symptoms and the possible causes. When two or more receivers are connected to one transmitter, similar symptoms occur. If all the receivers act up, the trouble is in the transmitter or the main bus. If the trouble appears in one receiver only, check that unit and its connection. The angles shown do not apply to systems using differentials, or to systems whose units are not zeroed. (See table 21-2.)

Table 21-1.—Synchro troubles.

General

Preliminary actions: Be sure TR is not jammed physically.
Turn TX slowly in one direction and observe TR.

| Symptoms | Trouble |
|---|---|
| Overload indicator lights. Units hum at all TX settings. One unit overheats. TR follows smoothly but reads wrong. | Rotor circuit open or shorted. See below. |
| Overload indicator lights. Units hum at all except two opposite TX settings. Both units overheat. TR stays on one reading half the time, then swings abruptly to the opposite one. TR may oscillate or spin. | Stator circuit shorted. See table 21-2 |
| Overload indicator lights. Units hum on two opposite TX settings. Both units get warm. TR turns smoothly in one direction, then reverses | Stator circuit open. See table 21-2. |
| TR reads wrong, or turns backward, follows TX smoothly. | Unit interconnections wrong. Unit not zeroed. See table 21-2. |

Open or Shorted Rotor

Preliminary action: Set TX to 0° and turn rotor smoothly counterclockwise.

| Symptoms | Trouble |
|---|---|
| TR turns counterclockwise from 0 to 180° in a jerky or erratic manner, and gets hot. | TX rotor open. |
| TR turns counterclockwise from or 180° in a jerky or erratic manner. TX gets hot. | TR rotor open. |
| TR turns counterclockwise from 90 or 270°, torque is about normal, motor gets hot, TX fuses blow. | TX rotor shorted. |
| TR turns counterclockwise from 90 or 270° torque is about normal, TX gets hot, TR fuses blow. | TR rotor shorted. |

Table 21-2.—Synchro troubles, stator.

### Shorted Stator

| Symptoms | | Trouble |
|---|---|---|
| Setting or conditions | Indication | |
| When TX is on 120° or 300° but | Overload Indicator goes out and TR reads correctly. | Stator circuit shorted from $S_1$ to $S_2$. |
| When TX is between 340° and 80°, or between 160° and 260° | Overload Indicator lights, units get hot and hum, and TR stays on 120° or 300° or may swing suddenly from one point to the other. | |
| When TX is on 60° or 240° but | Overload Indicator goes out and TR reads correctly. | Stator circuit shorted from $S_2$ to $S_3$. |
| When TX is between 280° and 20°, or between 100° and 200° | Overload Indicator lights, units get hot and hum, and TR stays on 60° or 240° or may swing suddenly from one point to the other. | |
| When TX is on 0° or 180° but | Overload Indicator goes out and TR reads correctly. | Stator circuit shorted from $S_1$ to $S_3$. |
| When TX is between 40° and 140°, or between 220° and 320° | Overload Indicator lights, units get hot and hum, and TR stays on 0° or 180° or may swing suddenly from one point to the other. | |
| All positions of TX. | Overload Indicator on continuously both units get very hot and hum, and TR does not follow at all or spins. | All three stator leads shorted together. |

### Open Stator

| Symptoms | | Trouble |
|---|---|---|
| Seetting or conditions | Indication | |
| When TX is on 150° or 330° When TX is held on 0° | TR reverses or stalls and Overload Indicator lights. TR moves between 300° and 0° in a jerky or erratic manner. | $S_1$ stator circuit open. |
| When TX is on 90° or 270° When TX is held on 0° | TR reverses or stalls and Overload Indicator lights. TR moves to 0° or 180°, with fairly normal torque. | $S_2$ stator circuit open. |

Table 21-2.—Synchro troubles, stator—Continued

## Open Stator

| Symptoms | | Trouble |
|---|---|---|
| Setting or conditions | Indication | |
| When TX is on 30° or 210°<br>When TX is held on 0° | TR reverses or stalls and Overload Indicator lights<br>TR moves between 0° and 60° in a jerky or erratic manner. | $S_3$ stator circuit open. |
| When TX is set at 0°, and then moved smoothly counter-clockwise | TR does not follow, no Overload Indication, no hum or over-heating. | Two or three stator leads open or both rotor circuits open. |

## Wrong Stator Connections, Rotor Wiring Correct

These problems must be worked using a stationary pointer on the chassis and a compass card mounted on the rotor.

| Setting or conditions | Indication | Trouble |
|---|---|---|
| TX set to 0° and rotor turned smoothly counterclockwise | TR indication is wrong, turns clockwise from 240°. | $S_1$ and $S_2$ stator connections are reversed. |
| | TR indication is wrong, turns clockwise from 120°. | $S_2$ and $S_3$ stator connections are reversed. |
| | TR indication is wrong, turns clockwise from 0°. | $S_1$ and $S_3$ stator connections are reversed. |
| | TR indication is wrong, turns counterclockwise from 120°. | $S_1$ is connected to $S_2$, $S_2$ is connected to $S_3$ and $S_3$ is connected to $S_1$. |
| | TR indication is wrong, turns counterclockwise from 240°. | $S_1$ is connected to $S_3$, $S_2$ is connected to $S_1$, and $S_3$ is connected to $S_2$. |

## Wrong Stator and/or Reversed Rotor Connections

These problems must be worked using a stationary pointer on the chassis and a compass card mounted on the rotor.

440

Table 21-2.—Synchro troubles, stator—Continued

| Setting or conditions | Indication | Trouble |
|---|---|---|
| TX is set to 0° and rotor turned smoothly counterclockwise | TR indication is wrong, turns counterclockwise from 180° | Stator connections are correct, but rotor connections are reversed. |
| | TR indication is wrong, turns clockwise from 60° | Stator connections $S_1$ and $S_2$ are reversed, and rotor connections are reversed. |
| | TR indication is wrong, turns clockwise from 300° | Stator connections $S_2$ and $S_3$ are reversed, and rotor connections are reversed. |
| | TR indication is wrong, turns clockwise from 180° | Stator connections $S_1$ and $S_3$ are reversed, and rotor connections are reversed. |
| | TR indication is wrong, turns counterclockwise from 300° | $S^1$ is connected to $S_2$, $S_2$ is connected to $S_3$, $S_3$ is connected to $S^1$, and rotor connections are reversed. |
| | TR indication is wrong, turns counterclockwise from 60° | $S_1$ is connected to $S_3$, $S_2$ is connected to $S_1$, $S_3$ is connected to $S_2$, and rotor connections are reversed. |

## SERVOMECHANISMS

In the operation of electrical and electronic equipment, it is often necessary to operate a mechanical load that is remotely located from its source of control. Examples of this are the movement of heavy radar antennas or small indicator motors. The mechanical load may require either high or low torque movement with variable speed and direction.

It can be stated that a servomechanism is an electromechanical device that positions an object in accordance with a variable signal. The signal source may be capable of supplying only a small amount of power. A servomechanism operates to reduce difference (error) between two quantities. These quantities are usually the CONTROL DEVICE position and the LOAD position.

The essential components of a servomechanism system are the INPUT CONTROLLER and OUTPUT CONTROLLER.

The input controller provides the means whereby the human operator may actuate or operate the remotely located load. This may be achieved either mechanically or electrically. Electrical means are most commonly used. Synchro systems and bridge circuits are most widely used for input control of servomechanism systems.

The output controller of a servomechanism system is the component or components in which power amplification and conversion occur. This power is usually amplified by vacuum-tube amplifiers or magnetic amplifiers. In many applications a combination of these are used. The power from the amplifier is then converted by the servomotor into mechanical motion of the direction required to produce the desired function. This sequence of functions is shown in figure 21-30. The figure shows a simplified block diagram of a simple servomechanism system.

## OPERATION OF BASIC SERVOMECHANISM

The block diagram of figure 21-30, is shown schematically in figure 21-31. Its basic principles of operation are as follows:

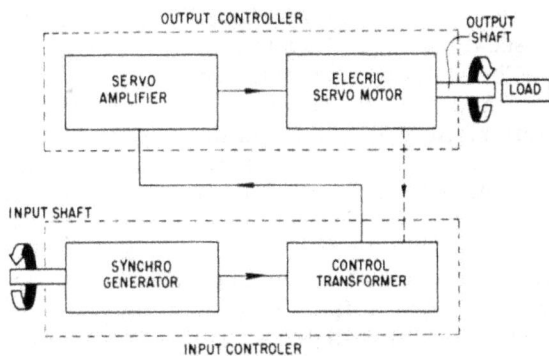

BE. 428

Figure 21-30.—Simplified block diagram of a servomechanism.

The INPUT CONTROLLER, which consists of a synchro transmitter (TX) and a synchro controller transformer (CT), originates, or commands, the system movement. The TX's rotor is attached to a shaft, which is turned by hand or by a controlling mechanical device. Movement of the TX rotor from its electrical zero will cause an unbalance of voltages in the stator windings $S_2$, $S_1$, and $S_3$. (This was explained under the heading, Control Synchro Systems.) A voltage is induced in the rotor of the control transformer. Its MAGNITUDE depends on the AMOUNT of displacement of the TX rotor; its PHASE RELATIONSHIP depends on the DIRECTION of displacement from electrical zero. This voltage represents the ERROR voltage. Since a control transformer is not designed to furnish enough power to drive a load of any significant size, the error voltage must be amplified before it is powerful enough to drive the servomotor, and thus move the load.

The servoamplifier, as stated before, may be an electronic type or magnetic type. In many cases, it is a combination of both. In the combination type, the electronic section amplifies the ERROR VOLTAGE, and the error voltage in turn controls a magnetic amplifier. The POWER amplification occurs in the magnetic amplifier. The various ways that a magnetic amplifier does this were explained in chapter 20, Magnetic Amplifiers.

Regardless of the type of amplifier used, the output of the power amplifier is connected to the control field of the servomotor. In this illustration, a single-phase induction motor is used. The fixed field is energized at all times but cannot turn the motor shaft unaided. The control field is energized only when an error voltage appears at the control transformer when the control field is energized, the motor operates.

The direction of rotation of the motor is determined by the phase relationship of the voltage applied to the fixed field to that of the control field. Varying phase relationships occur in the control field only, because its voltage (magnitude and phase relationship) is determined by the direction of displacement of the control transmitter rotor.

Once the input controller in the system just discussed produces an error voltage, the servomotor continues to turn until the rotor of the transmitter is again in its electrical zero (null) position. This is called an OPEN LOOP control system, because the servo output has no means of changing the input during operation.

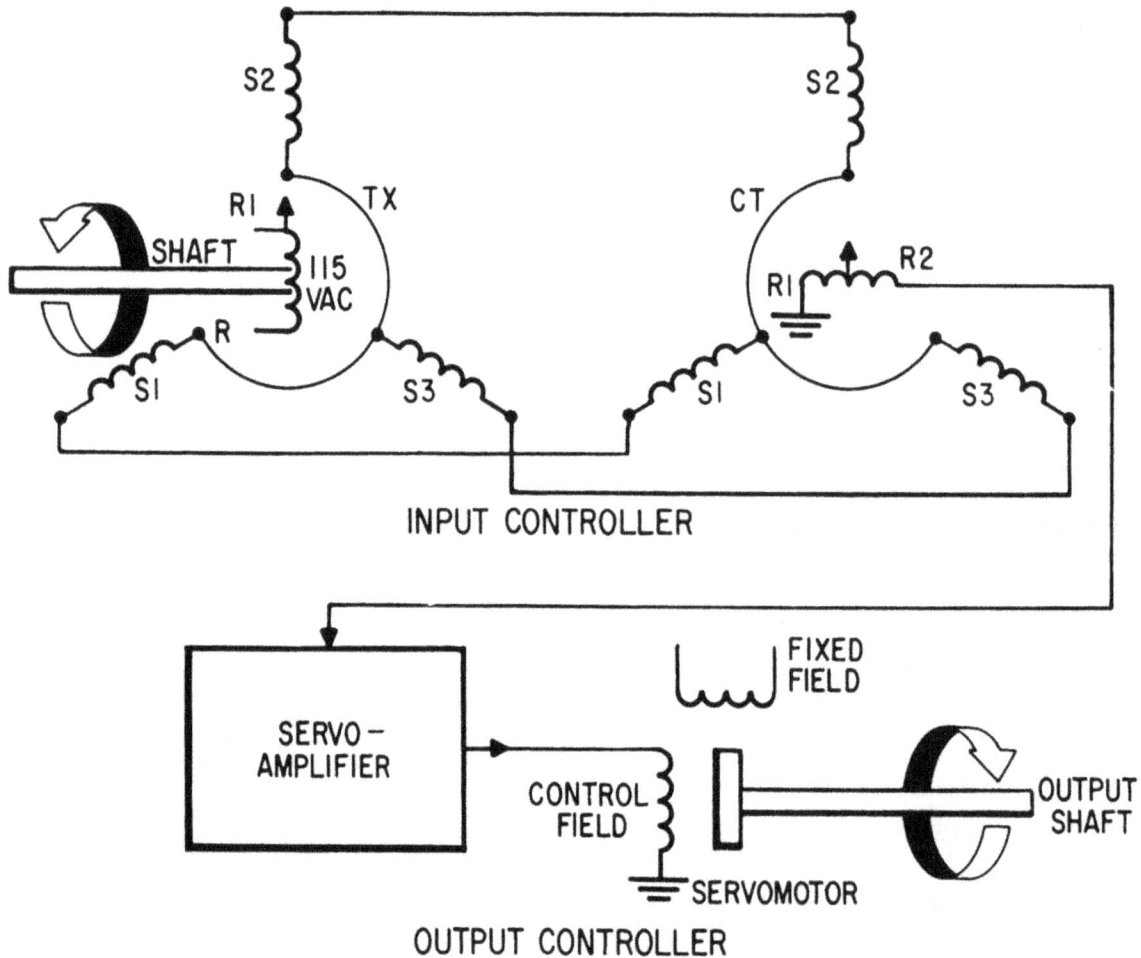

BE.429

Figure 21-31.—Schematic diagram of figure 21-30.

Figure 21-32 shows the simplified block diagram of a servomechanism system that is much more widely used. It is the CLOSED-LOOP control system. Note that it is similar to the open-loop system, except that it contains an additional followup signal stage between the servo output and the amplifier input.

In the closed-loop control system, the servomotor can be made to stop at any position, without returning the rotor of the transmitter to its electrical zero position. To accomplish this, an error voltage controlled by the servo is necessary. This error voltage is called a FOLLOWUP voltage.

To provide this followup voltage, there is usually a bridge element or a control transformer driven by the servo's output shaft. The electrical connection is such that the bridge or

control transformer generates a voltage opposing the error voltage of the input controller. Figure 21-33 shows a schematic diagram of the closed-loop control system.

The followup control transformer is basically the same in construction as those discussed earlier. Note that the stator is excited by the same a-c voltage as the rotor of the control transmitter. This voltage is connected only across two of the stator windings. The third stator winding is not used. With a voltage applied across the two windings 120° apart, a resultant magnetic field appears midway (60°) between the two windings. If the rotor of the followup control transmitter is placed so that its axis is 90° from the field axis, the voltage induced in the rotor from the two stator windings is zero. It can be seen that if the rotor

443

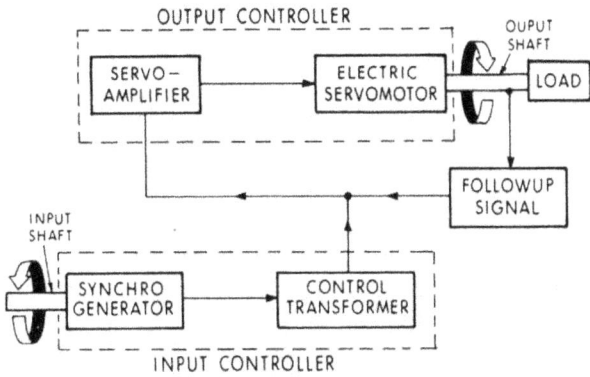

OUTPUT CONTROLLER

INPUT CONTROLLER

BE.430

Figure 21-32.—Closed-loop servomechanism.

is moved from this zero-voltage position, a voltage will be induced into the rotor. The magnitude and phase-relationship of this voltage is dependent upon the amount of displacement, and the direction of rotation of the rotor from its zero position.

When a followup control transformer is used in a servomechanism system, as illustrated in figure 21-33, the rotor is usually driven through a gear train by the servomotor's output shaft. The gear ratio will be such that the output shaft will turn many times before the rotor is turned any appreciable amount. Ratios of 1,500 to 1 are common. The servomotor output shaft may either be connected directly to its load, or geared up or down.

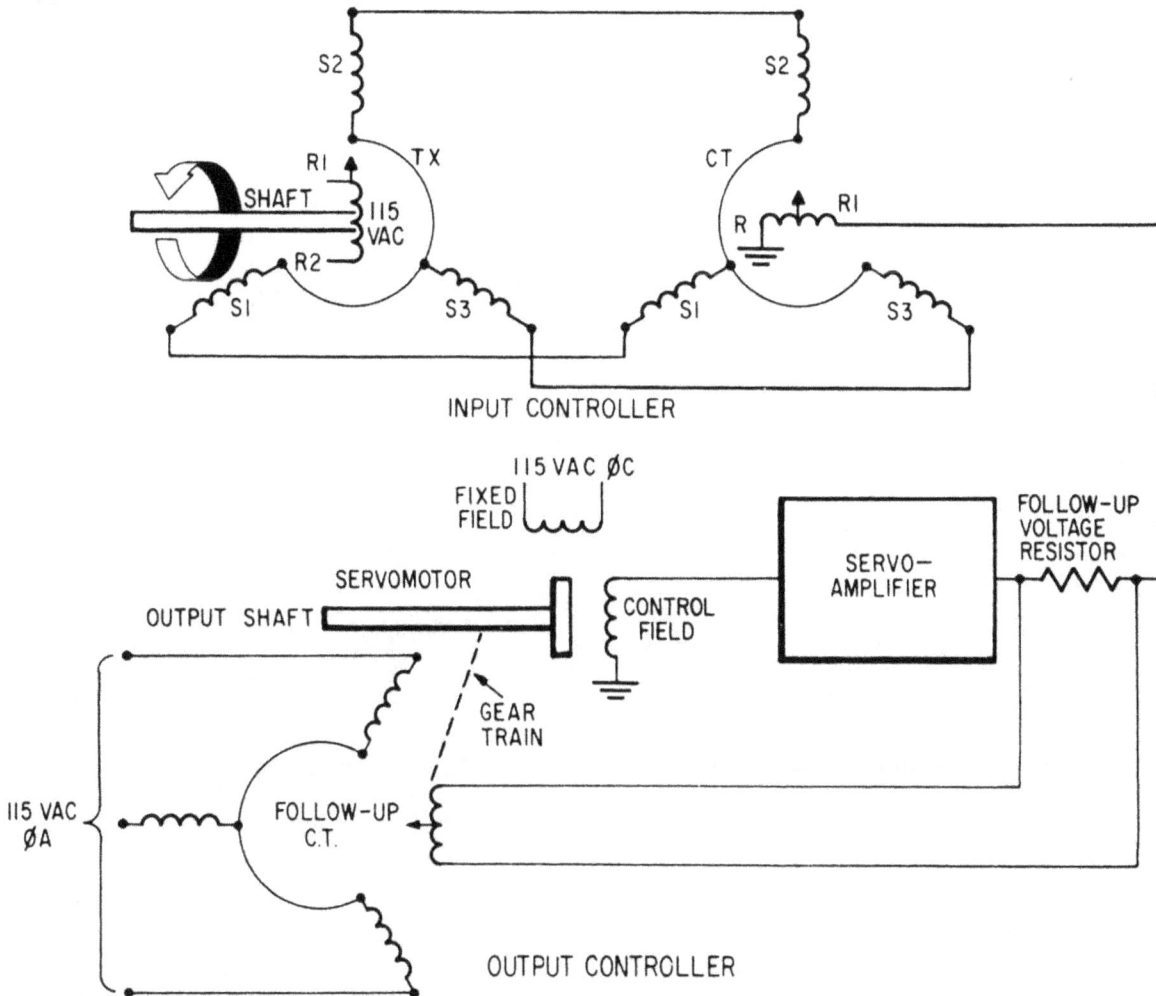

BE.431

Figure 21-33.—Closed-loop control system.

444

The principle of operation of the closed-loop servosystem is as follows:

At the static position (electrical zero) the output voltage of the INPUT CONTROLLER will be zero, and the output voltage of the FOLLOWUP TRANSFORMER will also be zero. Thus, no error signal is sent to the amplifier, and the voltage to the control field of the servomotor is zero.

When the input controller's shaft is rotated (either left or right) a voltage is induced into the rotor of the control transformer. This voltage is connected through the followup resistor to the amplifier, where it is sent through the necessary stages of amplification, and drives the servomotor. The direction of rotation of the servomotor shaft is determined by the direction of the TX rotor displacement, which controls the phase relationship between the motor's two fields. The variable control field will always lead or lag the fixed control field by 90°. As the servomotor shaft turns, it drives the load in the direction commanded. It also drives the gear train of the followup control transformer rotor. Driving the followup CT rotor from its electrical zero position causes a voltage to be induced into the rotor by the excited stator. This voltage will always be 180° OUT OF PHASE with the voltage generated by the INPUT CONTROLLER. (This is done by the gear arrangement which drives the followup CT rotor). The followup voltage is developed across the same followup voltage resistor through which command signals are sent to the amplifier. As the servomotor rotates, the followup voltage increases. When the followup voltage becomes equal to the command voltage, being opposite, the error voltage sensed by the amplifier will be zero. The servomotor stops rotating. Note that it is unnecessary to reposition the transmitter's rotor (input controller) to electrical zero in order to stop rotation of the servomotor. The servomotor will rotate to move the load to whatever position is commanded by the control transmitter.

## WARD-LEONARD SYSTEM

As previously stated, one of the basic requirements of a servomechanism is that the drive motor has variable speed and reversible rotation capabilities. Since the ordinary single-phase or 3-phase a-c motor is inherently a constant-speed device, the direct current motor is commonly used for controlled drives. The

direction of rotation can be changed readily by reversing either the armature current or the field current.

The speed can be controlled by the voltage on the armature.

The circuit shown in figure 21-34, commonly known as the Ward-Leonard system, accomplishes this result. The d-c motor is this circuit is fed directly from a d-c generator which is operated at a constant speed. The d-c field supply to the generator is variable in both magnitude and polarity by means of a rheostat and reversing switch as shown. Therefore, the motor armature is supplied by a generator having smoothly varying voltage output from zero to full-load value. The motor field is supplied with a constant voltage from the same source as that supplying the generator fields. The generator drive power could be from a single-phase or the 3-phase a-c motor, from an engine, or from any other constant speed source. In the same way the d-c field supply can be supplied from a rectifier, from an exciter on the end of the generator shaft, or from any other suitable d-c source.

The advantage of the Ward-Leonard system is that by means of the variation of a small field current, a smooth, flexible, yet stable control can be maintained over the speed and direction of rotation of a d-c motor. Such systems are applicable to ship propulsion, hoists, and elevators, diesel electric equipment and to the rotation of gun turrets, radar antennas, and similar heavy equipment. The action of the system is very much like that of an amplifier since a very small amount of power is used to control greatly increased power.

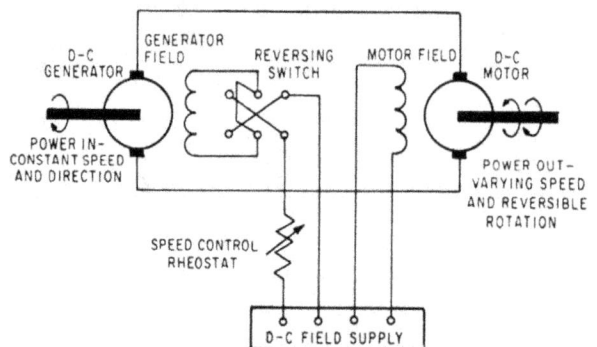

BE.432

Figure 21-34.—Ward-Leonard drive.

A simple Ward-Leonard drive for a radar antenna is shown in figure 21-35. The d-c generator in this system is driven by a 230-volt single-phase a-c motor. The same a-c line supplies a rectifier which furnishes field supply for both the d-c generator and the d-c motor fields. However, the generator field is connected to a potentiometer in such a way that the magnitude and polarity of the applied voltage can be varied. By varying the setting of the potentiometer control knob, the antenna can be rotated in either direction and at any speed from zero to full rate.

The system shown is an open-loop type and is applicable to a search type radar where the speed and direction of the antenna rotation is under direct control of the operator. This system could be modified for use as a closed-loop system by providing feedback from the antenna drive mechanism to the speed and direction control potentiometer so that the voltage supplied to the field of the d-c generator would be reduced to zero when the antenna position was in correspondence with the input order. One method of providing feedback for the system shown in figure 21-35 is by use of a simple transmitter-receiver system and a mechanical differential, illustrated in figure 21-36.

## AMPLIDYNES

As has been stated earlier in this chapter, SYNCHROS are used for the transmission of angular motion without developing a large amount of torque.

DIFFERENTIAL SYNCHROS are used for combining, in the desired manner angular motion from two different sources, again without developing a large amount of torque.

SYNCHRO CONTROL TRANSFORMERS are used to produce an output voltage that is proportional to the angular difference between the input and output shafts of a servomechanism.

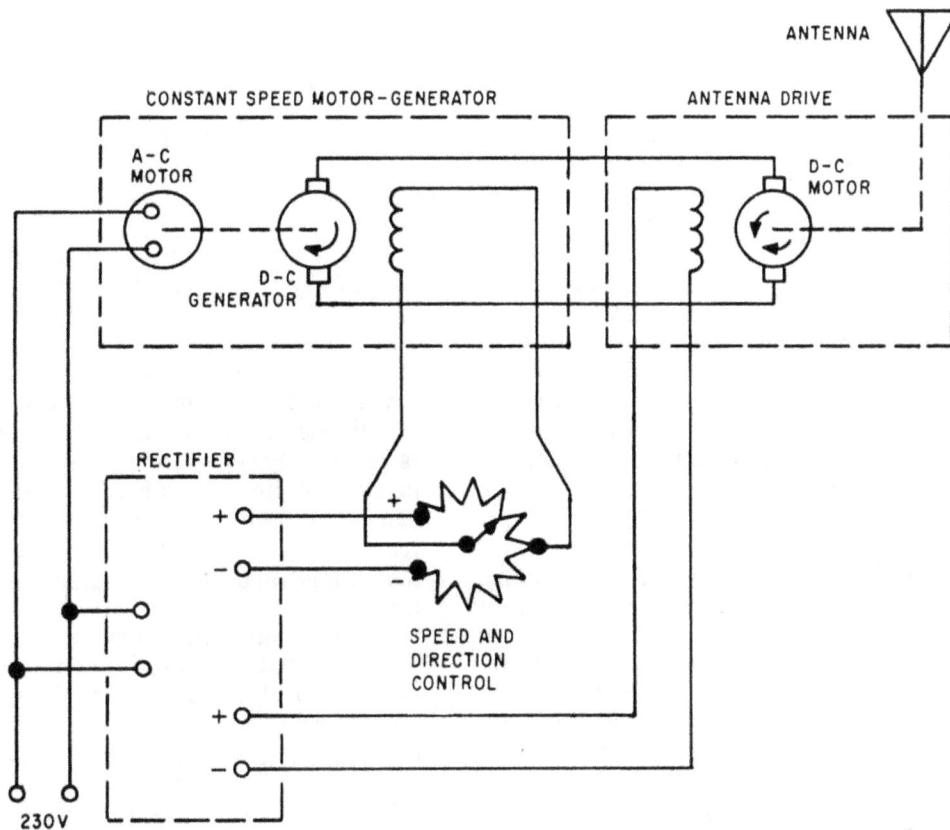

Figure 21-35.—Simple radar antenna drive.

BE.433

446

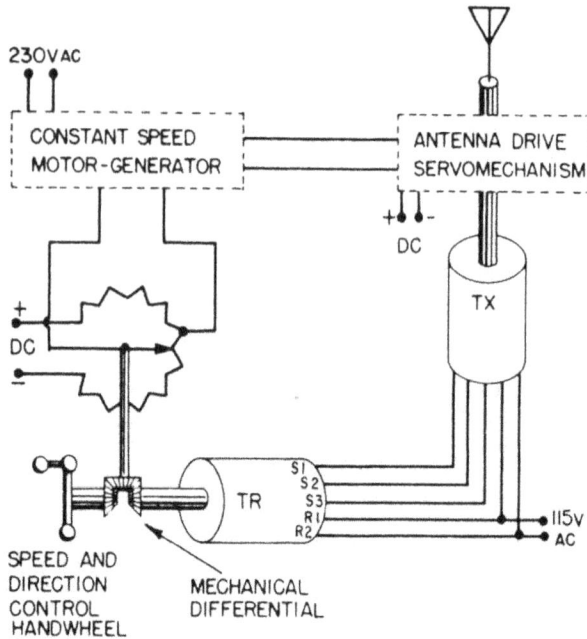

BE.434

Figure 21-36.—A feedback system for Ward-Leonard type servo control.

The error voltage from the control transformer is fed to a CONTROL AMPLIFIER (amplifies the output of the control transformer) which increases the amplitude of the error signal. In an amplidyne power drive the output of the control amplifier supplies the field coils of an amplidyne generator. A small variation in the strength of the current in the field coils of the amplidyne generator causes a great variation in its output power. In this respect the amplidyne is a d-c amplifier. Thus, the signal developed in the control amplifier can cause the amplidyne generator to supply enough output power to operate the d-c servomotor which has sufficient torque to move a heavy load.

The amplidyne power drive, as commonly used, consists of an amplidyne generator, a driving motor, and a d-c servomotor, as shown in simplified form in figure 21-37. The amplidyne generator has two sets of brushes on the commutator. One set is short-circuited and the other set supplies the armature of the servomotor. The field of the servomotor and the control field of the amplidyne are separately excited. The amplidyne control field has a split winding, one for each polarity of the applied signal. The series compensating winding is

discussed later in the text. The amplidyne drive motor is ordinarily a 3-phase a-c induction motor.

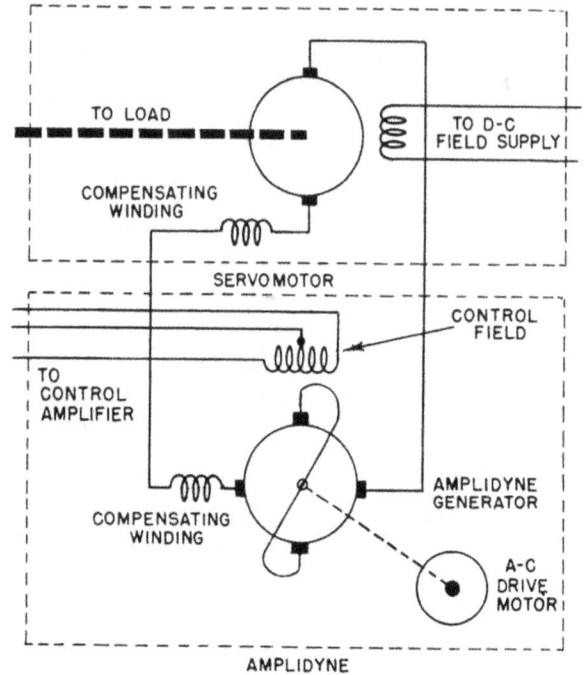

BE.435

Figure 21-37.—Basic amplidyne drive.

Operation

The amplidyne generator is a small separately excited direct-current generator having a control field that requires a low power input. The armature output voltage may be 100 volts and the output load current 100 amperes, giving an output power of 10 kilowatts (kw.). The amplidyne is essentially a control device that is extremely sensitive. For example, an increase in control field power from zero to 1 watt can cause the generator output power to increase from zero to 10 kw. It is thus a power amplifier that increases the power by an amplification factor of 10,000. The following analysis shows how this high amplification is accomplished.

In an ordinary 2-pole generator (fig. 21-38) the separately excited control field may take an input power of 100 watts in order to establish normal magnetism in the field poles and a normal armature voltage of 100 volts. When this voltage is impressed on a 1-ohm load the armature will deliver 100 amperes, or an output power of 10 kw. The circles with dots represent armature conductors carrying current toward

447

the observer; those with crosses represent electron flow away from the observer. The control field winding develops a north pole (N) on the left-hand field pole and a south pole (S) on the right-hand field pole.

The relative strength of the control field is represented by vector OA in figure 21-38 (B). The armature acting as an electromagnet establishes a quadrature, or cross magnetomotive force having an axis that coincides with that of the brushes. The relative strength and position of this armature cross magnetizing force is indicated by vector OB in figure 21-38 (B). It has about the same length as the control field vector, OA, and is perpendicular to it.

If the strength of the control field in figure 21-38 (A) is reduced to 1 percent of normal, the armature generated voltage will fall to 1 percent of normal, or to about 1 volt in this example. The output current from the armature will fall to 1 ampere through the 1-ohm load.

Normal armature current may be restored by short-circuiting the brushes, as shown in figure 21-39 (A). One volt acting through an internal armature resistance of 0.01 ohm will circulate 100 amperes between the brushes. The control field strength is still 1 percent of its normal value (vector OA, fig. 21-39 (B)), but the armature crossfield mmf has been restored to its normal value, as indicated by vector OB.

The armature crossfield is cut by the armature conductors and may be regarded as being responsible for normal load voltage, which is obtained across an additional pair of brushes, as shown in figure 21-40 (A). The axis of these brushes is perpendicular to that of the shorted brushes. The load is connected in series with these new brushes and a compensating winding, shown schematically around the south field pole, S.

Armature load current is toward the observer around the upper half of the armature windings and away from the observer around the lower half. The armature acts like an electromagnet, which creates an mmf directly opposed (in this case) to the control field mmf and in the same axis as that of the control field.

The purpose of the compensating winding is to create an opposing mmf along this same axis to counterbalance the armature load current mmf. This counterbalance is made automatic for any degree of load by connecting the compensating winding in series with the load brushes and the load. Vector OC (fig. 21-40 (B)) rep-

resents the relative magnitude and direction of the compensating winding mmf with respect to the control field mmf OA; the armature load current mmf OD; and the armature crossfield mmf OB.

The main field poles are shown slotted along the load-brush (horizontal) axis to indicate a method of achieving satisfactory commutation in the coils being shorted by these brushes. A restriction in the size of the amplidyne is the self-induced voltage in the coils being commutated and the resultant sparking at the commutator.

Because any residual magnetism along the axis of the control field would have an appreciable effect on the amplidyne output, it is necessary to demagnetize the core material when the control field winding is deenergized. This demagnetization is accomplished by means of a small a-c magneto generator (mounted on the amplidyne frame) which supplies a suitably placed demagnetizing winding, known as a KILLER WINDING (fig. 21-40 (A)).

The action of the amplidyne is summarized as follows:

1. A very low power (1 watt) input to the control field creates an armature short-circuit current of 100 amperes, thus producing a relatively strong armature crossfield. This is responsible, in turn, for the generation of a normal output voltage of 100 volts across the output load brushes. The output load of 1 ohm is supplied with a current of 100 amperes and an output power of 10,000 watts.

2. In larger size amplidyne generators, a control field input power of 4 watts will develop an output voltage of 200 volts and an armature load current of 200 amperes, or 40,000 watts output.

3. Because of the high power amplification, any residual magnetism in the poles would greatly interfere with the proportion between the input and output. Residual magnetism is removed by means of a small a-c magneto generator which supplies a demagnetizing winding having the same axis as the control field winding.

4. The laminated armature and stator cores are worked at a low flux density in order to maintain a straight-line proportion between input and output.

5. Amplidyne genrators are driven at relatively high speeds (1,800 rpm to 4,000 rpm in order that they may be of small size and light weight.

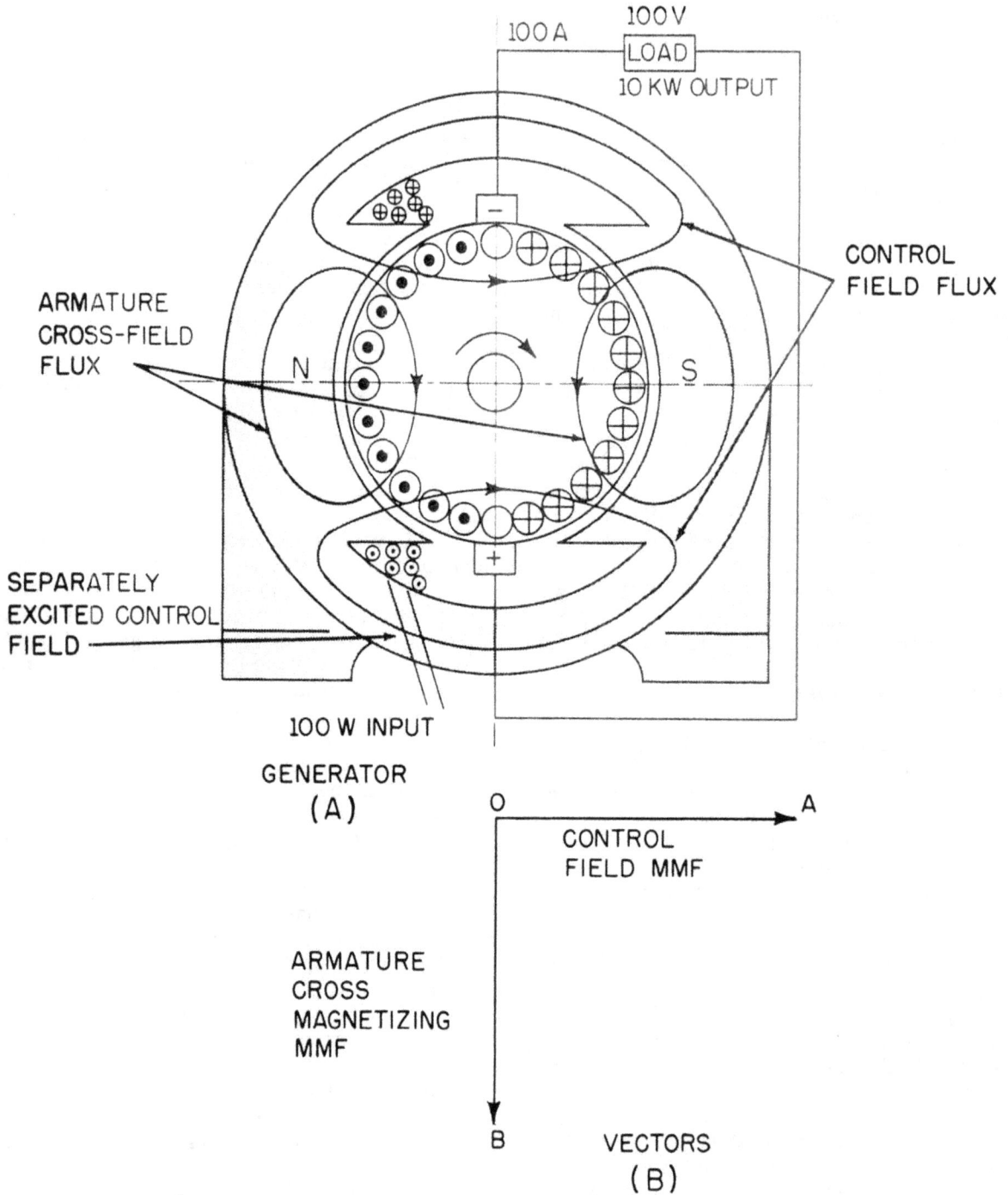

Figure 21-38.—Ordinary d-c generator.

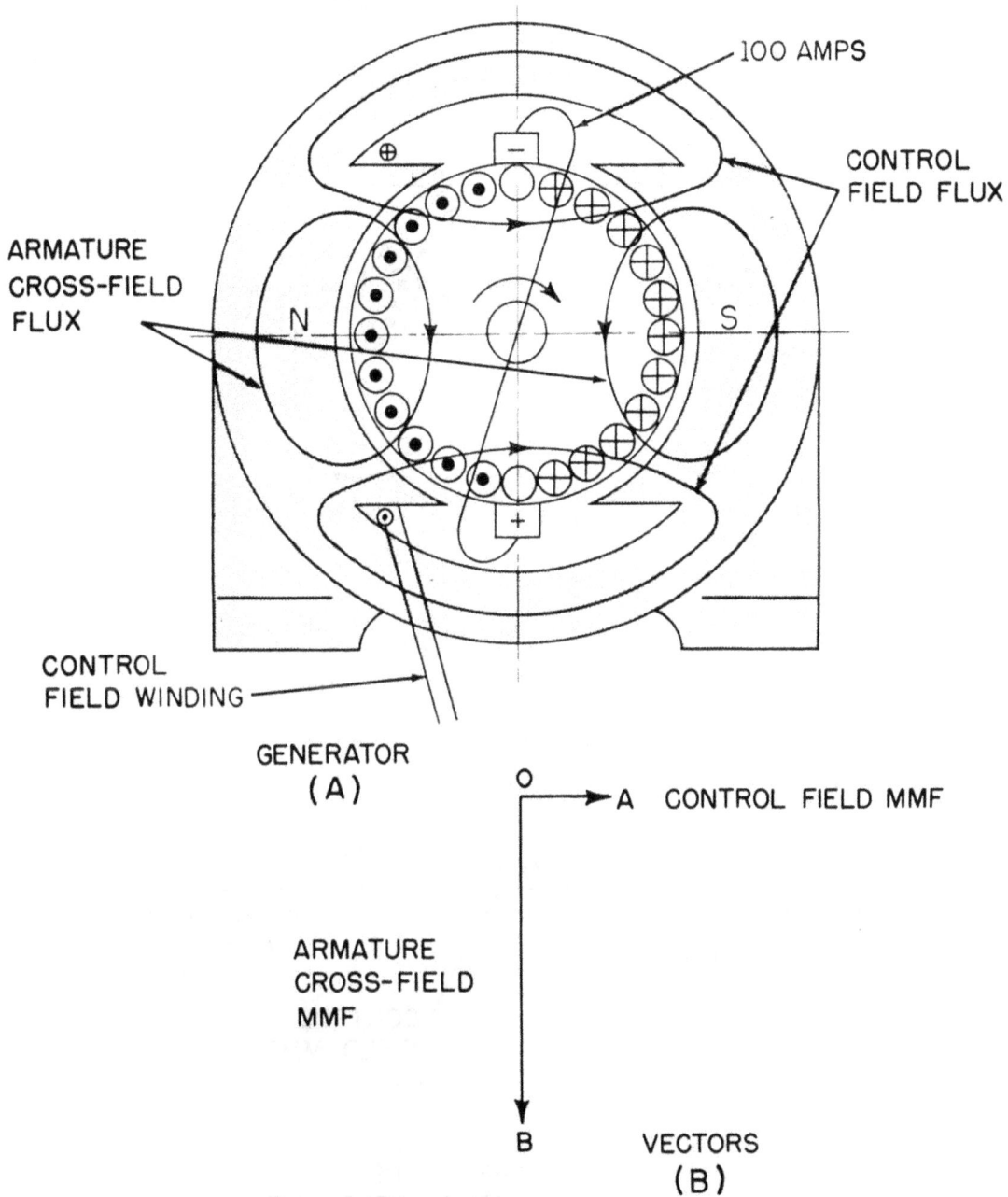

Figure 21-39.—Generator with short-circuited brushes.

6. The amplidyne generator has a quick response to changes in control field current. The lapse in time between a change in the magnitude of the control field current and the load output response is about 0.1 second.

ANTIHUNT CONSIDERATIONS

One problem in the application of the servo-system is that of hunting. Hunting refers to the tendency of a mechanical system to oscillate

450

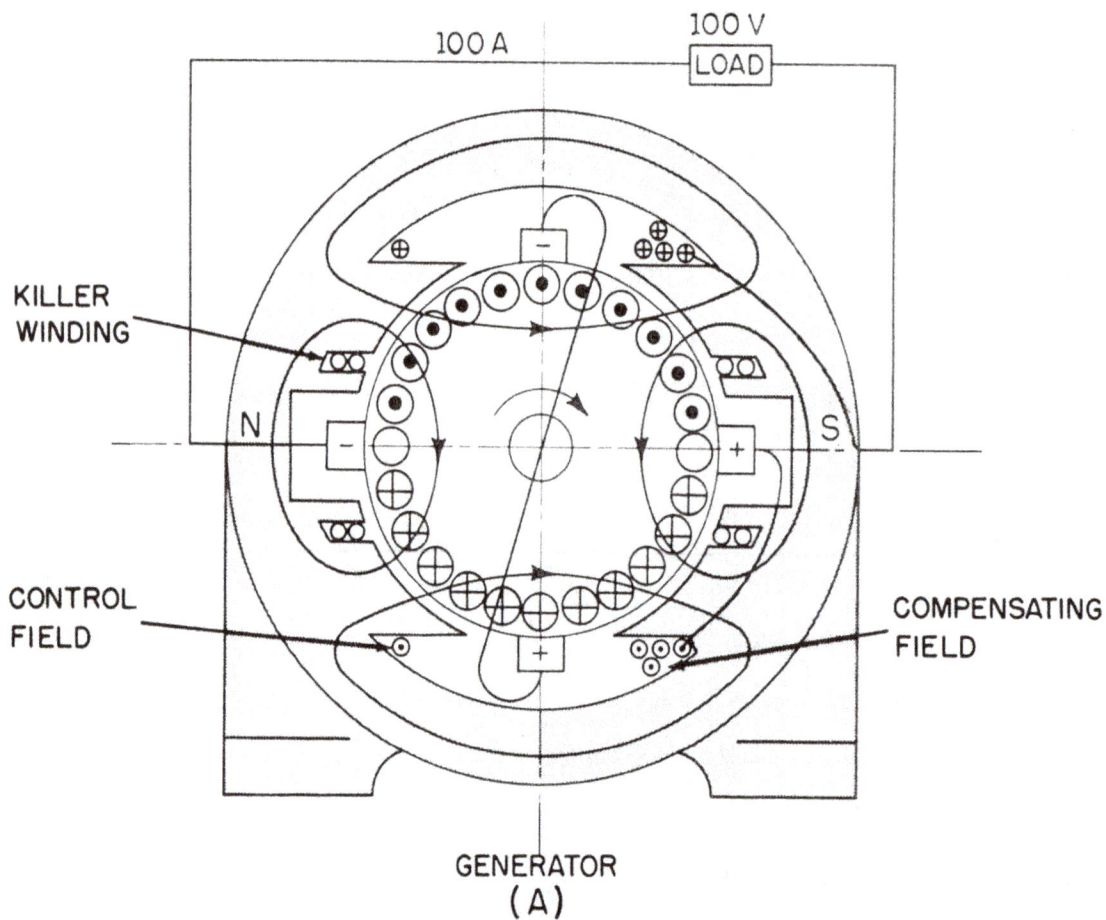

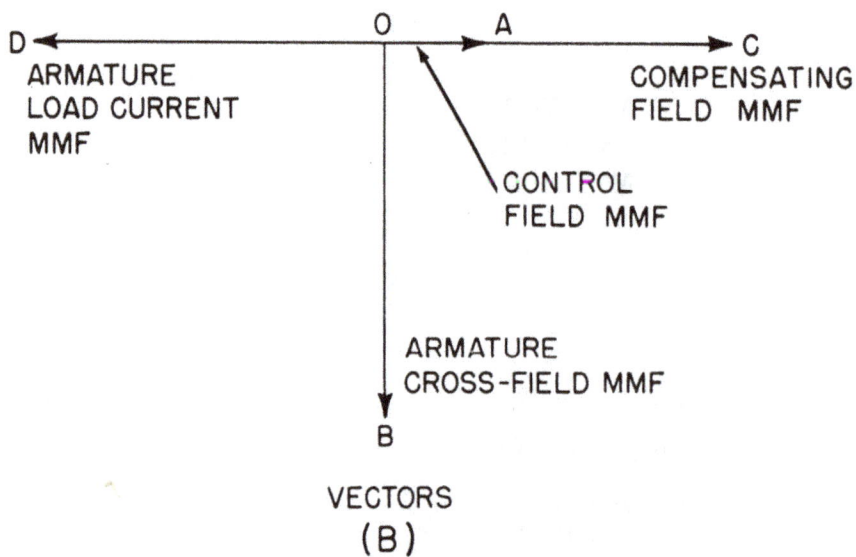

Figure 21-40.—Amplidyne generator.

about a normal position. Thus, in figure 21-41 the steel ball, if depressed from its normal position and suddenly released, oscillates vertically because of its inertia and the forces exerted by the springs. In time, the oscillation is damped out by the frictional losses in the oscillating system.

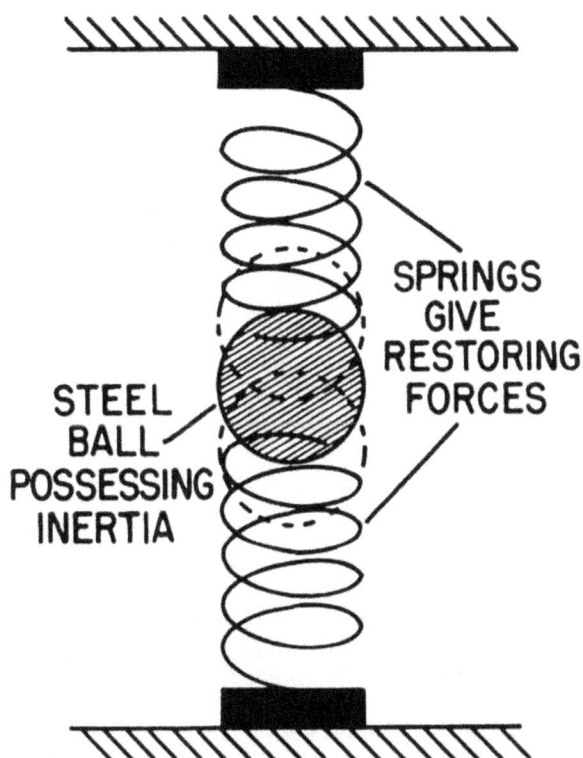

STEEL BALL POSSESSING INERTIA

SPRINGS GIVE RESTORING FORCES

BE.439

Figure 21-41.—Mechanical demonstration of hunting.

Although other factors are involved, a somewhat similar effect is observed in radar fire control, and other equipments employing servo drive. Because the connection between the handwheel or other device controlling the position of the equipment (antenna, gun, etc., to be moved) and the equipment itself is somewhat elastic (because of the action of the electrical and magnetic circuits involved), the inertia of the equipment causes it to overtravel its required position. An error voltage is developed in the servosystem in the opposite direction and the

equipment reverses. Successive overtravels by the equipment would be less and less, and the mechanical oscillation would die out except for one factor: there may be a time lag in the servosystem, which causes reinforced oscillations. In such a case, the equipment would continue to oscillate or HUNT about its normal position. This action would cause harmful mechanical vibration of the entire rotating system.

In order to eliminate hunting, an ANTIHUNT device or circuit is used. This device commonly consists of arrangements to slow up the drive motor as the equipment to be moved approaches its final position. If the drive power is reduced soon enough, the inertia of the moving part causes it to coast into its final position without any overtravel.

The antihunt circuit commonly connects the servo or followup motor to the control amplifier. A feedback signal is supplied by means of this circuit to the control amplifier where the hunting problem is handled electrically.

SERVOMOTORS

The output drive motor in a servosystem should have the required power, be easily reversible, and be capable of speed control over a fairly wide range. The d-c motor has characteristics that make its use advantageous. However, under certain circumstances the a-c motor has distinct advantages.

D-C Servomotor

The d-c servomotor used with the amplidyne generator must have a constant magnetic field if its output is to be proportional to the voltage applied across its armature. This voltage is supplied by the controlled output of the amplidyne generator. A constant field can be produced by a field winding supplied by a constant-voltage d-c source, or supplied by means of a suitable rectifier and an a-c source. A simpler method is to use permanent magnets that are capable of establishing the necessary flux. These are restricted to small-size, low-horsepower motors.

A disadvantage in using permanent magnets is that they may become demagnetized by armature reaction. However, this condition may be counteracted by the use of a compensating winding. This winding is connected in series with the armature and so placed around the pole faces that the armature reaction effect is completely eliminated.

## A-C Servomotor

As has been stated, the output drive motor in a servosystem should be easily reversible and should be capable of speed control over a fairly wide range. Ordinarily, an a-c motor cannot fulfill the requirements of a servo drive motor as completely as a d-c motor because the range of speed control is less. However, the use of an a-c motor may provide a much simpler drive system, especially where a-c power is the only source available and where some sacrifice in range of speed control can be made.

An a-c motor that can be adapted for servosystem use is the single-phase induction motor. This motor contains two stator windings. They are the starting and running windings and are displaced 90°. Their currents are displaced in time phase to produce a revolving magnetic field. The rotor may be either a WOUND rotor or a CAGE rotor. The latter type is the most common. It consists of heavy conducting bars embedded in the armature slots and connected at the ends by conducting rings. The revolving magnetic field originating in the stator cuts the rotor windings when power is applied at start and induces voltages in them. The resulting rotor currents react with the field to start the motor in the direction of the revolving field.

The starting winding is frequently designed to operate continuously with a capacitor to increase the phase displacement between the currents at start and to improve the motor power factor when the motor comes up to speed. When this arrangement is used it is called a CAPACITOR MOTOR. In order to reduce the size of the phase-splitting capacitor required for induction motor applications (and to produce essentially the same result), a high voltage capacitor of smaller capacitance and a small autotransformer are used. The operation of the capacitor motor and the method of employing the capacitor are explained in chapter 17.

In order to give the necessary range of speed control and higher starting torque for certain applications, it is possible to modify the single-phase motor in other ways. These include increasing the resistance of the rotor bars and end rings and the use of a tapped autotransformer, one arrangement of which is shown in figure 21-43. In this circuit the phase-splitting capacitor and autotransformer are placed in parallel with one coil, and the combination is placed in series with the second coil. The current through coil 1 is the vector sum of the lagging current through coil 2 and the leading current through the primary of the autotransformer. Because the current through the capacitor is leading the current through coil 2, the total current through coil 1 leads that through coil 2. The capacitor is chosen to give approximately a 90° phase shift between the currents in coils 1 and 2. Thus, the desired rotating magnetic field is produced. However, the efficiency of this type of motor is relatively poor because of the high resistance rotor.

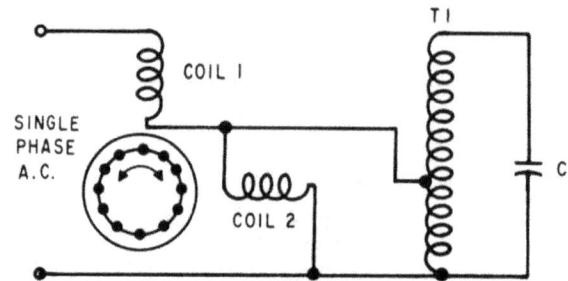

BE.440

Figure 21-42.—Single-phase capacitor motor with series stator coil connections.

The DIRECTION of rotation of the capacitor motor is reversed either by reversing the connections to one stator coil or by shifting the capacitor from one coil to the other. The SPEED of the motor is varied over a limited range by changing the voltage applied to the motor. The voltage may be changed by placing a variable impedance in series with one or both phases. The effect of such an impedance is lower the voltage, and hence the current, input to the windings. This action weakens the field and lowers the induced voltage in the rotor bars, hence reduces the motor torque and speed. The basic requirements of this system are: (1) a means of using an error signal to vary the impedance in series with the motor in order to control the speed, and (2) a means of comparing the phase of the error signal with a reference voltage in order to control the direction of rotation of the motor.

The block diagram of a servosystem in which a capacitor motor is used to provide the output power is shown in figure 21-43. A synchro control transformer is used to provide an error signal that is proportional to the difference between the actual antenna position and the desired position, represented by the position of the handwheel. The rotor of the control transformer is geared to the load so that rotation of the load turns the rotor toward the position in which the error voltage is zero. The antenna is rotated by turning the handwheel connected to the rotor of the synchro transmitter. The turning of the rotor shifts the position of the stator field of the control transformer and thus causes an error voltage to be induced in its rotor. The error voltage is fed to the control amplifier where it is amplified and used to lower the impedance in series with the a-c motor in order to start the motor, and to control (within limits) the speed of rotation. The error voltage is simultaneously fed to another section of the amplifier where its phase is compared with the phase of a reference voltage in order to control the direction of rotation. The output of this section of the amplifier is fed to a relay which selects the stator winding with which the phase-splitting capacitor is to be connected in series.

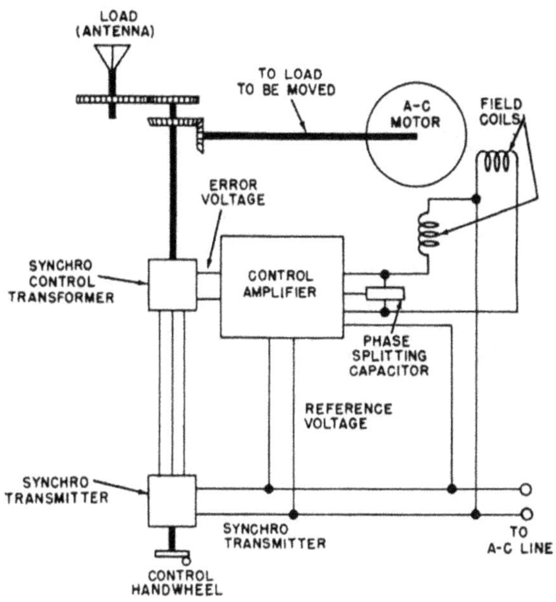

BE.441

Figure 21-43.—Servosystem using capacitor motor.

# APPENDIX I
# GREEK ALPHABET

| Name | Capital | Lower Case | Designates |
|------|---------|------------|------------|
| Alpha. . . . . . | A | $\alpha$ | Angles. |
| Beta. . . . . . | B | $\beta$ | Angles, flux density. |
| Gamma. . . . . | $\Gamma$ | $\gamma$ | Conductivity. |
| Delta . . . . . . | $\Delta$ | $\delta$ | Variation of a quantity, increment. |
| Epsilon. . . . . | E | $\epsilon$ | Base of natural logarithms (2.71828). |
| Zeta. . . . . . | Z | $\zeta$ | Impedance, coefficients, coordinates. |
| Eta . . . . . . . | H | $\eta$ | Hysteresis coefficient, efficiency, magnetizing force. |
| Theta. . . . . . | $\Theta$ | $\theta$ | Phase angle. |
| Iota . . . . . . . | I | $\iota$ | |
| Kappa . . . . . | K | $\kappa$ | Dielectric constant, coupling coefficient, susceptibility. |
| Lambda . . . . | $\Lambda$ | $\lambda$ | Wavelength. |
| Mu. . . . . . . . | M | $\mu$ | Permeability, micro, amplification factor. |
| Nu. . . . . . . . | N | $\nu$ | Reluctivity. |
| Xi . . . . . . . . | $\Xi$ | $\xi$ | |
| Omicron. . . . | O | $o$ | |
| Pi . . . . . . . . | $\Pi$ | $\pi$ | 3.1416 |
| Rho . . . . . . . | P | $\rho$ | Resistivity. |
| Sigma. . . . . . | $\Sigma$ | $\sigma$ | |
| Tau . . . . . . . | T | $\tau$ | Time constant, time-phase displacement. |
| Upsilon. . . . . | $\Upsilon$ | $\upsilon$ | |
| Phi . . . . . . . | $\Phi$ | $\phi$ | Angles, magnetic flux. |
| Chi . . . . . . . | X | $\chi$ | |
| Psi . . . . . . . | $\Psi$ | $\psi$ | Dielectric flux, phase difference. |
| Omega . . . . . | $\Omega$ | $\omega$ | Ohms (capital), angular velocity ($2\pi f$). |

# APPENDIX II

# COMMON ABBREVIATIONS AND LETTER SYMBOLS

| Term | Abbreviation or Symbol |
|------|------------------------|
| alternating current (noun) | a.c. |
| alternating-current (adj.) | a-c |
| ampere | a |
| audiofrequency (noun) | AF |
| capacitance | C |
| capacitive reactance | $X_c$ |
| centimeter | cm |
| conductance | G |
| coulomb | Q |
| counterelectromotive force | cemf |
| current (d-c or rms value) | I |
| current (instantaneous value) | i |
| dielectric constant | K,k |
| difference in potential (d-c or rms value) | E |
| difference in potential (instantaneous value) | e |
| direct current (noun) | d.c. |
| direct-current (adj.) | d-c |
| electromotive force | emf |
| frequency | f |
| henry | h |
| hertz | Hz |
| horsepower | hp |
| impedance | Z |
| inductance | L |
| inductive reactance | $X_L$ |
| kilovolt | kv |
| kilovolt-ampere | kva |
| kilowatt | kw |
| kilowatt-hour | kwhr |
| magnetic field intensity | H |
| magnetomotive force | m.m.f. |
| microampere | μa |
| microfarad | μf |
| microhenry | μh |
| microvolt | μv |
| microwatt | μw |
| milliampere | ma |
| millihenry | mh |
| millivolt | mv |
| milliwatt | mw |
| mutual inductance | M |
| picofarad | pf |
| power | P |
| resistance | R |
| revolutions per minute | rpm |
| root mean square | rms |
| time | t |
| torque | T |
| volt | v |
| volt-ampere | va |
| watt | w |

For computing the current or the horsepower of a d-c motor, use the formula given for single phase, but omit the power factor.

To find the value of an alternative current when voltage, power (in watts) and power factor are known, use the formulas:

$$\text{amp} = \frac{\text{watts}}{\text{volts times P.F.}} \text{ , or}$$

$$\text{amp} = \frac{\text{watts}}{\text{volts times 1.73 times P.F.}} \text{ ,}$$

according to whether the equipment is single phase or three phase.

# APPENDIX III

# ELECTRICAL TERMS

AGONIC.—An imaginary line of the earth's surface passing through points where the magnetic declination is 0°; that is, points where the compass points to true north.

AMMETER.—An instrument for measuring the amount of electron flow in amperes.

AMPERE.—The basic unit of electrical current.

AMPERE-TURN.—The magnetizing force produced by a current of one ampere flowing through a coil of one turn.

AMPLIDYNE.—A rotary magnetic or dynamo-electric amplifier used in servomechanism and control applications.

AMPLIFICATION.—The process of increasing the strength (current, power, or voltage) of a signal.

AMPLIFIER.—A device used to increase the signal voltage, current, or power, generally composed of a vacuum tube and associated circuit called a stage. It may contain several stages in order to obtain a desired gain.

AMPLITUDE.—The maximum instantaneous value of an alternating voltage or current, measured in either the positive or negative direction.

ARC.—A flash caused by an electric current ionizing a gas or vapor.

ARMATURE.—The rotating part of an electric motor or generator. The moving part of a relay or vibrator.

ATTENUATOR.—A network of resistors used to reduce voltage, current, or power delivered to a load.

AUTOTRANSFORMER.—A transformer in which the primary and secondary are connected together in one winding.

BATTERY.—Two or more primary or secondary cells connected together electrically. The term does not apply to a single cell.

BREAKER POINTS.—Metal contacts that open and close a circuit at timed intervals.

BRIDGE CIRCUIT.—The electrical bridge circuit is a term referring to any one of a variety of electric circuit networks, one branch of which, the "bridge" proper, connects two points of equal potential and hence carries no current when the circuit is properly adjusted or balanced.

BRUSH.—The conducting material, usually a block of carbon, bearing against the commutator or sliprings through which the current flows in or out.

BUS BAR.—A primary power distribution point connected to the main power source.

CAPACITOR.—Two electrodes or sets of electrodes in the form of plates, separated from each other by an insulating material called the dielectric.

CHOKE COIL.—A coil of low ohmic resistance and high impedance to alternating current.

CIRCUIT.—The complete path of an electric current.

CIRCUIT BREAKER.—An electromagnetic or thermal device that opens a circuit when the current in the circuit exceeds a predetermined amount. Circuit breakers can be reset.

CIRCULAR MIL.—An area equal to that of a circle with a diameter of 0.001 inch. It is used for measuring the cross section of wires.

COAXIAL CABLE.—A transmission line consisting of two conductors concentric with and insulated from each other.

COMMUTATOR.—The copper segments on the armature of a motor or generator. It is cylindrical in shape and is used to pass power into or from the brushes. It is a switching device.

CONDUCTANCE.—The ability of a material to conduct or carry an electric current. It is the reciprocal of the resistance of the material, and is expressed in mhos.

CONDUCTIVITY.—The ease with which a substance transmits electricity.

CONDUCTOR.—Any material suitable for carrying electric current.

CORE.—A magnetic material that affords an easy path for magnetic flux lines in a coil.

COUNTER EMF.—Counter electromotive force; an emf induced in a coil or armature that opposes the applied voltage.

CURRENT LIMITER.—A protective device similar to a fuse, usually used in high amperage circuits.

CYCLE.—One complete positive and one complete negative alternation of a current or voltage.

DIELECTRIC.—An insulator; a term that refers to the insulating material between the plates of a capacitor.

DIODE.—Vacuum tube—a two element tube that contains a cathode and plate; semiconductor—a material of either germanium or silicon that is manufactured to allow current to flow in only one direction. Diodes are used as rectifiers and detectors.

DIRECT CURRENT.—An electric current that flows in one direction only.

EDDY CURRENT.—Induced circulating currents in a conducting material that are caused by a varying magnetic field.

EFFICIENCY.—The ratio of output power to input power, generally expressed as a percentage.

ELECTROLYTE.—A solution of a substance which is capable of conducting electricity. An electrolyte may be in the form of either a liquid or a paste.

ELECTROMAGNET.—A magnet made by passing current through a coil of wire wound on a soft iron core.

ELECTROMOTIVE FORCE (emf).—The force that produces an electric current in a circuit.

ELECTRON.—A negatively charged particle of matter.

ENERGY.—The ability or capacity to do work.

FARAD.—The unit of capacitance.

FEEDBACK.—A transfer of energy from the output circuit of a device back to its input.

FIELD.—The space containing electric or magnetic lines of force.

FIELD WINDING.—The coil used to provide the magnetizing force in motors and generators.

FLUX FIELD.—All electric or magnetic lines of force in a given region.

FREE ELECTRONS.—Electrons which are loosely held and consequently tend to move at random among the atoms of the material.

FREQUENCY.—The number of complete cycles per second existing in any form of wave motion; such as the number of cycles per second of an alternating current.

FULL-WAVE RECTIFIER CIRCUIT.—A circuit which utilizes both the positive and the negative alternations of an alternating current to produce a direct current.

FUSE.—A protective device inserted in series with a circuit. It contains a metal that will melt or break when current is increased beyond a specific value for a definite period of time.

GAIN.—The ratio of the output power, voltage, or current to the input power, voltage, or current, respectively.

GALVANOMETER.—An instrument used to measure small d-c currents.

GENERATOR.—A machine that converts mechanical energy into electrical energy.

GROUND.—A metallic connection with the earth to establish ground potential. Also, a common return to a point of zero potential.

HERTZ.—A unit of frequency equal to one cycle per second.

HENRY.—The basic unit of inductance.

HORSEPOWER.—The English unit of power, equal to work done at the rate of 550 foot-pounds per second. Equal to 746 watts of electrical power.

HYSTERESIS.—A lagging of the magnetic flux in a magnetic material behind the magnetizing force which is producing it.

IMPEDANCE.—The total opposition offered to the flow of an alternating current. It may consist of any combination of resistance, inductive reactance, and capacitive reactance.

INDUCTANCE.—The property of a circuit which tends to oppose a change in the existing current.

INDUCTION.—The act or process of producing voltage by the relative motion of a magnetic field across a conductor.

INDUCTIVE REACTANCE.—The opposition to the flow of alternating or pulsating current caused by the inductance of a circuit. It is measured in ohms.

INPHASE.—Applied to the condition that exits when two waves of the same frequency pass through their maximum and minimum values of like polarity at the same instant.

INVERSELY.—Inverted or reversed in position or relationship.

ISOGONIC LINE.—An imaginary line drawn through points on the earth's surface where the magnetic variation is equal.

JOULE.—A unit of energy or work. A joule of energy is liberated by one ampere flowing for one second through a resistance of one ohm.

KILO.—A prefix meaning 1,000.

LAG.—The amount one wave is behind another in time; expressed in electrical degrees.

LAMINATED CORE.—A core built up from thin sheets of metal and used in transformers and relays.

LEAD.—The opposite of LAG. Also, a wire or connection.

LINE OF FORCE.—A line in an electric or magnetic field that shows the direction of the force.

LOAD.—The power that is being delivered by any power producing device. The equipment that uses the power from the power producing device.

MAGNETIC AMPLIFIER.—A saturable reactor type device that is used in a circuit to amplify or control.

MAGNETIC CIRCUIT.—The complete path of magnetic lines of force.

MAGNETIC FIELD.—The space in which a magnetic force exists.

MAGNETIC FLUX.—The total number of lines of force issuing from a pole of a magnet.

MAGNETIZE.—To convert a material into a magnet by causing the molecules to rearrange.

MAGNETO.—A generator which produces alternating current and has a permanent magnet as its field.

MEGGER.—A test instrument used to measure insulation resistance and other high resistances. It is a portable hand operated d-c generator used as an ohmmeter.

MEGOHM.—A million ohms.

MICRO.—A prefix meaning one-millionth.

MILLI.—A prefix meaning one-thousandth.

MILLIAMMETER.—An ammeter that measures current in thousandths of an ampere.

MOTOR-GENERATOR.—A motor and a generator with a common shaft used to convert line voltages to other voltages or frequencies.

MUTUAL INDUCTANCE.—A circuit property existing when the relative position of two inductors causes the magnetic lines of force from one to link with the turns of the other.

NEGATIVE CHARGE.—The electrical charge carried by a body which has an excess of electrons.

NEUTRON.—A particle having the weight of a proton but carrying no electric charge. It is located in the nucleus of an atom.

NUCLEUS.—The central part of an atom that is mainly comprised of protons and neutrons. It is the part of the atom that has the most mass.

NULL.—Zero.

OHM.—The unit of electrical resistance.

OHMMETER.—An instrument for directly measuring resistance in ohms.

OVERLOAD.—A load greater than the rated load of an electrical device.

PERMALLOY.—An alloy of nickel and iron having an abnormally high magnetic permeability.

PERMEABILITY.—A measure of the ease with which magnetic lines of force can flow through a material as compared to air.

PHASE DIFFERENCE.—The time in electrical degree by which one wave leads or lags another.

POLARITY.—The character of having magnetic poles, or electric charges.

POLE.—The section of a magnet where the flux lines are concentrated; also where they enter and leave the magnet. An electrode of a battery.

POLYPHASE.—A circuit that utilizes more than one phase of alternating current.

POSITIVE CHARGE.—The electrical charge carried by a body which has become deficient in electrons.

POTENTIAL.—The amount of charge held by a body as compared to another point or body. Usually measured in volts.

POTENTIOMETER.—A variable voltage divider; a resistor which has a variable contact arm so that any portion of the potential applied between its ends may be selected.

POWER.—The rate of doing work or the rate of expending energy. The unit of electrical power is the watt.

POWER FACTOR.—The ratio of the actual power of an alternating or pulsating current, as measured by a wattmeter, to the apparent power, as indicated by ammeter and voltmeter readings. The power factor of an inductor, capacitor, or insulator is an expression of their losses.

PRIME MOVER.—The source of mechanical power used to drive the rotor of a generator.

PROTON.—A positively charged particle in the nucleus of an atom.

RATIO.—The value obtained by dividing one number by another, indicating their relative proportions.

REACTANCE.—The opposition offered to the flow of an alternating current by the inductance, capacitance, or both, in any circuit.

RECTIFIERS.—Devices used to change alternating current to unidirectional current. These may be vacuum tubes, semiconductors such as germanium and silicon, and dry-disk rectifiers such as selenium and copper-oxide.

RELAY.—An electromechanical switching device that can be used as a remote control.

RELUCTANCE.—A measure of the opposition that a material offers to magnetic lines of force.

RESISTANCE.—The opposition to the flow of current caused by the nature and physical dimensions of a conductor.

RESISTOR.—A circuit element whose chief characteristic is resistance; used to oppose the flow of current.

RETENTIVITY.—The measure of the ability of a material to hold its magnetism.

RHEOSTAT.—A variable resistor.

SATURABLE REACTOR.—A control device that uses a small d-c current to control a large a-c current by controlling core flux density.

SATURATION.—The condition existing in any circuit when an increase in the driving signal produces no further change in the resultant effect.

SELF-INDUCTION.—The process by which a circuit induces an e.m.f. into itself by its own magnetic field.

SERIES-WOUND.—A motor or generator in which the armature is wired in series with the field winding.

SERVO.—A device used to convert a small movement into one of greater movement or force.

SERVOMECHANISM.—A closed-loop system that produces a force to position an object in accordance with the information that originates at the input.

SOLENOID.—An electromagnetic coil that contains a movable plunger.

SPACE CHARGE.—The cloud of electrons existing in the space between the cathode and plate in a vacuum tube, formed by the electrons emitted from the cathode in excess of those immediately attracted to the plate.

SPECIFIC GRAVITY.—The ratio between the density of a substance and that of pure water at a given temperature.

SYNCHROSCOPE.—An instrument used to indicate a difference in frequency between two a-c sources.

SYNCHRO SYSTEM.—An electrical system that gives remote indications or control by means of self-synchronizing motors.

TACHOMETER.—An instrument for indicating revolutions per minute.

TERTIARY WINDING.—A third winding on a transformer or magnetic amplifier that is used as a second control winding.

THERMISTOR.—A resistor that is used to compensate for temperature variations in a circuit.

THERMOCOUPLE.—A junction of two dissimilar metals that produces a voltage when heated.

TORQUE.—The turning effort or twist which a shaft sustains when transmitting power.

TRANSFORMER.—A device composed of two or more coils, linked by magnetic lines of force, used to transfer energy from one circuit to another.

TRANSMISSION LINES.—Any conductor or system of conductors used to carry electrical energy from its source to a load.

VARS.—Abbreviation for volt-ampere, reactive.

VECTOR.—A line used to represent both direction and magnitude.

VOLT.—The unit of electrical potential.

VOLTMETER.—An instrument designed to measure a difference in electrical potential, in volts.

WATT.—The unit of electrical power.

WATTMETER.—An instrument for measuring electrical power in watts.

# APPENDIX IV

# COMPONENT COLOR CODE

| COLOR | 1ST DIGIT | 2ND DIGIT | MULTIPLIER | TOLERANCE (percent) |
|-------|-----------|-----------|------------|---------------------|
| Black | 0 | 0 | 1 | |
| Brown | 1 | 1 | 10 | |
| Red | 2 | 2 | 100 | |
| Orange | 3 | 3 | 1,000 | |
| Yellow | 4 | 4 | 10,000 | |
| Green | 5 | 5 | 100,000 | |
| Blue | 6 | 6 | 1,000,000 | |
| Violet | 7 | 7 | 10,000,000 | |
| White | 9 | 9 | 1,000,000,000 | |
| Gold | | | .1 | 5 |
| Silver | | | .01 | 10 |
| No color | | | | 20 |

**(A)**

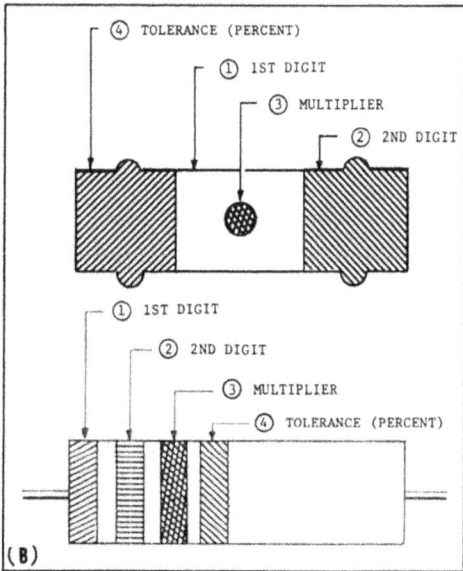

**(B)**

Resistor color code.

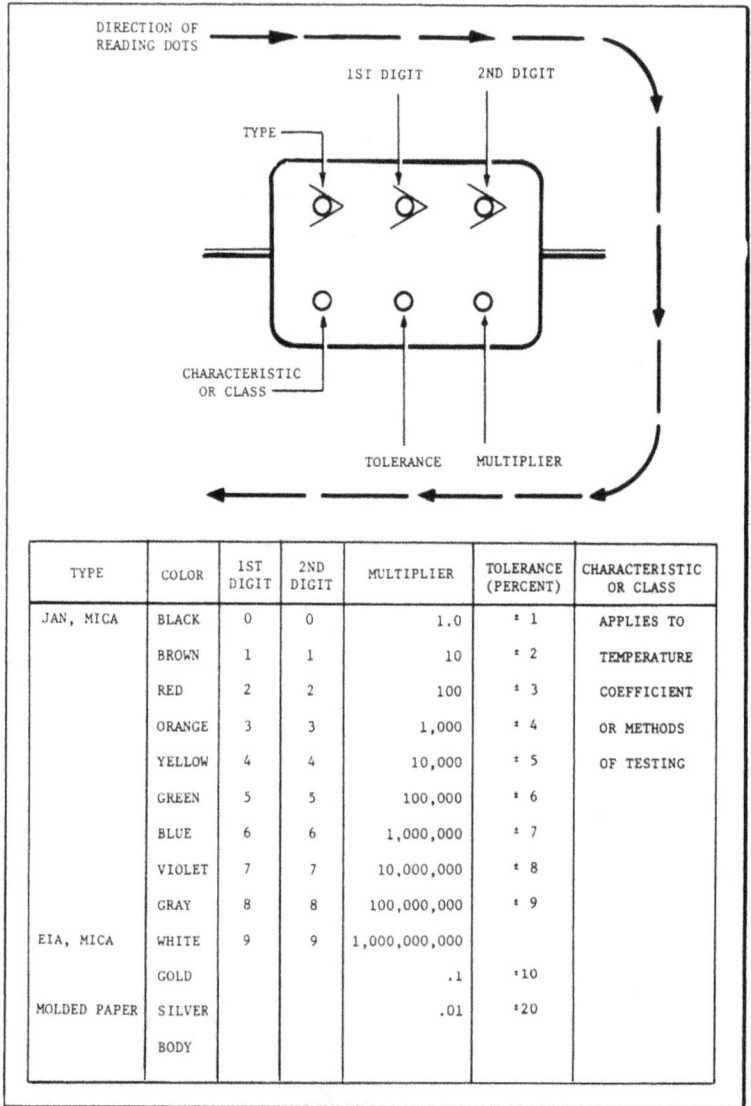

| TYPE | COLOR | 1ST DIGIT | 2ND DIGIT | MULTIPLIER | TOLERANCE (PERCENT) | CHARACTERISTIC OR CLASS |
|------|-------|-----------|-----------|------------|---------------------|-------------------------|
| JAN, MICA | BLACK | 0 | 0 | 1.0 | ± 1 | APPLIES TO |
| | BROWN | 1 | 1 | 10 | ± 2 | TEMPERATURE |
| | RED | 2 | 2 | 100 | ± 3 | COEFFICIENT |
| | ORANGE | 3 | 3 | 1,000 | ± 4 | OR METHODS |
| | YELLOW | 4 | 4 | 10,000 | ± 5 | OF TESTING |
| | GREEN | 5 | 5 | 100,000 | ± 6 | |
| | BLUE | 6 | 6 | 1,000,000 | ± 7 | |
| | VIOLET | 7 | 7 | 10,000,000 | ± 8 | |
| | GRAY | 8 | 8 | 100,000,000 | ± 9 | |
| EIA, MICA | WHITE | 9 | 9 | 1,000,000,000 | | |
| | GOLD | | | .1 | ±10 | |
| MOLDED PAPER | SILVER | | | .01 | ±20 | |
| | BODY | | | | | |

6-Dot color code for mica and molded paper capacitors.

463

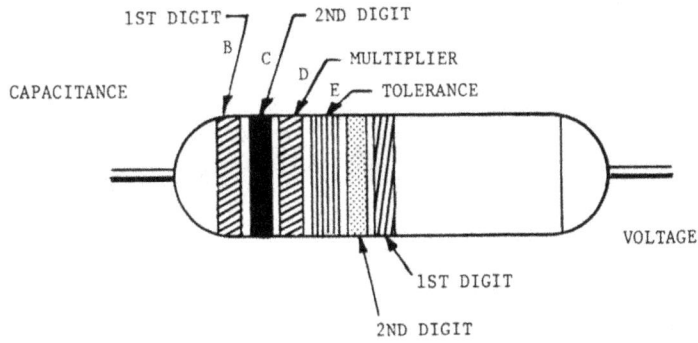

6-Band color code for tubular paper dielectric capacitors.

| COLOR | CAPACITANCE | | | TOLERANCE (PERCENT) | VOLTAGE RATING | |
|---|---|---|---|---|---|---|
| | 1ST DIGIT | 2ND DIGIT | MULTIPLIER | | 1ST DIGIT | 2ND DIGIT |
| BLACK | 0 | 0 | 1 | ±20 | 0 | 0 |
| BROWN | 1 | 1 | 10 | | 1 | 1 |
| RED | 2 | 2 | 100 | | 2 | 2 |
| ORANGE | 3 | 3 | 1,000 | ±30 | 3 | 3 |
| YELLOW | 4 | 4 | 10,000 | ±40 | 4 | 4 |
| GREEN | 5 | 5 | 100,000 | ± 5 | 5 | 5 |
| BLUE | 6 | 6 | 1,000,000 | | 6 | 6 |
| VIOLET | 7 | 7 | | | 7 | 7 |
| GRAY | 8 | 8 | | | 8 | 8 |
| WHITE | 9 | 9 | | ±10 | 9 | 9 |

## Appendix IV — COMPONENT COLOR CODE

B – A – TEMPERATURE COEFFICIENT

B – 1ST DIGIT

C – 2ND DIGIT

D – MULTIPLIER

E – TOLERANCE

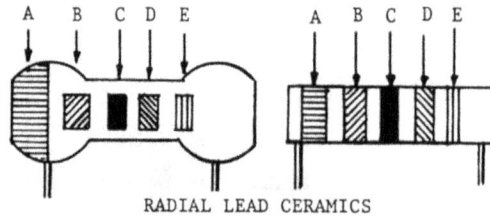

RADIAL LEAD CERAMICS

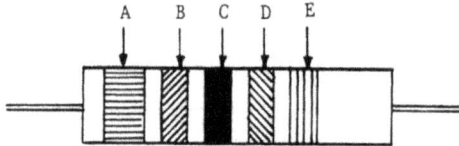

AXIAL LEAD CERAMIC

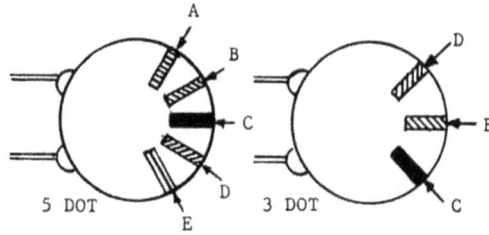

5 DOT          3 DOT

CERAMIC DISC CAPACITOR MARKING

| COLOR | 1ST DIGIT | 2ND DIGIT | MULTIPLIER | TOLERANCE | | TEMPERATURE COEFFICIENT * |
| | | | | MORE THAN 10 pf (IN PERCENT) | LESS THAN 10 pf (IN pf) | |
|---|---|---|---|---|---|---|
| BLACK | 0 | 0 | 1.0 | ±20 | ±2.0 | 0 |
| BROWN | 1 | 1 | 10 | ± 1 | | –30 |
| RED | 2 | 2 | 100 | ± 2 | | –80 |
| ORANGE | 3 | 3 | 1,000 | | | –150 |
| YELLOW | 4 | 4 | 10,000 | | | –220 |
| GREEN | 5 | 5 | | ± 5 | ±0.5 | –330 |
| BLUE | 6 | 6 | | | | –470 |
| VIOLET | 7 | 7 | | | | –750 |
| GRAY | 8 | 8 | .01 | | ±0.25 | +30 |
| WHITE | 9 | 9 | .1 | ±10 | ±1.0 | +120 TO –750 (EIA) |
| | | | | | | +500 TO –330 (JAN) |
| SILVER | | | | | | +100 (JAN) |
| GOLD | | | | | | BYPASS OR COUPLING (EIA) |

*PARTS PER MILLION PER DEGREE CENTIGRADE.

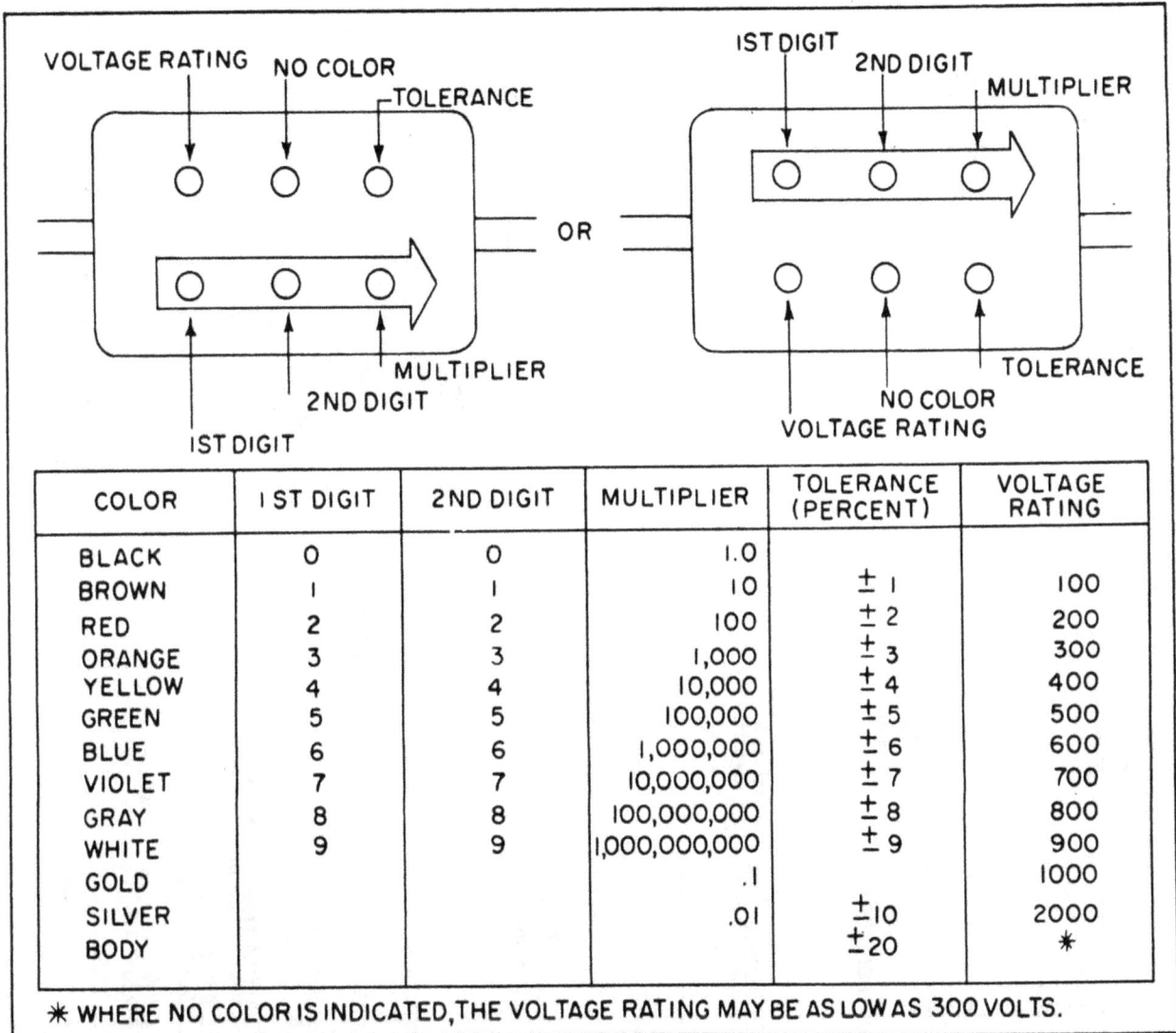

| COLOR | 1ST DIGIT | 2ND DIGIT | MULTIPLIER | TOLERANCE (PERCENT) | VOLTAGE RATING |
|---|---|---|---|---|---|
| BLACK | 0 | 0 | 1.0 | | |
| BROWN | 1 | 1 | 10 | ± 1 | 100 |
| RED | 2 | 2 | 100 | ± 2 | 200 |
| ORANGE | 3 | 3 | 1,000 | ± 3 | 300 |
| YELLOW | 4 | 4 | 10,000 | ± 4 | 400 |
| GREEN | 5 | 5 | 100,000 | ± 5 | 500 |
| BLUE | 6 | 6 | 1,000,000 | ± 6 | 600 |
| VIOLET | 7 | 7 | 10,000,000 | ± 7 | 700 |
| GRAY | 8 | 8 | 100,000,000 | ± 8 | 800 |
| WHITE | 9 | 9 | 1,000,000,000 | ± 9 | 900 |
| GOLD | | | .1 | | 1000 |
| SILVER | | | .01 | ± 10 | 2000 |
| BODY | | | | ± 20 | ✳ |

✳ WHERE NO COLOR IS INDICATED, THE VOLTAGE RATING MAY BE AS LOW AS 300 VOLTS.

466

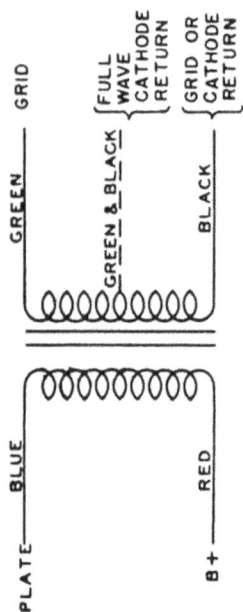

INTERSTAGE AUDIO
TRANSFORMERS

PLATE — BLUE
GREEN — GRID
GREEN & BLACK — FULL WAVE CATHODE RETURN
B+ — RED
BLACK — GRID OR CATHODE RETURN

STANDARD COLORS USED IN CHASSIS WIRING FOR THE
PURPOSE OF CIRCUIT IDENTIFICATION OF THE EQUIPMENT
ARE AS FOLLOWS:

| CIRCUIT | COLOR |
|---|---|
| GROUNDS, GROUNDED ELEMENTS, AND RETURNS | BLACK. |
| HEATERS OR FILAMENTS, OFF GROUND. | BROWN. |
| POWER SUPPLY B PLUS. | RED. |
| SCREEN GRIDS. | ORANGE. |
| CATHODES. | YELLOW. |
| CONTROL GRIDS. | GREEN. |
| PLATES. | BLUE |
| POWER SUPPLY, MINUS. | VIOLET (PURPLE). |
| A C POWER LINES. | GRAY. |
| MISCELLANEOUS, ABOVE OR BELOW GROUND RETURNS, A V C, ETC. | WHITE. |

FOR OTHER ELECTRICAL AND ELECTRONIC SYMBOLS
REFER TO MILITARY STANDARD, MIL-STD-15-IA

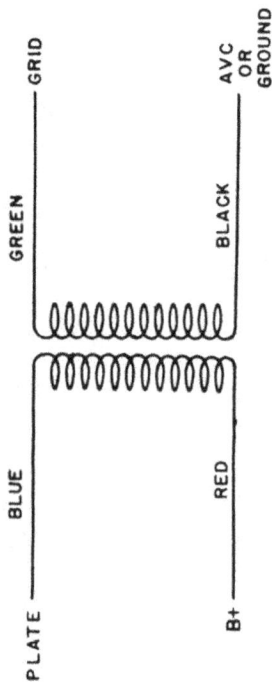

I F TRANSFORMERS.

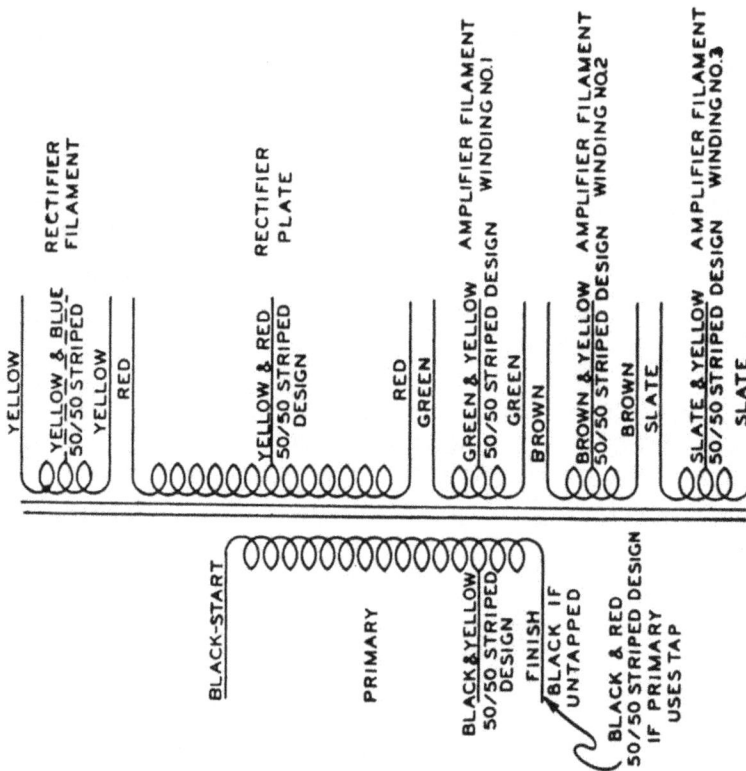

PLATE — BLUE
GREEN — GRID
B+ — RED
BLACK — AVC OR GROUND

POWER TRANSFORMERS.

YELLOW
YELLOW & BLUE 50/50 STRIPED — RECTIFIER FILAMENT
YELLOW
RED
YELLOW & RED 50/50 STRIPED DESIGN — RECTIFIER PLATE
RED
GREEN
GREEN & YELLOW 50/50 STRIPED DESIGN — AMPLIFIER FILAMENT WINDING NO.1
GREEN
BROWN
BROWN & YELLOW 50/50 STRIPED DESIGN — AMPLIFIER FILAMENT WINDING NO.2
BROWN
SLATE
SLATE & YELLOW 50/50 STRIPED DESIGN — AMPLIFIER FILAMENT WINDING NO.3
SLATE

BLACK-START
PRIMARY
BLACK & YELLOW 50/50 STRIPED DESIGN
FINISH
BLACK IF UNTAPPED
BLACK & RED 50/50 STRIPED DESIGN IF PRIMARY USES TAP

467

# APPENDIX V

# CABLE TYPE AND SIZE DESIGNATIONS

All cables described herein are identified by NAVY CABLE TYPE AND SIZE DESIGNATIONS, which consists of a letter followed by numerals.

I. The TYPE DESIGNATION indicates the type or construction of the cable and and consists of the first letters of the words used in describing the cable:
Example: Type SHFA—Single, Heat and Flame-resistance, Armored.

II. The SIZE DESIGNATION relates to the copper conductors. No set rule has been established for application of SIZE DESIGNATIONS since the numerals used may indicate one or more of the following:
(1) Size of copper conductor.
(2) Stranding of conductors.
(3) Number of copper conductors.
(4) Number of twisted pairs.

Examples of the most common uses of SIZE DESIGNATIONS are given below:
(1) To indicate approximate cross-sectional area of the conductor expressed in thousands of circular mils (abbreviated C.M.).
Example: The conductor area of SHFA-3 is approximately 3000 C.M. (exactly 2828 C.M.)
(2) To indicate approximate cross-sectional area with stranding shown in parentheses.
Example: The conductor area of SRI-2-1/2 (26) is approximately 2500 C.M. (exactly 2613 C.M.), and consists of 26 individual wire strands.

(3) To indicate number of conductors.
Example: MCOP-7 is a cable comprised of seven conductors.

(4) To indicate number of twisted pairs.
Example: TTOP-10 is a 20-conductor cable comprised of 10 twisted pairs.

The following are typical examples of type designations and constructions:

FLA—Four conductor, Lighting, Armored. Varnished cambric or rubber insulation, rubber hose jacket, armored.

MCMB—Multiple Conductor, Marker-Buoy. Rubber insulated identifying colors, steel supporting cable, tape over assembly, tough rubber sheath.

SHFA—Single conductor, Heat and Flame-resistant, Armored. Asbestos-varnished cambric-asbestos insulation, impervious sheath, armored.

THFR—Triple conductor, Heat and Flame-resistant, Radio. Synthetic resin-varnished cambric and felted asbestos insulation, asbestos belt, impervious sheath, armored.

TTHFA—Twisted pairs, Telephone, Heat and Flame-resistant, Armored. Solid conductor, textile wrap or synthetic resin insulation, twisted pair, felted asbestos belt, impervious sheath, armored.

# APPENDIX VI

# CURRENT-CARRYING CAPACITIES OF CONDUCTORS

Based on the National Electrical Code, 1951

| Size AWG or Thousand Cir. Mils | Current Capacity in Amperes | | |
|---|---|---|---|
| | Rubber or Thermoplastic Insulation | Paper or Varnished Cambric Insulation | Impregnated Asbestos Insulation |
| 14 | 15 | 25 | 30 |
| 12 | 20 | 30 | 40 |
| 10 | 30 | 40 | 50 |
| 8 | 40 | 50 | 65 |
| 6 | 55 | 70 | 85 |
| 4 | 70 | 90 | 115 |
| 3 | 80 | 105 | 130 |
| 2 | 95 | 120 | 145 |
| 1 | 110 | 140 | 170 |
| 0 | 125 | 155 | 200 |
| 00 | 145 | 185 | 230 |
| 000 | 165 | 210 | 265 |
| 0000 | 195 | 235 | 310 |
| 250 | 215 | 270 | 335 |
| 300 | 240 | 300 | 380 |
| 350 | 260 | 325 | 420 |
| 400 | 280 | 360 | 450 |
| 500 | 320 | 405 | 500 |
| 600 | 355 | 455 | 545 |
| 700 | 385 | 490 | 600 |
| 750 | 400 | 500 | 620 |
| 800 | 410 | 515 | 640 |
| 900 | 435 | 555 | ... |
| 1,000 | 455 | 585 | 730 |
| 1,250 | 495 | 645 | ... |
| 1,500 | 520 | 700 | ... |
| 1,750 | 545 | 735 | ... |
| 2,000 | 560 | 775 | ... |

Notes:
1. The ratings are for not more than three conductors in a cable or raceway, with a room temperature of 30° C or 86° F.
2. The above tables are for copper wires. The ratings for aluminum wires are 84 percent of these values.
3. Consult the National Electrical Code for more information. For example, higher ratings are allowed for single conductors in free air.

# FULL LOAD CURRENTS OF MOTORS

[The following data are approximate full-load currents for motors of various types, frequencies, and speeds. They have been compiled from average values for representative motors of their respective classes. Variations of 10 percent above or below the values given may be expected.]

**Amperes—Full-load current**

Column groups: **Direct-current motors** (115, 230, 550 volt); **Single-phase motors** (110, 220 volt); **Alternating-current motors** — **Squirrel-cage induction motors** (Two-phase: 110/220/440/550/2,200 volt; Three-phase: 110/220/440/550/2,200 volt) and **Slip-ring induction motors** (Two-phase: 110/220/440/550/2,200 volt; Three-phase: 110/220/440/550/2,200 volt).

| Hp | DC 115 | DC 230 | DC 550 | 1φ 110 | 1φ 220 | SqC 2φ 110 | SqC 2φ 220 | SqC 2φ 440 | SqC 2φ 550 | SqC 2φ 2200 | SqC 3φ 110 | SqC 3φ 220 | SqC 3φ 440 | SqC 3φ 550 | SqC 3φ 2200 | SlpR 2φ 110 | SlpR 2φ 220 | SlpR 2φ 440 | SlpR 2φ 550 | SlpR 2φ 2200 | SlpR 3φ 110 | SlpR 3φ 220 | SlpR 3φ 440 | SlpR 3φ 550 | SlpR 3φ 2200 |
|---|---|---|---|---|---|---|---|---|---|---|---|---|---|---|---|---|---|---|---|---|---|---|---|---|---|
| ¼ | | | | 4.8 | 2.4 | | | | | | | | | | | | | | | | | | | | |
| ½ | 4.5 | 2.3 | | 7 | 3.5 | 5.0 | 2.2 | 1.1 | 0.9 | | 5.0 | 2.5 | 1.3 | 1.0 | | | | | | | | | | | |
| ¾ | 6.5 | 3.3 | 1.4 | 9.4 | 4.7 | 5.4 | 2.4 | 1.2 | 1.0 | | 5.4 | 2.8 | 1.4 | 1.1 | | 6.2 | 3.1 | 1.6 | 1.3 | | 7.2 | 3.6 | 1.8 | 1.5 | |
| 1 | 8.4 | 4.2 | 1.7 | 11 | 5.5 | 6.6 | 2.9 | 1.4 | 1.2 | | 6.6 | 3.3 | 1.7 | 1.3 | | 6.7 | 3.4 | 1.7 | 1.4 | | 7.8 | 3.9 | 2.0 | 1.6 | |
| 1½ | 12.5 | 6.3 | 2.6 | 15.2 | 7.6 | 9.4 | 4.0 | 2 | 1.6 | | 9.4 | 4.7 | 2.4 | 2.0 | | 11.7 | 5.9 | 3.0 | 2.3 | | | | | | |
| 2 | 16.1 | 8.3 | 3.4 | 20 | 10 | 12.0 | 5 | 3 | 2.0 | | 12.0 | 6 | 3.0 | 2.4 | | 12.5 | 6.3 | 3.1 | 2.5 | | 14.4 | 7.2 | 3.6 | 2.9 | |
| 3 | 23 | 12.3 | 5.0 | 28 | 14 | | 8 | 4 | 3.0 | | | 9 | 4.5 | 4.0 | | | 8.7 | 4.3 | 3.5 | | 20.2 | 10 | 5.0 | 4 | |
| 5 | 40 | 19.8 | 8.2 | 46 | 23 | | 13 | 7 | 6 | | | 15 | 7.5 | 6.0 | | | 13.0 | 6.5 | 5.2 | | | 15 | 7.5 | 6 | |
| 7½ | 58 | 28.7 | 12 | 68 | 34 | | 19 | 9 | 7 | | | 22 | 11 | 9.0 | | | 20.0 | 10.0 | 7.6 | | | 25 | 13 | 10 | |
| 10 | 75 | 38 | 16 | 86 | 43 | | 24 | 12 | 10 | | | 27 | 14 | 11 | | | 24.3 | 12.1 | 10.0 | | | 28 | 14 | 11 | |
| 15 | 112 | 56 | 23 | | | | 33 | 16 | 13 | | | 38 | 19 | 15 | | | 39 | 19.5 | 15.6 | | | 45 | 23 | 18 | |
| 20 | 140 | 74 | 30 | | | | 45 | 23 | 19 | 5.7 | | 52 | 26 | 21 | 5.7 | | 49 | 24.7 | 19.8 | | | 56 | 28 | 22 | |
| 25 | 185 | 92 | 38 | | | | 55 | 28 | 22 | 6 | | 64 | 32 | 26 | 7 | | 60 | 30.0 | 24.0 | 6.4 | | 67 | 34 | 27 | 7.5 |
| 30 | 220 | 110 | 45 | | | | 67 | 34 | 27 | 7 | | 77 | 39 | 31 | 8 | | 72 | 36.0 | 28.8 | 7.8 | | 82 | 41 | 33 | 9 |
| 40 | 294 | 146 | 61 | | | | 88 | 44 | 35 | 9 | | 101 | 51 | 40 | 10 | | 93 | 46.5 | 37.3 | 9.5 | | 106 | 53 | 42 | 11 |
| 50 | 364 | 180 | 75 | | | | 108 | 54 | 43 | 11 | | 125 | 63 | 50 | 13 | | 113 | 57 | 45 | 12.1 | | 128 | 64 | 51 | 14 |
| 60 | 436 | 215 | 90 | | | | 129 | 65 | 52 | 13 | | 149 | 75 | 60 | 15 | | 135 | 68 | 54 | 14.0 | | 150 | 75 | 60 | 16 |
| 75 | 540 | 268 | 111 | | | | 156 | 78 | 62 | 16 | | 180 | 90 | 72 | 19 | | 164 | 82 | 65 | 17.3 | | 188 | 94 | 75 | 19 |
| 100 | | 357 | 146 | | | | 212 | 106 | 85 | 22 | | 246 | 123 | 98 | 25 | | 214 | 108 | 87 | 21.7 | | 246 | 123 | 99 | 25 |
| 125 | | 443 | 184 | | | | 268 | 134 | 108 | 27 | | 310 | 155 | 124 | 32 | | 267 | 134 | 108 | 27 | | 310 | 155 | 124 | 31 |
| 150 | | | 220 | | | | 311 | 155 | 124 | 31 | | 360 | 180 | 144 | 36 | | 315 | 158 | 127 | 32 | | 364 | 182 | 145 | 37 |
| 175 | | | | | | | | | | | | | | | | | | | | | | | | | |
| 200 | | | 295 | | | | 415 | 208 | 166 | 43 | | 480 | 240 | 195 | 49 | | 430 | 216 | 173 | 44 | | 490 | 245 | 196 | 52 |

470

# APPENDIX VIII
# FORMULAS

## Ohm's Law for D-C Circuits

$$I = \frac{E}{R} = \frac{P}{E} = \sqrt{\frac{P}{R}}$$

$$R = \frac{E}{I} = \frac{P}{I^2} = \frac{E^2}{P}$$

$$E = IR = \frac{P}{I} = \sqrt{PR}$$

$$P = EI = \frac{E^2}{R} = I^2R$$

## Resistors in Series

$$R_T = R_1 + R_2 + \ldots$$

## Resistors in Parallel

### Two resistors

$$R_T = \frac{R_1 R_2}{R_1 + R_2}$$

### More than two

$$\frac{1}{R_T} = \frac{1}{R_1} + \frac{1}{R_2} + \frac{1}{R_3} + \ldots$$

## RL Circuit Time Constant

$$\frac{L \text{ (in henrys)}}{R \text{ (in ohms)}} = t \text{ (in seconds), or}$$

$$\frac{L \text{ (in microhenrys)}}{R \text{ ( in ohms)}} = t \text{ (in microseconds)}$$

## RC Circuit Time Constant

R (ohms) x C (farads) = t (seconds)

R (megohms) x C (microfarads) = t (seconds)

R (ohms) x C (microfarads) = t (micro-seconds)

R (megohms) x C (micromicrofarads) = t (microseconds)

## Capacitors in Series

### Two capacitors

$$C_T = \frac{C_1 C_2}{C_1 + C_2}$$

### More than two

$$\frac{1}{C_T} = \frac{1}{C_1} + \frac{1}{C_2} + \frac{1}{C_3} + \ldots$$

## Capacitors in Parallel: $C_T = C_1 + C_2 + \ldots$

## Capacitive Reactance: $X_C = \frac{1}{2\pi fC}$

## Impedance in an RC Circuit (Series)

$$Z = \sqrt{R^2 + (X_C)^2}$$

## Inductors in Series

$$L_T = L_1 + L_2 + \ldots \text{ (No coupling between coils)}$$

## Inductors in Parallel

### Two inductors

$$L_T = \frac{L_1 L_2}{L_1 + L_2} \text{ (No coupling between coils)}$$

### More than two

$$\frac{1}{L_T} = \frac{1}{L_1} + \frac{1}{L_2} + \frac{1}{L_3} + \ldots \text{ (No coupling between coils)}$$

## Inductive Reactance

$$X_L = 2\pi fL$$

## Q of a Coil

$$Q = \frac{X_L}{R}$$

Impedance of an RL Circuit (Series)

$$Z = \sqrt{R^2 + (X_L)^2}$$

Impedance with R, C, and L in Series

$$Z = \sqrt{R^2 + (X_L - X_C)^2}$$

Parallel Circuit Impedance

$$Z = \frac{Z_1 Z_2}{Z_1 + Z_2}$$

Sine-Wave Voltage Relationships

Average value

$$E_{ave} = \frac{2}{\pi} \times E_{max} = 0.637 E_{max}$$

Effective or r.m.s. value

$$E_{eff} = \frac{E_{max}}{\sqrt{2}} = \frac{E_{max}}{1.414} = 0.707 E_{max}$$
$$= 1.11 E_{ave}$$

Maximum value

$$E_{max} = \sqrt{2}(E_{eff}) = 1.414 E_{eff}$$
$$= 1.57 E_{ave}$$

Voltage in an a-c circuit

$$E = IZ = \frac{P}{I \times P.F.}$$

Current in an a-c circuit

$$I = \frac{E}{Z} = \frac{P}{E \times P.F.}$$

Power in A-C Circuit

Apparent power: $P = EI$

True power: $P = EI \cos \theta = EI \times P.F.$

Power Factor

$$P.F. = \frac{P}{EI} = \cos \theta$$

$$\cos \theta = \frac{\text{true power}}{\text{apparent power}}$$

Transformers

Voltage relationship

$$\frac{E_p}{E_s} = \frac{N_p}{N_s} \text{ or } E_s = E_p \times \frac{N_s}{N_p}$$

Current relationship

$$\frac{I_p}{I_s} = \frac{N_s}{N_p}$$

Induced voltage

$$E_{eff} = 4.44 \times BAfN \times 10^{-8}$$

Turns ratio

$$\frac{N_p}{N_s} = \sqrt{\frac{Z_p}{Z_s}}$$

Secondary current

$$I_s = I_p \times \frac{N_p}{N_s}$$

Secondary voltage

$$E_s = E_p \times \frac{N_s}{N_p}$$

Three-Phase Voltage and Current Relationships

With wye connected windings

$$E_{line} = \sqrt{3}(E_{coil}) = 1.732 E_{coil}$$

$$I_{line} = I_{coil}$$

With delta connected windings

$$E_{line} = E_{coil}$$

$$I_{line} = 1.732 I_{coil}$$

With **wye** or delta connected winding

$$P_{coil} = E_{coil}I_{coil}$$

$$P_t = 3P_{coil}$$

$$P_t = 1.732E_{line}I_{line}$$

(To convert to true power multiply by $\cos \theta$ )

Synchronous Speed of Motor

$$r.p.m. = \frac{120 \text{ x frequency}}{\text{number of poles}}$$

## Comparison of Units in Electric and Magnetic Circuits

|  | Electric circuit | Magnetic circuit |
|---|---|---|
| Force............ | Volt, E, or e.m.f. | Gilberts, F, or m.m.f. |
| Flow ............ | Ampere, I | Flux, $\Phi$ , in maxwells |
| Opposition......... | Ohms, R | Reluctance, $\mathcal{R}$ |
| Law............. | Ohm's law, $I = \frac{E}{R}$ | Rowland's law, $\Phi = \frac{F}{\mathcal{R}}$ |
| Intensity of force ..... | Volts per cm. of length. | $H = \frac{1.257IN}{L}$, gilberts per centimeter of length. |
| Density........... | Current density—for example, amperes per cm$^2$. | Flux density—for example, lines per cm.$^2$ , or **gausses**. |

# APPENDIX IX

# TRIGONOMETRIC FUNCTIONS

In a right triangle, there are several relationships which always hold true. These relationships pertain to the length of the sides of a right triangle, and the way the lengths are affected by the angles between them. An understanding of these relationships, called trigonometric functions, is essential for solving many problems in a-c circuits such as power factor, impedance, voltage drops, and so forth.

To be a RIGHT triangle, a triangle must have a "square" corner; one in which there is exactly 90° between two of the sides. Trigonometric functions do not apply to any other type of triangle. This type of triangle is shown in figure IX-1.

By use of the trigonometric functions, it is possible to determine the UNKNOWN length of one or more sides of a triangle, or the number of degrees in UNKNOWN angles, depending on what is presently known about the triangle. For instance, if the lengths of any two sides are known, the third side and both angles $\theta$ (theta) and $\Phi$ (phi) may be determined. The triangle may also be solved if the length of any one side and one of the angles ($\theta$ or $\Phi$ in fig. IX-1) are known.

The first basic fact to accept, regarding triangles, is that IN ANY TRIANGLE, THE SUM OF THE THREE ANGLES FORMED INSIDE THE TRIANGLE MUST ALWAYS EQUAL 180°. If one angle is always 90° (a right angle) then the sum of the other two angles must always be 90°.

$$\theta + \Phi = 90°$$

and

$$90° + \theta + \Phi = 180°$$

or

$$90° + 90° = 180°$$

thus, if angle $\theta$ is known, $\Phi$ may be quickly determined.

For instance, if $\theta$ is 30°, what is $\Phi$ ?

$$90° + 30° + \Phi = 180°$$

Transposing $\quad \Phi = 180° - 90° - 30°$

$$\Phi = 60°$$

Also, if $\Phi$ is known, $\theta$ may be determined in the same manner.

The second basic fact you must understand is that FOR EVERY DIFFERENT COMBINATION OF ANGLES IN A TRIANGLE, THERE IS A DEFINITE RATIO BETWEEN THE LENGTHS OF THE THREE SIDES. Consider the triangle in figure IX-2, consisting of the base, side B; the altitude, side A; and the hypotenuse, side C. (The hypotenuse is always the longest side, and is always opposite the 90° angle.) If angle $\theta$ is 30°, $\Phi$ must be 60°. With $\theta$ equal to 30°, the ratio of the length of side B to side C is 0.866 to 1. That is, if the hypotenuse is 1 inch long, the side adjacent to $\theta$, side B, is 0.866 inch long. Also, with $\theta$ equal to 30°, the ratio of side A to side C is 0.5 to 1. That is. with the hypotenuse 1 inch long, the side opposite to $\theta$ (side A) is 0.5 inch long. With $\theta$ still at 30°, side A is 0.5774 of the length of B. With the combination of angles given (30° - 60° - 90°) these are the ONLY ratios of lengths that will "fit" to form a right triangle.

Note that three ratios are shown to exist for the given value of $\theta$ : the ratio $\frac{B}{C}$, which is always referred to as the COSINE ratio of $\theta$, the ratio $\frac{A}{C}$, which is always the SINE ratio of $\theta$, and the ratio $\frac{A}{B}$, which is always the TANGENT ratio of $\theta$. If $\theta$ changes, all three ratios

$$\theta + \Phi = 90°$$

$$90° + \theta + \Phi = 180°$$

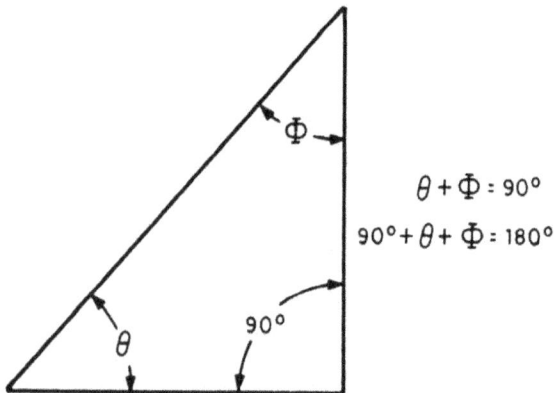

Figure IX-1.—A right triangle.

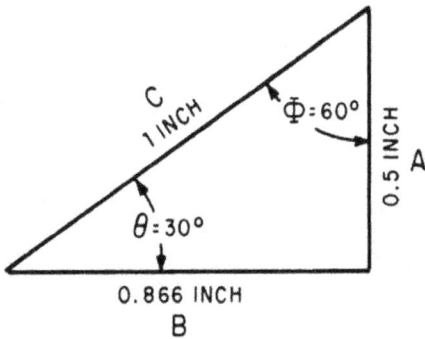

Figure IX-2.—A 30°-60°-90° triangle.

change, because the lengths of the sides (base and altitude) change. There is a set of ratios for every increment between 0° and 90°. These angular ratios, or sine, cosine, and tangent functions, are listed for each degree and tenth of degree in a table at the end of this appendix. In this table, the length of the hypotenuse of a triangle is considered fixed. Thus, the ratios of length given refer to the manner in which sides A and B vary with relation to each other and in relation to side C, as angle $\theta$ is varied from 0° to 90°.

The solution of problems in trigonometry (solution of triangles) is much simpler when the table of trigonometric functions is used properly. The most common ways in which it is used will be shown by solving a series of exemplary problems.

Problem 1: If the hypotenuse of the triangle (side C) in figure IX-3 (A) is 10 inches long, and angle $\theta$ is 33°, how long are sides B and A?

Solution: The ratio $\dfrac{B}{C}$ is the cosine function. By checking the table of functions, you will find that

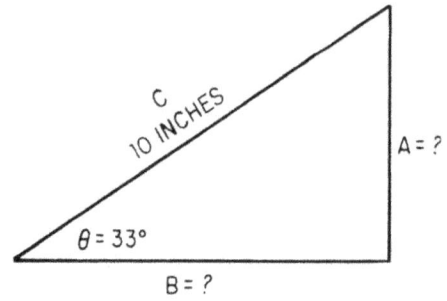

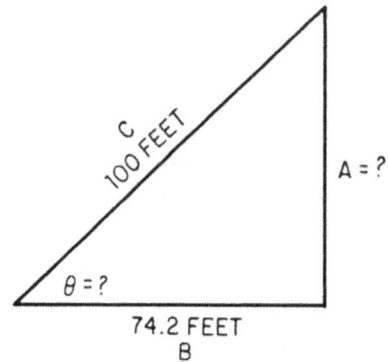

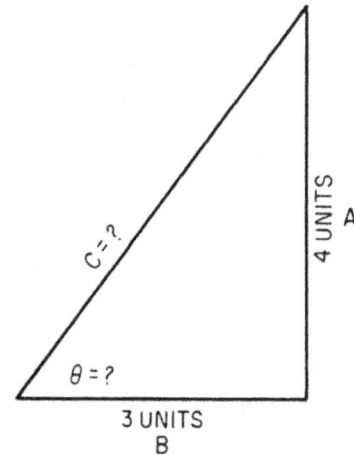

Figure IX-3.—Trigonometric problems.

the cosine of 33° is 0.8387. This means that the length of B is 0.8387 the length of side C. If side C is 10 inches long, then side B must be 10 x 0.8387, or 8.387 inches in length. To determine the length of side A, use the sine function, the ratio $\dfrac{A}{C}$. Again consulting the table of functions, you will find that the sine of 33° is 0.5446. Thus, side A must be 10 x 0.5446, or 5.446 inches in length.

Problem 2: The triangle in figure IX-3 (B) has a base 74.2 feet long, and hypotenuse 100 feet long. What is $\theta$, and how long is side A? Solution: When no angles are given, you must always solve for a known angle first. The ratio $\frac{B}{C}$ is the cosine of the unknown angle $\theta$; therefore $\frac{74.2}{100}$, or 0.742, is the cosine of the unknown angle. Locating 0.742 as a cosine value in the table, you find that it is the cosine of 42.1°. That is, $\theta = 42.1°$. With $\theta$ known, side A is solved for by use of the sine ratio $\frac{A}{C}$. The sine of 42.1°, according to the table, is 0.6704. Therefore, side A is 100 x 0.6704, or 67.04 feet long.

Problem 3: In the triangle in figure IX-3 (C), the base is 3 units long, and the altitude is 4 units. What is $\theta$, and how long is the hypotenuse? Solution: With the information given, the tangent of $\theta$ may be determined. $\text{Tan } \theta = \frac{A}{B} = \frac{4}{3} = 1.33$

Locating the value 1.33 as a tangent value in the table of functions, you find it to be the tangent of 53.1°. Therefore, $\theta = 53.1°$. Once $\theta$ is known, either the sine or cosine ratio may be used to determine the length of the hypotenuse. The cosine of 53.1° is 0.6004. This indicates that the base of 3 units is 0.6004, the length of the hypotenuse. Therefore, the hypotenuse is $\frac{3}{0.6004}$, or 5 units in length. Using the sine ratio, the hypotenuse is $\frac{4}{0.7997}$, or 5 units in length.

In the foregoing explanations and problems, the sides of triangles were given in inches, feet, and units. In applying trigonometry to a-c circuit problems, these units of measure will be replaced by such measurements as ohms, ampres, volts, and watts. Angle $\theta$ will often be referred to as the phase angle. However, the solution of these a-c problems is accomplished in exactly the same manner as the foregoing problems. Only the units and some terminology are changed.

## NATURAL SINES, COSINES, AND TANGENTS

0°–14.9°

| Degs. | Function | 0.0° | 0.1° | 0.2° | 0.3° | 0.4° | 0.5° | 0.6° | 0.7° | 0.8° | 0.9° |
|---|---|---|---|---|---|---|---|---|---|---|---|
| 0 | sin | 0 0000 | 0 0017 | 0 0035 | 0 0052 | 0 0070 | 0.0087 | 0.0105 | 0.0122 | 0 0140 | 0 0157 |
| | cos | 1 0000 | 1 0000 | 1 0000 | 1 0000 | 1 0000 | 1.0000 | 0.9999 | 0 9999 | 0.9999 | 0 9999 |
| | tan | 0 0000 | 0 0017 | 0 0035 | 0 0052 | 0.0070 | 0 0087 | 0.0105 | 0.0122 | 0 0140 | 0 0157 |
| 1 | sin | 0 0175 | 0.0192 | 0.0209 | 0.0227 | 0 0244 | 0.0262 | 0 0279 | 0 0297 | 0 0314 | 0 0332 |
| | cos | 0 9998 | 0 9998 | 0.9998 | 0.9997 | 0 9997 | 0 9997 | 0 9996 | 0 9996 | 0 9995 | 0 9995 |
| | tan | 0 0175 | 0 0192 | 0.0209 | 0.0227 | 0.0244 | 0 0262 | 0 0279 | 0 0297 | 0 0314 | 0 0332 |
| 2 | sin | 0 0349 | 0 0366 | 0.0384 | 0 0401 | 0 0419 | 0.0436 | 0.0454 | 0 0471 | 0 0488 | 0 0506 |
| | cos | 0 9994 | 0 9993 | 0 9993 | 0 9992 | 0 9991 | 0 9990 | 0 9990 | 0 9989 | 0 9988 | 0 9987 |
| | tan | 0 0349 | 0.0367 | 0.0384 | 0.0402 | 0.0419 | 0.0437 | 0.0454 | 0 0472 | 0.0489 | 0 0507 |
| 3 | sin | 0 0523 | 0 0541 | 0.0558 | 0.0576 | 0.0593 | 0.0610 | 0.0628 | 0.0645 | 0.0663 | 0 0680 |
| | cos | 0 9986 | 0 9985 | 0 9984 | 0 9983 | 0 9982 | 0 9981 | 0 9980 | 0 9979 | 0 9978 | 0 9977 |
| | tan | 0 0524 | 0 0542 | 0.0559 | 0.0577 | 0.0594 | 0.0612 | 0.0629 | 0.0647 | 0.0664 | 0 0682 |
| 4 | sin | 0 0698 | 0 0715 | 0.0732 | 0.0750 | 0.0767 | 0.0785 | 0.0802 | 0.0819 | 0 0837 | 0 0854 |
| | cos | 0 9976 | 0 9974 | 0 9973 | 0 9972 | 0 9971 | 0 9969 | 0 9968 | 0 9966 | 0 9965 | 0 9963 |
| | tan | 0 0699 | 0 0717 | 0.0734 | 0.0752 | 0.0769 | 0.0787 | 0.0805 | 0.0822 | 0.0840 | 0 0857 |
| 5 | sin | 0 0872 | 0 0889 | 0.0906 | 0.0924 | 0.0941 | 0.0958 | 0.0976 | 0.0993 | 0.1011 | 0 1028 |
| | cos | 0 9962 | 0 9960 | 0 9959 | 0.9957 | 0.9956 | 0.9954 | 0.9952 | 0.9951 | 0.9949 | 0 9947 |
| | tan | 0 0875 | 0 0892 | 0.0910 | 0.0928 | 0.0945 | 0.0963 | 0.0981 | 0.0998 | 0.1016 | 0 1033 |
| 6 | sin | 0.1045 | 0 1063 | 0.1080 | 0.1097 | 0.1115 | 0.1132 | 0.1149 | 0.1167 | 0.1184 | 0 1201 |
| | cos | 0 9945 | 0 9943 | 0.9942 | 0.9940 | 0 9938 | 0.9936 | 0.9934 | 0.9932 | 0.9930 | 0 9928 |
| | tan | 0 1051 | 0 1069 | 0.1086 | 0.1104 | 0.1122 | 0.1139 | 0.1157 | 0.1175 | 0.1192 | 0.1210 |
| 7 | sin | 0 1219 | 0 1236 | 0.1253 | 0.1271 | 0.1288 | 0.1305 | 0.1323 | 0.1340 | 0.1357 | 0 1374 |
| | cos | 0 9925 | 0 9923 | 0.9921 | 0.9919 | 0.9917 | 0.9914 | 0.9912 | 0.9910 | 0.9907 | 0 9905 |
| | tan | 0.1228 | 0.1246 | 0.1263 | 0.1281 | 0.1299 | 0.1317 | 0.1334 | 0.1352 | 0.1370 | 0 1388 |
| 8 | sin | 0 1392 | 0 1409 | 0 1426 | 0 1444 | 0 1461 | 0 1478 | 0 1495 | 0 1513 | 0 1530 | 0 1547 |
| | cos | 0 9903 | 0 9900 | 0 9898 | 0 9895 | 0 9893 | 0 98v0 | 0 9888 | 0 9885 | 0 9882 | 0 9880 |
| | tan | 0 1405 | 0 1423 | 0.1441 | 0 1459 | 0.1477 | 0.1495 | 0.1512 | 0.1530 | 0 1548 | 0 1566 |
| 9 | sin | 0 1564 | 0 1582 | 0 1599 | 0 1616 | 0 1633 | 0 1650 | 0.1668 | 0 1685 | 0 1702 | 0 1719 |
| | cos | 0 9877 | 0 9874 | 0 9871 | 0 9869 | 0 9866 | 0 9863 | 0 9860 | 0.9857 | 0.9854 | 0 9851 |
| | tan | 0 1584 | 0 1602 | 0.1620 | 0 1638 | 0.1655 | 0.1673 | 0.1691 | 0.1709 | 0.1727 | 0 1745 |
| 10 | sin | 0 1736 | 0 1754 | 0.1771 | 0 1788 | 0.1805 | 0.1822 | 0 1840 | 0 1857 | 0.1874 | 0 1891 |
| | cos | 0 9848 | 0 9845 | 0.9842 | 0 9839 | 0.9836 | 0.9833 | 0 9829 | 0 9826 | 0 9823 | 0 9820 |
| | tan | 0 1763 | 0.1781 | 0.1799 | 0 1817 | 0.1835 | 0.1853 | 0.1871 | 0.1890 | 0.1908 | 0 1926 |
| 11 | sin | 0 1908 | 0 1925 | 0 1942 | 0 1959 | 0.1977 | 0.1994 | 0 2011 | 0 2028 | 0 2045 | 0 2062 |
| | cos | 0 9816 | 0 9813 | 0 9810 | 0 9806 | 0.9803 | 0 9799 | 0 9796 | 0 9792 | 0 9789 | 0 9785 |
| | tan | 0 1944 | 0 1962 | 0.1980 | 0.1998 | 0.2016 | 0.2035 | 0 2053 | 0 2071 | 0.2089 | 0.2107 |
| 12 | sin | 0 2079 | 0 2096 | 0 2113 | 0 2130 | 0 2147 | 0 2164 | 0 2181 | 0 2198 | 0 2215 | 0 2232 |
| | cos | 0 9781 | 0 9778 | 0 9774 | 0 9770 | 0 9767 | 0 9763 | 0 9759 | 0 9755 | 0 9751 | 0 9748 |
| | tan | 0 2126 | 0 2144 | 0 2162 | 0 2180 | 0 2199 | 0 2217 | 0 2235 | 0 2254 | 0 2272 | 0 2290 |
| 13 | sin | 0 2250 | 0 2267 | 0 2284 | 0 2300 | 0 2318 | 0 2334 | 0 2351 | 0 2368 | 0 2385 | 0 2402 |
| | cos | 0 9744 | 0 9740 | 0 9736 | 0 9732 | 0 9728 | 0 9724 | 0 9720 | 0 9715 | 0 9711 | 0 9707 |
| | tan | 0 2309 | 0 2327 | 0 2345 | 0 2364 | 0 2382 | 0 2401 | 0 2419 | 0 2438 | 0 2456 | 0 2475 |
| 14 | sin | 0 2419 | 0 2436 | 0 2453 | 0 2470 | 0 2487 | 0 2504 | 0 2521 | 0 2538 | 0 2554 | 0 2571 |
| | cos | 0 9703 | 0 9699 | 0 9694 | 0 9690 | 0 9686 | 0 9681 | 0 9677 | 0 9673 | 0 9668 | 0 9664 |
| | tan | 0 2493 | 0 2512 | 0 2530 | 0 2549 | 0 2568 | 0 2586 | 0 2605 | 0 2623 | 0 2642 | 0 2661 |
| Degs. | Function | 0' | 6' | 12' | 18' | 24' | 30' | 36' | 42' | 48' | 54' |

Natural Sines, Cosines, and Tangents

15°–29.9°

| Degs. | Function | 0.0° | 0.1° | 0.2° | 0.3° | 0.4° | 0.5° | 0.6° | 0.7° | 0.8° | 0.9° |
|---|---|---|---|---|---|---|---|---|---|---|---|
| 15 | sin | 0.2588 | 0.2605 | 0.2622 | 0.2639 | 0.2656 | 0.2672 | 0.2689 | 0.2706 | 0.2723 | 0.2740 |
|  | cos | 0.9659 | 0.9655 | 0.9650 | 0.9646 | 0.9641 | 0.9636 | 0.9632 | 0.9627 | 0.9622 | 0.9617 |
|  | tan | 0.2679 | 0.2698 | 0.2717 | 0.2736 | 0.2754 | 0.2773 | 0.2792 | 0.2811 | 0.2830 | 0.2849 |
| 16 | sin | 0.2756 | 0.2773 | 0.2790 | 0.2807 | 0.2823 | 0.2840 | 0.2857 | 0.2874 | 0.2890 | 0.2907 |
|  | cos | 0.9613 | 0.9608 | 0.9603 | 0.9598 | 0.9593 | 0.9588 | 0.9583 | 0.9578 | 0.9573 | 0.9568 |
|  | tan | 0.2867 | 0.2886 | 0.2905 | 0.2924 | 0.2943 | 0.2962 | 0.2981 | 0.3000 | 0.3019 | 0.3038 |
| 17 | sin | 0.2924 | 0.2940 | 0.2957 | 0.2974 | 0.2990 | 0.3007 | 0.3024 | 0.3040 | 0.3057 | 0.3074 |
|  | cos | 0.9563 | 0.9558 | 0.9553 | 0.9548 | 0.9542 | 0.9537 | 0.9532 | 0.9527 | 0.9521 | 0.9516 |
|  | tan | 0.3057 | 0.3076 | 0.3096 | 0.3115 | 0.3134 | 0.3153 | 0.3172 | 0.3191 | 0.3211 | 0.3230 |
| 18 | sin | 0.3090 | 0.3107 | 0.3123 | 0.3140 | 0.3156 | 0.3173 | 0.3190 | 0.3206 | 0.3223 | 0.3239 |
|  | cos | 0.9511 | 0.9505 | 0.9500 | 0.9494 | 0.9489 | 0.9483 | 0.9478 | 0.9472 | 0.9466 | 0.9461 |
|  | tan | 0.3249 | 0.3269 | 0.3288 | 0.3307 | 0.3327 | 0.3346 | 0.3365 | 0.3385 | 0.3404 | 0.3424 |
| 19 | sin | 0.3256 | 0.3272 | 0.3289 | 0.3305 | 0.3322 | 0.3338 | 0.3355 | 0.3371 | 0.3387 | 0.3404 |
|  | cos | 0.9455 | 0.9449 | 0.9444 | 0.9438 | 0.9432 | 0.9426 | 0.9421 | 0.9415 | 0.9409 | 0.9403 |
|  | tan | 0.3443 | 0.3463 | 0.3482 | 0.3502 | 0.3522 | 0.3541 | 0.3561 | 0.3581 | 0.3600 | 0.3620 |
| 20 | sin | 0.3420 | 0.3437 | 0.3453 | 0.3469 | 0.3486 | 0.3502 | 0.3518 | 0.3535 | 0.3551 | 0.3567 |
|  | cos | 0.9397 | 0.9391 | 0.9385 | 0.9379 | 0.9373 | 0.9367 | 0.9361 | 0.9354 | 0.9348 | 0.9342 |
|  | tan | 0.3640 | 0.3659 | 0.3679 | 0.3699 | 0.3719 | 0.3739 | 0.3759 | 0.3779 | 0.3799 | 0.3819 |
| 21 | sin | 0.3584 | 0.3600 | 0.3616 | 0.3633 | 0.3649 | 0.3665 | 0.3681 | 0.3697 | 0.3714 | 0.3730 |
|  | cos | 0.9336 | 0.9330 | 0.9323 | 0.9317 | 0.9311 | 0.9304 | 0.9298 | 0.9291 | 0.9285 | 0.9278 |
|  | tan | 0.3839 | 0.3859 | 0.3879 | 0.3899 | 0.3919 | 0.3939 | 0.3959 | 0.3979 | 0.4000 | 0.4020 |
| 22 | sin | 0.3746 | 0.3762 | 0.3778 | 0.3795 | 0.3811 | 0.3827 | 0.3843 | 0.3859 | 0.3875 | 0.3891 |
|  | cos | 0.9272 | 0.9265 | 0.9259 | 0.9252 | 0.9245 | 0.9239 | 0.9232 | 0.9225 | 0.9219 | 0.9212 |
|  | tan | 0.4040 | 0.4061 | 0.4081 | 0.4101 | 0.4122 | 0.4142 | 0.4163 | 0.4183 | 0.4204 | 0.4224 |
| 23 | sin | 0.3907 | 0.3923 | 0.3939 | 0.3955 | 0.3971 | 0.3987 | 0.4003 | 0.4019 | 0.4035 | 0.4051 |
|  | cos | 0.9205 | 0.9198 | 0.9191 | 0.9184 | 0.9178 | 0.9171 | 0.9164 | 0.9157 | 0.9150 | 0.9143 |
|  | tan | 0.4245 | 0.4265 | 0.4286 | 0.4307 | 0.4327 | 0.4348 | 0.4369 | 0.4390 | 0.4411 | 0.4431 |
| 24 | sin | 0.4067 | 0.4083 | 0.4099 | 0.4115 | 0.4131 | 0.4147 | 0.4163 | 0.4179 | 0.4195 | 0.4210 |
|  | cos | 0.9135 | 0.9128 | 0.9121 | 0.9114 | 0.9107 | 0.9100 | 0.9092 | 0.9085 | 0.9078 | 0.9070 |
|  | tan | 0.4452 | 0.4473 | 0.4494 | 0.4515 | 0.4536 | 0.4557 | 0.4578 | 0.4599 | 0.4621 | 0.4642 |
| 25 | sin | 0.4226 | 0.4242 | 0.4258 | 0.4274 | 0.4289 | 0.4305 | 0.4321 | 0.4337 | 0.4352 | 0.4368 |
|  | cos | 0.9063 | 0.9056 | 0.9048 | 0.9041 | 0.9033 | 0.9026 | 0.9018 | 0.9011 | 0.9003 | 0.8996 |
|  | tan | 0.4663 | 0.4684 | 0.4706 | 0.4727 | 0.4748 | 0.4770 | 0.4791 | 0.4813 | 0.4834 | 0.4856 |
| 26 | sin | 0.4384 | 0.4399 | 0.4415 | 0.4431 | 0.4446 | 0.4462 | 0.4478 | 0.4493 | 0.4509 | 0.4524 |
|  | cos | 0.8988 | 0.8980 | 0.8973 | 0.8965 | 0.8957 | 0.8949 | 0.8942 | 0.8934 | 0.8926 | 0.8918 |
|  | tan | 0.4877 | 0.4899 | 0.4921 | 0.4942 | 0.4964 | 0.4986 | 0.5008 | 0.5029 | 0.5051 | 0.5073 |
| 27 | sin | 0.4540 | 0.4555 | 0.4571 | 0.4586 | 0.4602 | 0.4617 | 0.4633 | 0.4648 | 0.4664 | 0.4679 |
|  | cos | 0.8910 | 0.8902 | 0.8894 | 0.8886 | 0.8878 | 0.8870 | 0.8862 | 0.8854 | 0.8846 | 0.8838 |
|  | tan | 0.5095 | 0.5117 | 0.5139 | 0.5161 | 0.5184 | 0.5206 | 0.5228 | 0.5250 | 0.5272 | 0.5295 |
| 28 | sin | 0.4695 | 0.4710 | 0.4726 | 0.4741 | 0.4756 | 0.4772 | 0.4787 | 0.4802 | 0.4818 | 0.4833 |
|  | cos | 0.8829 | 0.8821 | 0.8813 | 0.8805 | 0.8796 | 0.8788 | 0.8780 | 0.8771 | 0.8763 | 0.8755 |
|  | tan | 0.5317 | 0.5340 | 0.5362 | 0.5384 | 0.5407 | 0.5430 | 0.5452 | 0.5475 | 0.5498 | 0.5520 |
| 29 | sin | 0.4848 | 0.4863 | 0.4879 | 0.4894 | 0.4909 | 0.4924 | 0.4939 | 0.4955 | 0.4970 | 0.4985 |
|  | cos | 0.8746 | 0.8738 | 0.8729 | 0.8721 | 0.8712 | 0.8704 | 0.8695 | 0.8686 | 0.8678 | 0.8669 |
|  | tan | 0.5543 | 0.5566 | 0.5589 | 0.5612 | 0.5635 | 0.5658 | 0.5681 | 0.5704 | 0.5727 | 0.5750 |
| Degs. | Function | 0' | 6' | 12' | 18' | 24' | 30' | 36' | 42' | 48' | 54' |

NATURAL SINES, COSINES, AND TANGENTS

30°-44.9°

| Degs. | Function | 0.0° | 0.1° | 0.2° | 0.3° | 0.4° | 0.5° | 0.6° | 0.7° | 0.8° | 0.9° |
|---|---|---|---|---|---|---|---|---|---|---|---|
| 30 | sin | 0.5000 | 0.5015 | 0.5030 | 0.5045 | 0.5060 | 0.5075 | 0.5090 | 0.5105 | 0.5120 | 0.5135 |
|  | cos | 0.8660 | 0.8652 | 0.8643 | 0.8634 | 0.8625 | 0.8616 | 0.8607 | 0.8599 | 0.8590 | 0.8581 |
|  | tan | 0.5774 | 0.5797 | 0.5820 | 0.5844 | 0.5867 | 0.5890 | 0.5914 | 0.5938 | 0.5961 | 0.5985 |
| 31 | sin | 0.5150 | 0.5165 | 0.5180 | 0.5195 | 0.5210 | 0.5225 | 0.5240 | 0.5255 | 0.5270 | 0.5284 |
|  | cos | 0.8572 | 0.8563 | 0.8554 | 0.8545 | 0.8536 | 0.8526 | 0.8517 | 0.8508 | 0.8499 | 0.8490 |
|  | tan | 0.6009 | 0.6032 | 0.6056 | 0.6080 | 0.6104 | 0.6128 | 0.6152 | 0.6176 | 0.6200 | 0.6224 |
| 32 | sin | 0.5299 | 0.5314 | 0.5329 | 0.5344 | 0.5358 | 0.5373 | 0.5388 | 0.5402 | 0.5417 | 0.5432 |
|  | cos | 0.8480 | 0.8471 | 0.8462 | 0.8453 | 0.8443 | 0.8434 | 0.8425 | 0.8415 | 0.8406 | 0.8396 |
|  | tan | 0.6249 | 0.6273 | 0.6297 | 0.6322 | 0.6346 | 0.6371 | 0.6395 | 0.6420 | 0.6445 | 0.6469 |
| 33 | sin | 0.5446 | 0.5461 | 0.5476 | 0.5490 | 0.5505 | 0.5519 | 0.5534 | 0.5548 | 0.5563 | 0.5577 |
|  | cos | 0.8387 | 0.8377 | 0.8368 | 0.8358 | 0.8348 | 0.8339 | 0.8329 | 0.8320 | 0.8310 | 0.8300 |
|  | tan | 0.6494 | 0.6519 | 0.6544 | 0.6569 | 0.6594 | 0.6619 | 0.6644 | 0.6669 | 0.6694 | 0.6720 |
| 34 | sin | 0.5592 | 0.5606 | 0.5621 | 0.5635 | 0.5650 | 0.5664 | 0.5678 | 0.5693 | 0.5707 | 0.5721 |
|  | cos | 0.8290 | 0.8281 | 0.8271 | 0.8261 | 0.8251 | 0.8241 | 0.8231 | 0.8221 | 0.8211 | 0.8202 |
|  | tan | 0.6745 | 0.6771 | 0.6796 | 0.6822 | 0.6847 | 0.6873 | 0.6899 | 0.6924 | 0.6950 | 0.6976 |
| 35 | sin | 0.5736 | 0.5750 | 0.5764 | 0.5779 | 0.5793 | 0.5807 | 0.5821 | 0.5835 | 0.5850 | 0.5864 |
|  | cos | 0.8192 | 0.8181 | 0.8171 | 0.8161 | 0.8151 | 0.8141 | 0.8131 | 0.8121 | 0.8111 | 0.8100 |
|  | tan | 0.7002 | 0.7028 | 0.7054 | 0.7080 | 0.7107 | 0.7133 | 0.7159 | 0.7186 | 0.7212 | 0.7239 |
| 36 | sin | 0.5878 | 0.5892 | 0.5906 | 0.5920 | 0.5934 | 0.5948 | 0.5962 | 0.5976 | 0.5990 | 0.6004 |
|  | cos | 0.8090 | 0.8080 | 0.8070 | 0.8059 | 0.8049 | 0.8039 | 0.8028 | 0.8018 | 0.8007 | 0.7997 |
|  | tan | 0.7265 | 0.7292 | 0.7319 | 0.7346 | 0.7373 | 0.7400 | 0.7427 | 0.7454 | 0.7481 | 0.7508 |
| 37 | sin | 0.6018 | 0.6032 | 0.6046 | 0.6060 | 0.6074 | 0.6088 | 0.6101 | 0.6115 | 0.6129 | 0.6143 |
|  | cos | 0.7986 | 0.7976 | 0.7965 | 0.7955 | 0.7944 | 0.7934 | 0.7923 | 0.7912 | 0.7902 | 0.7891 |
|  | tan | 0.7536 | 0.7563 | 0.7590 | 0.7618 | 0.7646 | 0.7673 | 0.7701 | 0.7729 | 0.7757 | 0.7785 |
| 38 | sin | 0.6157 | 0.6170 | 0.6184 | 0.6198 | 0.6211 | 0.6225 | 0.6239 | 0.6252 | 0.6266 | 0.6280 |
|  | cos | 0.7880 | 0.7869 | 0.7859 | 0.7848 | 0.7837 | 0.7826 | 0.7815 | 0.7804 | 0.7793 | 0.7782 |
|  | tan | 0.7813 | 0.7841 | 0.7869 | 0.7898 | 0.7926 | 0.7954 | 0.7983 | 0.8012 | 0.8040 | 0.8069 |
| 39 | sin | 0.6293 | 0.6307 | 0.6320 | 0.6334 | 0.6347 | 0.6361 | 0.6374 | 0.6388 | 0.6401 | 0.6414 |
|  | cos | 0.7771 | 0.7760 | 0.7749 | 0.7738 | 0.7727 | 0.7716 | 0.7705 | 0.7694 | 0.7683 | 0.7672 |
|  | tan | 0.8098 | 0.8127 | 0.8156 | 0.8185 | 0.8214 | 0.8243 | 0.8273 | 0.8302 | 0.8332 | 0.8361 |
| 40 | sin | 0.6428 | 0.6441 | 0.6455 | 0.6468 | 0.6481 | 0.6494 | 0.6508 | 0.6521 | 0.6534 | 0.6547 |
|  | cos | 0.7660 | 0.7649 | 0.7638 | 0.7627 | 0.7615 | 0.7604 | 0.7593 | 0.7581 | 0.7570 | 0.7559 |
|  | tan | 0.8391 | 0.8421 | 0.8451 | 0.8481 | 0.8511 | 0.8541 | 0.8571 | 0.8601 | 0.8632 | 0.8662 |
| 41 | sin | 0.6561 | 0.6574 | 0.6587 | 0.6600 | 0.6613 | 0.6626 | 0.6639 | 0.6652 | 0.6665 | 0.6678 |
|  | cos | 0.7547 | 0.7536 | 0.7524 | 0.7513 | 0.7501 | 0.7490 | 0.7478 | 0.7466 | 0.7455 | 0.7443 |
|  | tan | 0.8693 | 0.8724 | 0.8754 | 0.8785 | 0.8816 | 0.8847 | 0.8878 | 0.8910 | 0.8941 | 0.8972 |
| 42 | sin | 0.6691 | 0.6704 | 0.6717 | 0.6730 | 0.6743 | 0.6756 | 0.6769 | 0.6782 | 0.6794 | 0.6807 |
|  | cos | 0.7431 | 0.7420 | 0.7408 | 0.7396 | 0.7385 | 0.7373 | 0.7361 | 0.7349 | 0.7337 | 0.7325 |
|  | tan | 0.9004 | 0.9036 | 0.9067 | 0.9099 | 0.9131 | 0.9163 | 0.9195 | 0.9228 | 0.9260 | 0.9293 |
| 43 | sin | 0.6820 | 0.6833 | 0.6845 | 0.6858 | 0.6871 | 0.6884 | 0.6896 | 0.6909 | 0.6921 | 0.6934 |
|  | cos | 0.7314 | 0.7302 | 0.7290 | 0.7278 | 0.7266 | 0.7254 | 0.7242 | 0.7230 | 0.7218 | 0.7206 |
|  | tan | 0.9325 | 0.9358 | 0.9391 | 0.9424 | 0.9457 | 0.9490 | 0.9523 | 0.9556 | 0.9590 | 0.9623 |
| 44 | sin | 0.6947 | 0.6959 | 0.6972 | 0.6984 | 0.6997 | 0.7009 | 0.7022 | 0.7034 | 0.7046 | 0.7059 |
|  | cos | 0.7193 | 0.7181 | 0.7169 | 0.7157 | 0.7145 | 0.7133 | 0.7120 | 0.7108 | 0.7096 | 0.7083 |
|  | tan | 0.9657 | 0.9691 | 0.9725 | 0.9759 | 0.9793 | 0.9827 | 0.9861 | 0.9896 | 0.9930 | 0.9965 |
| Degs. | Function | 0' | 6' | 12' | 18' | 24' | 30' | 36' | 42' | 48' | 54' |

NATURAL SINES, COSINES, AND TANGENTS

45°-59.9°

| Degs. | Function | 0.0° | 0.1° | 0.2° | 0.3° | 0.4° | 0.5° | 0.6° | 0.7° | 0.8° | 0.9° |
|---|---|---|---|---|---|---|---|---|---|---|---|
| 45 | sin | 0.7071 | 0.7083 | 0.7096 | 0.7108 | 0.7120 | 0.7133 | 0.7145 | 0.7157 | 0.7169 | 0.7181 |
|    | cos | 0.7071 | 0.7059 | 0.7046 | 0.7034 | 0.7022 | 0.7009 | 0.6997 | 0.6984 | 0.6972 | 0.6959 |
|    | tan | 1.0000 | 1.0035 | 1.0070 | 1.0105 | 1.0141 | 1.0176 | 1.0212 | 1.0247 | 1.0283 | 1.0319 |
| 46 | sin | 0.7193 | 0.7206 | 0.7218 | 0.7230 | 0.7242 | 0.7254 | 0.7266 | 0.7278 | 0.7290 | 0.7302 |
|    | cos | 0.6947 | 0.6934 | 0.6921 | 0.6909 | 0.6896 | 0.6884 | 0.6871 | 0.6858 | 0.6845 | 0.6833 |
|    | tan | 1.0355 | 1.0392 | 1.0428 | 1.0464 | 1.0501 | 1.0538 | 1.0575 | 1.0612 | 1.0649 | 1.0686 |
| 47 | sin | 0.7314 | 0.7325 | 0.7337 | 0.7349 | 0.7361 | 0.7373 | 0.7385 | 0.7396 | 0.7408 | 0.7420 |
|    | cos | 0.6820 | 0.6807 | 0.6794 | 0.6782 | 0.6769 | 0.6756 | 0.6743 | 0.6730 | 0.6717 | 0.6704 |
|    | tan | 1.0724 | 1.0761 | 1.0799 | 1.0837 | 1.0875 | 1.0913 | 1.0951 | 1.0990 | 1.1028 | 1.1067 |
| 48 | sin | 0.7431 | 0.7443 | 0.7455 | 0.7466 | 0.7478 | 0.7490 | 0.7501 | 0.7513 | 0.7524 | 0.7536 |
|    | cos | 0.6691 | 0.6678 | 0.6665 | 0.6652 | 0.6639 | 0.6626 | 0.6613 | 0.6600 | 0.6587 | 0.6574 |
|    | tan | 1.1106 | 1.1145 | 1.1184 | 1.1224 | 1.1263 | 1.1303 | 1.1343 | 1.1383 | 1.1423 | 1.1463 |
| 49 | sin | 0.7547 | 0.7559 | 0.7570 | 0.7581 | 0.7593 | 0.7604 | 0.7615 | 0.7627 | 0.7638 | 0.7649 |
|    | cos | 0.6561 | 0.6547 | 0.6534 | 0.6521 | 0.6508 | 0.6494 | 0.6481 | 0.6468 | 0.6455 | 0.6441 |
|    | tan | 1.1504 | 1.1544 | 1.1585 | 1.1626 | 1.1667 | 1.1708 | 1.1750 | 1.1792 | 1.1833 | 1.1875 |
| 50 | sin | 0.7660 | 0.7672 | 0.7683 | 0.7694 | 0.7705 | 0.7716 | 0.7727 | 0.7738 | 0.7749 | 0.7760 |
|    | cos | 0.6428 | 0.6414 | 0.6401 | 0.6388 | 0.6374 | 0.6361 | 0.6347 | 0.6334 | 0.6320 | 0.6307 |
|    | tan | 1.1918 | 1.1960 | 1.2002 | 1.2045 | 1.2088 | 1.2131 | 1.2174 | 1.2218 | 1.2261 | 1.2305 |
| 51 | sin | 0.7771 | 0.7782 | 0.7793 | 0.7804 | 0.7815 | 0.7826 | 0.7837 | 0.7848 | 0.7859 | 0.7869 |
|    | cos | 0.6293 | 0.6280 | 0.6266 | 0.6252 | 0.6239 | 0.6225 | 0.6211 | 0.6198 | 0.6184 | 0.6170 |
|    | tan | 1.2349 | 1.2393 | 1.2437 | 1.2482 | 1.2527 | 1.2572 | 1.2617 | 1.2662 | 1.2708 | 1.2753 |
| 52 | sin | 0.7880 | 0.7891 | 0.7902 | 0.7912 | 0.7923 | 0.7934 | 0.7944 | 0.7955 | 0.7965 | 0.7976 |
|    | cos | 0.6157 | 0.6143 | 0.6129 | 0.6115 | 0.6101 | 0.6088 | 0.6074 | 0.6060 | 0.6046 | 0.6032 |
|    | tan | 1.2799 | 1.2846 | 1.2892 | 1.2938 | 1.2985 | 1.3032 | 1.3079 | 1.3127 | 1.3175 | 1.3222 |
| 53 | sin | 0.7986 | 0.7997 | 0.8007 | 0.8018 | 0.8028 | 0.8039 | 0.8049 | 0.8059 | 0.8070 | 0.8080 |
|    | cos | 0.6018 | 0.6004 | 0.5990 | 0.5976 | 0.5962 | 0.5948 | 0.5934 | 0.5920 | 0.5906 | 0.5892 |
|    | tan | 1.3270 | 1.3319 | 1.3367 | 1.3416 | 1.3465 | 1.3514 | 1.3564 | 1.3613 | 1.3663 | 1.3713 |
| 54 | sin | 0.8090 | 0.8100 | 0.8111 | 0.8121 | 0.8131 | 0.8141 | 0.8151 | 0.8161 | 0.8171 | 0.8181 |
|    | cos | 0.5878 | 0.5864 | 0.5850 | 0.5835 | 0.5821 | 0.5807 | 0.5793 | 0.5779 | 0.5764 | 0.5750 |
|    | tan | 1.3764 | 1.3814 | 1.3865 | 1.3916 | 1.3968 | 1.4019 | 1.4071 | 1.4124 | 1.4176 | 1.4229 |
| 55 | sin | 0.8192 | 0.8202 | 0.8211 | 0.8221 | 0.8231 | 0.8241 | 0.8251 | 0.8261 | 0.8271 | 0.8281 |
|    | cos | 0.5736 | 0.5721 | 0.5707 | 0.5693 | 0.5678 | 0.5664 | 0.5650 | 0.5635 | 0.5621 | 0.5606 |
|    | tan | 1.4281 | 1.4335 | 1.4388 | 1.4442 | 1.4496 | 1.4550 | 1.4605 | 1.4659 | 1.4715 | 1.4770 |
| 56 | sin | 0.8290 | 0.8300 | 0.8310 | 0.8320 | 0.8329 | 0.8339 | 0.8348 | 0.8358 | 0.8368 | 0.8377 |
|    | cos | 0.5592 | 0.5577 | 0.5563 | 0.5548 | 0.5534 | 0.5519 | 0.5505 | 0.5490 | 0.5476 | 0.5461 |
|    | tan | 1.4826 | 1.4882 | 1.4938 | 1.4994 | 1.5051 | 1.5108 | 1.5166 | 1.5224 | 1.5282 | 1.5340 |
| 57 | sin | 0.8387 | 0.8396 | 0.8406 | 0.8415 | 0.8425 | 0.8434 | 0.8443 | 0.8453 | 0.8462 | 0.8471 |
|    | cos | 0.5446 | 0.5432 | 0.5417 | 0.5402 | 0.5388 | 0.5373 | 0.5358 | 0.5344 | 0.5329 | 0.5314 |
|    | tan | 1.5399 | 1.5458 | 1.5517 | 1.5577 | 1.5637 | 1.5697 | 1.5757 | 1.5818 | 1.5880 | 1.5941 |
| 58 | sin | 0.8480 | 0.8490 | 0.8499 | 0.8508 | 0.8517 | 0.8526 | 0.8536 | 0.8545 | 0.8554 | 0.8563 |
|    | cos | 0.5299 | 0.5284 | 0.5270 | 0.5255 | 0.5240 | 0.5225 | 0.5210 | 0.5195 | 0.5180 | 0.5165 |
|    | tan | 1.6003 | 1.6066 | 1.6128 | 1.6191 | 1.6255 | 1.6319 | 1.6383 | 1.6447 | 1.6512 | 1.6577 |
| 59 | sin | 0.8572 | 0.8581 | 0.8590 | 0.8599 | 0.8607 | 0.8616 | 0.8625 | 0.8634 | 0.8643 | 0.8652 |
|    | cos | 0.5150 | 0.5135 | 0.5120 | 0.5105 | 0.5090 | 0.5075 | 0.5060 | 0.5045 | 0.5030 | 0.5015 |
|    | tan | 1.6643 | 1.6709 | 1.6775 | 1.6842 | 1.6909 | 1.6977 | 1.7045 | 1.7113 | 1.7182 | 1.7251 |
| Degs. | Function | 0' | 6' | 12' | 18' | 24' | 30' | 36' | 42' | 48' | 54' |

## Natural Sines, Cosines, and Tangents

60°–74.9°

| Degs. | Function | 0.0° | 0.1° | 0.2° | 0.3° | 0.4° | 0.5° | 0.6° | 0.7° | 0.8° | 0.9° |
|---|---|---|---|---|---|---|---|---|---|---|---|
| 60 | sin | 0.8660 | 0.8669 | 0.8678 | 0.8686 | 0.8695 | 0.8704 | 0.8712 | 0.8721 | 0.8729 | 0.8738 |
|    | cos | 0.5000 | 0.4985 | 0.4970 | 0.4955 | 0.4939 | 0.4924 | 0.4909 | 0.4894 | 0.4879 | 0.4863 |
|    | tan | 1.7321 | 1.7391 | 1.7461 | 1.7532 | 1.7603 | 1.7675 | 1.7747 | 1.7820 | 1.7893 | 1.7966 |
| 61 | sin | 0.8746 | 0.8755 | 0.8763 | 0.8771 | 0.8780 | 0.8788 | 0.8796 | 0.8805 | 0.8813 | 0.8821 |
|    | cos | 0.4848 | 0.4833 | 0.4818 | 0.4802 | 0.4787 | 0.4772 | 0.4756 | 0.4741 | 0.4726 | 0.4710 |
|    | tan | 1.8040 | 1.8115 | 1.8190 | 1.8265 | 1.8341 | 1.8418 | 1.8495 | 1.8572 | 1.8650 | 1.8728 |
| 62 | sin | 0.8829 | 0.8838 | 0.8846 | 0.8854 | 0.8862 | 0.8870 | 0.8878 | 0.8886 | 0.8894 | 0.8902 |
|    | cos | 0.4695 | 0.4679 | 0.4664 | 0.4648 | 0.4633 | 0.4617 | 0.4602 | 0.4586 | 0.4571 | 0.4555 |
|    | tan | 1.8807 | 1.8887 | 1.8967 | 1.9047 | 1.9128 | 1.9210 | 1.9292 | 1.9375 | 1.9458 | 1.9542 |
| 63 | sin | 0.8910 | 0.8918 | 0.8926 | 0.8934 | 0.8942 | 0.8949 | 0.8957 | 0.8965 | 0.8973 | 0.8980 |
|    | cos | 0.4540 | 0.4524 | 0.4509 | 0.4493 | 0.4478 | 0.4462 | 0.4446 | 0.4431 | 0.4415 | 0.4399 |
|    | tan | 1.9626 | 1.9711 | 1.9797 | 1.9883 | 1.9970 | 2.0057 | 2.0145 | 2.0233 | 2.0323 | 2.0413 |
| 64 | sin | 0.8988 | 0.8996 | 0.9003 | 0.9011 | 0.9018 | 0.9026 | 0.9033 | 0.9041 | 0.9048 | 0.9056 |
|    | cos | 0.4384 | 0.4368 | 0.4352 | 0.4337 | 0.4321 | 0.4305 | 0.4289 | 0.4274 | 0.4258 | 0.4242 |
|    | tan | 2.0503 | 2.0594 | 2.0686 | 2.0778 | 2.0872 | 2.0965 | 2.1060 | 2.1155 | 2.1251 | 2.1348 |
| 65 | sin | 0.9063 | 0.9070 | 0.9078 | 0.9085 | 0.9092 | 0.9100 | 0.9107 | 0.9114 | 0.9121 | 0.9128 |
|    | cos | 0.4226 | 0.4210 | 0.4195 | 0.4179 | 0.4163 | 0.4147 | 0.4131 | 0.4115 | 0.4099 | 0.4083 |
|    | tan | 2.1445 | 2.1543 | 2.1642 | 2.1742 | 2.1842 | 2.1943 | 2.2045 | 2.2148 | 2.2251 | 2.2355 |
| 66 | sin | 0.9135 | 0.9143 | 0.9150 | 0.9157 | 0.9164 | 0.9171 | 0.9178 | 0.9184 | 0.9191 | 0.9198 |
|    | cos | 0.4067 | 0.4051 | 0.4035 | 0.4019 | 0.4003 | 0.3987 | 0.3971 | 0.3955 | 0.3939 | 0.3923 |
|    | tan | 2.2460 | 2.2566 | 2.2673 | 2.2781 | 2.2889 | 2.2998 | 2.3109 | 2.3220 | 2.3332 | 2.3445 |
| 67 | sin | 0.9205 | 0.9212 | 0.9219 | 0.9225 | 0.9232 | 0.9239 | 0.9245 | 0.9252 | 0.9259 | 0.9265 |
|    | cos | 0.3907 | 0.3891 | 0.3875 | 0.3859 | 0.3843 | 0.3827 | 0.3811 | 0.3795 | 0.3778 | 0.3762 |
|    | tan | 2.3559 | 2.3673 | 2.3789 | 2.3906 | 2.4023 | 2.4142 | 2.4262 | 2.4383 | 2.4504 | 2.4627 |
| 68 | sin | 0.9272 | 0.9278 | 0.9285 | 0.9291 | 0.9298 | 0.9304 | 0.9311 | 0.9317 | 0.9323 | 0.9330 |
|    | cos | 0.3746 | 0.3730 | 0.3714 | 0.3697 | 0.3681 | 0.3665 | 0.3649 | 0.3633 | 0.3616 | 0.3600 |
|    | tan | 2.4751 | 2.4876 | 2.5002 | 2.5129 | 2.5257 | 2.5386 | 2.5517 | 2.5649 | 2.5782 | 2.5916 |
| 69 | sin | 0.9336 | 0.9342 | 0.9348 | 0.9354 | 0.9361 | 0.9367 | 0.9373 | 0.9379 | 0.9385 | 0.9391 |
|    | cos | 0.3584 | 0.3567 | 0.3551 | 0.3535 | 0.3518 | 0.3502 | 0.3486 | 0.3469 | 0.3453 | 0.3437 |
|    | tan | 2.6051 | 2.6187 | 2.6325 | 2.6464 | 2.6605 | 2.6746 | 2.6889 | 2.7034 | 2.7179 | 2.7326 |
| 70 | sin | 0.9397 | 0.9403 | 0.9409 | 0.9415 | 0.9421 | 0.9426 | 0.9432 | 0.9438 | 0.9444 | 0.9449 |
|    | cos | 0.3420 | 0.3404 | 0.3387 | 0.3371 | 0.3355 | 0.3338 | 0.3322 | 0.3305 | 0.3289 | 0.3272 |
|    | tan | 2.7475 | 2.7625 | 2.7776 | 2.7929 | 2.8083 | 2.8239 | 2.8397 | 2.8556 | 2.8716 | 2.8878 |
| 71 | sin | 0.9455 | 0.9461 | 0.9466 | 0.9472 | 0.9478 | 0.9483 | 0.9489 | 0.9494 | 0.9500 | 0.9505 |
|    | cos | 0.3256 | 0.3239 | 0.3223 | 0.3206 | 0.3190 | 0.3173 | 0.3156 | 0.3140 | 0.3123 | 0.3107 |
|    | tan | 2.9042 | 2.9208 | 2.9375 | 2.9544 | 2.9714 | 2.9887 | 3.0061 | 3.0237 | 3.0415 | 3.0595 |
| 72 | sin | 0.9511 | 0.9516 | 0.9521 | 0.9527 | 0.9532 | 0.9537 | 0.9542 | 0.9548 | 0.9553 | 0.9558 |
|    | cos | 0.3090 | 0.3074 | 0.3057 | 0.3040 | 0.3024 | 0.3007 | 0.2990 | 0.2974 | 0.2957 | 0.2940 |
|    | tan | 3.0777 | 3.0961 | 3.1146 | 3.1334 | 3.1524 | 3.1716 | 3.1910 | 3.2106 | 3.2305 | 3.2506 |
| 73 | sin | 0.9563 | 0.9568 | 0.9573 | 0.9578 | 0.9583 | 0.9588 | 0.9593 | 0.9598 | 0.9603 | 0.9608 |
|    | cos | 0.2924 | 0.2907 | 0.2890 | 0.2874 | 0.2857 | 0.2840 | 0.2823 | 0.2807 | 0.2790 | 0.2773 |
|    | tan | 3.2709 | 3.2914 | 3.3122 | 3.3332 | 3.3544 | 3.3759 | 3.3977 | 3.4197 | 3.4420 | 3.4646 |
| 74 | sin | 0.9613 | 0.9617 | 0.9622 | 0.9627 | 0.9632 | 0.9636 | 0.9641 | 0.9646 | 0.9650 | 0.9655 |
|    | cos | 0.2756 | 0.2740 | 0.2723 | 0.2706 | 0.2689 | 0.2672 | 0.2656 | 0.2639 | 0.2622 | 0.2605 |
|    | tan | 3.4874 | 3.5105 | 3.5339 | 3.5576 | 3.5816 | 3.6059 | 3.6305 | 3.6554 | 3.6806 | 3.7062 |
| Degs. | Function | 0′ | 6′ | 12′ | 18′ | 24′ | 30′ | 36′ | 42′ | 48′ | 54′ |

Natural Sines, Cosines, and Tangents.

75°–89.9°

| Degs. | Function | 0.0° | 0.1° | 0.2° | 0.3° | 0.4° | 0.5° | 0.6° | 0.7° | 0.8° | 0.9° |
|---|---|---|---|---|---|---|---|---|---|---|---|
| 75 | sin | 0.9659 | 0.9664 | 0.9668 | 0.9673 | 0.9677 | 0.9681 | 0.9686 | 0.9690 | 0.9694 | 0.9699 |
|  | cos | 0.2588 | 0.2571 | 0.2554 | 0.2538 | 0.2521 | 0.2504 | 0.2487 | 0.2470 | 0.2453 | 0.2436 |
|  | tan | 3.7321 | 3.7583 | 3.7848 | 3.8118 | 3.8391 | 3.8667 | 3.8947 | 3.9232 | 3.9520 | 3.9812 |
| 76 | sin | 0.9703 | 0.9707 | 0.9711 | 0.9715 | 0.9720 | 0.9724 | 0.9728 | 0.9732 | 0.9736 | 0.9740 |
|  | cos | 0.2419 | 0.2402 | 0.2385 | 0.2368 | 0.2351 | 0.2334 | 0.2317 | 0.2300 | 0.2284 | 0.2267 |
|  | tan | 4.0108 | 4.0408 | 4.0713 | 4.1022 | 4.1335 | 4.1653 | 4.1976 | 4.2303 | 4.2635 | 4.2972 |
| 77 | sin | 0.9744 | 0.9748 | 0.9751 | 0.9755 | 0.9759 | 0.9763 | 0.9767 | 0.9770 | 0.9774 | 0.9778 |
|  | cos | 0.2250 | 0.2232 | 0.2215 | 0.2198 | 0.2181 | 0.2164 | 0.2147 | 0.2130 | 0.2113 | 0.2096 |
|  | tan | 4.3315 | 4.3662 | 4.4015 | 4.4374 | 4.4737 | 4.5107 | 4.5483 | 4.5864 | 4.6252 | 4.6646 |
| 78 | sin | 0.9781 | 0.9785 | 0.9789 | 0.9792 | 0.9796 | 0.9799 | 0.9803 | 0.9806 | 0.9810 | 0.9813 |
|  | cos | 0.2079 | 0.2062 | 0.2045 | 0.2028 | 0.2011 | 0.1994 | 0.1977 | 0.1959 | 0.1942 | 0.1925 |
|  | tan | 4.7046 | 4.7453 | 4.7867 | 4.8288 | 4.8716 | 4.9152 | 4.9594 | 5.0045 | 5.0504 | 5.0970 |
| 79 | sin | 0.9816 | 0.9820 | 0.9823 | 0.9826 | 0.9829 | 0.9833 | 0.9836 | 0.9839 | 0.9842 | 0.9845 |
|  | cos | 0.1908 | 0.1891 | 0.1874 | 0.1857 | 0.1840 | 0.1822 | 0.1805 | 0.1788 | 0.1771 | 0.1754 |
|  | tan | 5.1446 | 5.1929 | 5.2422 | 5.2924 | 5.3435 | 5.3955 | 5.4486 | 5.5026 | 5.5578 | 5.6140 |
| 80 | sin | 0.9848 | 0.9851 | 0.9854 | 0.9857 | 0.9860 | 0.9863 | 0.9866 | 0.9869 | 0.9871 | 0.9874 |
|  | cos | 0.1736 | 0.1719 | 0.1702 | 0.1685 | 0.1668 | 0.1650 | 0.1633 | 0.1616 | 0.1599 | 0.1582 |
|  | tan | 5.6713 | 5.7297 | 5.7894 | 5.8502 | 5.9124 | 5.9758 | 6.0405 | 6.1066 | 6.1742 | 6.2432 |
| 81 | sin | 0.9877 | 0.9880 | 0.9882 | 0.9885 | 0.9888 | 0.9890 | 0.9893 | 0.9895 | 0.9898 | 0.9900 |
|  | cos | 0.1564 | 0.1547 | 0.1530 | 0.1513 | 0.1495 | 0.1478 | 0.1461 | 0.1444 | 0.1426 | 0.1409 |
|  | tan | 6.3138 | 6.3859 | 6.4596 | 6.5350 | 6.6122 | 6.6912 | 6.7720 | 6.8548 | 6.9395 | 7.0264 |
| 82 | sin | 0.9903 | 0.9905 | 0.9907 | 0.9910 | 0.9912 | 0.9914 | 0.9917 | 0.9919 | 0.9921 | 0.9923 |
|  | cos | 0.1392 | 0.1374 | 0.1357 | 0.1340 | 0.1323 | 0.1305 | 0.1288 | 0.1271 | 0.1253 | 0.1236 |
|  | tan | 7.1154 | 7.2066 | 7.3002 | 7.3962 | 7.4947 | 7.5958 | 7.6996 | 7.8062 | 7.9158 | 8.0285 |
| 83 | sin | 0.9925 | 0.9928 | 0.9930 | 0.9932 | 0.9934 | 0.9936 | 0.9938 | 0.9940 | 0.9942 | 0.9943 |
|  | cos | 0.1219 | 0.1201 | 0.1184 | 0.1167 | 0.1149 | 0.1132 | 0.1115 | 0.1097 | 0.1080 | 0.1063 |
|  | tan | 8.1443 | 8.2636 | 8.3863 | 8.5126 | 8.6427 | 8.7769 | 8.9152 | 9.0579 | 9.2052 | 9.3572 |
| 84 | sin | 0.9945 | 0.9947 | 0.9949 | 0.9951 | 0.9952 | 0.9954 | 0.9956 | 0.9957 | 0.9959 | 0.9960 |
|  | cos | 0.1045 | 0.1028 | 0.1011 | 0.0993 | 0.0976 | 0.0958 | 0.0941 | 0.0924 | 0.0906 | 0.0889 |
|  | tan | 9.5144 | 9.6768 | 9.8448 | 10.02 | 10.20 | 10.39 | 10.58 | 10.78 | 10.99 | 11.20 |
| 85 | sin | 0.9962 | 0.9963 | 0.9965 | 0.9966 | 0.9968 | 0.9969 | 0.9971 | 0.9972 | 0.9973 | 0.9974 |
|  | cos | 0.0872 | 0.0854 | 0.0837 | 0.0819 | 0.0802 | 0.0785 | 0.0767 | 0.0750 | 0.0732 | 0.0715 |
|  | tan | 11.43 | 11.66 | 11.91 | 12.16 | 12.43 | 12.71 | 13.00 | 13.30 | 13.62 | 13.95 |
| 86 | sin | 0.9976 | 0.9977 | 0.9978 | 0.9979 | 0.9980 | 0.9981 | 0.9982 | 0.9983 | 0.9984 | 0.9985 |
|  | cos | 0.0698 | 0.0680 | 0.0663 | 0.0645 | 0.0628 | 0.0610 | 0.0593 | 0.0576 | 0.0558 | 0.0541 |
|  | tan | 14.30 | 14.67 | 15.06 | 15.46 | 15.89 | 16.35 | 16.83 | 17.34 | 17.89 | 18.46 |
| 87 | sin | 0.9986 | 0.9987 | 0.9988 | 0.9989 | 0.9990 | 0.9990 | 0.9991 | 0.9992 | 0.9993 | 0.9993 |
|  | cos | 0.0523 | 0.0506 | 0.0488 | 0.0471 | 0.0454 | 0.0436 | 0.0419 | 0.0401 | 0.0384 | 0.0366 |
|  | tan | 19.08 | 19.74 | 20.45 | 21.20 | 22.02 | 22.90 | 23.86 | 24.90 | 26.03 | 27.27 |
| 88 | sin | 0.9994 | 0.9995 | 0.9995 | 0.9996 | 0.9996 | 0.9997 | 0.9997 | 0.9997 | 0.9998 | 0.9998 |
|  | cos | 0.0349 | 0.0332 | 0.0314 | 0.0297 | 0.0279 | 0.0262 | 0.0244 | 0.0227 | 0.0209 | 0.0192 |
|  | tan | 28.64 | 30.14 | 31.82 | 33.69 | 35.80 | 38.19 | 40.92 | 44.07 | 47.74 | 52.08 |
| 89 | sin | 0.9998 | 0.9999 | 0.9999 | 0.9999 | 0.9999 | 1.000 | 1.000 | 1.000 | 1.000 | 1.000 |
|  | cos | 0.0175 | 0.0157 | 0.0140 | 0.0122 | 0.0105 | 0.0087 | 0.0070 | 0.0052 | 0.0035 | 0.0017 |
|  | tan | 57.29 | 63.66 | 71.62 | 81.85 | 95.49 | 114.6 | 143.2 | 191.0 | 286.5 | 573.0 |
| Degs. | Function | 0' | 6' | 12' | 18' | 24' | 30' | 36' | 42' | 48' | 54' |

482

# APPENDIX X

# SQUARE AND SQUARE ROOTS

| N | $N^2$ | $\sqrt{N}$ | N | $N^2$ | $\sqrt{N}$ | N | $N^2$ | $\sqrt{N}$ |
|---|---|---|---|---|---|---|---|---|
| 1 | 1 | 1.000 | 41 | 1681 | 6.4031 | 81 | 6561 | 9.0000 |
| 2 | 4 | 1.414 | 42 | 1764 | 6.4807 | 82 | 6724 | 9.0554 |
| 3 | 9 | 1.732 | 43 | 1849 | 6.5574 | 83 | 6889 | 9.1104 |
| 4 | 16 | 2.000 | 44 | 1936 | 6.6332 | 84 | 7056 | 9.1652 |
| 5 | 25 | 2.236 | 45 | 2025 | 6.7082 | 85 | 7225 | 9.2195 |
| 6 | 36 | 2.449 | 46 | 2116 | 6.7823 | 86 | 7396 | 9.2736 |
| 7 | 49 | 2.646 | 47 | 2209 | 6.8557 | 87 | 7569 | 9.3274 |
| 8 | 64 | 2.828 | 48 | 2304 | 6.9282 | 88 | 7744 | 9.3808 |
| 9 | 81 | 3.000 | 49 | 2401 | 7.0000 | 89 | 7921 | 9.4340 |
| 10 | 100 | 3.162 | 50 | 2500 | 7.0711 | 90 | 8100 | 9.4868 |
| 11 | 121 | 3.3166 | 51 | 2601 | 7.1414 | 91 | 8281 | 9.5394 |
| 12 | 144 | 3.4641 | 52 | 2704 | 7.2111 | 92 | 8464 | 9.5917 |
| 13 | 169 | 3.6056 | 53 | 2809 | 7.2801 | 93 | 8649 | 9.6437 |
| 14 | 196 | 3.7417 | 54 | 2916 | 7.3485 | 94 | 8836 | 9.6954 |
| 15 | 225 | 3.8730 | 55 | 3025 | 7.4162 | 95 | 9025 | 9.7468 |
| 16 | 256 | 4.0000 | 56 | 3136 | 7.4833 | 96 | 9216 | 9.7980 |
| 17 | 289 | 4.1231 | 57 | 3249 | 7.5498 | 97 | 9409 | 9.8489 |
| 18 | 324 | 4.2426 | 58 | 3364 | 7.6158 | 98 | 9604 | 9.8995 |
| 19 | 361 | 4.3589 | 59 | 3481 | 7.6811 | 99 | 9801 | 9.9499 |
| 20 | 400 | 4.4721 | 60 | 3600 | 7.7460 | 100 | 10000 | 10.0000 |
| 21 | 441 | 4.5826 | 61 | 3721 | 7.8102 | 101 | 10201 | 10.0499 |
| 22 | 484 | 4.6904 | 62 | 3844 | 7.8740 | 102 | 10404 | 10.0995 |
| 23 | 529 | 4.7958 | 63 | 3969 | 7.9373 | 103 | 10609 | 10.1489 |
| 24 | 576 | 4.8990 | 64 | 4096 | 8.0000 | 104 | 10816 | 10.1980 |
| 25 | 625 | 5.0000 | 65 | 4225 | 8.0623 | 105 | 11025 | 10.2470 |
| 26 | 676 | 5.0990 | 66 | 4356 | 8.1240 | 106 | 11236 | 10.2956 |
| 27 | 729 | 5.1962 | 67 | 4489 | 8.1854 | 107 | 11449 | 10.3441 |
| 28 | 784 | 5.2915 | 68 | 4624 | 8.2462 | 108 | 11664 | 10.3923 |
| 29 | 841 | 5.3852 | 69 | 4761 | 8.3066 | 109 | 11881 | 10.4403 |
| 30 | 900 | 5.4772 | 70 | 4900 | 8.3666 | 110 | 12100 | 10.4881 |
| 31 | 961 | 5.5678 | 71 | 5041 | 8.4261 | 111 | 12321 | 10.5357 |
| 32 | 1024 | 5.6569 | 72 | 5184 | 8.4853 | 112 | 12544 | 10.5830 |
| 33 | 1089 | 5.7447 | 73 | 5329 | 8.5440 | 113 | 12769 | 10.6301 |
| 34 | 1156 | 5.8310 | 74 | 5476 | 8.6023 | 114 | 12996 | 10.6771 |
| 35 | 1225 | 5.9161 | 75 | 5625 | 8.6603 | 115 | 13225 | 10.7238 |
| 36 | 1296 | 6.0000 | 76 | 5776 | 8.7178 | 116 | 13456 | 10.7703 |
| 37 | 1369 | 6.0828 | 77 | 5929 | 8.7750 | 117 | 13689 | 10.8167 |
| 38 | 1444 | 6.1644 | 78 | 6084 | 8.8318 | 118 | 13924 | 10.8628 |
| 39 | 1521 | 6.2450 | 79 | 6241 | 8.8882 | 119 | 14161 | 10.9087 |
| 40 | 1600 | 6.3246 | 80 | 6400 | 8.9443 | 120 | 14400 | 10.9545 |

For numbers up to 120. For larger numbers divide into factors smaller than 120.

# APPENDIX XI

# LAWS OF EXPONENTS

The International Symbols Committee has adopted prefixes for denoting decimal multiples of units. The National Bureau of Standards has followed the recommendations of this committee, and has adopted the following list of prefixes:

| Numbers | Powers of ten | Prefixes | Symbols |
|---|---|---|---|
| 1, 000, 000, 000, 000 | $10^{12}$ | tera | T |
| 1, 000, 000, 000 | $10^9$ | giga | G |
| 1, 000, 000 | $10^6$ | mega | M |
| 1, 000 | $10^3$ | kilo | k |
| 100 | $10^2$ | hecto | h |
| 10 | 10 | deka | da |
| .1 | $10^{-1}$ | deci | d |
| .01 | $10^{-2}$ | centi | c |
| .001 | $10^{-3}$ | milli | m |
| .000001 | $10^{-6}$ | micro | u |
| .000000001 | $10^{-9}$ | nano | n |
| .000000000001 | $10^{-12}$ | pico | p |
| .000000000000001 | $10^{-15}$ | femto | f |
| .000000000000000001 | $10^{-18}$ | atto | a |

To multiply like (with same base) exponential quantities, add the exponents. In the language of alegebra the rule is $a^m \times a^n = a^{m+n}$

$$10^4 \times 10^2 = 10^{4+2} = 10^6$$

$$0.003 \times 825.2 = 3 \times 10^{-3} \times 8.252 \times 10^2$$

$$= 24.756 \times 10^{-1} = 2.4756$$

To divide exponential quantities, subtract the exponents. In the language of algebra the rule is

$$\frac{a^m}{a^n} = a^{m-n} \text{ or}$$

$$10^8 \div 10^2 = 10^6$$

$$3, 000 \div 0.015 = (3 \times 10^3) \div (1.5 \times 10^{-2})$$

$$= 2 \times 10^5 = 200,000$$

To raise an exponential quantity to a power, multiply the exponents. In the languague of algebra $(x^m)^n = x^{mn}$.

$$(10^3)^4 = 10^{3 \times 4} = 10^{12}$$

$$2, 500^2 = (2.5 \times 10^3)^2 = 6.25 \times 10^6 = 6, 250, 000$$

Any number (except zero) raised to the zero power is one. In the language of algebra $x^0 = 1$

$$x^3 \div x^3 = 1$$

$$10^4 \div 10^4 = 1$$

Any base with a negative exponent is equal to 1 divided by the base with an equal positive exponent. In the language of algebra $x^{-a} = \frac{1}{x^a}$

$$10^{-2} = \frac{1}{10^2} = \frac{1}{100}$$

$$5a^{-3} = \frac{5}{a^3}$$

$$(6a)^{-1} = \frac{1}{6a}$$

To raise a product to a power, raise each factor of the product to that power.

$$(2 \times 10)^2 = 2^2 \times 10^2$$

$$3, 000^3 = (3 \times 10^3)^3 = 27 \times 10^9$$

To find the nth root of an exponential quantity, divide the exponent by the index of the root. Thus, the nth root of $a^m = a^{m/n}$.

$$\sqrt{x^6} = x^{6/2} = x^3$$

$$\sqrt[3]{64 \times 10^3} = 4 \times 10 = 40$$

# INDEX

www.ingramcontent.com/pod-product-compliance
Lightning Source LLC
Chambersburg PA
CBHW082121210326
41599CB00031B/5832